T0326348

Soil Remediation and Plants

Soil Remediation and Plants
Prospects and Challenges

Khalid Rehman Hakeem
Faculty of Forestry, Universiti Putra Malaysia,
Serdang, Selangor, Malaysia

Muhammad Sabir
Institute of Soil and Environmental Sciences,
University of Agriculture, Faisalabad, Pakistan;
School of Plant Biology,
University of Western Australia,
Crawley, Australia

Münir Öztürk
Department of Botany, Ege University, Izmir, Turkey;
Faculty of Forestry, Universiti Putra Malaysia,
Serdang, Selangor, Malaysia;
ICCBS, Karachi University, Pakistan

Ahmet Ruhi Mermut
Department of Soil Sciences, University of Saskatchewan, Canada;
Harran University, Agriculture Faculty, Soil Science
Department, Şanlıurfa, Turkey

ELSEVIER

AMSTERDAM • BOSTON • HEIDELBERG
LONDON • NEW YORK • OXFORD • PARIS • SAN DIEGO
SAN FRANCISCO • SINGAPORE • SYDNEY • TOKYO

Academic Press is an imprint of Elsevier

Academic Press is an imprint of Elsevier
32 Jamestown Road, London NW1 7BY, UK
225 Wyman Street, Waltham, MA 02451, USA
525 B Street, Suite 1800, San Diego, CA 92101-4495, USA
The Boulevard, Langford Lane, Kidlington, Oxford OX5 1GB, UK

Notices
Knowledge and best practice in this field are constantly changing. As new research and experience broaden our understanding, changes in research methods, professional practices, or medical treatment may become necessary.

Practitioners and researchers must always rely on their own experience and knowledge in evaluating and using any information, methods, compounds, or experiments described herein. In using such information or methods they should be mindful of their own safety and the safety of others, including parties for whom they have a professional responsibility.

To the fullest extent of the law, neither the Publisher nor the authors, contributors, or editors, assume any liability for any injury and/or damage to persons or property as a matter of products liability, negligence or otherwise, or from any use or operation of any methods, products, instructions, or ideas contained in the material herein.

British Library Cataloguing-in-Publication Data
A catalogue record for this book is available from the British Library

Library of Congress Cataloging-in-Publication Data
A catalog record for this book is available from the Library of Congress

ISBN: 978-0-12-799937-1

For information on all Academic Press publications
visit our website at http://store.elsevier.com/

Typeset by TNQ Books and Journals
www.tnq.co.in

Printed and bound in the United States of America

Working together
to grow libraries in
developing countries

www.elsevier.com • www.bookaid.org

Contents

Preface   xix
Foreword   xxi
Contributors   xxv

1. Phytoremediation of Soils: Prospects and Challenges

Orooj Surriya, Sayeda Sarah Saleem, Kinza Waqar and Alvina Gul Kazi

Introduction   1
Types of Soil Pollutants   1
The Global Scenario of Soil Pollution   2
Effects of Soil Pollution on Human Health and Environment   4
Technologies for Site Remediation   5
Electroremediation   5
Soil Flushing   5
Soil Vapour Extraction   5
Stabilization   6
Soil Washing   7
Bio-piles   7
Phytoremediation   7
Phytoremediation System Design   8
Economics of Phytoremediation   11
Considerations for Waste Disposal   12
Phytoremediation Technologies   12
Heavy Metal Soil Pollutants and Use of Phytoremediation   14
Approach to the Removal of Heavy Metal Soil Pollutants   14
Ideal Plant Characteristics for Phytoremediation   15
Basic Process of Hyperaccumulation   15
Arsenic   16
The Health Hazards of Arsenic   16
Use of Phytoremediation to Remove Arsenic   17
Lead   18
The Health Hazards Caused by Lead   18
Use of Phytoremediation to Clean Up Lead   19
Zinc   20
The Health Hazards of Zinc   20
Use of Phytoremediation to Clean Up Zinc   20
Copper   21
The Health Hazards of Copper   21
Use of Phytoremediation to Clean Up Copper   21

Cadmium   23
The Health Hazards of Cadmium   23
Use of Phytoremediation to Clean Up Cadmium   23
Mercury   24
The Health Hazards of Mercury   24
Use of Phytoremediation to Clean Up Mercury   24
Prospects for Phytoremediation   26
Phytoremediation Is Solar Energy–Driven and Cost-Effective   27
Phytoremediation Is Environmental Friendly   27
Phytoremediation Can Help Mining Industries   27
Phytoremediation Cleans Up Contaminants in Soil and Water   27
The Reduction of Noise Pollution and Genetically Engineered
    Phytoremediation Plants   28
Challenges of Phytoremediation   28
The Contaminant Can Be Extracted from as Deep as the Plant's Roots   29
Slow Growth Cycle of Phytoremediation Plants   29
Suitable Climatic Conditions and the Availability of Space for
    Phytoremediation   29
Less Tolerance in Plants for Contaminant Uptake   29
Accumulation of Contaminants in Plant Tissues   30
Genetically Modified Plants in Phytoremediation: An Unexpected or
    Unknown Threat?   30
Accessibility of Sites for Phytoremediation   30
Techniques for Genetic Improvement of Plants Used for
    Phytoremediation   30
Example: Hyperaccumulation of Metal and Plant Response after
    Genetic Modification   31
Use of Recombinant DNA Technology in Genetic Improvement for
    Phytoremediation   32
Introduction of Genes Capable of Altering the Oxidation State of
    Heavy Metals   32
Use of Phytochelatins to Capture Metal Ions   32
Conclusion   33
References   33

2. Soil Contamination with Metals: Sources, Types and
    Implications
    Waqar Ahmad, Ullah Najeeb and Munir Hussain Zia

Introduction   37
Heavy Metals   39
Sources of Heavy Metals in Soil   39
Effects of Heavy Metals in Soils and Plants   43
Build-Up of Heavy Metals in Soil   43
Toxicity Symptoms in Plants   43
Risk Assessment Using Bioavailability and Bio-Accessibility Techniques   48
Control Measures   51
Removal of Heavy Metals from Soils   51

Phytoextraction of Heavy Metals   51
Chelator-Assisted Heavy Metals Phytoremediation   52
Biochar Application as an Emerging Tool for Reducing Heavy
  Metal Impacts   53
Conclusions   55
References   55

3. Phytoremediation: A Promising Strategy on the Crossroads of Remediation

Arif Tasleem Jan, Arif Ali and Qazi Mohd. Rizwanul Haq

Introduction   63
Metal Pollutants and Human Health   64
Microbial-Based Remediation   66
Enhancing Bioremediation Through Genetic Engineering   67
Surface Expression of Novel Metal Binding Peptides and Proteins   68
Phytoremediation   68
Phytoextraction   69
Rhizoremediation   73
Phytostabilization   74
Insights into Genetic Engineering Approaches for Phytoremediation   75
Conclusions   78
References   78

4. Phytoremediation: Mechanisms and Adaptations

Muhammad Sabir, Ejaz Ahmad Waraich, Khalid Rehman Hakeem,
Münir Öztürk, Hamaad Raza Ahmad and Muhammad Shahid

Introduction   85
Phytoremediation and Mechanisms   86
Phytoextraction   86
Phytostabilization   93
Phytovolatilization   96
Conclusions   97
References   97

5. Phytoremediation: An Eco-Friendly Green Technology for Pollution Prevention, Control and Remediation

Tanveer Bilal Pirzadah, Bisma Malik, Inayatullah Tahir,
Manoj Kumar, Ajit Varma and Reiaz Ul Rehman

Introduction   107
Background   108
Plants' Response to Heavy Metals   109
Metal Excluders   110
Metal Indicators   110
Metal Accumulating Plant Species   110

Factors Affecting Phytoremediation   115
Availability of Metals in Soil   115
Plant Uptake and Translocation   117
Plant Microbe Interactions   117
Role of Metal Chelators   118
Mechanism for Metal Detoxification   119
Conclusions and Future Perspectives   121
References   122

6. Recent Trends and Approaches in Phytoremediation

Bisma Malik, Tanveer Bilal Pirzadah, Inayatullah Tahir,
Tanvir ul Hassan Dar and Reiaz Ul Rehman

Introduction   131
Phytoremediation Technologies   132
Phytoextraction   132
Phytovolatilization   134
Phytostabilization   135
Rhizofiltration   136
Genetic Engineering to Improve Phytoremediation   137
Conclusions and Future Perspectives   142
References   142

7. Evaluation of Four Plant Species for Phytoremediation of Copper-Contaminated Soil

Parisa Ahmadpour, Fatemeh Ahmadpour, SeyedMousa Sadeghi,
Farhad Hosseini Tayefeh, Mohsen Soleimani and Arifin Bin Abdu

Introduction   147
General Background   147
Problem Statement   148
Objectives   148
Literature Review   149
Environmental Pollution and Sources of Contamination   149
Soil Contamination by Heavy Metals   149
Toxicity of Heavy Metals in Plants   151
Uptake and Translocation of Copper by Plant Parts (Leaves,
    Stems and Roots)   153
Remediation of Heavy Metals   153
Criteria for Metal Accumulation in Plants   156
Materials and Methods   159
Description of Study Area   159
Planting Materials   159
Experimental Design and Treatments   160
Plant Species and Planting   160
Data Collection   160
Laboratory Analysis   161

Evaluation of Heavy Metals Uptake Using Removal Efficiency (RE),
 Bioconcentration Factor (BCF) and Translocation Factor (TF)   163
Statistical Analysis   164
Results and Discussion   **164**
Physico-Chemical Properties of the Control Media   164
Summary, General Conclusion and Recommendation for
 Future Research   **195**
Conclusions   195
Recommendations for Future Research   197
References   **197**

8. Role of Phytoremediation in Radioactive Waste Treatment

L.F. De Filippis

Introduction   **207**
Radioactive Material and Safety   **215**
Classification and Categories   **217**
Management and Disposal   **219**
Transportation and Responsibility   **220**
Phytoremediation and Non-Plant Methods   **220**
Engineering-Based Technologies   220
In Situ Biological Remediation   221
Phytoremediation and Hyperaccumulation   **221**
Methods in Phytoremediation   **223**
Phytoextraction   223
Phytodegradation   231
Phytostabilization   232
Phytovolatilization   232
Rhizofiltration   233
Rhizodegradation   234
Vegetation Caps and Buffer Strips   234
Hydraulic Control   235
Tolerance and Extraction   **235**
Uptake and Distribution   **236**
Wetlands and Aquatic Phytoremediation   **237**
Constructed Wetlands   237
Combinations of Phytoremediation Methods   238
Treatment, Evaluation and Objectives   **238**
Ground Water   239
Surface Water and Waste Water   239
Soil, Sediment and Sludge   240
Air   240
Costs and Economics   **240**
Transgenic Phytoremediation   **244**
Conclusions and Future Directions   **245**
References   **247**

9. Plant–Microbe Interactions in Phytoremediation
   Ibrahim Ilker Ozyigit and Ilhan Dogan

   Definition of Phytoremediation   255
   Accumulator/Hyperaccumulator Plants   256
   Phytoremediation Applications   257
   Interactions Between Plants and Microbes in Phytoremediation   260
   Rhizosphere Microbiome   266
   Stimulation of Plant Growth by Microbial Communities   268
   Acknowledgements   271
   References   272

10. Soil Pollution in Turkey and Remediation Methods
    Hatice Dağhan and Münir Öztürk

    Introduction   287
    Land Use of Turkish Soils   289
    Sources of Soil Pollution in Turkey   291
    Erosion   292
    Parent Material   293
    Agricultural Activities   294
    Industrial Activities   299
    Urbanization   302
    Mining   303
    Remediation Methods for Polluted Soils   304
    Remediation Studies in Turkey   305
    Radioactive Pollution   307
    Conclusion   307
    References   308

11. Soil Pollution Status and Its Remediation in Nepal
    Anup K.C. and Subin Kalu

    Introduction   313
    Soil Characteristics   315
    Soils of Nepal   316
    Alluvial Soil   316
    Sandy and Alluvial Soil   316
    Gravelly Soil   317
    Residual Soil   317
    Glacial Soil   317
    Entisols   318
    Ustifluvents   318
    Ustorthents   318
    Fluvaquents   318
    Inceptisols   318

Haplaquents 318
Dystrochrepts 319
Ustochrepts 319
Cryumbrepts 319
Haplumbrepts 320
Cryochrepts 320
Eutrochrepts 320
Spodosols 320
Mollisols 320
Haplustolls 320
Cryoborolls 321
Alfisols 321
Ultisols 321
Aridosols 321
Nutrient and Heavy Metal Status in the Soils of Nepal 322
Remediation of Toxicity from Soil 325
Remediation Studies on Removal of Toxicity in Soil of Nepal 326
Conclusions 326
References 327

12. Transfer of Heavy Metals and Radionuclides from Soil to Vegetables and Plants in Bangladesh

Mahfuza S. Sultana, Y.N. Jolly, S. Yeasmin, A. Islam, S. Satter and Safi M. Tareq

Introduction 331
Materials and Methods 333
Geology and Geomorphology of the Study Area 333
Sampling 334
Analysis of Heavy Metals and Radioactivity 335
Data Analysis 338
Results and Discussion 341
Concentrations of Heavy Metals in the Top Soil 341
Pollution Index Assessment of Soil 342
Levels of Heavy Metals in Plants and Vegetables 344
Pollution Index Assessment of Different Crops 348
Comparative Study of the Heavy Metal Contents in Soil
    and Plants 349
Heavy Metal Transfer from Soil to Food Crops 350
Health Risk Assessment of Heavy Metals in Vegetables 354
Radioactivity in Soil 360
Radioactivity in Vegetables 361
Transfer Factor 362
Conclusions 362
Acknowledgements 363
References 364

13. Remediating Cadmium-Contaminated Soils by Growing Grain Crops Using Inorganic Amendments

Muhammad Zia-ur-Rehman, Muhammad Sabir, Muhammad Rizwan, Saifullah, Hamaad Raza Ahmed and Muhammad Nadeem

Introduction	367
Natural Cadmium Levels in Soil	368
Sources of Cadmium Contamination of Agricultural Soils	369
Bioassessment of Cadmium in Soils	370
Factors Influencing the Accumulation of Cadmium in Crops	370
Soil Factors	370
Microorganisms	372
Climatic Factors	372
Cadmium Uptake and Accumulation in Plants	372
Effects on Seed Germination	374
Effects on Plant Growth and Biomass	375
Effects on Mineral Nutrients	376
Effects on Photosynthetic Pigments	376
Cd-Induced Oxidative Stress in Plants	377
Miscellaneous Toxic Effects	377
Plant Response To Cd Concentrations	378
Threshold Bio-Available Concentration of Cd	378
Remediation of Cd-Contaminated Soils	379
Metals Response to Calcium Containing Inorganic Amendments	379
Metals Response to Phosphorus Containing Amendments	381
Metals Behaviour in Response to Ammonium Containing Amendments	384
Metals Behaviour in Response to Sulphur-Containing Amendments	384
Conclusions	385
References	386

14. Phytoremediation of Pb-Contaminated Soils Using Synthetic Chelates

Saifullah, Muhammad Shahid, Muhammad Zia-Ur-Rehman, Muhammad Sabir and Hamaad Raza Ahmad

Introduction	397
The Problem of Pb	399
Chelating Agents	399
Ethylene Diamine Tetraacetic Acid (EDTA)	400
Ethylene Diamine Disuccinic Acid	403
Nitrilotriacetic Acid (NTA)	406
Comparison of Synthetic Chelating Agents	407
Conclusions	407
References	408

15. Spatial Mapping of Metal-Contaminated Soils

H.R. Ahmad, T. Aziz, Z.R. Rehman and Saifullah

Introduction	415
Geophysical Techniques to Assess Spatial Variability	417
Global Positioning System (GPS)	417
Remote Sensing	417
Geographic Information System	418
Histogram	419
The QQPlot	420
Semivariogram	420
Inverse Weighted Distance	421
Krigging	421
Creating a Database File for GIS Environment	423
Conclusions	427
References	428

16. Arsenic Toxicity in Plants and Possible Remediation

*Mirza Hasanuzzaman, Kamrun Nahar, Khalid Rehman Hakeem,
Münir Öztürk and Masayuki Fujita*

Introduction	433
Environmental Chemistry of Arsenic	435
Sources of Arsenic Contamination in Soil and Environment	436
Status of Arsenic Toxicity in the World	438
Arsenic Hazard: A Bangladesh Perspective	441
Arsenic Uptake and Transportation in Plants	448
Plant Responses to Arsenic Toxicity	451
Seed Germination	451
Growth	455
Water Relations	456
Photosynthesis	457
Yield	459
Oxidative Stress	460
Antioxidant Defence in Plants in Response to Arsenic Stress	461
Remediation of Arsenic Hazards	468
Agronomic Management	468
Phytoremediation of Arsenic-Contaminated Soils	470
Presumed Phytochelatin-Mediated Detoxification of As Toxicity	483
Role of Biotechnology in Remediation of Arsenic Toxicity	485
Conclusion and Future Perspectives	485
Acknowledgements	486
References	486

17. **Phytoremediation of Metal-Contaminated Soils Using Organic Amendments: Prospects and Challenges**

Muhammad Sabir, Muhammad Zia-ur-Rehman, Khalid Rehman Hakeem and Saifullah

Background	503
Sources of Metals	505
Role of OM in Phytoavailability of Metals	506
Organic Amendments and Phytoavailability of Metals in Contaminated Soils	509
Manures	509
Compost	510
Activated Carbon / Biochar	511
Pressmud	512
Effect of Time on Decomposition of Organic Amendments and Metal Phytoavailability	512
Residual Effect of Organic Amendments on Metal Phytoavailability	514
Organic Acids and Metal Phytoavailability	515
Phytoremediation with Organic Amendments: Conclusion and Future Thrust	515
References	516

18. **Soil Contamination, Remediation and Plants: Prospects and Challenges**

M.S. Abdullahi

Introduction	525
Sources of Heavy Metals in Soil	527
Fertilizers	528
Pesticides	528
Biosolids and Manures	528
Wastewater	529
Metal Mining, Milling Processes and Industrial Wastes	530
Airborne Sources	530
Potential Risk of Heavy Metals to Soil	531
Soil Concentration Ranges and Regulatory Guidelines for Some Heavy Metals	531
Remediation of Contaminated Soil by Heavy Metals	532
Prevention of Heavy Metal Contamination	536
Traditional Remediation of Contaminated Soil	536
Management of Contaminated Soil	537
The Use of Plants for Environmental Clean-Up	538
The Use of Plants for Treating Metal-Contaminated Soils	538
Example of Disposal	539
Preventive Steps	539
Classification of Heavy Metals	540
Sources of Heavy Metals in the Environment	540
Benefits of Heavy Metals to Plants	541

Future Prospects   542
Challenges   542
Conclusions   543
References   543

19. Improving Phytoremediation of Soil Polluted with Oil Hydrocarbons in Georgia

Gia Khatisashvili, Lia Matchavariani and Ramaz Gakhokidze

Introduction   547
Characterization of Soil Types   549
Selection of Microorganisms   550
Selection of Plants   551
Determination of the Degree of Oxidative Degradation of
    Hydrocarbons   557
Revelation of Plant–Microbial Interaction   559
Model Experiments   559
References   567

20. Remediation of Cd-Contaminated Soils: Perspectives and Advancements

Syed Hammad Raza, Fahad Shafiq, Umer Rashid, Muhammad Ibrahim and Muhammad Adrees

Background and Introduction   571
Cadmium Emissions   571
Soil Dynamics, Retention and Availability of Metals   573
Dynamics of Cadmium in Soils   574
Influence of the Associated Cations and Anions on Cadmium
    Bioavailability in Soil   574
Response of Cd Towards Natural Elemental Inorganic Amendments   575
Calcium (Ca)   575
Nitrogen (N)   576
Sulphur (S)   576
Zinc (Zn)   576
Phosphorus (P)   577
Iron (Fe)   577
Manganese, Silicon and Chloride   578
Organic Amendments Versus Cadmium-Contaminated Soils   578
Natural Organic Additives   579
Root Exudates and the Concept of Organic Acids as Natural Chelators   579
Low-Molecular-Weight Organic Acids and Cadmium Chelation   580
Efficacy of Synthetic Organic Chelating Agents Towards Cadmium   581
Recent Presented Reports Regarding Grain Crops   582
Conclusions and the Concept of Coupled Phytoremediation as
    a Future Perspective   586
References   587

21. Phytoremediation of Radioactive Contaminated Soils
Muhammad Ibrahim, Muhammad Adrees, Umer Rashid, Syed Hammad Raza and Farhat Abbas

Introduction	599
Scope and Limitations	600
Major Sources of Radioactive Contaminants to Soil and Environment	600
Nuclear Weapons' Testing	602
Production of Nuclear Weapons	602
The Nuclear Fuel Cycle	602
Industrial Processes / Techniques Involving Radionuclides	603
Research Activities	603
Phytoremediation	604
Possible Roles of Phytoremediation	604
Phytomanagement	604
Phytoextraction	605
The Potential for Phytoextraction	605
Associated Risks	608
Important Radionuclides	608
Caesium (^{137}Cs)	608
Strontium (^{90}Sr)	609
Uranium (^{238}U)	609
Risks and Potential	613
Rhizofiltration	615
Possibilities and Prospects	615
Non-Food Crops / Alternative Crops	615
Forestry	616
Biofuel / Biodiesel Crops	616
Steps Involved in Remediation Programme Management	617
Major Steps in the Management of a Remediation Programme	617
Planning for Remediation	617
Site Characterization	618
Remediation Criteria	618
Remediation Strategy	618
Soil Solution Uranium (^{238}U)	619
Agricultural and Forested Zones and Their Remediation	620
Phytostabilization of Radionuclide Contaminated Soils	621
Remediation Actions Implementation	621
Conducting Post-Remediation Activities	621
Special Considerations	622
Remediation of Areas of Extensive Surface Contamination	622
References	623

22. Heavy Metal Accumulation in Serpentine Flora of Mersin-Findikpinari (Turkey) – Role of Ethylenediamine Tetraacetic Acid in Facilitating Extraction of Nickel

Nurcan Koleli, Aydeniz Demir, Cetin Kantar, Gunsu Altindisli Atag, Kadir Kusvuran and Riza Binzet

Introduction 629
Materials and Methods 632
Field Study: Site Description, Soil Characterization, Plant Analysis
 and Plant Selection 632
Greenhouse Study: Soil Characterization, Artificial Soil
 Contamination, Pot Experiment and Plant Analysis 633
Statistical Analysis 635
Results and Discussion 635
Field Study 635
Greenhouse Study 645
Conclusion 654
Acknowledgements 655
References 655

23. Phytomanagement of Padaeng Zinc Mine Waste, Mae Sot District, Tak Province, Thailand

M.N.V. Prasad, Woranan Nakbanpote, Abin Sebastian, Natthawoot Panitlertumpai and Chaiwat Phadermrod

Introduction 661
Phytomanagement of A Zinc-Mine-Industry-Ravaged Ecosystem 665
Phytomanagement for Sustainable Agriculture in the Vicinity of
 Mae Sot Zinc Mine 668
Feasible Options for the Management of Arable Lands Mine
 Tailing Water 675
Soil Remediation 677
Reduction of Cd in Crop Produce 681
Conclusions 682
Acknowledgements 682
References 683

24. Effect of Pig Slurry Application on Soil Organic Carbon

Ibrahim Halil Yanardağ, Asuman Büyükkılıç Yanardağ, Angel Faz Cano and Ahmet Ruhi Mermut

Introduction: Importance of Soil Organic Matter 689
Pig Slurry Application 690
Effect of Pig Slurry Application on Soil Organic Carbon 691
Black Carbon 692
Soluble Carbon 693
Microbial Biomass Carbon 695

Soil Respiration   697
Carbon Functional Groups   698
Stable Carbon Isotope   699
Conclusions   **701**
References   **701**

Index   **707**

Preface

For a layman, soil is the homogenous material also called 'dirt', 'mud' and 'ground' or 'earth'. We have disposed of our wastes through the soil and neglected its uses for as long as we have been living on mother earth. The living soils are not just the foundation of our food chain, but of our civilization in general. Knowing that the soils are a source of our food, it becomes imperative that we keep it healthy. It has always been considered as a disposable commodity. Our future will be dark if we do not look after our soils. Currently, every second, fertile soil equivalent to fill nearly 30 soccer fields is being destroyed, and around 52 million ha are being affected by different soil contaminants. Approximately 10 million ha of farmland is lost every year and about 1.6 billion ha of the world's best, most productive lands are used to grow crops at present. Unsustainable farming practices are leading to water and wind erosion, loss of organic matter, compaction of the soils, salinization and pollution, and nutrient loss. A New Paradigm for Agriculture, released earlier this year by FAO has been 'Save and Grow'; this save the soils campaign will run until December 2017, the international year of soils. As world population and food production demands rise, keeping our soil healthy and productive is of paramount importance. By focusing more attention on soil health and by educating ourselves about the positive impact healthy soils can have on productivity and conservation, we can save this living and life-giving substance, without which we would perish.

This thin layer is suffering from pollution, a common observation in our present day lives, which originates from our activities and our wastes which are full of chemicals lacking in nature and end up with the pollution of our soils. Nature itself produces wastes such as dead plants, dead animals, spoiled fruits and vegetables, but they increase the fertility of the soil. Our ancestors used the wastes to fertilize it, the populations then were small and this was no problem. Today demographic outburst is putting a pressure on the soil ecosystem to produce more and more food.

All through the 20th century the additions of fertilizer, pesticides and herbicides have created problems for us and all species. Industrialization has been the greatest contributor contaminating by the release of xenobiotics, which enter our ecosystems and degrade the soils. The toxic chemicals decrease the fertility, making the land unsuitable for the growth of natural plant cover or agriculture. The air pollutants return back to us as acid rains dissolving some of the soil nutrients, posing major problems for our health in the long run, because all the plants consumed by us absorb much of the pollution from the soils polluted by us. This is the cause for the sudden increase in some terminal illnesses, but long-term exposures can prove detrimental to our genetic make-up, resulting in congenital illnesses and chronic health problems.

Pollutants added to the soil may adhere to the soil matrix, become immobile, but may still be available to plant roots. They may be taken up into the plant structure and enter the food chain, or can percolate from the soil surface into our groundwaters and cause problems like nitrate loading in drinking waters. Whatever agriproduct we will cultivate on polluted soils, that will lack the quality and may even contain poisonous compounds causing serious health problems. The soil contaminants can have significant deleterious consequences for ecosystems. There are radical soil chemistry changes which can arise from the presence of many hazardous chemicals even at low concentrations of the contaminant species. We thus have to use very expensive primary accredited methods with proven high precision like X-Ray Fluorescence (XRF), ASV (Anodic Stripping Voltammetry), Colorimetric test kits, Atomic absorption and ICP (Inductive Coupled Plasma) to look at the pollution status of our soils. However, prevention is better than cure; therefore, we can restore 720 million hectares of lightly degraded soils on our planet simply by using sustainable and agro-ecological farming techniques. We can use several principal strategies for remediation like excavating the soil and disposing of it away from the human or sensitive ecosystem contact, but it will not serve the purpose in the long run. Phytoextraction, phytodegradation, phytovolatilization, rhizofiltration, evapotranspiration, phytostabilization and hydraulic control techniques are also available which can be applied. These remediations are becoming a subject of public/scientific interest as cost-effective technologies for cleaning our soils. The efficiency, however, depends on the physical and chemical properties of the soil, bioavailability of the pollutant, and the ability of the plants for uptake, accumulation, translocatation, sequesteration and detoxification. They are a type of ecological engineering, intermediate between engineering and natural attenuation. Genetic approaches have also been started to solve these problems.

In this book, an attempt has been made to provide a common platform to the biologists, agricultural engineers, environmental scientists and chemists, working with a common aim of finding sustainable solutions to various environmental issues. The volume addresses the topics crucial for understanding the ecosystem approaches for a sustainable development. Contributions from different authors present an overview of soil contaminants as well as the significant role of plants. The book provides an overview of ecosystem approaches and phytotechnologies and their cumulative significance in relation to various environmental problems and solutions. It will be a useful asset to students, researchers, practitioners and policy makers alike.

We are thankful to the contributors for their collaboration that made the present volume possible. We also extend our appreciation to Elsevier for their exceptional kind support, which made our efforts successful.

The Editors

Foreword

Land is a precious natural resource and base for agricultural sustainability and human civilization. Population growth, particularly the development of high-density urban populations leads to global industrialization and places major burdens on our environment, thereby considerably threatening environment sustainability. Contamination of soil mainly occurs due to release of industrial, urban and agricultural wastes generated by human activities. Controlled and uncontrolled solid discharge from industries, vehicle exhaustion, soluble salts, insecticides, pesticides, excessive use of fertilizers and heavy metals from organic and inorganic sources are environmental contaminants. These have resulted in build-up of chemical and biological containments throughout the biosphere, but most notably in soil and sediments. In addition to human-induced contamination of the environment, natural mineral deposits containing heavy metals are the major contaminants present in many regions of the globe.

Biological wastes and contaminants include raw and digested sewage, animal manures and vegetable wastes. However, microorganisms degrade or recycle these biological wastes into soil for agricultural benefits but increasing urbanization and continued expansion of cities require disposal of these materials far from cities. Traditional land disposal practices of biological wastes are often rendered uneconomic because of high transport costs. An additional problem regarding biological wastes is the risk of spread of infectious diseases when infected materials are applied to soil. Infected soil can facilitate disease transmission to plants, animals or humans who are directly or indirectly in contact with the soil.

Recent spreading out of the petroleum and chemical industries has resulted in the production of a wide range of organic and inorganic chemicals, which are considered major environmental pollutants. Among the chemical contaminants, inorganic contaminants being enriched with heavy metals are the most problematic for plants and humans. Industrial activities have also led to considerable contamination of soil and other media by enriching them with heavy metals, which have proven toxicity to both humans and animals. Contamination of soil and solid wastes with highly active radionuclides is another environmental risk with the potential for these metals to be radiotoxic to all life forms. Mention should also be made here of unwarranted concentrations of undesired chemicals mixed with commonly available inorganic fertilizers, such as nitrates, ammonia, phosphates, etc., which accumulate or contaminate water courses through run-off or air through volatilization. Although, several metals at their low levels are essential for normal functioning of metabolism, all metals are toxic to plants and other organisms when at higher concentrations. These heavy metals can easily replace essential metals associated with different pigments or enzymes, causing impairment of their functions.

Several legislative protocols have been framed aiming at reducing soil contaminants, but they are not so effective in controlling the contaminants. Many uncontrolled historical events like disposal of polychlorinated biphenyls by Hooker Company to the Love Canal area in Niagara Falls, and the dioxin crisis in Belgium, Italy and Bhopal caused miscarriages and birth abnormalities among the residents of affected areas. In addition, many chemicals have a great tendency to transfer from solid media to aqueous media and to be absorbed by plants or aquatic species.

The common remediation approaches being employed throughout the world to render soil enriched with toxic metals fit for use are based on the use of organic and inorganic chemicals. Some of the most common approaches are retention of toxicants within affected areas, degradation of organic contaminants by physico-chemical or biological means, and removal of contaminants from the soil. These approaches are being applied by an organization engaged in remediation to transform the contaminated soil into cultivable soil. Unfortunately, application of the above-stated strategies requires extensive earth moving and expensive machinery and infrastructure. For example, excavation of contaminated soil with heavy metals and offsite burial in landfill is not a suitable alternative, because it is just a shifting of the contamination problem somewhere else. Furthermore, non-biological processes have to bear heavy costs to remediate the entire known hazardous waste site worldwide. However, cost could be reduced substantially by using plants which are effective in phytoremediation.

Phytoremediation is the use of green plants to remove contaminants from contaminated sources such as soil, water, air and sediments. Generally, phytoremediation entails five processes of decontamination, for example rhizofiltration, phytostabilization, phytoextraction, phytovolatilization and phytodegradation. However, all these processes are meant for elimination of contaminants from soil and water, though to variable extents. Furthermore, these are the cost-effective and friendly techniques for cleaning the environment. Although thousands of species have been identified as heavy metal accumulators, there is a need to identify plants which can effectively phytoremediate the contaminated environment under the current scenario of climate change. The skill of selecting plant species, which can accumulate great amounts of heavy metals and are resistant to heavy metals, would facilitate reclamation of contaminated soils.

Phytoremediation is a hot topic being vigorously researched these days. Researchers, teachers and scholars engaged in the field of soil science, agronomy, ecology, botany, plant physiology, forestry, environmental chemistry, irrigation agriculture and biochemistry can greatly benefit from the detailed knowledge described in this book. Graduate and undergraduate students interested in phytoremediation may find this book to be a mandatory reference for their practical and theoretical study. Course instructors engaged with phytoremediation will find this book an adequate means to provide a fundamental background on the subject. I reviewed the book and a brief description is given below.

There are 24 chapters in the book written by the authors from 11 different countries. These cover topics like soil contamination with metals, different aspects of phytoremediation, evaluation of plants in phytoremediation, radioactive waste treatment and radionuclides in plants, plant–microbe interactions, pollution status of soils from different countries, heavy metal remediation, spatial mapping of metal-contaminated soils, organic amendments, soils polluted with oil hydrocarbons, role of serpentine flora and soil organic carbon.

A discussion on the different mechanisms plants adapt for remediation of metal-contaminated soils has been outlined at the start stressing the fact that an understanding of the inherently complex mechanisms is a prerequisite for developing suitable remediation techniques. This is followed by potential risks of heavy metals in soils, the role of plants in remediation, main limitations of phytoremediation and future prospects, and the sources and types of metal contamination in agricultural soils and the implications for the biosphere. Arsenic (As) is one of the oldest and most important poisons in the global environment and is becoming a serious threat for crop production. The chapter on this subject has summarized the work on As toxicity in relation to plants and environment. Authors have also discussed the progress made during last few decades to remediate the toxicity in soil, water, plant and food chains through different remediation technologies. Similarly studies on cadmium (Cd), being a promising ecotoxic metal, have been discussed at length because it poses inhibitory effects on plant metabolism, biodiversity, soil biological activity, and human and animal health. The strategies for the restriction of Cd entrance in grain crops by using different chemicals have been outlined. The role of Copper (Cu) as one of the most hazardous pollutants, particularly at higher concentrations has been included in the book, followed by the studies carried out in Malaysia which present results of assessing the phytoremediation potential of *Jatropha curcas*, *Acacia mangium*, *Dyera costulata* and *Hopea odorata* for Cu-contaminated soil. Soil contamination due to lead (Pb) warrants special attention because of its long-term retention in soil and hazardous effects on plant and human health. The role of synthetic chelators in the remediation of Pb-contaminated soils is given with a critical assessment of the risks and limitations associated with this technology. A chapter discussing the mechanisms of and factors affecting phytoremediation has also been included in the book.

A review of the role of phytoremediation in radioactive waste treatment has been presented as well as radionuclides in plants. Innovative new methodologies in this field and the different categories of phytoremediation techniques which may treat and control radioactive contaminated waste are addressed. The other chapter elaborates the scope and limitations of phytoremediation for radioactive contaminated soils. The authors have reviewed major sources of radioactive contaminants to soil and environment, possible role of phytoremediation, and post-remediation activities. This chapter concludes with implications for remediation of areas of extensive surface contamination. The transfer of heavy metals and radionuclides from soil to vegetables and plants in terms of transfer factor has been presented in a separate chapter. This factor is commonly used to estimate the food chain transfer of different elements and possible phytoremediation measures.

The extent of soil pollution in Turkey and the possible remediation measures and effective phytoremediation technology for cleaning soils polluted with oil hydrocarbons in Georgia have been discussed at length. The technology mentioned in the chapter from Georgia can be used to eliminate the pollution caused by accidental oil spills, which can be the result of oil transportation. Another chapter deals with the current status of soil pollution and nutrient deficiencies in the soils of Nepal. It has been concluded that plant-growth-promoting *Rhizobacteria* are beneficial for combating heavy metal stress. A chapter reviewing plant–microbe interactions in phytoremediation with a particular reference to the microbial dynamics in the rhizosphere of plants grown on contaminated soils has been added. Sufficient information has been given on the recent approaches in phytoremediation and how genetic engineering can play its role in improving the potential of phytoremediation in plants. The role of organic amendments to immobilize metals, improve plant growth and subsequent release of metals due to decomposition of organic matter has been discussed. Finally the spatial mapping of metal-contaminated soils using GIS techniques has been discussed and the fate of pig slurry application in a temperate region and its implications for sustainable agriculture and control of soil pollution evaluated.

Professor M. Ashraf (*PhD, DSc UK*)
Distinguished National Professor,
Professor & Chairman, Department of Agronomy,
University College of Agriculture & Director – Quality Enhancement Cell,
University of Sargodha, Sargodha, Pakistan.

Contributors

Farhat Abbas Department of Environmental Sciences, Government College University, Faisalabad, Pakistan

Arifin Bin Abdu Department of Forest Production, Faculty of Forestry, Universiti Putra Malaysia, Serdang, Selangor DarulEhsan, Malaysia

M.S. Abdullahi Department of Chemistry, Federal College of Education, Kontagora, Nigeria

Muhammad Adrees Department of Environmental Sciences, Government College University, Faisalabad, Pakistan

Waqar Ahmad Department of Environmental Sciences, Faculty of Agriculture and Environment, The University of Sydney, NSW, Australia

Hamaad Raza Ahmad Institute of Soil and Environmental Sciences, University of Agriculture, Faisalabad, Pakistan

Parisa Ahmadpour Ports and Martime Organization (PMO), Boushehr Maritime Rescue and Environmental Protection Department, Boushehr, Iran; Department of Forest Production, Faculty of Forestry, Universiti Putra Malaysia, Serdang, Selangor DarulEhsan, Malaysia

Fatemeh Ahmadpour Pars Special Economic Energy Zone, Pseez, National Iranian Oil Co, NIOC, Boushehr, Iran

Arif Ali Department of Biosciences, Jamia Millia Islamia, New Delhi, India

K.C. Anup Department of Environmental Science, Amrit Campus, Tribhuvan University, Thamel, Kathmandu, Nepal

Muhammad Ashraf Atta-ur-Rehman School of Applied Bio-sciences, National University of Science and Technology

Gunsu Altindisli Atag Alata Horticultural Research Station Directorate, Erdemli, Turkey

T. Aziz Institute of Soil and Environmental Sciences, University of Agriculture, Faisalabad, Pakistan

Riza Binzet Mersin University, Faculty of Engineering, Department of Environmental Engineering, Mersin, Turkey

Asuman Büyükkılıç Yanardağ Sustainable Use, Management and Reclamation of Soil and Water Research Group, Agrarian Science and Technology Department, Technical University of Cartagena, Cartagena, Murcia, Spain

Angel Faz Cano Sustainable Use, Management and Reclamation of Soil and Water Research Group, Agrarian Science and Technology Department, Technical University of Cartagena, Cartagena, Murcia, Spain

Hatice Dağhan Eskisehir Osmangazi University, Agricultural Faculty, Department of Soil Science and Plant Nutrition, Eskisehir, Turkey

Tanvir ul Hassan Dar Department of Bioresources, University of Kashmir, Srinagar, India

L.F. De Filippis School of the Environment, University of Technology, Sydney, NSW, Australia

Aydeniz Demir Mersin University, Faculty of Engineering, Department of Environmental Engineering, Mersin, Turkey

Ilhan Dogan Izmir Institute of Technology, Faculty of Science, Department of Molecular Biology and Genetics, Urla, Izmir, Turkey

Masayuki Fujita Laboratory of Plant Stress Responses, Department of Applied Biological Science, Faculty of Agriculture, Kagawa University, Miki-cho, Kita-gun, Kagawa, Japan

Ramaz Gakhokidze Department of Bioorganic Chemistry, Faculty of Exact & Natural Sciences, Tbilisi State University of Iv. Javakhishvili, Tbillisi, Georgia

Khalid Rehman Hakeem Faculty of Forestry, Universiti Putra Malaysia, Serdang, Selangor, Malaysia

Mirza Hasanuzzaman Department of Agronomy, Faculty of Agriculture, Sher-e-Bangla Agricultural University, Sher-e-Bangla Nagar, Dhaka, Bangladesh

Muhammad Ibrahim Department of Environmental Sciences, Government College University, Faisalabad, Pakistan

A. Islam Department of Environmental Sciences, Jahangirnagar University, Savar, Dhaka, Bangladesh

Arif Tasleem Jan Department of Biosciences, Jamia Millia Islamia, New Delhi, India

Y.N. Jolly Chemistry and Health Physics Division, Atomic Energy Centre, Dhaka, Bangladesh

Subin Kalu Central Department of Environmental Science, Tribhuvan University, Kirtipur, Kathmandu, Nepal

Cetin Kantar Canakkale Onsekiz Mart University, Faculty of Engineering and Architecture, Department of Environmental Engineering, Canakkale, Turkey

Alvina Gul Kazi Atta-ur-Rahman School of Applied Biosciences, National University of Sciences and Technology

Gia Khatisashvili Durmishidze Institute of Biochemistry and Biotechnology at Agricultural University of Georgia, Laboratory of Biological Oxidation, Tbillisi, Georgia

Nurcan Koleli Mersin University, Faculty of Engineering, Department of Environmental Engineering, Mersin, Turkey

Manoj Kumar Amity Institute of Microbial Technology, Amity University, Noida, Uttar Pradesh, India

Kadir Kusvuran Alata Horticultural Research Station Directorate, Erdemli, Turkey

Bisma Malik Department of Bioresources, University of Kashmir, Srinagar, India

Lia Matchavariani Department of Soil Geography Faculty of Exact & Natural Sciences, Tbilisi State University of Iv. Javakhishvili, Tbillisi, Georgia

Ahmet Ruhi Mermut Department of Soil Sciences, University of Saskatchewan, Canada; Harran University, Agriculture Faculty, Soil Science Department, Şanlıurfa, Turkey

Muhammad Nadeem Department of Environmental Sciences, COMSATS Institute of Information Technology (CIIT), Vehari, Pakistan

Kamrun Nahar Laboratory of Plant Stress Responses, Department of Applied Biological Science, Faculty of Agriculture, Kagawa University, Miki-cho, Kita-gun, Kagawa, Japan; Department of Agricultural Botany, Faculty of Agriculture, Sher-e-Bangla Agricultural University, Sher-e-Bangla Nagar, Dhaka, Bangladesh

Ullah Najeeb Department of Plant and Food Sciences, Faculty of Agriculture and Environment, The University of Sydney, NSW, Australia

Woranan Nakbanpote Department of Biology, Faculty of Science, Mahasarakham University, Khamriang, Kantarawichi, Mahasarakham, Thailand

Münir Öztürk Department of Botany, Ege University, Izmir, Turkey; Faculty of Forestry, Universiti Putra Malaysia, Selangor, Malaysia; ICCBS, Karachi University, Pakistan

Ibrahim Ilker Ozyigit Marmara University, Faculty of Science & Arts, Department of Biology, Goztepe, Istanbul, Turkey

Natthawoot Panitlertumpai Department of Biology, Faculty of Science, Mahasarakham University, Khamriang, Kantarawichi, Mahasarakham, Thailand

Chaiwat Phadermrod Padaeng Industry Public Co. Ltd, Phratad Padaeng, Mae Sot, Tak, Thailand

Tanveer Bilal Pirzadah Department of Bioresources, University of Kashmir, Srinagar, India

M.N.V. Prasad Department of Plant Sciences, University of Hyderabad, Hyderabad, India

Umer Rashid Institute of Advanced Technology, Universiti Putra Malaysia, Serdang, Selangor, Malaysia

Syed Hammad Raza Department of Botany, Government College University, Faisalabad, Pakistan

Reiaz Ul Rehman Department of Bioresources, University of Kashmir, Srinagar, India

Z.R. Rehman Institute of Soil and Environmental Sciences, University of Agriculture, Faisalabad, Pakistan

Muhammad Rizwan Department of Environmental Sciences, Government College University, Faisalabad, Pakistan

Qazi Mohd. Rizwanul Haq Department of Biosciences, Jamia Millia Islamia, New Delhi, India

Muhammad Sabir Institute of Soil and Environmental Sciences, University of Agriculture, Faisalabad, Pakistan; School of Plant Biology, University of Western Australia, Crawley, WA, Australia

SeyedMousa Sadeghi   Faculty of Forestry, Universiti Putra Malaysia, Serdang, Selangor DarulEhsan, Malaysia

Saifullah   Institute of Soil and Environmental Sciences, University of Agriculture, Faisalabad, Pakistan

Sayeda Sarah Saleem   Atta-ur-Rahman School of Applied Biosciences, National University of Sciences and Technology

S. Satter   Department of Environmental Sciences, Jahangirnagar University, Savar, Dhaka, Bangladesh

Abin Sebastian   Department of Plant Sciences, University of Hyderabad, Hyderabad, India

Fahad Shafiq   Department of Botany, Government College University, Faisalabad, Pakistan

Muhammad Shahid   Department of Environmental Sciences, COMSATS Institute of Information Technology, Vehari, Pakistan

Mohsen Soleimani   Department of Natural Resources, Isfahan University of Technology, Isfahan, Iran

Mahfuza S. Sultana   Department of Environmental Sciences, Jahangirnagar University, Savar, Dhaka, Bangladesh

Orooj Surriya   Atta-ur-Rahman School of Applied Biosciences, National University of Sciences and Technology

Inayatullah Tahir   Department of Bioresources, University of Kashmir, Srinagar, India

Safi M. Tareq   Department of Environmental Sciences, Jahangirnagar University, Savar, Dhaka, Bangladesh

Farhad Hosseini Tayefeh   Faculty of Forestry, Universiti Putra Malaysia, Serdang, Selangor DarulEhsan, Malaysia

Ajit Varma   Amity Institute of Microbial Technology, Amity University, Noida, Uttar Pradesh, India

Kinza Waqar   Atta-ur-Rahman School of Applied Biosciences, National University of Sciences and Technology

Ejaz Ahmad Waraich   Department of Crop Physiology, University of Agriculture, Faisalabad, Pakistan

Ibrahim Halil Yanardağ   Sustainable Use, Management and Reclamation of Soil and Water Research Group, Agrarian Science and Technology Department, Technical University of Cartagena, Cartagena, Murcia, Spain

S. Yeasmin   Chemistry and Health Physics Division, Atomic Energy Centre, Dhaka, Bangladesh

Munir Hussain Zia   Research and Development Section, Fauji Fertilizer Company Limited, Rawalpindi, Pakistan

Muhammad Zia-ur-Rehman   Institute of Soil and Environmental Sciences, University of Agriculture, Faisalabad, Pakistan

Chapter 1

Phytoremediation of Soils: Prospects and Challenges

Orooj Surriya, Sayeda Sarah Saleem, Kinza Waqar and Alvina Gul Kazi

Atta-ur-Rahman School of Applied Biosciences, National University of Sciences and Technology

INTRODUCTION

The rampant increase in the human population with every passing year has led to the clearing of different land forms to make space for urbanization. The gargantuan development of urbanization in turn has led to the aggravation of land, water and air pollution. Land pollution, also commonly referred to as soil pollution, is one of the greatest hazardous concerns of the twenty-first century. Soil is a non-renewable resource, interconnecting numerous anthropogenic activities with environmental ones by acting as a platform. The degree of chemical contact of industrial and other harmful effluents with the soil determines the level of disruption caused to either the surface soil or underground soil.

Types of Soil Pollutants

Fertilizers and Pesticides

About 11% of Earth's total land area is considered arable land for the cultivation of crops. To meet the inevitable increasing food demands made by the growing human population, fertilizers are used to increase crop yields. Most fertilizers contain potassium compounds, phosphorus compounds and ammonium nitrate, which may harbour minute traces of non-degradable metals such as lead and cadmium. These metals accumulate in large quantities and settle in soil particles. This accumulation greatly reduces the levels of vitamin C and carotene in fruit and vegetable crops harvested from the contaminated soil.

Similarly, large exudates of the pesticides sprayed on plant crops for weed and insect removal accumulate on the soil surface, where they have a greater chance of entering the human gastrointestinal (GI) tract via infected crop plants. Common pesticides that contaminate the soil are dichlorodiphenyltrichloroethane, chlorinated hydrocarbons, malathion, aldrin, furadon and organophosphates (Wokocha and Ihenko, 2010). In China, food crops such as groundnuts,

Soil Remediation and Plants. http://dx.doi.org/10.1016/B978-0-12-799937-1.00001-2

coca, rice and mustard were found to contain augmented levels of copper and zinc, most likely contributed by the widespread use of pesticides containing these compounds (Luo et al., 2009).

Municipal Waste

Solid wastes from domestic use, industry and various other commercial practices are dumped on large areas of ground with partially fertile soil, serving as host to many toxic volatile compounds such as chlorofluorocarbons. This toxicity disseminates to neighbouring soils, thereby spreading the contamination.

Heavy Metals

Transition metals, lanthanoids, actinoids and metalloids are all high-density metals belonging to the group of heavy metals. In high concentrations, heavy metals render serious toxicity to plant roots. Industrial practices that contribute to heavy metal soil accumulation include mining, petroleum gas leaks and leaded paints.

Flying Ash

Large quantities of metallic pollutants from anthropogenic activities are released into the clean atmosphere. These pollutants, referred to as flying ash, are derived from vehicle exhausts, coal burning, mining and electrical power and contribute greatly to soil contamination via absorption and settlement on the soil surface. In a study conducted in China, it was discovered that 43–85% of metallic pollutants, such as lead, arsenic, zinc, mercury and cadmium, resulted from atmospheric deposition (Luo et al., 2009).

Wastewater Irrigation

Water pollution is concomitant with land and air pollution owing to the increase in areas of commercial development. Polluted water acquired from rivers, tributaries and the water table for irrigation purposes also leads to soil contamination. Sewage dumping alongside river banks and oceans in densely populated countries brings toxic materials into contact with the soil, which find their way through alternative routes into human consumption. In northern Greece major rivers and their tributary systems were tested for 3 years and the results showed contamination with the elements Silver(Ag), Cadmium(Cd), Boron(B), Arsenic(As), Barium(Ba), Mercury(Hg), Copper(Cu), Nickel(Ni), Iron(Fe), Lead(Pb), Manganese(Mn), Selenium(Se), Zinc(Zn) (Farmaki and Thomaidis, 2008).

The Global Scenario of Soil Pollution

Europe

Pollution of both land and water caused by contamination with persistent heavy metals is a growing threat in both developed and developing countries. Studies in Western Europe have shown that approximately 1,400,000 land areas have been found to contain heavy metals; of this total area, 300,000 sites were identified as contaminated (McGrath et al., 2001). Recent reports and surveys have gathered numerous data on various European countries with large contaminated areas. The United Kingdom, Germany, Spain, Denmark, Belgium, Finland, Italy and The Netherlands are European countries with at least 400,000 contaminated land sites (Perez, 2012), whereas France, Sweden, Hungary, Austria and Slovakia belong to the group of European countries with at least 200,000 polluted land areas. Poland and Greece had more than 10,000 polluted land sites, whereas Portugal and Ireland were reported to have fewer than 10,000 polluted sites (Perez, 2012).

Asia

Disposing of untreated sewage and industrial waste in nearby drains is a common practice in countries such as India, Pakistan, Bangladesh and Sri Lanka. There are not enough treatment plants and programmes to dispose of the harmful effluents safely, and as a consequence the practice of dumping solid wastes in clean water drains is widespread (Lone et al., 2008). In China, one-sixth of the total agricultural land area has been contaminated by heavy metals, and about 40% was reported to be disrupted by excessive deforestation and erosive activities (Liu et al., 2005). In a study conducted in China, it was calculated that cultivated croplands irrigated by contaminated water totalled 7.3% of the total irrigated land area. The amount of polluted water is also reported to have got out of control in China (Luo et al., 2009).

In recent years, rice paddy fields in Korea have been found to be contaminated with the heavy metals Zn, Pb, Cd and Cu in concentrations up to $0.11\,mg\,kg^{-1}$. Similarly, in Japan, concentrations in rice paddy fields contaminated with similar heavy metals were estimated to be $75.9\,mg\,kg^{-1}$, $3.71\,mg\,kg^{-1}$ and $22.9\,mg\,kg^{-1}$.

America

A study by McKeehan estimated that about 600,000 brownfield sites in the United States are contaminated with heavy metals. A report released by United States Department of Agriculture in 2003 highlighted that most of the water and land pollution in the United States was accounted for by the disposal of 'big waste' from poultry farms. In 2007 the number of chicken broilers exceeded 200 million, leading to an alarming increase in water and land pollution via unchecked chicken waste disposal.

Pacific Islands

Cook Islands

A study by Convard and Nancy in 2005 revealed that 9000 cubic meters of solid waste is dumped into nearby land sites and water drains, resulting in colossal land and water resources being heavily contaminated. Increasing tourism, if continued at the prevailing rate, will also increase pollution 10-fold in coastal areas (Convard et al., 2005).

Fiji, Kiribati, Nauru, Marshall Islands and Niue

Anthropogenic activities such as the use of fertilizers, herbicides, pesticides, fossil fuel combustion and tourism release huge amounts of solid wastes that are carelessly dumped into the marine environment. The contaminated water is then used for irrigating agricultural areas (Convard et al., 2005).

Effects of Soil Pollution on Human Health and Environment

All types of pollutants have an effect on animals, plants, humans and the environment. Heavy metals pose the greatest risks of harm to life. Since they are not biologically degradable, heavy metals can only be oxidized from one state to another. Thus their persistent existence in nature poses the most serious concern of all pollutant types.

Human consumption of heavy-metal-intoxicated plants can lead to carcinogenic disorders and other chronic diseases. Cadmium and zinc, when consumed in large quantities, can cause respiratory and GI disorders, as well as damage to heart, brain and kidneys. Heavy metals in high concentrations can also adversely affect plant crops. Stunted growth, poor yield and aberrations in metabolic functions such as respiration and photosynthesis can result from heavy metal toxicity in plants (Garbisu and Alkorta, 2001; Schwartz et al., 2003). Heavy metal contamination can also alter the microbial composition of the soil, which can eventually destroy its biochemical properties, such as fertility (Kozdrój and van Elsas, 2001; Kurek and Bollag, 2004) (Figure 1.1). Therefore, over recent decades continuous efforts have been dedicated to developing a technology that can sustain the intrinsic natural properties of soil with minimal economic and environmental damage.

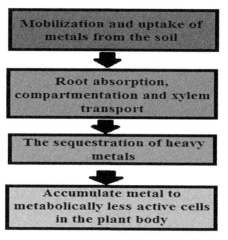

FIGURE 1.1 The hyperaccumulation of heavy metals involves bioactivation in the rhizophere, root absorption, xylem transport and finally distribution and sequestration.

TECHNOLOGIES FOR SITE REMEDIATION

Contaminated site(s) may require a series of different procedures in order to restore the natural integrity of the soil to its maximum degree. Therefore, over the decades soil remediation procedures have been split into three combinations: physical, chemical and biological.

Electroremediation

This technique is not in widespread use for the removal of contaminants, as in many areas it is still under development. It works on the principle of attracting charged particles within the soil. Two electrodes, a cathode and an anode, are inserted in the contaminated soil at two opposite sites and current is made to flow across the soil. The magnetic field that is generated propels the ionic heavy metals present in the soil towards the respective electrodes. Once the metal particles arrive at the electrodes, they are extracted via attachment to ion-exchange resins or adsorption to the electrodes themselves. However, if the target soil bears excessive heterogeneity, this might hinder the efficiency of the process. Electroremediation is successful in removing a wide range of heavy metals (Lindgren et al., 1994).

Soil Flushing

Soil flushing is an *in situ* chemical method of soil remediation. It involves the extraction of heavy metals via a fluid injected into the contaminated soil. The extraction fluid is pumped to the surface, which brings along the absorbed contaminants with itself. The extraction fluid is made by the liquefaction of various gases such as propane, carbon dioxide and butane. Soil flushing works on all types of soil pollutants, usually in combination with other techniques. Only soil types that contain spaces large enough to allow the extraction fluid to seep through the soil particles can be purified using this technique (Di Palma et al., 2003; Boulding, 1996). See Figure 1.2 for an outline of the basic procedure of soil flushing.

Soil Vapour Extraction

As its name implies, this technique is used for the removal of volatile organic compounds through evaporation. It works by building vertical or horizontal walls within the contaminated soil site through which a vacuum is blown into the contaminated areas, to permit the evaporation of volatile pollutants. An extraction well is placed at one end of these wells to extract the evaporated pollutants. The pollutant gases are treated before disposal into the atmosphere. Soil vapour extraction can only be applied to soil types that have considerable amounts of permeability. Heavy pollutants such as kerosene and diesel oils are very poorly – or in some cases never – removed via this technique (Barnes et al., 2002; Park and Zahn, 2003).

Stabilization

Reducing the toxicity, mobility and solubility of the contaminant in order to minimize its risk to neighbouring areas is known as stabilization (Anderson and Mitchell, 2003). Stabilization of waste pollutants can be achieved by both chemical and physical means.

Asphalt Batching

This is a chemical method that is used for the treatment of hydrocarbon contaminants. It involves adding a petroleum-containing soil to a hot bitumen mixture. The resultant mixture forms an aggregate that is further treated to extract soil contaminants. The aggregate is thermally treated, which causes the volatile compounds to volatize. The remaining asphalt mixture is cooled to restrict the dissemination of the left-over contaminants (Alpaslan and Yukselen, 2002).

Vitrification

Vitrification is a technique that involves the application of high temperatures (1600–2000°C) to melt the soil and the pollutants within it, thereby blocking the migration of harmful constituents to non-polluted areas (Khan et al., 2004). Vitrification of soil pollutants is carried out by three different procedures:

- Thermal: uses heat from an external source with a reactor.
- Electrical: uses the insertion of graphite electrodes to provide heat in the form of electrical energy.
- Plasma: highly favourable for achieving high temperatures to about 5000°C (Khan et al., 2004).

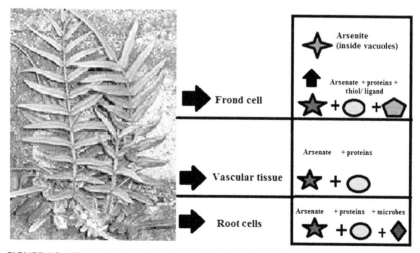

FIGURE 1.2 The mechanism of arsenic uptake in *Pterisvittata*. The arrows show movement and transformation of arsenic inside the cells.

Soil Washing

Soil washing refers to the use of solvents such as water to wash the contaminated soil so as to separate fine soil from its larger constituents such as gravel and sand. Studies have shown that hydrocarbons have a tendency to cling to smaller soil parts such as clay. Hence separating larger soil particles from the smaller ones can help achieve soil remediation (Riser-Roberts, 1998). The separated small soil particles are then treated to obtain complete purity. Solvents are chosen for their solubilizing ability and environmental effects.

Bio-piles

Bio-piles are a biological method of soil remediation that fall under the heading biodegradation. In this method, stacks or piles of soil containing concentrations of petroleum are separated. The separated heaps are aerated with microbes for biodegradation at an optimum temperature and pH. This is an effective method for removing petroleum, volatile organic compounds and pesticide chemicals, and is also easy to design. However, it is a short-term technology that may last only weeks or months. To achieve 100% efficacy it should be used in combination with the aforementioned techniques (Filler et al., 2001).

Over decades of using physical and chemical methods for soil remediation, experts and research groups around the world have unanimously agreed on several important drawbacks of these methods. Physical and chemical methods are considered not only cumbersome but extremely costly, not applicable in developing countries, and capable of inducing secondary damage, for example, to ecology and the economy (Saier and Trevors, 2010). Thus there was a deep-felt need for an alternative technology that would be cost-effective and environmentally friendly, one that could provide reliable efficacy and relatively fewer limitations to worry about. One such alternative still considered an emerging technology is phytoremediation.

PHYTOREMEDIATION

The term phytoremediation is derived by fusion of the Greek word *phyto* (meaning 'plant') with the Latin word *remedium* (meaning 'cure of evil') and was first coined in 1994 by Ilya Raskan (Vamerali et al., 2010). Phytoremediation is a technique that uses various plant species to facilitate soil or water reclamation (Ali et al., 2012). The method emerged as a result of remarkable material demonstrating the extracting, metabolizing and accumulating features of plants (Bollag et al., 1994). Studies have shown that plants react to the presence of soil contaminants in many ways, namely, by accumulating them, indicating their presence, and/or extruding them out to the surface (Baker et al., 1988). As autotrophs, plants store pollutants in their vegetative areas and aid in the removal of undesirable content from their environment (Cunningham et al., 1995). These salient features of plants that favour environmental safety as well as remediation efficiency have made phytoremediation the leading technology of choice for soil reclamation purposes (Chaudhry et al., 1998). Being cheaper than the physical and chemical methods, phytoremediation requires a technical design strategy by professionals that have ample in-field experience to guide in the selection of proper plant species to use according to the kind of metal, land and climate (Pollard et al., 2002).

Phytoremediation System Design

When designing an appropriate system of phytoremediation, different aspects of certain parameters must be thoroughly considered. These parameters include the choice of contaminant to be treated, the characteristics of the chosen contaminants, the selection of plant species and land type and other biotic and abiotic conditions affecting the process of phytoremediation (Mudgal, Madaan, & Mudgal, 2010).

Consideration of Type of Pollutants

Phytoremediation has a wide range in terms of removal of soil contaminants. Pollutants ranging from heavy metals to volatile organic and heavy organic compounds can be easily treated with this technology (Henry, 2000).

Organic Pollutants

The degree of hydrophobicity of an organic contaminant greatly alters the manner and efficiency of its uptake by the plant. Moderately hydrophobic pollutants are readily taken up and translocated by the plant (Cunningham et al., 1997).

Inorganic Pollutants

The ease and efficacy of removing one metal compared to a mixture of metals may be different for phytoremediation. Over the years a list of extractable metals has been generated in order of least to most easily achieved: Cr, Cd, Ni, Zn, Cu, Pb and Cr (Dushenkov et al., 1995).

Concentration of Pollutants

Before using plants to extract soil pollutants, the concentration of the pollutants must be estimated, because high levels can pose the risk of rendering healthy plants toxic and completely damaging them. In some laboratory specimen experiments high concentrations of contaminants were readily tolerated by plants, unlike microorganisms (Miller et al., 1985).

Pollutant Characteristics

Pollutant type is a serious consideration prior to plant type selection. Pollutants in the form of 'light non-aqueous phase' or 'dense non-aqueous phase' liquid mixtures can significantly reduce plant growth. Pollutant compounds that have been present in the soil for ages have very low bioavailability, and it is therefore difficult to extract them using phytoremediation (EPA, 2000).

Consideration and Selection of Plant Species

It requires geologists and plant engineers, rather than ordinary farmers, to design the set-up of a plant biological system suitable for phytoremediation in a given area. Ample information should be gathered regarding the candidate plant before it can be finalized for use. Certain characteristics must be checked during the process of plant selection.

Type of Root

There are two types of root systems present in plants: fibrous and tap root. Fibrous roots provide a greater contact area with the soil and hence a larger degree of pollutant extraction can be carried out. Tap roots involve a central large root extending towards the bottom of the soil. They may also be efficient in absorption (Schwab, 1998).

Depth of Roots

The depth of roots varies according to the species of plant. Certain intrinsic and external factors, such as soil structure, depth of soil water and cropping pressures, may also affect root depth (EPA, 2000). The desirable root depth for non-woody plants has been estimated to be 1–2 feet (30.48–60.96 cm), whereas for tree roots, a depth of less than 10–20 feet (304.8–609.6 cm) is effective (Gatliff, 1994).

Plant Growth Rate

The effect of growth rate is directly proportional to the efficacy of phytoreme-diation. However, it varies for different types of phytoremediation. Some types require a fast growth rate for the part of plant above the soil, whereas others require a fast growth rate for the root mass. A faster growth rate reduces the time taken by the plant roots to extract large masses of heavy metals (EPA, 2000).

Rate of Transpiration

Transpiration rate is an important factor in cases when contaminant uptake is involved during phytoremediation. Plant species, age, size, climatic conditions, size and surface area significantly affect the rate of transpiration (EPA, 2000).

Seed and Plant Source

Verifying the seed source is imperative before commencing the phytoremedia-tion process. The authenticity and whereabouts of the seed supplier must be thoroughly checked. Mostly seeds from local regions are preferred, because it is easier for the plant to adapt to the environment. The seeds purchased must be in good health. Seed quality should be thoroughly scrutinized. Any disease, infection or contamination in the purchased seed will only hamper the process rather than help in soil reclamation (EPA, 2000).

Allelopathy

The production of chemicals by one plant species to inhibit the growth of another plant species in its vicinity is known as allelopathy. When more than one type of plant species are grown adjacent to one another, it is important to scan and investigate the possible amounts of chemicals secreted by one species. Some plants exhibit allelopathic effects that tend to increase soil fertility; for instance, canola secretes residues from its leaves, stem and roots that obliterate the growth of wheat, corn and barley. This feature could be put to a wider use, as it can give an insight to which kinds of plants produce residues that facilitate the increase of soil fertility.

Plant Type

A number of plant types have been investigated that are commonly and efficiently used for phytoremediation purposes. More than 400 plant species worldwide have been identified that have the ability to extract heavy metals from the soil (Baker et al., 2000). Some commonly investigated plants suitable for phytoremediation are listed in Table 1.1.

Many plant species have been explored to filter out those capable of accumulating heavy metals. Among the most researched are *Thlaspi* sp., *Arabidopsis* sp. and *Sedum alfredi*. The genus *Thlaspi* is recognized for accumulating more than one type of heavy metal, for instance, *Thlaspi caerulescens* for Zn, Pb, Cd and Ni; *Thlaspi goesingense* for Zn and Ni; *Thlaspi ochroleucum* for Ni, Zn and T; and *Thlaspi rotundifolium* for Zn, Pb and Ni (Prasad and Freitas, 2003). Studies have shown that *Thlaspi caerulescens* possesses the remarkable ability to hyperaccumulate $8.4\,kg\,Cd\,ha^{-1}$ and $60\,kg\,Zn\,ha^{-1}$. It can also accumulate 2600×10^{-6} Zn with considerable tolerance.

Thlaspi caerulescens stores zinc residues in a soluble form inside the vacuoles of epidermal cells. In some plants, such as *Arabidopsis halleri,* zinc residues are sequestered in mesophyll cells.

Other plant types applicable for phytoremediation purposes include rice, sugarcane, tobacco and soybean (EPA, 2000). Minute plant characteristics are not specific in general, but differ for each phytoremediation type.

Climate Considerations

Climatic conditions such as temperature, weather, water availability from rainfall, sunlight and precipitation levels greatly influence seed germination and plant growth. Preparations for climatic variation are a crucial step prior to the establishment of the soil remediation process. Dry spells may require the setting

TABLE 1.1 List of Common Plants Known to Be Suitable for Phytoremediation

Trees	*Poplars* *Willows*
Grasses	*Prairie grasses* *Fescue*
Legumes	*Alfalfa*
Aquatic Plants	*Parrot feather* *Phragmites reeds* *Cattails*
Accumulators	*Sunflower*
Metal Accumulators	*Thlaspi caerulescens* *Brassica juncea*

up of perennial irrigation networks. Heavy winds may affect the evaporation rate of plants, and shade from nearby vegetation or buildings can reduce the amount of sunlight available to the growing plant. All such factors contributing to climatic changes must be kept in check for effective results from phytoremediation (EPA, 2000).

Economics of Phytoremediation

Phytoremediation is an emerging technology that has overshadowed earlier physical and chemical technologies used for soil reclamation, particularly because of its economic, industrial and commercial benefits. However, certain factors influence the fluctuating economics of phytoremediation. The system costs of phytoremediation can be divided into operation costs, design costs and installation costs.

In 1998, however, the total systems cost of phytoremediation was calculated and was shown to be 50–80% lower than that of previous methods used for soil reclamation. Although the total cost of each individual phytoremediation application is different from the others, a general total cost for phytoremediation compared to conventional techniques has been estimated. The cost estimate for the removal of lead from soil by conventional means is twice the amount spent on phytoremediation.

Considerations for Waste Disposal

Disposing of the accumulated waste in the cultivated area is essential to prevent circumstances that may block further reclamation of the soil. However, the removal of biomass depends greatly upon the type of phytoremediation applied (EPA, 2000). In systems where plants that require long growth periods are used, the periodic removal of biomass is not necessary (EPA, 2000). Damaged or diseased plant parts, and fallen plant parts such as leaves and twigs, have to be removed periodically in order to ensure proper functioning of the system. They may even have to be tested for the presence of any contamination and, if found to be contaminated, they are disposed of off-site (EPA, 2000).

If, after verification, the plant biomass proves not to be contaminated it can be reused as a cash crop (EPA, 2000). For proper disposal, disposal facilities are constructed nearby or at a considerable distance from the site.

Phytoremediation Technologies

The technology of phytoremediation is subcategorized into five types.

Phytoextraction

Phytoextraction was developed decades ago to translocate metal contaminants from the soil to the ground surface via the root systems of plants (Brooks, 1998), and is most efficiently used for the removal of heavy metals (EPA, 2000). Different species of plants show varying morphophysiological responses to soil contamination. The metal biological absorption coefficient (BAC) determines the capacity of a given plant to accumulate metal: it is the ratio of plant to soil metal concentration. An efficient translocation factor (i.e. shoot to root metal concentration ratio) alongside a good BAC can have a significant effect on the functioning of phytoextraction. Mostly plant species that are not tolerant and which have a BAC factor > 1 work efficiently for phytoextraction purposes.

A plant species which is eminently suitable for phytoextraction must have a number of desirable characteristics, such as rapid growth, a high translocation factor, high biomass, a good assimilation rate, high tolerance to large amounts of metals, a vast root system and easy growth and harvesting management.

Heavy metals that can be removed via phytoextraction include Cr, Cd, Cu, Co, Ag, Zn, Ni, Mo, Pb and Hg (EPA, 2000). The concentration of heavy metals in soils depends upon the site and pollution status of a country. Plants used for phytoextraction are commonly termed hyperaccumulators. Hyperaccumulators are famous for their slow growth, and a shallow root system and low biomass accompany this feature (EPA, 2000).

The plant families Asteraceae, Brassicaceae, Lamiaceae, Euphorbiaceae and Scrophulariaceae include plants that can be employed for phytoextraction (Baker et al., 1988).

Rhizofiltration

The technique of rhizofiltration is used for the remediation of waste water by aquatic or land plants (Henry, 2000). Pb, Cd, Cu, Ni, Zn and Cr can be extracted using rhizofiltration. A few examples of plants that are employed for rhizofiltration are sunflower, tobacco, spinach, rye and Indian mustard. Plant species besides hyperaccumulators can also be used, as the heavy metals need not be translocated to the shoots (Henry, 2000). Terrestrial plants are widely preferred for rhizofiltration as they have fibrous root systems with fast growth.

The rhizofiltration technique can be constructed either as floating rafts on ponds or as tank systems. One of the main disadvantages of rhizofiltration involves growing plants in a greenhouse first and then transferring them to the remediation site. Great care must be taken to maintain an optimum pH in the effluent solution.

Phytostabilization

Phytostabilization is the restriction of contaminant mobility and bioavailability in soil. The secondary purpose of phytostabilization is to suppress the migration of soil contaminants via water and wind erosion and leaching. The main function of this technique also revolves around root-zone microbiology and chemistry. Phytostabilization works on the principle of hindering metal mobility and altering metal solubility. It confers certain changes to the soil environment that leads to the insolubility of metals present in an oxidative state (Aggarwal and Goyal, 2007). Soil contaminated with As, Cd, Cu, Pb and Cr can be remediated by phytostabilization.

Phytodegradation

Phytodegradation is the term used to describe the breakdown of metal contaminants by plant enzymes following uptake from the soil (EPA, 2000). Before the contaminants can be degraded, however, they must first be taken up by the plants. Uptake of heavy metals depends largely on plant type, age and size, and the chemical characteristics of the soil. A wide variety of contaminants have been identified that can be metabolized. These include the herbicide atrazine, munitions trinitrotoluene (Thompson et al., 1998) and the chlorinated solvent trichloroethane.

The plant enzymes dehalogenase, nitrilase, phosphalase, nitroreductase and oxidoreductase are most commonly involved in phytodegradation (EPA, 2000). Plant species such as yellow poplar, black willow, live oak, river birch, bald cypress and cherry bark have been shown to exhibit phytodegradation of certain herbicides. Phytodegradation works best in soil areas with shallow contamination.

Phytovolatilization

Phytovolatilization is the use of plants to extract soil contaminants and then transform them into volatile substances out to the atmosphere. This technique converts the metal contaminant into its less toxic elemental form; for example, mercuric mercury is one of the chief metal compounds that are obliterated by phytovolatilization. Heavy metals such as Se and Hg and organic solvents such as carbon tetrachloride and trichloroethane have been eliminated using phytovolatilization. Since this technique involves the use of transpiration on parts of plants, its efficiency is greatly influenced by climatic conditions.

Many varieties of plant species have been studied for their ability to volatilize organic and inorganic pollutants. The species black locust, alfalfa, Indian mustard and canola have shown laudable performances in phytovolatilization (EPA, 2000).

HEAVY METAL SOIL POLLUTANTS AND USE OF PHYTOREMEDIATION

The main inorganic soil pollutants are the heavy metals, which cause significant areas of land to become contaminated. The main reason for this enhanced toxicity is the constant use of fertilizers and pesticides, sludge, car exhausts, smelting industries, the residues emerging from metal mines and emissions from incinerators (Garbisu and Alkorta, 2001; Halim et al., 2003; Yang et al., 2005). Even though Earth's crust contains diverse naturally occurring metals, they are toxic in high concentration, especially where they are unwanted, and can affect human health and the environment, leading to increased death rates and poor agricultural products (McIntyre, 2003; Yang et al., 2005).

Approach to the Removal of Heavy Metal Soil Pollutants

It is essential to remove all these dangerous and persistent contaminants effectively, thereby allowing polluted areas to be restored. There are many approaches consisting of physical, chemical and biological techniques. Mainly they are divided into two classes (Ghosh and Singh, 2005; Kotrba et al., 2009).

Ex Situ Method

This process helps in decontamination via the destruction of contaminants either chemically or physically. The pollutant is destroyed, stabilized, immobilized or solidified, and the treated soil is restored to its natural state.

In Situ Method

This process helps in decontamination without the need to dig the contaminated site. The pollutant is destroyed or transformed to reduce its bioavailability, and can thus can be separated from the bulk soil. These techniques are better than the previous ones due to low cost and environmental friendliness.

Phytoremediation is included in the *in situ* method of soil detoxification and has gained increasing attention from ecologists, evolutionary biologists and plant physiologists over the last 50 years. The use of plants to reduce the spread of heavy metals in the soil is beneficial because a plant is a solar energy–driven pump, and can grow in low-nutrient conditions to extract a definite number of specific heavy metals. The process of phytoremediation involves various techniques, viz. phytovolatization, phytostabilization, rhizofiltration and phytoextraction, as discussed above (Yang et al., 2005; Kotrba et al., 2009).

Hyperaccumulation is the ability of plants to take up heavy metals from the soil to a level higher than the substrate soil. It could be 50–500 times greater (Clemens, 2006) and at least 100 times greater (Brooks, 1998). About 450 plant species from 45 families having the property of hyperaccumulation are currently being used. The reason for hyperaccumulation is unclear, but it might be a protection mechanism by which plant tissues become poisonous to pathogens or herbivores. Usually plants that have this feature are slow-growing and hence produce less biomass, which does not help much to fulfill the requirement of the process (Boyd, 2007; Kotrba et al., 2009).

Ideal Plant Characteristics for Phytoremediation

An ideal plant for removing heavy metals from the soil should have the following properties:

- an inborn capacity for hyperaccumulation and tolerance of heavy metals;
- fast growth of plant and biomass;
- a well-developed root system that is widely branched;
- easy to develop and cultivate;
- an extensive geographical distribution;
- easy to harvest;
- susceptible to genetic manipulation.

Such plants are also called as metallophytes. Some well-established metallophytes were produced via genetic manipulation, and so appear to be good specimens of genetically engineered plants for phytoremediation, for example, *Helianthus annuus* (sunflower), *Nicotiana glaucum* and others (Kotrba et al., 2009).

Basic Process of Hyperaccumulation

The main process of hyperaccumulation of metals involves the following (see also Figure 1.1) (Kotrba et al., 2009):

- Mobilization and uptake of metals from the soil;
- Sequestration of heavy metals by the formation of metal complexes and their accumulation in the vacuoles in order to carry out detoxification in plant roots;

- The ability to translocate metals to shoots through the apoplast or symplast pathway, along with the efficacy of loading of xylem
- The distribution of plant organs and tissues above ground
- The sequestration of heavy metals inside the tissues, at the cellular level
- Finally, ejection of the accumulated metal to metabolically less active cells in plant body

The use of plants to extract heavy metals such as arsenic, lead, zinc, copper, cadmium and mercury is explained in the next part of this chapter.

ARSENIC

Arsenic is known as 'the king of poisons', as it has affected human history in ways no other element has ever done, and caused the death of a significant number of human beings. It is the cause of poisoning epidemics all around the world.

The Health Hazards of Arsenic

Arsenic is a carcinogen that contaminates the environment when present in high concentrations. The level of toxicity depends on the chemical form and solubility of arsenic. It has two important inorganic forms, arsenate and arsenite; these are the main contaminants as stated by United States Environmental Protection Agency (U.S. EPA). In the past it was seen that about 52,000–112,000 tons per annum of arsenic was contaminating the soil. This resulted in other problems, the most important being leaching of arsenic into drinking water, causing a health hazard to humans. The main reason for the increasing amount of arsenic is anthropogenic activities. Table 1.2 shows some of the leading sources of arsenic contamination (Rathinasabapathi et al., 2006; Butcher, 2009).

Mild arsenic exposure causes diarrhoea, vomiting and abdominal pain, whereas long exposure is responsible for skin darkening, corns, cardiac and respiratory effects (oral poisoning). It was seen that the cancer-causing ability of arsenic is due to dimethyl-arsinous acid and mono-methyl-arsonous acid, which are genotoxic,

TABLE 1.2 Commercial Products Having a Large Content of Arsenic

Arsenic compound	Uses
Sodium arsenite Monosodium methanearsonate (MMSA) Cacodylic acid (CA)	Herbicides
Lead arsenate Calcium arsenate Zinc arsenate etc.	Pesticides
Arsenic pentoxide	Weed killers
Arsenic disulphide	Leather industry, paints
Poultry feed additive	Roxarsone

leading to the inhibition of DNA repair systems. The concentration of arsenic on the surface of soil is greater than in the deep layers. Therefore, strategies are needed to extract it from the soil in order to negate its harmful effects (Butcher, 2009).

Use of Phytoremediation to Remove Arsenic

The most useful and cost-effective way to remove arsenic in the soil is phytoremediation, in which plants are used as weapons to solve the contamination problem as other methods are quite expensive. The two most effective methods of carrying out this process are phytoextraction and phytostabilization. Of all the plants, the Chinese fern *Pterisvittata* is significant as it has the ability to hyperaccumulate 2500–22,630 mg kg^{-1}dry weight of arsenic. This plant doubles its biomass in a week if subjected to 100 ppm arsenic (Alkorta et al., 2004; Rathinasabapathi et al., 2006).

The mechanism of *Pterisvittata* for the transport of arsenic is shown in Figure 1.2. The transport of arsenic is aided by certain proteins and microbes present in the soil. A study also notes that the phosphate transport pathway might be involved in arsenic transport because of the structural similarity of the two elements (Alkorta et al., 2004). The arsenic is transported as arsenite using the particular transport protein. It is later reduced to arsenate, which is present in the form of a protein complex inside the vacuoles of plant cells to avoid reoxidation (Alkorta et al., 2004; Rathinasabapathi et al., 2006).

The efficiency of *Pterisvittata* can be further increased in three main ways. This can be achieved by understanding the factors that increase the rate of plant growth, the uptake of arsenic and its hyperaccumulation. The use of this plant helps in soil decontamination through the following steps (Rathinasabapathi et al., 2006; Butcher, 2009):

- the growth of *Pterisvittata* in the area where arsenic is contaminating the soil
- the collection of contaminated biomass from the plants
- the collected biomass is pre-treated, which involves compaction, composting and pyrolysis
- finally, direct disposal of the contaminant, or incineration.

These plants can also be used to extract arsenic from ground water, via optimization of the rhizofiltration process. This process involves the following steps, as shown in Figure 1.3; this is basically a suggested model (Alkorta et al., 2004).

- The ground water is pre-treated by various methods, e.g. the use of nano filtration membranes. The latest one is by using waste tea fungal biomass for the removal of arsenic from the contaminated ground water.
- Certain parameters are adjusted for the growth of *Pterisvittata* in the presence of ground water that needs to be decontaminated.
- The plant grows and takes up the arsenic. The arsenic is then recovered from the plant.
- Finally, the effluent is removed from the ground water, hence the water is purified.

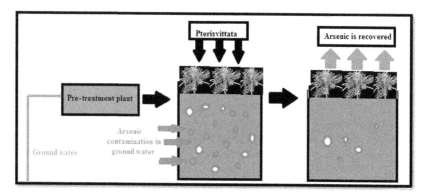

FIGURE 1.3 Optimized rhizofiltration process for the extraction of arsenic.

LEAD

The lead concentration in soil is $400\,\mathrm{mg\,kg^{-1}}$ in play zones and $1200\,\mathrm{mg\,kg^{-1}}$ in yards according to the recent U.S. EPA. This concentration is quite high, thus the level of lead should be controlled otherwise it can lead to serious health issue.

The Health Hazards Caused by Lead

Lead is a very dangerous element and one of the most common soil contaminants. It is widespread owing to its excessive use in fertilizers, pesticides (lead arsenate), fuels (leaded petrol), paints and industry. It has severe effects on human health which can lead to brain and kidney damage, as it is both a physiological and a neurological toxin. It is a non-essential metal, hence even the smallest concentration can have lethal effects.

The toxic effects of lead are related to its ability to react with other functional groups, such as carboxyl, amine and sulphydryl, resulting in loss of cellular activities. In order to control this a low-cost method is required, which nowadays is phytoremediation (Boonyapookana et al., 2005; Butcher, 2009).

The presence of high levels of lead in the soil can lead to serious health problems, as the food plants growing in the soil will take up lead in higher concentrations than required, which when consumed will act as a poison in the body – even in small concentrations. However, if the vegetables are eaten without their skin (e.g. potato without skin), this can help reduce the toxic effects by reducing the concentration of lead consumed. The concentration of lead in the skin of potatoes grown in moderately contaminated soil was $13–47\,\mathrm{mg\,kg^{-1}}$, whereas in heavily contaminated soil it was $72–226\,\mathrm{mg\,kg^{-1}}$. Figure 1.4 shows the concentration of lead in potatoes for both mildly and heavily contaminated soils.

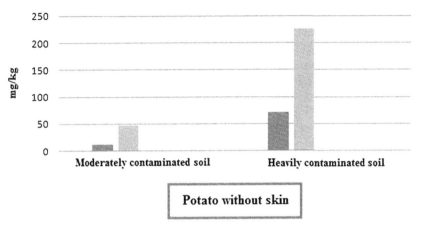

FIGURE 1.4   Bar graph showing the concentration of lead in potatoes grown in moderate and heavily contaminated soil.

Use of Phytoremediation to Clean Up Lead

Phytoextraction is the most common method used, which works on the principle of using plants that are hyperaccumulative for lead, harvesting them and selling the harvested plants to industries that extract the metal; the metal is then further processed to be recycled or disposed of as waste. Along with this process, phytofiltration and phytostabilization are also used. Five main species of plants have been identified as metallophytes for lead which are very commonly used; *Thlaspi alpestre, Polycarpae asynandra, Armeria martima, Alyssum wulfenianum* and *Thlaspi rotundifolium* (Butcher, 2009).

Another factor that is of great importance for metallophytes is the bioavailability of lead in the soil, so that the plants are able to take it up. This can be enhanced using certain chelating agents that help in the formation of soluble lead complexes. The most commonly used chelating agent is ethylene-diamine tetra-acetic acid (EDTA). This is a very efficient and effective agent which helps in the cleaning of lead, but it has two major drawbacks. First, it is persistent in the environment, and second, its biodegradability is low (Boonyapookana et al., 2005).

Three plant species – *N. tabacum, V. zizanioides* and *H. annuus* were tested for lead accumulation in a hydroponic culture containing lead nitrate, with and without EDTA. These plants were grown bare-rooted in long, narrow channels which were waterproof and through which nutrient was circulated using a pump. The nutrient was fed continuously to the plants and then collected back into the nutrient reservoir. It was then analysed to see which tissues accumulated the lead, and which of the three plants was the best to use for phytoremediation of lead. Also, it was seen that EDTA increases the uptake of lead. Therefore, in order to use EDTA for phytoremediation it is essential to know the threshold value to be used, so that the environment is only minimally affected (Boonyapookana et al., 2005).

All the plants had the ability to be used as lead extractors, as more lead was stored in the leaves than in the stem, but *Helianthus annuus* was the best of them. It has a high potential to be used in the restoration of soil, mines and factory sites, where considerable amount of lead is present owing to industrialization and a greater consumption of leaded fuels, thus affecting human health (Boonyapookana et al., 2005; Butcher, 2009).

The process of phytoextraction can help in reducing and controlling lead contamination in tea. Moreover the use of lead in fuels, pesticides, fertilizers and industries should be minimized as they are the main reason of lead increment in the environment (Han et al., 2006).

ZINC

Zinc is required for various important physiological functions, such as reactions carried out by enzymes, cell signalling, transcription, regulation of pH, etc.

The Health Hazards of Zinc

Zinc is toxic when present in high concentration. Its mild side effects are abdominal pain, nausea, epigastric pain, vomiting and diarrhoea. High zinc intake can cause nervous system disorders, problems with cholesterol metabolism and iron balance disorders (Mertens et al., 2007; Lemire et al., 2008).

The main source of the toxic effects of zinc is industrial release, so it is very important to maintain a balanced level of zinc in the environment, otherwise it can lead to mitochondrial defects affecting hepatocytes owing to the production of less adenosine triphosphate (ATP). The recommended dietary intake of zinc is $15\,mg\,day^{-1}$, so if more than this is consumed it can have toxic effects (Lemire et al., 2008).

Use of Phytoremediation to Clean Up Zinc

Phytoremediation is a very significant technique to control zinc. Many plants that possess specific characters are being used, of which various poplar floras are some of the best candidates. Some transgenic changes were made in the poplar floras so that they would work more effectively against zinc. The most significant transformed poplar floras are those that show an overexpression of glutamyl cystein synthetase (g-EC), which is a bacterial gene introduced into the plant. This helps to protect the plant by working as an antioxidant against oxidative stress from the atmosphere (Bittsánszky et al., 2005; Mertens et al., 2007; Lemire et al., 2008).

In order to see the response of transgenic plants in the presence of zinc, two wild and two transgenic poplar floras, one having g-EC in the cytosol and the other having g-EC in the chloroplast, were examined. It was seen that zinc uptake was same in all the plants, whereas the accumulation of zinc was maximal in the transgenic plant having g-EC in the cytosol (Bittsánszky et al., 2005).

It was observed that zinc in phytotoxic concentrations induced more stress in wild plants than the transgenic ones. This is because the overexpression of g-EC in the transgenic plants provides more stress tolerance and allows the accumulation of zinc in the plants. The g-EC enzymes have peroxidase activity, which allows them to detoxify reactive oxygen species. Thus it was seen that these transgenic plants are really excellent candidates for phytoremediation of zinc-contaminated soils compared to the wild poplar floras (Bittsánszky et al., 2005).

COPPER

Copper is another essential metal for the body. It is required for many enzymatic reactions and is present in trace amount in human body.

The Health Hazards of Copper

Copper is advantageous for human health but high doses can prove toxic, as copper does not decompose biologically. Copper causes the formation of hydroxyl and reactive oxygen species, leading to damage to DNA, lipids and proteins in humans, animals and plants. The copper in the environment is mainly due to anthropogenic activities, being discharged as waste products from fertilizer, mining, paint and dye industries, etc. Elevated copper intake causes central nervous system irritation, GI disorders, haemolysis, and liver and kidney toxicity (Ariyakanon and Winaipanich 2006; Das et al., 2013).

Use of Phytoremediation to Clean Up Copper

Bearing the lethal effects of copper in mind, it is important to control its concentration in the environment using methods that are easy and economically possible. Phytoremediation using hyperaccumulating plants is the most feasible method for this. About 400 plants are currently used for this purpose, and some names of the terrestrial plants being used are shown in Figure 1.5 (Das et al., 2013).

Some of the unique bacteria that live in the rhizosphere of plants can be the source of tolerance to heavy metal accumulation. These endophytic bacteria assist in the production of phytohormones, siderophores, etc., which help in the synthesis of 1-aminocyclopropane-1-carboxylate (ACC) deaminase, which in turn increases ethylene in the flora, thus making plants tolerant to heavy metal storage.

There are two main copper-tolerant species, *C. communis* and *E. splendens,* which are used for phytoremediation; they are usually used at mines and in the soils where toxic level of copper is present. Both of these species are widely used in China to control the level of copper. These bacteria can be used by hyperaccumulating plants to provide copper resistance to floras and thereby allowing greater copper uptake (Sun et al., 2010).

Bidens alba and Brassica juncea are copper-tolerant hyperaccumulating plants that have other essential characteristics too, such as short lifecycle, ease of handling and greater shoot biomass. These two plants were tested in order to understand their efficiency of copper removal. The experiment was carried out in a greenhouse. It was seen that the ability of B. juncea to remove copper was 11 times greater than that of B. alba. The maximum copper removal was 1.61–0.14% of 150 mg copper kg⁻¹ soil. Hence both species were able to take up a significant amount of copper at a 99% confidence level. This can easily be seen in Figure 1.6 (Ariyakanon and Winaipanich 2006).

It was also seen that copper uptake by the roots was far more significant than that of the shoot areas of both plants. Also, in order to increase the efficiency of copper uptake by B. alba, chemical substances, organic acids

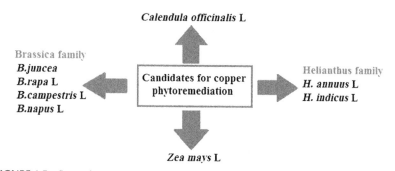

FIGURE 1.5   Some plants used as candidates for copper phytoremediation.

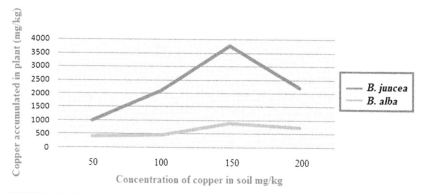

FIGURE 1.6   Copper accumulation by *Brassica juncea* and *Bidens alba*.

and chelates should be used to make more copper available to the plant by increasing its solubility (Ariyakanon and Winaipanich 2006).

CADMIUM

Cadmium is used in industry and in the making of nickel–cadmium batteries. It is also used in fertilizers and is an ingredient in cigars. When only traces are ingested it is not poisonous, but an excess of cadmium in the body can prove lethal.

The Health Hazards of Cadmium

Cadmium is a very toxic element, leading to toxic effects on farms, in agricultural areas and in industry. In large amounts it is a human health hazard, causing contamination of foods, low yields from plants, affecting the normal functioning of ecosystem. The major reason for cadmium excess in the environment is again because of anthropogenic activity. Excess intake of cadmium from the environment, either directly or indirectly, leads to toxicity that causes a disease, Itai-Itai (reported mainly in Japan), which leads to softening of the bones, anaemia, leading to problems in ovulation, endometrial cancer and reproductive cycle of the females and certain effects on kidney. It is one of the four major pollution diseases in Japan (Wei and Zhou, 2006; Ji et al., 2011).

Use of Phytoremediation to Clean Up Cadmium

Bearing all the above in mind, it is obvious that levels of cadmium should be controlled, especially in the soil, and for this purpose cadmium-specific hyperaccumulating plants are used. About 400 species of such plants are currently known. One significant candidate for cadmium uptake is *Rorippa globosa*, a newly discovered hyperaccumulator for cadmium which is very strongly tolerant of Cd accumulation (Wei and Zhou, 2006).

It was seen that when $25 \, mg \, kg^{-1}$ cadmium was given to *R. globosa* the biomass was 92.3% greater in the above-ground part of the plants at the flowering stage than in mature plants, at about 74.1%. Thus it was concluded that by using these two stages of planting in a year the removal of cadmium from the soil could be significantly improved (Wei and Zhou, 2006).

Another important Cd hyperaccumulator is *Solanum nigrum*. This plant was studied further in order to assess its potential for Cd uptake and effectiveness, as well as its ease of use in removing Cd. The research was based on important factors such as its cost-effectiveness, ease of use, its environmentally friendly effects and its effectiveness in phytoremediation.

Four strategies were used to analyse these plants: density variation, double harvesting, double cropping and fertilization (Ji et al., 2011). The outcome of the study showed that *S. nigrum* had the potential to be used for Cd hyperaccumulation, as it was able to carry out phytoextraction of Cd from contaminated soil.

The most important factor to consider for this plant is that it was able to take up Cd even when its concentration in the soil was low, thus helping keep soil concentrations always in check. Also, its biomass was much greater than other plants.

The double-cropping strategy helped to increase the total yield of biomass in the plant; therefore, for the best results this method is highly recommended (Ji et al., 2011).

MERCURY

Mercury is non-essential and one of the most toxic elements. Owing to its ability to be transported long distances in atmosphere, mercury is a leading cause of environmental pollution worldwide. The yearly emission of this element is between 4354 and 7530 t, owing to both natural and human activities. It is also responsible for hazards to human health and ecosystems. In the USA coal-fired power plants were seen to be responsible for emitting about 43.5 t of mercury per year (Meagher and Heaton 2005; Ruiz and Daniell, 2009; Pedron et al., 2011).

The Health Hazards of Mercury

The poisonous effects of mercury depend mainly on the form in which it is present in the surroundings. Organic forms are more deleterious than inorganic, the main reason being that inorganic forms bind very firmly to various soil components and are thus least available to the environment. Organic forms of mercury contribute towards toxicity owing to their hydrophobic nature, which allows them to accumulate in certain organelles that are membrane bound, leading to the inhibition of important pathways such as oxidation, photosynthesis, etc.

The organic forms of mercury are usually neurotoxins and are 90% absorbed by the intestine, leading to significant health hazards in the body. The organomercuric form also causes damage to the aquaporins (Ruiz and Daniell, 2009; Pedron et al., 2011).

Use of Phytoremediation to Clean Up Mercury

Plants are unable to convert mercury into less toxic forms, and so cannot detoxify it. Genetic engineering can help plants enhance their tolerance towards mercury thereby enhance the process of phytoremediation by integrating specific genes that lead to mercury removal and tolerance, thus reducing the toxicity of mercury in the environment (Ruiz and Daniell, 2009).

It was also seen that phytoremediation using hyperaccumulating plants was increased using ammonium thiosulphate and potassium iodide, as they allow metals to be biologically accessible to the plant (Ruiz and Daniell, 2009; Pedron et al., 2011). Methyl mercury is a neurotoxin. It is usually present in fish, where it is 990 times more toxic than elsewhere. It is taken up very effectively by humans. A detailed analysis of the uptake, transport, accumulation and steady state of mercury in hyperaccumulating plant tissues might greatly assist in the development of an improved phytoremediation process (Ruiz and Daniell, 2009).

Phytoremediation depends mainly on a combination of genes to enhance the process, for example mercury uptake, its translocation in the plant, chelation or conversion into less toxic substances, and finally allowing the release of less toxic mercury into the surroundings. Edible plants should not be used for phytoremediation, as these metals are toxic to both animals and humans (Ruiz and Daniell, 2009).

The use of genetic recombination in hyperaccumulating plants has led to the formation of recombinants having genes merA, merB and merC. The merA gene encodes for mercuric ion reductase, which helps detoxify mercury to its elemental form, as shown in Figure 1.7.

merB is basically a bacterial gene which helps cleave organic mercury to less toxic forms such as methane and ionic mercury. If both of these bacterial genes are expressed in a plant, it will have 100 times more tolerance towards mercury accumulation than the wild plant. The reactions are shown in Figure 1.8.

merC is also a significant gene which is a specialized mercury transporter, present at various sites in plant cell and organelles, which helps convert maximum mercury intake into a less toxic state (Figure 1.9) (Meagher and Heaton 2005; Ruiz and Daniell, 2009).

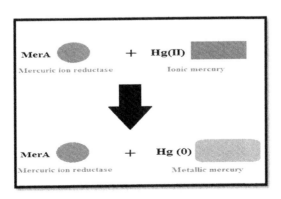

FIGURE 1.7   The reactions catalyzed by MerA.

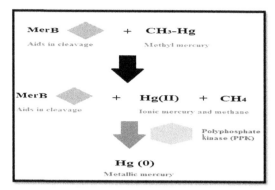

FIGURE 1.8 The use of MerB conversion of methyl mercury into a less toxic form using poly-phosphate kinase (PPK).

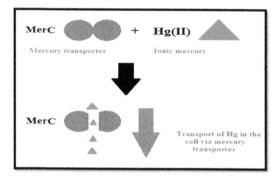

FIGURE 1.9 MerC use in a plant can assist with greater mercury uptake.

PROSPECTS FOR PHYTOREMEDIATION

Phytoremediation technology is becoming increasingly popular, and as a result ideas on how, when and where to use the process are being developed. Currently it is being used at various sites to dispose of hazardous waste in order to fulfill regulatory demands, as well as on sites listed as national priorities. There are a huge number of pollutants present in the environment to which it can be applied, such as chlorine solvents, insecticides, pesticides, heavy metals or metals, crude oil, explosives, etc. (Russell, 2005).

Phytoremediation is not only an area of interest in research centres and universities, but has also been seen to interest new business contractors as well as their consulting firms. The consultants for this process usually advise the stakeholders on whether or not this process will clean their site, and the contractors install the system at the site. The US phytoremediation market is growing very swiftly; the same trend is also being seen in Canada as well as other countries (Russell, 2005).

Phytoremediation Is Solar Energy–Driven and Cost-Effective

Phytoremediation has various benefits over other processes used to remove pollutants, as it is solar energy–driven, thereby using the natural processes of the plant to carry out all the work. This process requires little manpower compared to other physical and chemical techniques. The cost of equipment and operations is affordable, therefore the process is quite cost-effective. It can be used in conjunction with other techniques, so it works well as a multifunctional method compared to stand-alone processes (Kozdrój and van Elsas, 2001; Campos et al., 2008).

Phytoremediation Is Environmental Friendly

Phytoremediation is environmentally friendly as it reduces water and air emissions into the environment, and the secondary waste that results is not hazardous. It helps control soil erosion and the run-off that results from heavy rain or flooding. The process provides many wildlife habitats, thereby increasing biodiversity and aiding in the restoration of the ecosystem. It also acts as a carbon sink. Thus phytoremediation is a less destructive and less offensive process (Ghavzan and Trivedy, 2005; Russell, 2005).

It has also been seen that this technique usually results in about 50–80% cost savings compared to traditional processes such as physical or chemical methods. The USA Department of Energy also showed that the use of environment friendly processes provides important benefits to the ecosystem, either directly or indirectly. Henceforth, phytoremediation can be used in the petroleum industry, as the degradation of hydrocarbons can be controlled using plants, thereby also having an advantageous effect on the ecosystem (Ghavzan and Trivedy, 2005; Russell, 2005; Kozdrój and van Elsas, 2001).

Phytoremediation Can Help Mining Industries

The use of phytoremediation can help not only in the reduction and uptake of heavy metals from the soil, but also in the mining industry. For example, in the gold mining process the plants or hyperaccumulators are made specific for gold by adding ammonium thiocyanate to the substrate, and are later burnt to obtain a bio-ore which is harvested and sent to various industries to extract gold. Sometimes the metals extracted as pollutants can also be used in the respective mining industries, for example, gold, nickel, thallium, etc. (Figure 1.10). Therefore this process carries the possiblity of recovery and reuse of metals for beneficial purposes (e.g. in phytomining). The plants used can be easily monitored (Russell, 2005).

Phytoremediation Cleans Up Contaminants in Soil and Water

Phytoremediation helps preserve the texture of the soil during the extraction of heavy metals. It is also used for the extraction of pollutants from contaminated water.

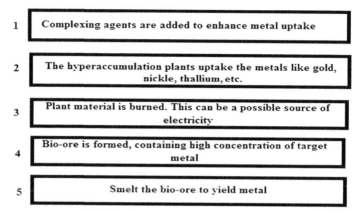

FIGURE 1.10   The basic steps of phytomining.

Often industries release heavy metals into their waste water. If regulatory measures are not being strictly adhered to, phytoremediation can be used to reduce the toxic levels of such waste (Ghavzan and Trivedy, 2005; Russell, 2005; Kozdrój and van Elsas, 2001).

The Reduction of Noise Pollution and Genetically Engineered Phytoremediation Plants

The process also reduces noise pollution in the environment, which is common when physical or chemical processes are used. The phytoremediation process can be further enhanced and optimized by finding genes that are able to provide the plant with useful characteristics, for example, increased tolerance to heavy metal accumulation, a short lifecycle, tolerance towards climatic change, less water intake, etc. These properties allow the plants to work more effectively and efficiently (Kozdrój and van Elsas, 2001).

The use of modern molecular tools in phytoremediation, via our knowledge of biochemistry, physiology and plant genetics, creates the expectation that methods of pollutant removal may be further enhanced by allowing the development of superior varieties of hyperaccumulating plants.

The selection of suitable plant species is a chief factor in achieving success in this technology. The poplar floras are important in this area, mainly because of their significant inherent properties. They have also proved to be of great importance with respect to genome sequencing, the ease and comfort of genetic transformation, and the development of systematic molecular techniques. These plants are already the first choice for organic pollution control using phytoremediation (Campos et al., 2008).

CHALLENGES OF PHYTOREMEDIATION

The use of phytoremediation is a novel modern technology that has many benefits over traditional methods of reducing environmental pollution. They have been used in both laboratory and greenhouse tests. For their use in a full-scale process it is essential that they are first extensively field tested in order to achieve the best results possible (Campos et al., 2008; Kulakow and Pidlisnyuk, 2009).

The Contaminant Can Be Extracted from as Deep as the Plant's Roots

Phytoremediation technology is expanding and progressing very effectively, enhancing the effects of its application, but there are still some unavoidable limitations to the process. The main problem is that in order to remove a pollutant or contaminant from the soil it is essential that it is available to the plants' roots. Briefly, the phytoremediation process is only successful when the contaminant is located shallow enough for the roots of plants to reach it; otherwise it must be brought close to the roots (Ghavzan and Trivedy, 2005; Russell, 2005; Kozdrój and van Elsas, 2001).

Slow Growth Cycle of Phytoremediation Plants

Another major drawback of the process is that the duration of lifecycle of most plants to reach maturity slows down the process of pollutant uptake, as the plant puts all its energy into growth. Thus phytoremediation becomes a slow process. Nowadays specific plants with short lifecycle are being identified in order to overcome this problem. Genetic modification in such plants can prove really helpful (Russell, 2005; Kozdrój and van Elsas, 2001).

Suitable Climatic Conditions and the Availability of Space for Phytoremediation

Some important factors to bear in mind with regard to phytoremediation are the local climate, the availability of space for the plants to be grown on contaminated sites and the season in which the plants grow. The plants can also transmit the pollutant to the environment. These factors are very challenging for phytoremediation, and a great deal of effort must be made to create suitable conditions for the plants. Research has shown a preference for plants that grow year-round, rather than being season specific, in order to reduce the problems associated with this technology (Russell, 2005; Kulakow and Pidlisnyuk, 2009).

Less Tolerance in Plants for Contaminant Uptake

Phytotoxicity is also a leading cause of failure of this process, as extremely high levels of contaminants in the plant do not permit it to grow or survive. Thus the process is seen to be effective only for lower concentrations of contaminant. Researchers have solved this problem by genetically modifying plants by inserting genes from other organisms or plants that are resistant to heavy metal accumulation, thereby allowing the process to work more swiftly and increasing the plants' uptake capacity (Russell, 2005; Kozdrój and van Elsas, 2001).

Accumulation of Contaminants in Plant Tissues

Phytoextraction can lead to the accumulation of contaminants in plant tissues, which can be the main cause of ecosystem exposure to contaminants; otherwise the plants will require harvesting to extract the contaminant. Another process known as phytovolatilization is responsible for removing toxic elements from the subsurface, thereby leading to an increase in airborne contamination (Kozdrój and van Elsas, 2001).

Genetically Modified Plants in Phytoremediation: An Unexpected or Unknown Threat?

Another important issue to be addressed is the introduction of an unnatural or genetically modified plant into the environment. This can lead to ecosystem problems which are either unexpected or unknown, thereby posing a threat to the natural environment.

Field trials are carried out for such plants, but there is a great deal of difference between open fields and controlled field conditions, and so there is a threat to environment via cross-pollination or via the negative effects of pollutants being introduced into the problem area by the presence of such plants (Ghavzan and Trivedy, 2005, Russell, 2005; Kozdrój and van Elsas, 2001).

Accessibility of Sites for Phytoremediation

Accessibility to the site is essential when carrying out this process, as these plants should not be consumed by livestock or the general public as they contain high accumulation of metals and can prove to be carcinogenic. This is why plants for remediation are not usually edible. If taken lightly, this factor can pose a serious threat. Moreover, if mixed contamination sites are to be treated – i.e. both organic and inorganic contaminants are present in a location – more than one phytoremediation process might be required to clean up the area (Ghavzan and Trivedy, 2005).

TECHNIQUES FOR GENETIC IMPROVEMENT OF PLANTS USED FOR PHYTOREMEDIATION

Phytoremediation is the natural ability of the plant to deal with hazardous material to which it is exposed, and all plants show different levels of activity depending upon their environment and genetic make-up. Genetic modification acts as an enrichment tool to enhance the phytoremediation ability of the plants. Advances in the field of plant biotechnology have created a large number of possibilities with the potential prospect to enhance phytoremediation activity (Shah and Nongkynrih, 2007). This creates a need for the identification and manipulation of beneficial genes which can provide plants with improved traits to withstand environmental stresses.

Example: Hyperaccumulation of Metal and Plant Response after Genetic Modification

Metal toxicity and metal hyperaccumulation are faced by plants in different regions around the world at different intensities. Different plants and some microorganisms that can withstand high metal concentrations maintain the metabolic integrity and unite at the common focal point of the set of genes that bestow these organisms with these traits (Danika and Norman, 2005).

With laborious work on genetic sequencing these sets of genes have been identified and can be extracted from the source organism and transferred to the intended plant. To increase the degree of active phytoremediation, these conserved genes can be transferred to rapidly growing plants (De Souza et al., 1998).

Copper Hyperaccumulation and Mimulus guttatus

In order to further elaborate the above, consider the example of *Mimulus guttatus* and copper (Cu) tolerance. When experiments were conducted on two lines of *Mimulus guttatus,* one with Cutolerant genes and one without, it was found that the plants with Cu tolerance genes were able to respond better to Cu stress, which was due to the presence of a special conserved set of tolerance genes (Smith and McNair, 1998).

Zinc Hyperaccumulation and Arabidopsis

Experiments on zinc hyperaccumulation were conducted on two species of *Arabidopsis:*

- A zinc hyperaccumulator, i.e., *Arabidopsis halleri*
- A zinc non-accumulator, i.e., *Arabidopsis petrea.*

The difference in plant tolerance of metal hyperaccumulation suggested the presence of the set of genes present in *Arabidopsis halleri* (McNair et al., 2000). *Arabidopsis halleri* has a set of genes that allows it to hyper accumulate zinc while these genes are absent in *Arabidopsis petrea* (non-hyper accumulator of zinc). With the help of plant biotechnology these efficient genes can be identified, isolated and transferred to other plant species to improve their tolerance abilities.

Use of Recombinant DNA Technology in Genetic Improvement for Phytoremediation

One potential way of improving plant genetics is recombinant DNA technology. This carries the possibility of providing improved traits in the form of new genes or the transfer of sets of pre-existing genes to plants with high functional abilities. It involves the following steps:

- isolation of the gene of interest;
- transfer to a vector;
- introduction of vector to host via physical or biological means;
- integration of the new gene (gene of interest) into host chromosomes;
- expression of the gene in cells;
- selection and culturing of these cells;
- tissue culturing to produce plants for further breeding.

Introduction of Genes Capable of Altering the Oxidation State of Heavy Metals

A change in the oxidation state of the metal ion changes its interaction power and means of exit from the plant; in this method sets of genes are introduced into the plants that encode for special enzymes which alter the oxidation state of heavy metal ions. The plants now use the process of phytovolatilization to dispose of the metal ions, which have become volatile after the change of oxidation state.

For example:

- the merA gene, which encodes for the enzyme mercuric oxide reductase (Rugh et al., 1996)
- genes that encode for enzymes that can methylate selenium into dimethylselenate (Hansen et al., 1998)

Use of Phytochelatins to Capture Metal Ions

In order to sequester the toxic heavy metal ions, plants secrete proteins, metal chelators and peptides. These help the plants to capture and finally dispose of the metal ions. They can be secreted in large amounts into the soil by the plant roots, or in the cells in the form of small molecules. The production and types of these proteins, enzymes, peptide molecules or metal chelators can be enhanced by genetic modification or gene transfer to the plant.

An example is the cad-1 gene of *Arabidopsis*, which provides a variable degree of cadmium sensitivity to mutant plants. Experiments showed that this gene is responsible for the formation of cadmium peptide complexes in the plants, and the plants with low response to the gene should have a negative response to cadmium tolerance.

CONCLUSION

Phytoremediation uses the innate ability of plants to protect themselves against metal toxicity and other stresses, such as high accumulations of toxic substances. Phytoremediation thus helps the plants to maintain their metabolic and structural integrity in harsh and unfavourable conditions. Advances in plant biotechnology have the potential to improve plant genetic traits and thus provide plants with better resistance to metal toxicity, using different methods of recombinant DNA technology. As very large numbers of plants all around the world are devastated by the problem of soil toxicity from the accumulation of hazardous materials, there is an urgent need for extensive research in the genetic improvement of plants.

REFERENCES

Aggarwal, H., Goyal, D., 2007. Phytoremediation of some heavy metals by agronomic crops. Dev. Environ. Sci. 5, 79–98.

Ali, H., Naseer, M., Sajad, M.A., 2012. Phytoremediation of heavy metals by Trifolium alexandrinum. Int. J. Environ. Sci. 2 (3), 1459–1469.

Alkorta, I., Hernández-Allica, J., Garbisu, C., 2004. Plants against the global epidemic of arsenic poisoning. Environ. Int. 30 (7), 949–951.

Alpaslan, B., Yukselen, M.A., 2002. Remediation of lead contaminated soils by stabilization/solidification. Water, Air, and Soil Pollut., 253–263.

Anderson, A., Mitchell, P., 2003. Treatment of mercury-contaminated soil, mine waste and sludge using silica micro-encapsulation. TMS Annual Meeting, Extraction and Processing Division, 265–274.

Ariyakanon, N., Winaipanich, B., 2006. Phytoremediation of copper contaminated soil by *Brassica juncea* (L.) Czern and *Bidens alba* (L.) DC. Var. *radiata*. J. Sci. Res. Chula Univ. 31, 49–56.

Baker, A.J.M., Brookes, R., Reeves, R.D., 1988. Growing for gold and copper and zinc. New sci. 1603, 44–48.

Baker, A.J.M., McGrath, S.P., Reeves, R.D., 2000. Metal hyperaccumulator plants: a review of the ecology and physiology of a biological resource for phytoremediation of metal-polluted soils. In: Terry, N., Bañuelos, G. (Eds.), Phytoremediation of Contaminated Soils and Waters. Lewis Publishers, Boca Raton, pp. 85–108.

Barnes, D.L., Cosden, E., Johnson, B., Johnson, K., Stjarnstrom, S., Johansson, K., Filler, D., 2002. Operation of soil vapor extraction in cold climates, cold regions engineering cold regions impacts on transportation and infrastructure. Proceedings of the Eleventh International Conference, 956–967.

Bittsánszky, A., Kömives, T., Gullner, G., Gyulai, G., Kiss, J., Heszky, L., Radimszky, L., Rennenberg, H., 2005. Ability of transgenic poplars with elevated glutathione content to tolerate zinc (2+) stress. Environ. Int. 31 (2), 251–254.

Bollag, J., Mertz, T., Otjen, L., 1994. Role of microorganisms in soil bioremediation. Bioremediation through Rhizosphere Technology. ACS Symposium Series, ACS Publications.

Boonyapookana, B., Parkpian, P., Techapinyawat, S., DeLaune, R., Jugsujinda, A., 2005. Phytoaccumulation of lead by sunflower (*Helianthus annuus*), tobacco (*Nicotiana tabacum*), and vetiver (*Vetiveri azizanioides*). J. Environ. Sci. Health 40 (1), 117–137.

Boulding, J.R., 1996. USEPA Environmental Engineering Sourcebook. Ann Arbor Press.

Boyd, R., 2007. The defense hypothesis of elemental hyperaccumulation: status, challenges and new directions. Plant Soil 293, 153–176.

Brooks, R.R., 1998. General introduction. In: Brooks, R.R. (Ed.), Plants that Hyperaccumulate Heavy Metals. CABI, Wallingford, pp. 1–14.

Butcher, D.J., 2009. Phytoremediation of arsenic: fundamental studies, practical applications, and future prospects. Appl. Spectrosc. Rev. 44 (6), 534–551.

Butcher, D.J., 2009. Phytoremediation of lead in soil: recent applications and future prospects. Appl. Spectrosc. Rev. 44 (2), 123–139.

Campos, V., Merino, I., Casado, R., Gómez, L., 2008. Review of phytoremediation of organic pollutants. Span. J. Agric. Res. 6, 38–47.

Chaudhry, T.M., Hayes, W.J., Khan, A.G., Khoo, C.S., 1998. Phytoremediation focusing on accumulator plants that remediate metal contaminated soils. Australasian J. Ecotoxicol 4, 37–51.

Clemens, S., 2006. Toxic metal accumulation, responses to exposure and mechanisms of tolerance in plants. Biochimie 88, 1707–1719.

Convard, Nancy S.A., Tomlinson, A., Welch, C., 2005. Strategies for preventing and mitigating land-based sources of pollution to trans-boundary water resources in the Pacific region, SPREP.

Cunningham, S.D., Berti, W.R., Huang, J.W., 1995. Phytoremediation of contaminated soils. Trends Biotechnol., 393–397.

Cunningham, S.D., Crowley, D., Anderson, T., 1997. Phytoremediation of water and soil. Presented at Phytoremediation Soil Water Contam., Washington, DC. ACS Symp. Ser. No. 664.

Danika, L., LeDuc Norman, T., 2005. Phytoremediation of toxic trace elements in soil and water. J. Ind. Microbiol.Biotechnol. 32, 514–520.

Das, S., Goswami, S., Talukdar, Das, 2013. Copper hyperaccumulating plants from Barak valley, South Assam, India for phytoremediation. Int. J. Toxicol. Pharmacol. Res. 5(1): 30–32.

De Souza, M.P., Pilon-Smits, E.A.H., Lytle, C.M., Hwang, S., Tai, J., Honma, T.S.U., Yeh, L., Terry, N., 1998. Rate-limiting steps in selenium assimilation and volatilization by Indian mustard. Plant Physiol. 117, 1487–1494.

Di Palma, L., Ferrantelli, P., Merli, C., Biancifiori, F., 2003. Recovery of EDTA and metal precipitation from soil flushing solutions. J. Hazard. Mater. 103 (1), 153–168.

Dushenkov, V., Nanda-Kumar, P.B.A., Motto, H., Raskin, I., 1995. Rhizofiltration: the use of plants to remove heavy metals from aqueous streams. Environ. Sci. Technol. 29, 1239–1245.

EPA, 2000. Introduction to phytoremediation. National Risk Management Research Laboratory.

Farmaki, E., Thomaidis, N., 2008. Current status of the metal pollution of the environment of Greece- a review. Global nest. Int. J. 10 (3), 366–375.

Filler, D.M., Lindstrom, J.E., Braddock, J.F., Johnson, R.A., Nickalaski, R., 2001. Integral biopile components for successful bioremediation in the Arctic. Cold Regions Science and Technology, 143–156.

Garbisu, C., Alkorta, I., 2001. Phytoextraction: A cost effective plant-based technology for the removal of metals from the environment. Biores. Technol. 77 (2001), 229–236.

Gatliff, E.G., 1994. Vegetative remediation process offers advantages over traditional pump-and-treat technologies. Remed. Summer 4 (3), 343–352.

Ghavzan, N.J., Trivedy, R., 2005. Environmental pollution control by using phytoremediation technology. Pollut. Res. 24 (4), 875.

Ghosh, M., Singh, S., 2005. A review on phytoremediation of heavy metals and utilization of its by-products. Appl. Ecol. Environ. Res. 3 (1), 1–18.

Halim, M., Conte, P., Piccolo, A., 2003. Potential availability of heavy metals to phytoextraction from contaminated soils induced by exogenous humic substances. Chemosphere 52 (1), 265–275.

Han, W.-Y., Zhao, F.-J., Shi, Y.-Z., Ma, L.-F., Ruan, J.-Y., 2006. Scale and causes of lead contamination in Chinese tea. Environ. Pollut. 139 (1), 125–132. http://dx.doi.org/10.1016/j.envpol.2005.04.025.

Hansen, D., Duda, P., Zayed, A.M., Terry, N., 1998. Selenium removal by constructed wetlands: role of biological volatilization. Environ. Sci.Technol. 32, 591–597.

Henry, J.R., 2000. An overview of phytoremediation of lead and mercury. U.S Environmental Protection Agency.

Ji, P., Sun, T., Song, Y., Ackland, M.L., Liu, Y., 2011. Strategies for enhancing the phytoremediation of cadmium-contaminated agricultural soils by Solanum nigrum L. Environ. Pollut. 159 (3), 762–768.

Khan, F.I., Husain, T., Hejazi, R., 2004. An overview and analysis of site remediation technologies. J. Environ. Manag. 71 (2), 95–122.

Kotrba, P., Najmanova, J., Macek, T., Ruml, T., Mackova, M., 2009. Genetically modified plants in phytoremediation of heavy metal and metalloid soil and sediment pollution. Biotechnol. Adv. 27 (6), 799–810. http://dx.doi.org/10.1016/j.biotechadv.2009.06.003.

Kozdrój, J., van Elsas, J.D., 2001. Structural diversity of microbial communities in arable soils of a heavily industrialized area determined by PCR-DGGE finger printing and FAME profiling. Appl. Soil Ecol., 31–42.

Kulakow, P.A., Pidlisnyuk, V.V., 2009. Application of phytotechnologies for cleanup of industrial, agricultural and wastewater contamination. Springer Verlag.

Kurek, E., Bollag, J.M., 2004. Microbial immobilization of cadmium released from CdO in the soil. Biogeochemistry, 69 (2), 227–239.

Lemire, J., Mailloux, R., Appanna, V.D., 2008. Zinc toxicity alters mitochondrial metabolism and leads to decreased ATP production in hepatocytes. J. Appl. Toxicol. 28 (2), 175–182.

Lindgren, E.R., Kozak, M.W., Mattson, E.D., 1994. Electrokinetic remediation of unsaturated soils. American Chemical Society Symposium Series on Emerging Technologies in Hazardous Waste Management IV, Washington, DC, pp. 33–50.

Liu, X.M., Wu, Q.T., Banks, M.K., 2005. Effect of simultaneous establishment of Sedum alfridii and Zea mays on heavy metal accumulation in plants. Int. J. Phytoremediation, 7, 43–53.

Lone Mohammad Iqbal, Z.L.H., Peter, J., Stoffella, E Yang, Xiao, 2008. Phytoremediation of heavy metal polluted soils and water: progresses and perspectives. J. Zhejiang Univ. Sci. B. 9 (3), 210–220.

Luo, L.M.Y., Zhang, S., Wei, D., Zhu, Y.G., 2009. Inventory of trace element inputs to agricultural soils in China. J. Environ. Manag. 90 (8), 2524–2530.

Luo, Y.M., Teng, Y., 2006. Status of soil pollution-caused degradation and countermeasures in China, 505–508.

Mcgrath, S.P., Zhao, F.J., Lombi, E., 2001. Plant and rhizosphere process involved in phytoremediation of metal-contaminated soils. Plant Soil, 207–214.

McIntyre, T., 2003. Phytoremediation of heavy metals from soils. Advance Biochemistry Engineering Biotechnology 78, 97–123.

McNair, M.R., Tilstone, G.H., Smith, S.S., 2000. The genetics of metal tolerance and accumulation in higher plants. In: Terry, N., Bañuelos, G. (Eds.), Phytoremediation of Contaminated Soil and Water. Lewis Publishers, Boca Raton, pp. 235–250.

Meagher, R.B., Heaton, A.C., 2005. Strategies for the engineered phytoremediation of toxic element pollution: mercury and arsenic. J. Ind. Microbiol. Biotechnol. 32 (11–12), 502–513.

Mertens, J., Degryse, F., Springael, D., Smolders, E., 2007. Zinc toxicity to nitrification in soil and soilless culture can be predicted with the same biotic ligand model. Environ. Sci. Technol. 41 (8), 2992–2997.

Miller, W., Peterson, J., Greene, J.C., Callahan, C.A., 1985. Comparative toxicology of laboratory organisms for assessing hazardous waste sites. J. Environ. Qual. 14 (4), 569–574.

Mudgal, V., Madaan, N., Mudgal, A., 2010. Heavy metals in plants: phytoremediation: Plants used to remediate heavy metal pollution. Agriculture and Biol. J. North Am. 1 (1), 40–46.

Park, E., Zhan, H., 2003. Hydraulics of horizontal wells in fractured shallow aquifer systems. J. Hydrol. 281 (1), 147–158.

Pedron, F., Petruzzelli, G., Barbafieri, M., Tassi, E., Ambrosini, P., Patata, L., 2011. Mercury mobilization in a contaminated industrial soil for phytoremediation. Commun. Soil Sci. Plant Anal. 42 (22), 2767–2777.

Perez, J., 2012. The soil remediation industry in Europe: the recent past and future perspectives. pp. 2–22.

Pollard, A.J., Powell, K.D., Harper, F.A., Smith, J.A.C., 2002. The genetic basis of metal hyperaccumulation in plants. Critical reviews in plant sciences. 21 (6), 539–566.

Prasad, M.N.V., Freitas, H., 2003. Metal hyperaccumulation in plants—biodiversity prospecting for phytoremediation technology. Electron J. Biotechnol. 6 (3), 285–321.

Rathinasabapathi, B., Ma, L.Q., Srivastava, M., 2006. Arsenic hyperaccumulating ferns and their application to phytoremediation of arsenic contaminated sites. In: Teixeira da Silva, J.A. (Ed.), Floriculture, Ornamental and Plant Biotechnology, vol. 3. Advances and Topical Issues, Global Science Books, pp. 305–311.

Riser-Roberts, E., 1998. Remediation of petroleum contaminated soil: biological, physical, and chemical processes. Lewis Publishers.

Rugh, C.L., Wilde, D., Stack, N.M., Thompson, D.M., Summers, A.O., Meagher, R.B., 1996. Mercuric ion reduction and resistance in transgenic *Arabidopsis thaliana* plants expressing a modified bacterial merA gene. Proc. Natl. Acad. Sci. USA 93, 3182–3187.

Ruiz, O.N., Daniell, H., 2009. Genetic engineering to enhance mercury phytoremediation. Curr. Opin. Biotechnol. 20 (2), 213–219.

Russell, K., 2005. The use and effectiveness of phytoremediation to treat persistent organic pollutants. Environ. Careers Organ. (for US EPA).

Saier, M., Trevors, J., 2010. Phytoremediation. Water, Air, & Soil Pollution, 205, 61–63.

Schwab, A.P., 1998. Phytoremediation of soils contaminated with PAHS and other petroleum compounds. Beneficial Eff. Vegetation in Contaminated Soils Workshop.

Schwartz, C., Echevarria, G., Morel, J.L., 2003. Phytoextraction of cadmium with *Thlaspi caerulescens*. Plant Soil, 27–35.

Shah, K., Nongkynrih, J., 2007. Metal hyperaccumulation and bioremediation. Biologia Plantarum 51 (4), 618–634.

Smith, S.E., McNair, M.R., 1998. Hypostatic modifiers cause variation in degree of copper tolerance in *Mimulus guttatus*. Heredity 80, 760–768.

Sun, L.-N., Zhang, Y.-F., He, L.-Y., Chen, Z.-J., Wang, Q.-Y., Qian, M., Sheng, X.-F., 2010. Genetic diversity and characterization of heavy metal-resistant-endophytic bacteria from two copper-tolerant plant species on copper mine wasteland. Bioresour. Technol. 101 (2), 501–509.

Thompson, P.L., Ramer, L.A., Schnoor, J.L., 1998. Uptake and transformation of TNT by hybrid poplar trees. Environ. Sci. Technol. 32, 975–980.

Vamerali, T., Bandiera, M., Mosca, G., 2010. Field crops for phytoremediation of metal-contaminated land. A review. Environ. Chem. Lett. 8, 1–17.

Wei, S., Zhou, Q.-X., 2006. Phytoremediation of cadmium-contaminated soils by *Rorippa globosa* using two-phase planting (5 pp). Environ. Sci. Pollut. Res. 13 (3), 151–155.

Wokocha, G.A., Ihenko, S., 2010. Effects of fertilizer on soil Ph in Rivers State. Am. J. Sci. Res. 2010 (12), 125–128.

Yang, X., Feng, Y., He, Z., Stoffella, P.J., 2005. Molecular mechanisms of heavy metal hyperaccumulation and phytoremediation. J. Trace. Elem. Med. Biol. 18 (4), 339–353. http://dx.doi.org/10.1016/j.jtemb.2005.02.007.

Chapter 2

Soil Contamination with Metals: Sources, Types and Implications

Waqar Ahmad,[*] Ullah Najeeb[†] and Munir Hussain Zia[‡]

*Department of Environmental Sciences, Faculty of Agriculture and Environment, The University of Sydney, NSW, Australia, †Department of Plant and Food Sciences, Faculty of Agriculture and Environment, The University of Sydney, NSW, Australia, ‡Research and Development Section, Fauji Fertilizer Company Limited, Rawalpindi, Pakistan

INTRODUCTION

Unlike organic pollutants, heavy metals are not degraded into harmless forms; therefore, they persist in the environment for the long term. The toxicity of heavy metals adversely affects soil quality, agricultural production, human health and the environment. Soil metal contamination is particularly important in the developing countries, where a large number of industries often discharge their untreated waste into the open environment (Wuana and Okieimen, 2011). Other than smelting and electroplating industries, city effluents (sewage and industrial) are also among the potential sources of metal pollution, used for growing crops, mostly vegetables and fodder, in the pre-urban agriculture (Kapungwe, 2013; Murtaza et al., 2010).

Application of urban and industrial wastes containing a variety of pollutants for cultivation is responsible for building up the higher concentrations of heavy metals such as iron (Fe), manganese (Mn), copper (Cu), zinc (Zn), lead (Pb), chromium (Cr), nickel (Ni), cadmium (Cd) and cobalt (Co) in agricultural soils (Kabir et al., 2012), which ultimately enter into the food chain and prove hazardous to humans and animals. Suruchi and Khanna (2011) reviewed crop production on heavy-metal-contaminated soils, and observed significantly high metal concentrations in the vegetable tissues, indicating a potential hazard to human health. However, due to limited availability of data on the issue, they suggested further studies for assessment of risk to humans on consumption of such types of food. The industrial units established two to three decades earlier inside metropolitan cities limits dispose of their effluents into a city sewerage system that might pollute the existing effluent. Untreated effluents released into rivers and canals ultimately end up in agricultural fields and become a serious

Soil Remediation and Plants. http://dx.doi.org/10.1016/B978-0-12-799937-1.00002-4

threat to biota. In many countries, due to shortage of river/canal water, the untreated effluents are now directly uplifted by the farmers for irrigation purposes. Such indiscriminate use of effluents may cause a serious hazard for cultivating lands due to the risk of the addition of harmful metals into human and animal food chains (Antil, 2012). Concentrations of heavy metals in air-dried forage above certain limits, viz. Zn (500–1000 mg/kg (ppm), Cd (0.5 ppm), Cu (25–80 ppm), Pb (30 ppm) and organic As (100 ppm), are considered toxic for domestic animals (NRC, 1980).

About 2 million tons of sewage, industrial and agricultural wastes are discharged into the land and water every day, which contributes to a total annual estimation of wastewater at about 1500 km^3 (UN-WWAP, 2003). In developed countries, the effluents are properly treated by drying; depending upon the legislation of the country, the dried sludge is either burnt or applied to land as manure. In the underdeveloped or developing countries, wastewater treatment plants are either absent or non-functional. Therefore, these plants are practically ineffective and do not produce any treated sewage sludge that can be used as manure for cultivation of crops. Comparing the water quality, Khan et al. (2007) observed significantly higher concentrations of heavy metals such as Cd, Cr, Fe, Pb and Mn in effluents compared to canal water. As discussed earlier, there are various potential sources of soil metal pollution, but most commonly it arises from the purification and processing of metals that can cast long-term adverse effects on the soil ecosystem. For example, mining of sulphide-containing rocks can cause acid mine drainage that, in turn, increases heavy metal release including arsenic (As) into the atmosphere by reacting with oxygen and water (Lawrence and Higgs, 1999). Long-term application of phosphate fertilizers and smelter dust deposition on agricultural soils can also increase levels of heavy metals in soils; agricultural soils close to metal-processing industries are more vulnerable to these risks. Long-term (10 years) sewage sludge application to agricultural soils significantly elevated heavy metal solubility and uptake (Bergkvist et al., 2003), where the relationship between soluble Cd concentrations in soils and grain became more linear after sludge application (Chaudri et al., 2001).

Several studies have reported higher metal concentrations (total and available) in the top layer of soils irrigated with contaminated water. Therefore, metal-contaminated crops especially the vegetables, which have the major network of their roots in the top soil layer, may be unfit for consumption if the heavy metals contents in consumable portions are above the permissible limits. Moderate to strong positive correlations between crop and soil in respect of Fe, Cd, Co and Pb metals have been reported in literature (Smolen et al., 2010). This demonstrates the toxic build up of many of the heavy metals in plant bodies via soil routes. Almost all the studies lead to a conclusion that growing crops, especially vegetables, should be restricted in wastewater-irrigated areas to avoid metal toxification to the consumers through the food chain.

HEAVY METALS

The term 'heavy metal' is applied to a large number of industrially and biologically important trace elements. Although not completely satisfactory from the chemistry point of view, it is still the most widely recognized term for a large group of elements with an atomic density over 6 gm cm^{-3} (Alloway, 1991). Heavy metals that have been identified in the polluted environment include As, Cu, Cd, Pb, Cr, Ni, mercury (Hg) and Zn (Corbett, 2002; Wong, 2003). Accumulation of these metals in soils may be harmful to soil and its boundary ecosystems (Gupta et al., 1996). Some of the heavy metals are essential in trace amounts, namely, Co, Cu, Fe, Mn, Mo and Zn for plants and in addition Cr and Ni for animals. Cd, Hg and Pb are essential neither for plants nor for animals. The subsequent transfer of these metals from soil to plants is of most increasing concern due to the possible adverse effects that they might have on plants, animals and human health. Some metals are abundant in the environment, and might be more toxic to plants and animals than others; therefore, in this chapter, we have focused on Cd, Pb, Ni and As only.

Sources of Heavy Metals in Soil

The major sources of the heavy metals include natural occurrence, derived from parent materials and human activities. Anthropogenic inputs are associated with urbanization, industrialization and agricultural practices, such as atmospheric deposition, waste disposal, waste incineration, urban effluent, traffic emissions, fertilizer application and long-term application of wastewater to agricultural lands (Koch and Rotard, 2001). Pesticides may be included as one of the sources of the heavy metals due to their unjudicious application, especially in the developing countries. The fate of heavy metals added to soils, including their mobility, reactions in soil and their subsequent uptake and distribution in plant bodies, is therefore of critical importance in relation to human health.

Agricultural-Based Sources

Pesticides

Pesticides are extremely toxic substances, and are hazardous to human health and the environment. Despite a slight reduction in pesticide application in agriculture over a period from 2000 to 2007, still 2.3 million tons of industrial pesticides are being used, a figure that is 50 times greater than the 1950s. The Center for Disease Control's (CDC) Fourth National Report on Human Exposure to Environmental Chemicals indicated the presence of neurotoxins in the bodies of most Americans, caused by excessive use of organophosphates (CDC, 2009). The widespread use of pesticides may cause more harm to the underdeveloped and developing countries, where proper legislations for pesticide application are either absent or not implemented. The incidence of pesticide poisoning in these countries may even be greater than reported due to under-reporting, lack of data and misdiagnosis (Tariq et al., 2007). In a study, Gimeno-García et al. (1996) showed that out of the three pesticides (two weedicides and a fungicide) used in paddy fields, highest levels of Fe, Mn, Zn, Pb and Ni were found in the herbicides.

Arsenical pesticides were largely used for controlling ticks in livestock from the early 1900s to 1955. This resulted in As contamination of surrounding soil in cattle-dip sites in many countries, including, Australia, the United States, Republic of South Africa and New Zealand (Okonkwo, 2007; Sarkar et al., 2007). Since the 1960s, there has been a significant reduction in agricultural application of As-based pesticides; however, some arsenical pesticides are still in use such as lead arsenate, copper acetoarsenite (Paris green), Bordeaux blue and organic mercury fumigants (Peryea, 1998). However, it is hard to give the exact list of chemicals (insecticides and fungicides) that include, but are not limited to, Cu, Hg, Mn, Pb or Zn being used on the farm over a wide variety of crops.

Fertilizers

More or less all the essential metals are present in the soil (Table 2.1); however, the availability of essential elements in soil and crops is limited. Thus, crop nutrients may be supplied either as a supplement to the soil or as a foliar spray. Heavy metals are present in several fertilizers as impurities, and their repeated application could cause metal accumulation in soils. The most commonly used

TABLE 2.1 Most Common Chemical Compositions of Soils

Sr. No.	Element	Range (mg kg⁻¹)	Reference
1	Mn	200–3000	Swaine, 1955
2	Zn	10–300	Kiekens, 1995
3	Cu	5–50	McLaren and Crawford, 1973
4	Mo	0.6–3.3	Swaine, 1955
5	B	20–200	Gupta, 1979
6	Co	1–40	Vanselow, 1966
7	Ni	1–450	Kabata-Pendias and Pendias, 1992
8	Cr	< 100	Alloway, 1968
9	Pb	2–200	Alloway, 1968
10	Fe	4 %	Bohn et al., 2010

phosphate fertilizers, single super phosphate, contains Cd and other radionu-clides, although the concentrations may vary depending upon the phosphate rock origin (Table 2.2).

Heavy metal mobility and availability in soil solution is regulated by various soil physico-chemical properties such as precipitation–dissolution, adsorption–desorption, ion exchange, pH and organic matter contents. The increased acidic fertilizer application could cause soil acidification, which changes the soil's physical and chemical properties and increases concentrations of heavy metals including Al^{3+}, Cd^{2+}, As^{3+} and Pb^{2+}. An inverse relationship has been proposed between soil pH and Cd uptake by plant roots (Yanai et al., 2006).

Depending on the crop nature and soil pH value, Zn and Fe deficiencies are common and therefore application of these nutrients is also recommended (Table 2.3).

Industrial-Based Sources

Aerial Emissions

All solid particles in smoke from fires and emissions from factory chimneys are eventually deposited either on the land or sea surface. Most forms of fos-sil fuel contain small quantities of heavy metals and are potential sources of

TABLE 2.2 Cadmium (Cd) Contents in Various Rock Phosphates

S. No	Origin	Name	Mean Cd (mg kg^{-1})
1	USSR	Kola phosphate	0.3
2	USA	Florida pebble	7
3	USA	North Carolina	36
4	Israel	Oron phosphate	3
5	Israel	NahalZin phosphate	20
6	Morocco	Khouribga phosphate	12
7	Morocco	Joussufia phosphate	4
8	Algeria	Algier phosphate	23
9	Spanish Sahara	Bu Craa phosphate	43
10	Tunisia	Gafsa phosphate	56
11	Togo	Togo phosphate	53
12	Senegal	Taiba phosphate	84

Adapted from Baechle H.T. and Wolsrein F. (1984) Cd compounds in minerals fertilizers. *The Fertil-izer Society, Proceeding No. 226,* London.

TABLE 2.3 Commonly Used Zinc Fertilizer Materials

	Formula	Zinc Content (%)
Inorganic Compounds		
Zinc sulphate monohydrate	$ZnSO_4.H_2O$	36–37
Zinc sulphate heptahydrate	$ZnSO_4.7H_2O$	22–23
Zinc oxysulphate	$xZnSO_4.xZnO$	20–50
Basic zinc sulphate	$ZnSO_4.4Zn(OH)_2$	55
Zinc oxide	ZnO	50–80
Zinc carbonate	$ZnCO_3$	50–56
Zinc chloride	$ZnCl_2$	50
Zinc nitrate	$Zn(NO_3)_2.3H_2O$	23
Zinc frits	Fritted glass	10–30
Ammoniated zinc sulphate solution	$Zn(NH_3)_4SO_4$	10
Organic Compounds		
Disodium zinc EDTA	$Na_2ZnEDTA$	8–14
Sodium zinc HEDTA	$NaZnHEDTA$	6–10
Sodium zinc EDTA	$NaZnEDTA$	9–13
Zinc polyflavonoid	—	5–10
Zinc lignosulphonate	—	5–8

Based on: (1) Mortvedt J.J. and Gilkes R.J. (1993) Zinc fertilisers. *In* A. D. Robson (Ed.), *Zinc in Soils and Plants*, pp. 33–44, Dordrecht: Kluwer Academic Publishers; (2) Martens D.C. and Westermann D.T. (1991) Fertilizer applications for correcting micronutrient deficiencies. *In* J.J. Mortvedt, F.R. Cox, L.M. Shuman, and R.M. Welch (Eds) *Micronutrients in Agriculture*, 2nd ed, pp. 549–592, Madison: Soil Science Society of America; (3) Srivastava P.C. and Gupta U.C. (1996) *Trace Elements in Crop Production*, p. 356, Science Pub. Inc., Lebanon.

contamination. Emission of Pb from petrol combustion contaminates the air and substantially contributes to the soil Pb content in urban areas and those adjacent to major roads. The contamination of agricultural soils by Cd and Pb is a persistent problem in many countries. Soil contamination with these toxic metals is primarily due to their release from mining areas. The United States Environmental Protection Agency (USEPA) has classified Cd and Pb as the most hazardous heavy metals in the environment due to their increased accumulation in the environment.

Raw Effluents

In many developing countries, application of sewage effluent for raising vegetables and fodders around the big cities is a well-established practice. The sewage effluents are likely to become more widely used sources of irrigation on agricultural lands due to their high contents of crop nutrients and shortage of irrigation water. Dense population in cities is responsible for the changes in the chemical composition of domestic sewage effluents with respect to fairly high N, P and K concentrations. Hussain, et al. (2006) conducted a study across Faisalabad, Pakistan, and observed higher Cr, Mn and Ni in all effluent samples, whereas only Cd contents were higher than permissible limits. All the metallic ions were above the safe limits in edible parts of vegetables. Thus, the untreated effluent irrigation would make the soil unproductive and may have adverse effects on human health through introducing toxic metals into the food chain.

EFFECTS OF HEAVY METALS IN SOILS AND PLANTS

Build-Up of Heavy Metals in Soil

The variable metal loadings from numerous anthropogenic activities coupled with some transmissivity factors such as heavy rainfall-runoff, adverse land topography and cropping systems are primarily responsible for the build-up of toxic heavy metals in the soils and their spatial variations due to horizontal movement. The absorption of metals from the soil by plants is influenced by a variety of factors, including pH, temperature, soil cation exchange capacity, soil organic matter (SOM) contents, the type and concentration of metals and plant species (Salim et al., 1993). The level of heavy metals in the soils builds up with the repeated application of amendments containing heavy metals. Crops would remove only small amounts, and presumably precipitation would not normally wash the metals down to soil profile, suggesting that there is no common natural mechanism for heavy metals removal from polluted soils.

Toxicity Symptoms in Plants

The phytotoxicity is normally associated with metal accumulation in plant tissues to a level that could affect their optimal growth and development. The occurrence of metal phytotoxicity is based on two evidences: (1) the injuries and observed abnormalities in plants are sustained (persist through maturity), and (2) the source of these alterations in the plants is not the plant growth malfunctioning. Native soil concentrations of a particular metal or background concentrations of the metal in applied irrigation water are responsible for some toxicities (Gupta and Gupta, 1998). For example, soils developed from serpentine deposits are considered high in Ni and Cr (Proctor and Woodell, 1975), and the mine tailings originate Hg-polluted soils (Harsh and Doner, 1981). In fact, based on the available literature, the phytotoxicity could be categorized into primary and secondary toxicities. The most common human-induced phytotoxicity is secondary and includes additions of metal sludge, non-ferrous metal smelters, coal fly ash and accidental applications of high levels of metals in addition to contribution from mineral weathering of parent material

(Gupta and Gupta, 1998); the polluted lands usually contain excessive levels of two or more of these metals. For example, Cd contamination is much less frequent and is usually accompanied by >100 times more Zn such that crops cannot accumulate excessive Cd before Zn phytotoxicity substantially reduces plant yield. Therefore, it is imperative to quantify the multiple metals in different plant parts to understand the interactive toxic effect of various metals. However, the nature of interactions could vary with soil properties, especially soil reaction and SOM contents (Wallace and Wallace, 1994).

Cadmium (Cd)

Among heavy metals, Cd is considered as one of the potentially important environmental pollutants. There is a growing concern about Cd contamination as being one of the most ecotoxic metals that exhibit highly adverse effects on soil biological activity, plant metabolism and the health of biota. Chemical composition of the parent rock is the main factor determining the Cd content of soil. As a result of weathering, Cd goes readily into the soil solution, where it may be present as Cd or it may form several complex ions and organic chelates. Among frequent human-impacted sources for Cd release are metal (Zn) mining and smelting, rock phosphate-derived fertilizers, atmospheric deposition and sewage sludges. Metal plating industries, batteries and pigments are also potential sources of Cd release into environments. Presence of Cd in plant tissues may inhibit cell division and cell proliferation, cause chromosomal aberrations, alteration in RNA synthesis (Liu et al., 1994), detrimental plasmolytic shrinkage and increased vacuole numbers (Daud et al., 2009b). There has been a wide variation in plant species for Cd tolerance (Grant et al., 1998); this tolerance mechanism may consist of Cd exclusion from cells or compartmentalization and detoxification of Cd-binding peptides (Sanita di Toppi and Gabbrielli, 1999).

Daud et al. (2009a, b) observed toxic effects of elevated Cd concentrations on cotton cultivars and found the slightly positive impact of low Cd concentration (10 and 100 µM) on seed germination; however, higher concentration of Cd (1000 µM) significantly inhibited the seed germination. The internal detoxification of lower Cd concentrations might be through apoplastic and symplastic binding. The authors further observed that higher Cd levels were responsible for damaging chloroplasts and mitochondria. Reduction in plant growth under Cd stress was also suggested by Wan et al. (2011), who observed significant reductions in photosynthesis and nutrient uptake in *Brassica napus* seedlings.

Cd-contaminated soils always have approximately 100–200-fold higher Zn concentrations than Cd. Zn toxicity limits the maximum Cd uptake by plants, even in calcareous soils. High chloride in a saline soil can promote Cd uptake and, therefore, the measures of the chloride need to be made before the start of any experimental work on Cd. Many scientists have also reported a small positive response on growth and yield with low levels of added Cd (Liu et al., 1994).

Some consider it a 'hormesis' effect, which only gives it a name. Sometimes the cause could be clarified by looking at the nutrient solution used. Sometimes, it seems real, but when researchers try to test for an essential response, they find plants could complete their cycle with no added Cd. This is an area where further research could help us understand several hidden aspects of Cd regarding plant physiology.

Nickle (Ni)

Ni holds a special place among the heavy metals, because it is a constituent of urease, and plays a role in hydrogenase metabolism. Its essentiality for plants is based on the numerous symptoms caused by deficiency, its physiological functions and critical roles in various enzymes (Muyssen et al., 2004). It is not as important for plant metabolism as Zn or Cu and is toxic in excessive concentrations. According to Singh et al. (1990), the toxicity of Ni is the most prevalent among the non-essential elements and toxicity symptoms cannot be easily described. In addition, the introduction of large quantities of Ni in the environment through increased industrialization has posed potential hazards to animals and human health. Higher Ni concentrations have been found in plants and soils in the areas of Ni mines and smelters, and toxic effects of Ni have been recorded on vegetation cultivated in those areas (Temple and Bisessar, 1981). Elevated Ni levels in the nearby farmland soil were found from a smelting facility in Port Colborne, Ontario (Canada).

Boreal ecosystems contaminated by emission of Ni through smelters depict severe Ni phytotoxicity, whereas natural serpentine ecosystems with significant amounts of Ni contamination and neutral pH soils contaminated with smelter emissions may not reflect any phytotoxicity. Identifying why these differences exist in ecosystems is important in order to remediate Ni-contaminated phytotoxic soils as well as for the development of environmental regulations. Many of the older published works (Beckett and Davis, 1978; Cunningham et al., 1975a, b) have relied on examining soil Ni risk assessments based on the methods, conditions or treatments that are now considered irrelevant and, therefore, helped very little in understanding a valid risk-assessment approach. For example, many scientists have relied on soluble salt addition studies only, and other toxicological approaches, ignoring the fact that animals can gain access to plants with Ni over about 50–75 ppm, and tolerate considerably higher levels. Similarly, Aboulroos et al. (1996) measured Ni activities in 11 alkaline and slightly acidic soils using the chelate method and evaluated the results against the stability diagram for different Ni solids. None really fit, but stretched to suggest that perhaps amorphous Ni-ferrite was controlling Ni activity in soils. The method can be applied to soils of much lower pH, so the important range of Ni activity cannot be tested by these methods. Moreover, all the soils were so low in Ni that plant uptake was limited by both the solubility and the total supply, leading to the analytical results being suspect in the first place. Further, the reaction of Ni-diethylene-trinitrilopentaacetic acid (DTPA) with the soils did not show ideal behaviour. The 'soil-Ni' line was estimated from these data and was a good fit to the data. Ni, $Ni(OH)_2$, $NiCO_3$, Ni_2SiO_4, $Ni_3(PO_4)_2$, $NiSL_2O_4$ and $NiFe_2O_4$ were in poor agreement with the apparent activity of Ni^{2+}. The authors gave little or no consideration to adsorption being the soil reaction which controls Ni^{2+} activity and strongly stressed the Ni-ferrite compound which is not known to be important in nature. Beckett and Davis (1978) examined the effect of Cu, Ni, Zn or Cd in the leaves and shoots of young barley on their yield of dry matter; however, several serious questions arise from the study: (1) displacement of Fe from the Fe-EDTA lowers the activity of free metals in the test solutions; severity is affected by pH; (2) Fe supply to grasses by Fe-EDTA is not a good idea because of phytosiderophores; (3) the truth is that each of the intersecting lines has a range of variance, or the upper and lower 95% confidence intervals. The points at which these intersect says something about the reliability of the results. The authors simply took the intersection points from different experiments, and estimated the variance on these numbers – not at all the correct way to handle these data. The data could be reinterpreted at the 25% yield reduction point from curves, and where some symptoms should have been noted. The authors did not report the symptoms of toxicity for any of the elements tested. This combination of confounding factors, coupled with the use of a point estimate of the upper critical level, probably caused the results to overstate the severity of metal toxicity in plants. Moreover, the use of sand culture, and Fe-EDTA, with a grass, and nitrate-N, make it difficult to use these findings; therefore, this research needs to be redone with modern methods.

Lead (Pb)

The average abundance of Pb in Earth's crust is estimated at about 15 ppm, mainly derived from magmatic rocks and argillaceous sediments. However, due to widespread Pb pollution, most soils are likely to be enriched with this metal, especially in the top horizon. In the soils, it is mainly associated with clay minerals, Mn oxides, Fe and Al hydroxides and SOM (Bradl, 2004). In soils, lead sulphide (PbS) is the most stable solid form within the soil matrix, and it is formed under reducing conditions in the presence of high sulphide concentrations. Frequent anthropogenic sources include metal mining, smelting and processing, traffic (leaded gasoline), incineration processes, disposed lead acid batteries (in particular metal scrap sites), paints and waste. Although organic Pb compounds are released from traffic and with gasoline spills, their transformation to inorganic Pb compounds is believed to be fast, and their significance is low. Pb is considered the least mobile of the heavy metals in soils. Moreover, its reactions vary widely among areas because it enters the soil in the form of various and complex compounds (Kabata-Pendias and Pendias, 2001). However, contradictory views are present in literature about the stability/mobility of Pb in soils.

Pb is seldom phytotoxic except immediately after salt addition to acidic soils. In the calcareous soil, it is unlikely that Pb toxicity would be observed unless the soil is deficient in P. Immediate cropping is useless. For an excellent review of the lack of phytotoxicity, readers are referred to Koeppe (1981). The key risk from Pb in soil is bioavailability to children who ingest Pb. In field trials in 1969–70, lead acetate applied at eight rates from 0 to 3.2 t Pb ha^{-1} had no effect on emergence, morphology, grain yield or grain Pb content of maize cv. De Kalb XL-45A. The Pb content (which increased with each increment of applied Pb) of whole young plants, leaves at tasselling and whole plants at grain harvest with 0 and 3.2 t applied Pb ha^{-1} were 2.4 and 37.8, 3.6 and 27.6, and 4.2 and 20.4 ppm, respectively (Baumhardt and Welch, 1972). Food-chain transfer could cause some crops, especially low-lying vegetables, to contain higher than normal Pb levels, but if children were exposed to the soil, the soil ingestion risk would be much greater than the food ingestion. Thus bio-accessibility testing would be more useful than all the other planned testing. Using the pH 2.2 extraction according to Zia et al. (2011) would be useful.

Arsenic (As)

Soil and water contamination with As is a major environmental problem due to its toxic and carcinogenic nature. This metal has been an issue of concern for the last few decades because of its serious impact on human health due to prolonged-period ingestion of water containing small As contents. Around 100 million people are at risk in Asia alone, and 14 million people are potentially affected in Latin America. Marine organisms are among the greatest bio-accumulators of As due to their tendency to replace N or P in several compounds, thus producing AsBet, AsChol, algal arsenosugars, etc. Interestingly, these marine bio-accumulators of As are harmless to the marine ecosystem (Saha et al., 1999).

Considering the intensity of health risks associated with As contamination, in 2002 the USEPA lowered the maximum contaminant level of As in drinking water from 50 to 10 µg l^{-1}, thereby making remediation of As-contaminated water an increasingly important and potentially expensive issue (Smith et al., 2002). In groundwaters of Pakistan, India, Bangladesh, Sri Lanka, Japan and other countries, As levels have been reported to surpass the maximum permissible limit of 50 µg l^{-1} as recommended by the World Health Organization. Ingestion of inorganic As could cause both non-cancer and cancer health effects for the human body. By contrast, organic arsenicals in soils are often less toxic and are exemplified by methyl and phenyl arsenates and compounds (Smith et al., 1998). The mechanism of the carcinogenic effect of As is not well understood, but presumably both genomic and epigenomic alterations associated with As-driven neoplastic processes and changes in other factors such as keratins, could explain the association between As and squamous cell carcinomas in humans (Martinez et al., 2011).

Historical use of As-containing pesticides has also resulted in soil contamination with high and variable concentrations of As. Until now, various strategies such as filtration, immobilization and precipitation have been developed to remediate As-contaminated soil environments. However, these techniques are often expensive and proven only in laboratory tests with little or no applicability under natural field conditions (Mahimairaja et al., 2005). After the first identification of an As hyperaccumulator *Pterisvittata* L (Chinese brake fern), by Ma et al. (2001) capable of accumulating up to 2.3% dry biomass of As, many As-hyperaccumulators have been reported, with large variations in their bioaccumulation potentials and tolerance levels (Meharg, 2003). It has been found that *P. calomelanos* var. *austroamericana*(Domin) Farw. (Pteridaceae) can accumulate As up to 16415 ppm in fronds (Kachenko et al., 2007). Recently, Niazi et al. (2011) found that gold dust fern (*P. calomelanos* var. *austroamericana*(Domin) Farw. (Pteridaceae) is a more efficient hyperaccumulator of As than the Chinese brake fern under natural field conditions in NSW, Australia. The higher accumulation of As (1262–3941 ppm) and bioaccumulation factor (BF = 3–5) in gold dust fern suggested that this species is more efficient with regards to As mining.

Literature shows that the phytoremediation using As-hyperaccumulating ferns could potentially be utilized as an environmentally friendly and low-cost remediation technology to remove As from contaminated soil and water. Nevertheless, this technology is still in its infancy, and it has yet to be used commercially to any extent in realistic field conditions. The commercialization of ferns as a sustainable remediation tool warrants understanding of the hyperaccumulation functionality of these ferns on a range of soils with contrasting mineralogy.

RISK ASSESSMENT USING BIOAVAILABILITY AND BIO-ACCESSIBILITY TECHNIQUES

The presence of heavy metal content in the environment increases the potential risk of their uptake by plants. Crops containing excessive amounts of these elements could cause toxicity to livestock and human beings upon consumption. However, the toxic effects on livestock depend on different factors such as species, metal type and concentrations (Table 2.4). Metal toxicity in humans has also been attributed to cases of overdose or therapeutic use. Therefore, it is essential to develop a detailed risk-assessment plan for estimating heavy metal accumulation in soils and water for judicious application of fertilizers, organic wastes and pesticides to agricultural soils (Papafilippaki et al., 2007). Bioavailability and bio-accessibility are complex issues that determine the possible adverse effects when the biota is exposed to contaminants. The determinants of bioavailability and bio-accessibility must be well understood to decipher the effects of metals. Assessment of soil heavy metal bioavailability may profoundly affect the extent of remediation required at contaminated sites by improving human exposure estimates. Metal bioavailability estimates can significantly alter risk assessments and remediation goals; therefore, convenient, rapid, reliable, and

TABLE 2.4 Heavy Metals Sources, Critical Levels in Soils with respect to Human Health and Concentrations Associated with Deficiency, Sufficiency, Phytotoxicity and Tolerance in Plants

Metal	Sources	Total (Acid Extractable)	AB-DTPA/DTPA Extractable	Deficient	Normal	Toxic	Tolerance	Phytotoxicity
Cd	Geogenic sources, anthropogenic activities (Nriagu and Pacyna, 1988); Metal smelting and refining, fossil fuel burning, application of phosphate fertilizers, sewage sludge (Alloway, 1995; Kabata-Pendias and Pendias, 2001)	0.5 (Rowell, 1994)	0.31 (MacLean et al., 1987)	—	0.05–0.20	5–30	3	>2
Cr	Electroplating industry, sludge, solid waste, tanneries (Knox et al., 1999)	200.0 (Huamain et al., 1999)	—	—	0.1–0 5	5–30	2	>5
Cu	Electroplating industry, smelting and refining, mining, biosolids (Liu et al., 2005)	20.0 (Rowell, 1994)	0.5 (Soltanpour, 1985)	2–2	5–30	20–100	50	20–100
Fe		—	5.00 (Soltanpour, 1985)	—	—	—	—	—
Mn		—	1.00 (Soltanpour, 1985)	10–30	30–300	100–400	300	—

Continued

TABLE 2.4 Heavy Metals Sources, Critical Levels in Soils with respect to Human Health and Concentrations Associated with Deficiency, Sufficiency, Phytotoxicity and Tolerance in Plants—cont'd

Metal	Sources	Total (Acid Extractable)	AB-DTPA/DTPA Extractable	Deficient	Normal	Toxic	Tolerance	Phytotoxicity
Ni	Volcanic eruptions, land fill, forest fire, bubble bursting and gas exchange in ocean, weathering of soils and geological materials (Knox et al., 1999)	20.0 (Rowell, 1994)	8.10 (MacLean et al., 1987)	—	0.1–5.0	10–100	50	>50
Pb	Mining and smelting of metalliferous ores, burning of leaded gasoline, municipal sewage, industrial wastes enriched in Pb, paints (Gisbert et al., 2003; Seaward and Richardson, 1990)	50.0 (Rowell, 1994)	13.00 (MacLean et al., 1987)	—	5–15	30–300	10	>20
Zn	Electroplating industry, smelting and refining, mining, biosolids (Liu et al., 2005)	80.0 (Rowell, 1994)	1.50 (Soltanpour, 1985)	10–20	27–150	100–400	300	150–200

Adapted from Kabata-Pendias and Pendias (2001).

inexpensive test methods are needed to determine soil contamination that does not necessarily mimic human / animal digestive tract.

In the above sections, we provided an overview of the current science related to Pb, Cd, Ni and As toxicity, and summarized the existing management actions, and additionally proposed how to reduce risks associated with exposure to these heavy metals. For example, estimation of the spatial distribution of As is imperative to precisely quantify its concentration at contaminated sites (Aelion et al., 2009; Smith et al., 1998). Such information can also assist in designing suitable remediation and management strategies of As-contaminated sites (Alary and Demougeot-Renard, 2009), as it can delineate the level and spread of As contamination.

CONTROL MEASURES

Removal of Heavy Metals from Soils

Studies have indicated that vegetables, particularly leafy crops, grown over metal-contaminated soils have higher concentrations of these toxic elements than those grown in normal soils (Guttormensen et al., 1995). Keeping in view the importance of heavy metal pollution for human and animal health, there is an urgent need to develop technologies for heavy metals removal from the contaminated soils and waters. The first, and most effective technique on a small scale, is soil removal and replacement with new. This brings obvious drawbacks of disposing of the soil and sourcing appropriate new soil. The cost could be very high. As an alternate, soil washing has also been proposed as a way to remove the heavy metals from soil in a water-based solvent that contains iron chloride. Again it is a time-consuming, labour-intensive and expensive technique, and is applicable only to small, usually urban sites.

Phytoextraction of Heavy Metals

Phytoremediation is an environment-friendly and cheaper technology for the remediation of metal-polluted environments. Plant-based heavy metal removal from the soils can be divided into different categories: (1) use of plants with capacity to uptake higher amount of metals from soils (hyperaccumulators); (2) use of fast-growing plant species coupled with chemical amendments; and (3) changes in chemical form of elements inside the plant body, making it less toxic (phytovolatilization) (McGrath et al., 2002). The term 'hyperaccumulators' was first coined by Brooks et al. (1977) for the plants having potential to accumulate Ni > 1000 ppm on a dry leaf basis in their natural habitats. A similar concentration (1000 ppm in plant shoot dry matter) is defined for other metals such as Co, Cu, Pb and Se, while it is 10,000 ppm for Zn and Mn and 100 ppm for Cd. More than 400 plant species have been identified worldwide that have potential to accumulate high metal concentrations from contaminated soils (Baker et al., 2000).

Removal of heavy metals from soils to plant tissues involves several steps, viz. metal entry to plant roots via root cell plasma membranes, and its translocation to above-ground parts through xylem tissues, where it is sequestered in specialized cells through compartmentation and complexation with legends (Najeeb et al., 2009).

Hyperaccumulating plants have relatively higher potential of translocating metal from roots to shoots, which make them an efficient metal harvester. A Zn hyperaccumulator ecotype of *Thlaspi caerulescens* is reported to have a 10 times higher translocation factor (ability to translocate metals to shoots) for Zn compared to its non-accumulator ecotype (Lasat et al., 1996). Similar results were obtained by Liu et al. (2008): they observed a threefold greater Pb translocation in the hyperaccumulating ecotype of *Sedum alfredii*.

Some of the wetland plants such as *Leersia hexandra*, *Juncus effusus* and *Equisetum ramosisti* have been found to be potential candidates for phytoremediation due to their ability to sequester high metal concentrations in their below-ground parts (Deng et al., 2004; Najeeb et al., 2011). These wetland plants can be ideal candidates for metal removal from contaminated sites due to their fast growth rate. Attempts have been made to improve translocation factors for wetland metal-accumulating plant species using chelating agents, and a degree of success has been achieved (Najeeb et al., 2009).

Provided soil and environmental conditions are favourable, metal-contaminated soil replacement and removal costs about $2 million per hectare for 30 cm soil depth. Similarly, in situ remediation costs could be estimated to be of the order of $3000–10,000 ha^{-1}. Opposite to this, phytoextraction, for example of soil Ni, could help farm managers earn more money from the recovery of soil Ni as *Alyssum murale* plant ash for sale to Ni recycling facilities than growing common crops on the remediated soils. The concept of Ni phytoextraction was first reported by Chaney (1983). One can phytoextract between 200 and 400 kg ha^{-1} Ni annually by growing Ni-hyperaccumulator plant species with high biomass yield potential (e.g. *Alyssum murale*). The research work conducted by United States Department of Agriculture - Agriculture Research Service (Chaney et al., 2007) in cooperation with Viridian LLC has licensed all of the patented Ni phytoextraction technologies. *A. murale* is perennial, and re-grows after cutting at flowering to harvest biomass. This species accumulated up to 25,000 ppm Ni in above-ground biomass from field plots in Oregon without yield reduction from the Ni (compared to the 100 ppm Ni which causes visible phytotoxicity in crop plants). Leaves contain about double the Ni level of stems plus petioles, so leaves of these plants can accumulate up to 40,000 ppm Ni (Chaney*et al.*, 2007).

Chelator-Assisted Heavy Metals Phytoremediation

Use of metal-mobilizing chemical agents is proposed for improving metal removal from polluted sites. Since most of the hyperaccumulating plants are slow growing and low biomass producing (McGrath et al., 2002), use of chelator-assisted phytoremediation could speed up the process by using non-hyperaccumulating plant species having higher biomass production potential (Najeeb et al., 2009).

Numerous synthetic chelators including ethylene diamine tetra acetic acid (EDTA), diethylenetrinitrilopentaacetic acid (DTPA) and nitrilotriacetic acid (NTA) have been used for increasing phytoremediation of metal-contaminated soils. These chelating agents increased not only the metal mobility and bioavailability but also its translocation to aerial plant parts (Luo et al., 2006; Tandy et al., 2006). For example, low solubility and mobility of Pb due to strong binding to soil particles restrict its uptake by plant roots. Pb uptake from polluted soils is as low as 50 ppm (Cunningham et al., 1995) with a minimum translocation factor among the phytotoxic metals. Using different chemical amendments such as EDTA and DTPA for Pb phytoremediation from contaminated soils, Huang et al. (1997) observed a 120-fold increase in net Pb translocation with 1.0 g of EDTA kg^{-1} soil compared to the control plants.

Saifullah et al. (2009), in a review, found EDTA as the most effective chelator for increasing Pb solubility in soil, while DPTA and EDTA were found to be more suitable for phytoextraction of Zn and Cd, and Ethylenediamine-N,N'-disuccinic acid, EDTA and DTPA for Cu. In addition, the synergetic effect of these chelators has also been reported for phytoremediation of heavy metals (Liu et al., 2007). López et al. (2005) found a 2800% and about a 600% increase in Pb contents of *Medicago sativa* leaves under the combined application of 100 μM indole-3-acetic acid (IAA) / 200 μM EDTA compared to treatments with Pb alone and Pb + EDTA, respectively.

However, low biodegradability and long persistence in soils make EDTA unsuitable for in situ application (Meers et al., 2005). Higher EDTA addition to soil causes micronutrient losses, approximately ⅓ of total applied EDTA leached down to the soil with other ions posing potential risk for groundwater pollution (Wu et al., 2004). Alternatively, low-molecular-weight organic acids (LMWOA) could be used to substitute EDTA due to their easy biodegradation as the natural products of root exudates, microbial secretions, and plant and animal residue decomposition in soils. Exploring phytoextraction potential of wetland plant *Juncus effuses*, Najeeb et al. (2009, 2011) observed the positive impact of LMWOA (citric acid) on growth and the ability of plants to remove heavy metals (Cd and Mn) from the nutrient media. Although EDTA was more efficient at accumulating metals, its higher concentrations damaged plant root and shoot cells. Conversely, citric acid protected plants from metal toxicity. Thus, LMWOA might be preferred over traditional synthetic chelators for a sustainable phytoremediation process.

Biochar Application as an Emerging Tool for Reducing Heavy Metal Impacts

Biochar is a carbon-rich, porous material (Atkinson et al., 2010), which is proposed to have various agronomic and environmental benefits once added to the soil (Kookana et al., 2011). It is produced by the thermal decomposition of organic biomass, through pyrolysis, which involves heating biomass under conditions of partial or complete absence of oxygen (Sohi et al., 2010). Biochar comprises mainly of stable aromatic forms of organic carbon (C) (Sohi et al., 2010); however, it is distinguished from charcoal with regards to its intended purpose of use (Kookana et al., 2011). The incorporation of biochar into the soil is expected to achieve agronomic and environmental benefits.

Biochar and Heavy Metal Pollution?

In the agricultural context, heavy metal contamination is a serious problem facing land managers. The adsorption of heavy metals from the soil environment plays a critical role in reducing ecological and human health risks associated with heavy metal contamination. There are various pathways for, heavy metals' entry into the soil environment, and once in the soil, these are readily taken up by plants (Evangelou et al., 2007). Numerous soil remediation methods have been proposed; of these, the most common involves immobilizing the toxic elements in soil to reduce uptake by plants (Namgay et al., 2010). Addition of SOM to soil offers a promising approach for immobilizing most trace elements in the soil (van Herwijnen et al., 2007); however, a caveat is that once the organic matter decomposes, the adsorbed metals are released (Hjortenkrans et al., 2007). Biochar is an alternative that may prove efficient in adsorbing heavy metals from the soil, by forming complexes with metal ions (Namgay et al., 2010). To be effective as a contaminant soil remediation strategy, biochar should fulfil two key criteria. Foremost, its high pollutant sorption capacity should be sustained over a long time period, under changing environmental conditions; and secondly, the sorption capacity should be maintained in the presence of soil (Hale et al., 2011). Due to the highly recalcitrant nature of biochar, it is likely that the biochar-trace elements formed would be more stable than those of organic matter (Namgay et al., 2010).

What Mechanisms Are Responsible for the Adsorption of the Metals?

Available evidence suggests that the organic functional groups and adsorption sites present on the surface of biochar may influence its cation exchange capacity (Cheng et al., 2008) and in turn play a crucial role in the formation of biochar–trace element complexes (Lee et al., 2010).

What Next?

Another fundamental issue concerning land managers is nutrient retention in the soils and their uptake by plants. It is known that biochar properties are influenced by various parameters, notably the pyrolysis conditions and the feedstock source (Singh et al., 2010). Furthermore, although biochar is regarded as being chemically and biologically recalcitrant in nature (Cross and Sohi, 2011), it undergoes 'aging' once added to the soil (Joseph et al., 2010), a weathering process whereby oxidization takes place (Cheng et al., 2008). Currently, a huge knowledge gap exists in our understanding of how ageing impacts the integrity of biochar, which in turn will have significant implications for the use of biochar for agronomic and environmental purposes (Uchimiya et al., 2010).

Previous studies have demonstrated the absorptive capabilities of biochar on fertilizers, pesticides and other organic compounds, pyrene and heavy metals. Based on the literature reviewed, there is evidence to suggest that the surface charge characteristics of biochar play a significant role in its interaction with other compounds. However, with the ageing process, how long it takes to return heavy metals into the environment again is yet to be explored.

CONCLUSIONS

Heavy-metal-polluted soil environments can be reclaimed through the application of chemicals, soil amendments or phytoaccumulators. Phytoremediation is a cost-effective and sustainable method of reclamation of heavy-metal-polluted environments. Certain plant species have the ability to uptake higher concentrations of metal ions into their harvestable parts. Selection of plants with high potential to remove heavy metals from polluted soils and engineering genes regulating metal accumulation can further assist the phytoremediation process. Synthetic chelators (EDTA, DTPA) accelerate the phytoextraction process by enhancing metal bioavailability and translocation. However, slow degradation and long persistence of the same in soil pose a threat to environmental sustainability; low-molecular-weight organic acids offer an alternative option. In addition, to supplement the phytoremediation work, exploration of new technologies such as the use of microorganisms, biochar, etc., is crucial for making land reclamation processes more sustainable. Another challenge before the scientific community is to put the phytoremediation theory into practice, (i.e. commercial phase), which is only valid in the case of nickel at the moment. Moreover, bioavailability-based phytoremediation efforts can save significant costs in time and space. The farming community should use only those rock phosphate-based fertilizers that comply with relevant fertilizer legislation as per specific soil requirement for specific crop rotation.

REFERENCES

Aboulroos, S.A., El-Falaky, A.A., Lindsay, W.L., 1996. Activity measurements of Ni^{2+} in soils. Z. Pflanzenernaehr. Bodenk 159, 399–404.

Aelion, C.M., Davis, H.T., Liu, Y., Lawson, A.B., McDermott, S., 2009. Validation of Bayesian Kriging of arsenic, chromium, lead, and mercury surface soil concentrations based on internode sampling. Environ. Sci. Technol. 43, 4432–4438.

Alary, C., Demougeot-Renard, H., 2009. Factorial Kriging analysis as a tool for explaining the complex spatial distribution of metals in sediments. Environ. Sci. Technol. 44, 593–599.

Alloway, B., 1995. Heavy metals in soils, third ed. Blackie Academic and Professional, London. p. 368.

Alloway, B.J., 1991. Heavy metals in soils. John Willey and Sons, Inc, N.Y. USA.

Alloway, W.H., 1968. Trace elements cycling. Adv. Agron. 20, 235–274.

Antil, R.S., 2012. In: Show, Kuan-Yeow (Ed.), Impact of sewage and industrial effluents on soil–plant health, Industrial Waste. ISBN: 978-953-51-0253-3, InTech, Available at http://www.intechopen.com/books/industrial-waste/impact-of-sewer-water-and-industrial-wastewaters-on-soilplant-health(Accessed 29.09.13.).

Atkinson, C., Fitzgerald, J., Hipps, N., 2010. Potential mechanisms for achieving agricultural benefits from biochar application to temperate soils: a review. Plant Soil 337, 1–18.

Baechle, H.T., Wolsrein, F., 1984. Cd compounds in minerals fertilizers. The Fertilizer Society. Proceeding No. 226, London.

Baker, A., McGrath, S., Reeves, R., Smith, J., 2000. Metals, plant strategies toward detoxification, metal ligands, sequestration oxygen donor ligands, sulfur donor ligands, nitrogen donor. Norman Terry.

Baumhardt, G., Welch, L., 1972. Lead uptake and corn growth with soil-applied lead. J. Environ. Qual. 1, 92–93.

Beckett, P.H.T., Davis, R.D., 1978. Upper critical levels of toxic elements in plants. New Phytol. 79, 95–106.

Bergkvist, P., Jarvis, N., Berggren, D., Carlgren, K., 2003. Long-term effects of sewage sludge applications on soil properties, cadmium availability and distribution in arable soil. Agric. Ecosys. Environ. 97, 167–179.

Bohn, H.L., McNeal, B.L., O, Cornor, G.A., 2010. Soil Chemistry, third ed. John Wiley and Sons, N.Y. p. 320.

Bradl, H.B., 2004. Adsorption of heavy metal ions on soils and soils constituents. J Colloid. Interf. Sci. 277, 1–18.

Brooks, R., Lee, J., Reeves, R., Jaffré, T., 1977. Detection of nickeliferous rocks by analysis of herbarium specimens of indicator plants. J. Geochem. Explor. 7, 49–57.

Centre for Disease Control and Prevention (CDC), 2009. Fourth National Report on Human Exposure to Environmental Chemicals. Available at http://www.cdc.gov/ExposureReport/pdf/Fourth Report.pdf (accessed 29.09.13.).

Chaney, R., 1983. Plant uptake of inorganic waste constituents. In: Land Treatment of Hazardous Wastes. Noyes Publications, p. 50.

Chaney, R.L., Angle, J.S., Li, Y.M., and Baker, A.J. (2007). Recovering metals from soil. In US Patent 7, 268–273.

Chaudri, A.M., Allain, C.M., Badawy, S., Adams, M.L., McGrath, S.P., Chambers, B.J., 2001. Cadmium content of wheat grain from a long-term field experiment with sewage sludge. J. Environ. Qual. 30, 1575–1580.

Cheng, C.H., Lehmann, J., Engelhard, M.H., 2008. Natural oxidation of black carbon in soils: Changes in molecular form and surface charge along a climosequence. Geochim. Cosmochim. Ac. 72, 1598–1610.

Corbett, J.J., 2002. Emissions from ships in the Northwestern United States. Environ. Sci. Technol. 36, 1299–1306.

Cross, A., Sohi, S.P., 2011. The priming potential of biochar products in relation to labile carbon contents and soil organic matter status. Soil Biol. Biochem. 43, 2127–2134.

Cunningham, J.D., Ryan, J.A., Keeney, D.R., 1975a. Phytotoxicity in and metal uptake from soil treated with metal-amended sewage sludge. J. Environ. Qual. 4, 455–460.

Cunningham, J.D., Ryan, J.A., Keeney, D.R., 1975b. Phytotoxicity and uptake of metals added to soils as inorganic salts or in sewage sludge. J. Environ. Qual. 4, 460–462.

Cunningham, S.D., Berti, W.R., Huang, J.W., 1995. Phytoremediation of contaminated soils. Tibtech 13, 393–397.

Daud, M., Sun, Y., Dawood, M., Hayat, Y., Variath, M., Wu, Y.X., Mishkat, U., Najeeb, U., Zhu, S., 2009a. Cadmium-induced functional and ultrastructural alterations in roots of two transgenic cotton cultivars. J. Hazard. Mater. 161, 463–473.

Daud, M.K., Variath, M.T., Ali, S., Najeeb, U., Jamil, M., Hayat, Y., Khan, M.D., Zaffar, M.I., Cheema, S.A., Tong, X.H., Zhu, S., 2009b. Cadmium-induced ultramorphological and physiological changes in leaves of two transgenic cotton cultivars and their wild relative. J. Hazard. Mater. 168, 614–625.

Deng, H., Ye, Z.H., Wong, M.H., 2004. Accumulation of lead, zinc, copper and cadmium by 12 wetland plant species thriving in metal-contaminated sites in China. Environ. Pollut. 132, 29–40.

Evangelou, M.W.H., Ebel, M., Schaeffer, A., 2007. Chelate assisted phytoextraction of heavy metals from soil. Effect, mechanism, toxicity, and fate of chelating agents. Chemosphere 68, 989–1003.

Gimeno García, E., Andreu, V., Boluda, R., 1996. Heavy metals incidence in the application of inorganic fertilizers and pesticides to rice farming soils. Environ. Pollut. 92, 19–25.

Gisbert, C., Ros, R., de Haro, A., Walker, D.J., Pilar, B.M., Serrano, R., Avino, J.N., 2003. A plant genetically modified that accumulates Pb is especially promising for phytoremediation. Biochem. Biophysic. Res. Commun. 303, 440–445.

Grant, C., Buckley, W., Bailey, L., Selles, F., 1998. Cadmium accumulation in crops. Can. J. Plant Sci. 78, 1–17.

Gupta, S.K., Vollmer, M.K., Krebs, R., 1996. The importance of mobile, mobilisable and pseudo total heavy metal fractions in soil for three-level risk assessment and risk management. Sci. Total Environ. 178, 11–20.

Gupta, U.C., 1979. Boron in nutrition of crops. In: Sparks, D.L. (Ed.). Advances in Agronomy, vol. 31. Academic Press, pp. 273–307.

Gupta, U.C., Gupta, S.C., 1998. Trace element toxicity relationships to crop production and livestock and human health: implications for management. Commun. Soil Sci. Plant Anal. 29, 1491–1522.

Guttormensen, G., Singh, B.R., Jeng, A.S., 1995. Cadmium concentration in vegetable crops grown in a sandy soil as affected by Cd levels in fertilizer and soil pH. Fertilizer Res. 41, 27–32.

Hale, S., Hanley, K., Lehmann, J., Zimmerman, A., Cornelissen, G., 2011. Effects of chemical, biological, and physical aging as well as soil addition on the sorption of pyrene to activated carbon and biochar. Environ. Sci. Technol. 45, 10445–10453.

Harsh, J.B., Doner, H., 1981. Characterization of mercury in a riverwash soil. J. Environ. Qual. 10, 333–337.

Hjortenkrans, D.S., Bergbäck, B.G., Häggerud, A.V., 2007. Metal emissions from brake linings and tires: case studies of Stockholm, Sweden 1995/1998 and 2005. Environ. Sci. Technol. 41, 5224–5230.

Huamain, C., Chunrong, Z., Cong, T., Yongguan, Z., 1999. Heavy metal pollution in soils of China: Status and counter-measures. Ambio. 28, 130–134.

Huang, J.W., Chen, J., Berti, W.R., Cunningham, S.D., 1997. Phytoremediation of lead-contaminated soils: role of synthetic chelates in lead phytoextraction. Environ. Sci. Technol. 31, 800–805.

Hussain, S.I., Ghafoor, A., Ahmad, S., Murtaza, G., Sabir, M., 2006. Irrigation of crops with raw sewage: Hazard assessment of effluent, soil and vegetables. Pak. J. Agric. Sci. 43, 97–102.

Joseph, S., Camps-Arbestain, M., Lin, Y., Munroe, P., Chia, C., Hook, J., Van Zwieten, L., Kimber, S., Cowie, A., Singh, B., 2010. An investigation into the reactions of biochar in soil. Soil Res. 48, 501–515.

Kabata-Pendias, A., Pendias, H., 1992. Trace elements in soils and plants, second ed. CRC Press, Boca Raton, FL. p. 365.

Kabata-Pendias, A., Pendias, H., 2001. Trace elements in soils and plants, third ed. CRC Press, FL USA.

Kabir, E., Ray, S., Kim, K.H., Yoon, H.O., Jeon, E.C., Kim, Y.S., Cho, Y.S., Yun, S.T., Brown, R.J.C., 2012. Current Status of Trace Metal Pollution in Soils Affected by Industrial Activities Scientific World Journal. 2012; 2012: 916705. http://dx.doi.org/10.1100/2012/916705.

Kachenko, A., Bhatia, N., Singh, B., Siegele, R., 2007. Arsenic hyperaccumulation and localization in the pinnule and stipe tissues of the gold-dust fern (*Pityrogramma calomelanos* (L.) Link var. austroamericana (Domin) Farw.) using quantitative micro-PIXE spectroscopy. Plant Soil 300, 207–219.

Kapungwe, E.M., 2013. Heavy metal contaminated water, soils and crops in peri urban wastewater irrigation farming in Mufulira and Kafue Towns in Zambia. J. Geogr. Geol. 5, 55–72.

Khan, J.M., Bhatti, A.U., Hussain, S., Wasiullah, 2007. Heavy metal contamination of soil and vegetable with industrial effluents from sugar mill and tanneries. Soil Environ. 26, 139–145.

Kiekens, L., 1995. Zinc. In: Alloway, B.J. (Ed.), Heavy metals in soils. Blackie Academic and Professional, London, pp. 284–305.

Knox, A.S., Gamerdinger, A.P., Adriano, D.C., Kolka, R.K., Kaplan, D.I., 1999. Sources and practices contributing to soil contamination. In: Adriano, D.C., Bollag, J.M., Frankenberg Jr, W.T., Sims, R.C. (Eds.), Bioremediation of the Contaminated Soils, pp. 53–87. Agronomy Series No. 37, ASA, CSSA, SSSA, Madison, Wisconsin, USA.

Koch, M., Rotard, W., 2001. On the contribution of background sources to the heavy metal content of municipal sewage sludge. Water Sci. Technol. 43, 67–74.

Koeppe, D.E., 1981. Lead: Understanding the minimal toxicity of lead in plants. In: Effect of Heavy Metal Pollution on Plants. Springer, pp. 55–76.

Kookana, R.S., Sarmah, A.K., Van Zwieten, L., Krull, E., Singh, B., 2011. Biochar Application to Soil: Agronomic and Environmental Benefits and Unintended Consequences. In: Sparks, D.L. (Ed.). Advances in Agronomy, vol. 112. Academic Press, pp. 103–143.

Lasat, M.M., Baker, A.J.M., Kochian, L.V., 1996. Physiological characterisation of root Zn^{2+} absorption and translocation to shoots in Zn hyperaccumulator and nonaccumulator species of Thlaspi. Plant Physiol. 112, 1715–1722.

Lawrence, R.W., Higgs, S.A.T.W., 1999. Removing and stabilizing as in acid mine water. JOM 51, 27–29.

Lee, J.W., Kidder, M., Evans, B.R., Paik, S., Buchanan- III, A.C., Garten, C.T., Brown, R.C., 2010. Characterization of biochars produced from cornstovers for soil amendment. Environ. Sci. Technol. 44, 7970–7974.

Liu, D., Jiang, W., Wang, W., Zhao, F., Lu, C., 1994. Effects of lead on root growth, cell division, and nucleolus of *Allium cepa*. Environ. Pollut. 86, 1–4.

Liu, D., Li, T., Yang, X., Islam, E., Jin, X., Mahmood, Q., 2007. Enhancement of lead uptake by hyperaccumulator plant species *Sedum alfredii* Hance using EDTA and IAA. Bull. Environ. Contam. Toxicol. 78, 280–283.

Liu, D., Li, T.Q., Jin, X.F., Yang, X.E., Islam, E., Mahmood, Q., 2008. Lead induced changes in the growth and antioxidant metabolism of the lead accumulating and non-accumulating ecotypes of *Sedum alfredii*. J. Integ. Plant Biol. 50, 129–140.

Liu, X.M., Wu, Q.T., Banks, M.K., 2005. Effect of simultaneous establishment of *Sedum alfridii* and *Zea mays* on heavy metal accumulation in plants. Int. J. Phytorem. 7, 43–53.

López, M.L., Peralta-Videa, J.R., Benitez, T., Gardea-Torresdey, J.L., 2005. Enhancement of lead uptake by alfalfa (*Medicago sativa*) using EDTA and a plant growth promoter. Chemosphere 61, 595–598.

Luo, C., Shen, Z., Lou, L., Li, X., 2006. EDDS and EDTA-enhanced phytoextraction of metals from artificially contaminated soil and residual effects of chelant compounds. Environ. Pollut. 144, 862–871.

Ma, L.Q., Komar, K.M., Tu, C., Zhang, W., Cai, Y., Kennelley, E.D., 2001. A fern that hyperaccumulates arsenic. Nature 409, 579.

MacLean, K.S., Robinson, A.R., MacConnell, H.M., 1987. The effect of sewage-sludge on the heavy metal content of soils and plant tissue. Commun. Soil Sci. Plant Anal. 18, 1303–1316.

Mahimairaja, S., Bolan, N., Adriano, D., Robinson, B., 2005. Arsenic contamination and its risk management in complex environmental settings. Adv. Agron. 86, 1–82.

Martinez, V.D., Becker-Santos, D.D., Vucic, E.A., Lam, S., Lam, W.L., 2011. Induction of human squamous cell-type carcinomas by arsenic. J. Skin Cancer. 2011, 1–9. http://dx.doi.org/10.1155/2011/454157.

McGrath, S.P., Zhao, J., Lombi, E., 2002. Phytoremediation of metals, metalloids, and radionuclides. In: Sparks, D.L. (Ed.). Advances in Agronomy, vol. 75. Academic Press, pp. 1–56.

McLaren, R.G., Crawford, D.V., 1973. Studies on soil copper. I. The fractionation of copper in soils. J. Soil Sci. 24, 172–181.

Meers, E., Ruttens, A., Hopgood, M.J., Samson, D., Tack, F.M.G., 2005. Comparison of EDTA and EDDS as potential soil amendments for enhanced phytoextraction of heavy metals. Chemosphere 58, 1011–1022.

Meharg, A.A., 2003. Variation in arsenic accumulation: hyperaccumulation in ferns and their allies. New Phytol. 157, 25–31.

Murtaza, G., Ghafoor, A., Qadir, M., Owens, G., Aziz, M.A., Zia, M.H., Saifullah, 2010. Disposal and use of sewage on agricultural lands in Pakistan: A review. Pedosphere 20, 23–34.

Muyssen, B.T., Brix, K., DeForest, D., Janssen, C., 2004. Nickel essentiality and homeostasis in aquatic organisms. Environ. Rev. 12, 113–131.

Najeeb, U., Jilani, G., Ali, S., Sarwar, M., Xu, L., Zhou, W., 2011. Insights into cadmium induced physiological and ultra-structural disorders in *Juncus effusus* L. and its remediation through exogenous citric acid. J. Hazard. Mater. 186, 565–574.

Najeeb, U., Xu, L., Ali, S., Jilani, G., Gong, H.J., Shen, W.Q., Zhou, W.J., 2009. Citric acid enhances the phytoextraction of manganese and plant growth by alleviating the ultrastructural damages in *Juncus effusus* L. J. Hazard. Mater. 170, 1156–1163.

Namgay, T., Singh, B., Singh, B.P., 2010. Influence of biochar application to soil on the availability of As, Cd, Cu, Pb, and Zn to maize (*Zea mays* L.). Soil Res. 48, 638–647.

Niazi, N.K., Singh, B., Van Zwieten, L., Kachenko, A.G., 2011. Phytoremediation potential of *Pityrogramma calomelanos* var. austroamericana and *Pteris vittata* L. grown at a highly variable arsenic contaminated site. Int. J. Phytorem. 13, 912–932.

NRC, 1980. National Research Council. Subcommittee on Mineral Toxicity in Animals. Mineral tolerance of domestic animals. National Academies Press.

Nriagu, J.O., Pacyna, J.M., 1988. Quantitative assessment of worldwide contamination of air water and soils by trace metals. Nature 333, 134–139.

Okonkwo, J., 2007. Arsenic status and distribution in soils at disused cattle dip in South Africa. Bull. Environ. Contam. Toxicol. 79, 380–383.

Papafilippaki, A., Gasparatos, D., Haidouti, C., Stavroulakis, G., 2007. Total and bioavailable forms of Cu, Zn, Pb and Cr in agricultural soils: a study from the hydrological basin of Keritis, Chania, Greece. Global Nest J. 9, 201.

Peryea, F.J., August 1998. Historical use of lead arsenate insecticides, resulting soil contamination and implications for soil remediation [Online]: Proceedings of the 16th World Congress of Soil Science, 25th, Montpellier, France, pp. 24–25. http://natres.psu.ac.th/Link/SoilCongress/en/symt25.htm.

Proctor, J., Woodell, S.R., 1975. The ecology of serpentine soils. Adv. Ecol. Res. 9, 255–366.

Rowell, D.L., 1994. Pesticides and metals. In: Rowell, D.L. (Ed.), Soil Science: Methods and Applications. Longman Singapore Publisher Ltd, Singapore, pp. 303–327.

Saha, J., Dikshit, A., Bandyopadhyay, M., Saha, K., 1999. A review of arsenic poisoning and its effects on human health. Crit. Rev. Environ. Sci. Tech. 29, 281–313.

Saifullah, Meers, E., Qadir, M., de Caritat, P., Tack, F.M.G., Du Laing, G., Zia, M.H., 2009. EDTA-assisted Pb phytoextraction. Chemosphere 74, 1279–1291.

Salim, R., Al-Subu, M.M., Atallah, A., 1993. Effects of root and foliar treatments with lead, cadmium, and copper on the uptake distribution and growth of radish plants. Environ. Int. 19, 393–404.

Sanita di Toppi, L., Gabbrielli, R., 1999. Response to cadmium in higher plants. Environ. Exp. Bot. 41, 105–130.

Sarkar, D., Makris, K.C., Parra-Noonan, M.T., Datta, R., 2007. Effect of soil properties on arsenic fractionation and bioaccessibility in cattle and sheep dipping vat sites. Environ. In. 33, 164–169.

Seaward, M.R.D., Richardson, D.H.S., 1990. Atmospheric sources of metal pollution and effects on vegetation. In: Shaw, A.J. (Ed.), Heavy Metal Tolerance in Plants: Evolutionary Aspects. CRC Press, Florida, pp. 75–92.

Singh, B., Dang, Y.P., Mehta, S.C., 1990. Influence of nitrogen on the behaviour of nickel in wheat. Plant Soil 127, 213–218.

Singh, B., Singh, B.P., Cowie, A.L., 2010. Characterisation and evaluation of biochars for their application as a soil amendment. Soil Res. 48, 516–525.

Smith, A.H., Lopipero, P.A., Bates, M.N., Steinmaus, C.M., 2002. Arsenic epidemiology and drinking water standards. Science 296, 2145–2146.

Smith, E., Naidu, R., Alston, A.M., 1998. Arsenic in the soil environment: A review. In: Sparks, D.L. (Ed.). Advances in Agronomy, vol. 64. Academic Press, pp. 149–195.

Smolen, S., Sady, W., Ledwozyw-Smolen, I., 2010. Quantitative relations between the content of selected trace elements in soil extracted with 0.03 M CH_3COOH or 1 M HCL and its total concentration in lettuce and spinach. Acta Sci. Polonorum Hortorum Cultus 4, 13–23.

Sohi, S.P., Krull, E., Lopez-Capel, E., Bol, R., 2010. A review of biochar and its use and function in soil. In: Sparks, D.L. (Ed.). Advances in Agronomy, vol. 105. Academic Press, pp. 47–82.

Soltanpour, P.N., 1985. Use of AB-DTPA soil test to evaluate elemental availability and toxicity. Commun. Soil Sci. Plant Anal. 16, 323–338.

Suruchi, Khanna, P., 2011. Assessment of heavy metal contamination in different vegetables grown in and around urban areas. Res. J. Environ. Toxicol. 5, 162–179.

Swaine, D.J., 1955. The trace elements of soils. Soil Science Technology Communication No. 48. Herald Printing works, Coney St, York (England).

Tandy, S., Schulin, R., Nowack, B., 2006. The influence of EDDS on the uptake of heavy metals in hydroponically grown sunflowers. Chemosphere 62, 1454–1463.

Tariq, M.I., Afzal, S., Hussain, I., Sultana, N., 2007. Pesticides exposure in Pakistan: A review. Environ. Int. 33, 1107–1122.

Temple, P.J., Bisessar, S., 1981. Uptake and toxicity of nickel and other metals in crops grown on soil contaminated by a nickel refinery. J. Plant Nutri. 3, 473–482.

Uchimiya, M., Lima, I.M., Klasson, K.T., Wartelle, L.H., 2010. Contaminant immobilization and nutrient release by biochar soil amendment: Roles of natural organic matter. Chemosphere 80, 935–940.

UN-WWAP, 2003. United Nations World Water Assessment Programme. The World Water Development Report 1: Water for People, Water for Life. UNESCO, Paris, France.

van Herwijnen, R., Laverye, T., Poole, J., Hodson, M.E., Hutchings, T.R., 2007. The effect of organic materials on the mobility and toxicity of metals in contaminated soils. Appl. Geochem. 22, 2422–2434.

Vanselow, A.P., 1966. Cobalt. In: Chapman, H.D. (Ed.), Criteria for Plants and Soils. University of California, Division of Agricultural Sciences, pp. 142–156.

Wallace, G.A., Wallace, A., 1994. Lead and other potentially toxic heavy metals in soil. Commun. Soil Sci. Plant Anal. 25, 137–141.

Wan, G., Najeeb, U., Jilani, G., Naeem, M.S., Zhou, W., 2011. Calcium invigorates the cadmium-stressed *Brassica napus* L. plants by strengthening their photosynthetic system. Environ. Sci. Pollut.Res. 18, 1478–1486.

Wong, M.H., 2003. Ecological restoration of mine degraded soils, with emphasis on metal contaminated soils. Chemosphere 50, 775–780.

Wu, L.H., Luo, Y.M., Xing, X.R., Christie, P., 2004. EDTA-enhanced phytoremediation of heavy metal contaminated soil with Indian mustard and associated potential leaching risk. Agric. Ecosys. Environ. 102, 307–318.

Wuana, R.A., Okieimen, F.E., 2011. Heavy metals in contaminated soils: a review of sources, chemistry, risks and best available strategies for remediation. International Scholarly Research Network http://dx.doi.org/10.5402/2011/102617. Article ID 402647, 20 pages.

Yanai, J., Zhao, F.J., McGrath, S.P., Kosaki, T., 2006. Effect of soil characteristics on Cd uptake by the hyperaccumulator *Thlaspi caerulescens*. Environ. Pollut. 139, 167–175.

Zia, M.H., Codling, E.E., Scheckel, K.G., Chaney, R.L., 2011. In vitro and in vivo approaches for the measurement of oral bioavailability of lead (Pb) in contaminated soils: a review. Environ. Pollut. 159, 2320–2327.

Chapter 3

Phytoremediation: A Promising Strategy on the Crossroads of Remediation

Arif Tasleem Jan, Arif Ali and Qazi Mohd. Rizwanul Haq
Department of Biosciences, Jamia Millia Islamia, New Delhi, India

INTRODUCTION

In the era of industrialization, developmental progress marked with the technological advancement has raised serious environmental challenges as they bypassed the safety demands of the natural environment (Bennett et al., 2003). Although, a number of natural processes (volcanoes, forest fires, etc.) release different pollutants into the environment, anthropogenic activities are believed to be the major cause of environmental pollution. Pollutants of various kinds, particularly heavy metals that arise as a result of accidental or process spillage, disposal of untreated industrial waste and by sludge application to agricultural soils, have contributed significantly to the deterioration of land and water resources, as is evident from changes in ecosystem processes. Amid their distribution along with potential deleterious effects on human health, their contamination presents a black picture of a complex industrialized hi-tech society. Being unable to degrade by soil microbiota, their evacuation from the polluted site generally relies on immobilization or physical removal. Besides being expensive and non-specific, use of physicochemical techniques for the treatment of soil pollutants has rendered land useless for the growth of plants as the population of microbes associated with various biological activities such as nitrogen fixation are greatly affected during the process of decontamination.

Continued from the past, progressive deterioration of environmental quality resulting from the discharge of untreated industrial effluents and sewage has severely threatened human and environmental health. Metal pollutants that present a continuing threat to mankind and to its surroundings need to be addressed in order to reverse their effects so as to maintain the balance of the system. Sensing chemicals in the environment and responding to changes in their concentrations so as to bring the system back into balance is the fundamental principle behind

Soil Remediation and Plants. http://dx.doi.org/10.1016/B978-0-12-799937-1.00003-6

the process of bioremediation. In the simplest sense, the goal of bioremediation (use of biological agents) is to attenuate pollutants by transforming them to less toxic or to innocuous products. Over the years, the success of bioremediation technology has emerged as the most advantageous clean-up technique for the treatment of pollutants that have polluted soil and water resources, with bacteria being the most important organism in the context of immobilization or the transformation of metal pollutants (Malik, 2004; Fu and Wang, 2011).

METAL POLLUTANTS AND HUMAN HEALTH

Since the commencement of the industrial revolution, redistribution of metals in the environment is gradually ending with the build-up in terrestrial and aquatic habitats. They are not just part of our long industrial heritage, however; but are finding increasing uses in areas as diverse as medicine, electronics, catalysis and the generation of nuclear power. Being non-degradable and thus persistent, metal pollutants with known effects on human health represent widespread pollutants of great concern. Toxicity of metals is generally represented in terms of their concentration, inability to undergo degradation completely by any methods, ability to get transformed to more toxic forms under certain environmental conditions and by their ability to get accumulated in the food chain that hamper normal physiological activities and ultimately endanger human life (McLaughlin et al., 1999; Nair et al., 2008). Given our long and intimate association with metals, they present a great concern both at personal and at public health levels as their disturbed hazardous concentrations result in toxicity and sometimes even in lethal effects as they escape control mechanisms such as homeostasis, transport, compartmentalization and binding to specific cell constituents. Toxicity of metal compounds is accredited to their prevalence in close proximity to humans which facilitates their access into living systems.

Mercury (Hg), a transition metal belonging to group 12 (IIB) of the periodic table, exists in three valence states: elemental or metallic form (Hg^0), inorganic (Hg^{2+} or Hg^+) and organic forms ($R-Hg^+$ or $R-Hg-X$). Despite the fact that all forms are toxic, the organic form is considered to be the most toxic and the elemental the least toxic. The detrimental fate of organomercurial compounds in humans is endorsed to its hydrophobic character, a property which it employs in crossing the placental and blood–brain barriers (Jan et al., 2009; Singh et al., 2011). Transport of mercury within the body of humans is attributed to molecular mimicry, a mechanism by which metal compounds compete with endogenous substrates in getting entry into different cells and tissues. Having high affinity for thiol groups of enzymes and proteins, both organic and inorganic mercury cross membranes with different abilities, which leads to their inactivation (Hajela et al., 2002; Gupta and Ali, 2004). Compared to organic mercury, inorganic mercury exerts its toxic consequences mainly on getting methylated to methylmercury compounds that lead to increases in the intracellular calcium level rapidly from the extracellular calcium pools, thereby hampering production of neurotransmitters and creating serious imbalances in the development of the brain (Pedersen et al., 1999).

Arsenic (As), a group 15 (V) member of the periodic table, exists as metalloid arsenic (As^0), arsenite (As^{3+}), arsenate (As^{5+}) and as an arsine gas (As^{3-}). However, arsenic compounds are generally categorized as organic (non-toxic) as well as inorganic (toxic) forms. Inorganic arsenic predominantly exists in trivalent (As^{3+}) and pentavalent (As^{5+}) forms, with trivalent compounds having the property to get absorbed more rapidly due to their high lipid solubility and, thus, being more toxic than pentavalent ones (Kaur et al., 2009). They are interconvertible depending on the redox status of the environment; with pentavalent arsenic more stable and predominant under aerobic conditions and trivalent species under anaerobic conditions (Duker et al., 2005). Arsenic exerts its toxic consequences soon after binding to proteins such as haemoglobin, whereby it gets redistributed to organs such as liver, kidney, lung, spleen and intestines. However, unlike other metals (mercury, cadmium, etc.) that get biomagnified through the food chain, arsenic does not possess the property to get biomagnified. Arsenate (As^{5+}), a phosphate analogue competing with phosphates for adenosine triphosphate, results in adenosine diphosphate monoarsine formation by substituting arsenic for phosphate. Soon after its formation, it interferes with essential cellular processes such as oxidative phosphorylation by hampering adenosine triphosphate (ATP) synthesis (Jomova et al., 2011). Compared to arsenate, toxicity of arsenite (As^{3+}) is accredited to its propensity to bind sulphydryl groups of dihydrolipoic acid (a pyruvate dehydrogenase cofactor associated with the conversion of pyruvate to acetyl coenzyme A), thereby resulting in the reduction in citric acid cycle activity. In addition to that, binding of arsenite (As^{3+}) to sulphydryl groups is associated with the inhibition of gluconeogenesis and production of glutathione, responsible for protecting cells against oxidative stress (Jomova et al., 2011).

Lead (Pb) is a persistent and common environmental contaminant. Inorganic Pb can be absorbed following inhalation, oral and dermal exposure through the respiratory tract, as well as through the gastrointestinal tract primarily in the duodenum (ATSDR, 2007). Nutritional status of iron and calcium contributes as a risk factor for Pb intoxication, as their deficiency increases retention of ingested Pb (Ruff et al., 1996). By inhibiting the activities of enzymes involved in haem biosynthesis, particularly δ-aminolevulinic acid dehydratase, it is known to amend the haematological system. Increased circulating aminolevulinic acid, a weak γ-aminobutyric acid (GABA) agonist, decreases GABA release by presynaptic inhibition (Hernberg and Nikkanen, 1970; Millar et al., 1970). In blood, Pb is primarily found in red blood cells, where it destabilizes the cellular membranes of red blood cells (RBC). By decreasing the cell membrane fluidity and increasing the rate of erythrocyte haemolysis, it leads to the development of anaemia (Needleman, 2004; Bellinger and Bellinger, 2006). Ability of lead to mimic calcium or perturb calcium homeostasis makes lead a powerful neurotoxin having diverse impacts on the central nervous system (ATSDR, 2007).

In addition to interfering with cellular functions, metal intoxication also involves generation of reactive oxygen (ROS) and nitrogen species (RNS) that induces a complex series of downstream adaptive and reparative events, driven by oxidative and nitrative stress. ROS includes free radicals such as hydroxyl (OH·), superoxide (O_2·⁻), peroxyl (RO_2·) and alkoxyl (RO·), along with certain non-radical species such as peroxynitrite (ONOO⁻), H_2O_2, etc., which are either oxidizing agents or get easily converted into radicals. It is plausible that an increase in ROS/RNS production due to exogenous stimuli alters cellular functions through direct modifications of biomolecules (proteins, nucleic acids and lipids) (Jan et al., 2011). In view of the essentiality of redox homeostasis, induction of oxidative stress due to fluctuations in the defence machinery appears to be a remarkable phenomenon by which metals cause toxicity in living organisms.

MICROBIAL-BASED REMEDIATION

Besides being responsible for redistribution to the aquatic and terrestrial ecosystems, metals from anthropogenic sources contribute more towards a wide range of toxic effects in microbial communities. Their toxic consequences preferentially vary with the species available for interactions with the microflora in their natural environment. Despite being toxic, microbes flourish in metal-polluted locations apparently by adapting themselves through a variety of resistance mechanisms that are both active as well as incidental, with frequencies ranging from a few per cent in pristine environments to nearly 100% in heavily polluted environments (Gadd and Griffith, 1978; Gadd, 2010). Microbes with their symbiotic associations to each other and higher organisms, contribute actively to ecological phenomena that involve metal transformations having beneficial or detrimental consequences with respect to the human context (Gadd, 2010). Natural attenuation that, as a whole, depends on natural processes for transformation or immobilization of metal contaminants is believed to be the major process involved in the reduction/transformation of contaminant concentrations, spurred by the addition of nutrients (biostimulation) or through the addition of bacterial strains with desired catalytic capabilities (bioaugmentation) that enhance the resident microbial population's ability to transform contaminants (Rosenberg et al., 1992; Tyagi et al., 2011). However, for deciding whether a particular approach is appropriate for the clean-up of a contaminated site, it is necessary to have knowledge of degradative capabilities of native as well as the organisms that could be added against the particular contaminants so as to enhance their remediation from the site.

Microbes being prime mediators of biochemical cycling are ubiquitous and, as such, are well adapted to exist on substances that are often present in the environment at toxic concentrations. They interact with metals, alter their physical and chemical state by making changes in metal speciation that lead to a decrease or increase in their mobility, so as to enhance their survival in metal-contaminated sites (Gadd, 1992). Among various strategies through which bacteria resist toxicity of metals, important ones include: (1) toxic metal may get sequestered on cell wall components or by intracellular metal binding proteins and peptides such as metallothioneins, phytochelatins, etc.; (2) changing uptake pathways which block uptake of a metal; (3) enzymatic conversion of the metal to a less toxic form and (4) lowering the intracellular concentration of the metal by a specific efflux system or by intracellular compartmentalization (Mowll and Gadd, 1984; White and Gadd, 1998; Gadd, 2004).

As microorganisms inhabiting native habitats evolved towards ecological fitness rather than to biotechnological competence, cleaning contaminated sites by bacteria through the process of natural selection is a challenging task. In order to broaden the horizon of capacity for biological systems towards creation of superior microbial chemical factories, it seems necessary to go beyond the genetic confines of microorganisms for the assembly of biosynthetic pathways for both natural and unnatural compounds so as to exploit them for bioremediation purposes (Doong et al., 1998; Mohan and Pittman, 2007; Chauhan and Jain, 2010; Patel et al., 2010). Recent developments in the genetic engineering approach that have begun to uncover the endogenous detoxification system in microbes, along with the ways to genetically engineer microbes, can be employed in making inroads successfully into this environmental clean-up technology.

ENHANCING BIOREMEDIATION THROUGH GENETIC ENGINEERING

Considerable interest in exploring metal resistance mechanisms in bacteria serves to highlight the credentials that are particularly advantageous for their practical application in dealing with unfavourable metal burdens in nature. As the efficacy of a desired in situ catalytic activity (biodegradation) depends on the presence of microflora possessing desired catalytic activity at the target site, it is equally manifested that one key enzyme may or may not be present and, if present, might be harboured by only a small proportion of the microbial population. Therefore, with the commencement of the genetic era, a better understanding of microorganisms' natural transformation abilities at the genetic level has accelerated the progress to genetically engineer microbes so as to achieve the factual expression of genes that are necessary for the detoxification of toxic pollutants. Natural processes that involve mutations, DNA shuffling, horizontal gene transfers, etc. have a significant contribution in the formation of hybrid genes and metabolic operons (Van der Meer, 1997). However, compared to natural attenuation, engineered bioremediation relies on accelerating the growth of microorganisms as well as on optimizing the environment in which they carry out the detoxification reactions. Using the genetic engineering approach, bacterial surface display of metal-binding peptides and proteins to remove metal pollutants from water and waste water had a significant impact in making it an increasingly active research area with a wide range of biotechnological and industrial applications.

Surface Expression of Novel Metal Binding Peptides and Proteins

Besides offering higher metal-binding capacity and/or specificity and selectivity for different metal ions, surface expression of novel metal binding peptides and proteins has enabled rapid binding and, as such, their immobilization for improved metal resistance of growing cultures (Pazirandeh et al., 1998). Surface exposure of metal binding peptides and proteins is fortuitous and relatively efficient as it improves metal binding properties of microorganisms that are employed in various systems for their removal from the environment. Depending on the degrading capability of the indigenous microbes and the extent of contamination of the site to be treated, the decision to implement either or both of these techniques for bioremediation seems essential.

From the previous figures, it is quite apparent that compared to fast growers that result in a large heap of unwanted biomass, an optimal clean-up agent for use in bioremediation would be one that displays maximum catalytic ability with a minimum cell mass. It has readily been found that unwelcome biomass produced by fast growers provides a palatable niche for protozoa to flourish which prevents normal microflora from growing beyond a certain level (Iwasaki et al., 1993). One major issue pertaining to bacterial species being propagated as monoculture under optimized controlled environment for use in the bioremediation process is that they may not be the ones performing the bulk of biodegradation in natural niches. To be exploited in the process of bioremediation, it is necessary that they might establish themselves, interact with other bacteria in unknown ways as they are expected to be hindered in their activities by exposure to a multitude of poorly controllable external factors, some of which place them under considerable stress (Timmis and Pieper, 1999; Sayler and Ripp, 2000). Knowing that genetically modified organisms (GMOs) are generally exposed to unfavourable conditions that often place them under stress, conceptualization to genetically engineer bacteria for characteristics such as stability without selection, minimal physiological burden, small size, non-antibiotic selection markers, minimal lateral transfer of cloned genes to indigenous organisms and traceability of specific genes seems necessary so as to enhance their exploitation for use in bioremediation in order to meet the growing demands for their eventual applications in the field.

PHYTOREMEDIATION

Despite the fact that whether the introduced strains are fast growers or slow growers, the introduction of GMOs, particularly those which are raised in laboratory conditions with modified metabolic machinery at the start point, faces problems in colonization, niche specificity and poor adaptation of its transcriptional machinery. Whether fast growers or slow growers, introduction of foreign microbes, no matter whether they are recombinant or not, into an unfamiliar territory in itself poses a threat of ecological imbalance (Lindow et al., 1989). Keeping in view the difficulties that might arise on introducing GMOs, expression of bacterial genes in plant carriers seems a quite reasonable approach that is believed to open a new perspective as a promising alternative aimed at environmental clean-up. The use of genetically modified plants generated by ectopic expression of bacterial genes to achieve remediation of polluted sites has several advantages over microbial-based remediation: spontaneous delivery of catalysts well below the soil surface through roots and methods for the extensive agricultural dispersion of seeds are already optimized and importantly chances for the horizontal transfer of genes seem less probable.

Owing to its 'green' approach, exploiting plants to remediate environmentally toxic compounds is expanding its dimensions by gaining a significant amount of public attention. Plants possessing traits such as being fast growing, having a deep root system, having a high biomass, being easy to harvest, tolerant and having the ability to accumulate a wide range of heavy metals in its aerial and harvestable parts have commonly been exploited for use in phytoremediation. However, to date, no plant has been found that fulfils all the above criteria as most hyperaccumulator plants have slow growth rates and are low biomass producers; whereas, plants that show rapid growth have been found to be sensitive to metals and as such are known to accumulate only low concentrations in their shoots. Researchers have found that plants that generally grow in areas commonly contaminated with metal pollutants, belonging to Brassicaceae and Fabaceae families, possess the characteristics of hyperaccumulators (Gleba et al., 1999). To date, 400 hyperaccumulator plant species possessing the characteristics of hyperaccumulators from 22 families have been identified, with the Brassicaceae family alone containing 87 hyperaccumulating species with the widest range of metal accumulation properties from 11 genera (Baker and Brooks, 1989).

Based on the method by which remediation of metal contaminants is achieved, phytoremediation has been broadly categorized into the following areas:

Phytoextraction: use of metal accumulators to remove metals upon concentrating them in the harvestable parts from the soil.

Rhizoremediation: use of microorganisms that are associated with plants to eliminate the contaminant from the soil.

Phytostabilization: reducing the environmental burden of pollutants by stabilizing them through the use of plants.

Phytovolatilization: conversion of toxic pollutants to innocuous products by employing genetic engineering of genes from microbial origin.

Phytoextraction

Over the last few decades, remediation of sites contaminated with metal pollutants by plants has gained much attention over the microbial-based remediation strategies (Kramer, 2005; Pilon-Smits, 2005; Doty, 2008). Although not exclusive, various phytoremediation strategies can be employed simultaneously in order to achieve remediation against the particular pollutant at a given time. However, for any remediation strategy to be effective in eliminating the toxic pollutant, basic criteria settles on the type and form of the contaminant present at that site. Despite the fact that soluble fractions of various pollutants are generally present as ions or unionized organometallic complexes, heavy metals also exist in colloidal, ionic, particulate and dissolved phases, with high affinity for humic acids, organic clays and oxides coated with organic matter (Connell and Miller, 1984; Elliot et al., 1986). Preferentially, their solubility in the environment (soil and groundwater) is controlled by pH, cation exchange capacity, oxidation state and the amount of metal present at that site along with the redox potential of the system (Connell and Miller, 1984; Elliot et al., 1986; Baker and Walker, 1990; McNeil and Waring, 1992; Martinez and Motto, 2000). In the rhizospheric region, bioavailability of metals is affected up to a greater extent by various plant and/or microbial activities that affect their uptake by plants via altering their mobility and bioavailability (Ma et al., 2011; Miransari, 2011; Aafi et al., 2012; Yang et al., 2012). Presence of lipophilic compounds in plant exudates or lysates has an effect either in increasing their solubility in water or in promoting growth of biosurfactant-producing microbial populations. Owing to the beneficial roles that they are known to play in metal mobilization, use of beneficial microbes possesses several advantages over chemical amendments such as the fact that metabolites produced by microbes are biodegradable, less toxic and can be produced in situ at rhizosphere soils. Furthermore, their bioavailability can be enhanced by chelators such as siderophores, organic acids and phenolics of plant and microbial origin associated with the release of metal cations from soil particles followed by its uptake by plants. Besides contributing to plant growth, microbial processes and/or activities in the rhizosphere soils enhance the effectiveness of phytoremediation either by: (1) enhancing metal translocation (facilitate phytoextraction) or by reducing metal bioavailability in the rhizosphere (phytostabilization) and (2) by indirect promotion of phytoremediation achieved either by conferring resistance to plants and/or enhancing their biomass production so as to achieve remediation of pollutants up to a greater extent (Glick, 2010; Kuffner et al., 2010; Rajkumar et al., 2010; Babu and Reddy, 2011).

Despite being permissive, a metal after mobilization into soil solution has to be captured by the cells of the root. Following adherence, transport of metal ions is guided by transporters, particularly channel proteins and/or by H^+-coupled carrier proteins that are embedded in the membrane. On getting inside, metals undergo precipitation to carbonate, sulphate or phosphate precipitates for their immobilization through extracellular (apoplastic) and intracellular (symplastic) compartments (Raskin et al., 1997). Flow of substances through the apoplastic pathway is relatively unregulated as substances flow without crossing membranes. However, subsequent to sequestration inside root cells, symplastic transport of ions involves movement into the stele followed by release into the xylem on crossing the casparian strip. On gaining entry into the root, metals are either stored in the root or translocated to the shoots. Metal ions are transported actively as free ions or as metal–chelate complexes across the tonoplast

(Cataldo and Wildung, 1978). Being important with respect to metal ion storage, metals are often chelated either by organic acid or phytochelatins. Inside the vacuole, formation of insoluble precipitates followed by their compartmentalization occurs as a mechanism of resistance against the damaging effect of metals (Cunningham et al., 1995).

Among the various approaches that are employed to achieve phytoremediation, phytoextraction of metals and metalloids is considered to be the most challenging due to the high quantity of pollutants present. Phytoextraction is broadly categorized into: chelate-assisted phytoextraction (*induced phytoextraction*) and long-term phytoextraction (*continuous phytoextraction*), of which chelate-assisted phytoextraction is more acceptable and presently being implemented commercially.

Chelate-Assisted Phytoextraction

Chelate-assisted phytoextraction involves the addictive use of metal chelating agents in combination with non-accumulator plants, bearing far less ability to accumulate metal than hyperaccumulators, but with high biomass potential so as to increase the soluble metal fraction. It involves the additive function of two basic processes: transport of metals to the harvestable shoot upon release into soil solution. It has been observed that use of fast-growing tree species fulfils the need for high biomass production aided by chemicals in order to achieve high metal accumulation. Chelates such as ethylenediaminetetraacetic acid (EDTA) are accredited with the function to increase the formation of soluble metal-EDTA within the plants that allows their movement from root system towards shoot, where actual accumulation of metals occurs as metal–EDTA complex. Transport of metal–chelate complexes that occur through xylem to shoots appears to play an important role in the accumulation of chelate-assisted metal complexes in plants (Blaylock et al., 1997; Huang et al., 1997). Inside the plants, movement of metal–chelate complex, besides relying on efficient capillary plumbing system, relies on the high-surface-area collection system provided by the roots. Movement of metals as a metal–chelate complex to shoots is followed by the retention of metal–chelate complex as water evaporates via the transpiration stream.

In contrast to soils that possess only a trace amount of various elements, metalliferous soils containing high concentrations of elements have been found to vary widely in their effects on different plant species. The total amount of metal removed from a site is generally calculated as a product concentration of metals in harvested plant material to total harvested biomass. Despite the fact that certain plant species posses an inherent capacity to sequester high concentrations of metals in the above-ground tissues, addition of chelates have been found to have a profound effect in enhancing metal accumulation up to a greater extent. Studies carried by Huang et al. (1997) and Blaylock et al. (1997) reported rapid accumulation of lead (> 1% of shoot dry biomass) in the shoots. Based on the observation that high biomass crops (Indian mustard, corn and sunflower) could be 'induced' to accumulate high concentrations of lead, it is generally apprehended that these strategies would lead the way in designing strategies that can be employed to achieve phytoextraction of other toxic metals using appropriate chelates (Huang and Cunningham, 1996; Blaylock et al., 1997; Huang et al., 1997). Meers et al. (2007) reported that the use of ethylene diamine disuccinate, a naturally occurring chelator with a high degree of biodegradability, on soil sediments contaminated with Zn and Cd, enhances metal uptake to a significant extent by willows grown on the contaminated soils (Meers et al., 2007). On comparing the effects of different chelates, it was observed that metal accumulation efficiency is directly related to the affinity of chelate that is used to achieve the decontamination of available metal. From the results of various studies, it is concluded that for efficient phytoextraction, chelates possessing a high affinity for the metal of interest should be used, such as EDTA for lead, ethylene glycol tetraacetic acid for cadmium and citrate for uranium (Blaylock et al., 1997). Following harvest, metal-enriched plant residue can either be disposed off as hazardous material or, if economically feasible, used for metal recovery.

Continuous Phytoextraction

Metal chelate uptake in plants is correlated with stress, which is sometimes followed by plant death. However, it is not clear whether stress is necessary for induction or whether it simply reflects the accumulation of high concentrations of synthetic chelate in the plant. Despite the beneficial effect in terms of increasing the efficiency of phytoextraction, it has been observed that chemical amendments (e.g. EDTA) are not only phytotoxic, but also exert their toxic consequences on beneficial soil microorganisms known to play important roles in plant growth and development (Ultra et al., 2005; Evangelou et al., 2007; Muhlbachova, 2009). In an alternative approach, accumulation of metals can be achieved based on the specialized physiological processes occurring over the complete growth cycle in plants. Unlike induced metal uptake, continuous phytoextraction is achieved under the genetic and physiological capacity of plants specialized to accumulate, translocate and resist high amounts of metals. There are reports where the removal of metals such as zinc, cadmium and nickel along with arsenic and chromium by hyperaccumulating plants that grow on soils rich in heavy metals has been achieved by continuous phytoextraction (Baker and Brooks, 1989). Dissecting the mechanism that is employed to achieve hyperaccumulation of metals will help in understanding the biological mechanisms by which plants become superior for the remediation of soils. As a type of management strategy, phytoextraction for metal contaminated sediment is thought to address efficiency, duration, leaching, food chain contamination and cost benefits. However, its application is limited by low biomass production and insufficiently high metal uptake into plant tissue along with the lack of knowledge regarding minimum amendments that are required (e.g. compost, irrigation) to support long-term plant establishment.

Rhizoremediation

As an alternative strategy, degradation of contaminants that is achieved by microorganisms inhabiting the rhizosphere holds great potential for the remediation of contaminated soil (Kuiper et al., 2004). As an essentiality for rhizoremediation, the association commonly referred to as the 'rhizosphere effect' shows great potential where nutrients in the form of root exudates, oxygen and favourable redox conditions to soil microorganisms provided by plants results in increased bacterial diversity, population density and activity compared with bulk soil (Molina et al., 2000).

Rhizofiltration

Rhizofiltration that involves the use of plants (both terrestrial and aquatic) to absorb, concentrate and precipitate contaminants from polluted aqueous sources with low contaminant concentration in their roots finds its application in dealing with effluents discharged from industries and agricultural run-off as well as acid mine drainage. It is more commonly used for lead (Pb), cadmium (Cd), copper (Cu), nickel (Ni), zinc (Zn) and chromium (Cr), which are primarily retained within the roots (Chaudhry et al., 1998). As a cost-effective technology, it has been used for the treatment of surface or groundwater containing significant proportions of heavy metals such as Cr, Pb and Zn (Kumar et al., 1995; Ensley, 2000). Its commercialization is driven by its technical advantages in its applicability to many metals, besides possessing the ability to treat high volumes with less need for toxic chemicals, reduced volume of secondary waste and the possibility of recycling (Dushenkov et al., 1995; Kumar et al., 1995). Being advantageous in its ability for use in in situ as well as ex situ applications, studies performed on plants such as sunflower, Indian mustard, tobacco, rye, spinach and corn have highlighted its application in the removal of lead from effluent, with sunflower having the greatest ability.

Rhizodegradation

Among the various strategies that are employed to achieve remediation of polluted sites, rhizodegradation involves use of plants to stimulate the microbial community near the root–soil interface, in order to enhance the degradation of recalcitrant compounds in the soil. In this type of association, plants are known for their contribution in making contaminants of the soil more bioavailable through release of a series of low molecular weight organic acids, carbon and nitrogen compounds to nourish microbes in the rhizosphere and by releasing exudates that enhance degradation of soil contaminants through induction of biochemical pathways within bacteria (Leigh et al., 2002; White et al., 2003). Kuiper and colleagues (2004) described a two-step enrichment approach where bacteria after harvesting from the roots of plants growing in contaminated sites are enhanced for their biodegradation potential and are then made to re-colonize plant roots.

Successful rhizosphere colonization depends not only on interactions between the plants and the microorganisms of interest but also on interactions with other rhizospheric microorganisms and the environment. In the past two decades, several techniques such as visualization of bacteria that carry the *luxAB* genes encoding bacterial luciferase, the green fluorescent protein and/or another reporter gene along with in situ hybridization assays using fluorescent probes have been devised to track colonization of roots by bacteria (Broek et al., 1998; Tombolini et al., 1999; Ramos et al., 2000; Sharma et al., 2013). These techniques have been found to be highly advantageous in illustrating the introduced microorganisms that are often unable to compete with indigenous microorganisms or are unable to establish in high numbers at the rhizosphere region (Rattray et al., 1995; Lubeck et al., 2000). In a major boost to this strategy, some bacteria have emerged with the strategies to out-compete other microorganisms by delivering toxins, using extremely efficient nutrient utilization systems or by physical exclusion (Lugtenberg et al., 1991).

Phytostabilization

At times, when there seems no immediate effort to clean up sites polluted by metals as they are not of high priority on a remediation agenda, traditional means such as in-place inactivation, a remediation technique that employs use of soil amendments to immobilize or fix metals, are employed to reduce metal toxicity (Berti and Cunningham, 2000). Phytostabilization (also known as phytorestoration) is a plant-based remediation technique that is aimed at reducing the risk of metal pollutants by stabilizing them through formation of a vegetative cap at the plant rhizosphere, where sequestration (binding and sorption) processes immobilize metals so as to make them unavailable for livestock, wildlife and human exposure (Munshower, 1994; Cunningham et al., 1995; Wong, 2003). Unlike other phytoremediative techniques, the goal of phytostabilization is not to remove metal contaminants from a site, but rather to stabilize them and reduce the risk to human health and the environment. Phosphate fertilizers, organic matter, iron or manganese oxyhydroxides, natural or artificial clay minerals being the most prominent soil amendments fulfil the criteria of fixing metals rapidly following incorporation and/or making chemical alterations that are very long lasting, if not permanent. Besides being poor translocators of metal contaminants to above-ground plant tissues that could be consumed by humans or animals, characteristics of plants that are employed for phytostabilization include easy to establish and care for, quickly grown, produce dense canopies and root systems and be tolerant of metal contaminants and other site conditions. In being less expensive, less environmentally evasive and easy to implement, phytostabilization is considered to be more advantageous than other soil-remediation practices (Berti and Cunningham, 2000). For more heavily contaminated soils, phytostabilization aims at stabilizing the contaminated sites with tolerant plants in order to reduce the risk of erosion and leaching of these pollutants to water bodies.

Phytovolatilization

Phytovolatilization is aimed at transforming organic and inorganic contaminants that are taken along with water to volatile gaseous species within the plant followed by their eventual release into the atmosphere at comparatively low concentrations (Mueller et al., 1999). To date, the phytovolatilization approach has been used primarily for the removal of mercury, where Hg^{2+} is transformed into less toxic Hg^0. Recent efforts to achieve remediation of mercury were aimed at inserting bacterial Hg ion reductase genes into plants such as *Arabidopsis thaliana* L. and tobacco (*Nicotiana tabacum* L.) so as to achieve phytovolatilization of mercury to a greater extent (Rugh et al., 1996; Heaton et al., 1998; Rugh et al., 1998; Bizily et al., 1999). Although this remediation approach has added benefits of minimal site disturbance, less erosion and no need to dispose of contaminated plant material, it is still considered as the most controversial of all phytoremediation technologies as release of mercury into the environment is likely to be recycled by precipitation and then redeposit back into the ecosystem (Henry, 2000).

Insights into Genetic Engineering Approaches for Phytoremediation

For any plant to be employed in the process of phytoremediation, it needs to fulfil the basic criteria of being tolerant to various metal pollutants. Since the beginning of the genetic era, there is now ample data available regarding the expression of various genes that are believed to impart resistance through various mechanisms of plants and, more particularly, bacteria. In order to enhance the metal accumulation property of plants, current remediation strategies are particularly aimed at enhancing the metal accumulation property following its transport from the soil to the roots and translocation to shoots in plants. For any particular strategy to be effective, it is necessary to have a thorough understanding of the mechanisms that are operating within the hyperaccumulators that are employed in acquisition and translocation, followed by accumulation so as to achieve remediation of polluted sites. Feasibility in employing any strategy to genetically engineer a rapidly growing non-accumulator in order to enhance it for the property that hyperaccumulators generally harbour requires quite good understanding of the biological processes that are operating within the hyperaccumulators. Unfortunately, apart from the mechanisms of tolerance, these biological processes are not well understood. For phytoremediation to succeed, current research needs to be targeted at improving existing lines of phytoextracting plants for high metal acquisition, high-biomass and hyperaccumulation both in hyperaccumulating as well as in nonaccumulating plants.

Metal Sequestering Proteins and Peptides

Following mobilization, metal uptake and its accumulation in plants is affected by concentration and affinities of chelating molecules as well as by the presence and selectivity of transport activities. For enhancing metal uptake and, as such, its accumulation, three approaches to genetically engineer plants have been envisioned: increasing the number of uptake sites, altering the specificity of uptake systems so as to reduce the competition by unwanted cations and increasing the number of intracellular high-affinity binding sites so as to enhance the sequestering capacity of the plants. Among them, increasing the number of intracellular binding sites drives the passage of transition metal ions across the plasma membrane followed by its sequestration that enhances accumulation. Metallothioneins (MT) are cysteine-rich, low-molecular-weight metal-binding proteins endowed with a wide range of functional capabilities in a biosystem (Hamer, 1986; Sousa et al., 1998). They possess high metal content (predominantly Zn, Cu or Cd), bound by sulphur atoms in thiolate clusters, besides having highly conserved cysteine residues (18–23) that bind metal ions and sequester them in the biologically inactive form (Vasak, 2005; Bell and Vallee, 2009). Since establishing its role in metal detoxification, plant metallothioneins are grouped into four subfamilies, MT1, MT2, MT3, MT4, with different expressions for fulfilling different functions during development in plant tissues (Cobbett and Goldsbrough, 2002). It was observed that expression of MT1a and MT1b in roots of non-accumulator plants, like *A. thaliana* and *poplar* were high during exposure to Cd, Cu and Zn (Garcia-Hernandez et al., 1988; Zhou and Goldsbrough, 1994; Kohler et al., 2004), while as in *Thlaspi caerulescens*, levels of expression of MT1 mRNA were constitutively higher in leaves than in roots (Roosens et al., 2005). Although MT2b is not considered as the major player in the process of metal tolerance, increased expression of MT2b was found to be associated with Cu tolerance and increased As translocation from root to shoot in *A. thaliana* (Schat et al., 1996; Grispen et al., 2009). In non-accumulator plants, expression of MT3 genes increases on exposure to Cu and as such is known to play an important role in the maintenance of Cu homeostasis under conditions of high Cd and Zn in the cytoplasm (Guo et al., 2003; Kohler et al., 2004; Roosens et al., 2004). This suggestion is supported by the discovery of a small cavity that is adapted for Cd chelation in the MT3 protein by *A. thaliana* as compared with the analogous protein from *T. caerulescens* (Roosens et al., 2004, 2005). In *A. thaliana*, it is reported that MT4 plays an important role in metal homeostasis during seed development and seed germination rather than in metal decontamination (Guo et al., 2003).

Research performed on microorganisms has resulted in providing valuable models to understand and engineer plants for tolerance to metal as the number of potential gene targets in microorganisms is much higher and more diverse than in plants (Silver and Misra, 1988). Among the various mechanisms that are involved in detoxification and transformation of metals, engineering plants for proteins and peptides that possess the potential to efficiently immobilize metals has emerged as a technique that is useful for their removal from the environment (Beveridge and Murray, 1980). Transgenic plants expressing metallothionein genes usually exhibit an increased metal tolerance as was reported in transgenic tobacco expressing the MT gene from *Silene vulgaris* (Zimeri et al., 2005;

Zhigang et al., 2006; Gorinova et al., 2007). In another study, expression of MT1 from *Brassica rapa* L. in the chloroplasts of *A. thaliana* showed increased tolerance to Cd and to oxidative stress (Kim et al., 2007). These observations were complementary to observations in *A. thaliana* mutants lacking MT1a and MT2b; the double mutant had normal Cu tolerance but Cu accumulation in roots was reduced by 30% (Guo et al., 2008). As overexpression of MTs result in enhanced metal accumulation, it offers a promising strategy to achieve remediation of metals at contaminated sites.

Similarly, Histidine (His), a free amino acid present at high concentration in the roots of hyperaccumulators, forms stable complexes with Ni, Zn and Cd (Kramer et al., 1996; Haydon and Cobbett, 2007). Compared to non-accumulator *Alyssum montanum*, hyperaccumulators such as *Alyssum lesbiacum* show a larger pool of His available for chelation in roots. In the hyperaccumulator *A. lesbiacum*, concentration of His is increased by Ni treatment without inducing the expression of the genes involved in the biosynthetic pathway (Persans et al., 1999; Ingle et al., 2005). Overexpression of ATP-phosphoribosyltransferase, the first enzyme of the His biosynthetic pathway in transgenic *A. thaliana* plants, increases tolerance but not the accumulation (Ingle et al., 2005).

Engineering for Improved Enzyme Activities

One major challenge to overcome the major hurdles associated with phytoremediation is to construct genetically modified plants that, besides possessing the ability to tolerate high shoot metal concentrations, should produce large biomass in order to keep the required number of cropping cycles to a minimum. In addition to the enhancement of the metal accumulation properties via genetic engineering, a thorough understanding of the biological processes that are involved in metal transformation after acquisition from soil seems equally essential. As metal tolerance is a key prerequisite for phytoremediation, the cellular mechanism for/behind metal tolerance that involves the detoxification of metal ions, via chelation or transformation to a less toxic or easier to handle form followed by compartmentalization, is apprehended as essential to achieve the desired fate. It is generally estimated that microorganisms that act as valuable models for understanding and engineering metal tolerance in higher plants can provide valuable information regarding the number of known potential gene targets that can be engineered into plants to achieve decontamination of the target site (Silver and Misra, 1988). With the exception of mammalian metallothionein and phytochelatin synthase, plants have broadly been engineered with genes of microbial origin for their tolerance via accumulation and decontamination against the wide range of contaminants (Mejare and Bulow, 2001). In view of the possibility to increase the remediation potential of plants, research carried out by Meagher's group has demonstrated enhanced mercury (Hg) tolerance and phytovolatilization in transgenic plants expressing *E. coli* mercuric ion reductase (*mer*A) and organomercury lyase (*mer*B) genes (Meagher and Heaton, 2005). In the last decade, more progress as a step towards enhancement of remediation potential includes transformation of aquatic plants for Hg phytovolatilization in wetlands, transformation in the fast-growing tree eastern cottonwood (*Populus deltoides*) as well as in the chloroplast genome of tobacco, with both *mer*A and *mer*B genes (Ruiz et al., 2003; Czako et al., 2006; Lyyra et al., 2007). Chloroplast transformation is much preferred over the nuclear transformation as it prevents escape of transgene via pollen to related species, high levels of foreign gene expression and engineering multiple genes or pathways in a single transformation event.

CONCLUSIONS

The metals being toxic at higher concentrations posses the ability to replace essential metals in pigments or enzymes, thereby disrupting their function, in addition to oxidative stress via the formation of free radicals. An increase in the concentration of metal pollutants is on the rise as the paradigm shifts more towards industrialization. Rapid industrialization with faulty waste disposal, increased anthropogenic activities and modern agricultural practices have resulted in an increase in the pollution of the environment. Recognition of the ecological and human health hazards has necessitated the urge for management and remediation of contaminated sediments so as to minimize ecological risks and risks associated with food chain contamination. Compared to the conventional treatment methods, the employment of various phytoremediation strategies to clean up soil and water pollution is gaining a lot of importance because of its cost effectiveness and high efficiency of detoxification and because it is environmentally friendly. On the contrary to microbial-based remediation strategies, plant-based remediation technology appears to be a promising alternative in its applicability to address the challenges in terms of minimum requirements for an external energy source for growth, in possessing an extensive root system that reaches every crevice and covers a large area in the soil, where it functions as a pump driven by solar energy to extract and concentrate elements from the soil. Besides that, this technology also offers the flexibility for developing non-destructive desorption techniques for biomass regeneration and/or quantitative metal recovery. Despite being a promising alternative in the management and remediation of contaminated sediments, phytoremediation still requires further development and optimization in order to improve traits that will lead to improvement in its applicability in the restoration of polluted sites.

REFERENCES

Aafi, N.E., Brhada, F., Dary, M., Maltouf, A.F., Pajuelo, E., 2012. Rhizostabilization of metals in soils using *Lupinus luteus* inoculated with the metal resistant rhizobacterium *Serratia* sp. MSMC 541. Int. J. Phytorem. 14, 261–274.

ATSDR, 2007. Toxicological profile for Lead. http://www.atsdr.cdc.gov/toxprofiles/tp13.html.

Babu, A.G., Reddy, S., 2011. Dual inoculation of *Arbuscular mycorrhizal* and phosphate solubilising fungi contributes in sustainable maintenance of plant health in fly ash ponds. Water Air Soil Poll. 219, 3–10.

Baker, A.J.M., Brooks, R.R., 1989. Terrestrial higher plants which hyperaccumulate metallic elements. A review of their distribution, ecology and phytochemistry. Biorecovery 1, 81–126.

Baker, A.J.M., Walker, P.L., 1990. Heavy metal tolerance in plants: Evolutionary aspects. In: Shaw, A.J. (Ed.). CRC Press, Boca Raton, pp. 155–177.

Bell, S.G., Vallee, B.L., 2009. The metallothionein/thionein system: An oxidoreductive metabolic zinc link. ChemBioChem. 10, 55–62.

Bellinger, D.C., Bellinger, A.M., 2006. Childhood lead poisoning: The torturous from science to policy. J. Clin. Invest. 116, 853–857.

Bennett, L.E., Burkhead, J.L., Hale, K.L., Tery, N., Pilon, M., Pilon-Smits, E.A.H., 2003. Analysis of transgenic Indian mustard plants for phytoremediation of metal contaminated mine tailings. J. Environ. Qual. 32, 432–440.

Berti, W.R., Cunningham, S.D., 2000. Phytostabilization of metals. In: Raskin, I., Ensley, B.D. (Eds.), Phytoremediation of toxic metals: Using plants to clean-up the environment. John Wiley & Sons, Inc, New York, pp. 71–88.

Beveridge, T.J., Murray, R.G.E., 1980. Sites of metal deposition in the cell walls of Bacillus subtilis. J. Biotechnol. 141, 876–887.

Bizily, S.P., Rugh, C.L., Summers, A.O., Meagher, R.B., 1999. Phytoremediation of methylmercury pollution: *Mer B* expression in *Arabidopsis thaliana* confers resistance to organomercurials. Proc. Natl. Acad. Sci. USA 96, 6808–6813.

Blaylock, M.J., Salt, D.E., Dushenkov, S., Zakharova, O., Gussman, C., et al., 1997. Enhanced accumulation of lead in Indian mustard by soil applied chelating agents. Environ. Sci. Technol. 31, 860–865.

Broek, A.V., Lambrecht, M., Vanderleyden, J., 1998. Bacterial chemotactic motility is important for the initiation of wheat root colonization by *Azospirillum brasilense*. Microbiology 144, 2599–2606.

Cataldo, D.A., Wildung, R.E., 1978. Soil and plant factors influencing the accumulation of heavy metals by plants. Environ. Health perspec. 27, 149–159.

Chaudhry, T.M., Hayes, W.J., Khan, A.G., Khoo, C.S., 1998. Phytoremediation – focusing on accumulator plants that remediate metal contaminated soils. Austr. J. Ecotoxicol. 4, 37–51.

Chauhan, A., Jain, R.K., 2010. Biodegradation: Gaining insight through proteomics. Biodegradation 21, 861–879.

Cobbett, C., Goldsbrough, P., 2002. Phytochelatins and metallothioneins: Role in heavy metal detoxification and homeostasis. Ann. Rev. Plant Biol. 53, 159–182.

Connell, D.W., Miller, G.J., 1984. Chemistry and Ecotoxicology of Pollution. John Wiley & Sons, NY. 444.

Cunningham, S.D., Berti, W.R., Huang, J.W.W., 1995. Phytoremediation of contaminated soils. Trends Biotechnol. 13, 393–397.

Czako, M., Feng, X., He, Y., Liang, D., Marton, L., 2006. Transgenic *Spartina alterniflora* for phytoremediation. Environ. Geochem. Health 28, 103–110.

Doong, R.A., Wu, Y.W., Lei, W.G., 1998. Surfactant enhanced remediation of cadmium contaminated soils. Water Sci. Tech. 37, 65–71.

Doty, S.L., 2008. Enhancing phytoremediation through the use of transgenics and endophytes. New Phytol. 179, 318–333.

Duker, A.A., Carranza, E.J.M., Hale, M., 2005. Arsenic geochemistry and health. Environ. Int. 31, 631–641.

Dushenkov, V., Kumar, P.B.A.N., Motto, H., Raskin, I., 1995. Rhizofiltration: The use of plants to remove heavy metals from aqueous streams. Environ. Sci. Technol. 29, 1239–1245.

Elliot, H.A., Liberali, M.R., Huang, C.P., 1986. Competitive adsorption of heavy metals by soils. J. Environ. Qual. 15, 214–219.

Ensley, B.D., 2000. Rationale for use of phytoremediation. In: Raskin, I., Ensley, B.D. (Eds.), Phytoremediation of toxic metals. Using Plants to Clean up the Environment. Wiley, New York, pp. 3–12.

Evangelou, M.W.H., Bauer, U., Ebel, M., Schaeffer, A., 2007. The influence of EDDS and EDTA on the uptake of heavy metals of Cd and Cu from soil with tobacco *Nicotiana tabacum*. Chemosphere 68, 345–353.

Fu, F., Wang, Q., 2011. Removal of heavy metal ions from wastewaters – A review. J. Environ. Manag. 92, 407–418.

Gadd, G.M., 1992. Metals and microorganisms: A problem of definition. FEMS Microbiol. Lett. 100, 197–204.

Gadd, G.M., 2004. Microbial influence on metal mobility and application for bioremediation. Geoderma. 122, 109–119.

Gadd, G.M., 2010. Metals, Minerals and Microbes: Geomicrobiology and bioremediation. Microbiol. 156, 609–643.

Gadd, G.M., Griffith, A.J., 1978. Microorganism and heavy metals. Microb. Ecol. 4, 303–317.

Garcia-Hernandez, M., Murphy, A., Taiz, L., 1988. Metallothioneins 1 and 2 have distinct but overlapping expression patterns in Arabidopsis. Plant Physiol. 118, 387–397.

Gleba, D., Borisjuk, N.V., Borisjuk, L.G., Kneer, R., Poulev, A., Skarzhinskaya, M., Dushenkov, S., Logendra, S., Gleba, Y.Y., Raskin, I., 1999. Use of plant root for phytoremediation and molecular farming. Proc. Natl. Acad. Sci. USA 96, 5973–5977.

Glick, B.R., 2010. Using soil bacteria to facilitate phytoremediation. Biotechnol. Adv. 28, 367–374.

Gorinova, N., Nedkovska, M., Todorovska, E., Simova-Stoilova, L., Stoyanova, Z., Georgieva, K., Demirevaska-Kepova, K., Atanassov, A., Herzig, R., 2007. Improved phytoaccumulation of cadmium by genetically modified tobacco plants (*Nicotiana tabacum* L.). Physiological and biochemical response of the transformants to cadmium toxicity. Environ. Pollut. 145, 161–170.

Grispen, V.M.J., Irtelli, B., Hakvoort, H.W.J., Vooijs, R., Bliek, T., ten Bookum, W.M., Verkleij, J.A.C., Schat, H., 2009. Expression of the Arabidopsis metallothionein 2b enhances arsenite sensitivity and root to shoot translocation in tobacco. Environ. Exp. Bot. 66, 69–73.

Guo, W.J., Bundithya, W., Goldsbrough, P.B., 2003. Characterization of the Arabidopsis metallothionein gene family: Tissue-specific expression and induction during senescence and in response to copper. New Phytol. 159, 369–381.

Guo, W.J., Meetam, M., Goldsbrough, P.B., 2008. Examining the specific contributions of individual Arabidopsis metallothioneins to copper distribution and metal tolerance. Plant Physiol. 146, 1697–1706.

Gupta, N., Ali, A., 2004. Mercury volatilization by R factor systems in *Escherichia coli* isolated from aquatic environments of India. Curr. Microbiol. 48, 88–96.

Hajela, N., Murtaza, I., Qamri, Z., Ali, A., 2002. Molecular intervention in the abatement of mercury pollution. In: Environmental education. Anmol Publication Pvt. Ltd, New Delhi, pp. 22–144.

Hamer, D.H., 1986. Metallothionein. Ann. Rev. Biochem. 55, 913–951.

Haydon, M.J., Cobbett, C.S., 2007. Transporters of ligands for essential metal ions in plants. New Phytol. 174, 499–506.

Heaton, A.C.P., Rugh, C.L., Wang, N., Meagher, R.B., 1998. Phytoremediation of mercury and methylmercury polluted soils using genetically engineered plants. J. Soil Contam. 7, 497–510.

Henry, 2000. In: An overview of phytoremediation of lead and mercury. NNEMS Report, Washington, D.C, pp. 3–9.

Hernberg, S., Nikkanen, J., 1970. Enzyme inhibition by lead under normal urban conditions. Lancet 10, 63–64.

Huang, J.W.W., Cunningham, S.D., 1996. Lead phytoextraction: Species variation in lead uptake and translocation. New Phytol. 134, 75–84.

Huang, J.W.W., Chen, J.J., Berti, W.R., Cunningham, S.D., 1997. Phytoremediation of lead contaminated soils: Role of synthetic chelates in lead phytoextraction. Environ. Sci. Technol. 31, 800–805.

Ingle, R.A., Mugford, S.T., Rees, J.D., Campbell, M.M., Smith, J.A.C., 2005. Constitutively high expression of the histidine biosynthetic pathway contributes to nickel tolerance in hyperaccumulator plants. Plant Cell. 17, 2089–2106.

Iwasaki, K., Yoshikawa, G., Sakurai, K., 1993. Fractionation of zinc in greenhouse soils. Soil Sci. Plant Nutr. 39, 507–515.

Jan, A.T., Ali, A., Haq, Q.M.R., 2011. Glutathione as an antioxidant in inorganic mercury induced nephrotoxicity. J Postgrad. Med. 57, 72–77.

Jan, A.T., Murtaza, I., Ali, A., Haq, Q.M.R., 2009. Mercury pollution: An emerging problem and potential bacterial remediation strategies. World. J. Microbiol. Biotech. 25, 1529–1537.

Jomova, K., Jenisova, Z., Feszterova, M., Baros, S., Liska, J., Hudecova, D., Rhodes, C.J., Valko, M., 2011. Arsenic: Toxicity, oxidative stress and human disease. J. Appl. Toxicol. 31, 95–107.

Kaur, S., Kamli, M.R., Ali, A., 2009. Diversity of arsenate reductase genes (arsC genes) from arsenic resistant environmental isolates of E. coli. Curr. Microbiol. 59, 288–294.

Kim, S.H., Lee, H.S., Song, W.Y., Choi, K.S., Hur, Y., 2007. Chloroplast-targeted BrMT1 (Brassica rapa Type-1 metallothionein) enhances resistance to cadmium and ROS in transgenic Arabidopsis plants. J. Plant Biol. 50, 1–7.

Kohler, A., Blaudez, D., Chalot, M., Martin, F., 2004. Cloning and expression of multiple metallothioneins from hybrid poplar. New Phytol. 164, 83–93.

Kramer, 2005. Phytoremediation: Novel approaches to cleaning up polluted soils. Curr. Opin. Biotechnol. 2, 133–141.

Kramer, U., Cotter-Howells, J.D., Charnock, J.M., Baker, A.J.M., Smith, J.A.C., 1996. Free histidine as a metal chelator in plants that accumulate nickel. Nature 379, 635–638.

Kuffner, M., De Maria, S., Puschenreiter, M., Fallmann, K., Wieshammer, G., Gorfer, M., et al., 2010. Culturable bacteria from Zn and Cd accumulating Salix caprea with differential effects on plant growth and heavy metal availability. J. Appl. Microbiol. 108, 1471–1484.

Kuiper, I., Lagendijk, E.L., Bloemberg, G.V., Lugtenberg, B.J.J., 2004. Rhizoremediation: A beneficial plant–microbe interaction. Mol. Plant Microb. Interact. 17, 6–15.

Kumar, P.B.A.N., Dushenkov, V., Motto, H., Raskin, I., 1995. Phytoextraction: The use of plants to remove heavy metals from soils. Environ. Sci. Technol. 29, 1232–1238.

Leigh, M.B., Fletcher, J.S., Fu, X., Schmitz, F.J., 2002. Root turnover: An important source of microbial substances in rhizosphere remediation of recalcitrant contaminants. Environ. Sci. Technol. 36, 1579–1583.

Lindow, S.E., Panopoulos, N.J., Mcfarland, B.L., 1989. Genetic engineering of Bacteria from managed and natural habitats. Science 244, 1300–1307.

Lubeck, P.S., Hansen, M., Sorensen, J., 2000. Simultaneous detection of the establishment of seed-inoculated Pseudomonas fluorescens strain DR54 and native soil bacteria on sugar beet root surfaces using fluorescence antibody and in situ hybridization techniques. FEMS Microbiol. Ecol. 33, 11–19.

Lugtenberg, B.J.J., de Weger, L.A., Bennett, J.W., 1991. Microbial stimulation of plant growth protection from disease. Curr. Biotechnol. 2, 457–465.

Lyyra, S., Meagher, R.B., Kim, T., Heaton, A., Montello, P., Balish, R.S., Merkle, S.A., 2007. Coupling two mercury resistance genes in Eastern cottonwood enhances the processing of organomercury. Plant Biotechnol J. 5, 254–262.

Ma, Y., Prasad, M.N.V., Rajkumar, M., Freitas, H., 2011. Plant growth promoting rhizobacteria and endophytes accelerate phytoremediation of metalliferous soils. Biotechnol. Adv. 29, 248–258.

Malik, A., 2004. Metal bioremediation through growing cells. Environ. Inter. 30, 261–278.

Martinez, C.E., Motto, H.L., 2000. Solubility of lead, zinc, and copper added to mineral soils. Environ. Poll. 107, 153–158.

McLaughlin, M.J., Parker, D.R., Clarke, J.M., 1999. Metals and micronutrients – food safety issues. Field Crops Res. 60, 143–163.

McNeil, K.R., Waring, S., 1992. Contaminated Land Treatment Technologies. In: Rees, J.F. (Ed.), Society of chemical industry. Elsevier Applied Sciences, London, pp. 143–159.

Meagher, R.B., Heaton, A.C.P., 2005. Strategies for the engineered phytoremediation of toxic element pollution: Mercury and arsenic. J. Ind. Microbiol. Biotechnol. 32, 502–513.

Meers, E., Vandecasteele, B., Ruttens, A., Vangronsveld, J., Tack, F.M.G., 2007. Potential of five willow species (*Salix* spp.) for phytoextraction of heavy metals. Environ. Exp. Bot. 60, 57–68.

Mejare, M., Bulow, L., 2001. Metal binding proteins and peptides in bioremediation and phytoremediation of heavy metals. Trends Biotechnol. 19, 67–75.

Millar, J.A., Battistini, V., Cumming, R.L.C., Carswell, F., Goldberg, A., 1970. Lead and δ-aminolevulinic acid dehydratase levels in mentally retarded children and in lead-poisoned suckling rats. Lancet 3, 695–698.

Miransari, M., 2011. Hyperaccumulators, arbuscular mycorrhizal, fungi and stress of heavy metals. Biotechnol. Adv. 29, 645–653.

Mohan, D., Pittman Jr, C.U., 2007. Arsenic removal from water/wastewater using adsorbents – a critical review. J. Hazard. Mat. 142, 1–53.

Molina, L., Ramos, C., Duque, E., Ronchel, M.C., Garcia, J.M., Wyke, L., Ramos, J.L., 2000. Survival of *Pseudomonas putida* KT2440 in soil and in the rhizosphere of plants under greenhouse and environmental conditions. Soil Biol. Biochem. 32, 315–321.

Mowll, J.L., Gadd, G.M., 1984. Cadmium uptake by *Aureobasidium pullulans*. J. Gen. Microbiol. 130, 279–284.

Mueller, B., Rock, S., Gowswami, D., Ensley, D., 1999. Phytoremediation decision tree. Prepared by – Interstate Technology and Regulatory Cooperation Work Group. 1–36.

Muhlbachova, G., 2009. Microbial biomass dynamics after addition of EDTA into heavy metal contaminated soils. Plant Soil Environ. 55, 544–550.

Munshower, F.F., 1994. Practical handbook of disturbed land revegetation. Lewis Publishing, Boca Raton, FL.

Nair, A., Juwarkar, A.A., Devotta, S., 2008. Study of speciation of metals in an industrial sludge and evaluation of metal chelators for their removal. J. Hazard. Mater. 152, 545–553.

Needleman, H., 2004. Lead poisoning. Ann. Rev. Med. 55, 209–222.

Patel, J., Zhang, Q., Michael, R., McKay, L., Vincent, R., Xu, Z., 2010. Genetic engineering of *Caulobacter crescentus* for removal of cadmium from water. Appl. Biochem. Biotechnol. 160, 232–243.

Pazirandeh, M., Wells, B., Ryan, R.L., 1998. Development of bacterium-based heavy metal biosorbents: Enhanced uptake of cadmium and mercury by *Escherichia coli* expressing a metal binding motif. Appl. Environ. Microbiol. 64, 4068–4072.

Pedersen, M.B., Hansen, J.C., Mulvad, G., Pedersen, S.H., Gregersen, M., Danscher, G., 1999. Accumulation in brains from population exposed to high and low dietary levels of methylmercury. Int. J. Circumpolar Health 58, 96–107.

Persans, M.W., Yan, X., Patnoe, J.M.M.L., Kraemer, U., Salt, D.E., 1999. Molecular dissection of the role of histidine in nickel hyperaccumulation in *Thlaspi goesingense* (Halacsy). Plant Physiol. 121, 1117–1126.

Pilon-Smits, E., 2005. Phytoremediation. Ann. Rev. Plant Biol. 56, 15–39.

Rajkumar, M., Ae, N., Prasad, M.N.V., Freitas, H., 2010. Potential of siderophore-producing bacteria for improving heavy metal phytoextraction. Trends Biotechnol. 28, 142–28, 149.

Ramos, C., Molina, L., Molbak, L., Ramos, J.L., Molin, S., 2000. A bioluminescent derivative of *Pseudomonas putida* KT2440 for deliberate release into the environment. FEMS Microbiol. Ecol. 34, 91–102.

Raskin, I., Smith, R.D., Salt, D.E., 1997. Phytoremediation of metals: Using plants to remove pollutants from the environment. Curr. Opin. Biotechnol. 8, 221–226.

Rattray, E.A.S., Prosser, J.I., Glover, L.A., Killham, K., 1995. Characterization of rhizosphere colonization by luminescent *Enterobacter cloacae* at the population and single-cells levels. Appl. Environ. Microbiol. 61, 2950–2957.

Roosens, N.H., Bernard, C., Leplae, R., Verbruggen, N., 2004. Evidence for copper homeostasis function of metallothionein (MT3) in the hyperaccumulator. *Thlaspi caerulescens*. FEBS Lett. 577, 9–16.

Roosens, N.H., Leplae, R., Bernard, C., Verbruggen, N., 2005. Variations in plant metallothioneins: The heavy metal hyperaccumulator *Thlaspi caerulescens* as a case study. Planta. 222, 716–729.

Rosenberg, E., Lagmann, R., Kushmaro, A., Taube, R., Adler, R., Ron, E.Z., 1992. Petroleum bioremediation – a multiphase problem. Biodegradation 3, 337–350.

Ruff, H.A., Markowitz, M.E., Bijur, P.E., Rosen, J.F., 1996. Relationships among blood lead levels, iron deficiency, and cognitive development in two-year-old children. Environ. Health Perspect. 104, 180–185.

Rugh, C.L., Gragson, G.M., Meagher, R.B., Merkle, S.A., 1998. Toxic mercury reduction and remediation using transgenic plants with a modified bacterial gene. Hortscience 33, 618–621.

Rugh, C.L., Wilde, H.D., Stacks, N.M., Thompson, D.M., Summers, A.O., Meagher, R.B., 1996. Mercuric ion reduction and resistance in transgenic *Arabidopsis thaliana* plants expressing a modified bacterial merA gene. Proc. Natl. Acad. Sci. USA 93, 3182–3187.

Ruiz, O.N., Hussein, H.S., Terry, N., Daniell, H., 2003. Phytoremediation of organomercurial compounds via chloroplast genetic engineering. Plant Physiol. 132, 1344–1352.

Sayler, G.S., Ripp, S., 2000. Field applications of genetically engineered microorganisms for bioremediation processes. Curr. Opin. Biotech. 11, 286–289.

Schat, H., Vooijs, R., Kuiper, E., 1996. Identical major gene loci for heavy metal tolerances that have independently evolved in different local populations and subspecies of *Silene vulgaris*. Evolution 50, 1888–1895.

Sharma, P., Asad, S., Ali, A., 2013. Bioluminescent bioreporter for assessment of arsenic contamination in water samples of India. J. Biosci. 38, 251–258.

Silver, S., Misra, T.K., 1988. Plasmid-mediated heavy metal resistances. Ann. Rev. Microbiol. 42, 717–743.

Singh, J.S., Abhilash, P.C., Singh, H.B., Singh, R.P., Singh, D.P., 2011. Genetically engineered bacteria: An emerging tool for environmental remediation and future research perspectives. Gene 480, 1–9.

Sousa, C., Kotrba, P., Ruml, T., Cebolla, A., de Lorenzo, V., 1998. Metalloadsorption by *Escherichia coli* cells displaying yeast and mammalian metallothioneins anchored to the outer membrane protein LamB. J Bacteriol. 180, 2280–2284.

Timmis, K.N., Pieper, D.H., 1999. Bacteria designed for bioremediation. Tibtech 17, 201–204.

Tombolini, R., van der Gaag, D.J., Gerhardson, B., Jansson, J.K., 1999. Colonization pattern of the biocontrol strain *Pseudomonas chlororaphis* MA 342 on barley seeds visualized by using green fluorescent protein. Appl. Environ. Microbiol. 65, 3674–3680.

Tyagi, M., da Fonseca, M.M., de Carvalho, C.C., 2011. Bioaugmentation and biostimulation strategies to improve the effectiveness of bioremediation processes. Biodegradation 22, 231–241.

Ultra, V.U., Yano, A., Iwasaki, K., Tanaka, S., Kang, Y.M., Sakurai, K., 2005. Influence of chelating agent addition on copper distribution and microbial activity in soil and copper uptake by brown mustard (*Brassica juncea*). Soil Sci. Plant Nutr. 51, 193–202.

Van der Meer, J.R., 1997. Evolution of novel metabolic pathways for the degradation of chloroaromatic compounds. Antonie Van Leeuwenhoek 71, 159–178.

Vasak, M., 2005. Advances in metallothionein structure and function. J. Trace Elem. Med. Biol. 19, 13–17.

White, C., Gadd, G.M., 1998. Accumulation and effects of cadmium on sulphate-reducing bacterial biofilms. Microbiology 144, 1407–1415.

White Jr., P.M., Wolf, D.C., Thoma, G.J., Reynolds, C.M., 2003. Influence of organic and inorganic soil amendments on plant growth in crude oil-contaminated soil. Int. J. Phytoremed. 5, 381–397.

Wong, M.H., 2003. Ecological restoration of mine degraded soils, with emphasis on metal contaminated soils. Chemosphere 50, 775–780.

Yang, Q., Tu, S., Wang, G., Liao, X., Yan, X., 2012. Effectiveness of applying arsenate reducing bacteria to enhance arsenic removal from polluted soils by *Pteris vittata* L. Int. J. Phytorem. 14, 89–99.

Zhigang, A., Cuijie, L., Yuangang, Z., Yejie, D., Wachter, A., Gromes, R., Rausch, T., 2006. Expression of BjMT2, a metallothionein 2 from *Brassica juncea*, increases copper and cadmium tolerance in *Escherichia coli* and *Arabidopsis thaliana*, but inhibits root elongation in *Arabidopsis thaliana* seedlings. J. Exp. Bot. 57, 3575–3582.

Zhou, J., Goldsbrough, P.B., 1994. Functional homologs of fungal metallothionein genes from. Arabidopsis. Plant Cell 6, 875–884.

Zimeri, A.M., Dhankher, O.P., McCaig, B., Meagher, R.B., 2005. The plant MT1 metallothioneins are stabilized by binding cadmium and are required for cadmium tolerance and accumulation. Plant Mol. Biol. 58, 839–855.

Chapter 4

Phytoremediation: Mechanisms and Adaptations

Muhammad Sabir,*,† Ejaz Ahmad Waraich,‡ Khalid Rehman Hakeem,§ Münir Öztürk,¶ Hamaad Raza Ahmad* and Muhammad Shahid‡‡

*Institute of Soil and Environmental Sciences, University of Agriculture, Faisalabad Pakistan, †School of Plant Biology, University of Western Australia, Crawley, Australia, ‡Department of Crop Physiology, University of Agriculture, Faisalabad, Pakistan, §Faculty of Forestry, Universiti Putra Malaysia, Serdang, Selangor, Malaysia, ¶Department of Botany, Ege University, Izmir, Turkey, ‡‡Department of Environmental Sciences, COMSATS Institute of Information Technology, Vehari, Pakistan

INTRODUCTION

Metal contamination of soils is ubiquitous around the globe. Metals enter the soil due to anthropogenic activities such as the use of sewage sludge, urban composts, fertilizers, pesticides, sewage irrigation, incineration of municipal waste, auto-vehicle exhausts, industrial emissions and metal mining and smelting (Hussain et al., 2006; McGrath et al., 2001; Murtaza et al., 2010). These metals include iron (Fe), copper (Cu), manganese (Mn), zinc (Zn), cadmium (Cd), lead (Pb), chromium (Cr), mercury (Hg) and nickel (Ni) (McIntyre, 2003). Metals accumulate in the soil to toxic levels that may lead to accumulation of metals in plants to unacceptable levels. Metal accumulation is a subject of serious concern due to threat to plant growth, soil quality, animal and human health (McGrath et al., 2001). Cleaning up soils to remove metals is a sign of the times, but it is a challenging task. Different technologies being used nowadays are ex situ which lead to destruction of soil structure, thus leaving it unusable with poor vegetative cover (He and Yang, 2007). Growing plants to clean up the soils is a cost-effective and environmentally friendly alternative (Yang et al., 2005). Phytoremediation seems attractive due to non-invasive and non-destructive technologies which leave the soil intact and biologically productive (Wenzel, 2009). Plants respond differentially to metal contamination in soils and can be classified into different categories, depending upon their responses to metal contamination in their rooting medium. Plants can be classified into accumulators, indicators or excluders depending upon absorption and translocation of metals by the plants to above-ground parts (Baker, 1981). Accumulators can survive by maintaining high concentration of metals in their tissues. Indicator plants are reported to have mechanisms that control transloca-tion of metals from roots to shoots and excluders restrict the entry of metals into plants at root level (Chaudhry et al., 1998). Plants use different adaptive mecha-nisms to accumulate or exclude metals and thus maintain their growth. Accumula-tion and tolerance of metals by the plants is a complex phenomenon. Movement of metals across the root membrane, loading and translocation of metals through xylem and sequestration and detoxification of metals at the cellular and whole-plant levels are important mechanisms adopted by accumulator plants (Lombi et al., 2002). Indicator plants absorb the metals from the soils and then restrict their movement to the shoots while excluders restrict the entry of metals into the plant roots. Understanding the mechanisms involved in phytoremediation is nec-essary to effectively use this technique on metal-contaminated soils. This chapter discusses different mechanisms adopted by plants for remediation of metal-con-taminated soils, which are briefly given in Table 4.1.

PHYTOREMEDIATION AND MECHANISMS

Heavy metals degrade soil and water resources and thus pose a serious threat to human and animal health. This threat is further aggravated due to the persistent and non-biodegradable nature of metals (Gisbert et al., 2003). Accumulation of metals in the bodies of animals and humans after entering the food chain has seri-ous implications for health as some metals are known to damage DNA and cause cancer due to their mutagenic abilities (Steinkellner et al., 1998). Remediating the soils contaminated with metals is thus necessary for safe use of such soils and sev-eral in situ and ex situ technologies are used for this purpose. Phytoremediation is considered environmentally friendly, non-invasive and cost-effective technology to clean up the metal-contaminated soils. Plants adopt different mechanisms to grow in the metal-contaminated soils without adverse effects on their growth. Some plants exclude the metals from metabolically active sites by restricted uptake or root to shoot transfer of metals (Küpper et al., 1999). Some other plants can tolerate high metal concentrations in their tissues through binding of metals with organic compounds, metal compartmentalization at cellular and sub-cellular levels and metabolic alterations (Küpper et al., 1999; Peng et al., 2006; Wei et al., 2005). Heavy metals tolerance in plants may be defined as the ability of plants to survive in a soil that is toxic to other plants (Macnair et al., 1999). Phytoremedia-tion can be classified into phytoextraction, phytostabilization and phytovolatiliza-tion (Alkorta et al., 2004; Raskin et al., 1997), in addition to various other classes of phytoremediation which are beyond the scope of this chapter.

Phytoextraction

Removal of the metals from the soil by growing plants is known as phytoextrac-tion. Metal-extracting plants absorb metals from the soils, transport and con-centrate them in the above-ground parts of plants. The above-ground parts of

TABLE 4.1 Mechanisms Adopted by Plants for Remediation of Metal-Contaminated Soils

Plant	Type of phytoremediation	Metal	Mechanism	Reference
Silene vulgaris	Phytostabilization	Fe, Ni, Cu, Al, Sn, Zn	Binding with a protein with oxalate oxidase activity in cell wall Accumulation in cell well as silicates	Bringezu et al., 1999
Sedum alfredii H	Phytostabilization	Pb, Cd	Induction of glutathione biosynthesis that bind metals in roots	Anjum et al., 2012; Gupta et al., 2010; Sun et al., 2007; Zhang et al., 2008
Imperata cylindrical, Miscanthus floridulus	Phytostabilization	Cd, Zn, Cu, Pb	Fibrous root system retaining the metals	Peng et al., 2006
Lupinus albus	Phytostabilization	As, Cd	Metal accumulation in root nodules Increasing the pH in rhizosphere by citrate release	Vázquez et al., 2006
Athyrium wardii	Phytostabilization	Cd, Pb	Root retention of metals	Zhang et al., 2012; Zou et al., 2011
Salicornia bigelovii	Phytovolatilization	Se	Volatilization as dimethyle selenide	Lin et al., 2000
Sedum alfredii	Phytoextraction	Pb, Cd	Induction and accumulation of phytochelatin that binds metals in above-ground parts	Zhang et al., 2008
Ceratophyllum demersum	Phytoextraction	Cd	Production of phytochelatin for metal binding in shoots Activation of cysteine synthase, glutathione-S-transferase glutathione,	Mishra et al., 2009

Continued

TABLE 4.1 Mechanisms Adopted by Plants for Remediation of Metal-Contaminated Soils—cont'd

Brassica juncea	Phytoextraction	Cd	Synthesis of phytochelatins (PCs), glutathione reductase, non-protein thiols and glutathione for metal binding in shoots	Seth et al., 2008
Thlaspi caerulescens Thlaspi ochroleucum	Phytoextraction	Zn, Cd, Cr, Cu, Ni, Pb	Lowering the pH of rhizosphere; thus enhancing metal solubilization	McGrath et al., 1997
Cynodon dactylon	Phytostabilization	As, Zn, Pb	Binding with hyphae of mycorrhizae Release of organic acids	Leung et al., 2007
Pteris vittata	Phytoextraction	As	Increased colonization Exploring more soil	Leung et al., 2007
Thlaspi goesingense	Phytoextraction	Ni	Lowering the soil pH Release of ligands into rhizosphere	Puschenreiter et al., 2003; Wenzel et al., 2003a
Sedum alfredii	Phytoextraction	Zn	Metals loaded into leaf sections and protoplast	Yang et al., 2006
Arabidopsis halleri	Phytoextraction	Cd, Zn	Accumulation in trichomes and mesophyll cells	Küpper et al., 2000
Alyssum Species, Brassica juncea	Phytoextraction	Ni	Binding of the metals with histidine for detoxification	Kerkeb and Krämer, 2003; Krämer et al., 1996

the plants are harvested and can be safely processed for disposal or recycling of metals (Ali et al., 2013; Garbisu and Alkorta, 2001). Plants used for phytoextraction must not only be metal tolerant but they must be fast growing with the potential to produce high biomass. However, most of the metal-accumulating plants are slow growing with low biomass production (Evangelou et al., 2007). These characteristics of metal-accumulating plants have made the process of phytoextraction of metals very slow as phytoextraction is a function of tissue metals concentration and biomass produced (Chaney et al., 2007). Such metal-accumulator plants having capacity to accumulate 100 mg kg^{-1} of cadmium (Cd), 1000 mg kg^{-1} of arsenic (As), cobalt (Co), copper (Cu), lead (Pb) or nickel (Ni) or > 10,000 mg kg^{-1} of manganese (Mn) and zinc (Zn), are classified as hyperaccumulator plants. Hyperaccumulation of heavy metals by plants depends upon several steps, including absorption and transportation of metals across the membranes of root cells, loading of metals into xylem and translocation to the shoots and sequestration and detoxification of metals within plant tissues (Yang et al., 2005). Epidermis, trichomes and cuticle are the preferred sites of metal detoxification, and in many cases subsidiary and stomatal cells are protected against metal toxicity (Rascio and Navari-Izzo, 2011). Metal detoxification or sequestration traits are controlled by expression of genes encoding the protein responsible for exclusion of metals from cytoplasm and transfer across tonoplast and plasma membranes. Cation Diffusion Facilitator (CDF) family members like metal transporter proteins present in the tonoplast are overexpressed in Zn and Ni hyperaccumulators and these transporters are also reported to be involved in Ni accumulation by Ni hyperaccumulators (Gustin et al., 2009; Hammond et al., 2006; Persans et al., 2001; Rascio and Navari-Izzo, 2011). About 400 plants have been identified as hyperaccumulators which constitute only < 0.2% of higher plants (McGrath and Zhao, 2003). Low biomass production of hyperaccumulators discourages their adoption on a commercial scale for phytoextraction. However, some high-biomass-producing plants capable of tolerating metals can be effectively used for phytoextraction on a commercial scale (Saifullah et al., 2009). However, these species have an inherently low ability to absorb metals but can accumulate higher concentrations of metals if grown in the soils treated with chemical amendments to increase metal phytoavailability and plant uptake (Meers et al., 2005).

Cellular Detoxification of Metals

Mechanisms of phytoextraction and hyperaccumulation have been studied considerably but still a lot more research is required to fully understand the mechanisms. Plants adopt different mechanisms to maintain their growth in metal-contaminated soil environments. Plants may immobilize, exclude, chelate compartmentalize metal ions or may release ethylene or stress proteins (Cobbett, 2000b). Phytoextraction of metals is a function of two factors; biomass and metal bioconcentration factor. Bioconcentration factor is the ratio of metal concentration in the shoots of plants to that in the soil and thus indicates the uptake and translocation of metals by the plants to the shoots (McGrath and Zhao, 2003). Most plants have bioconcentration factors < 1 which make them unsuitable for use in phytoextraction, irrespective of biomass produced. However, most hyperaccumulators are known to have bioconcentration factors of > 1 (and in some cases the biocentration factors reach up to that of 50–100); the ratio of metal concentration between shoot and root is greater than one signifying efficient root to shoot translocation and having enhanced metal tolerance due to internal metal detoxification (McGrath and Zhao, 2003; Zhao et al., 2003). A study focusing on the genetic basis of hyperaccumulation reveals that zinc (Zn) hyperaccumulation and tolerance are independent traits controlled by separate genes (Macnair et al., 1999). Enhanced transfer of metals from roots to shoots (through decreased metal sequestration in roots or enhanced xylem loading) is an important component of metal hyperaccumulation and is evidenced by enhanced loading of histidine into xylem of *Alyssum lesbiacum* upon exposure to Ni (Krämer et al., 1996; Lasat et al., 1998). Hypertolerance is necessary for hyperaccumulation which is achieved through internal detoxification mediated by metal compartmentation and complexation. Metals are generally sequestered in leaf vacuoles and this is achieved through enhanced tonoplast transport of metals into vacuoles being controlled by a gene (Bert et al., 2003; Küpper et al., 2000; Macnair et al., 1999; Vázquez et al., 1994; Zhao et al., 2000). The vacuole serves as the dumping site for most of the complexed metals in plants and yeast; but against the yeast, movement of Ni into vacuoles is not a pH-gradient-dependent phenomenon in plants (Ramsay and Gadd, 1997; Salt et al., 1995). *Thlaspi goesingense* (Ni-hyperaccumulator) accumulates Ni by compartmentalizing intracellular Ni into leaf vacuoles (Krämer et al., 2007). The Metal ion transporter gene TgMTP1 is found to be involved in transport and accumulation of Ni in shoot vacuoles (Freeman et al., 2004). Vacuoles are the site of accumulation for large numbers of heavy metals including Cd and Zn (De, 2000), and Zn treatment was reported to enhance vacuolation in root meristematic cells of *Festuca rubra* L. (Davies et al., 1991). Zinc, Al and Cu were found in the cell walls, vacuoles and mucilage vesicles and higher concentrations of oxygen were found in the areas where Zn, Al and Cu were localized, indicating sequestration of metals as oxides (Volland et al., 2011). Among different mechanisms, metal chelation with ligands is an important mechanism adopted by plants to detoxify the metals within plant tissues (Cobbett, 2000b; Cobbett and Goldsbrough, 2002). As the metals enter the cytosol of plant cells, metal ions are bound with organic molecules, thus protecting the metabolically active sites from the toxic effects of metals (Zenk, 1996). These ligands include organic acids, amino acids, peptides and polypeptides. Phytochelatins (PCs) and metallothioneins (MTs) are the two best metal-binding polypeptide ligands in the plant cells and bind metals through thiolate coordination (Clemens et al., 2002; Cobbett, 2000a; Gupta et al., 1999). A mutant *Arabidopsis thaliana* is very sensitive to Cd because it does not have the enzyme 'phytochelatin synthase' but it grows well as the wild type plant at normal Zn and Cu concentrations, which are two essential metal ions, indicating that phytochelatin is only involved in resistance to metal poisoning (Cobbett et al., 1998). Cadmium is complexed with citrate in the vacuole when the cell is exposed to high Cd levels (Wagner, 1993). Glutathione (GSH) is one source of non-protein thiols found in variety of cell components like cytosol, chloroplast, endoplasmic reticulum, vacuole and mitochondria (Yadav, 2010). The thiol group forms mercaptide bonds with metals, thus making them an important biochemical molecule in the protection of plants against stresses caused by metals, exogenous and endogenous organic chemicals and oxidative stress (Mullineaux and Rausch, 2005; Rausch et al., 2007). Exposure of plants to metals triggers the formation of reactive oxygen species and free radicals in the plants which damage metabolic components of cells. Excess reactive oxygen species in the plant cells initiate oxidation of amino acids, proteins, membrane lipids and DNA resulting in decreased growth and development (Ogawa and Iwabuchi, 2001). GSH decreases the level of reactive species in the plants cells and thus prevents the damage caused thereby (Foyer and Noctor, 2005). GSH causes cellular detoxification of metals and xenobiotics by conjugating with such molecules through S-transferase and conjugates are transported to vacuoles (Dixon et al., 2002; Edwards and Dixon, 2005; Klein et al., 2006; Yazaki, 2006). In addition to its protective role, GSH is the precursor for biosynthesis of phytochelatins which are excellent heavy-metal-binding peptides (Grill et al., 1988). Nickel hyperaccumulator *Thlaspi* contained higher concentrations of GSH which is strongly correlated with Ni hyperaccumulation (Freeman et al., 2004). It was observed by Freeman et al. (2004) that elevated levels of GSH in hyperaccumulators is due to enhanced activity of Ser acetyltransferase which provides plants with the ability to detoxify non-sequestered Ni. It was reported that there was a significant increase in histidine concentration in Ni hyperaccumulator *Alyssum* when exposed to high Ni concentration. Moreover, the supply of histidine to non-accumulators caused hyperaccumulation of Ni and enhanced transport towards shoots (Krämer et al., 1996). Metallothioneins (MT) are low-molecular-weight cysteine-rich metal-binding proteins and are synthesized from mRNA (Memon and Schröder, 2009). Metallothioneins are reported to be involved in metal homeostasis in different ecotypes of Arabidopsis evidenced by correlation between MT RNA and metal tolerance level of Arabidopsis (Guo et al., 2003, 2008). Metal detoxifications, development of plants and resistance against abiotic stress are some of the functions listed for metallothioneins in plants (Domènech et al., 2006; Roosens et al., 2005; Zhou et al., 2005).

Rhizosphere Changes and Root Adaptations

The rhizosphere represents the soil volume around the plant root which is directly influenced by root activity (Hinsinger et al., 2005). Plants change the bioavailability of metals in the rhizosphere due to changes in elemental concentration, soil reaction (pH), partial pressure of carbon dioxide (pCO_2) and oxygen (pO_2), redox potential, organic ligand concentration and microbial biomass

(Kidd et al., 2009; Wenzel et al., 1999). Plant species differ in their effects on the rhizosphere due to the nature of root exudates, nutrient absorption strategies, root architecture, soil type and properties (Hinsinger et al., 2008). Plants release about 10–20% of photosynthetic carbon from roots into the rhizosphere in the form of different rhizodepositions (Singer et al., 2003). Root exudates are involved in weathering of soil parent material, mobilization and enhancement of nutrient uptake and stress resistance to toxic metals. Plants can enhance metal bioavailability by altering the rhizosphere. Some plants can mobilize sparingly soluble nutrients like iron (Fe) and phosphorus (P) in soil by releasing chelators into the rhizosphere. However, evidences are very rare regarding alteration of rhizosphere by hyperaccumulators for absorption of less soluble fractions of metals in the soil. It has been reported that root exudates of *Thlaspi caerulescens* cannot mobilize heavy metals in soil; rather, roots proliferate in pursuance of Zn in the soils but the mechanism involved in sensing Zn in soil is still unknown. Others have reported the release of metal-solubilizing phytosiderophores and hydrogen ions which can chelate metals or acidify the rhizosphere and increasing metals in soil solution (Lone et al., 2008; Thangavel and Subbhuraam, 2004). Additionally, rhizomicrobes like rhizobacteria and *Mycorrhizal fungi* can increase the bioavailability of heavy metals in soil (Vamerali et al., 2011). Several researchers have reported that the roots of hyperaccumulators proliferate in response to metal contamination in the soil to accumulate more metals while non-hyperaccumulators restrict their root growth as a result of metal contamination in soils (Schwartz et al., 1999; Whiting et al., 2001). These results indicate the presence of some mechanism in the roots triggering its preferred growth towards metal-rich areas of soil which still needs to be uncovered. Hyperaccumulator plants absorb excessive metals due to high-affinity transport systems across the plasma membranes of root tissues (Lasat, 2002; Wenzel, 2009). Hyperaccumulators like *Thlaspi caerulescens* grow their roots efficiently into metal-enriched areas for excessive absorption of metals (Schwartz et al., 1999), while some other researchers have reported the development of dense root systems with a large proportion of fine roots by hyperaccumulators in metal-contaminated soils which enhance metal uptake (Himmelbauer et al., 2005; Keller et al., 2003). Conversely, roots and root debris can adsorb cationic metals/metalloids and thus can decrease their bioavailability (Keller et al., 2003). Plant root can avoid or actively fetch heavy metals from polluted soils (Keller et al., 2003; Schwartz et al., 2003). Plant roots can increase metal solubility by changing speciation including acidification/alkalinization, modification of the redox potential, exudation of metal chelants and organic ligands (in particular low molecular organic acids and phytosiderophores) that compete with anionic species for binding sites (Fitz and Wenzel, 2002; Puschenreiter et al., 2005). Increased metal solubility is not necessarily related to increased absorption of metals by the plants (Shenker et al., 2001). Hyperaccumulator plants represent the main model for phytoextraction of metals from the soils but the main question of metal mobilization by these plants through rhizosphere changes still needs to be answered. Alternately, pH in the rhizosphere of these plants can be manipulated by using amendments. Plants can mobilize the soil-bound metals by adopting different strategies which are briefly explained below. Plants can secrete metal-chelating substances into rhizosphere such as phytosiderophores for solubilizing Fe. Largely phytosiderophores have been studied for Fe mobilization but these compounds can also mobilize other metals. Mugineic and deoxymugeneic acids from barley and corn and avenic acid from oats are considered the best studied plant phytosiderophores (Kinnersley, 1993). Histidine is reported to be used by Ni hyperaccumulator, *Alyssum lesbiacum*, for the acquisition and transport of Ni (Krämer et al., 1996). Roots can release metal reductase for the reduction and solublization of soil-bound metals (Raskin et al., 1997). Some plants deficient in Fe or Cu can reduce these metals by releasing certain organic compounds and thereby increase their availability and absorption by the plants (Marschner and Römheld, 1994; Mejáre and Bulow, 2001; Raskin et al., 1994). Alternately, some plants solubilize metals by acidifying their rhizospheres by releasing protons (Crowley et al., 1991). Mycorrhizal association with roots or root colonization by bacteria can enhance the availability of metals. Hyperaccumulator *Thlaspi* species depleted labile and EDTA extractable fractions of the metals at high concentrations, whereas labile metal pools increase at low metal concentration (Puschenreiter et al., 2003). It was concluded that the rhizosphere of Ni hyperaccumulator *Thlaspi goesingense* contained higher levels of dissolved organic carbon and some sugars than that of bulk soils (Wenzel et al., 2003b). This signifies the ligand-induced solublization of metals in rhizosphere which helps in hyperaccumulation of metals by this species. Efficiency of phytoextraction depends on the availability of metals in the root zone which is affected by soil factors like cation exchange capacity, pH or organic matter contents (Evangelou et al., 2007; Felix, 1997; Schmidt, 2003). Growing high-biomass crops is a key factor for the success of phytoextraction; however, the problem of low mobility of metals in the soil can be overcome by enhancing the mobility through application of chelating agents (Evangelou et al., 2007). The chelant-assisted mobilization of metals could be environmentally unsafe due to leaching of mobilized metals into the ground water as these overcome the absorption capacity of most of the plants.

Phytostabilization

Some soils are highly contaminated to the extent that phytoextraction of metals from such soils would take a considerably longer period of time which is neither economical nor suitable. If such soils are not remediated, these could be a major source of metal dispersion into the environment. The risk posed by such soils can be decreased by using plants to stabilize the metals in the soil (Marques et al., 2009). Phytostabilizing plants can grow in metal-contaminated soils by keeping the metals in below-ground parts, immobilizing in the rhizosphere through various mechanisms. Plants provide litter, vegetative cover, decrease the leaching losses from soil, control erosion and release organic matter to the soil and thus bind metals (Pulford and Watson, 2003; Robinson et al., 2006). In phytostabilization, plants immobilize the metals in the rhizosphere thereby leaving them less bioavailable and less toxic to plants, animals and humans or retain the metals in the roots by restricting their translocation to above-ground parts (Mendez and Maier, 2008; Wong, 2003). Additionally, plants provide vegetative cover to reduce the aeolian dispersion of metals into the environment.

Mechanisms of Phytostabilization

Plants sequester the metals in the rhizosphere through adsorption and precipitation of metals into less soluble forms like carbonates and sulphides of metals, metal complexes with organic compounds, metal adsorption on root surfaces and metal accumulation in root tissues (Mendez and Maier, 2008; Wong, 2003). The presence of plants in metal-contaminated soils promotes heterotrophic microbial communities which may, in turn, promote plant growth and participate in metal stabilization. Metal-tolerant plants with the capacity to keep the metals out of metabolic sites (shoots) are the best candidates for phytostabilization. Although such plants have developed mechanisms to restrict the metals in the rhizosphere or roots, even then concentration of metals in shoots must be monitored (Mendez and Maier, 2008). The *Cynodon dactylon* was found to be the best accumulator of As in roots and thus a promising candidate for phytostabilization and have wide adaptations in Pb- and Zn-contaminated soils (Leung et al., 2007). Mycorrhizae play an important role in stabilization by binding the metals with hyphae and some mycorrhizae like ericoid and *Ectomycorrhizal fungi* colonizing in *Cynodon dactylon* can modify the rhizosphere by excreting organic acids and thus stabilizing metals in the rhizosphere (Meharg, 2003). Hyphae of *Mycorrhizal fungi* contain polyphosphate which can bind heavy metals up to saturation and >60% metals are reported to be retained in apoplast cell walls (Bücking and Heyser, 1999, 2000; Yang et al., 2005). Some plants can detoxify the metals in the rhizosphere by releasing organic acids thus tendering the metals less available (Brunner et al., 2008; Qin et al., 2007). Another process of metal detoxification is immobilization of metals in fine roots through binding with pectins in the cell walls and to the negatively charged cytoplasm-membrane surfaces due to their strong electrochemical potential (Kochian et al., 2005; Rengel and Zhang, 2003). Lupin (*Lupinus albus* L.) stabilized As and Cd in contaminated soil by increasing pH due to release of citrate and accumulation in root nodules (Vázquez et al., 2006). Some plants have the ability to reduce the valence of metals by releasing redox enzymes and thus toxic metals are converted into less toxic forms (Ali et al., 2013). Transformation of tetravalent chromium (more toxic) to trivalent chromium (less toxic) is the best-studied example of this strategy being adopted by the plants (Bluskov et al., 2005). Some plants accumulate metals to high concentrations in their roots and restrict their translocation to shoots and thus become good candidates for phytostabilization (Pignattelli et al., 2012). Phytostabilization has very promising results for stabilization of chromium and lead in soils. Hexavalent chromium (Cr^{6+}) is highly toxic and is transformed into less soluble and less toxic trivalent chromium (Cr^{3+}) by deep-rooted plants (Chaney et al., 1997; James, 1996). Lead (Pb) is present in the soil in different species which are mostly bioavailable but the Pb-phosphate mineral, chloropyromorphite, is insoluble and non-bioavailable (Chaney et al., 1997; Cotter-Howells, 1996). Formation of chloropyromorphite is induced by roots of *Agrostis capillaris* growing in highly contaminated Pb/Zn mine waste soils (Cotter-Howells and Caporn, 1996). Norway spruce (*Picea abies*) and poplar (*Populus tremula*) accumulate 10 to 20 times more Cu and Zn in their roots compared to those plants grown in uncontaminated soils and accumulation was restricted to fine root, cell walls and epidermis (Brunner et al., 2008). Plants which can survive in metal-contaminated soils without affecting growth and maintain low concentrations of metals in aerial parts, even though concentration of metals is very high in the roots, are known as metal excluder plants (Baker, 1981; Krämer, 2010; Wei *et al.*, 2005). Several plants with the potential to exclude metals from aerial parts have been identified. These include Ni-excluders such as *Silene vulgaris, Zea mays* L., Cu excluder *Hyparrhenia hirta* and Co excluder *Armeria maritima* (Brewin et al., 2003; Poschenrieder et al., 2001; Seregin et al., 2003; Wenzel et al., 2003a). Although, excluder plants can grow in metal-contaminated soils without affecting their growth and keeping metal concentration in aerial parts at minimum levels, it is most important that metal concentration should not exceed standards for agricultural products (Wei *et al.*, 2005). Plants use different strategies to exclude metals from the tissues and these may include the role of mycorrizae, cell walls and plasma membranes (Hall, 2002). Mycorrhizae can play an effective role in amelioration of metal toxicity in plants. Mycorrhizae generally adopt the same mechanisms as those are adopted by higher plants like binding to extracellular materials or sequestration in the vacuolar compartment (Hall, 2002; Tam, 1995). In relation to the role of ectomycorrhizae in metal tolerance by the host plant, most mechanisms that have been proposed involve various exclusion processes that restrict metal movement to the host roots (Jentschke and Godbold, 2000). These have been extensively reviewed and assessed and include absorption of metals by the hyphal sheath, reduced access to the apoplast due to the hydrophobicity of the fungal sheath, chelation by fungal exudates and adsorption onto the external mycelium (Jentschke and Godbold, 2000). Clearly, from the variation between species described above, these different exclusion mechanisms are likely to vary in significance between different plant-fungal interactions. Discovering the mechanisms of metal exclusion and genes responsible for metal exclusion and their ultimate induction in field crops is very important and thus can lead to the safe use of metal-contaminated soils (Wei *et al.*, 2005). Several hypotheses have been suggested regarding the mechanisms of metal exclusion. These include metal binding in cell walls, exudation of metal-chelating ligands and formation of redox and pH barriers at the plasma membrane (Taylor, 1987). There are contradictory reports about the role of the cell wall in metal tolerance of plants (Hall, 2002). Some researchers found that the cell wall plays a very minor role in metal tolerance, whereas others have found heavy metals accumulated in the cell wall as bound with protein or as silicates (Bringezu et al., 1999). The cell membrane is the first living structure of the plant which is likely be damaged by metals. Metal toxicity causes leaky behaviour of the plasma membrane due to oxidation of protein thiols, inhibition of HATPase, alternation of composition and fluidity of membrane lipids (Astolfi et al., 2005; Devi and Prasad, 1999; Hall, 2002). Thus protection of the plasma membrane against metal toxicity damage is the key to metal tolerance in plants. Metal-tolerant plants opt for homeostasis to sustain the high concentration of metals due to their inability to tolerate reactive oxygen species or free radicals (Dietz et al., 1999; Sharma and Dietz, 2009; Panda et al., 2003). Aluminium tolerance in wheat is initiated by extracellular chelation of Al with citrate and malate (Delhaize and Ryan, 1995) and release of organic acids from roots has also been reported in Al-resistant Arabidopsis (Larsen et al., 1998). Phytostabilization is considered a very good alternative for those soils which cannot be immediately remediated through phytoextraction. Efficiency of phytostabilization can be enhanced by involving soil amendments like zeolites, beringite, steel shot and hydroxyapatite (Lothenbach et al., 1998). *Vetiveria zizanioides, Sesbania rostrata,* herb legume and *Leucaena leucocephala* have been successfully grown in metal-contaminated soils for metal stabilization (Shu et al., 2002; Zhang et al., 2001).

Phytovolatilization

Transformation of toxic metals and metalloids like mercury (Hg), selenium (Se) and arsenic (As) into less toxic and volatile forms released through foliage by plants into the atmosphere is known as phytovolatilization (Malik and Biswas, 2012; Marques et al., 2009). In phytostabilization metals are assimilated into organic compounds which are volatile in nature and ultimately released into atmosphere as biomolecules (Marques et al., 2009). *Brassica Juncea* has been shown to volatilize Se into the atmosphere through assimilation of Se from the soil into organic seleno-amino acids, seleno-cysteine and seleno-methionine which later can be biomethylated to form the volatile compound dimethylselenide (Banuelos et al., 1993; Banuelos and Meek, 1990; Terry et al., 2000). A gene responsible for reducing mercuric ion into elemental mercury through enzyme mercury reductase has been introduced into *Arabidopsis thaliana* which ultimately volatilizes large amounts of Hg into the atmosphere (Rugh et al., 1996). The practical application of phytovolatilization is questioned due to the release of toxic volatile compounds into the atmosphere and a risk assessment should be done. However, some researchers have reported that volatile compounds released into the atmosphere are dispersed and diluted in the atmosphere and pose no environmental risk (Lin et al., 2000; Meagher et al., 2000).

Arsenic was successfully volatilized in a frond of *Pteris vittata* in the form of arsenic compounds, arsenite and arsenate (Sakakibara et al., 2010).

CONCLUSIONS

Metal contamination of soils is a widespread problem around the globe with varying intensities and magnitudes in different regions. Several remediation technologies have already been discussed in detail elsewhere with each one carrying a wide range of merits and demerits. Every technology is aimed at the safe use of metal-contaminated soils for environmental quality and safe food chains. Among all remediation methods, phytoremediation is considered to be environmentally friendly, non-disruptive and low in cost. At the same time, adoption of phytoremediation technology on a commercial scale warrants serious consideration of issues of being slow and time consuming and the fate of the plants being used. A variety of plants have been identified which are capable of accumulating high concentrations of metals in their aerial parts (phytoextraction), retaining the metals in roots or stabilizing the metals in soils and thus restricting their translocation to the shoots (phytostabilization) and removing the metals from the soil through synthesis of volatile compounds (phytovolatilization). Each of the aforementioned technologies involves distinct mechanisms which are already explained in detail. The choice of phytoremediation technology to be employed for remediation of metal-contaminated sites depends on soil type, metal type, degree and extent of contamination and environmental disturbance involved. An understanding of the different mechanisms involved would really improve the decision making in the adoption of a specific technology. Among different phytoremediation technologies, phytoextraction is being widely used and a wide range of hyperaccumulator plants capable of accumulating high concentrations of metals have been identified. However, most of hyperaccumulators are slow growing and low-biomass-producing plants which make this technology a slow process. Chelant-assisted phytoextraction through fast-growing and high-biomass-producing plants is an alternative option, but leaching or run-off of solubilized metals into surface and sub-surface water bodies is a serious issue. Identification and induction of genes responsible for hyperaccumulation in hyperaccumulator plants into those plants which are capable of accumulating metals and producing high biomass could revolutionize the phytoremediation technology.

REFERENCES

Ali, H., Khan, E., Sajad, M.A., 2013. Phytoremediation of heavy metals—Concepts and applications. Chemosphere, 91, 869–881.

Alkorta, I., Hernández-Allica, J., Becerril, J., Amezaga, I., Albizu, I., Garbisu, C., 2004. Recent findings on the phytoremediation of soils contaminated with environmentally toxic heavy metals and metalloids such as zinc, cadmium, lead, and arsenic. Rev. Environ. Sci. Biotechnol. 3, 71–90.

Anjum, N.A., Ahmad, I., Mohmood, I., Pacheco, M., Duarte, A.C., Pereira, E., Umar, S., Ahmad, A., Khan, N.A., Iqbal, M., 2012. Modulation of glutathione and its related enzymes in plants' responses to toxic metals and metalloids—a review. Environ. Exp. Bot. 75, 307–324.

Astolfi, S., Zuchi, S., Passera, C., 2005. Effect of cadmium on H+ATPase activity of plasma membrane vesicles isolated from roots of different S-supplied maize *Zea mays* L plants. Plant Science 169, 361–368.

Baker, A.J., 1981. Accumulators and excluders-strategies in the response of plants to heavy metals. J. Plant Nutr. 3, 643–654.

Banuelos, G., Cardon, G., Mackey, B., Ben-Asher, J., Wu, L., Beuselinck, P., Akohoue, S., Zambrzuski, S., 1993. Boron and selenium removal in boron-laden soils by four sprinkler irrigated plant species. J. Environ. Qual. 22, 786–792.

Banuelos, G., Meek, D., 1990. Accumulation of selenium in plants grown on selenium-treated soil. J. Environ. Qual. 19, 772–777.

Bert, V., Meerts, P., Saumitou-Laprade, P., Salis, P., Gruber, W., Verbruggen, N., 2003. Genetic basis of Cd tolerance and hyperaccumulation in *Arabidopsis halleri*. Plant and Soil 249, 9–18.

Bluskov, S., Arocena, J., Omotoso, O., Young, J., 2005. Uptake, distribution, and speciation of chromium in *Brassica juncea*. Int. J. Phytoremediation 7, 153–165.

Brewin, L., Mehra, A., Lynch, P., Farago, M., 2003. Mechanisms of copper tolerance by *Armeria maritima* in Dolfrwynog Bog, North Wales-Initial Studies. Environmental geochemistry and health 25, 147–156.

Bringezu, K., Lichtenberger, O., Leopold, I., Neumann, D., 1999. Heavy metal tolerance of *Silene vulgaris*. J. Plant Physiol. 154, 536–546.

Brunner, I., Luster, J., Günthardt-Goerg, M.S., Frey, B., 2008. Heavy metal accumulation and phytostabilisation potential of tree fine roots in a contaminated soil. Environ. Pollut. 152, 559–568.

Bücking, H., Heyser, W., 1999. Elemental composition and function of polyphosphates in *Ectomycorrhizal fungi*—an X-ray microanalytical study. Mycol. Res. 103, 31–39.

Bücking, H., Heyser, W., 2000. Subcellular compartmentation of elements in non-mycorrhizal and mycorrhizal roots of *Pinus sylvestris*: An X-ray microanalytical study. I. The distribution of phosphate. New Phytol. 145, 311–320.

Chaney, R.L., Angle, J.S., Broadhurst, C.L., Peters, C.A., Tappero, R.V., Sparks, D.L., 2007. Improved understanding of hyperaccumulation yields commercial phytoextraction and phytomining technologies. J. Environ. Qual. 36, 1429–1443.

Chaney, R.L., Malik, M., Li, Y.M., Brown, S.L., Brewer, E.P., Angle, J.S., Baker, A.J.M., 1997. Phytoremediation of soil metals. Curr. Opin. Biotechnol. 8, 279–284.

Chaudhry, T., Hayes, W., Khan, A., Khoo, C., 1998. Phytoremediation—focusing on accumulator plants that remediate metal-contaminated soils. Australas. J. Ecotoxicol. 4, 37–51.

Clemens, S., Palmgren, M.G., Krämer, U., 2002. A long way ahead: Understanding and engineering plant metal accumulation. Trends Plant Sci. 7, 309–315.

Cobbett, C., Goldsbrough, P., 2002. Phytochelatins and metallothioneins: Roles in heavy metal detoxification and homeostasis. Annu. Rev. Plant Biol. 53, 159–182.

Cobbett, C.S., 2000a. Phytochelatin biosynthesis and function in heavy-metal detoxification. Curr. Opin. Plant Biol. 3, 211–216.

Cobbett, C.S., 2000b. Phytochelatins and their roles in heavy metal detoxification. Plant Physiol. 123, 825–832.

Cobbett, C.S., May, M.J., Howden, R., Rolls, B., 1998. The glutathione-deficient, cadmium-sensitive mutant, cad2-1, of Arabidopsis thaliana is deficient in γ-glutamylcysteine synthetase. Plant J. 16, 73–78.

Cotter-Howells, J., 1996. Lead phosphate formation in soils. Environ. Pollut. 93, 9–16.

Cotter-Howells, J., Caporn, S., 1996. Remediation of contaminated land by formation of heavy metal phosphates. Appl. Geochem. 11, 335–342.

Crowley, D., Wang, Y., Reid, C., Szaniszlo, P., 1991. Mechanisms of iron acquisition from siderophores by microorganisms and plants. Plant and Soil 130, 179–198.

Davies, K., Davies, M., Francis, D., 1991. The influence of an inhibitor of phytochelatin synthesis on root growth and root meristematic activity in *Festuca rubra* L. in response to zinc. New Phytol. 118, 565–570.

De, D., 2000. Plant Cell Vacuoles: An Introduction. Csiro Publishing.

Delhaize, E., Ryan, P.R., 1995. Aluminum toxicity and tolerance in plants. Plant Physiology 107, 315.

Devi, S.R., Prasad, M., 1999. Membrane lipid alterations in heavy metal exposed plants, Heavy Metal Stress in Plants. Springer, 99–116.

Dietz, K.-J., Baier, M., Krämer, U., 1999. Free radicals and reactive oxygen species as mediators of heavy metal toxicity in plants, Heavy metal stress in plants. Springer, 73–97.

Dixon, D.P., Lapthorn, A., Edwards, R., 2002. Plant glutathione transferases. Genome Biol 3 3004 3001–3004.3010.

Domènech, J., Mir, G., Huguet, G., Capdevila, M., Molinas, M., Atrian, S., 2006. Plant metallothionein domains: Functional insight into physiological metal binding and protein folding. Biochimie 88, 583–593.

Edwards, R., Dixon, D.P., 2005. Plant glutathione transferases. Methods Enzymol. 401, 169–186.

Evangelou, M.W., Ebel, M., Schaeffer, A., 2007. Chelate assisted phytoextraction of heavy metals from soil. Effect, mechanism, toxicity, and fate of chelating agents. Chemosphere 68, 989–1003.

Felix, H., 1997. Field trials for in situ decontamination of heavy metal polluted soils using crops of metal-accumulating plants. Zeitschrift für Pflanzenernährung und Bodenkunde 160, 525–529.

Fitz, W.J., Wenzel, W.W., 2002. Arsenic transformations in the soil–rhizosphere–plant system: Fundamentals and potential application to phytoremediation. J. Biotechnol. 99, 259–278.

Foyer, C.H., Noctor, G., 2005. Redox homeostasis and antioxidant signaling: A metabolic interface between stress perception and physiological responses. The Plant Cell Online 17, 1866–1875.

Freeman, J.L., Persans, M.W., Nieman, K., Albrecht, C., Peer, W., Pickering, I.J., Salt, D.E., 2004. Increased glutathione biosynthesis plays a role in nickel tolerance in *Thlaspi nickel hyperaccumulators*. The Plant Cell Online 16, 2176–2191.

Garbisu, C., Alkorta, I., 2001. Phytoextraction: A cost-effective plant-based technology for the removal of metals from the environment. Bioresour. Technol. 77, 229–236.

Gisbert, C., Ros, R., De Haro, A., Walker, D.J., Pilar Bernal, M., Serrano, R., Navarro-Aviñó, J., 2003. A plant genetically modified that accumulates Pb is especially promising for phytoremediation. Biochem. Biophys. Res. Commun. 303, 440–445.

Grill, E., Thumann, J., Winnacker, E., Zenk, M., 1988. Induction of heavy-metal binding phytochelatins by inoculation of cell cultures in standard media. Plant Cell Rep. 7, 375–378.

Guo, W.-J., Meetam, M., Goldsbrough, P.B., 2008. Examining the specific contributions of individual *Arabidopsis metallothioneins* to copper distribution and metal tolerance. Plant Physiol. 146, 1697–1706.

Guo, W.J., Bundithya, W., Goldsbrough, P.B., 2003. Characterization of the *Arabidopsis metallothionein* gene family: Tissue-specific expression and induction during senescence and in response to copper. New Phytol. 159, 369–381.

Gupta, D., Huang, H., Yang, X., Razafindrabe, B., Inouhe, M., 2010. The detoxification of lead in *Sedum alfredii* H. is not related to phytochelatins but the glutathione. J. Hazard. Mater. 177, 437–444.

Gupta, M., Tripathi, R., Rai, U., Haq, W., 1999. Lead induced synthesis of metal binding peptides (Phytochelatins) in submerged macrophyte *Vallisneria spiralis* L. Physiol. Mol. Biol. Plants 5, 173–180.

Gustin, J.L., Loureiro, M.E., Kim, D., Na, G., Tikhonova, M., Salt, D.E., 2009. MTP1-dependent Zn sequestration into shoot vacuoles suggests dual roles in Zn tolerance and accumulation in Zn-hyperaccumulating plants. Plant J. 57, 1116–1127.

Hall, J., 2002. Cellular mechanisms for heavy metal detoxification and tolerance. J. Exp. Bot. 53, 1–11.

Hammond, J.P., Bowen, H.C., White, P.J., Mills, V., Pyke, K.A., Baker, A.J., Whiting, S.N., May, S.T., Broadley, M.R., 2006. A comparison of the *Thlaspi caerulescens* and *Thlaspi arvense* shoot transcriptomes. New Phytol. 170, 239–260.

He, Z.-l., Yang, X.-e., 2007. Role of soil rhizobacteria in phytoremediation of heavy metal contaminated soils. J. Zhejiang Univ. Sci. B 8, 192–207.

Himmelbauer, M.L., Puschenreiter, M., Schnepf, A., Loiskandl, W., Wenzel, W.W., 2005. Root morphology of *Thlaspi goesingense* Hálácsy grown on a serpentine soil. J. Plant Nutr. Soil Sci. 168, 138–144.

Hinsinger, P., Courchesne, F., Violante, A., Huang, P., Gadd, G., 2008. Biogeochemistry of metals and metalloids at the soil–root interface. John Wiley & Sons, Hoboken, NJ.

Hinsinger, P., Gobran, G.R., Gregory, P.J., Wenzel, W.W., 2005. Rhizosphere geometry and heterogeneity arising from root-mediated physical and chemical processes. New Phytol. 168, 293–303.

Hussain, S.I., Ghafoor, A., Ahmad, S., Murtaza, G., Sabir, M., 2006. Irrigation of crops with raw sewage: Hazard assessment of effluent, soil and vegetables. Pak. J. Agric. Sci. 43, 97–102.

James, B.R., 1996. Peer reviewed: The challenge of remediating chromium-contaminated soil. Environ. Sci. Technol. 30, 248A–251A.

Jentschke, G., Godbold, D., 2000. Metal toxicity and ectomycorrhizas. Physiologia Plantarum 109, 107–116.

Keller, C., Hammer, D., Kayser, A., Richner, W., Brodbeck, M., Sennhauser, M., 2003. Root development and heavy metal phytoextraction efficiency: Comparison of different plant species in the field. Plant and Soil 249, 67–81.

Kerkeb, L., Krämer, U., 2003. The role of free histidine in xylem loading of nickel in *Alyssum lesbiacum* and *Brassica juncea*. Plant Physiol. 131, 716–724.

Kidd, P., Barceló, J., Bernal, M.P., Navari-Izzo, F., Poschenrieder, C., Shilev, S., Clemente, R., Monterroso, C., 2009. Trace element behaviour at the root–soil interface: Implications in phytoremediation. Environ. Exp. Bot. 67, 243–259.

Kinnersley, A.M., 1993. The role of phytochelates in plant growth and productivity. Plant Growth Regul. 12, 207–218.

Klein, M., Burla, B., Martinoia, E., 2006. The multidrug resistance-associated protein (MRP/ABCC) subfamily of ATP-binding cassette transporters in plants. FEBS Lett. 580, 1112–1122.

Kochian, L.V., Pineros, M.A., Hoekenga, O.A., 2005. The physiology, genetics and molecular biology of plant aluminum resistance and toxicity. Plant and Soil 274, 175–195.

Krämer, U., 2010. Metal hyperaccumulation in plants. Annu. Rev. Plant Biol. 61, 517–534.

Krämer, U., Cotter-Howells, J.D., Charnock, J.M., Baker, A.J., Smith, J.A.C., 1996. Free histidine as a metal chelator in plants that accumulate nickel. Nature, 379, 635–638.

Krämer, U., Talke, I.N., Hanikenne, M., 2007. Transition metal transport. FEBS Lett. 581, 2263–2272.

Küpper, H., Zhao, F.J., McGrath, S.P., 1999. Cellular compartmentation of zinc in leaves of the hyperaccumulator *Thlaspi caerulescens*. Plant physiol. 119, 305–312.

Küpper, H., Lombi, E., Zhao, F.-J., McGrath, S.P., 2000. Cellular compartmentation of cadmium and zinc in relation to other elements in the hyperaccumulator *Arabidopsis halleri*. Planta 212, 75–84.

Larsen, P.B., Degenhardt, J., Tai, C.-Y., Stenzler, L.M., Howell, S.H., Kochian, L.V., 1998. Aluminum-resistant Arabidopsis mutants that exhibit altered patterns of aluminum accumulation and organic acid release from roots. Plant Physiology 117, 9–17.

Lasat, M.M., 2002. Phytoextraction of toxic metals. J. Environ. Qual. 31, 109–120.

Lasat, M.M., Baker, A.J., Kochian, L.V., 1998. Altered Zn compartmentation in the root symplasm and stimulated Zn absorption into the leaf as mechanisms involved in Zn hyperaccumulation in *Thlaspi caerulescens*. Plant Physiol. 118, 875–883.

Leung, H., Ye, Z., Wong, M., 2007. Survival strategies of plants associated with arbuscular *Mycorrhizal fungi* on toxic mine tailings. Chemosphere 66, 905–915.

Lin, Z.-Q., Schemenauer, R., Cervinka, V., Zayed, A., Lee, A., Terry, N., 2000. Selenium Volatilization from a Soil–Plant System for the Remediation of Contaminated Water and Soil in the San Joaquin Valley. J. Environ. Qual. 29, 1048–1056.

Lombi, E., Tearall, K.L., Howarth, J.R., Zhao, F.-J., Hawkesford, M.J., McGrath, S.P., 2002. Influence of iron status on cadmium and zinc uptake by different ecotypes of the hyperaccumulator *Thlaspi caerulescens*. Plant Physiol. 128, 1359–1367.

Lone, M.I., He, Z.-l., Stoffella, P.J., Yang, X.-e, 2008. Phytoremediation of heavy metal polluted soils and water: Progresses and perspectives. J. Zhejiang Univ. Sci. B 9, 210–220.

Lothenbach, B., Krebs, R., Furrer, G., Gupta, S., Schulin, R., 1998. Immobilization of cadmium and zinc in soil by Al-montmorillonite and gravel sludge. Eur. J. Soil Sci. 49, 141–148.

Macnair, M.R., Bert, V., Huitson, S.B., Saumitou-Laprade, P., Petit, D., 1999. Zinc tolerance and hyperaccumulation are genetically independent characters. Proceedings of the Royal Society of London. Series B: Biological Sciences 266, 2175–2179.

Malik, N., Biswas, A., 2012. Role of higher plants in remediation of metal contaminated sites. Sci. Rev. Chem. Commun. 2, 141–146.

Marques, A.P., Rangel, A.O., Castro, P.M., 2009. Remediation of heavy metal contaminated soils: Phytoremediation as a potentially promising clean-up technology. Crit. Rev. Environ. Sci. Technol. 39, 622–654.

Marschner, H., Römheld, V., 1994. Strategies of plants for acquisition of iron. Plant and Soil 165, 261–274.

McGrath, S., Shen, Z., Zhao, F., 1997. Heavy metal uptake and chemical changes in the rhizosphere of *Thlaspi caerulescens* and *Thlaspi ochroleucum* grown in contaminated soils. Plant and Soil 188, 153–159.

McGrath, S., Zhao, F., Lombi, E., 2001. Plant and rhizosphere processes involved in phytoremediation of metal-contaminated soils. Plant and Soil 232, 207–214.

McGrath, S.P., Zhao, F.-J., 2003. Phytoextraction of metals and metalloids from contaminated soils. Curr. Opin. Biotechnol. 14, 277–282.

McIntyre, T., 2003. Phytoremediation of heavy metals from soils, Phytoremediation. Springer. pp. 97–123.

Meagher, R., Rugh, C., Kandasamy, M., Gragson, G., Wang, N., 2000. Engineered phytoremediation of mercury pollution in soil and water using bacterial genes. Phytoremediation of Contaminated Soil and Water. Lewis Publishers, Boca Raton, FL. 201–219.

Meers, E., Ruttens, A., Hopgood, M., Samson, D., Tack, F., 2005. Comparison of EDTA and EDDS as potential soil amendments for enhanced phytoextraction of heavy metals. Chemosphere 58, 1011–1022.

Meharg, A.A., 2003. The mechanistic basis of interactions between mycorrhizal associations and toxic metal cations. Mycol. Res. 107, 1253–1265.

Mejáre, M., Bülow, L., 2001. Metal-binding proteins and peptides in bioremediation and phytore-mediation of heavy metals. Trends. Biotechnol. 19, 67–73.

Memon, A.R., Schröder, P., 2009. Implications of metal accumulation mechanisms to phytoremediation. Environ. Sci. Pollut. Res. 16, 162–175.

Mendez, M.O., Maier, R.M., 2008. Phytostabilization of mine tailings in arid and semiarid environments—an emerging remediation technology. Environ. Health Perspect. 116, 278.

Mishra, S., Tripathi, R., Srivastava, S., Dwivedi, S., Trivedi, P.K., Dhankher, O., Khare, A., 2009. Thiol metabolism play significant role during cadmium detoxification by *Ceratophyllum demersum* L. Bioresour. Technol. 100, 2155–2161.

Mullineaux, P.M., Rausch, T., 2005. Glutathione, photosynthesis and the redox regulation of stress-responsive gene expression. Photosynth. Res. 86, 459–474.

Murtaza, G., Ghafoor, A., Qadir, M., Owens, G., Aziz, M., Zia, M., 2010. Disposal and use of sewage on agricultural lands in Pakistan: A review. Pedosphere 20, 23–34.

Ogawa, K.i., Iwabuchi, M., 2001. A mechanism for promoting the germination of *Zinnia elegans* seeds by hydrogen peroxide. Plant and cell physiol. 42, 286–291.

Panda, S., Chaudhury, I., Khan, M., 2003. Heavy metals induce lipid peroxidation and affect antioxidants in wheat leaves. Biologia Plantarum 46, 289–294.

Peng, K., Li, X., Luo, C., Shen, Z., 2006. Vegetation composition and heavy metal uptake by wild plants at three contaminated sites in Xiangxi area, China. J. Environ. Sci. Health. A 41, 65–76.

Persans, M.W., Nieman, K., Salt, D.E., 2001. Functional activity and role of cation-efflux family members in Ni hyperaccumulation in *Thlaspi goesingense*. Proc. Natl. Acad. Sci. 98, 9995–10,000.

Pignattelli, S., Colzi, I., Buccianti, A., Cecchi, L., Arnetoli, M., Monnanni, R., Gabbrielli, R., Gonnelli, C., 2012. Exploring element accumulation patterns of a metal excluder plant naturally colonizing a highly contaminated soil. J. Hazard. Mater. 227, 362–369.

Poschenrieder, C., Bech, J., Llugany, M., Pace, A., Fenés, E., Barceló, J., 2001. Copper in plant species in a copper gradient in Catalonia (North East Spain) and their potential for phytoremediation. Plant and soil 230, 247–256.

Pulford, I., Watson, C., 2003. Phytoremediation of heavy metal-contaminated land by trees—a review. Environ. Int. 29, 529–540.

Puschenreiter, M., Schnepf, A., Millan, I.M., Fitz, W.J., Horak, O., Klepp, J., Schrefl, T., Lombi, E., Wenzel, W.W., 2005. Changes of Ni biogeochemistry in the rhizosphere of the hyperaccumulator *Thlaspi goesingense*. Plant and Soil 271, 205–218.

Puschenreiter, M., Wieczorek, S., Horak, O., Wenzel, W.W., 2003. Chemical changes in the rhizosphere of metal hyperaccumulator and excluder Thlaspi species. J. Plant Nutr. Soil Sci. 166, 579–584.

Qin, R., Hirano, Y., Brunner, I., 2007. Exudation of organic acid anions from poplar roots after exposure to Al, Cu and Zn. Tree Physiol. 27, 313–320.

Ramsay, L.M., Gadd, G.M., 1997. Mutants of *Saccharomyces cerevisiae* defective in vacuolar function confirm a role for the vacuole in toxic metal ion detoxification. FEMS Microbiol. Lett. 152, 293–298.

Rascio, N., Navari-Izzo, F., 2011. Heavy metal hyperaccumulating plants: How and why do they do it? And what makes them so interesting? Plant Sci. 180, 169–181.

Raskin, I., Kumar, P., Dushenkov, S., Salt, D.E., 1994. Bioconcentration of heavy metals by plants. Curr. Opin. Biotechnol. 5, 285–290.

Raskin, I., Smith, R.D., Salt, D.E., 1997. Phytoremediation of metals: Using plants to remove pollutants from the environment. Curr. Opin. Biotechnol. 8, 221–226.

Rausch, T., Gromes, R., Liedschulte, V., Müller, I., Bogs, J., Galovic, V., Wachter, A., 2007. Novel insight into the regulation of GSH biosynthesis in higher plants. Plant Biol. 9, 565–572.

Rengel, Z., Zhang, W.H., 2003. Role of dynamics of intracellular calcium in aluminium-toxicity syndrome. New Phytol. 159, 295–314.

Robinson, B., Schulin, R., Nowack, B., Roulier, S., Menon, M., Clothier, B., Green, S., Mills, T., 2006. Phytoremediation for the management of metal flux in contaminated sites. Forest Snow and Landscape Res. 80, 221–224.

Roosens, N.H., Leplae, R., Bernard, C., Verbruggen, N., 2005. Variations in plant metallothioneins: The heavy metal hyperaccumulator *Thlaspi caerulescens* as a study case. Planta 222, 716–729.

Rugh, C.L., Wilde, H.D., Stack, N.M., Thompson, D.M., Summers, A.O., Meagher, R.B., 1996. Mercuric ion reduction and resistance in transgenic *Arabidopsis thaliana* plants expressing a modified bacterial merA gene. Proc. Natl. Acad. Sci. 93, 3182–3187.

Saifullah, Meers, E., Qadir, M., de Caritat, P., Tack, F.M.G., Du Laing, G., Zia, M.H., 2009. EDTA-assisted Pb phytoextraction. Chemosphere 74, 1279–1291.

Sakakibara, M., Watanabe, A., Inoue, M., Sano, S., Kaise, T., 2010. Phytoextraction and phytovolatilization of arsenic from As-contaminated soils by *Pteris vittata*. Proceedings of the Annual International Conference on Soils, 26. Sediments, Water and Energy.

Salt, D.E., Blaylock, M., Kumar, N.P., Dushenkov, V., Ensley, B.D., Chet, I., Raskin, I., 1995. Phytoremediation: A novel strategy for the removal of toxic metals from the environment using plants. Nat. Biotechnol. 13, 468–474.

Schmidt, U., 2003. Enhancing phytoextraction. J. Environ. Qual. 32, 1939–1954.

Schwartz, C., Echevarria, G., Morel, J.L., 2003. Phytoextraction of cadmium with *Thlaspi caerulescens*. Plant and Soil 249, 27–35.

Schwartz, C., Morel, J.L., Saumier, S., Whiting, S.N., Baker, A.J., 1999. Root development of the zinc-hyperaccumulator plant *Thlaspi caerulescens* as affected by metal origin, content and localization in soil. Plant and Soil 208, 103–115.

Seregin, I., Kozhevnikova, A., Kazyumina, E., Ivanov, V., 2003. Nickel toxicity and distribution in maize roots. Russian Journal of Plant Physiology 50, 711–717.

Seth, C.S., Kumar Chaturvedi, P., Misra, V., 2008. The role of phytochelatins and antioxidants in tolerance to Cd accumulation in *Brassica juncea* L. Ecotoxicol. environ. saf. 71, 76–85.

Sharma, S.S., Dietz, K.-J., 2009. The relationship between metal toxicity and cellular redox imbalance. Trends in plant science 14, 43–50.

Shenker, M., Fan, T.-M., Crowley, D., 2001. Phytosiderophores influence on cadmium mobilization and uptake by wheat and barley plants. J. Environ. Qual. 30, 2091–2098.

Shu, W., Xia, H., Zhang, Z., Lan, C., Wong, M., 2002. Use of vetiver and three other grasses for revegetation of Pb/Zn mine tailings: Field experiment. Int. J. Phytoremediation 4, 47–57.

Singer, A.C., Crowley, D.E., Thompson, I.P., 2003. Secondary plant metabolites in phytoremediation and biotransformation. Trends. Biotechnol. 21, 123–130.

Steinkellner, H., Mun-Sik, K., Helma, C., Ecker, S., Ma, T.H., Horak, O., Kundi, M., Knasmüller, S., 1998. Genotoxic effects of heavy metals: Comparative investigation with plant bioassays. Environ. Mol. Mutagen. 31, 183–191.

Sun, Q., Ye, Z.H., Wang, X.R., Wong, M.H., 2007. Cadmium hyperaccumulation leads to an increase of glutathione rather than phytochelatins in the cadmium hyperaccumulator *Sedum alfredii*. J. plant physiol. 164, 1489–1498.

Tam, P.C., 1995. Heavy metal tolerance by ectomycorrhizal fungi and metal amelioration by *Pisolithus tinctorius*. Mycorrhiza 5, 181–187.

Taylor, G.J., 1987. Exclusion of metals from the symplasm: A possible mechanism of metal tolerance in higher plants. J. Plant Nutr. 10, 1213–1222.

Terry, N., Zayed, A., De Souza, M., Tarun, A., 2000. Selenium in higher plants. Annu. rev. plant biol. 51, 401–432.

Thangavel, P., Subbhuraam, C., 2004. Phytoextraction: Role of hyperaccumulators in metal contaminated soils. Proc. Indian Natl. Sci. Acad. B 70, 109–130.

Vamerali, T., Bandiera, M., Mosca, G., 2011. In situ phytoremediation of arsenic- and metal-polluted
pyrite waste with field crops: Effects of soil management. Chemosphere 83, 1241–1248.

Vázquez, M., Poschenrieder, C., Barcelo, J., Baker, A., Hatton, P., Cope, G., 1994. Compartmentation of zinc in roots and leaves of the zinc hyperaccumulator *Thlaspi caerulescens* J. & C. Presl.
Bot. Acta 107, 243–250.

Vázquez, S., Agha, R., Granado, A., Sarro, M., Esteban, E., Peñalosa, J., Carpena, R., 2006. Use
of white lupin plant for phytostabilization of Cd and As polluted acid soil. Water, Air, and Soil
Pollut. 177, 349–365.

Volland, S., Andosch, A., Milla, M., Stöger, B., Lütz, C., Lütz-Meindl, U., 2011. Intracellular metal
compartmentalization in the green algal model system *Micrasterias denticulata* (streptophyta)
measured by transmission electron microscopy–coupled electron energy loss spectroscopy1.
J. Phycol. 47, 565–579.

Wagner, G.J., 1993. Accumulation of cadmium in crop plants and its consequences to human health.
Adv. Agronomy 51, 173–212.

Wei, S., Zhou, Q., Wang, X., 2005. Identification of weed plants excluding the uptake of heavy metals. Environment International 31, 829–834.

Wenzel, W., Bunkowski, M., Puschenreiter, M., Horak, O., 2003a. Rhizosphere characteristics of
indigenously growing nickel hyperaccumulator and excluder plants on serpentine soil. Environ.
Pollut. 123, 131–138.

Wenzel, W., Lombi, E., Adriano, D., 1999. Biogeochemical processes in the rhizosphere: Role in
phytoremediation of metal-polluted soils, Heavy metal stress in plants. Springer. 273–303.

Wenzel, W.W., 2009. Rhizosphere processes and management in plant-assisted bioremediation
(phytoremediation) of soils. Plant and Soil 321, 385–408.

Wenzel, W.W., Bunkowski, M., Puschenreiter, M., Horak, O., 2003b. Rhizosphere characteristics of
indigenously growing nickel hyperaccumulator and excluder plants on serpentine soil. Environ.
Pollut. 123, 131–138.

Whiting, S.N., Leake, J.R., McGrath, S.P., Baker, A.J., 2001. Assessment of Zn mobilization in the
rhizosphere of *Thlaspi caerulescens* by bioassay with non-accumulator plants and soil extraction. Plant and Soil 237, 147–156.

Wong, M.H., 2003. Ecological restoration of mine degraded soils, with emphasis on metal contaminated soils. Chemosphere 50, 775–780.

Yadav, S.K., 2010. Heavy metals toxicity in plants: An overview on the role of glutathione and phytochelatins in heavy metal stress tolerance of plants. S. Afr. J. Bot. 76, 167–179.

Yang, X., Feng, Y., He, Z., Stoffella, P.J., 2005. Molecular mechanisms of heavy metal hyperaccumulation and phytoremediation. J. Trace Elem. Med. Biol. 18, 339–353.

Yang, X., Li, T., Yang, J., He, Z., Lu, L., Meng, F., 2006. Zinc compartmentation in root, transport
into xylem, and absorption into leaf cells in the hyperaccumulating species of Sedum alfredii
Hance. Planta 224, 185–195.

Yazaki, K., 2006. ABC transporters involved in the transport of plant secondary metabolites. FEBS
Lett. 580, 1183–1191.

Zenk, M.H., 1996. Heavy metal detoxification in higher plants—a review. Gene 179, 21–30.

Zhang, S., Li, T., Huang, H., Zou, T., Zhang, X., Yu, H., Zheng, Z., Wang, Y., 2012. Cd accumulation
and phytostabilization potential of dominant plants surrounding mining tailings. Environ. Sci.
Pollut. Res. 19, 3879–3888.

Zhang, Z., Gao, X., Qiu, B., 2008. Detection of phytochelatins in the hyperaccumulator *Sedum
alfredii* exposed to cadmium and lead. Phytochemistry 69, 911–918.

Zhang, Z., Shu, W., Lan, C., Wong, M., 2001. Soil seed bank as an input of seed source in revegetation of lead/zinc mine tailings. Restoration Ecol. 9, 378–385.

Zhao, F., Lombi, E., Breedon, T., 2000. Zinc hyperaccumulation and cellular distribution in *Arabidopsis halleri*. Plant. Cell Environ. 23, 507–514.

Zhao, F., Lombi, E., McGrath, S., 2003. Assessing the potential for zinc and cadmium phytoremediation with the hyperaccumulator *Thlaspi caerulescens*. Plant and Soil 249, 37–43.

Zhou, G.-K., Xu, Y.-F., Liu, J.-Y., 2005. Characterization of a rice class II metallothionein gene: Tissue expression patterns and induction in response to abiotic factors. J. Plant Physiol. 162, 686–696.

Zou, T., Li, T., Zhang, X., Yu, H., Luo, H., 2011. Lead accumulation and tolerance characteristics of *Athyrium wardii* (Hook.) as a potential phytostabilizer. J. Hazard. Mater. 186, 683–689.

Chapter 5

Phytoremediation: An Eco-Friendly Green Technology for Pollution Prevention, Control and Remediation

Tanveer Bilal Pirzadah,* Bisma Malik,* Inayatullah Tahir,*
Manoj Kumar,† Ajit Varma† and Reiaz Ul Rehman*
*Department of Bioresources, University of Kashmir, Srinagar, India; †Amity Institute of
Microbial Technology, Amity University, Noida, Uttar Pradesh, India

INTRODUCTION

Currently the quality of soil has been degraded considerably due to continuous utilization of huge quantities of synthetic fertilizers and agricultural malpractices. In addition, other factors such as natural disasters and industrialization are also responsible for discharging toxic metals into the soil and thus degrading its quality (Raven and Leoppert, 1997). Both organic and inorganic pollutants are responsible for the deterioration in the health of soil. Among organic pollutants, trichloroethylene, herbicides (Burken and Schnoor, 1997), explosives such as trinitrotoluene (Hughes et al., 1997), hydrocarbons (Schnoor et al., 1995) and fossil fuels and methyl tertiary butyl ether are the leading polluting agents (Hong et al., 2001). As these xenobiotic toxic compounds persist in the ecosystem due to their non-biodegradable nature, a severe threat is posed to the environment. However, these toxic compounds can only be changed from one oxidation state to another (Marques et al., 2009) and, without intervention, remain in the soil for a long period of time. Generally, these toxicants exist in distinct states, viz. colloidal state (hydroxides, oxides and silicates) (Lytle et al., 1998), metalloids (As and Se) (Rathinasabapathi and Srivastava, 2006) radionuclides (U/Sr/Cs/Co/Ra) (Dushenkov, 2003), nutrients, pesticides, herbicides and petroleum derivatives (Macek et al., 2000). These hazardous metallic elements possess a high degree of toxicity and thus have a negative impact on the whole ecosystem. Due to their mutagenic nature, these heavy metals cause DNA damage in animals as well as humans (Baudouin et al., 2002). For sustainable development, it is mandatory to restore these contaminated sites and to minimize the entry of toxicants into the food chain. Previously, soil was restored by using various classical engineering approaches. However, due to certain limitations such as expensive technology, labour-intensive methods, environmental incompatibility and certain technical barriers, conventional techniques are not fully acceptable (Salt et al., 1995). There has been an increasing interest over the last 15 years in developing a novel plant-based technology to decontaminate the soil from heavy metals (Raskin et al., 1994). Scientists now focus on the remediation technologies that are sustainable, cost-effective and eco-friendly in nature (Gupta and Sinha, 2007). The plant-based technology known as phytoremediation has emerged as a novel low-tech, cost-effective technology that utilizes the potential of plants and their associated microbial flora for environmental clean-up (Salt et al., 1998). Phytoremediation technology is simple, cost-effective, sustainable, compatible, eco-friendly and constitutes one of the main components of green technology. Plants have the natural ability to degrade these heavy metals by means of various processes such as bioaccumulation, translocation and pollutant storage/degradation. Phytoremediation is about 10-fold cheaper than classical engineering approaches since it is performed *in situ*, is solar driven, and can function with minimal maintenance once established (Nascimento and Xing, 2006). It has been reported that plants possess a greater tolerance to heavy metal pollution without being seriously harmful, indicating that this property of plants can be exploited to detoxify contaminants using novel agricultural and genetic engineering approaches. Some plants possess the natural ability to degrade numerous recalcitrant xenobiotics and are thus regarded as 'green livers', acting as an essential sink for environmentally obnoxious chemicals (Schwitzguébel, 2000). Nature has bestowed on plants an excellent capacity to neutralize these toxic elements within the growth matrix, be it soil or water.

Background

Phytoremediation technology is not new to science. It has been reported that plants were used for the treatment of waste water some 300 years ago (Hartman, 1975). The two plant species, viz. *Thlaspi caerulescens* and *Viola calaminaria*, were the first to be documented to accumulate high concentrations of heavy metals in leaves (Baumann, 1885). Analysis of the dry shoot biomass of the genus *Astragalus* has revealed that it has the capacity to accumulate selenium (Se) concentration up to 0.6% (Byers, 1935). In 1948, more plants were identified as accommodating Nickel (Ni) concentration up to the level of 1% (Minguzzi and Vergnano, 1948). It has been recently reported that *Thlaspi caerulescens* possesses a high resistance to Zn and accumulates high Zn concentration in its shoots (Rascio, 1977). Even though there have been successive reports claiming the identification of Cobalt (Co), Copper (Cu) and Manganese (Mn) hyperaccumulating plant species, some plant species possess tremendous potential to accumulate and scavenge toxic metals other than Cadmium (Cd), Ni, Se and Zinc (Zn) and thus need proper attention from the scientific community in order to explore their phytoremediation potential (Salt et al., 1995). During the dawn of the nuclear era, few semi-aquatic plants used for treating radionuclide-contaminated waters existed in Russia (Salt et al., 1995). Plants growing on metalliferous soils adapted themselves to accumulate high concentrations of heavy metals in their distinct parts without showing any significant ill effects (Reeves and Brooks, 1983). Chaney (1983) was the first to recommend the use of 'hyperaccumulators' for the phytoremediation of metal polluted areas and the first field trial on Zn and Cd phytoremediation was conducted in 1991 (Baker et al.,1991). Generally these hyperaccumulators are seeded or transplanted in contaminated sites to extract these heavy metals by using various agronomic practices. Usually plants absorb these elements from the soil with the help of the dense root system and later transport them to the shoots. Finally, the above-ground parts of the plant are harvested and metal is recovered from the plant biomass; this phenomenon of recovering metals from the metal-contaminated sites by employing the use of plants is referred as phytomining. Phytomining is one of the important components of 'green' technologies for the production of bio-ore (Robinson et al., 2009). It not only helps to produce bio-ore, but also finds great application to completely eradicate toxicants from the contaminated sites. Currently it is of paramount importance to explore the biology of metal phytoextraction. Although we have achieved some progress in this field, our knowledge of the dynamics of plant–metal interaction is still emerging. It needs a multi-interdisciplinary approach in order to commercialize this technology.

PLANTS' RESPONSE TO HEAVY METALS

The phenomenon of phytoremediation, i.e. utilization of plants to evacuate the harmful metals from the soil, appears to be a cost-effective substitute for traditional methods for clean-up of soil adulterated by heavy metals. It utilizes the capacity of plants to clear, decompose and immobilize toxic chemicals. In the late 1980s, the concept of phytoremediation originated from the study of heavy metal resistance in plants. A significant portion of the world is adulterated with different types of pollutants (organic and inorganic) which also include a wide range of heavy metals (Ensley, 2000). Various human activities like mining, transportation, agricultural activities, waste disposal and military actions subsequently liberate a large amount of these harmful inorganic pollutants into the environment. Due to their impregnable origin, metals are a group of adulterators that are of much concern. The effect of harmful metals is intensified by their imprecise tenacity in the environment. In many regions of South Asia, inorganic (heavy metals) and organic pollutant adulteration at present pose a serious threat to human health and ecosystems (World Health Organization, 2003). In Sri Lanka, only a small number of published reports are available on the origin of soil and water adulteration (Bandara, 2003; Dissanayake et al., 2007). Nevertheless, the majority of soil area and water possesses extreme amounts of heavy metals such as Zn, Ni, Co, Cu and Cd, including other pollutants, due to several human activities (Dissanayake et al., 2002). Cd is present in significant quantities in water, soils, plant and animal biomass in the dry zone (Bandara, 2006). It has been found that discharge of industrial waste, agricultural waste, domestic waste water and solid waste makes a huge contribution to heavy metal adulteration on land as well as in surface- and ground-water resources (Bandara, 2003). Plants have three underlying strategies for their growth on metal-adulterated soil (Raskin et al., 1994).

Metal Excluders

Metal excluders block metal so that it can not enter their aerial parts or maintain low and constant metal levels over a wide range of metal concentrations in soil; they mostly confine metal in their roots. The plant may change or modify their membrane permeabilities, change metal binding abilities of cell walls or secrete some natural chelating agents (Cunningham, 1995).

Metal Indicators

Metal indicators are species which functionally build-up metals in their aerial tissues and mainly speculate metal concentration in the soil. They resist the existing concentration level of metals by secreting intracellular chelating compounds, or by modifying the metal compartmentalization pattern by depositing metals in their non-sensitive parts.

Metal Accumulating Plant Species

Metal accumulating plant species can concentrate metals in their aerial parts, to an extent that is greater than that in the soil. Hyperaccumulators are the plants that can accumulate greater amounts of adulterants present in the soil either in their roots, shoot or leaves (Baker et al., 1994; Raskin et al., 1994; Cunningham, 1995). Baker and Brooks (1989) have described metal hyperaccumulators as plants that contain more than or up to 0.1% (1000 mg g^{-1}) of Ni, Cd, Cu, Pb, Cr, Co or 1% (> 10,000 mg g^{-1}) of Zn or Mn by dry weight. For Cd and other rare metals, it is > 0.01% by dry weight. Investigators have found the hyperaccumulating species by collecting them from their natural environments or areas where the soil exhibits higher concentrations than normal levels of metals, as in the case of adulterated areas or geographically sound areas in a specific metal (Gleba et al., 1999). The identification of hyperaccumulating plants, which exhibit a large amount of heavy metals, that would be extremely harmful to other plants, give rise to the concept of using these plant species for extracting metals from the soil, and also help to clean up the soil for other less resistant plants. It is currently critically important to engineer plants in such a way so that they accumulate large amounts of heavy metals. The development of the hyperaccumulator attributes is believed to take place as a feature that is advantageous to the plant (Baker et al., 2000). It is a known fact that the biological mechanisms of metal build-up in plants are complex and diverse. However, a necessity for hyperaccumulation is the capacity to efficiently resist the extreme levels of metals in aerial tissues, particularly the leaf tissue (Reeves and Baker, 2000). Furthermore, the ecological role of metal hyperaccumulation is still dubious. However, various studies have described how metal-accumulating plant species have the ability to protect themselves from insect, fungal as well as herbivore attacks (Boyd, 2004). Another assumption for the advantage of metal hyperaccumulation is 'elemental allelopathy', which means that it stops the growth of other plants in the area underneath hyperaccumulating plant species (Zhang et al., 2007). It is because of this that current research on phytoremediation is more focused on crop species and plants such as *Avena sativa* (oat), *Hordeum vulgare* (barley), *Brassica juncea* (Indian mustard), *Thlaspi caerulescens* (alpine penny-cress) and *Phaseolus vulgaris* (garden bean) which have been shown to accumulate Zn and Pb (Whitting et al., 2003). It has been estimated that 400 hyperaccumulating plant species belonging to 22 families have been recognized. A vast number of hyperaccumulators belong to the family *Brassicaceae* with a broad range of metals that include 87 species from 11 genera (Baker and Brooks, 1989). Distinct genera of the family *Brassicaceae* are known to concentrate heavy metals such as Zn, Ni, Cd and Pb (Robinson et al., 2009). Various assumptions have been made to describe the underlying system of metal hyperaccumulation and the evolutionary benefit of this approach, as discussed below.

Complex Formation and Compartmentation

It is believed that hyperaccumulators integrate chelators that lead to the detoxification of metal ions by the formation of certain complexes. The less-toxic and soluble organic–metal complex is then allocated to different cell components with a minimum metabolic activity where it is deposited in the form of a stable inorganic or organic complex (Barceló and Poschenrieder, 2003).

Deposition Hypothesis

Metals are separated from the roots by the hyperaccumulators and then concentrated in plant parts that are abscised (old leaves), leached by rain (epidermis and hairs) or burnt. Being an inadvertent form of uptake, it is also believed that hyperaccumulation of the metal is the derivative of an adaptive system to protect against various hostile soil characteristics (e.g. Serpentinophytes shows Ni hyperaccumulation).

Hyperaccumulation as a Defense System against Various Environmental Stresses

There are several reports on the influence of metals against certain pathogenic fungi, bacteria and herbivores that feed on leaves (Boyd, 1998). Phloem parasites (Ernst et al., 1990), however, are not greatly influenced, presumably because the metals have less phloem mobility. Due to the presence of high levels of metals in the leaves, they can act as feeding restraints or, after ingestion, they may decrease the reproduction rate in herbivores or be lethal for them. Another advantage that is bestowed by the metal hyperaccumulation mechanism is the trade-off of organic defenses (Tolrà et al., 2001). It was recommended that during drought conditions, extreme metal accumulation in leaves can be used in osmotic adjustment (Poschenrieder and Barceló, 1999). Broad studies in *Alyssum murale* (Ni hyperaccumulator) and *Thlaspi caerulescens* (Zn-hyperaccumulator) did not establish this hypothesis (Whitting et al., 2003).

Noteworthy attention is paid towards Indian mustard by present day researchers, geneticists and particularly plant breeders, because of the presence of the novel polyploid genome. *Brassica juncea* is called as Allotetraploid plant because of the presence of the genome which is made up of entire diploid genomes of both the parents (*B. nigra* and *B. campestris*) and in the current breeding programmes, selection of *B. juncea* is based on extensive study of a number of important characters. A significant aim of the plant breeders is the enhancement of oil and meal quality either by excluding nutritionally unwanted components or by altering the fatty-acid composition of oil (Banga, 1997). Other selections are based on insect (Andrahennadi and Gillott, 1998) and disease resistance (Pang and Halloran, 1996) and several temperature adaptations (Banga, 1997). Currently there has been an interest in selecting Indian mustard lines by plant breeders which is based on their capacity to resist and concentrate heavy metals. Various accessions of *B. juncea* have been classified as mild accumulators of metallic elements and are maintained by the United States Department of Agriculture–Agriculture Research Service (USDA-ARS) Plant Introduction Station at Iowa State University, Ames, Iowa. The advantage of employing *B. juncea* seed from the plant introduction station is that the genetic integrity of the accessions is maintained through appropriate breeding methodology. It has been found that those experiments which employ these seeds are more exact than those which are carried out with seeds that are available from commercial sources. Exactness is also higher, because present day researchers can obtain the same accessions and the seeds (metal accumulators) for their experiments as the USDA-ARS Plant Introduction Station maintains and distributes them to public and private research institutions free of cost. Currently, Indian mustard (*B. juncea*) is one of the most expectable members used for the phytoextraction of various metals, comprising Zn, Cd, Ni, Pb, Cu and Cr (IV) (Zhu et al., 1999). Kumar et al. (1995) performed an experiment to examine various rapid *Brassica* taxa for their capability to resist and concentrate heavy metals, including. *B juncea* L. (Indian mustard), *B. nigra* Koch (black mustard), *B. campestris* L. (turnip), *B. napus* L. (rape) and *B. oleracea* L. (kale). However, it has been found that all *Brassica* taxa concentrate heavy metals, but *B. juncea* exhibited greater ability to concentrate heavy metals in roots and then allocated Pb, Cr (IV), Zn and Cd to shoots for build-up (Jiang et al., 2000). Several experiments have also been carried out by Kumar et al. (1995) to examine reasonable genetic variations in distinct *B. juncea* accessions in order to find the metals which have the highest phytoextraction potential. In addition, the application of chelating agents such as EDTA to adulterated soils has been shown to induce the uptake of metals by plants (Salido et al., 2003). It has also been documented that the addition of EDTA significantly intensified Pb concentration in *B. juncea* (Indian mustard). Another technology to remove Pb using Indian mustard from adulterated soil has been described using the combination of electric potential and EDTA (Lim et al., 2004). A recent study has observed that there is a two- to fourfold increase in Pb accumulation in shoots of *B. juncea* which is achieved by applying EDTA plus a DC electric potential rather than an AC electric potential with six graphite electrodes (Lim et al., 2012). Another study describes the mercury (Hg) uptake and phytotoxicity (antioxidative responses) by two varieties of ferns (*P. vittata* and *N. exaltata)* and Indian mustard by using anatomical, histochemical and biochemical approaches grown in a hydroponic system. The results show that mercury exposure results in severe phytotoxicity followed by lipid peroxidation and rapid deposition of hydrogen peroxide (H_2O_2) in ferns and Indian mustard. Indian mustard (*B. juncea*) efficiently produces an enzymatic antioxidant defense system to scavenge H_2O_2 which results in less H_2O_2 in shoots with huge mercury concentrations. The Indian mustard (*B. juncea*) is confirmed as an effective metabolic defense and adaption system to mercury-induced oxidative stress (Su et al., 2009). Another recent study investigates the application of adulterated soil with nitrogen fertilizer at the rate of 0, 2 and 4 g m^{-2} on Cu and Pb uptake by Indian mustard and Amaranth. Total Cu uptake (15.7–21.4 mg kg^{-1}) was considerably greater than the lead uptake (12.7–16.9 mg kg^{-1}) in both the plant species with the application of nitrogen in the soil. Adulterated soil containing nitrogen fertilizer showed highest shoot accumulation of Pb (10.1–11.6 mg kg^{-1}) and Cu (11.6–14.3 mg kg^{-1}), which is two- to fourfolds greater than its roots (2.6–5.7 mg kg^{-1} for Pb; 3.5–4.7 mg kg^{-1} for Cu). Results showed that both the plant species removed an appreciable amount of Pb and Cu from the contaminated soil with the utilization of nitrogen fertilizer. The results also recommend that nitrogen fertilizer is much more efficient for phytoextraction of Pb and Cu from adulterated soils (Rahman et al., 2013). It was also suggested that the proper plant nutrition has the capacity to be an efficient, cost-effective agronomic practice for increasing the phytoextraction of heavy metals by plants; however, much study is needed before fertilizers can be employed efficiently for this purpose.

Buckwheat (Fagopyrum esculentum) as Hyperaccumulator

Aluminium toxicity is detrimental to agricultural production especially in soils with low pH, i.e. acidic soils which constitute about 40% of the world's total arable land area (Foy et al., 1978). Common buckwheat is a prime candidate for an aluminium phytoremediation program. The mechanisms of aluminium tolerance in buckwheat are quite complex, as they exhibit both exclusionary resistance mechanisms and internal detoxification mechanisms. Ma et al. (1997) reported that the roots of the buckwheat secrete oxalic acid in response to aluminium stress. Besides, buckwheat leaves show high aluminium concentration without any toxicity. Oxalic acid is considered to be a strong aluminium chelator (Hue et al., 1986), and thus renders buckwheat a high aluminium-resistant crop. In buckwheat, gene encoding transporters for Al-induced secretion of organic acids such as oxalic acid have recently been identified and characterized (Ma, 2007). Al^{3+} detoxification takes place inside vacuoles by forming Al-complexes with organic acids or other chelating agents which subsequently undergo sequestration (Ma and Hiradate, 2000). Under exposure to Al^{3+}, the root tips of buckwheat immediately produce oxalates. These oxalates in turn chelate Al^{3+} resulting in the formation of stable, non-phytotoxic complexes of Al-oxalate. During translocation of Al^{3+} from root to leaves, Al-oxalate is converted into Al-citrate inside the xylem. This Al-citrate gets converted back into Al-oxalate during translocation from xylem to leaf cells (Ma and Hiradate, 2000). In addition, buckwheat accumulates a high concentration of Al^{3+} in its leaves (Ma et al., 1997). The majority (60%) of the Al^{3+} is present in the cells of buckwheat roots as Al-oxalates. It has also been reported that the suppression of transpiration rate declines the accumulation of Al^{3+} inside the leaves (Shen and Ma, 2001). It seems very interesting that buckwheat accumulates Al^{3+} in the leaves and not in the seeds (Shen et al., 2006). All these facts are important due to the potential use of buckwheat plants in phytoremediation of metal-adulterated soils (Barceló and Poschenrieder, 2003). Besides remediating Al-rich soils, buckwheat has also been recognized as a potential lead (Pb) hyperaccumulator, which is defined as a plant constituting over $1000\,mg\,kg^{-1}$ of Pb in its shoots on a dry-weight (DW) basis. Tamura et al. (2005) reported that the resistance of the buckwheat could be enhanced by more than five times without any side effects on treating with a biodegradable chelator ($20\,mmol\,kg^{-1}$ methylglycinediacetic acid trisodium salt). Common buckwheat possesses a great potential to accumulate high Pb concentration in its leaves ($8000\,mg\,kg^{-1}$ DW), stem ($2000\,mg\,kg^{-1}$ DW) and roots ($3300\,mg\,kg^{-1}$ DW DW) without serious damage (Tamura et al., 2005). This type of study reveals that buckwheat has enormous potential to decontaminate soils with

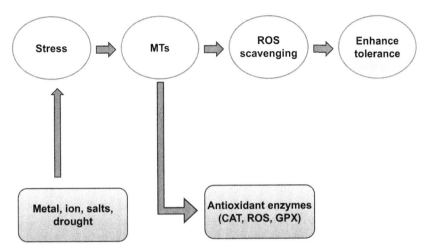

FIGURE 5.1   Model mechanism of MTs action in plants.

high Pb concentrations. A buckwheat cDNA clone (pBM 290) coding for metallothionein (MT)-like proteins was isolated from its seed cDNA library and the reduced amino acid sequence contains the maximum homology with the MT3-like protein from *A. thaliana* (Brkljačić et al., 1999). A study of expression analysis revealed that buckwheat MT3 mRNAs are apparent in the root and leaf tissue and during seed development in natural conditions. The expression of MT3-like proteins in both leaves and seeds is greatly influenced by heavy metal stress; therefore, identifying novel MT3 family members in order to explicate their possible role in plant defense systems during specific stressed environmental conditions (Brkljačić et al., 2004). MTs belong to the diverse family of small cysteine-rich proteins present throughout the plant and animal kingdom, besides being also present in some prokaryotes. These MTs upon exposure to heavy metals quench their toxicity under normal physiological conditions as they bind to metal ions by using thiol groups of their cysteine residues (Kagi and Vallee, 1960). Thus, these MTs play a pivotal role in maintaining homeostasis of important transition metals, subside the toxic effects of heavy metals and provide a shield against distinct abiotic stresses (Wong et al., 2004) (Figure 5.1).

FACTORS AFFECTING PHYTOREMEDIATION

There are various factors that are responsible for the achievement of phytoremediation as an environmental restoration technology which include: metal availability in soil, uptake by plants, transport and metal concentration in shoots and plant–microbe interactions. Knowledge of the difficulties that lie beneath the biological mechanisms of plant detoxification processes is inadequate and the majority of the information is still unknown.

Availability of Metals in Soil

In order to remediate adulterants by plants in relation to microbes, they should be in touch with them and also have the capability to operate on them; availability of adulterant in the soil is essential for its remediation. The availability of adulterants is influenced by many factors such as chemical attributes of the adulterant, soil attributes, environmental conditions and biological activity. Clay has a smaller particle size than sandy soil, and hence it contains more water as well as more binding sites particularly for cations (cation exchange capacity; CEC) (Taiz and Zeiger 2002). The amount of humus in the soil is correlated with the CEC since humus mostly contains dead plant material, as the cell walls of the plants possess negatively charged ions that bind cations, and lignin that binds hydrophobic components (Burken, 2003). Metals are hydrophilic in nature because of the presence of negatively (anions) and positively (cations) charged ions. In the presence of lower soil pH, there is the substitution of the H^+ ions on the cations of soil CEC sites which results in the enhanced cation bioavailability (Taiz and Zeiger, 2002). Another factor which influences the bioavailability of ions is the oxidation–reduction reaction. The majority of land soil exists in the oxidizing state and the elements of distinct oxidation states are present to a greater degree in their oxidized form, while water soil exists in the reducing state and hence possesses more reduced elemental forms. The bioavailability of the metal such as its solubility, metal uptake capability of the plant and also its toxicity may be influenced by its oxidation state and other influencing factors such as temperature and moisture. Usually elevated temperatures increase the physical, chemical and biological processes. Other major factors such as precipitation trigger the usual plant growth and greater soil moisture elevates movement of water-soluble adulterants. Metal chelators that are liberated by plants in the rhizosphere enhance the bioavailability of metals while as bacterial chelating agents like siderophores, organic acids and phenolic compounds help to detach metal ions that are bound to soil particles and thus make these metal ions available for plant uptake. In addition to this, plants exude H^+ ions through ATPases, which substitute cations at soil CEC sites, resulting in an increase in metal bioavailability (Taiz and Zeiger 2002). By comprehending the mechanism underlying adulterant availability the phytoremediation capabilities can be increased. It has been found that aged soils contain adulterants that are available in lower amounts (i.e. recalcitrant in nature) and so it is harder to phytoremediate aged soil than soil that has recently being adulterated. So in these cases, the soil could be altered to produce metal cations that are easily available for uptake by plants. For example, cation bioavailability can be increased by the addition of citric and malic acids (natural organic acids) that reduce the pH of soil particles. Bhatia et al. (2005) classified malate as a ligand for nickel to maintain decontamination / translocation and deposition of this massive metal in *Stackhousiatryonii*. It has been found that the synthetic EDTA metal chelator is highly effective in liberating metals from soil. This postulate is utilized in chelate-assisted phytoextraction where, before harvesting the plants, a small amount of EDTA is added to the soil which highly enhances the uptake of metal by the plants (Salt et al., 1998). A current investigation describes how adding EDTA to adulterated soil can enhance the potential and capability of phytoextraction of Cu by *Elsholtzia splendens* (Jiang and Yang, 2004). It has been observed that the incorporation of EDTA and citric acid (Turgut et al., 2004) as well as EDTA and N-(2-hydroxyethyl)-ethylenediaminetriacetic acid (HEDTA) together in *Helianthus annus* (Chen and Cutright, 2001) can increase the uptake of Ni, Cr and Cd. Before utilizing any phytoremediation technology, it is mandatory to monitor the risk assessment to find out the future consequences of the chelator on metal leaching while, on the other hand, it may be advantageous to reduce the bioavailability of metals if they exist at phytotoxic levels or in phytostabilization. In such conditions, lime is blended with soil in order to elevate the pH making it suitable for the phytoextraction process (Brown et al., 2003).

Plant Uptake and Translocation

Once plants are exposed to heavy metals, the root tips get stimulated and secrete ligand molecules in order to absorb these heavy metal ions, and later translocate them to other tissues. Plants generally employ two distinct pathways to transport these heavy metal ions which include: (1) passive transport across membranes and (2) inducible substrate-specific and energy dependent transport across membranes (Williams et al., 2000). Either by means of passive or active uptake, root cells act as the main players to grab metals from soil and later transport them across the membrane. Due to the lipophilic nature of the membrane, the charged metal ions become a hindrance to diffuse freely across the membrane into cytosol. Therefore, some energy-driven pumps like the ATP-dependent protein pump play a pivotal role in transporting these metal ions across the membrane. In one of the model planta (*A. thaliana*) there exist about 150 distinct cation transporters (Axelsen and Palmgren, 2001) and 14 transporters only for sulphate (Hawkesford, 2003). These energy-driven pumps (transporter proteins) possess some unique characteristics with respect to transport rate, substrate specificity and affinity, and they also obey Michaelis–Menton kinetics (Marschner, 1995). These properties can be regulated by subjecting them to metabolic levels or regulatory proteins (e.g. kinases). Moreover, these transporter proteins generally vary depending upon the type of tissue and some environmental factors that may be regulated at the transcription level or through endocytosis. As a result, the uptake and transport of metal ions in plant species are complex and dependent upon the process conditions (Pilon-Smits, 2005).

Plant Microbe Interactions

The metal ions present in the soil are not readily available to plants because of their reduced solubility in water and strong affinity with soil particles, and this restricts their deposition and translocation by plants. However, soil is enriched with some beneficial microorganisms (root colonizing bacteria and mycorrhiza) that have the ability to degrade these toxicants so that they become easily available to the plant (Chaudhry et al., 1999). Besides, some bacteria also ooze out biosurfactants (rhamnolipids) that possess the ability to make adulterants more water soluble (Volkering et al., 1998). Chemicals secreted by these microorganisms not only make these heavy metals easily available to the plant but also help to facilitate absorption of a number of metal ions such as iron (Burkal et al., 2000), manganese (Barber and Lee, 1974) and lead (Salt et al., 1995). Moreover, roots in association with microorganisms secrete some natural ligands (citric acid) that enhance the ion mobility or promote biosurfactant-producing microbial populations. Due to the complex feedback mechanisms between roots and rhizosphere microorganisms, they adapted themselves to the changing environmental conditions. For instance, plants growing in soils deficient in phosphorus usually secrete some chemicals (citric acid) that help them to mobilize any phosphorous component present in the soil. The mutual association between plants and fungi is very important not only to increase root absorption up to 47-fold (Smith and Reed, 1997), but also to stimulate the acquisition of plant nutrients including metal ions (Khan et al., 2000). This symbiotic relationship also plays an important role in maintaining homeostasis in plants, i.e. increases metal uptake in plants when these are present in low concentration and vice versa (Frey et al., 2000).

Role of Metal Chelators

As previously specified, plants exude some important natural chelating factors from their complex roots that influence solubility of the adulterant and uptake by the plants. In the plant tissues, the presence of chelator agents plays an important role in resistance, separation and translocation (Ross, 1994). It has been also found that phytosiderophores are chelator agents that stimulate the uptake of iron and possibly other metals in grasses (Higuchi et al., 1999). Some of the notable organic acids such as citric, malic and acetic also play a vital role in the stimulation of metal uptake by the roots and also facilitate processes like translocation, separation and metal resistance (Von Wiren et al., 1999). In the resistance and decontamination mechanism, metal chelators are eliminated from the cytoplasm and segregated in the vacuolar compartments that keep them out from the cellular sites where processes such as cell division and respiration occur and, therefore, contribute towards efficient protective mechanisms (Chaney et al., 1997; Hall, 2002). Decontamination of Cd and

Zn in *Thlaspi caerulescens* is achieved by vacuolar compartmentalization (Ma et al., 2005). The function of cysteine-rich MTs (Cobbett and Goldsbrough, 2000) and thiol-rich phytochelatins (PCs) as metal-chelating agents is also well defined. It has been observed that production of MTs in animals and some fungi takes place in reaction to metal uptake (Hamer, 1986), and in plants like wheat (Lane et al., 1987) and *Arabidopsis* (Murphy et al., 1997) MTs proteins have been recognized. However, identification of plant MTs proteins has been difficult because it is evident from the information that they have a greater affinity to bind metal cations such as Zn, Cu and Cd (Cobbett and Goldsbrough, 2002). Several investigations have revealed that MT proteins play a vital function in heavy metal decontamination as overexpression of Fe or Cu chelating MTs and ferritin defends plants against metal-induced oxidative damage (Fabisiak et al., 1999). Phytochelatins (PCs) that belong to the family of metal complexing peptides are rapidly induced on overexposure to metals or metalloids in plants, animals and some yeasts (Cobbett and Goldsbrough, 2002). Phytochelatins (PCs) have been demonstrated to be mainly involved in Cd and Cu resistance (Rauser, 1995; Ow 1996) but Schmoger et al. (2000) displayed that phytochelatins (PCs) may also play a role in arsenic decontamination as they exhibit greater affinity to sulphydryl groups, and after chelation by phytochelatins (PCs), the metal chelate complex is operatively translocated to the vacuole, where it is further complexed by sulphide (Cobbett and Goldsbrough, 2000). However, further studies of the function of these distinct chelators in translocation and decontamination processes of metals are yet to be fully established.

MECHANISM FOR METAL DETOXIFICATION

Despite the fact that micronutrients are essential for the growth and development of plants, the excess concentration of these metals becomes a hindrance which ultimately leads to cytotoxic effects. In order to counteract the hazardous effects of heavy metals, plants adapt themselves to such environmental conditions and develop a mechanism that will regulate the homeostasis of intracellular metal ions. These include regulation of ion influx and elimination of intracellular ions back into the extraneous solution. However, the intensity of hyperaccumulation depends upon various factors and can fluctuate from species to species (Roosens et al., 2003). Hyperaccumulation generally depends on three basic criteria that distinguish hyperaccumulators from non-hyperaccumulator species. These traits include: (1) excellent efficiency to absorb heavy metal ions from the soil; (2) rapid and powerful root-to-shoot translocation of heavy metal ions and (3) higher potency to detoxify and sequester these toxicants in their leaves. By means of proteomic and genomic approaches, significant improvements have been made to better understand the mechanism governing metal hyperaccumulation in plants, especially in *T. caerulescens* and *A. halleri*, which have become model plants for this study (Frérot et al., 2010). Most of the indispensable steps involved in the hyperaccumulation pathway do not depend on novel genes, but rely on genes common to both hyperaccumulators as well as non-hyperaccumulators, which are differently expressed and controlled in the two kinds of plants (Verbruggen et al., 2009). However, the plants capable of detoxifying the heavy metals at a very high concentration bear some supplementary mechanisms. For instance, in *T. goesingense*, a Ni hyperaccumulator, high resistance was due to Ni complexation by histidine which rendered the metal inactive (Krämer et al., 1997). Lasat et al. (1998) reported that the resistance of Zn-hyperaccumulator *T. caerulescens* is due to sequestration of Zn ions in the vacuole. Zn detoxification takes place inside the vacuoles by means of several mechanisms, viz. precipitation (Zn-phytate) (Van Steveninck et al., 1990) and binding to low-molecular-weight organic acids (Salt et al., 1999). Ni resistance in plants is also due to the complex formation with organic acids secreted by the plants (Lee et al., 1977). In some plant species phytochelatins exist that belong to the family of thiol (SH)-rich peptides, which play an important role in detoxifying Cd heavy metal (Rauser, 1990; Steffens, 1990). There are some important players such as CDF (Cation Diffusion Facilitator) family members, also named MTPs (Metal Transporter Proteins) that mediate bivalent cation efflux from the cytosol. In the leaves of Zn/Ni hyperaccumulator species one of the MTPs, i.e. MTP1, a gene encoding a protein bound at tonoplast, is highly overexpressed (Gustin et al., 2009). It has also been revealed that MTP1-gene not only provides resistance to Zn but also enhances the ability to accumulate high Zn concentration. In *T. goesingense* shoots, MTPs are responsible for mediating the Ni storage in its vacuoles (Persant et al., 2001). During the translocation of Zn metal ions inside the vacuoles a systemic Zn deficiency response is initiated that includes the enhancement of the heavy metal uptake and translocation via the increased expression of ZIP transporters in hyperaccumulator plants (Gustin et al., 2009). Kim et al. (2004) reported that the presence of MTP1 at both vacuolar as well as plasma membrane serves to operate in both Zn and Ni efflux from cytoplasm to the cell wall. The P_{1B}-type ATPase, also termed as the Heavy Metal transporting ATPase (HMAs), plays an essential role in carrying metal ions against their electrochemical gradient using the energy provided by the ATP hydrolysis. HMAs is a special class of ATPase enzymes found in all living organisms ranging from archaea to humans including yeast and plants. Usually these HMAs are classified into various classes, viz. monovalent cations (Cu/Ag-group) and those transporting divalent cations (Zn/Co/Cd/Pb-group) (Axelsen and Palmgrem, 2001). In the case of prokaryotes, all types of P_{1B}-type ATPase are present (Rensing et al., 1999), whereas in the case of non-plant eukaryotes only two classes such as Cu/Ag P_{1B}-type ATPase have been found to date. Baxter et al. (2003) reported eight HMA genes in *Arabidopsis thaliana* and *Oryza sativa* suggesting that these enzymes play a crucial role in the translocation of metal ions in plants. With the help of genomics and other bioinformatic tools, a complete genome sequence analysis of *A. thaliana* was carried out that revealed the distribution of the eight HMAs in two classes, viz. HMA1–4 which are responsible for the transport of Zn/Co/Cd/Pb and HMA5–8 for Cu and Ag. Among different classes of HMAs, HMA4 was the first plant P_{1B}-type ATPase to be cloned and characterized in *A. thaliana* (Mills et al., 2003). In transgenic Arabidopsis HMA4 plays many important functions such as detoxification of Cd metal ions, Zn homeostasis and also helps in transporting these toxic metal ions from root to shoot (Verret et al., 2005). In the Cd/ Zn hyperaccumulators, HMA4 is more highly expressed in both roots and shoots compared with Cd/Zn-sensitive close relatives and thus the overexpression of HMA4 in two distinct species (Zn/Cd hyperaccumulator) that evolved independently, strongly supports the idea that HMA4 plays a pivotal role in providing resistance to both metals (Courbot et al., 2007). Overexpression of HMA3 that codes for a vacuolar P_{1B}-ATPase, probably involved in Zn compartmentation, and that of CAX genes encoding members of a cation exchanger family that seems to mediate Cd sequestration, have been noticed in *T. caerulescens* and *A. halleri* and are supposed to be involved in heavy metal hyperaccumulation (van de Mortel et al., 2008). In the stalks of hyperaccumulating ferns arsenic is usually stored in the form of inorganic arsenite in its vacuoles, although the transport system located at the tonoplast has not been identified yet (Zhao et al., 2009). Besides, some small ligands, such as organic acids possess a major function as detoxifying agents. These small ligands either act as barriers to prevent incorporation of heavy metals inside cytoplasm or bind with these toxic metal ions to form chelate complexes or thus subside their toxic effects. Citrate is the main ligand of Ni in leaves of *T. goesingense* (Krämer et al., 2000), while citrate and acetate bind Cd in leaves of *Solanum nigrum* (Sun et al., 2006). Moreover, most Zn in *A. halleri* and Cd in *T. caerulescens* are complexed with malate (Sarret et al., 2002). Similarly, in *Fagopyrum esculentum* oxalic acid is an important source of ligand secreted by root tips which forms a chelate complex with aluminium ions and thus provides resistance to the plant against aluminium element (Figure 5.2).

CONCLUSIONS AND FUTURE PERSPECTIVES

In this current era of industrialization, the complications associated with heavy metal pollution are continuously deteriorating the quality of soil. A concept of eradication through a utilization technique should be developed. Future strategies should be developed in such a way that they will not only help to rejuvenate soil health but the biofeedstock could be utilized for biofuel production. Nowadays phytoremediation has gained worldwide attention due to the unique feature of plants growing on metalliferous soils and accumulating heavy metal concentration several 100-folds higher than other plants. Phytoremediation in combination with biofuel technology might provide a sustainable option for the better management of other contaminated areas, especially those heavily impacted by mining activities, and thus these contaminated areas can be exploited as potential biomass feedstock reserves for biofuel production. Understanding

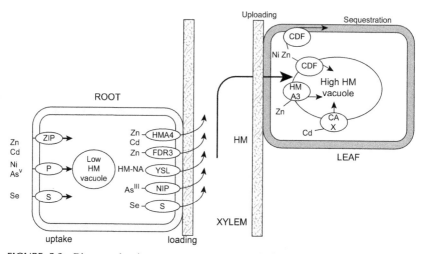

FIGURE 5.2 Diagram showing transport systems constitutively overexpressed and/or with enhanced affinity for heavy metals, which are thought to be involved in the uptake, root-to-shoot translocation and heavy metal sequestration traits of hyperaccumulator plants. CAX, Cation Exchangers; CDF, Cation Diffusion Facilitators; FDR3, a member of the Multidrug and Toxin Efflux family; HM, Heavy Metals; HMA, Heavy Metal transporting ATPases; NA, Nicotinamine; NIP, Nodulin 26-like Intrinsic Proteins; P, Phosphate transporters; S, Sulphate transporters; YSL, Yellow Strip 1-Like Proteins; ZIP, Zinc-regulated transporter Iron-regulated transporter Proteins).

the molecular biology of the hyperaccumulating plant species is of paramount importance in order to engineer them and enhance their metal-chelating and quenching activities. Finally, it should be kept in mind that phytoremediation is an interdisciplinary area of research where plant biology, microbiology, soil science, genetic engineering and environmental modelling converge. In addition, there is a novel phenomenon known as phytomining, one of the aspects of 'green' technology that utilizes metal-hyperaccumulating plant species to extract metal ores generally termed as bio-ore from soil (Robinson et al., 2009). Currently phytomining has gained a lot of attention especially in developed countries to remediate heavily contaminated soils as it causes permanent elimination of heavy metals from the contaminated sites. Current research should be focused on rhizosphere and soil–microbial interaction in order to identify microorganisms associated with metal solubility or chelation. Genes responsible for metal detoxification or chelation should be identified and characterized for genetic manipulation.

REFERENCES

Andrahennadi, R., Gillott, C., 1998. Resistance of *Brassica*, especially *B. juncea* (L.) Czern, genotypes to the diamondback moth, *Plutellaxylostella* (L.). Crop Prot. 17 (1), 85–94.

Axelsen, K.B., Palangren, M.G., 2001. Inventory of the superfamily of P-type ion pumps in Arabidopsis. Plant Physiol. 120, 696–706.

Baker, A.J.M., Brooks, R.R., 1989. Terrestrial higher plants which hyperaccumulate metalic elements. A review of their distribution, ecology and phytochemistry. Biorecovery 1, 81–126.

Baker, A.J.M., McGrath, S.P., Reeves, R.D., Smith, J.A.C., 2000. Metal hyperaccumulator plants: a review of the ecology and physiology of a biological resource for phytoremediation of metal polluted soils. In: Terry, N., Bañuelos, G. (Eds.), Phytoremediation of Contaminated Soils and Waters. CRC Press LLC, Boca Raton, FL, USA, pp. 85–107.

Baker, A.J.M., McGrath, S.P., Sidoli, C.M.D., Reeves, R.D., 1994. The possibility of in situ heavy metal decontamination of polluted soils using crops of metal-accumulating plants. Resour. Conserv. Recycl. 11, 41–49.

Baker, A.J.M., Reeves, R.D., McGrath, S.P., 1991. *In situ* decontamination of heavy metal polluted soils using crops of metal-accumulating plants—a feasibility study. In: Hinchee R.E., Olfenbuttel, R.F. (Eds.), *In situ* bioreclamation. Butterworth-Heinemann, Stoneham MA, pp. 539–544.

Bandara, N.I.G.J., 2003. Water and waste water issues in Sri Lanka. Water Sci. Technol. 47, 305–312.

Bandara, S., 2006. "Rajarata Goveen Wakugadu Roogeen Karana Pohora". Lakbima. December 10, 21.

Banga, S.S., 1997. Genetics and breeding in *Brassica* oilseed crops. In: Thomas, G., Monteiro, A.A. (Eds.), Proceedings of the International Symposium on *Brassicas*. Acta Hortic. No. 459. ISHS, Rennes, France, pp. 389–395.

Barber, A., Lee, R.B., 1974. The effect of microorganisms on the absorption of manganese by plants. New Phytol. 73, 97–106.

Barceló, J., Poschenrieder, C., 2003. Phytoremediation: Principles and perspectives. Contribut. Sci. 2, 333–344.

Baudouin, C., Charveron, M., Tarrouse, R., Gall, Y., 2002. Environmental pollutants and skin cancer. Cell Biol. Toxicol. 18, 341–348.

Baumann, A., 1885. Das Verhalten von Zinksatzengegen Pflanzen und im Boden. Landwirtsch. Vers.-Statn. 31, 1–53.

Baxter, I., Tchieu, J., Sussman, M.R., Boutry, M., Palmgren, M.G., Gribskov, M., Harper, J.E., Axelsen, K.B., 2003. Genomic comparison of P-type ATPase ion pumps in Arabidopsis and rice. Plant Physiol. 132, 618–628.

Bhatia, N.P., Walsh, K.B., Baker, A.J., 2005. Detection and quantification of ligands involved in nickel detoxification in a herbaceous Ni hyperaccumulator Stackhousiatryonii Bailey. J. Exp. Bot. 56, 1343–1349.

Boyd, R.S., 1998. Hyperaccumulation as a plant defensive strategy. In: Brooks, R.R. (Ed.), Plants that Hyperaccumulate Heavy Metals. Their Role in Phytoremediation, Microbiology, Archaeology, Mineral Exploration and Phytomining. CAB International, Wallingford, UK, pp. 181–201.

Boyd, R.S., 2004. Ecology of metal hyperaccumulation. New Phytol. 162, 563–567.

Brkljacic, J.M., Maksimovic, V.R., Radovic, S.R., Savic, A.P., 1999. Isolation of metallothionein-like cDNA clone from buckwheat. J. Plant Physiol. 154, 802–804.

Brkljacic, J.M., Samardzic, J.T., Timotijevic, G.S., Maksimovic, V.R., 2004. Expression analysis of buckwheat (*Fagopyrum esculentum* Moench) metallothionein-like gene (MT3) under different stress and physiological conditions. J. Plant Physiol. 161 (6), 741–746.

Brown, S.L., Henry, C.L., Chaney, R., Compton, H., DeVolder, P.M., 2003. Using municipal biosolids in combination with other residuals to restore metal contaminated mining areas. Plant Soil 249, 203–215.

Burken, J.G., 2003. Uptake and metabolism of organic compounds: Green-livermodel. In: McCutcheon, S.C., Schnoor, J.L. (Eds.), Phytoremediation: Transformation and Control of Contaminants. Wiley, New York, pp. 59–84.

Burken, J.G., Schnoor, J.L., 1997. Uptake and metabolism of atrazine by poplar trees. Environmental Science and Technology 31, 1399–1406.

Byers, H.G., 1935. Selenium occurrence in certain soils in the United States, with a discussion of the related topics. US Dept. Agric. Technol. Bull. 482, 1–47.

Chaney, R.L., 1983. Plant uptake of inorganic waste. In: Parr, J.E., Marsh, P.B., Kla, J.M. (Eds.), Land Treatment of Hazardous Waste. Noyes Data Corp., Park Ridge Il, pp. 50–76.

Chaney, R.L., Malik, M., Li, Y.M., Brown, S.L., Brewer, E.P., Angle, J.S., Baker, A.J.M., 1997. Phytoremediation of soil metals. Curr. Opin. Biotechnol. 8, 279–284.

Chaudhry, T.M., Hill, L., Khan, A.G., Keuk, C., 1999. Colonization of iron and zinc contaminated dumped filter cake waste by microbes, plants and associated mycorrhizae. In: Wong, M.H., Baker, A.J.M. (Eds.), Remediation and Management of Degraded Land. CRC Press, Boca Raton, FL, pp. 275–283.

Chen, H., Cutright, T., 2001. EDTA and HEDTA effects on Cd, Cr, and Ni uptake by Helianthus annuus. Chemosphere 45 (1), 21–28.

Cobbett, C., Goldsbrough, P., 2002. Phytochelatins and metallothioneins: Roles in heavy metal detoxification and homeostasis. Ann. Rev. Plant Physiol. Plant Mol. Biol. 53, 159–182.

Cobbett, C.S., Goldsbrough, P.B., 2000. Mechanisms of metal resistance: Phytochelatins and metallothioneins. In: Raskin, I., Ensley, B.D. (Eds.), Phytoremediation of Toxic Metals using Plants to Clean Up the Environment. Wiley, New York, pp. 247–271.

Courbot, M., Willems, G., Motte, P., Arvidsson, S., Roosens, N., Saumitou-Laprade, P., Verbruggen, N., 2007. A major quantitative trait locus for cadmium tolerance in Arabidopsis halleri colocalizes with HMA4, a gene encoding a heavy metal ATPase. Plant Physiol. 144, 1052–1065.

Cunningham, S., 1995. In: Proceedings / Abstracts of the Fourteenth Annual Symposium, Current Topics in Plant Biochemistry, Physiology, and Molecular Biology Columbia, pp. 47–48.

Dissanayake, P., Clemett, A., Jayakody, P., Amarasinghe, P., 2007. Report on Water Quality Survey and Pollution in Kurunegala, Sri Lanka.

Dissanayake, U.S., Tennakoon, K.U., Priyantha, N., 2002. Potential of two invasive plant species, *Lantana camara* L. and *Wedeliatrilobata* L. for selective heavy metal uptake. Ceylon J. Sci. (Bio. Sci.) 29, 1–11.

Dushenkov, S., 2003. Trends in phytoremediation of radionuclides. Plant and Soil 249, 167–175.

Ensley, B.D., 2000. Phytoremediation for toxic metals – using plants to clean-up the environment. In: Raskin, I., Ensley, B.D. (Eds.), Rational for Use of Phytoremediation. John Wiley & Sons Inc., pp. 3–13.

Ernst, W.H.O., Schat, H., Verkleij, J.A.C., 1990. Evolutionary biology of metal resistance in *Silene vulgaris*. Evol. Trends in Plants 4, 45–51.

Fabisiak, J.P., Pearce, L.L., Borisenko, G.G., Tyhuriana, Y.Y., Tyurin, V.A., Razzack, J., Lazo, J.S., Pitt, B.R., Kagan, K.E., 1999. Bifunctional anti prooxidant potential of metallothionein: Redox signaling of copper binding and release. Antioxid. Redox. Signal. 1, 349–364.

Foy, C.D., Chaney, R.L., White, M.C., 1978. The physiology of metal toxicity in plants. Annu. Rev. Plant Physiol. 29, 511–566.

Frérot, H., Faucon, M.-P., Willems, G., Godé, C., Courseaux, A., Darracq, A., Verbruggen, N., Saumitou-Laprade, P., 2010. Genetic architecture of zinc hyperaccumulation in Arabidopsis halleri: The essential role of QTLx environment interactions. New Phytol. 187, 355–367.

Frey, B., Zierold, K., Brunner, I., 2000. Extracellular complexation of Cd in the Hartig net and cytosolic Zn sequestration in the fungal mantle of *Picea abies Hebeloma crustuliniforme* ectomycorrhizas. Plant Cell Environ. 23, 1257–1265.

Gleba, D., Borisjuk, N.V., Borisjuk, L.G., Kneer, R., Poulev, A., Skarzhinskaya, M., Dushenkov, S., Logendra, S., Gleba, Y.Y., Raskin, I., 1999. Use of plant root for phytoremediation and molecular farming. Proc. Natl. Acad. Sci. USA 96, 5973–5977.

Gupta, A.K., Sinha, S., 2007. Phytoextraction capacity of the plants growing on tannery sludge dumping sites. Bioresour. Technol. 98, 1788–1794.

Gustin, J.L., Loureiro, M.E., Kim, D., Na, G., Tikhonova, M., Salt, D.E., 2009. MTP1-dependent Zn sequestration into shoot vacuoles suggests dual roles in Zn tolerance and accumulation in Zn hyperaccumulating plants. Plant J. 57 (6), 1116–1127.

Hall, J.L., 2002. Cellular mechanisms for heavy metal detoxification and tolerance J. Exp. Bot. 53, 1–11.

Hamer, D.H., 1986. Metallothioneins. Ann. Rev. Biochem. 55, 913–951.

Hartman Jr., W.J., 1975. An evaluation of land treatment of municipal wastewater and physical siting of facility installations. US Department of Army, Washington DC.

Hawkesford, M.J., 2003. Transporter gene families in plants: The sulphate transporter gene family—redundancy or specialization? Physiol. Plant 117, 155–163.

Higuchi, K., Suzuki, K., Nakanishi, H., Yamaguchi, H., Nishizawa, N.K., Mori, S., 1999. Cloning of nicotianamine synthase genes, novel genes involved in the biosynthesis of phytosiderophores. Plant Physiol. 119, 471–479.

Hong, M.S., Farmayan, W.F., Dortch, I.J., Chiang, C.Y., McMillan, S.K., Schnoor, J.L., 2001. Phytoremediation of uranium-contaminated soils: Role of organic acids in triggering uranium hyperaccumulation in plants. Environ. Sci. Technol. 35, 1231–1239.

Hue, N.V., Craddock, G.R., Adams, F., 1986. Effect of organic acids on aluminum toxicity in subsoils. Soil Sci. Soc. Am. J. 50, 28–34.

Hughes, J.B., Shanks, J., Vanderford, M., Lauritzen, J., Bhadra, R., 1997. Transformatiom of TNT by aquatic plants and plant tissue cultures. Environ. Sci. Technol. 31, 266–271.

Jiang, L.Y., Yang, X.E., 2004. Chelators effect on soil Cu extractability and uptake by Elsholtzi asplendens. J. Zhejiang Univ. Sci. 5 (4), 450–456.

Jiang, W., Liu, D., Hou, W., 2000. Hyperaccumulation of lead by roots, hypocotyls, and shoots of *Brassica juncea*. Biologia Plantarum 43 (4), 603–606.

Khan, A.G., Keuk, C., Chaudhry, T.M., Khoo, C.S., Hayes, W.J., 2000. Role of plants, mycorrhizae and phytochelators in heavy metal contaminated landremediation. Chemosphere 41, 197–207.

Kim, D., Gustin, J.L., Lahner, B., Persans, M.W., Baek, D., Yun, D.J., Salt, D.E., 2004. The plant CDF family member TgMTP1 from the Ni/Zn hyperaccumulator *Thlaspi goesingense* acts to enhance efflux of Zn at the plasma membrane when expressed in *Saccharomyces cerevisiae*. Plant J. 39 (2), 237–251.

Krämer, U., Pickering, I.J., Prince, R.C., Raskin, I., Salt, D.E., 2000. Subcellular localization and speciation of nickel in hyperaccumulator and non-accumulator *Thlaspi* species. Plant Physiol. 122 (4), 1343–1354.

Krämer, U., Smith, R.D., Wenzel, W., Raskin, I., Salt, D.E., 1997. The role of metal transport and tolerance in nickel hyperaccumulation by *Thlaspi goesingense* Halacsy. Plant Physiol. 115, 1641–1650.

Kumar, P.B.A.N., Dushenkov, V., Motto, H., Raskin, I., 1995. Phytoextraction: The use of plants to remove heavy metals from soils. Environ. Sci. Tech. 29, 1232–1238.

Lane, B., Kajioka, R., Kennedy, T., 1987. The wheat germ Ec protein is a zinc containing metallothionein. Biochem. Cell. Biol. 65, 1001–1005.

Lasat, M.M., Baker, A.J.M., Kochian, L.V., 1998. Altered Zn compartmentation in the root symplasm and stimulated Zn absorption into the leaf as mechanisms involved in Zn hyperaccumulation in *Thlaspi caerulescens*. Plant Physiol. 118, 875–883.

Lee, J., Reeves, R.D., Brooks, R.R., Jaffré, T., 1977. Isolation and identification of a citratocomplex of nickel from nickel-accumulating plants. Phytochemistry 16, 1502–1505.

Lim, J.M., Jin, B., Butcher, D.J., 2012. A comparison of electrical stimulation for electrodic and EDTA-enhanced phytoremediation of lead using Indian mustard (Brassica juncea). Korean Chem. Soc. vol. 33. No. 82737. http://dx.doi.org/10.5012/bkcs.2012.33.8.2737.

Lim, J.M., Salido, A.L., Butcher, D.J., 2004. Phytoremediation of lead using Indian mustard (*Brassica juncea*) with EDTA and Electrodics. Microchem. J. 76, 3–9.

Lytle, C.M., Lytle, F.W., Yang, N., Qian, H., Hansen, D., Zayed, A., Terry, N., 1998. Reduction of Cr(VI) to Cr (III) by wetland plants: Potential for in situ heavy metal detoxification. Environ. Sci. Technol. 32, 3087–3093.

Ma, J.F., 2007. Syndrome of aluminum toxicity and diversity of aluminum resistance in higher plants. Int. Rev. Cytol. 264, 225–252.

Ma, J.F., Hiradate, S., 2000. Form of aluminum for uptake and translocation in buckwheat (*Fagopyrum esculentum* Moench). Planta 211, 355–360.

Ma, J.F., Zheng, S.J., Hiradate, S., Matsumoto, H., 1997. Detoxifying aluminum with buckwheat. Nature 390, 569–570.

Ma, X., Richter, A.R., Albers, S., Burken, J.G., 2005. Phytoremediation of MTBE with hybrid poplar trees. Int. J. Phytoremediation 6 (2), 157–167.

Macek, T., Mackova, M., Kas, J., 2000. Exploitation of plants for the removal of organics in environmental remediation. Biotechnol. Adv. 18, 23–34.

Marques, A., Rangel, A.O.S.S., Castro, P.M.L., 2009. Remediation of heavy metal contaminated soils: Phytoremediation as a potentially promising clean-up technology. Crit. Rev. Environ. Sci. Technol. 39, 622–654.

Marschner, H., 1995. Mineral nutrition of higher plants. Academic Press, San Diego. 889.

Mills, R.F., Krijger, G.C., Baccarini, P.J., Hall, J.L., Williams, L.E., 2003. Functional expression of AtHMA4, a P-1B type ATPase of the Zn/Co/Cd/Pb subclass. Plant J. 35, 164–176.

Minguzzi, C., Vergnano, O., 1948. Il contento di nichel nelliceneri di Alyssum bertlonii Desv. Atti della Società Toscana di Scienze Naturali. Mem. Ser. A. 55, 49–77.

Murphy, A., Zhou, J., Goldsbrough, P.B., Taiz, L., 1997. Purification and immunological identification of metallothioneins 1 and 2 from *Arabidopsis thaliana*. Plant Physiol. 113 (4), 1293–1301.

Nascimento, C.W.A., Xing, B., 2006. Phytoextraction: A review on enhanced metal availability and plant accumulation. Sci. Agric. 63, 299–311.

Ow, D.W., 1996. Heavy metal tolerance genes—prospective tools for bioremediation. Res. Conserv. Recycling 18, 135–149.

Pang, E.C.K., Halloran, G.M., 1996. The genetics of adult plant blackleg (*Leptosphaeriamaculans*) resistance from *B. juncea* in *B. napus*. Theor. Appl. Genet. 92, 382–387.

Persant, M.W., Nieman, K., Salt, D.E., 2001. Functional activity and role of cation efflux family members in Ni hyperaccumulation in *Thlaspi goesingense*. Plant Biol. 98 (17), 9995–10000.

Pilon-Smits, E.A., 2005. Phytoremediation. Ann. Rev. Plant Biol. 56, 15–39.

Poschenrieder, C., Barceló, J., 1999. Water relations in heavy metal stressed plants. In: Prasad, M.N.V., Hagemeyer, J. (Eds.), Heavy Metal Stress in Plants. From Molecules to Ecosystems. Springer, Berlin, pp. 207–229.

Rahman, M., Tan, P.J., Faruq, G., Sofian, A.M., Rosli, H., Boyce, A.N., 2013. Use of Amaranth (*Amaranthuspaniculatus*) and Indian Mustard (*Brassica juncea*) for phytoextraction of lead and copper from contaminated soil. Int. J. Agriculture and Biol. 15, 903–908.

Rascio, W., 1977. Metal accumulation by some plants growing on Zn mine deposits. Oikos 29, 250–253.

Raskin, I., Kumar, P.B.A.N., Dushenkov-S, Salt D.E., 1994. Bioconcentration of heavy metals by plants. Curr. Opin. Biotechnol. 5, 285–290.

Rathinasabapathi, B., Srivastava, M., 2006. Arsenic hyperaccumulating ferns and their application to phytoremediation of arsenic contaminated sites. In: Texeira da Silva, J.A. (Ed.), Floriculture, Ornamental and Plant Biotechnology. Global Science Books, London, pp. 304–311.

Rauser, W.E., 1990. Phytochelatins. Ann. Rev. Biochem. 59, 61–86.

Rauser, W.E., 1995. Phytochelatins and related peptides. Plant Physiol. 109, 1411–1419.

Raven, K.P., Leoppert, R.H., 1997. Trace element composition of fertilizers and soil amendments. J. Environ. Qual. 26, 551–557.

Reeves, R.D., Brooks, R.R., 1983. Hyperaccumulation of lead and zinc by two metallophytes from a mining area of Central Europe. Environ. Pollut. Series A 31, 277–287.

Reeves, R.D., Baker, A.J.M., 2000. Metal Accumulating Plants. In: Raskin, I., Ensley, B.D. (Eds.), Phytoremediation of Toxic Metals—Using Plants to Clean Up the Environment. John Wiley & Sons, New York USA, pp. 193–229.

Rensing, C., Ghosh, M., Rosen, B.P., 1999. Families of soft metal ion transporting ATPase. J. Bacteriol. 181, 5891–5897.

Robinson, B.H., Banuelos, G., Conesa, H.M., Evangelon, W.H., Schulin, R., 2009. The phytomanagement of trace elements in soil. Crit. Rev. Plant Sci. 28 (4), 240–266.

Roosens, N., Verbruggen, N., Meerts, P., Ximenez-Embun, P., Smith, J.A.C., 2003. Natural variation in cadmium tolerance and its relationship to metal hyperaccumulation for seven populations of *Thlaspi caerulescens* from Western Europe. Plant Cell Environ. 26, 1657–1672.

Ross, S.M., 1994. Toxic metals in soil–plant systems. Chichester, Wiley, England.

Salido, A.L., Hasty, K.L., Lim, J.M., Butcher, D.J., 2003. Phytoremediation of arsenic and lead in contaminated soil using Chinese brake ferns (*Pteris vittata*) and Indian mustard (*Brassica juncea*). Int. J. Phytoremediat. 5, 89–103.

Salt, D.E., Blaylock, M., Kumar, N.P.B.A., Dushenkov, V., Ensley, D., Chet, I., Raskin, I., 1995. Phytoremediation: A novel strategy for the removal of toxic metals from the environment using plants. Biotechnol. 13, 468–474.

Salt, D.E., Prince, R.C., Baker, A.J.M., Raskin, I., Pickering, I.J., 1999. Zinc ligands in the metal hyperaccumulator *Thlaspi caerulescens* as determined using X-ray absorption spectroscopy. Environ. Sci. Technol. 33, 713–717.

Salt, D.E., Smith, R.D., Raskin, I., 1998. Phytoremediation. Annu. Rev. Plant Physiol. Plant Mol. Biol. 49, 643–668.

Sarret, G., Saumitou-Laprade, P., Bert, V., Proux, O., Hazemann, J.L., Traverse, A., Matthew, A., Manceau, Manceau, A., 2002. Forms of zinc accumulated in the hyperaccumulator *Arabidopsis halleri*. Plant Physiol. 130 (4), 1815–1826.

Schmoger, M.E., Oven, M., Grill, E., 2000. Detoxification of arsenic by phytochelatins in plants. Plant Physiol. 122, 793–801.

Schnoor, J.L., Light, L.A., Mccutcheon, S.C., Wolfe, N.L., Carreira, L.H., 1995. Phytoremediation of organic and nutrient contaminants. Environ. Sci. Technol. 29, 318–323.

Schwitzguébel, J., 2000. Potential of phytoremediation, an emerging green technology. In: Ecosystem Service and Sustainable Watershed Management in North China. Proceedings of International Conference, Beijing, P.R. China, p. 5. August 23–25, 2000.

Shen, R., Ma, J.F., 2001. Distribution and mobility of aluminum in an Al-accumulating plant, *Fagopyrum esculentum* Moench. J. Exp. Bot. 52, 1683–1687.

Shen, R.F., Chen, R.F., Ma, J.F., 2006. Buckwheat accumulates aluminum in leaves but not in seeds. Plant and Soil 284, 265–271.

Smith, S.E., Reed, D.J., 1997. Mycorrhizal Symbiosis. Academic Press, London. 589.

Steffens, J.C., 1990. The heavy metal-binding peptides of plants. Annu. Rev. Plant Physiol. Plant Mol. Biol. 41, 553–575.

Su, Y., Han, F.X., Chen, J., Shiyab, S., Monts, D.L., 2009. Phytotoxicity and Phytoremediation Potential of Mercury in Indian Mustard and Two Ferns with Mercury Contaminated Water and Oak Ridge Soil – 9241 Phoenix AZ March 1–5.

Sun, R., Zhou, Q., Jin, C., 2006. Cadmium accumulation in relation to organic acids in leaves of *Solanum nigrum* L. as a newly found cadmium hyperaccumulator. Plant and Soil 285 (1–2), 125–134.

Taiz, L., Zeiger, E., 2002. Plant Physiology. Sinauer, Sunderland, MA. 690.

Tamura, H., Honda, M., Sato, T., Kamachi, H., 2005. Pb hyperaccumulation and tolerance in common buckwheat (*Fagopyrum esculentum* Moench). J. Plant Res. 118, 355–359.

Tolrà, R.P., Poschenrieder, C., Alonso, R., Barceló, D., Barceló, J., 2001. Influence of zinc hyperaccumulation on glucosinolates in *Thlaspi caerulescens*. New Phytol. 151, 621–626.

Turgut, C., Katie Pepe, M., Cutright, T.J., 2004. The effect of EDTA and citric acid on phytoremediation of Cd, Cr and Ni from soil using Helianthus annuus. Environ. Pollut. 131 (1), 147–154.

Vallee, Kagi, 1960. Metallothionein: A cadmium- and zinc-containing protein from equine renal cortex. J. Biol. Chem. 235, 3460–3465.

van de Mortel, J.E., Schat, H., Moerland, P.D., Van Themaat, E.V.L., Van der Ent, S., Blankestijn, H., Ghandilyan, A., Tsiatsiani, S., Aarts, M.G.M., 2008. Expression differences for genes involved in lignin, glutathione and sulphate metabolism in response to cadmium in *Arabidopsis thaliana* and the related Zn/Cd hyperaccumulator *Thlaspi caerulescens*. Plant Cell Environ. 31, 301–324.

Van Steveninck, R.F.M., Van Steveninck, M.E., Wells, A.J., Fernando, D.R., 1990. Zinc tolerance and the binding of zinc as zinc phytate in *Lemna minor*. X-ray microanalytical evidence. J. Plant Physiol. 137, 140–146.

Verbruggen, V., Hermans, C., Schat, H., 2009. Molecular mechanisms of metal hyperaccumulation in plants. New Phytol. 181, 759–776.

Verret, F., Gravot, A., Auroy, P., Preveral, S., Forestier, C., Vavasseur, A., Richaud, P., 2005. Heavy metal transport by AtHMA4 involves the N-terminal degenerated metal binding domain and the C-terminal His (II) stretchs. FEBS Lett. 579, 1515–1522.

Volkering, F., Breure, A.M., Rulkens, W.H., 1998. Microbiological aspects of surfactant use for biological soil remediation. Biodegradation 8, 401–417.

Von Wiren, N., Klair, S., Bansal, S., Briat, J.F., Khodr, H., Shiori, T., 1999. Nicotianamine chelates both Fe III and Fe II. Implications for metal transport in plants. Plant Physiol. 119, 1107–1114.

Whitting, S.N., Neumann, P.M., Baker, A.J.M., 2003. Nickel and zinc hyperaccumulation by *Alyssum murale* and *Thlaspi caerulescens*(Brassicaceae) do not enhance survival and whole plant growth under drought stress. Plant Cell Environ. 26, 351–360.

Williams, L.E., Pittman, J.K., Hall, J.L., 2000. Emerging mechanisms for heavy metal transport in plants. Biochim. Biophys. Acta 1465 (1), 104–126.

Wong, H.L., Sakamoto, T., Kawasaki, T., Umemura, K., Shimamoto, K., 2004. Down-regulation of metallothionein, a reactive oxygen scavenger, by the small GTPase OsRac1 in rice. Plant Physiol. 135, 1447–1456.

World Health Organization, 2003. Health as a cross-cutting issue in dialogues on water for food and the environment. Report for an international workshop. http://whqlibdoc.who.int/2004/WHO_SDE_WSH_04.02.pdf.

Zhang, L., Angle, J.S., Chaney, R.L., 2007. Do high-nickel leaves shed by the Ni-hyperaccumulator Alyssum murale inhibit seed germination of competing plants? New Phytol. 173, 509–516.

Zhao, F.J., Ma, J.F., Meharg, A.A., McGrath, S.P., 2009. Arsenic uptake and metabolism in plants. New Phytol 181, 777–794.

Zhu, Y., Pilon-Smits, E.A.H., Jouanin, L., Terry, N., 1999. Overexpression of glutathione synthetase in *Brassica juncea* enhances cadmium tolerance and accumulation. Plant Physiol. 119, 73–79.

Chapter 6

Recent Trends and Approaches in Phytoremediation

Bisma Malik, Tanveer Bilal Pirzadah, Inayatullah Tahir, Tanvir ul Hassan Dar and Reiaz Ul Rehman

Department of Bioresources, University of Kashmir, Srinagar, India

INTRODUCTION

The tremendous progress in scientific technology has evolved human life globally and these developments have raised new issues which pose challenges to the environment, particularly protection and conservation (Bennett et al., 2003). Further, in industrialized society exposure to toxic chemicals and metals becomes unavoidable. Various anthropogenic activities lead to heavy metal enrichment, especially in developing countries where it has become a serious threat to the environment (Wang et al., 2001) by affecting the health of animals as well as humans. The pollution by heavy metals degrades the quality of water, soil and air, thus having deleterious effects on agricultural production. The governments around the world are duty bound to their citizens to advocate a safe environment free from pollution. However, these environmental issues are outweighed by other concerns for the countries' economic, agricultural and industrial development for ever-increasing populations. Thus this prioritization in a particular direction becomes the driving force actually responsible for environmental pollution (Ikhuoria and Okieimen, 2000). Plants play a pivotal role in transforming solar energy to green energy and have tremendous potential to remediate soil from heavy metals. The conventional methods available currently for the remediation of heavy-metal-contaminated soils are expensive and are not necessarily eco-friendly. However, phytoremediation is a novel 'green technology' which utilizes plants' potential to restore the health of the environment. The conclusions drawn on the basis of fundamental and applied research is that the plants possess tremendous potential for eliminating, degrading or neutralizing a variety of heavy metal toxicants. The wonderful nature of this technology is in its cost effectiveness, simplicity, sustainability, environmental compatibility and the fact that it is more aesthetically attractive than the conventional classical technologies. It can be

Soil Remediation and Plants. http://dx.doi.org/10.1016/B978-0-12-799937-1.00006-1

implemented in situ to remediate large expanses of contaminated ground or to treat large volumes of dilute wastewater. To enhance the efficiency of phytoremediation it is necessary to understand the biology of the plant kingdom at the molecular level. This review is an attempt to summarize the recent approaches in phytoremediation and how genetic engineering plays an important role in phytoremediation technology.

PHYTOREMEDIATION TECHNOLOGIES

Phytoremediation is a plant-based technology broadly classified into four classes, each having its own mechanism of degrading heavy metals. Figure 6.1. These include the following.

Phytoextraction

Phytoextraction or phytomining is generally an innovative technology to produce bio-ore and restore the health of the soil. This technology involves planting hyperaccumulator species especially in heavy-metal-contaminated soils and later harvesting the biomass leading to the permanent heavy metals removal from the soil. This process can be termed as 'concentration technology' which produces smaller mass for disposal in comparison to conventional engineering including excavation and landfilling. Phytoremediation evaluation by Superfund Innovative Technology Evaluation (SITE) should demonstrate its amenability for recovery and recycling of contaminants. As compared to conventional soil remediation technologies, the cost involved in phytoextraction is more than

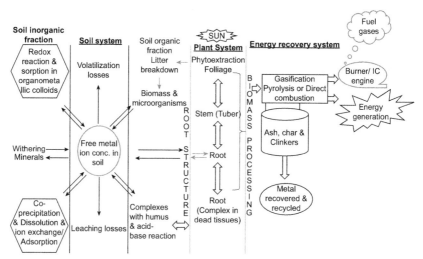

FIGURE 6.1 Soil, plant and green energy recovery system depicting the key components associated with the mass transfer and dynamics of phytoextraction.

10-times less per hectare and is therefore considered to be a cost-effective technology (Salt et al., 1995). This technology not only helps in producing bio-ore, but also controls leaching and soil erosion. Besides, the concentration of heavy metals can be reduced after successive cropping and harvesting (Vandenhove et al., 2001). The efficiency of this plant-based technology usually depends on following criteria: (1) site selection is of fundamental importance, i.e. it should be conducive; (2) it must possess the property to accumulate and scavenge bulk quantity of heavy metals (McGrath, 1998); (3) plants must possess fast growth rate and produce large amounts of biomass; (4) bioavailability of heavy metals so that it gets easily absorbed by the plants; (5) acceptable land topography and cheap agronomic practices; (6) capacity to resist hardy soil conditions like soil pH, salinity, soil texture, water content, etc.; root depth also determines the efficiency of effective phytoextraction. Currently, the two main approaches of phytoextraction used include: (i) chelate-assisted phytoextraction or induced phytoextraction – this approach involves the use of artificial chelates in order to enhance the mobility of heavy metal ions so that they become amenable to plants; (ii) continuous phytoextraction – this approach involves the natural capability of plants to uptake and scavenge the toxicants (Salt et al., 1997). Currently research is ongoing to improve the efficiency of phytoextraction. Researchers all over the world target the novel hyperaccumulating plant species and strive to understand their biological pathways. Some plant families are reported to accumulate high concentrations of heavy metals which include *Asteraceae, Brassicaceae, Euphorbiaceae, Fabaceae, Flacourtiaceae* and *Violaceae* (Kumar et al., 1995). Among the *Brasicaceae* family, some species possess enormous potential to scavenge heavy metal ions and are thus considered as potential candidates for phytoextraction of metals. Generally, this family is known to scavenge metals like Pb, Cd, Zn and Ni (Robinson et al., 2009). Although *B. juncea* that belongs to the family *Brassicaceae* possess one-third of the concentration of zinc in its tissues, it is more efficient to remediate zinc than *T. caerulescens* (known zinc hyperaccumulator) and the reason behind this fact is that the biomass production of *B. juncea* is about 10-times more than *T. caerulescens* (Ebbs and Kochian, 1997). Experimental analysis was carried out in different Brassica species in order to check their ability to resist and accumulate heavy metals. These include: Indian mustard (*B. juncea* L.), kale (*B. oleracea* L.), turnip (*B. campestris* L.), black mustard (*B. nigra* Koch) and rape (*B. napus* L.) (Kumar et al., 1995). Among different *Brassica* species, Indian mustard (*B. juncea*) is the leading candidate crop to remediate numerous heavy metals from the soil which include: cadmium (Cd), chromium-IV (Cr^{IV}), caesium-137 (^{137}Cs) copper (Cu), nickel (Ni), lead (Pb), uranium (U) and zinc (Zn) (Jiang et al., 2000). In another study, scientists at the University of Florida have discovered that the Chinese brake fern, *P. vittata*, possesses efficient capacity to hyperaccumulate arsenic metal. This fern, when planted at a wood-preserving site containing soil loaded with arsenic (18.8–1603 ppm), exhibits the highest accumulation of arsenic heavy metal (3280–4980 ppm) in its tissues (Ma et al., 2001). Some plant species also have the potential to decontaminate the soil from radionuclides and the best example is Sunflower (*H. annus*). Experimental analysis revealed that the level of radionuclides such as caesium-137 and strontium-90 in water declines by 90% within a 2-week time period; besides, the highest concentration of these radionuclides was found in roots (8000 times) rather than water when flowers were planted as a demonstration of phytoremediation in a pond contaminated with these radioactive elements (Chernobyl nuclear disaster in the Ukraine). Another study carried out by Phytotech for the US Department of Energy, shows that *H. annus* has tremendous potential to diminish the uranium level at the site from 350 parts per billion to 5 parts per billion, thus achieving about 95% reduction within 24 hours (Schnoor, 1997). Currently very few plants are known for phytoextraction although they possess few desirable traits, but with the help of biotechnological approaches transgenic plants can be developed which possess all the traits essential for phytoextraction technology.

Phytovolatilization

Phytovolatilization generally involves biomethylation of some obnoxious and heavy metals such as selenium, arsenic and mercury and transforming them into volatile molecules that can easily be lost into the atmosphere. This technology is generally applied to remediate groundwater, soil, sediments and sludges. Formerly only microorganisms were known to play this role (Karlson and Frankenberger, 1989), but recently it has been discovered that plants (*B. juncea, B. napus*) also possess an excellent ability to perform the same function (Terry et al., 1992). A few aquatic plant species, viz. Azolla, rabbit foot grass, rice, cattail (*Typha latifolia* L.) and pickle weed, also have the ability to act as the best volatilizers of selenium (Zayed et al., 2000). Nowadays research is focussed on selenium volatilization due to its deleterious effects in many parts of the world. Generally, phytovolatilization has been utilized to decontaminate soil from mercury as it converts mercuric ions into less toxic elemental mercury. The mechanism behind selenium phytoremediation involves assimilation of inorganic Se and later converting it into organic forms such as selenoaminoacids, selenocysteine (SeCys) and selenomethionine (SeMet). These compounds are further processed and can be methylated to form dimethylselenide (DMSe), which is volatile in nature (Terry et al., 2000). Microorganisms have the ability to transform mercury (II) the known toxic element into organomercurials which, in turn, changes into elemental mercury (less toxic) under optimal conditions (Bizily et al., 2000) and later enters into the global biogeochemical cycle upon volatilization. As the process of volatilization does not require plant harvesting and disposal, it becomes an attractive and cheap remediation technology (Rugh et al., 2000). Meagher et al. (2000) carried out a risk assessment study for volatile selenium and mercury and the results revealed that these pollutants were dispersed and diluted in such a way that they did not pose a threat to the ecosystem. Although the pace of the phytovolatilization process is very slow, it can be enhanced by using novel plant species with high rates of transpiration, overexpression of enzymes such as cystathionine-V-synthase that mediates S/Se volatilization (Van Huysen et al., 2003) and also by means of transgenic technology to produce transgenic plants with efficient selenium volatilization (Le Duc et al., 2004). Recently, a bacterial Hg ion reductase gene was incorporated into plants to produce a transgenic plant which shows excellent mercury volatilization. In addition, it has also been reported that bacterial organomecurial lyase (*merB*) and mercuric reductase (*merA*) genes were incorporated into model plants such as *A. thaliana* and *N. tabacum*; the resulting transgenic plants have the potential to absorb elemental mercury (II) as well as methyl mercury from the soil and transform it into a volatile form (Hg^0) (Heaton et al., 1998). Transgenic yellow poplar (*Liriodendron tulipifera*) plantlets were produced which exhibited resistance to and grew well in normally toxic levels of ionic mercury, and the transformed plantlets volatilized about 10-times more elemental mercury than non-transgenic plantlets (Rugh et al., 1998). Moreover it has also been revealed that the addition of Hg^0 into the atmosphere would not contribute significantly to the atmospheric pool. Higher plants release the volatile selenium and it was first reported by Lewis et al. (1966). Few members of *Brassicaceae* family are known to release 40 g of selenium ha^{-1}/day as a volatile compound. Volatilization of these toxic metals is essential in various aspects such as the volatile form of selenium (dimethylselenide) is less toxic than inorganic forms of selenium found in soil (DeSouza et al., 2000). Furthermore, it also involves the permanent removal of these toxicants (Atkinson et al., 1990). It is also recommended that phytovolatilization is not applicable for the sites near population centres or at places with unique meteorological conditions as it promotes the rapid deposition of volatile compounds (Rugh et al., 2000). Although this technology is a promising tool for the remediation of selenium and mercury, at the same time it has some demerits like loss of control of volatile compounds over their migration to other areas. Currently by means of this technology, tritium (^3H), a radioactive isotope of hydrogen is decayed to stable helium with a half-life of about 12 years as reported Dushenkov et al. (1995).

Phytostabilization

Phytostabilization, also known as phytorestoration or in-place inactivation, is generally used for the remediation of soil, sediment and sludges (United States Protection Agency, 2000). One of the important applications of this technology is to stabilize the toxicants and prevent exposure to wind and water erosion. In addition, it provides hydraulic control to suppress the migration of these toxicants into the water table and thus physiochemically immobilizes contaminants by root sorption, precipitation, complex action or metal valence reduction with different soil amendments (Berti and Cunningham, 2000). The main aim of this technique is not to restore sites from toxicants but rather to stabilize them and reduce the risks to the ecosystem. It is usually meant for those soils contaminated with Zn/As/Cr/Cd/Pb/Cu. Generally the role of plants in this technique is to reduce the amount of water percolating through the soil matrix that will ultimately lead to the formation of toxicants (hazardous leachates) and avoid soil erosion and transport of toxic metals to distinct areas. The plants chosen for this technology should obey certain criteria: (1) efficiency of translocation of heavy metals from root-to-shoot system should be low; (2) they should have a fast growth rate and resistance to heavy metals and (3) there should be cost-effective management. Practically this technique is not only applicable at sites with high organic load and porosity but is also effective for a wide range of surface contamination sites (Berti and Cunningham, 2000). One of the drawbacks of this technique is that it is not applicable to those areas which are heavily contaminated because such conditions become a hindrance in plant growth and development (Berti and Cunningham, 2000). Besides, the contaminant remains in soil as it is, and therefore requires regular monitoring.

Rhizofiltration

This phytoremediation technique is usually intended to restore aquatic ecosystems. It involves the use of both aquatic as well as terrestrial plants to treat industrial discharge, agricultural runoff or acid mine drainage. It is mainly used for the removal of Ni, Cd, Cu, Zn, Cr and Pb which are primarily retained within the roots (United States Protection Agency, 2000). Formerly some chemical- and mechanical-based techniques such as precipitation, flocculation followed by sedimentation were utilized for the removal of industrial pollutants. However, there are certain limitations to this conventional technology as it is not eco-friendly and cost-effective. Thus, the rhizofiltration technique which is sustainable, eco-friendly and cost-effective has gained worldwide importance to restore contaminated sites. In this process, plants are raised hydroponically and later transplanted into contaminated sites in order to absorb and concentrate the toxicants in their roots and shoots (Zhu et al., 1999b). It has also been reported that some chemicals ooze out of the roots that lead to a change in rhizosphere pH which ultimately causes metals to precipitate on to the root surface. Once roots get saturated with these toxicants, either only the roots or the whole plants are harvested for further processing (Zhu et al., 1999b). This process is not feasible in those areas where metal concentration is high in water because the contaminants should be in the solution form in order to be sorbed into the plant system. Other complications associated with this technique involve drying, composting or incineration. For efficient removal of contaminants from the site, plants must possess efficient and fast-growing root systems with the ability to scavenge these toxicants and thus neutralize their harmful effects. Low maintenance costs, easy handling and resistance of plants to heavy metals are other important criteria for effective removal of heavy metals from a particular area. A number of aquatic plant species show a better response in the removal of toxicants from water. These include: water hyacinth (*Eichhornia crassipes*

Mart.; Zhu et al., 1999c), pennywort (*Hydrocotyle umbellata* L; Dierberg et al., 1987) and duckweed (*Lemna minor* L; Mo et al., 1989). However, there are certain limitations to these plants as they have limited potential for rhizofiltration, because of their small, slow-growing root systems (Dushenkov et al., 1995). Zhu et al. (1999c) reported that water hyacinth is a potential candidate for the removal of trace elements from waste streams. A number of plants such as tobacco, corn, rye, spinach, Indian mustard and sunflower have been studied for their ability to remove Pb from effluents and the results revealed that sunflower has the greatest potential to detoxify Pb content. Experimental evidences show that terrestrial plants with dense and fibrous root systems are suitable for this technique as they possess greater metal-absorbing powers, and the leading examples include Sunflower (*Helianthus annuus* L.) and Indian mustard (*Brassica juncea* Czern.). Indian mustard is known to remove a wide concentration of Pb (4–500 mg l^{-1}) (Raskin and Ensley, 2000). These terrestrial plants are also being used to remove various metals such as Cd, Zn, Cu, Ni and Cr (Dushenkov et al., 1995), uranium (Dushenkov et al., 1997a) and ^{90}Sr (Dushenkov et al., 1997b) from hydroponic solutions. Currently, researchers are in the process of improving the efficiency of rhizofiltration technology in order to decontaminate the soil. Salt et al. (1995) carried out an experimental analysis and concluded that young seedlings exhibit better abilities to remove heavy metals from water. This technique of using young plant seedlings in order to remove toxicants (heavy metals) from water is known as blastofiltration (blasto is 'seedling' in Greek) and is considered to be second-generation, plant-based water treatment technology. In this technique there is a dramatic enhancement in surface to volume ratio that usually occurs after germination and some germinating seedlings also ab/adsorb huge quantities of toxic metal ions; this is why young seedlings are suitable for restoring water quality. Indian mustard is considered to be a potential crop for blastofiltration as it is effective in sorbing divalent cations of toxic metals (Salt et al., 1997) due to some unique features of Indian mustard such as fast growth rate and resistance to heavy metals or microbial infection. Through data analysis, it has been reported that for a few metals, blastofiltration techniques are more potent and economical than rhizofiltration; however the main advantage of rhizofiltration is that it can be utilized both in situ as well as ex situ and species other than hyperaccumulators can also be taken into account.

GENETIC ENGINEERING TO IMPROVE PHYTOREMEDIATION

Genetic engineering is one of the important approaches that can be used as an alternative to improve the phytoremediation potential of plant species with high biomass production. It was very hard to develop general plant productivity, which is regulated by several genes, by the insertion of only one gene. Biotechnological approaches (genetic engineering) are used to insert more effective accumulator genes into other plants, and these methods have been recommended by various authors (Chaney et al., 2000). This approach of inserting many effective accumulator genes from taller plants into other natural plants enhanced the final biomass. Zhu et al. (1999a) studied the transgenic *Brassica juncea* for rate-limiting factors by inserting gshl-gene from *Escherichia coli* for glutathione and phytochelatin production. The γ-ECS genetically engineered seedlings exhibited improved resistance to Cd and contained greater amounts of phytochelatins, γ-GluCys, glutathione and totally non-proteinthiols when compared to wild-type seedlings. There are several factors that restrict the achievement of genetic engineering and one of the limiting factors is the anatomical restraint (Ow, 1996). Commercial utilization of phytoremediation can be increased by transforming metal-scavenging properties of hyperaccumulating plants, such as *T. caerulescens,* to high-biomass generating species, such as Indian mustard (*Brassica juncea*) or maize (*Zea mays*) (Brown et al., 1995). At present, genetic engineering is successfully utilized to manipulate metal uptake and stress resistance properties in various species. For example, enhanced metal resistance in tobacco (*Nicotiana tabacum*) was achieved by expressing the mammalian metallothionein metal binding proteins (Maiti et al., 1991). Currently, genetic engineering plays a pivotal role in the production of transgenic plant species that have tremendous potential to remediate soil contaminated with methyl mercury (a neurotoxic agent). Examples are transgenic Tobacco and Arabidopsis which express bacterial genes *merB* and *merA*, and have the capacity to remove mercury from the soil. The *merB* gene present in these transgenic plants has the capacity to carry out protonolysis of the carbon–mercury bond and thus liberate Hg^{2+}, a reduced mobile mercury species, and the *MerA* gene converts Hg^{2+} to Hg^0 (elemental-Hg) a less toxic, volatile element which is liberated into the atmosphere (Rugh et al., 1996). Despite the fact that regulatory concerns have confined the application of genetically engineered plants transformed with *merA* and *merB,* this work demonstrates the enormous capability of genetic engineering for restoration of the environment. In order to declaim these regulatory concerns related to phytovolatilization of mercury, Bizily et al. (1999) illustrated genetically engineered plants that express *MerBpe* (an organomercurial lyase under the control of a plant promoter) employed to decompose methyl mercury and consequently take-off ionic mercury through extraction. In spite of current progress in biotechnology, less is known about the eugenics of metal hyperaccumulation in plants. Specifically, the genetics pertinent to plant mechanisms, such as metal transport and storage (Lasat et al., 2000) and metal resistance (Ortiz et al., 1992) must be clearly explained. Chaney et al. (1999) recently suggested the application of conventional breeding processes for enhancing metal-hyperaccumulating species and bioengineering by inducing some important traits such as metal resistance and uptake characteristics into high-biomass-yielding plants. Complete success has not been reported in the literature. For example, Brewer et al. (1997) in an investigation attempted to correct the small size of hyperaccumulating plants by producing somatic hybrids between

T. caerulescens (a Zn hyperaccumulator) and *Brassica napus* (canola) followed by hybrid selection for Zn resistance. High biomass hybrids with extreme Zn resistance were obtained. These researchers have also made an effort to collect and conserve germplasm of hyperaccumulating species. Comprehensively, any dicotyledonous plant species can be modified by employing the Agrobacterium vector system, while as the majority of the monocotyledonous plants can be reconstructed employing particle gun or electroporation methods. Several feasible methods have been developed that describe greater resistance, concentration and/or degradation capabilities of several adulterants. These were carried out either by the overproduction of citrate (metal chelating molecule) (de la Fuente et al., 1997), phytochelatins (PCs) (Zhu et al., 1999a, b), metallothioneins (MTs) (Hasegawa et al., 1997) or ferritin (Goto et al., 1999) or by overexpression of metal transporter proteins (Curie et al., 2001). Using these bioengineering techniques in order to enhance the metal concentration in plants in turn increases the phytoremediation capability by the same factor. However, it has not been fully investigated how appropriate these genetically modified plants are for environmental restoration, because only a little field work has been done so far, including one study using genetically modified Indian mustard plant that overproduces enzymes that catalyse the sulphate/selenate reduction process (Pilon-Smits et al., 1999). Also, it has been suggested by many researchers that the consequences of the genetically modified plants should be regularly observed, such as the competition between altered plants and wild ones, the influence on birds, insects, etc., that might feed on plant biomass exhibiting extreme amount of noxious metals, and also there is a probability of gene transfer to other plants through pollination. Bioengineering of the chloroplast genome put forward a unique alternative to acquire high expression and reduce the risk of gene transfer through pollination (Ruiz et al., 2003). Table 6.1 lists some of the transgenic plants, genes incorporated, gene sources and the enhanced trait of modified plants (Vinita, 2007). Recently, the transcriptomic studies on hyperaccumulators have provided novel insights into molecular mechanisms underlying metal resistance and concentration, as well as reaching into the classification of a huge pattern of genes which are constitutively (in the absence of excess of metallic ions) overexpressed in the hyperaccumulation trait (Filatov et al., 2006). Classic genetic studies and recent Quantitative trait loci (QTL) analysis of hyperaccumulators support the existence of only a few genes that are responsible for tolerance to accumulation (Willems et al., 2007). Genetic engineering of metal transporters also plays a vital role in phytoremediation, as these are extremely complex and diverse. HMA4 [heavy metal transporting *adenosine triphosphatase* (ATPase)] was the first plant P_{1B}- ATPase of the divalent transport group to be cloned and characterized in *A. thaliana* (Mills et al., 2003). The function of HMA4 has been described in *Arabidopsis thaliana* for Zn homeostasis, removal of Cd and in the transportation of these metals from the root to the shoot (Verret et al., 2005). In *A. halleri* and *T. caerulescens*, which are

TABLE 6.1 Some Genetically Engineered Plants with the Respective Gene Transferred, Gene Product, Gene Source and Improved Trait

S. no	Transgenic plant	Transgene and its product	Gene source	Trait enhanced	References
1.	S. cerevisiae	Cup1 gene: MT's	A. thaliana	Zn & Cu tolerance	Robinson et al., 1996
2.	S. cerevisiae	TaPCSI gene: PC's	Wheat	Cd tolerance	Vatamaniuk et al., 1999
3.	N. tobaccum and Liriodendron tulipifera	merA: Hg(II) reductase	Gram–ve bacteria	Hg tolerance and volatilization	Heaton et al., 1998
4.	A. thaliana	merA: Hg(II) reductase merB: organomercurial lyase	Gram–ve bacteria	Hg tolerance and volatilization	Rugh et al., 1996
5.	Brassica juncea	APs gene: ATP-sulphurylase	A. thaliana	Se hyperaccumulation	Banuelos et al., 2005
6.	Lycopersicon esculentum	ACC gene: 1-amino-cyclopropane 1-carboxilic acid deaminase	Bacteria	Cd, Co, Cu, Mg, Ni, Pb, Zn tolerance	Gricheko et al., 2000
7.	A. thaliana	PsMTA gene: MTs	Peas	Cu tolerance	Evans et al., 1992
8.	Brassica juncea	gshl gene: g-glutamylcysteine	E.coli	Cd tolerance	Zhu et al., 1999a, b
9.	N. tobaccum	MT-1 gene: MT's	Mouse	Cd tolerance	Thomine et al., 2000
10.	N. Tobaccum Carcia papaya	Citrate synthase gene	Pseudomonas aeruginosa	High levels of citrate	de la Fuente et al., 1997
11.	N. tobaccum	FRE1, FRE2: Ferric reductase	S. cerevisiae	Ni hyperaccumulation	Samuelsen et al., 1998

Adapted from Vinita (2007).

considered to be Cd/Zn hyperaccumulators, the HMA4 shows overexpression both in roots and shoots in comparison to Cd-/Zn-sensitive close relatives (Courbot et al., 2007). Two distinct autonomously produced Zn/Cd hyperaccumulating plant species with increased expression of HMA4 have been found, powerfully supporting the concept that HMA4 plays a vital role in resistance to both metals. In *A. halleri* this presumption is supported by the QTL analysis of Cd and Zn resistance indicating the co-localization of major QTLs of Cd and Zn resistance with HMA4 gene (Courbot et al., 2007). At present there is genetic evidence for only one gene, i.e. HMA4, which has a function in both Zn and Cd resistance. HMA4 is involved in the detoxification of roots by transporting Zn and Cd to the shoots as observed in AtHMA4-overexpressing lines (Verret et al., 2004), which translocate more Zn and Cd to the shoot and also show resistance to hyperaccumulation of Zn and Cd. The enhanced expression of HMA4 in *A. halleri* or *T. Caerulescens* may be related to the enhanced resistance to Zn and Cd and the hyperaccumulation phenotype of these species (Willems et al., 2007). Several ZIP (zinc regulated transporters, iron regulated transporter proteins) members are overexpressed in *A. halleri and T. caerulescens* (ZIP 3, 4, 6, 9 and 10 and IRT3) and are supposed to function by increasing root metal uptake of those Zn hyperaccumulators (Kramer et al., 2007). IRT1, one of the first-discovered members of the large ZIP family, is involved in the uptake of Fe^{2+} from the soil. Plants overexpressing IRT1 engineered to improve Fe nutrition also accumulated more Cd (Cannoly et al., 2002). Also some MATE (multi-drug and toxin extrusion) member proteins, e.g. FRD3, are constitutively overexpressed in *A. halleri* compared to its close relative and play an important role in Zn translocation (Talke et al., 2006). Overexpression of ZAT (zinc transporter of Arabidopsis Thaliana) in *A. thaliana* described extreme Zn resistance and a two-fold enhanced Zn concentration in roots (Van der Zaal et al., 1999). Manganese-resistant *Strylosanthes hamate* (a tropical legume) includes ShMTP in the vacuolar storage of manganese and when overexpressed in *A. thaliana* confirmed greater resistance and concentration of manganese (Delhaize et al., 2003). Using CAX (CAtioneXchanger) under the influence of the constitutive 35S CaMV promoter it has been revealed that the expression of AtCAX2 and AtCAX4 in root vacuoles of *A. thaliana* resulted in greater deposition of Cd (Korenkov et al., 2007). Another study described that the overexpression of SMT (Se cysMethylTransferase) from Se-hyperaccumulator *A. bisulcatus* in *A. thaliana* and Indian mustard resulted in the considerable enhancement of Se resistance, concentration and volatilization in both species (Le Duc et al., 2006). Interestingly, constitutive expression of existing genes seems to be a general process in the acclimatization of plants into the worst environmental conditions (Taji et al., 2004). Novel powerful tools to control the timing, the location and the level of the transgene expression are expected from the characterization of native promoters of hyperaccumulators. Future studies require more information from fieldwork and risk assessment monitoring along with biotechnological techniques; this will directly result in the commercializing of phytoremediation on a large scale.

CONCLUSIONS AND FUTURE PERSPECTIVES

Phytoremediation is currently an emerging technology and is a fast-growing field involving the approaches corresponding with green technology. Scientists as well as the general masses around the world are concerned about the changing environmental conditions, and since the last decade we have witnessed the initiation of many field applications including the phytoremediation of organic, inorganic and radionuclides. Phytoremediation, being a cost-effective, sustainable and eco-friendly technology, is one of the feasible alternatives, especially in developing countries. This technology can be applied in situ to immediate shallow soil, ground water and runoff water bodies. Currently, phytoremediation technology is in its infancy and there are many technical hindrances that need to be addressed. However, with the help of modern innovative agronomic practices and genetic engineering approaches to its efficiency, this green technology would be enhanced. Many hyperaccumulator plants remain to be discovered and there is an urgent need to explore more about their physiology and molecular biology. Optimization of the process, proper understanding of the dynamics of plant–heavy metal interactions, plant–microbe interactions and proper disposal of biomass produced is still needed.

REFERENCES

Atkinson, R., Aschmann, S.M., Hasegawa, D., Eagle-Thompson, E.T., Frankenberger Jr., W.T., 1990. Kinetics of the atmospherically important reactions of dimethyl selenide. Environ. Sci. Technol. 24, 1326–1332.

Banuelos, Terry, D.N., Leduc, D.L., Pilon Smits, E.A., Mackey, B., 2005. Field trial of transgenic Indian mustard plants shows enhanced phytoremediation of selenium contaminated sediment. Environ. Sci. Technol. 39 (6), 1771–1777.

Bennett, L.E., Burkhead, J.L., Hale, K.L., Terry, N., Pilon, M., Pilon-smits, E.A.H., 2003. Analysis of transgenic Indian Mustard plants for phytoremediation of metals-contaminated mine tailings. J. Environ. Qual. 32, 432–440.

Berti, W.R., Cunningham, S.D., 2000. Phytostabilization of metals. In: Raskin, I., Ensley, B.D. (Eds.), Phytoremediation of toxic metals – using plants to clean-up the environment. John Wiley & Sons, Inc, New York, pp. 71–88.

Bizily, S.P., Clayton, L.R., Summers, A.O., Meagher, R.B., 1999. Phytoremediation of methylmercury pollution: *merB* expression in *Arabidopsis thaliana* confers resistance to organomercurials. Proc. Natl. Acad. Sci. 96, 6808–6813.

Bizily, S.P., Rugh, C.L., Meagher, R.B., 2000. Phytodetoxification of hazardous organomercurials by genetically engineered plants. Nat. Biotechnol. 18, 213–217.

Brewer, E.P., Saunders, J.A., Angle, J.S., Chaney, R.L., McIntosh, M.S., 1997. Somatic hybridization between heavy metal hyperaccumulating *Thlaspi caerulescens* and canola. Agron Abstr. 154.

Brown, S.L., Chaney, R.L., Angle, J.S., Baker, A.J.M., 1995. Zinc and cadmium uptake by hyperaccumulator *Thlaspi caerulescens* grown in nutrient solution. Soil Sci. Soc. Am. J. 59, 125–133.

Cannoly, E.L., Fett, J.P., Guerinot, M.L., 2002. Expression of the IRT1 metal transporter is controlled by metals at the levels of transcript and protein accumulation. Plant Cell 14, 1347–1357.

Chaney, R.L., Li, Y.M., Angle, J.S., Baker, A.J.M., Reeves, R.D., Brown, S.L., Homer, F.A., Malik, M., Chin, M., 1999. Improving metal hyperaccumulators wild plants to develop commercial phytoextraction systems: Approaches and progress. In: Terry, N., Bañuelos, G.S. (Eds.), Phytoremediation of Contaminated Soil and Water. CRC Press, Boca Raton, FL.

Chaney, R.L., Li, Y.M., Angle, J.S., Baker, A.J.M., Reeves, R.D., Brown, S.L., Homer, F.A., Malik, M., Chin, M., 2000. In: Terry, N., Banelos, G. (Eds.), Phytoremediation of contaminated soil and water. Lewis Publishers, Boca Raton, FL, pp. 129–158.

Courbot, M., Willems, G., Motte, P., Arvidsson, S., Roosens, N., Saumitou-laprade, P., Verbrugger, N., 2007. A major quantitative trait locus for cadmium tolerance in *Arabidopsis halleri* colocalizes with HMA4, a gene encoding a heavy metal ATPase. Plant Physiol. 144, 1052–1065.

Curie, C.Z., Panaviene, C., Loulergue, S.L., Dellaporta, J.F., Briat, Walker EL., 2001. Maize yellow stripe 1 encodes a membrane protein directly involved in Fe(III) uptake. Nature 409, 346–349.

de la Fuente, J.M., Ramírez-Rodriguez, V., Cabrera-Ponce, J.L., Herrera-Estrella, L., 1997. Aluminum tolerance in transgenic plants by alteration of citrate synthesis. Science 276, 1566–1568.

Delhaize, E., Kataoka, T., Hebb, D.M., White, R.G., Ryan, P.R., 2003. Genes encoding proteins of the cation diffusion facilitator family that confer manganese tolerance. Plant Cell 15, 1131–1142.

DeSouza, M.P., Pilon-Smits, E.A.H., Terry, N., 2000. The physiology and biochemistry of selenium volatilization by plants. In: Raskin, I., Ensley, B.D. (Eds.), Phytoremediation of Toxic Metals – Using Plants to Clean-Up the Environment. John Wiley & Sons, Inc., New York, pp. 171–190.

Dierberg, F.E., DeBusk, T.A., Goulet Jr., N.A., 1987. In: Reddy, K.R., Smith, W.H. (Eds.), Aquatic Plants for Water Treatment and Resource Recovery. Magnolia Publishing, Orlando, FL, pp. 497–504.

Dushenkov, V., Kumar, P.B.A.N., Motto, H., Raskin, I., 1995. Rhizofiltration: The use of plants to remove heavy metals from aqueous streams. Environ. Sci. Technol. 29, 1239–1245.

Dushenkov, S., Vasudev, D., Kapulnik, Y., Gleba, D., Fleisher, D., Ting, K.C., Ensley, B., 1997a. Removal of uranium from water using terrestrial plants. Environ. Sci. Technol. 31, 3468–3474.

Dushenkov, S., Vasudev, D., Kapolnik, Y., Gleba, D., Fleisher, D., Ting, K.C., Ensley, B., 1997b. Environ. Sci. Technol. 31 (12), 3468–3476.

Ebbs, S.D., Kochian, L.V., 1997. Toxicity of zinc and copper to *Brassica* species: Implications for phytoremediation. J. Environ. Qual. 26, 776–781.

Evans, K.M., Gatehouse, J.A., Lindsay, W.P., Shi, J., Tommey, A.M., Robinson, N.J., 1992. Expression of the pea metallothionein like gene PsMTA in Escherichia coli and Arabidopsis thaliana and analysis of trace metal ion accumulation: implications for gene PsMTA function. Plant Mol. Biol. 20, 1019–1028.

Filatov, V., Dowdle, J., Smirnoff, N., Ford-Llyod, B., Newburry, H.J., Macnain, M.M., 2006. Comparison of genes expression in segregating families identifies genes and genomic regions involved in a novel adaptation, zinc hyperaccumulation. Mol. Ecol. 15, 3045–3059.

Goto, F., Yoshihara, T., Shigemoto, N., Toki, S., Takaiwa, F., 1999. Iron fortification of rice seed by the *soybean ferritin gene*. Nature Biotechnol. 17, 282–286.

Gricheko, V.P., Filby, B., Glick, B.R., 2000. Increased ability of transgenic plants expressing the bacterial enzyme ACC deaminase to accumulate Cd, Co, Cu, Ni, Pb and Zn. J. Biotech. 81, 45–53.

Hasegawa, I., Terada, E., Sunairi, M., Wakita, H., Shinmachi, F., 1997. Genetic improvement of heavy metal tolerance in plants by transfer of the yeast metallothionein gene (CUP1). Plant Soil 196, 277–281.

Heaton, A.C.P., Rugh, C.L., Wang, N., Meagher, R.B., 1998. Phytoremediation of mercury and methyl mercury polluted soils using genetically engineered plants. J. Soil Contam. 7 (4), 497–509.

Ikhuoria, E.U., Okieimen, F.E., 2000. Scavenging Cadmium, copper, lead, nickel and zinc ions from aqueous solution by modified cellulosic sorbent. Int. J. Environ. Stud. 57 (4), 401.

Jiang, W., Liu, D., Hou, W., 2000. Hyperaccumulation of lead by roots, hypocotyls, and shoots of *Brassica juncea*. Biol. Plantarum 43 (4), 603–606.

Karlson, U., Frankenberger, W.T., 1989. Accelerated rates of selenium volatilization from California soils. Soil Sci. Soc. Am. J. 53, 749–753.

Korenkov, V., Park, S.H., Chang, N.H., Sreevidya, C., Jlachmansingh, J., Hirschi, K., Wagner, G.J., 2007. Enhanced Cd^{2+} selective root-transport-transport in tobaccos expressing Arabidopsis cation exchangers. Planta 225, 403–411.

Kramer, U., Talke, I., Hanikenne, M., 2007. Transition metal transport. FEBS Lett. 581, 2263–2272.

Kumar, P.B.A.N., Dushenkov, V., Motto, H., Raskin, I., 1995. Phytoextraction: The use of plants to remove heavy metals from soils. Environ. Sci. Technol. 29, 1232–1238.

Lasat, M.M., Pence, N.S., Garvin, D.F., Ebbs, S.D., Kochian, L.V., 2000. Molecular physiology of zinc transport in the Zn hyperaccumulator Thlaspi caerulescens. J. Exp. Bot. 51 (342), 71–79.

Le Duc, D.L., Tarun, A.S., Montes-Bayon, M., Meija, J., Malit, M.F., 2004. Overexpression of selenocysteine methyltransferase in Arabidopsis and Indian mustard increases selenium tolerance and accumulation. Plant Physiol. 135, 377–383.

LeDuc, D.L., Abdelsamie, M., Montes-Bayon, M., Wu, C.P., Reisinger, S.J., Terry, N., 2006. Overexpressing both ATPsulfurylase and selenocysteine methyl transferase enhances selenium phytoremediation traits in Indian mustard. Environ. Pollut. 144, 70–76.

Lewis, B.G., Johnson, C.M., Delwiche, C.C., 1966. Release of volatile selenium compounds by plants: Collection procedures and preliminary observations. J. Agric. Food Chem. 14, 638–640.

Ma, L.Q., Komar, K.M., Tu, C., 2001. A fern that accumulates arsenic. Nature 409, 579.

Maiti, I.B., Wagner, G.J., Hunt, A.G., 1991. Light inducible and tissue specific expression of achimeric mouse metallothionein cDNA gene in tobacco. Plant Sci. 76, 99–107.

McGrath, S.P., 1998. Phytoextraction for soil remediation. In: Brooks, R.R. (Ed.), Plants that hyperaccumulate heavy metals: Their role in phytoremediation, microbiology, archeology, mineral exploration and phytomining. CAN International, New York, pp. 261–288.

Meagher, R.B., Rugh, C.L., Kandasamy, M.K., Gragson, G., Wang, N.J., 2000. Engineered phytoremediation of mercury pollution in soil and water using bacterial genes. In: Terry, N., Banuelos, G. (Eds.), Phytoremediation of contaminated soil and water. Lewis, Boca Raton, pp. 201–221.

Mills, R.F., Krijger, G.C., Baccarini, P.J., Hall, J.L., Williams, L.E., 2003. Functional expression of AtHMA4, a p-13-type ATPase of the Zn/Co/Cd/Pb subclass. Plant J. 35, 164–176.

Mo, S.C., Choi, D.S., Robinson, J.W., 1989. Uptake of mercury from aqueous solution by duckweed: The effect of pH, copper, and humic acid. J. Environ. Health Sci. A. 24, 135–146.

Ortiz, D.F., Kreppel, L., Speiser, D.M., Scheel, G., McDonald, G., Ow, D.V., 1992. Heavy metal tolerance in the fission yeast requires an ATP-binding cassette-type vacuolar membrane transporter. EMBO J. 11, 3491–3499.

Ow, D.W., 1996. Heavy metal tolerance genes—prospective tools for bioremediation. Res. Conserv. Recycling 18, 135–149.

Pilon-Smits, E.A.H., de Souza, M.P., Hong, G., Amini, A., Bravo, R.C., 1999. Selenium volatilization and accumulation by twenty aquatic plant species. J. Environ. Qual. 28, 1011–1017.

Raskin, I., Ensley, B.D., 2000. Phytoremediation of Toxic Metals: Using Plants to Clean Up the Environment. John Wiley & Sons, Inc., New York. 53–70.

Robinson, N.J., Wilson, J.R., Turner, J.S., 1996. Expression of the type 2 metallotienin-like gene MT2 from *Arabidopsis thaliana* in Zn2+- metallothionein deficient *Synechococcus* PCC 7942: Putative role for MT2 in Zn2+-metabolism. Plant Mol. Biol. 30, 1169–1179.

Robinson, B.H., Banuelos, G., Conesa, H.M., Evangelon, W.H., Schulin, R., 2009. The phytomanagement of trace elements in soil. Crit. Rev. Plant Sci. 28 (4), 240–266.

Rugh, C.L., Bizily, S.P., Meagher, R.B., 2000. Phytoreduction of environmental mercury pollution. In: Raskin, I., Ensley, B.D. (Eds.), Phytoremediation of Toxic Metals – Using Plants to Clean-Up the Environment. John Wiley & Sons, Inc., New York, NY, pp. 151–170.

Rugh, C.L., Gragson, G.M., Meagher, R.B., Merkle, S.A., 1998. Toxic mercury reduction and remediation using transgenic plants with a modified bacterial gene. Hort. Sci. 33 (4), 618–621.

Rugh, C.L., Wilde, H.D., Stack, N.M., Thompson, D.M., Summers, A.O., Meagher, R.B., 1996. Mercuric ion reduction and resistance in transgenic *Arabidopsis thaliana* plants expressing a modified bacterial *merA* gene. Proc. Natl. Acad. Sci. USA 93, 3182–3187.

Ruiz, O.N., Hussein, H.S., Terry, N., Daniell, H., 2003. Phytoremediation of organomercurial compounds via chloroplast genetic engineering. Plant Physiol. 132, 1344–1352.

Salt, D.E., Blaylock, M., Kumar, N.P.B.A., Dushenkov, V., Ensley, B.D., Chet, I., Raskin, I., 1995. Phytoremediation: A novel strategy for the removal of toxic metals from the environment using plants. Biotechnol. 13, 468–475.

Salt, D.E., Pickering, I.J., Prince, R.C., Gleba, D., Dushenkov, S., Smith, R.D., Raskin, I., 1997. Metal accumulation by aquacultured seedlings of Indian Mustard. Environ. Sci. Technol. 31 (6), 1636–1644.

Samuelsen, A.I., Martin, R.C., Mok, D.W.S., Machteld, C.M., 1998. Expression of the yeast FRE genes in transgenic tobacco. Plant Physiol. 118, 51–58.

Schnoor, J.L., 1997. Phytoremediation. University of Lowa, Department of Civil and Engineering. 1, 62.

Taji, T., Seki, M., Satou, M., Sakurai, T., Kobayashi, M., Ishiyama, K., Narusaka, M., Zhu, J.K., Shinozaki, K., 2004. Comparative genomics in salt tolerance between Arabidopsis and Arabidopsis- Related halophyte salt cress using Arbidopsis microarray. Plant Physiol. 135, 1697–1709.

Talke, I., Hanikenne, M., Kramer, U., 2006. Zinc dependent global transcriptional control, transcriptional de-regulation and higher gene copy number for genes in metal homeostasis of the hyperaccumulator *Arabidopsis halleri*. Plant Physiol. 142, 148–167.

Terry, N., Carlson, C., Raab, T.K., Zayed, A., 1992. Rates of selenium volatilization among crop species. J. Environ. Qual. 21, 341–344.

Terry, N., Zayed, A.D., de Souza, M.P., Tarun, A.S., 2000. Selenium in higher plants. Annu. Rev. Plant Physiol. Plant Mol. Biol. 51, 401–432.

Thomine, S., Wang, R., Ward, J.M., Crawford, N.M., Schroeder, J.I., 2000. Cadmium and iron transport by members of a plant metal transporter family in *Arabidopsis* with homology to *Nramp* genes. Proc. Natl. Acad. Sci. USA 97, 4991–4996.

United States Protection Agency (USPA), 2000. Introduction to Phytoremediation. EPA 600/R-99/107. U.S. Environmental Protection Agency, Office of Research and Development, Cincinnati, OH.

Van der Zaal, B.J., Neuteboom, L.W., Pinas, J.E., Chardonnens, A.N., Schat, H., 1999. Overexpression of a novel Arabidopsis gene related to putative zinc-transporter genes from animals can lead to enhanced zinc resistance and accumulation. Plant Physiol. 119, 1047–1055.

Van Huysen, T., Abdel Ghany, S., Hale, K.L., Le Duc, D., Terry, N., Pilon Smits, E.A.H., 2003. Overexpression of cystathionine synthase enhances selenium volatilization in *Brassica juncea*. Planta 218, 71–78.

Vandenhove, H., van Hees, M., van Winkel, S., 2001. Feasibility of phytoextraction to clean up low-level uranium-contaminated soil. Int. J. Phytoremediation 3, 301–320.

Vatamaniuk, O.K., Mari, S., Lu, Y.P., Rea, Y.A., 1999. AtPCS1, a phytochelatin synthase from *Arabidopsis:* Isolation and in vitro reconstruction. Proc. Natl. Acad. Sci. USA 96, 7110–7115.

Verret, F., Gravot, A., Auroy, P., Leonhardt, N., David, P., Nussaume, L., Vavasseur, A., Richaud, P., 2004. Overexpression of AtHMA4 enhances root-to-shoot translocation of zinc and cadmium and plant heavy metal tolerance. FEBS Lett. 576, 306–312.

Verret, F., Gravot, A., Auroy, P., Preveral, S., Forestier, C., Vavasseur, A., Richaud, P., 2005. Heavy metal transport by AtMHA4 involves N-terminal degenerated metal binding domain and the C-terminal HIS (II) stretch. FEBS Lett. 579, 1515–1522.

Vinita, H., 2007. Phytoremediation of toxic metals from soil and waste water. J. Environ. Biol. 28 (2), 367–376.

Wang, H., Kimberley, M.O., Schlegelmilch, M., 2001. Biosolids derived nitrogen mineralization and transformation in forest soils. J. Environ. Qual. 32, 1851–1856.

Willems, G., Gode, C., Verbrugger, N., Saumitou-laprade, P., 2007. Quantitative trait loci mapping zinc tolerance in the metallophyte *Arabidopsis halleri* spp. *halleri*. Genetics 176, 659–674.

Zayed, A., Pilon Smits, E., DeSouza, M., Lin, Z.Q., Terry, N., 2000. Remediation of selenium polluted soils and waters by phytovolatilization. In: Terry, N., Bañuelos, G. (Eds.), Phytoremediation of Contaminated Soil and Water. Lewis, Boca Raton, pp. 61–83.

Zhu, Y.L., Pilon-Smits, E.A.H., Jouanin, L., Terry, N., 1999a. Overexpression of gluthathione synthetase in *Brassica juncea* enhances cadmium tolerance and accumulation. Plant Physiol. 119, 73–79.

Zhu, Y., Pilon Smits, E.A.H., Tarun, A., Weber, S.U., Jouanin, L., Terry, N., 1999b. Cadmium tolerance and accumulation in Indian mustard is enhanced by overexpressing glutamyl cysteine synthetase. Plant Physiol. 121, 1169–1177.

Zhu, Y.L., Zayed, A.M., Quian, J.H., de Souza, M., Terry, N., 1999c. Phytoaccumulation of trace elements by wetland plants: II. Water hyacinth. J. Environ. Qual. 28, 339–344.

Chapter 7

Evaluation of Four Plant Species for Phytoremediation of Copper-Contaminated Soil

Parisa Ahmadpour,*,† Fatemeh Ahmadpour,‡ SeyedMousa Sadeghi,§
Farhad Hosseini Tayefeh,§ Mohsen Soleimani†† and Arifin Bin Abdu†
*Ports and Maritime Organization (PMO), Boushehr Maritime Rescue and Environmental
Protection Department, Boushehr, Iran; †Department of Forest Production, Faculty of Forestry,
Universiti Putra Malaysia, Serdang, Selangor DarulEhsan, Malaysia; ‡Pars Special Economic
Energy Zone, Pseez, National Iranian Oil Co, NIOC, Boushehr, Iran; ††Department of Natural
Resources, Isfahan University of Technology, Isfahan, Iran; §Faculty of Forestry, Universiti Putra
Malaysia, Serdang, Selangor DarulEhsan, Malaysia

INTRODUCTION

General Background

The environment has been contaminated with organic and inorganic pollutants. Organic pollutants are largely anthropogenic and are introduced to the environment in many ways, such as through the use of solvents, agricultural practices, industrial activities and fuel spills. Excess amounts of fertilizers, such as phosphates, nitrates, micronutrients of copper (Cu), iron (Fe), molybdenum (Mo), manganese (Mn) and zinc (Zn), and the nonessential elements [such as cadmium (Cd), lead (Pb) and arsenic (As)] are classified as inorganic pollutants (Rajakaruna et al., 2006).

Soil is one of the most important resources that are being contaminated with heavy metals. Soil contamination with toxic metals, such as Cd and Cu, as a result of worldwide industrialization has increased noticeably within the past few years (Manousaki et al., 2008). Therefore, it is necessary to clean the contaminated areas to remediate polluted soils and to reduce the transfer of toxic metals to the food chain. Many countries are trying to control the heavy metal contamination and to find an appropriate remediation method. There are some conventional remediation technologies to clean polluted areas, specifically soils contaminated with metals. In spite of being efficient, these methods are expensive, time-consuming and environmentally devastating (Liu et al., 2010). In addition to their high cost, these methods create soil deterioration

Soil Remediation and Plants. http://dx.doi.org/10.1016/B978-0-12-799937-1.00007-3

(Manousaki et al., 2008). Therefore, it is necessary to develop new technologies that are low-cost and environmentally friendly (Garbisu and Alkorta, 2001). Recently, a cost effective and environmentally friendly technology has been developed by scientists and engineers in which biomass/microorganisms or live plants are used to remediate the polluted areas (Jadia and Fulekar, 2009). A simple, unique and 'green' technology, phytoremediation came into existence to remediate contaminated areas (Shuhe et al., 2005). The name 'phytoremediation' is based on a Greek and a Latin word: the Greek word *phyton* (plant) and the Latin word *remediate* (to remedy) describe the purpose of this green technology (Karami and Shamsuddin, 2010). It can be categorized into various applications, including phytofiltration, phytostabilization, phytoextraction and phytodegradation.

Some specific plants, such as herbs and woody species, have been proven to have noticeable potential to absorb toxic metals. These plants are known as hyperaccumulators. Researchers are trying to find new plant species that are suitable to be used in removing heavy metals from contaminated soils.

Problem Statement

Since the beginning of the industrial revolution, heavy metal contamination of the biosphere has increased considerably and has become a serious environmental concern. Contamination by heavy metals can be considered to be one of the most critical threats to soil and water resources as well as to human health (Yoon et al., 2006). During the past decades, the annual widespread release of heavy metals reached 22,000 t (metric ton) for Cd, 939,000 t for Cu, 1,350,000 t for Zn and 738,000 t for Pb (Singh et al., 2003). Cu is non-degradable and is known to be the most dangerous pollutant, especially at higher rates.

Despite being an essential element, Cu has a toxic effect on the environment at higher concentrations and is harmful to humans, animals and plants (Khellaf and Zerdaoui, 2009). Contamination of soil with Cu has become a serious environmental problem. The persistence of this metal in the soil at higher rates causes a reduction in crop yield and threatens human health by entering the food chain (Ma et al., 2008). Cu threatens human health by contaminating soil and water bodies (Chiew, 2007). Therefore, soil contamination needs to be cleaned to ensure a safe environment.

Objectives

Although many studies have been carried out on phytoremediation of contaminated soils using weeds, leafy wild vegetables and ornamental plants, information is lacking regarding the potential of tropical plant species to remediate Cu-contaminated soils. Therefore, four tropical plant species such as [Jarakpagar (*Jatropha curcas*), Acacia (*Acacia mangium*), Jelutong (*Dyera costulata*) and Merawan Siput Jantan (*Hopea odorata*)] were tested on Cd- and Cu-contaminated soil. The objectives of this study were:

(1) To assess the growth performance of the four tropical plant species and absorption of heavy metals in Cu-contaminated soil.
(2) To evaluate the phytoremediation potential of tropical plant species and to determine the most suitable species for phytoremediation of Cu-contaminated soils.

LITERATURE REVIEW

Environmental Pollution and Sources of Contamination

Environmental pollution is the concentration of chemicals at poisonous levels in land, water and air. Pollution can be defined as an accidental or deliberate contamination of the environment with waste generated by human activities.

Our environment has been contaminated with organic and inorganic pollutants, because pollutants are released into the environment through many different ways. Soil, water and air have been contaminated as a result of industrial activities and the unmanageable growth of large cities. Metals such as Cd, Cu, chromium (Cr), nickel (Ni), zinc (Zn) and lead (Pb) are known to be serious environmental pollutants. Sources of metal contamination include anthropogenic and geological activities. Industrial pollutants, smelting, mining, military activities, fuel production and agricultural chemicals are some of the anthropogenic activities that cause metal contamination (Jadia and Fulekar, 2009).

Soil Contamination by Heavy Metals

The two different ways that heavy metals enter the environment are from natural and anthropogenic sources. Natural sources of heavy metals contamination usually result from the weathering of mines, which are themselves created anthropogenically (Wei et al., 2008). Heavy metal is defined as any element with metallic characteristics, such as density, conductivity, stability as cations and an atomic number greater than 20 (Raskin et al., 1994). Heavy metal pollution is a crucial environmental concern throughout the world. It occurs in the soil, water, living organisms and at the bottom of the sediment. Unlike organic matter, these metals cannot be altered by microorganisms. The toxicity of heavy metals is a very serious issue, because they have a long persistence in the environment. The half-life of these toxic elements is more than 20 years (Ruiz et al., 2009). According to the United States Environmental Action Group (USEAG), this environmental problem has threatened the health of more than 10 million people in many countries (Environmental News Service, 2006). Environmental contamination by heavy metals as a result of industrial and mining activities became widespread in the late nineteenth and early twentieth centuries (Benavides et al., 2005). In fact, heavy metal pollution has spread throughout the world. Fifty-three elements are classified as heavy metals. Their densities exceed $5\,g\,cm^{-3}$, and they are known as universal pollutants in industrial areas (Sarma, 2011). Heavy metals, including Cd, Cu, Cr, Zn, Ni and Pb as critical pollutants, have an adverse effect on the environment, specifically at high concentrations in areas with severe anthropogenic activities (United States Protection Agency, 1997). Although they are natural components of the Earth's crust, heavy metals' biochemical equivalence and geochemical cycles have changed noticeably due to human activities (Baccio et al., 2003). These metals are just being transformed from one form to another, because of their inability to degrade naturally. The heavy metals namely Cu, Fe, Zn, Mo and Mn are micronutrients and are considered to be essential to maintaining life in biological systems. However, at higher concentrations, these metals become highly toxic and threaten the health of animals and humans by influencing the quality of crops, water and atmosphere. The heavy metals Cd, Cu, Ni and mercury (Hg) create greater phytotoxicity than Zn and Pb (Raskin et al., 1994).

The pollution of soil is a crucial matter that has attracted considerable public attention over the past few decades. A large proportion of land has become hazardous and non-arable for humans and animals, because of extensive pollution. It is unusual to have soils without at least traces of heavy metals, and the levels of these elements become more toxic due to anthropogenic or natural activities that are harmful for living systems (Turan and Esringu, 2007). Organic pollutants are anthropogenic and degrade in the soil compared to heavy metals, which are non-degradable and occur naturally in the environment (Garbisu and Alkorta, 2001). Heavy metals can be developed by industrial activities, volcanic operations and parent material. Generally, depending on the type of element and its location, the concentration of metals in the soil ranges from traces levels to as high as $100,000\,mg\,kg^{-1}$ (Blaylock and Huang, 2000).

Malaysia is a developing country with the goal of becoming an industrial country by the year 2020. Most of the urban areas and industries are situated on the west coast of Peninsular Malaysia. Malaysia's Department of Environment (DOE) reported that the sources of heavy metals in the west coast of Peninsular Malaysia are agricultural and animal management, manufacturing industries, urbanization practices and agro-based industries (DOE, 1998). Moreover, the production and application of toxic chemicals, such as trace elements, in Malaysia has been increased by the country's rapid economic development. In the west coast of Peninsular Malaysia, 113,750 ha (hectare) of the soil surface is coated with tailings, which are a source of heavy metals, such as Pb, Cd, As and Hg (Ang et al., 2003). In addition, the town of Sungai Lembing abandoned their tin mine, which during its operation produced dangerous waste with high heavy metal content. Heavy metals, such as Cu, As, Cr, Ni, Zn and Pb, present in the tailings are mobilized and cause extensive contamination in the soils and ground water (Alshaebi et al., 2009). Furthermore, the straits of Malacca, an important international shipping lane and a centre of agricultural and industrial practices as well as urbanization on the west coast of Peninsular Malaysia, have contributed to various types of pollution (Abdullah et al., 1999).

Malaysia does not currently have definite regulations on the remediation and control of contaminated soils and ground waters. Many areas of contamination have drawn considerable attention, and the DOE has undertaken some studies to improve the standards for remediation of contaminated lands (Mohamed et al., 2009). Moreover, Malaysia's only copper mine is situated in Sabah, East Malaysia. The mine was activated in 1975 and stopped its operation in 1999. The acid mine drainage pollution has become noticeable over a long period of time. Throughout this time, the open-cast mine produced more than 100 Mt of tailings and about 250 Mt of waste rocks and overburden (Jopony and Tongkul, 2009).

Copper Contamination in the Soil

Contamination of soil with the heavy metal Cu has become a serious problem throughout the world, causing the reduction of agricultural yield and harmful effects on human health by entering the food chain. Despite being essential for plants, Cu is a toxic heavy metal at higher concentrations and causes soil to be contaminated recurrently in the environment. In fact, one of the most complicated issues for environmental engineering is to detoxify soils polluted with Cu and other heavy metals (Jiang et al., 2004). Cu is introduced into the soil through various products of human activity, including pesticides, fertilizers, municipal compost, sludge and car exhausts, emissions from municipal wastes incinerators, smelting industries, mining and residues from metalliferous mines (Yang et al., 2002; Wei et al., 2008). In Malaysia, a copper mine located in Mamut, Sabah sprung a leak in its main pipe, releasing hazardous material waste rocks into Mamut, Lohn and Bambangan Rivers (Chiew, 2007).

The concentration of Cu in the soil solution is naturally very low, between 0.01 and 0.6 µM. However, the total Cu concentration in soil ranges from 25 to 40 mg kg^{-1}. Cu can form strong complexes with organic matter. This strong binding limits the mobility of this element in the soil (Shorrocks and Alloway, 1988). Kabata-Pendias and Pendias (1984) reported that the ranges of critical soil data are included: 8 mg kg^{-1} (Cd), 125 mg kg^{-1} (Cu), 3000 mg kg^{-1} (Mn), 400 mg kg^{-1} (Pb and Zn) and 100 mg kg^{-1} (Ni and Cr). In general, at least one metal appears in concentrations higher than the critical levels in most soils.

Toxicity of Heavy Metals in Plants

Heavy metals can be poisonous for organism and microorganism via direct influence on the biochemical and physiological procedures, reducing growth, deteriorating cell organelles and preventing photosynthesis. Regarding the transportation of metals from roots to the aerial parts of the plants, some metals (especially Pb) tend to be accumulated in roots more than in aerial parts, because of some barriers that prevent their movement. However, other metals, such as Cd, moves easily in plants (Garbisu and Alkorta, 2001). Generally, all plants are able to accumulate essential elements, such as Cu, Fe, Zn, Ca, K, Mg and Na, from soil solutions for growth and development. However, during this process, plants also accumulate some non-essential elements, such as Cd, As, Cr, Al and Pb that have no biological activity.

Copper Toxicity in Plants

Cu occupies 0.1% of the Earth's crust and is an essential element for plants. It occurs in two oxidation states: Cu^+ (cuprous ion) and Cu^{2+} (cupric ion). As an essential metal in many enzymatic reactions, Cu acts as an electron donor or acceptor (Gambling et al., 2004; Mehta et al., 2006). It acts as a co-factor in many enzymes, including superoxide dismutase, alcohol dehydrogenase, peroxidases, phosphatases and catalases (Brazeau et al., 2004; MacPherson and Murphy, 2007). Despite being an essential element and acting as a co-factor in numerous enzymes, at higher concentrations, Cu is highly toxic to plants, microorganisms and invertebrates. It creates disorders in physiological process, such as unexpected leaf fall, chlorotic spots and interference with root growth, because roots are the first plant parts to be damaged when exposed to toxic concentrations of Cu (Atienzar et al., 2001; Babu et al., 2001; Merian et al., 2004; Ke et al., 2007a). The normal concentration of Cu in plant material ranges from 5 to $25\,mg\,kg^{-1}$ (Ariyakanon and Winaipanich, 2006). However, the critical concentrations of Cu in plant tissue at 10% reduction of dry weight production ranged from 5 to $30\,mg\,kg^{-1}$, depending on different crop species (Yang et al., 2002). Shorrocks and Alloway (1988) found that Cu toxicity in crops occurred when the concentration in the shoots was more than $20\,mg\,kg^{-1}$ in the dry matter. However, Evangelou (2007) reported that plants exposed to Cu showed low toxicity, because of different protective activities, such as interaction between Cu and zinc and binding to definite Cu proteins closely as a prosthetic element that decreases thermodynamic activity. Generally, Cu contaminates soil frequently as a result of its extensive use in agricultural and industrial activities (Jiang et al., 2004). Cu is introduced to the soils and ground water from many sources, including poultry manures, pig wastes, metal processing and microelectronics by-products, smelting, discarding sewage sludge and agricultural activities (such as using cheap fertilizers, fungicides and pesticides) (Brun et al., 2003; Ariyakanon and Winaipanich, 2006; Yruela, 2009).

Benimeli et al. (2009) investigated the accumulation of Cu^{2+} by different parts of *Zea mays* (roots, shoots and leaves) in various concentrations of Cu sulphate (10^{-4}–10^{-2} M) and found that the growth of *Z. mays* was not stunted by exposure to 10^{-4}–10^{-2} M Cu^{2+}. The root was the only part of this species to accumulate Cu^{2+} when exposed to 10^{-2} M Cu^{2+} (three times more than the control plant). Moreover, the authors reported that *Z. mays* has the potential to accumulate Cu^{2+} without showing any adverse effects. Khellaf and Zerdaoui (2009) investigated the effect of different concentrations of Cu, Ni, Cd and Zn on growth response of duckweed (*Lemna minor*). The result showed that the growth of *Lemna* fronds was increased at $0.2\,mg\,l^{-1}$ Cu and $0.5\,mg\,l^{-1}$ Ni. The growth of these plants was affected adversely at higher concentrations of heavy metals. The order of toxicity in duckweed, from greatest to least, was Cu followed by Cd, Ni and Zn. Duckweed was the most sensitive plant to Cu and Cd in polluted soils.

Uptake and Translocation of Copper by Plant Parts (Leaves, Stems and Roots)

The uptake and translocation of metals can be affected strongly by soil properties, such as pH, organic matter and soil mineralogy (Yanai et al., 2006). Most metals in soil are bound to organic matter and inorganic (clay) soil components and present in an insoluble form. Therefore, plants have to mobilize the soil-bound metals into the soil solution to accumulate them in different plant parts. The plant roots can acidify the soil environment with proton expelled from the roots and thus solubilize the heavy metals (Raskin et al., 1994). The chemical mobility and uptake of metal ions by the plants can be affected by microbial activities, rhizospheres and roots (Hinsinger and Courchesne, 2007). There are some chelator compounds released by plant roots that can affect the solubility and uptake of pollutants. These compounds inside plant tissue play an important role in the toleration, sequestration and transportation of contaminants (Ross, 1994).

In fact, uptake of heavy metals by the plant happens accidentally, because plants cannot distinguish between definite essential and non-essential elements. However, the translocation and absorption of essential trace elements occurs in plants actively (Meers et al., 2005). Moreover, heavy metals are detoxified by plants in two ways: the plant is either an excluder or an accumulator. The excluder plants are able to detoxify heavy metals in their roots, whereas the accumulators can carry the heavy metals to their above-ground parts (shoots), accumulate them in their leaf cells and subsequently discard these ions during seasonal leaf drops (Kuzovkina et al., 2004). Baker (1981) reported that plants are classified as accumulators if heavy metal concentration ratio (shoots–roots) is > 1 and as excluders if this ratio is < 1. Many studies have shown that tree species have considerable potential to remediate soil contaminated with heavy metals, such as Cd, Cu, Zn, Pb and Ni (Li, 2006; Ang et al., 2010).

Liu et al. (2010) investigated the differences between 40 Chinese cabbage cultivars in translocation and accumulation of Cd when exposed to different concentrations of Cd (1.0, 2.5 and 5.0 mg kg^{-1}). The cultivars had low accumulation of Cd in the edible parts, and Cd concentrations in the shoots differed significantly ($p \leq 0.05$). The Chinese cabbage was also found to have a high tolerance for Cd-stressed soil.

Remediation of Heavy Metals

Current conventional methods to remediate heavy-metal-contaminated soil and water, such as ex situ excavation, landfill of the top contaminated soils

(Zhou and Song, 2004), detoxification (Ghosh and Singh, 2005) and physico-chemical remediation, are expensive (Danh et al., 2009), time consuming, labour exhaustive, increase the mobilization of contaminants and destroy the biotic structure of the soil. Therefore, these remediation techniques are not techni-cally or financially suitable for large contaminated areas (Baccio et al., 2003; Soleimani et al., 2010). Bioremediation was developed as a technology to degrade pollutants into a low toxic level by using microorganisms. However, the use of this technology to remediate contaminated areas by applying living organisms was less successful for extensive metal and organic pollutants. Plants are able to metabolize substances produced in natural ecosystems (Vidali, 2001). Phytoreme-diation is an approach in which plants are applied to detoxify contaminated areas (Garbisu and Alkorta, 2001; Mangkoedihardjo and Surahmaida, 2008).

Definition and General Types of Phytoremediation

Phytoremediation is a promising new technology that uses plants to clean up contaminated areas. It is a low-cost, long-term, environmentally and aestheti-cally friendly method of immobilizing/stabilizing, degrading, transferring, removing or detoxifying contaminants, such as metals, pesticides, hydrocar-bons and chlorinated solvents (Susarla et al., 2002; Jadia and Fulekar, 2008a; Zhang et al., 2010).

Over the past two decades, it has become a highly accepted means of detoxi-fying contaminated water and soil (US EPA, 2001). Historically, phytoremedia-tion has been considered a natural process, first identified and proved more than 300 years ago (Lasat, 2000). The specific plant and wild species that are used in this technique are effective at accumulating increasing amounts of toxic heavy metals (Ghosh and Singh, 2005; Brunet et al., 2008). These plants are known as accumulators. They accumulate heavy metals at higher concentrations (≥ 100 times) above ground than do non-hyperaccumulators growing in the same con-ditions, without showing any observable symptoms in their tissues (Barceló and Poschenrieder, 2003). Phytoremediation can be applied to detoxify areas with trivial pollution of metal, nutrients, organic matter or contaminants. Nagaraju and Karimulla (2002) described that some species, including *J. curcas* (from *Euphorbiaceae*), *Dodonaea viscosa* (from *Sapindaceae*) and *Cassia auriculata* (from *Fabaceae*), have potential for remediation of soils polluted with different kinds of trace and major elements.

Phytoremediation can be classified into different applications, such as (1) phytofiltration or rhizofiltration, (2) phytostabilization, (3) phytovolatilization, (4) phytodegradation (Long et al., 2002) and (5) phytoextraction (Jadia and Fulekar, 2009).

1. Phytofiltration or rhizofiltration is the removal by plant roots of contaminants in wastewater, surface water or extracted ground water (Pivetz, 2001). Abhilash et al. (2009) investigated the potential of *Limnocharis flava* (L.) Buchenau grown for phytofiltration of Cd in polluted water with low concentrations of

Cd in a hydroponic experiment. They spiked 45-day-old seedlings of *L. flava* with different concentrations of Cd (0.5, 1, 2 and $4\,mg\,l^{-1}$). The concentration of Cd in different parts of the plant was highest in the roots followed by leaves and peduncle. This suggested that *L. flava* is a suitable species for phytofiltration of Cd in water with low concentrations of Cd.

2. Phytostabilization is a simple, cost-effective and less environmentally invasive approach to stabilize and reduce the bioavailability of contaminants by using plants. In fact, this approach uses plant roots to restrict the mobility and bioavailability of contaminants in the soil (Jadia and Fulekar, 2009). Plants can reduce the future adverse effects of pollutants in the environment by keeping them from entering the ground water or spreading in the air. This method is applicable when there is no prompt action to detoxify contaminated areas (e.g. if a responsible company only exists for a short time, or if an area is not of high concern on a remediation agenda) (Garbisu and Alkorta, 2001). In this approach, the chemical and biological characteristics of polluted soils are amended by increasing the organic matter content, cation-exchange capacity (CEC), nutrient level and biological actions (Alvarenga et al., 2008). In phytostabilization, plants are responsible for reducing the percolation of water within the soil matrix, which may create a hazardous leachate, inhibiting direct contact with polluted soil by acting as a barrier, and interfering with soil erosion, which results in the spread of toxic metals to the other sites (Raskin and Ensley, 2000). Phytostabilization is a suitable technique to remediate Cd, Cu, As, Zn and Cr. Alvarenga et al. (2009) investigated the effect of three organic residues, sewage sludge, municipal solid waste compost and garden waste compost, on the phytostabilization of an extremely acidic metal-contaminated soil. The plant species used in this experiment was perennial ryegrass (*Lolium perenne* L.). The organic residues were used at 25, 50 and $100\,Mg\,ha^{-1}$ (dry weight basis). These reagents immobilize and decrease the mobile fraction of Cu, Pb and Zn. It was inferred that ryegrass had the potential to be used in phytostabilization for mine-polluted soil and municipal solid waste compost and, to a lesser extent, sewage sludge, used at $50\,Mg\,ha^{-1}$ and that it is efficient in the in situ immobilization of metals, developing the chemical properties of the soil, and greatly enhancing the plant biomass.

3. Phytovolatilization is the use of green plants to extract volatile contaminants, such as Hg and selenium (Se), from polluted soil and to ascend them into the air from their foliage (Karami and Shamsuddin, 2010). Gray Banuelos of USDS Agricultural Research Service perceived that some plants are able to transform Se in the form of dimethylselenide and dimethyldiselenide in high-selenium media (Bañuelos et al., 2000).

4. Phytodegradation is the use of plants and microorganisms to uptake, metabolize and degrade the organic contaminant. In this approach, plant roots are used in association with microorganisms to detoxify soil

contaminated with organic compounds (Garbisu and Alkorta, 2001). It is also known as phytotransformation. Some plants are able to decontaminate soil, sludge, sediment and ground and surface water by producing enzymes. This approach involves organic compounds, including herbicides, insecticides, chlorinated solvents and inorganic contaminants (Pivetz, 2001).

5. Phytoextraction is a phytoremediation technique that uses plants to remove heavy metals, such as Cd, from water, soil and sediments (Yanai et al., 2006; Van Nevel et al., 2007). It is an ideal method for removing pollutants from soil without adversely affecting the soil's properties. Furthermore, in this approach, metals accumulated in harvestable parts of the plant can be simply restored from the ash that is produced after drying, ashing and composting these harvestable parts (Garbisu and Alkorta, 2001). Phytoextraction has also been called phytomining or biomining (Pivetz, 2001). This technology is a more advanced form of phytoremediation, in which high-biomass crops grown in the contaminated soil are used to bioharvest and recover heavy metals. It can be applied in the mineral industry to commercially produce metals by cropping (Sheoran et al., 2009).

The ability of plants to transport and uptake heavy metals from the soil into their above-ground shoots and the harvestable parts of their under-ground roots is the key to successful phytoextraction (Garbisu and Alkorta, 2001; Chen et al., 2003). Robinson et al. (2006) stated that a few field experiments and commercial exercises have been done in the past decade to investigate successful phytoextraction. Moreover, for phytoextraction to be considered successful, the contaminated areas need to be detoxified to a level specified by environmental rules and for a lower cost than conventional techniques (Kos and Le tan, 2003). Nascimento and Xing (2006) expressed that phytoextraction may be considered as a commercial technology in the future.

Criteria for Metal Accumulation in Plants

All plant species have the ability to absorb metals; however, some can accumulate greater amounts of metals (100 times more than the average plant in the same condition without showing any adverse effects). The woody or herbaceous plants that accumulate and tolerate heavy metals in an amount greater than the toxic levels in their tissue are known as hyperaccumulators (Baker et al., 2000; Barceló and Poschenrieder, 2003; Zhou and Song, 2004). In recent years, the use of hyperaccumulators for remediation of contaminated sites due to their capacity to take up heavy metals from polluted soil and accumulate them in their shoots has been receiving a great deal of attention from researchers (Sun et al., 2007a, 2009). The main criteria for hyperaccumulators are (1) accumulating capability, (2) tolerance capability, (3) removal efficiency based on plant biomass, (4) bioconcentration factor (BCF) index and (5) translocation factor (TF) index.

1. Accumulating capability is the natural capacity of plants to accumulate metals in their above-ground parts (the threshold concentration) in amounts greater than 100 $mg\,kg^{-1}$ for Cd (Zhou and Song, 2004; Soleimani et al., 2010), 1000 $mg\,kg^{-1}$ for Cu, Cr, Pb and cobalt (Co), 10 $mg\,kg^{-1}$ for Hg (Baker et al., 2000) and 10,000 $mg\,kg^{-1}$ dry weight of shoots for Ni and Zn (Lasat, 2002).
2. Tolerance capability is the ability of plants to grow in heavy-metal-contaminated sites and to have considerable tolerance to heavy metals without showing any adverse effects, such as chlorosis, necrosis, whitish-brown colour or reduction in the above-ground biomass (or at least not a significant reduction) (Sun et al., 2009).
3. Removal efficiency based on plant biomass is the total concentration of metal and dry biomass of plants compared to the total loaded metal in the growth media (Soleimani et al., 2010).
4. BCF index is the ratio of heavy metal concentration in plant roots to that in the soil (Yoon et al., 2006). Cluis (2004) reported that the BCF for hyperaccumulators is > 1 and in some cases can go up to 100.
5. TF is the capability of plants to take up heavy metals in their roots and to translocate them from the roots to their above-ground parts (shoots). Therefore, it is the ratio of heavy metal concentration in aerial parts of the plant to that in its roots (Mattina et al., 2003; Liu et al., 2010). This specific criterion for hyperaccumulators should reach > 1 to indicate that the concentration of heavy metals above ground is greater than that below ground (roots). Therefore, it can be concluded that this criterion is more crucial in phytoextraction, where harvesting the aerial parts of the plant is the most important objective (Wei and Zhou, 2004; Karami and Shamsuddin, 2010). Baker and Whiting (2002) reported that excluders can be identified by a TF < 1, whereas accumulators are characterized by a TF > 1.

McGrath and Zhao (2003) reported that BCFs and TFs are > 1 in hyperaccumulators. More than 400 species from 45 families all over the world have been classified as hyperaccumulators (Sun et al., 2009). Sarma (2011) reported the latest number of metal hyperaccumulators. According to his report, more than 500 plant species consisting of 101 families are classified as metal hyperaccumulators, including *Euphorbiaceae, Violaceae, Poaceae, Lamiaceae, Flacourtiaceae, Cunoniaceae, Asteraceae, Brassicaceae, Caryophyllace* and *Cyperaceae*. Zhou and Song (2004) reported that the hyperaccumulation of Cd and As occurs rarely in the plant families. They found that because hyperaccumulators produce low shoot biomass with long periods of maturity and long growing seasons, there are only a few plants with high metal accumulation ability and high biomass. However, Baker et al. (2000) found many species that can be classified as hyperaccumulators based on their capacity to tolerate toxic concentrations of metals, such as Cd, Cu, As, Co, Mn, Zn, Ni, Pb and Se (Table 7.1).

Bidens pilosa L. was grown in Cd-contaminated media to examine the growth conditions and physiological processes at the seedling period, and Cd accumulation

Soil Remediation and Plants

TABLE 7.1 Some Hyperaccumulator Species of Metals

Species	Metal	References
Clerodendrum infortunatum	Cu	Rajakaruna and Böhm, 2002
Croton bonplandianus	Cu	Rajakaruna and Böhm, 2002
Thordisa villosa	Cu	Rajakaruna and Böhm, 2002
Pityrogramma calomelanos	As	Dembitsky and Rezanka, 2003
Pistia stratiotes	Zn, Pb, Ni, Hg, Cu, Cd, Ag and Cr	Odjegba and Fasidi, 2004
Alyssum lesbiacum	Ni	Cluis, 2004
Helicotylenchus indicus	Pb	Sekara et al., 2005
Bidens pilosa	Cd	Sun et al., 2009
Thlaspi caerluescens	Cd, Zn and Pb	Cluis, 2004; Banasova et al., 2008
Lonicera japonica	Cd	Liu et al., 2009a
Solanum nigrum L.	Cd	Sun et al., 2008
Sedum alfredii	Cd	Sun et al., 2007a
Brassica junceae	Ni and Cr	Saraswat and Rai, 2009

at the flowering and mature steps. Results manifested that this species has the potential to be used as Cd hyperaccumulator (Sun et al., 2009). Moreover, the tolerance of *Lonicera japonica* to Cd stress was investigated by applying various concentrations of Cd. The plants did not show any adverse effects when treated with 5 and 10 mg l^{-1} Cd, and even with higher concentrations (50 mg l^{-1}) there was no significant difference ($p \leq 0.05$) between treated plants and control plants in terms of height or production of dry biomass (shoots and roots). Hence, they classified *L. japonica* as a species with a high tolerance to Cd and a potential for Cd hyperaccumulation (Liu et al., 2009a). Shuhe et al. (2005) suggested *Solanumnigrum* L. as a newly discovered Cd hyperaccumulator and applied Cd in various concentrations (10, 25, 50, 100 and 200 mg kg^{-1}) in a concentration-gradient experiment. The height and shoot biomass showed no significant reduction under two concentrations of Cd (10 and 25 mg kg^{-1}). It was revealed that this species has great tolerance when exposed to these concentrations of Cd and could be a Cd hyperaccumulator, because of its high tolerance and accumulation of this metal. Sun et al. (2007b) also suggested *S. nigrum* L. as a new Cd hyperaccumulator.

Rajakaruna et al. (2006) stated that the hyperaccumulation of Cu occurs rarely. However, Rajakaruna and Böhm (2002) discovered five Cu hyperaccumulators: *Geniosporum tenuifelorum* (from family *Lamiaceae*), with a total Cu accumulation of 2266 ppm; *Waltheria indica* (from family *Sterculiaceae*), with a total of 1504 ppm; *Clerodendrum infortunatum* (from family *Verbenaceae*), with a total of 2278 ppm; *Tephrosia villosa* (from family *Fabaceae*), with a total of 1858 ppm; and *Croton bonplandianus* (from family *Euphorbiaceae*), with a total of 2163 ppm. Rajakaruna and Baker (2004) also identified other species that can be considered as Cu hyperaccumulators, including *Cassia auriculata* and *Phyllanthus* sp.

MATERIALS AND METHODS

Description of Study Area

The study was conducted in the greenhouse, Faculty of Forestry, Universiti Putra Malaysia (2°59′18.24″N latitude and 101°42′45.45″E longitude), Serdang, Selangor, Malaysia, from February to June 2010. The average temperature at the greenhouse ranged from 27°C in the morning, to 36°C in the afternoon and to 32°C in the evening. The relative humidity of the greenhouse was 65%.

Planting Materials

Seedlings

Healthy seedlings of the same age were collected from the Department of Agriculture, Serdang, Selangor. The plants were selected based on their ability to produce a high biomass in a very short time as well as their performing high growth rate. Plants were grown under greenhouse conditions. They were kept in ambient conditions and watered periodically. Each pot was marked by non-toxic paint to differentiate each treatment level.

Growth Media for Copper Contamination

Soil was taken from the field, Faculty of Agriculture, Universiti Putra Malaysia. It was air-dried until it could be crushed to pass through a 4-mm sieve. A stainless steel sieve was used to supply a homogenous soil composite to be used as the growing media. Ten kilogrammes of soil was transferred to each polyethylene pot. $CuSO_4 \cdot 5H_2O$ was applied to provide different levels of Cu. The different types of growth media prepared by mixing soil with different levels of Cu including Cu_0 (control, soil), Cu_1 (soil + 50 mg kg^{-1} Cu), Cu_2 (soil + 100 mg kg^{-1} Cu), Cu_3 (soil + 200 mg kg^{-1} Cu), Cu_4 (soil + 300 mg kg^{-1} Cu) and Cu_5 (soil + 400 mg kg^{-1} Cu). The Cu salts were dissolved in distilled water and added to the soil. The treated soils were saturated, mixed thoroughly and air-dried. The wetting–drying mixing technique was continued for 1 month to equilibrate adsorption of the added Cu salt, as explained in the first experiment.

Experimental Design and Treatments

A randomized complete block design (RCBD) with four replications and six different levels of treatment was used in a factorial arrangement. Two experiments were conducted. Experiment 1 consisted of four plant species with six levels of Cd, and Experiment 2 consisted of four plant species with six levels of Cu. Treatment combinations of Cd and Cu on different plant species are presented in Table 7.2.

Plant Species and Planting

After filling the pots with the growth media, seedlings of *J. curcas, A. mangium, D. costulata* and *H. odorata* (all of which were of the same age) were transplanted into the plastic pots (32.0 cm height, 106.0 cm upper diameters and 69.0 cm lower diameters). They were transplanted gently, without injury to the root system. One seedling was planted into each pot (Figure 7.1). The pot soil was fertilized with NPK fertilizer during transplanting and at 60 days after planting to supply nutrients for optimum growth and development.

Intercultural operations, such as watering and weeding, were accomplished when necessary to ensure normal growth of the plants. The pots were watered based on 75% of the soil field capacity to prevent leaching.

Data Collection

Measurement of Basal Stem Diameter, Height and Number of Leaves

The growth variables such as basal stem diameter, height and number of leaves were measured once a month for 5 months. The basal stem diameter was measured 5 cm above the ground using a vernier caliper whereas the height was measured with diameter tape. A specific point on the stem was marked with yellow paint to indicate the point used for basal stem diameter measurement.

TABLE 7.2 Treatment Combination of Plant Species and Different Levels of Copper

Plant Species	Cu Levels ($mg\,kg^{-1}$)					
	Control	Cu_1	Cu_2	Cu_3	Cu_4	Cu_5
Jatropha curcas	0	50	100	200	300	400
Acacia mangium	0	50	100	200	300	400
Dyera costulata	0	50	100	200	300	400
Hopea odorata	0	50	100	200	300	400

FIGURE 7.1   An overview of the greenhouse.

Dry Biomass of Plants

Plants were pulled out carefully from the pot to avoid any damage to the roots. The plants were then cleaned thoroughly with tap water and rinsed in distilled water. The samples were oven-dried at 70°C for 48 h (Wu et al., 2009) for *H. odorata* and *A. mangium* and for 72 h for *J. curcas* and *D. costulata* where the temperature decreased to 50°C after 48 h. Dry weight biomass was measured by weighing each part of the dried plant (leaves, stems and roots) separately.

Laboratory Analysis

Physical Analysis

Soil samples were analyzed for texture and electrical conductivity. Soil texture was determined using a pipette gravimetric method (Tan, 2005; Yanai et al., 2006).

The calculation associated with the pipette method was as follows:

$$\% \text{ pipette fraction} = \frac{W-D}{S} \times \frac{1000}{50} \times 100$$

where W = weight of pipetted fraction, D = weight of dispersing agent in 50 ml, S = weight of soil on an oven-dry basis.

Soil electrical conductivity (solid:deionized water = 1:2, w/v) was determined by using an electric conductivity metre (Yadav et al., 2010).

Chemical Analysis

Soil chemical properties, including heavy metal concentrations, pH, total carbon, total nitrogen, CEC, exchangeable calcium (Ca), magnesium (Mg) and potassium (K), exchangeable Al and hydrogen (H) and available phosphorous (P), were determined following standard laboratory methods. Heavy metal concentrations were determined by using the aqua regia method (Karaca, 2004). Dry soil (0.5 g) was weighed into the digestion tube. Aqua regia (3 HCl:1 HNO_3) solution was prepared, and 4 ml of this solution was added and digested at 110°C. After cooling, 10 ml of 1.2% HNO_3 was added and reheated at 80°C for 30 min. After that, the solution was made up to just under 20 ml with deionized water and reheated for another 30 min at 80°C. The solution made up to 20 ml with deionized water. Finally, the sample was whirl mixed and filtered through Whatman no. 42 filter paper into a plastic vial. The Cd and Cu concentrations were measured by flame atomic absorption spectrometry (Perkin Elmer, A-Analyst 200).

Soil pH was determined in a suspension of 1:2.5 soil:water (w/v) by using a glass electrode pH meter (Zhang et al., 2009). Total soil C and N were determined by dry combustion using a LECO CNS 2000 analyzer (Leco, St Joseph, MI) (Moore et al., 2010). Soils with weight ranges from 0.1000 to 0.1010 g were measured using a balance and were put on a boat for each sample. The CEC and exchangeable cations (Ca, Mg and K) were determined by using the leaching method with 1 M ammonium acetate at pH 7 (Ariyakanon and Winaipanich, 2006). Ten grammes of air-dried soil were leached with 100 ml of 1 M ammonium acetate (NH_4OAc) solution for 6 h. The leached solution collected was made up to volume with ammonium acetate and was applied for determination of exchangeable cations (Ca, Mg and K). The exchangeable cations were measured by atomic absorption spectrometry (Perkin Elmer, A-Analyst 200). The soil was then washed with 95% ethanol to measure CEC for another 5–6 h through the leaching process and again leached with 100 ml of 0.1 M K_2SO_4 for 5–6 h. CEC was measured using an auto-analyzer.

The exchangeable Al and H were determined by the NaOH titration method (Perez et al., 2009). The filtrate obtained from a pH-KCl suspension was used. Ten millilitres of the filtrate was pipetted into a 250 ml Erlenmeyer flask. Phenolphthalein (1%) was used as an indicator. After that, 0.01 M NaOH was added to the filtrate to reach the first permanent pink endpoint for about 2–3 min. The volume of NaOH solution used, recorded as (x). Five millilitres of 4% NaF, was added subsequently. This solution was titrated with 0.01 M HCl until the pink colour permanently disappeared. The volume of HCl solution used was recorded as (y). The calculation was as follows:

$$y = \text{Exchange Al and}$$

$$x - y = \text{Exchange H.}$$

Available P (ppm) was determined by using the Bray and Kurtz II method with a mixture of ammonium fluoride (0.03 M NH$_4$F) and hydrochloric acid (0.1 M HCl) (Akbar et al., 2010). One gramme of air-dried soil (2.0 mm) was weighed into a test tube and then poured into 20 ml of Bray II extracting reagent (0.1 M HCl + 0.03 M NH$_4$F). After that, it was shaken well for 1 min. The solution was filtered with Watman no. 42 filter paper, and the filtrate was collected. Finally, the filtrate was sent to an auto analyzer for determination.

Plant Tissue Analysis

The plants were cut into the required parts (roots, stems and leaves). The fresh samples were weighed, oven-dried at 70°C for about 48 h and then the dry weight of samples was weighed. The oven-dried samples were shredded into small sections. A stainless steel Fritch pulverization mill was used to grind the dried plant samples into less than 1 mm (Yadav et al., 2009). Dried and well-ground plant samples (0.5 g) were digested with 5 ml HNO$_3$ (65% w/w, Merck, Darmstadt, Germany) and heated at 110°C for 2 h. The solution was left to cool, and then 1 ml H$_2$O$_2$ (30% w/w, Merck, Darmstadt, Germany) was added, and the solution was boiled for 1 h (Lim and Salido, 2004).

Triple deionized water was used to dilute the clear digests to 50 ml. Finally, the solution was filtered through a Whatman no. 42 filter paper, and the heavy metal concentration in plant parts was determined by using flame atomic absorption spectrometry.

Evaluation of Heavy Metals Uptake Using Removal Efficiency (RE), Bioconcentration Factor (BCF) and Translocation Factor (TF)

In order to evaluate the potential of plant species, three indicators were used: BCF (metal concentration ratio of plant roots to soil), TF (metal concentration ratio of plant shoots to roots) and removal efficiency based on total dry biomass (RE; total concentrations of metal and dry biomass of plants to total loaded metal in growth media). BCF, TF and RE were calculated as follows:

$$BCF = \left[\frac{\text{Metal Concentration in Root } \left(\text{mg kg}^{-1} \right)}{\text{Metal Concentration in Soil } \left(\text{mg kg}^{-1} \right)} \right]$$

$$TF = \left[\frac{\text{Metal Concentration in Shoot } \left(\text{mg kg}^{-1} \right)}{\text{Metal Concentration in Root } \left(\text{mg kg}^{-1} \right)} \right]$$

$$RE\,(\%) = \left[\frac{\text{Metal Concentration in Shoot } \left(\text{mg kg}^{-1} \right) \times \text{Shoot Biomass (kg)} +}{\text{Total Added Metal per Pot (mg)}} \right] \cdots$$

$$\left[\frac{\text{Metal Concentration in Root} \left(\text{mg kg}^{-1}\right) \times \text{Root Biomass (kg)}}{\text{Total Added Metal per Pot (mg)}} \right]$$

Statistical Analysis

The data were statistically analyzed using the Statistical Analysis Software (SAS) program (Release 9.2). The analysis of variance (ANOVA) for growth and heavy metals concentrations in the growth media and plants parts were performed, and the mean values were adjusted post hoc by the Duncan's multiple range test (DMRT) ($p \leq 0.05$). Correlation analysis was done for each species between Cu concentrations in the growth media with dry biomass production and total Cu concentration in plant parts.

RESULTS AND DISCUSSION

Physico-Chemical Properties of the Control Media

The physico-chemical properties of the control media are presented in Table 7.3. The soil used in this study had a sandy clay texture. The proportions of sand, silt and clay were 57.88%, 5.25% and 36.87%, respectively. Total N, C, P and K were 0.03%, 0.74%, 0.03% and 0.1%, respectively. The soil was acidic, with pH 4.62. The available P was 9.17 mg kg^{-1}, CEC and electrical conductivity were 14.03 cmol$_c$ kg^{-1} and 0.3 dS m^{-1}, respectively. The concentration of exchangeable cations for K$^+$, Mg^{2+} and Ca^{2+} were 0.005, 0.004 and 0.046 cmol$_c$ kg^{-1}, respectively. These values were regarded as low compared to the values of exchangeable Al (0.75 cmol$_c$ kg^{-1}) and H (0.13 cmol$_c$ kg^{-1}). High concentrations of exchangeable Al and H were the main source of soil acidity. Judging the content of total C and N, CEC and exchangeable bases, the plant nutrient status in this growth medium was relatively low due to the acidic nature of the soil. The concentrations of Cd, Cu, Zn, Fe and Mn were 2.6, 9.93, 46.75, 479.4 and 30.6 mg kg^{-1}, respectively.

Copper Concentration in the Growth Media

Copper Concentration in the Growth Media before Planting

As shown in Table 7.4, the Cu concentration in the growth media before planting ranged from 9.37 ± 0.28 to 313.48 ± 0.48 mg kg^{-1} in all Cu levels. The highest concentration (313.48 ± 0.48 mg kg^{-1}) was found in Cu$_5$, and the lowest (9.37 ± 0.28 mg kg^{-1}) was in control media. Cu concentration in the growth media was significantly different ($p \leq 0.05$) between various Cu levels applied to the growth media before planting.

Copper Concentration in the Growth Media after Harvest

Cu concentration in the growth media was varied between species after harvest and was less than the Cu concentration before planting. In *J. curcas*,

TABLE 7.3 Selected Physico-Chemical Properties of the Control Media

Properties	Values
Texture	Sandy clay
Sand (%)	$57.88 \pm 1.97^*$
Silt (%)	5.25 ± 0.42
Clay (%)	36.87 ± 1.86
Field capacity (%)	28.92 ± 0.46
Total N (%)	0.03 ± 0.03
Total C (%)	0.74 ± 0.05
Total P (%)	0.03 ± 0.002
Total K (%)	0.1 ± 0.003
pH (1:2.5 soil: water)	4.62 ± 0.16
Available P (mg kg^{-1})	9.17 ± 1.12
CEC (cmol$_c$kg^{-1})	14.03 ± 1.77
EC (dS m^{-1})	0.3 ± 0.13
Exchangeable Cations (cmol$_c$ kg^{-1})	
K (cmol$_c$ kg^{-1})	0.005 ± 0.001
Mg (cmol$_c$ kg^{-1})	0.004 ± 0.001
Ca (cmol$_c$ kg^{-1})	0.046 ± 0.005
Al (cmol$_c$ kg^{-1})	0.75 ± 0.13
H (cmol$_c$ kg^{-1})	0.13 ± 0.06
Total Heavy Metal (mg kg^{-1})	
Cd	2.6 ± 0.21
Cu	9.93 ± 0.31
Zn	46.75 ± 4.55
Fe	479.4 ± 22.87
Mn	30.6 ± 3

*SE (standard errors) values after ±.

the highest reduction (19.72%) was recorded in Cu_0 (9.68 ± 0.62 mg kg^{-1} before planting to 7.8 ± 1.13 mg kg^{-1} after harvest), and the lowest (5.55%) was found in Cu_4 (236.6 ± 2.70 mg kg^{-1} before planting and decreased to 223.48 ± 10.55 mg kg^{-1} after harvest) (Figure 7.2a). In *A. mangium*, the highest

TABLE 7.4 Copper Concentrations in the Growth Media before Planting

Treatment	Cu Concentration (mg kg^{-1})
Cu_0	9.37 ± 0.28^f
Cu_1	44.56 ± 1.37^e
Cu_2	97.92 ± 0.87^d
Cu_3	158.51 ± 4.76^c
Cu_4	245.47 ± 5.26^b
Cu_5	313.48 ± 0.48^a

Note: $Cu_0 = 100\%$ soil, $Cu_1 = 50$ ppm Cu, $Cu_2 = 100$ ppm Cu, $Cu_3 = 200$ ppm Cu and $Cu_5 = 400$ ppm Cu. Values are means ± standard errors. Different letters within a column represent significant difference among means at 5% level of DMRT ($p \leq 0.05$).

reduction (12.82%) was recorded in Cu_1 (45.8 ± 2.64 mg kg^{-1} before planting to 39.93 ± 0.48 mg kg^{-1} after harvest), and the lowest (4.38%) was in Cu_4 (242.88 ± 4.62 mg kg^{-1} before planting to 232.23 ± 6.63 mg kg^{-1} after harvest) (Figure 7.2b). In *D. costulata*, the highest reduction (12.49%) was found in Cu_2 (98.23 ± 2.36 mg kg^{-1} before planting to 85.96 ± 5.58 mg kg^{-1} after harvest), and the lowest (3.46%) was recorded in Cu_4 (226.73 ± 8.5 mg kg^{-1} before planting to 218.88 ± 4.55 mg kg^{-1} after harvest) (Figure 7.2c). In *H. odorata*, the highest reduction (18.47%) was recorded in Cu_2 (97.83 ± 1.46 mg kg^{-1} before planting to 79.76 ± 5.48 mg kg^{-1}), and the lowest (5.77%) was observed in Cu_5 (313.88 ± 0.69 mg kg^{-1} before planting to 295.78 ± 11.06 mg kg^{-1} after harvest) (Figure 7.2d).

Of all the species, *J. curcas* showed the highest reduction (14.02%) in all Cu levels, because it was the most effective species at removing Cu from contaminated soil. Similarly, Justin et al. (2011) evaluated the heavy metal uptake and translocation abilities of *A. mangium* for phytoremediation of Cd in contaminated soils and found that concentrations of Cd, Cu and Zn in growth media after harvest were decreased.

The Cu concentration in the growth media after harvest was significantly different ($p \leq 0.05$) among plant species grown under different Cu levels (Figure 7.3). Cu concentration in the growth media after harvest increased with increase in the Cu levels added to the growth media. The Cu concentration in the growth media at harvest ranged from 7.75 to 8.45 mg kg^{-1} in control media. The highest Cu concentration (8.45 ± 0.71 mg kg^{-1}) was recorded in *A. mangium,* and the lowest (7.75 ± 0.72 mg kg^{-1}) was in *J. curcas*. The Cu concentration also varied among plant species grown in the growth media treated with Cu_5, ranging from 279.13 to 295.78 mg kg^{-1}.

H. odorata showed the highest concentration (295.78 ± 10.86 mg kg^{-1}), followed by *A. mangium* (292.7 ± 10.86 mg kg^{-1}), *D. costulata* (287.08 ± 15.99 mg kg^{-1})

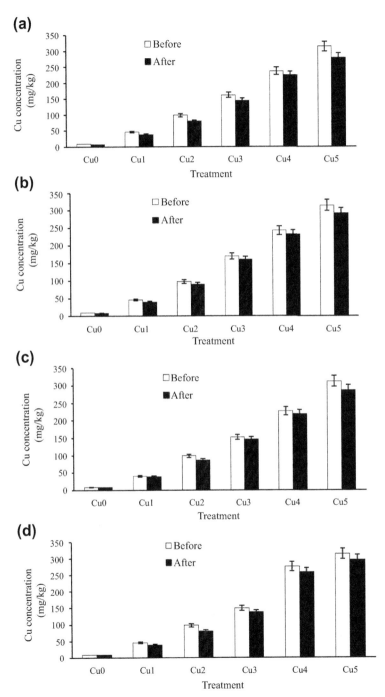

FIGURE 7.2 Copper concentrations in the growth media before and after cultivation of *Jatropha curcas* (a), *Acacia mangium* (b), *Dyera costulata* (c) and *Hopea odorata* (d), as influenced by different copper levels.

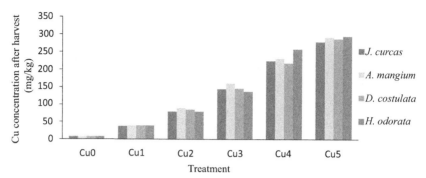

FIGURE 7.3 Copper concentrations in the growth media at harvest, as influenced by different copper levels.

and *J. curcas* ($279.13 \pm 10.46 \, \text{mg kg}^{-1}$) (Figure 7.3). The Cu concentration in soils generally ranges from 2 to $250 \, \text{mg kg}^{-1}$, and in healthy plant tissues ranges from 20 to $30 \, \mu\text{g g}^{-1}$ dry weight (Khatun et al., 2008). However, at higher concentrations, Cu causes toxicity via oxidative stress, because of an increase in the level of reactive oxygen species in subcellular compartments (Khatun et al., 2008). The critical range of Cu in soil is between 60 and $125 \, \text{mg kg}^{-1}$ (Gardea-Torresdey et al., 2005). However, Jiang et al. (2004) reported that the critical range of total Cu in soils ranged from 140 to $180 \, \text{mg kg}^{-1}$ for different vegetable crops.

Soil pH before Planting and after Harvest

Soil pH ranged from 4.41 to 4.63 before planting, having the highest (4.63) in Cu_2 and the lowest (4.41) in Cu_3. There was no significant difference ($p \leq 0.05$) between Cu levels in soil pH before planting; however, a significant difference ($p \leq 0.05$) was observed between Cu levels in pH after harvest, ranging from 4.44 to 4.76, having the highest (4.76) in Cu_1 and the lowest (4.44) in control media (Table 7.5). However, there were no significant differences ($p \leq 0.05$) between Cu levels added to the growth media in soil pH after harvest. Soil pH increased at harvest compared to before planting, with the highest increase (4.41 to 4.73) in Cu_3.

One of the most important factors that control heavy metal uptake is pH (Amini et al., 2005). In fact, solubility and bioavailability of heavy metals increased with a decrease in the soil pH, resulting in an increased metal uptake by the plants (Fässler et al., 2010). The solubility and mobility of Cu is affected by soil pH, and in soils in which the pH reached 5, Cu^{2+} showed toxicity effects (Lin et al., 2003). Cu is found in two ionic forms: cuprous and cupric ions. Cuprous ions are more soluble than cupric ions, and when soil pH < 7, Cu is found in cuprous form (Ariyakanon and Winaipanich, 2006). Similarly, Justin et al. (2011) reported that soil pH increased significantly when *A. mangium* was grown in contaminated soils, which was in line with the results of our study.

TABLE 7.5 pH of Soil with Various Copper Concentrations, before Planting and after Harvest, as Influenced by Different Copper Levels

Treatment	Soil pH	
	Before Planting	After Harvest
Cu_0	4.50 ± 0.10^a	4.44 ± 0.06^b
Cu_1	4.46 ± 0.06^a	4.76 ± 0.04^a
Cu_2	4.63 ± 0.11^a	4.74 ± 0.04^a
Cu_3	4.41 ± 0.05^a	4.73 ± 0.05^a
Cu_4	4.60 ± 0.08^a	4.72 ± 0.03^a
Cu_5	4.51 ± 0.05^a	4.63 ± 0.04^a

Note: Cu_0 = 100% soil, Cu_1 = 50 ppm Cu, Cu_2 = 100 ppm Cu, Cu_3 = 200 ppm Cu, Cu_4 = 300 ppm Cu and Cu_5 = 400 ppm Cu. Values are means ± standard errors. Different letters within a column represent significant difference among means at 5% level of DMRT ($p \leq 0.05$).

Growth Performance of Plant Species under Different Levels of Copper

Growth Performance of *Jatropha curcas* under Different Levels of Copper

The increment of basal stem diameter and height, and the total number of leaves are depicted in Figure 7.4. The results of ANOVA showed a significant difference ($p \leq 0.05$) in basal stem diameter, height and number of leaves between growth media with different Cu levels, at harvest. The increment of basal stem diameter ranged from 1.03 to 3.39 mm. The highest increment of basal stem diameter (3.39 ± 0.29 mm) was observed in seedlings grown in the growth media treated with Cu_2, the next highest (3.33 ± 0.62 mm) was in seedlings grown in the media treated with Cu_1, and the lowest (1.03 ± 0.20 mm) was in seedlings grown in the media treated with Cu_5 (Figure 7.4a).

The plant height increment ranged from 6.30 to 20.7 cm. The greatest increment of height (20.7 ± 4.45 cm) was in seedlings grown in control media, the second greatest (18.48 ± 1.78 cm) was found in seedlings grown in the growth media treated with Cu_1, and the lowest height increment (6.30 ± 0.62 cm) was in seedlings grown in the growth media treated with Cu_5 (Figure 7.4b).

As depicted in Figure 7.4c, the number of leaves ranged from 18 to 30 under different Cu levels. There was significant difference ($p \leq 0.05$) between Cu levels in number of leaves, at harvest. The number of leaves decreased with increase in the Cu level added to the growth media. Seedlings grown in control media showed the highest number of leaves (30 ± 1.83), seedlings grown in Cu_1 exhibited the next highest (29 ± 1.19), and seedlings grown in Cu_5 showed the lowest (18 ± 1.04).

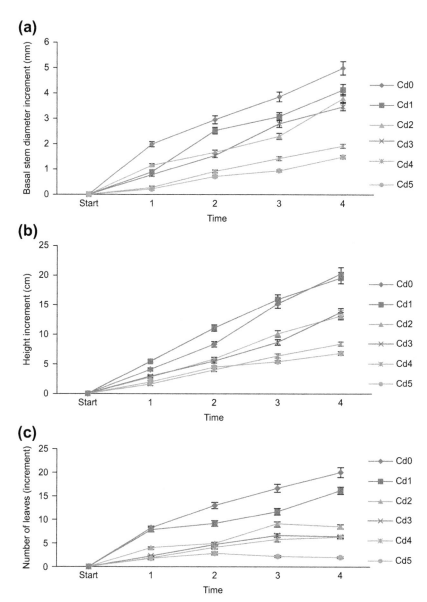

FIGURE 7.4 Increment (increase per month) of basal stem diameter (a) and height (b) and the total number of leaves (c) of *Jatropha curcas*, as influenced by different copper levels. For a colour version of this figure, please see the section at the end of this book.

  Khatun et al. (2008) reported that the growth variables of *Withania som-nifera* decreased significantly when exposed to the different concentrations of $CuSO_4$ (0, 10, 25, 50, 100 and 200 µM), indicating that higher concentrations of Cu had a toxic effect on plant growth. Dengyi and Youbao (2002) also reported that the growth of wheat improves under low Cu concentrations ($\leq 80 \, mg \, l^{-1}$), whereas higher Cu concentrations ($> 80 \, mg \, l^{-1}$) interfered with seedling growth and germination, possibly because of plant characteristics and heavy metals.

Growth Performance of *Acacia mangium* under Different Levels of Copper

The effect of Cu levels on the increment of the growth variables is depicted in Figure 7.5. There was a significant difference ($p \leq 0.05$) between Cu levels in the increment of basal stem diameter, height and the total number of leaves, at harvest. The increment of basal stem diameter ranged from 1.05 to 3.77 mm. Seedlings grown in the media treated with Cu_2 exhibited the highest increment of basal stem diameter ($3.77 \pm 0.59 \, mm$), which was closely followed by seedlings grown in the media treated with Cu_1 ($3.40 \pm 0.23 \, mm$). Seedlings grown in the media treated with Cu_5 had the lowest increment of basal stem diameter ($1.05 \pm 0.06 \, mm$) (Figure 7.5a).

As shown in Figure 7.5b, the increment of height ranged from 10.75 to 23 cm at harvest. The maximum increment ($23 \pm 1.47 \, cm$) was found in Cu_1. The second highest increment ($22.25 \pm 2.46 \, cm$) was in the control media. Cu_5 showed the lowest increment of height ($10.75 \pm 1.31 \, cm$). This implies that Cu at higher levels is detrimental to plant growth.

Based on the results shown in Figure 7.5c, the Cu levels added to the growth media had an inconsistent effect on the number of leaves, which ranged from 18 to 36 at harvest. The highest number of leaves (36 ± 4.30) was found in seedlings exposed to the growth media treated with Cu_1 and the lowest (18 ± 0.91) was in seedlings grown in the media treated with Cu_5. The number of leaves significantly ($p \leq 0.05$) decreased at higher Cu levels, from Cu_3 (34 ± 2.29) to Cu_5 (18 ± 0.91).

Metals, such as Cu, are essential for plant growth but have a serious toxic effect at higher concentrations, and inhibition of growth is the general response of plants when exposed to toxic concentrations of heavy metals. Jiang et al. (2004) reported that the shoot growth of *Elsholtzia splendens* increased with an increase in the Cu concentration in the growth media up to $62.6 \, mg \, kg^{-1}$ and then showed a slight reduction.

Growth Performance of *Dyera costulata* under Different Levels of Copper

The effect of Cu levels on the increment of basal stem diameter and height, and the total number of leaves is shown in Figure 7.6. The basal stem diameter, height and number of leaves were significantly different ($p \leq 0.05$) between growth media with different Cu levels, at harvest. Control media showed the greatest increment of basal stem diameter ($3.62 \pm 0.45 \, mm$), followed by Cu_1 with the next greatest ($3.49 \pm 0.52 \, mm$), while Cu_5 exhibited the lowest ($1.29 \pm 0.23 \, mm$) (Figure 7.6a).

The increment of height ranged from 6.50 to 20.88 cm, at harvest. The trend of the increment of height was not very dependent on different Cu levels. Seedlings exposed to Cu_2 showed the greatest increment of height ($20.88 \pm 0.75 \, cm$),

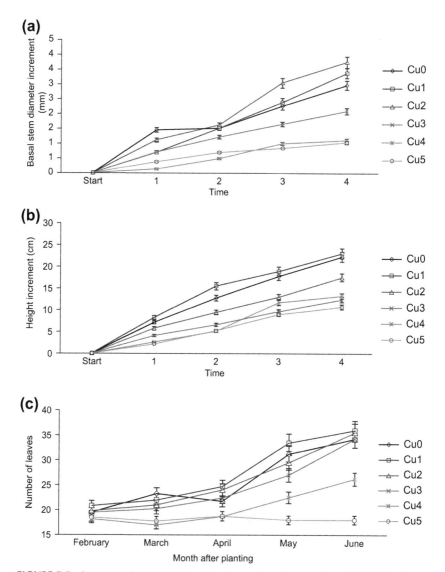

FIGURE 7.5   Increment (increase per month) of basal stem diameter (a) and height (b) and the total number of leaves (c) of *Acacia mangium*, as influenced by different copper levels. For a colour version of this figure, please see the section at the end of this book.

followed by Cu_1 (20 ± 5.72 cm) with the second greatest, and then control media (17.25 ± 2.44 cm). Cu_5 showed the lowest increment of height (6.50 ± 0.89 cm) (Figure 7.6b).

As shown in Figure 7.6c, the number of leaves also varied between Cu levels at harvest, ranging from 15 to 28. Seedlings grown in media treated with Cu_1

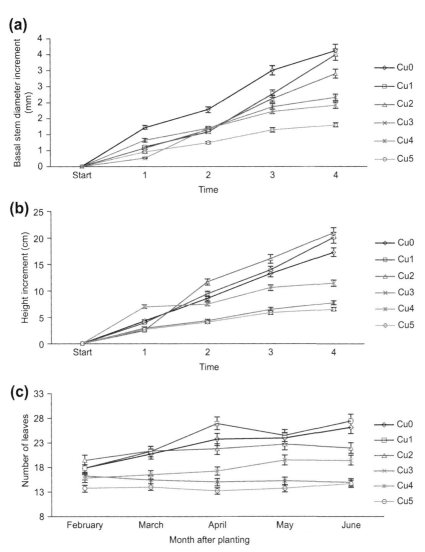

FIGURE 7.6   Increment (increase per month) of basal stem diameter (a) and height (b) and the total number of leaves (c) in *Dyera costulata*, as influenced by different copper levels. For a colour version of this figure, please see the section at the end of this book.

showed the highest number of leaves (28 ± 0.65), seedlings grown in the control media exhibited the second highest (26 ± 0.48) and seedlings grown in Cu_5 showed the lowest number of leaves (15 ± 1.89). The number of leaves slightly increased in seedlings grown in control media up to media treated with Cu_1; however, further increase in the Cu level applied to the growth media significantly decreased ($p \leq 0.05$) the total number of leaves.

Ariyakanon and Winaipanich (2006) reported that *Brassica juncea* (L.) Czern and *Bidens alba* (L.) DC. var. *radiata* grew well in pot soil spiked with 50, 100 and 150 mg kg^{-1} Cu. However, the results of Yan and Pan (2002) on the bioaccumulation of Cu (50, 68 and 200 µg l^{-1}) in three green microalgal species (*Closterium lunula, Scenedesmus obliquus* and *Chlorella pyrenoidosa*) showed that when the concentration of applied Cu increased from 50 to 200 µg l^{-1}, the growth of *C. lunula* was restricted.

Cu is an important nutrient in photosynthetic and respiratory electron transport and is known as a co-factor for many enzymes. However, at higher concentrations it has an adverse effect on plant growth and causes chlorosis and ion leakage and prevents root growth (Clemens, 2001; Ke et al., 2007b). Schmidt (2003) found that higher concentrations of heavy metals in soil can result in increased plant uptake and have an adverse effect on plant growth. In fact, higher concentrations of metals inhibit metabolic processes and interfere with plant growth.

Growth Performance of *Hopea odorata* under Different Levels of Copper

The effect of different Cu levels added to the growth media on the increment of basal stem diameter and height and the total number of leaves is depicted in Figure 7.7. The basal stem diameter and number of leaves were significantly different ($p \leq 0.05$) between different Cu levels applied to the growth media, at harvest. However, no significant difference ($p \leq 0.05$) was observed between Cu levels in the height increment. The increment of basal stem diameter ranged from 1.55 to 3.78 mm. Seedlings exposed to Cu$_1$ showed the highest increment of basal stem diameter (3.78 ± 0.60 mm), followed by seedlings grown in the control media with the next highest (3.61 ± 0.28 mm) (Figure 7.7a).

The increment of seedling height ranged from 15.25 to 29.53 cm, at harvest. Seedlings grown in the control media showed the highest increment of height (29.53 ± 6.48 cm), which was followed by Cu$_2$ (26.25 ± 0.75 cm) with the next highest and seedlings exposed to Cu$_5$ exhibited the lowest (15.25 ± 2.10 cm) (Figure 7.7b).

The number of leaves ranged from 49 to 92 (Figure 7.7c). Seedlings grown in the control media showed the greatest number of leaves (92 ± 1.55), whereas seedlings exposed to Cu$_5$ exhibited the lowest (49 ± 1.70). The seedling response to Cu levels in the growth media was steady and significantly decreased ($p \leq 0.05$) with increase in the Cu levels added to the growth media.

In spite of being essential for plant growth, Cu can cause toxicity when plant shoots and roots accumulate more than 20 mg kg^{-1} of this element (Borkert et al., 1998). Ait Ali et al. (2002), observed the inhibition of shoot and root growth in maize (*Z. mays* L.) when that plant was exposed to the higher concentrations of Cu which corroborated the findings of our study. Moreover, Kopittke and Menzies (2006) reported that the root and shoot growth of Cowpea

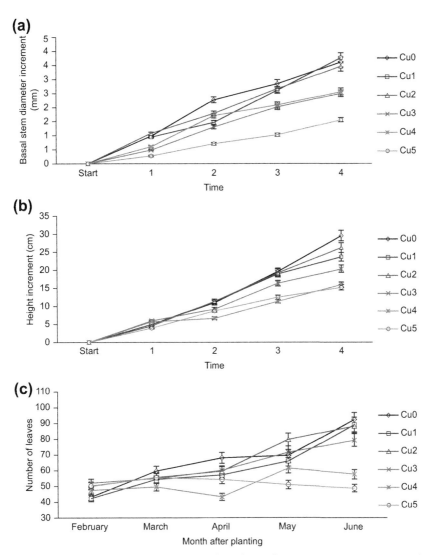

FIGURE 7.7   Increment (increase per month) of basal stem diameter (a), height (b) and the total number of leaves (c) of *Hopea odorata*, as influenced by different copper levels. For a colour version of this figure, please see the section at the end of this book.

(*Vigna unguiculata*) decreased with increase in solution of Cu^{2+} activity (1.3, 4.0, 5.0, 20, 45, 65, 75 and 85% Cu-form resin). Kabata-Pendias and Pendias (2001) found that Cu was presented in many physiological activities, such as photosynthesis, carbohydrate distribution, respiration and metabolism of nitrogen and the cell wall. However, at higher concentrations, this essential element can interfere with root growth and in some cases create leaf chlorosis (Baker and Brooks, 1989).

Production of Dry Biomass under Different Levels of Copper

Production of Dry Biomass in *Jatropha curcas* under Different Levels of Copper

The effect of Cu concentrations on dry biomass of plant parts in *J. curcas* is presented in Table 7.6. There was a significant difference ($p \leq 0.05$) between Cu levels in dry biomass of leaves, stems and roots. Seedlings grown in control media showed the highest dry biomass of leaves (21.85 ± 1.16 g) compared to the other Cu levels. Increase in the Cu level applied to the growth media significantly decreased the dry biomass of leaves, with the lowest dry biomass (7.98 ± 0.52 g) found in seedlings exposed to Cu_5.

The dry biomass of stems ranged from 15.88 to 26.7 g. Seedlings exposed to Cu_1 showed the highest dry biomass of stems (26.7 ± 1.14 g), followed by control (24.53 ± 1.85 g) and seedlings exposed to Cu_5 exhibited the lowest dry biomass of stems (15.88 ± 0.18 g). The dry biomass of stems increased from 24.53 ± 1.85 to 26.7 ± 1.14 g in seedlings grown in control media and in growth media treated with Cu_1, respectively. Increase in the Cu level added to the growth media significantly decreased ($p \leq 0.05$) the dry biomass of stems from 22.3 ± 1.19 to 15.88 ± 0.18 g when seedlings were exposed to Cu_2 and Cu_5, respectively, indicating the toxic effect of Cu at high concentrations (Table 7.6).

The dry biomass of roots ranged from 10.15 ± 0.18 to 18.48 ± 1.22 g. The root dry biomass significantly increased ($p \leq 0.05$) from 12.3 ± 0.81 to

TABLE 7.6 Dry Biomass of Leaves, Stems and Roots of *Jatropha curcas* at Harvest, as Influenced by Different Levels of Copper

Treatments	Plant Parts (g)			Total (g)
	Leaf	Stem	Root	
Cu_0	21.85 ± 1.16^a	24.53 ± 1.85^{ab}	12.3 ± 0.81^b	58.68
Cu_1	19.23 ± 0.38^b	26.70 ± 1.14^a	18.48 ± 1.22^a	63.71
Cu_2	15.88 ± 0.54^c	22.3 ± 1.19^{bc}	16.85 ± 0.56^a	55.03
Cu_3	14.15 ± 0.5^c	21.35 ± 0.51^{bc}	16.65 ± 0.43^a	52.15
Cu_4	9.53 ± 0.55^d	19.03 ± 1.34^{cd}	11.05 ± 0.12^{bc}	39.61
Cu_5	7.98 ± 0.52^d	15.88 ± 0.18^d	10.15 ± 0.18^c	34.01
Total	88.62	129.09	85.48	

Note: Cu_0 = 100% soil, Cu_1 = 50 ppm Cu, Cu_2 = 100 ppm Cu, Cu_3 = 200 ppm Cu, Cu_4 = 300 ppm Cu and Cu_5 = 400 ppm Cu. Values are means ± standard errors. Different letters within a column represent significant difference among means at 5% level of DMRT ($p \leq 0.05$).

18.48 ± 1.22 g when seedlings were grown in control media and in media treated with Cu_1, respectively. The dry biomass of roots decreased from 16.85 ± 0.56 to 10.15 ± 0.18 g in seedlings exposed to Cu_2 and Cu_5, respectively. The dry biomass of different plant parts exposed to Cu_5, were highest in stems, followed by leaves and roots (stems > leaves > roots) (Figure 7.8).

In general, the ideal plant species for phytoremediation should be fast-growing, tolerate the higher concentrations of heavy metals, accumulate the metal in its above-ground parts, produce great biomass, and should be harvested easily (Sereno et al., 2007). Ye et al. (2003) found that the dry weight of shoots and roots of *Phragmites australis* was significantly reduced when exposed to Cu (0.1 and 0.5 μg ml⁻¹) compared to the control. The reduction in biomass production could be caused by the interference of cell elongation and division by heavy metals (Khatun et al., 2008). Heavy metals can interfere with the synthesis of proteins, such as phosphoenolpyruvate carboxylase (Chen et al., 2011).

Production of Dry Biomass in *Acacia mangium* under Different Levels of Copper

The dry biomass in *A. mangium* was varied in different plant parts exposed to various levels of Cu, as shown in Table 7.7. The dry biomass of leaves, stems and roots was significantly different ($p \leq 0.05$) between Cu levels applied to the growth media. Seedlings grown in media treated with Cu_2 exhibited the highest dry biomass of leaves (33.23 ± 1.52 g), followed by Cu_1 (31.53 ± 1.68 g) and seedlings exposed to Cu_5 showed the lowest dry biomass of leaves (14.48 ± 0.65 g). The dry biomass of leaves increased from 22.2 ± 1.26 to 33.23 ± 1.52 g when seedlings were grown in control media and in soil treated with Cu_2. However, there was no significant difference ($p \leq 0.05$) between seedlings exposed to Cu_1 and Cu_2 or Cu_3 and Cu_4. The dry biomass of leaves decreased significantly ($p \leq 0.05$) with increase in the Cu level added to the

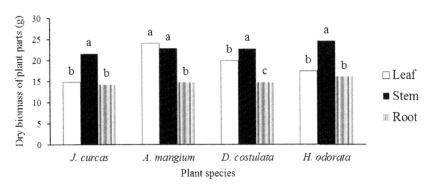

FIGURE 7.8   Dry biomass of leaves, stems and roots of four tested species at harvest.

TABLE 7.7 Dry Biomass of Leaves, Stems and Roots of *Acacia Mangium* at Harvest, as Influenced by Different Levels of Copper

Treatments	Plant Parts (g)			Total (g)
	Leaf	Stem	Root	
Cu_0	22.2 ± 1.26^b	16 ± 1.69^d	11.80 ± 0.6^b	50
Cu_1	31.53 ± 1.68^a	30.13 ± 1.45^a	17.30 ± 0.82^a	78.95
Cu_2	33.23 ± 1.52^a	29.1 ± 2.23^a	17.83 ± 0.73^a	80.16
Cu_3	21.96 ± 0.88^b	23.53 ± 1.73^b	15.98 ± 0.79^a	61.47
Cu_4	20.93 ± 1.18^b	20.8 ± 1.35^{bc}	13.43 ± 0.49^b	55.16
Cu_5	14.48 ± 0.65^c	17.93 ± 0.49^{cd}	11.90 ± 0.78^b	44.31
Total	144.3	137.48	88.24	

Note: $Cu_0 = 100\%$ soil, $Cu_1 = 50\,ppm$ Cu, $Cu_2 = 100\,ppm$ Cu, $Cu_3 = 200\,ppm$ Cu, $Cu_4 = 300\,ppm$ Cu and $Cu_5 = 400\,ppm$ Cu. Values are means ± standard errors. Different letters within a column represent significant difference among means at 5% level of DMRT ($p \leq 0.05$).

growth media starting from Cu_3 to Cu_5, which may be attributed to the toxic effect of Cu on dry biomass.

The effect of Cu on the dry biomass of stems was also not steady. The dry biomass of stems ranged from 16 ± 1.69 to $30.13 \pm 1.45\,g$. The highest dry biomass of stems ($30.13 \pm 1.45\,g$) was found in seedlings exposed to Cu_1, and the lowest ($16 \pm 1.69\,g$) was in seedlings grown in control media. There was a significant, sharp increase ($p \leq 0.05$) in the dry biomass of stems, from 16 ± 1.69 to $30.13 \pm 1.45\,g$, when seedlings were grown in control media and in media treated with Cu_1, respectively. However, increase in the Cu levels applied to the growth media significantly decreased ($p \leq 0.05$) the dry biomass of stems. This probably occurred due to the adverse effect of Cu, especially at higher levels, on stem dry biomass.

As shown in Table 7.7, the plant response to the different Cu levels was also inconsistent in the case of dry biomass of roots. Seedlings grown in the growth media treated with Cu_2 showed the greatest dry biomass of roots ($17.83 \pm 0.73\,g$), followed by Cu_1 ($17.30 \pm 0.82\,g$), whereas seedlings grown in the control media showed the lowest dry biomass of roots ($11.80 \pm 0.6\,g$). The dry biomass of root decreased from 17.83 ± 0.73 to $11.90 \pm 0.78\,g$ when seedlings were grown under Cu_2 and Cu_5, respectively. This implied that lower concentrations of Cu improve the plant growth and dry biomass, whereas higher concentrations of this essential element could adversely affect plant growth. The dry biomass of plant parts exposed to different Cu levels was highest in leaves followed by stems and roots (leaves > stems > roots) (Figure 7.8).

Roots are first influenced by higher concentrations of Cu, which resulted in interference with root growth, enzyme passivation and damage to the plasma membrane (Ke et al., 2007a). Cu plays an important role in photosynthesis and respiratory electron transport, metabolism of the cell wall, ethylene sensing and oxidative stress protection. Therefore, deficiencies of this nutrient can create disorders in plant metabolism (Yruela, 2009). Production of high biomass and high Cu tolerance are required for plant species to be successful at phytoremediation. Similar results were reported by Yan and Pan (2002) regarding the bioaccumulation of Cu in three green microalgal species (*Chlorella pyrenoidosa*, *Scenedesmus obliquus* and *Closterium lunula*) exposed to 50, 68 and 200 μg l^{-1}. The researchers observed that the dry biomass of each species decreased with increase in Cu concentration.

Production of Dry Biomass in *Dyera costulata* under Different Levels of Copper

The dry biomass of leaves, stems and roots is shown in Table 7.8. The dry biomass of different parts was significantly different ($p \leq 0.05$) between Cu levels. The dry biomass of leaves ranged from 9.85 ± 0.52 to 31.2 ± 2.14 g. The seedling response to Cu level showed an inconsistent trend. The dry biomass of leaves increased significantly from 21.33 ± 2.86 to 31.2 ± 2.14 g in seedlings grown in control media and media treated with Cu_1, respectively. However, further increase in Cu level in the growth media significantly decreased ($p \leq 0.05$) the dry biomass of leaves starting from Cu_2 (30.9 ± 1.62 g) to Cu_5 (9.85 ± 0.52 g).

TABLE 7.8 Dry Biomass of Leaves, Stems and Roots of *Dyera costulata* at Harvest, as Influenced by Different Levels of Copper

Treatments	Plant Parts (g)			Total (g)
	Leaf	Stem	Root	
Cu_0	21.33 ± 2.86^b	21.75 ± 0.42^{bc}	16.08 ± 0.25^a	59.16
Cu_1	31.2 ± 2.14^a	28.38 ± 2.84^{ab}	17 ± 0.59^a	76.58
Cu_2	30.9 ± 1.62^a	29.68 ± 1.35^a	17.1 ± 1.01^a	77.68
Cu_3	14.25 ± 0.47^c	22.43 ± 0.47^{bc}	16.05 ± 0.61^a	52.73
Cu_4	11.8 ± 0.48^c	18.3 ± 1.4^c	11.75 ± 1.13^b	41.85
Cu_5	9.85 ± 0.52^c	16.13 ± 0.74^c	10.75 ± 0.73^b	36.73
Total	119.33	136.67	88.73	

Note: $Cu_0 = 100\%$ soil, $Cu_1 = 50$ ppm Cu, $Cu_2 = 100$ ppm Cu, $Cu_3 = 200$ ppm Cu, $Cu_4 = 300$ ppm Cu and $Cu_5 = 400$ ppm Cu. Values are means ± standard errors. Different letters within a column represent significant difference among means at 5% level of DMRT ($p \leq 0.05$).

The effect of Cu concentrations on the dry biomass of stems was also inconsistent. Seedlings grown in media treated with Cu_2 showed the highest dry biomass of stems (29.68 ± 1.35 g), followed by Cu_1 (28.38 ± 2.84 g) and those exposed to Cu_5 showed the lowest (16.13 ± 0.74 g). An increase in dry biomass of stems from 21.75 ± 0.42 to 29.68 ± 1.35 g was found in seedlings grown in control media and media treated with Cu_2, respectively. However, the dry biomass of stems significantly decreased ($p \leq 0.05$) with increase in the Cu level added to the growth media. There was no significant difference ($p \leq 0.05$) between seedlings exposed to Cu_4 and Cu_5.

The dry biomass of roots showed a similar trend to the dry biomass of stems. The dry biomass of roots increased from 16.08 to 17.1 g when seedlings were grown in control media and in the growth media treated with Cu_2, respectively. However, there was no significant difference ($p \leq 0.05$) between seedlings grown in control media and in media treated with Cu_1 and Cu_2. Increase in the Cu level in the growth media significantly decreased ($p \leq 0.05$) the dry biomass of roots starting from Cu_3 (16.05 ± 0.61 g) to Cu_5 (10.75 ± 0.73 g). There was also no significant difference ($p \leq 0.05$) in the dry biomass of roots between seedlings exposed to Cu_4 and Cu_5. The dry biomass of different plant parts was highest in stems, followed by leaves and roots (stems > leaves > roots) (Figure 7.8).

These results revealed that *D. costulata* could tolerate low Cu concentrations in the growth media, in which the dry biomass of stems and roots slightly increased. However, higher Cu concentrations had an adverse effect on dry biomass, with the lowest biomass found in seedlings grown in the growth media treated with Cu_5. In general, plants typically exhibited a reduction in growth and in the biomass of roots and shoots, as well as necrosis and chlorosis, when exposed to high Cu concentrations from 3 to 100 μM (Yruela, 2009). Higher Cu concentrations can create phytotoxicity in plants by creating reactive oxygen radicals that destroy cells or by causing disorders in enzymes and protein structures (Yruela, 2009). Luo et al. (2006) reported a 45% reduction of shoot biomass compared to the control in soil that was artificially contaminated with heavy metals (Cu, Cd, Zn and Pb).

Production of Dry Biomass in *Hopea odorata* under Different Levels of Copper

The effect of Cu concentrations on the dry biomass of plant parts is presented in Table 7.9. The dry biomass of leaves, stems and roots was significantly different ($p \leq 0.05$) between Cu levels applied to the growth media. The highest dry biomass of leaves (22.1 ± 1.64 g) was recorded in seedlings exposed to Cu_1. The dry biomass of leaves slightly increased from 22.03 ± 1.14 to 22.1 ± 1.64 g when seedlings were grown in control media and in spiked media with Cu_1, respectively. However, there was no significant difference ($p \leq 0.05$) between seedlings grown in control media and media treated with Cu_1. The dry biomass of leaves significantly decreased ($p \leq 0.05$) with increase in the Cu level added to

TABLE 7.9 Dry Biomass of Leaves, Stems and Roots of *Hopea odorata* at Harvest, as Influenced by Different Levels of Copper

Treatments	Plant Parts (g)			Total (g)
	Leaf	Stem	Root	
Cu_0	22.03 ± 1.14^a	23.45 ± 1.63^b	16.78 ± 1.21^{ab}	62.25
Cu_1	22.1 ± 1.64^a	25.93 ± 2.12^{ab}	17.25 ± 1.09^{ab}	65.28
Cu_2	17.2 ± 0.57^b	28.45 ± 0.61^a	19.3 ± 1.63^a	64.95
Cu_3	16.03 ± 0.59^{bc}	27.2 ± 1.19^{ab}	18.03 ± 0.58^a	61.26
Cu_4	14.1 ± 1.3^{cd}	23.55 ± 1.5^b	14.35 ± 0.87^b	52
Cu_5	13.35 ± 0.44^d	18.83 ± 0.74^c	10.75 ± 0.64^c	42.93
Total	104.8	147.41	96.46	

Note: $Cu_0 = 100\%$ soil, $Cu_1 = 50\,ppm$ Cu, $Cu_2 = 100\,ppm$ Cu, $Cu_3 = 200\,ppm$ Cu, $Cu_4 = 300\,ppm$ Cu and $Cu_5 = 400\,ppm$ Cu. Values are means ± standard errors. Different letters within a column represent significant difference among means at 5% level of DMRT ($p \leq 0.05$).

the growth media, with the lowest dry biomass of leaves ($13.35 \pm 0.44\,g$) being recorded in seedlings exposed to Cu_5.

The dry biomass of stems ranged from 18.83 ± 0.74 to $28.45 \pm 0.61\,g$. The effect of Cu levels on the dry biomass of stems was inconsistent. Seedlings grown in media treated with Cu_2 exhibited the highest dry biomass of stems ($28.45 \pm 0.61\,g$). The dry biomass of stems significantly increased ($p \leq 0.05$) when seedlings were grown in control media and in media spiked with Cu_2, respectively. However, the dry biomass of stems significantly decreased ($p \leq 0.05$) at higher Cu levels applied to the growth media, starting from Cu_3 ($27.2 \pm 1.19\,g$) to Cu_5 ($18.83 \pm 0.74\,g$), which may be due to Cu toxicity.

The dry biomass of roots ranged from 10.75 to 19.3 g, and showed a similar trend to the dry biomass of stems (Table 7.9). The highest dry biomass of roots ($19.3 \pm 1.63\,g$) was found in seedlings exposed to Cu_2, whereas the lowest ($10.75 \pm 0.64\,g$) was found in seedlings exposed to Cu_5. The dry biomass of different plant parts grown in media treated with various Cu levels was highest in stems, followed by leaves and roots (stems > leaves > roots) (Figure 7.8).

For successful phytoremediation, plants must be fast-growing and must produce a large biomass. Cu toxicity is one of the limiting factors in phytoremediation of polluted areas (Lombi et al., 2001). Cu competes with essential ion pathways and has a toxic effect on plant cells. Ke et al. (2007b) studied the effect of Cu [0.3 (control), 25 and $100\,\mu M$] on growth of two *Rumex japonicas* populations from a Cu mine and an uncontaminated field site, and found that at the higher Cu concentration ($100\,\mu M$), the shoot biomass of both populations decreased considerably.

Production of Total Dry Biomass of Plant Species

The effect of Cu concentrations on the total dry biomass of the four plants species is depicted in Figure 7.9. There was significant difference ($p \leq 0.05$) between plant species grown under different levels of Cu applied to the growth media in production of total dry biomass. The total dry biomass of plants grown in control media ranged from 50 to 62.25 g. *H. odorata* showed the highest total dry biomass (62.25 ± 2.01 g) and *A. mangium* exhibited the lowest total dry biomass (50 ± 3.12 g). Overall, *A. mangium* showed the highest total dry biomass in every Cu level applied to the growth media. However, there was no significant difference ($p \leq 0.05$) between *A. mangium* and *D. costulata* in total dry biomass when exposed to Cu_1 or Cu_2. *J. curcas* showed the lowest total dry biomass (64.4 ± 1.8, 55.03 ± 1.36, 52.15 ± 0.94, 39.6 ± 1.17 and 34 ± 0.33 g) in seedlings exposed to Cu_1, Cu_2, Cu_3, Cu_4 and Cu_5, respectively. There was also no significant difference ($p \leq 0.05$) between *A. mangium* and *H. odorata* when seedlings were exposed to Cu_3, Cu_4 or Cu_5. The total dry biomass of plant species exposed to the highest Cu level (Cu_5) was varied. *A. mangium* achieved the highest total dry biomass (44.3 ± 0.7 g), followed by *H. odorata* (42.93 ± 1.14 g), *D. costulata* (36.73 ± 1.62 g) and *J. curcas* (34 ± 0.33 g).

Sukganah et al. (2009) reported that tropical species of *Acacia* are fast-growing plant species. As such, *A. mangium* can be used for reforestation in degraded landscapes. Similar results were obtained by Liu et al. (2009b) on the Cu tolerance of biomass crops, including Elephant grass (*Pennisetumpur pureumSchumach*), Vetiver grass (*Vetiveria zizanioides*) and the upland reed (*P. australis*). The researchers found that the biomass of all three species decreased with increase in Cu levels added to growth media (50, 100, 500, 1500 and 3000 mg kg^{-1}). At low concentrations, Cu is an essential nutrient for metabolism in plants and animals and is a component in many enzymes, co-factors and proteins. At higher concentrations, however, this essential nutrient can be highly toxic and can cause destructive effects (Khellaf and Zerdaoui, 2010).

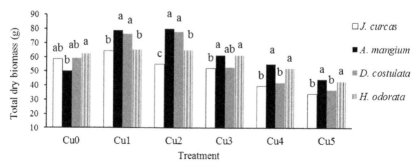

FIGURE 7.9 Total dry biomass of plant species at harvest, as influenced by different copper levels.

Copper Concentration in Plant Parts (Leaves, Stems and Roots)

Copper Concentration in Different Parts (Leaves, Stems and Roots) of Jatropha curcas

The Cu concentration in different plant parts is shown in Table 7.10. The Cu concentrations in leaves, stems and roots were significantly different ($p \leq 0.05$) under Cu levels applied to the growth media. The Cu concentrations in leaves ranged from 17.63 to 73.74 mg kg^{-1} and were not consistent under different Cu levels added to the growth media. The highest Cu concentration in leaves (73.74 ± 0.23 mg kg^{-1}) was found in seedlings exposed to Cu_3, and the lowest (17.63 ± 0.19 mg kg^{-1}) was in seedlings grown in control media. The Cu concentration in leaves significantly increased ($p \leq 0.05$) when seedlings were exposed to control media up to the media treated with Cu_3. However, the Cu concentration in leaves significantly decreased ($p \leq 0.05$) when seedlings were exposed to Cu_4 and Cu_5, indicating the toxic effect of Cu on the accumulation of this element in plants.

Cu concentration in stems showed a consistent trend when seedlings were exposed to the growth media with different Cu levels. The highest Cu concentration in stems (81.26 ± 0.7 mg kg^{-1}) was recorded in seedlings exposed to Cu_5, and the lowest (12.5 ± 0.19 mg kg^{-1}) was observed in seedlings grown in control media. The Cu concentration in stems significantly increased ($p \leq 0.05$) with an

TABLE 7.10 Copper Concentrations in Various Parts of Jatropha curcas, as Influenced by Different Copper Levels

Treatments	Plant Parts (mg kg^{-1})			Total (mg kg^{-1})
	Leaf	Stem	Root	
Cu_0	17.63 ± 0.19^e	12.5 ± 0.19^e	24.57 ± 0.55^e	54.7
Cu_1	36.26 ± 1.48^d	43.7 ± 3.66^d	190.46 ± 9.25^d	270.42
Cu_2	52.2 ± 3.19^b	48.35 ± 0.16^d	208.07 ± 0.24^d	308.62
Cu_3	73.74 ± 0.23^a	54.03 ± 1.53^c	262.66 ± 4.31^c	390.43
Cu_4	45.57 ± 1.12^c	75.34 ± 1.71^b	374.28 ± 11.6^b	495.19
Cu_5	32.78 ± 1.83^d	81.26 ± 0.7^a	551 ± 17.42^a	665.04
Total	258.18	315.18	1611.04	

Note: Cu_0 = 100% soil, Cu_1 = 50 ppm Cu, Cu_2 = 100 ppm Cu, Cu_3 = 200 ppm Cu, Cu_4 = 300 ppm Cu and Cu_5 = 400 ppm Cu. Values are means ± standard errors. Different letters within a column represent significant difference among means at 5% level of DMRT ($p \leq 0.05$).

increase in the Cu levels applied to the growth media, which indicates that the stems of *J. curcas* may tolerate Cu toxicity. However, the Cu concentrations in the stems did not differ significantly ($p \leq 0.05$) between seedlings exposed to Cu_1 and Cu_2.

The concentration of Cu in roots ranged from 24.57 ± 0.55 to $551 \pm 17.42\,mg\,kg^{-1}$ and showed a similar trend to the concentration of Cu in stems. Based on the results shown in Table 7.10, Cu concentration in roots increased with an increase in the Cu levels added to the growth media. The highest Cu concentration in roots ($551 \pm 17.42\,mg\,kg^{-1}$) was achieved by seedlings exposed to Cu_5, and the lowest ($24.57 \pm 0.55\,mg\,kg^{-1}$) was found in seedlings grown in control media. This may be attributed to the high tolerance of this species to Cu toxicity even at high concentrations of Cu in the growth media. Cu concentration between different plant parts was highest in roots, followed by stems and leaves (roots > stems > leaves).

In general, much of the Cu in soils is unavailable for plant uptake, because Cu binds strongly with organic matter and other colloids in soils; however, at higher concentrations, this metal becomes toxic and interferes with the plant growth and root elongation and destroys the root cell membrane and epidermal cells (Lin et al., 2003). The normal Cu concentration in plant tissue ranges from $5-25\,mg\,kg^{-1}$. Cu concentration in plants rarely exceeds $100\,mg\,kg^{-1}$, even when plants are exposed to higher concentrations of Cu in soils (Ariyakanon and Winaipanich, 2006). Jadia and Fulekar (2008b) reported that the uptake of heavy metals (Cd, Ni, Pb, Cu and Zn) by alfalfa (*M. sativa*) increased with increases in the concentration of these metals in contaminated soil. Chehregani et al. (2011) reported that *Euphorbia macroclada* Boiss., which comes from the same family as *J. curcas* (*Euphorbiaceae*), had the potential to be used as a hyperaccumulator of some heavy metals, such as Cu, Zn, Mn, Fe and Pb, when grown on a lead and zinc mine. Lorestani et al. (2011) also reported that *E. macroclada* Boiss. is the most suitable species to be used for phytostabilization of Cu and Fe when grown on soil contaminated with heavy metals in Cu mines in Iran.

Copper Concentration in Different Parts (Leaves, Stems and Roots) of *Acacia mangium*

The Cu concentrations in plant parts are presented in Table 7.11. The Cu concentration in leaves, stems and roots was significantly different ($p \leq 0.05$) under Cu levels applied to the growth media. The Cu concentrations in different plant parts showed a similar and steady trend. The Cu concentrations in different parts of seedlings increased with increase in Cu applied to the growth media, indicating that *A. mangium* had a high tolerance of Cu, even at high concentrations. The highest Cu concentrations in leaves, stems and roots ($38.5 \pm 1.51\,mg\,kg^{-1}$, $29.75 \pm 1.17\,mg\,kg^{-1}$ and $327.7 \pm 5.58\,mg\,kg^{-1}$, respectively) was found in seedlings exposed to Cu_5, whereas the lowest Cu concentrations in leaves, stems and roots ($17.39 \pm 0.84\,mg\,kg^{-1}$, $12.16 \pm 1.13\,mg\,kg^{-1}$ and $15.2 \pm 1.56\,mg\,kg^{-1}$, respectively) was found in seedlings grown in control

TABLE 7.11 Copper Concentrations in Various Parts of *Acacia mangium*, as Influenced by Different Copper Levels

Treatments	Plant Parts ($mg\,kg^{-1}$)			
	Leaf	Stem	Root	Total ($mg\,kg^{-1}$)
Cu_0	17.39 ± 0.84^e	12.16 ± 1.13^c	15.2 ± 1.56^f	44.75
Cu_1	22.71 ± 0.14^d	23.28 ± 0.32^b	85.36 ± 5.48^e	131.35
Cu_2	24.45 ± 0.16^d	23.39 ± 0.47^b	102.84 ± 3.39^d	150.68
Cu_3	29.58 ± 0.2^c	24.98 ± 0.8^b	178.16 ± 7.08^c	232.72
Cu_4	32.03 ± 0.86^b	29.6 ± 0.57^a	256.29 ± 5.56^b	317.92
Cu_5	38.5 ± 1.51^a	29.75 ± 1.17^a	327.7 ± 5.58^a	395.95
Total	164.66	143.16	965.55	

Note: $Cu_0 = 100\%$ soil, $Cu_1 = 50\,ppm$ Cu, $Cu_2 = 100\,ppm$ Cu, $Cu_3 = 200\,ppm$ Cu, $Cu_4 = 300\,ppm$ Cu and $Cu_5 = 400\,ppm$ Cu. Values are means ± standard errors. Different letters within a column represent significant difference among means at 5% level of DMRT ($p \leq 0.05$).

media. However, the Cu concentration in stems did not differ significantly ($p \leq 0.05$) between seedlings exposed to Cu_1, Cu_2 and Cu_3 or between seedlings grown in media treated with Cu_4 and Cu_5. The Cu concentration between different plant parts was highest in roots, followed by leaves and stems (roots > leaves > stems).

Lin et al. (2003) found that the Cu content in the roots, hypocotyls, cotyledons and leaves of sunflowers (*Helianthus annuus* L.) increased with increase in the Cu concentration in solution. Evangelou (2007) described some different protective activities, including interaction between Cu and Zn and close binding to definite Cu proteins as a prosthetic element to decrease the thermodynamic activity. Therefore, plants exposed to Cu show low toxicity. The uptake efficiency of plants depends on the plant species (Jadia and Fulekar, 2008a). Hussain and Khan (2010) reported that the critical level of Cu in plants is between 20 and $100\,mg\,kg^{-1}$.

Copper Concentration in Different Parts (Leaves, Stems and Roots) of *Dyera costulata*

The Cu concentration of different plant parts is presented in Table 7.12. The concentration of Cu in leaves, stems and roots was significantly different ($p \leq 0.05$) between Cu levels applied to the growth media. The Cu concentration in leaves showed an inconsistent trend. It was highest ($52.99 \pm 2.19\,mg\,kg^{-1}$) in seedlings exposed to Cu_3, whereas the lowest ($18.18 \pm 0.24\,mg\,kg^{-1}$) was found in seedlings grown in control media. The Cu concentration in leaves increased when grown in control media ($18.18 \pm 0.24\,mg\,kg^{-1}$) up to the media treated

TABLE 7.12 Copper Concentrations in Various Parts of *Dyera costulata,* as Influenced by Different Copper Levels

Treatments	Plant Parts ($mg\,kg^{-1}$)			Total ($mg\,kg^{-1}$)
	Leaf	Stem	Root	
Cu_0	18.18 ± 0.24^e	11.44 ± 0.11^f	18.73 ± 0.21^f	48.35
Cu_1	21.58 ± 0.5^d	21.68 ± 0.43^e	79.27 ± 4.67^e	122.53
Cu_2	23.09 ± 0.54^d	23.98 ± 0.38^d	132.06 ± 3.42^d	179.13
Cu_3	52.99 ± 2.19^a	33.18 ± 0.91^c	161.1 ± 6.84^c	247.27
Cu_4	32.42 ± 0.87^b	36.03 ± 1.29^b	341.29 ± 2.83^b	409.74
Cu_5	26.27 ± 0.4^c	47.65 ± 0.59^a	418.06 ± 9.59^a	491.98
Total	174.53	173.96	1150.51	

Note: $Cu_0 = 100\%$ soil, $Cu_1 = 50\,ppm$ Cu, $Cu_2 = 100\,ppm$ Cu, $Cu_3 = 200\,ppm$ Cu, $Cu_4 = 300$ ppm Cu and $Cu_5 = 400$ ppm Cu. Values are means ± standard errors. Different letters within a column represent significant difference among means at 5% level of DMRT ($p \leq 0.05$).

with Cu_3 ($52.99 \pm 2.19\,mg\,kg^{-1}$) and significantly decreased ($p \leq 0.05$) in seedlings exposed to Cu_4 ($32.42 \pm 0.87\,mg\,kg^{-1}$) and Cu_5 ($26.27 \pm 0.4\,mg\,kg^{-1}$). However, there was no significant difference ($p \leq 0.05$) between seedlings treated with Cu_1 and Cu_2.

The Cu concentration in stems and roots increased with an increase in the level of Cu added to the growth media. The highest Cu concentrations ($47.65 \pm 0.59\,mg\,kg^{-1}$ and $418.06 \pm 9.59\,mg\,kg^{-1}$) in stems and roots, respectively, were in seedlings exposed to Cu_5. The lowest Cu concentrations ($11.44 \pm 0.11\,mg\,kg^{-1}$ and $18.73 \pm 0.21\,mg\,kg^{-1}$) in stems and roots, respectively, were found in seedlings grown in control media. In general, the Cu concentration in plant parts was highest in roots, followed by leaves and stems (roots > leaves > stems) (Figure 7.10).

Similar to our results, Lin et al. (2008) found that the Cu concentration in shoots and roots of *Z. mays* L. increased with increase in the Cu levels (50, 100 and $500\,mg\,kg^{-1}$) in growth media. Some plant species can grow in Cu-contaminated soils without being influenced by higher concentration of this element. These plants are heavy metal-tolerant species (Ernst et al., 2000). Song et al. (2004) showed that the Cu concentrations in roots of *E. splendens* and *Silene vulgaris* were much higher than in the shoots, which is a typical aspect of heavy metal excluders. Similar results were also obtained by Benimeli et al. (2009), who found that the root was the only part of this species to accumulate Cu, which it did at three times higher than control, when exposed to $10^{-2}\,M\,Cu^{+2}$. Physiological features, such as the membrane

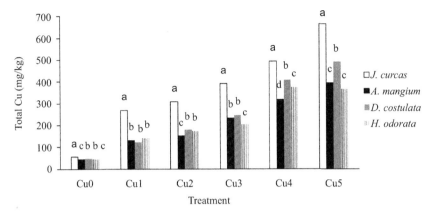

FIGURE 7.10 Total copper concentrations among plant species at harvest, as influenced by different copper levels.

and morphological features, including the length and diameter of roots and root hairs, are important factors in determining metal uptake from soil (Ghassemzadeh et al., 2008).

Copper Concentration in Different Parts (Leaves, Stems and Roots) of *Hopea odorata*

The Cu concentration in leaves, stems and roots of seedlings grown under different Cu levels applied to the growth media is shown in Table 7.13. The Cu concentration in plant parts was significantly different ($p \leq 0.05$) when seedlings were exposed to different levels of Cu added to growth media. The Cu concentrations in leaves, stems and roots followed similar consistent trends. The Cu concentration in different plant parts increased with increase in the Cu levels applied to the growth media. The highest Cu concentration in leaves and roots ($29.3 \pm 0.64\,\mathrm{mg\,kg^{-1}}$ and $310.98 \pm 5.04\,\mathrm{mg\,kg^{-1}}$, respectively) was found in seedlings exposed to Cu_5. The highest Cu concentration in stem ($51.76 \pm 2.56\,\mathrm{mg\,kg^{-1}}$) was in Cu_4. The lowest Cu concentration in leaves, stems and roots ($18.06 \pm 2.06\,\mathrm{mg\,kg^{-1}}$, $11.26 \pm 0.05\,\mathrm{mg\,kg^{-1}}$ and $17.32 \pm 0.37\,\mathrm{mg\,kg^{-1}}$, respectively) was found in seedlings grown in control media. However, no significant difference ($p \leq 0.05$) was observed between the Cu concentration in leaves of seedlings exposed to Cu_2 and Cu_3 or between the roots of seedlings exposed to Cu_4 and Cu_5. In general, the Cu concentration in different plant parts was highest in roots, followed by stems and leaves (roots > stems > leaves) (Figure 7.10).

These results indicate that *H. odorata* may tolerate Cu toxicity even at high concentrations of Cu in the growth media. Beladi et al. (2011) reported that the Cu absorption in the roots and aerial parts of Sainfoin (*Onobrychis viciifolia*) exposed to different concentrations of Cu (0, 150, 300, 450 mg kg⁻¹) increased with increase in the Cu content in the growth media. The accumulation and

TABLE 7.13 Copper Concentrations in Various Parts of *Hopea odorata*, as Influenced by Different Copper Levels

Treatments	Plant Parts (mg kg^{-1})			Total (mg kg^{-1})
	Leaf	Stem	Root	
Cu_0	18.06 ± 2.06^d	11.26 ± 0.05^e	17.32 ± 0.37^e	46.64
Cu_1	20.8 ± 0.03^c	24.60 ± 0.18^d	94.97 ± 2.98^d	140.37
Cu_2	24.67 ± 0.39^b	31.85 ± 1.45^c	117.8 ± 9.7^c	174.32
Cu_3	24.96 ± 0.12^b	41.59 ± 1.27^b	139.39 ± 3.1^b	205.94
Cu_4	26.84 ± 0.32^{ab}	51.76 ± 2.56^a	296.56 ± 7.94^a	375.16
Cu_5	29.3 ± 0.64^a	25.86 ± 0.38^d	310.98 ± 5.04^a	366.14
Total	144.63	186.92	977.02	

Note: Cu_0=100% soil, Cu_1=50 ppm Cu, Cu_2=100 ppm Cu, Cu_3=200 ppm Cu, Cu_4=300 ppm Cu and Cu_5=400 ppm Cu. Values are means±standard errors. Different letters within a column represent significant difference among means at a 5% level of DMRT ($p \leq 0.05$).

immobilization of heavy metals in roots is a mechanism of heavy metal tolerance in plant species (Ke et al., 2007b). Nwaichi and Onyeike (2010) evaluated the Cu tolerance and accumulation by *Centrosema pubescen* Benth and *Mucuna pruriens* VarPruriens. They found that the Cu concentrations in the roots of all *Centrosema* plants were much higher than those in the shoots. In addition, Yoon et al. (2006) had similar results with native plants, such as *Phyla nodiflora*, *Gentiana pennelliana* and *Cynodon dactylon* (L.), in which concentrations of Pb, Cu and Zn in roots were greater than those in shoots. The Cu concentration in roots of *C. dactylon* (L.) was 310 mg kg^{-1}, which was similar to the Cu concentration in roots of *H. odorata* exposed to Cu_5 in our study. In general, the results of our study support previous studies on high accumulation of Cu in roots and the immobility of this metal in plants (Zheljazkovet al., 2006, 2008). The Cu accumulation in leaves, stems and roots of the four tested species is shown in Figure 7.10.

Overall, none of the species in this study showed Cu concentration >1000 mg kg^{-1} (the criteria for Cu hyperaccumulator) in the shoots; therefore, none of them are hyperaccumulator according to the criteria of Baker and Brooks (1989). However, the ability of these plant species to accumulate and tolerate Cu may be useful for phytostabilization (Yoon et al., 2006).

Total Copper Concentration in Plant Species

As shown in Figure 7.10, total Cu concentration in different plant species was significantly different ($p \leq 0.05$) between Cu levels applied to the growth media. Among the plant species, *J. curcas* showed the highest total Cu concentrations (54.7 ± 0.64, 270.42 ± 10.18, 308.62 ± 2.96, 390.43 ± 4.75, 495.18 ± 12.19 and $665.044 \pm 0.3 \, \text{mg kg}^{-1}$) when grown in control media and in the media treated with Cu_1, Cu_2, Cu_3, Cu_4 and Cu_5, respectively. However, the species with the lowest total Cu concentration within each level of Cu applied to the growth media varied. *A. mangium* showed the lowest total Cu concentrations (44.75 ± 1.15, 150.68 ± 3.74 and $317.91 \pm 5.28 \, \text{mg kg}^{-1}$) in seedlings grown in control media and in media treated with Cu_2 and Cu_4, respectively. However, *D. costulata* showed the lowest total Cu concentration ($122.53 \pm 5.31 \, \text{mg kg}^{-1}$) in seedlings exposed to Cu_1. *H. odorata* showed the lowest total Cu concentrations (205.93 ± 3.74 and $366.14 \pm 4.71 \, \text{mg kg}^{-1}$) when seedlings were exposed to Cu_3 and Cu_5, respectively. The total Cu concentration between plant species exposed to the highest Cu level (Cu_5) added to the growth media was highest in *J. curcas* ($665.04 \pm 15.41 \, \text{mg kg}^{-1}$) followed by *D. costulata* ($491.98 \pm 9.44 \, \text{mg kg}^{-1}$), *A. mangium* ($395.95 \pm 4.15 \, \text{mg kg}^{-1}$) and *H. odorata* ($366.14 \pm 4.71 \, \text{mg kg}^{-1}$). However, there was no significant difference ($p \leq 0.05$) between *A. mangium* and *H. odorata* when seedlings were exposed to Cu_5.

Macnair (1987) reported that *J. curcas* has the potential to grow well in soil contaminated with high concentrations of heavy metals. In general, the maximum potential concentration of heavy metals in plants varies with plant species (Lorestani et al., 2011). All tree and shrub species have the potential to take up heavy metals, but each species may vary in its capacity to bioaccumulate them (Ang et al., 2010). The uptake of heavy metals by plants happens in two ways: passive uptake with the mass flow of water toward the roots or active transport within the plasma membrane of root epidermal cells (Yoon et al., 2006).

Murakami and Ae (2009) evaluated five cultivars of three species of rice (*Sativa* L.), soybeans (*G. max* [L.] Merr.) and maize (*Z. mays* L.) for phytoextraction of Cu, Pb and Zn. They found that the highest Cu concentration was in shoots of Milyang 23 rice, whereas the lowest was observed in two soybean cultivars. Gupta and Sinha (2007) studied the phytoremediation potential of the plants growing on tannery sludge dumping sites. They reported that the accumulation of metals, such as Cu, Cd, Co, Mn, Zn and Cr, was highest in *Calotropis procera* followed by *Sida acuta*, *Ricinus communis* and *Cassia fistula* (which were equal). It can be inferred that different plant species have different nutrient uptakes, which is due to their specific physiology (Abdu, 2005). Cu hyperaccumulation occurred rarely, which makes Cu a regular pollutant of many metal-contaminated habitats.

Relationship between Copper Concentration in the Growth Media and Dry Biomass of the Plant Species

There was a significant negative relationship between the total dry biomass of the plant and the total Cu concentration for *J. curcas*, *D. costulata*, and *H. odorata* and *A. mangium*. The negative correlation between total Cu in the growth media and dry biomass of the plant indicated that total dry biomass decreased with increase in the concentration of Cu in the growth media (Table 7.14). Higher concentrations of heavy metals can reduce plant growth and biomass production by affecting the physiology of the plant (Grifferty and Barrington, 2000). In addition, heavy metals can inhibit the uptake and transportation of essential nutrients and, therefore, can interrupt the mineral nutrition composition of plants (Ke et al., 2007b). Subsequently, heavy metals can create phytotoxic effects in plants by disturbing the metabolism of mineral nutrition (Monni et al., 2000). Cu is highly toxic to normal plants, and depending on the crop plant species, critical tissue Cu concentrations at 10% decrease of dry mass ranged from 5 to $30\,mg\,kg^{-1}$ (Jiang et al., 2004).

Relationship between Copper Concentration in the Growth Media and in the Plant Species

The total Cu concentration in the growth media and plant species was significantly correlated ($p \leq 0.01$). The positive correlation revealed that the total concentration of Cu in plants increased with increase in Cu level in the growth media (Table 7.15). Similar results were obtained by Xiong and Wang (2005) on Cu toxicity and bioaccumulation in Chinese cabbage (*Brassica pekinensis* Rupr.), in which the Cu concentration in shoots increased noticeably with increase in the Cu level in growth media (0, 0.2 and $1.0\,mM\,kg^{-1}$). Some plants have the ability to tolerate, endure and survive in heavy-metal-contaminated soils that are toxic for other plants. The tolerance of some species to high concentrations of heavy metals may be due to some potential mechanisms at the cellular level. These mechanisms seem to be incorporated in preventing the accumulation of toxic levels at sensitive areas inside the cell, thus avoiding harmful effects (Yruela, 2005).

TABLE 7.14 Correlation between Copper Concentration in the Growth Media and Total Plant Dry Biomass

Total Cu in Soil	n	R
Dry biomass of *J. curcas*	24	−0.92**
Dry biomass of *A. mangium*	24	−0.49*
Dry biomass of *D. costulata*	24	−0.76**
Dry biomass of *H. odorata*	24	−0.81**

*Pearson correlation is significant at the 0.05 level (two-tailed).
**Correlation is significant at the 0.01 level (two-tailed).

TABLE 7.15 Correlation between Copper Concentration in the Growth Media and in Plants

Total Cu in Soil	n	r
Total Cu in *J. curcas*	24	0.96**
Total Cu in *A. mangium*	24	0.98**
Total Cu in *D. costulata*	24	0.98**
Total Cu in *H. odorata*	24	0.96**

*Pearson correlation is significant at the 0.05 level (two-tailed).
**Correlation is significant at the 0.01 level (two-tailed).

Evaluation of Phytoremediation Potential of Tested Species by Removal Efficiency (RE), Bioconcentration Factor (BCF) and Translocation Factor (TF)

Evaluation of Phytoremediation Potential of Tested Species by Removal Efficiency (RE)

The Cu removal efficiency based on plant biomass is presented in Figure 7.11. The results showed that the removal efficiency was varied and significantly different ($p \leq 0.05$) between plant species under different Cu levels. In control media, the highest Cu removal ($1.07 \pm 0.03\%$) was seen in *D. costulata*, whereas the lowest ($0.84 \pm 0.1\%$) was observed in *A. mangium*. However, there was no significant difference ($p \leq 0.05$) between *J. curcas*, *A. mangium*, *D. costulata* and *H. odorata* in Cu removal when grown in control media. Among different plant species exposed to various Cu levels, *J. curcas* always attained the highest Cu removals ($1.18 \pm 0.14\%$, $0.55 \pm 0.02\%$, $0.41 \pm 0.01\%$, $0.25 \pm 0.009\%$ and $0.23 \pm 0.007\%$) when seedlings were exposed to Cu_1, Cu_2, Cu_3, Cu_4 and Cu_5, respectively. *A. mangium* showed the lowest Cu removals of all the species ($0.34 \pm 0.009\%$, $0.24 \pm 0.01\%$ and $0.2 \pm 0.005\%$) in seedlings grown in media treated with Cu_2, Cu_3 and Cu_4, respectively. The lowest Cu removals ($0.61 \pm 0.06\%$ and $0.13 \pm 0.007\%$) were found in *H. odorata* when seedlings were exposed to Cu_1 and Cu_5, respectively. However, there was no significant difference ($p \leq 0.05$) between *A. mangium*, *D. costulata* and *H. odorata* grown in media treated with Cu_1, Cu_2 and Cu_3, or between *D. costulata* and *H. odorata* when seedlings were exposed to different levels of Cu except Cu_5. The Cu removal between plant species grown under Cu_5 was highest in *J. curcas* ($0.23 \pm 0.007\%$) followed by *D. costulata* ($0.18 \pm 0.01\%$), *A. mangium* ($0.16 \pm 0.006\%$) and *H. odorata* ($0.13 \pm 0.007\%$).

As shown in Figure 7.11, *J. curcas* was the most efficient species at removing Cu from Cu-contaminated soils under all Cu levels applied to the growth media except for control media. The Cu removal of the four plant species decreased with increase in the Cu levels applied to the growth media, which may be due to the reduction of plant dry biomass when exposed to different Cu levels. Similarly,

Soil Remediation and Plants

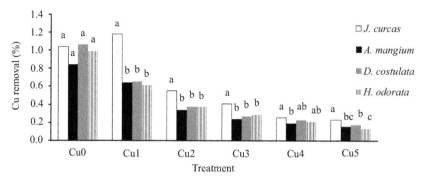

FIGURE 7.11 Total copper removals by plant species at harvest, as influenced by different copper levels.

Burke et al. (2000) investigated the effect of the application of soil fungicide Benomyl on maize root colonization by *arbuscular mycorrhizae*. They found that total Pb removal decreased as a result of reduction in shoot biomass which occurred due to the application of fungicide. Lasat (2000) reported that the biomass of harvested species and the metal concentration in the harvested biomass are important factors to estimate the rate of metal removal. He found that the contamination of soil with metals can significantly decrease plant biomass and rate of metal removal due the toxic effect of metals to plant species. Soleimani et al. (2010) evaluated the effect of endophytic fungi on bioaccumulation of Cd by *Festuca pratensis* and *Festuca arundinacea* grown under different levels of Cd (0, 5, 10 and 20 mg l^{-1}) in a hydroponic condition. The removal efficiency in plants was in the range of 4.4 to 31% and decreased with increase in Cd level in the growth media. Regardless of the control media (Cd$_0$), Cd$_5$ (5 mg l^{-1}) exhibited the highest removal due to the maximum production of biomass and lower Cd concentration in the growth media.

The result of our study was also in line with the finding of Ariyakanon and Winaipanich (2006) who reported that the Cu-removal efficiency of *B. juncea* (L.) Czern was 11 times higher than *Bidens alba* (L.) DC. var. *radiate* when grown in the Cu-contaminated soil. The highest removal efficiencies for *B. juncea* (L.) Czern (1.16%) and for *B. alba* (L.) DC. var. *radiate* (0.14%) were recorded in pot experiments with 150 mg kg^{-1} Cu concentration.

Successful phytoremediation depends on the genotype and phenotype of the plant; however, interaction between heavy metals and organic matter in soil should also be taken into consideration, because of the fixation of heavy metals to organic matter, clays and oxides, which reduces the solubility and mobility of metals for plant uptake (Chen et al., 2003).

Evaluation of Phytoremediation Potential of Tested Species by Bioconcentration Factor (BCF)

The BCFs of Cu between plant species grown in the growth media treated with various concentrations of Cu are presented in Figure 7.12. BCFs were

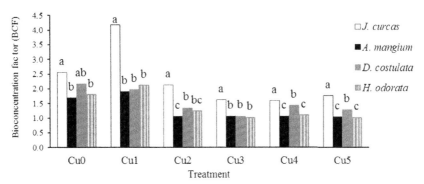

FIGURE 7.12   Bioconcentration factors of plant species at harvest, as influenced by different copper levels.

significantly different ($p \leq 0.05$) between plant species exposed to different Cu levels added to the growth media. The BCFs of Cu were > 1 within each level of Cu and in control media. *J. curcas* always showed the highest BCFs of all the species (2.55 ± 0.09, 4.19 ± 0.04, 2.13 ± 0.04, 1.62 ± 0.04, 1.58 ± 0.05 and 1.76 ± 0.05) in seedlings grown in control media and in the media treated with Cu_1, Cu_2, Cu_3, Cu_4 and Cu_5, respectively. *A. mangium* showed the lowest BCFs (1.68 ± 0.22, 1.89 ± 0.19, 1.06 ± 0.04 and 1.05 ± 0.01) when seedlings grown in control media and in media treated with Cu_1, Cu_2 and Cu_4, respectively. *H. odorata* showed the lowest BCFs (1.01 ± 0.14 and 1.02 ± 0.02) in seedlings exposed to Cu_3 and Cu_5, respectively. However, there was no significant difference ($p \leq 0.05$) between *A. mangium*, *D. costulata* and *H. odorata* in BCFs when seedlings were exposed to Cu_1 and Cu_3 or between *A. mangium* and *H. odorata* when seedlings grown under different levels of Cu. The BCFs between plant species exposed to the highest Cu level (Cu_5) added to the growth media were highest in *J. curcas* (1.76 ± 0.05) followed by *D. costulata* (1.29 ± 0.04), *A. mangium* (1.04 ± 0.01) and *H. odorata* (1.02 ± 0.03).

J. curcas showed the highest BCF of all the species (1.76 ± 0.05) when seedlings were exposed to Cu_5, which may be due to the highest Cu concentration in roots (Figure 7.12). Certain factors can affect the BCF value, such as type and concentration of heavy metals and the physiological and accumulative abilities of plants (Niu et al., 2007). Yoon et al. (2006) evaluated the potential of 36 plants from 17 species for accumulation of Cu, Zn and Pb in contaminated site. They observed that among the plant species, *Gentiana pennelliana* was the most efficient at phytostabilization of sites contaminated with Cu, Zn and Pb (with BCFs values of 22, 2.6 and 11, respectively). They suggested that plants with high BCFs and low TFs have the potential to be used in phytoremediation through phytostabilization rather than phytoextraction. Malik et al. (2010) also found that *Xanthium stromarium* L., *S. nigrum* L. and *Parthenium oleracea* L. showed higher BCFs (14.2, 14.8 and 25.4, respectively) for Cu compared to the other wild species (*Amaranthus viridis* L. and *Brachiaria reptans* L.) when grown on heavy metal-contaminated soil in industrial areas of Islamabad, Pakistan.

Evaluation of Phytoremediation Potential of Tested Species by Translocation Factor (TF)

The TFs of Cu among plant species grown under different Cu levels applied to the growth media are depicted in Figure 7.13. TFs differed significantly ($p \leq 0.05$) between plant species exposed to various Cu levels added to the growth media. Plant species grown in control media showed TFs > 1, whereas plant species grown in media treated with various levels of Cu exhibited very small TFs (< 1), which may imply the restriction of root–shoot transfer of Cu, due to the toxicity of this metal specifically at higher levels. In control media, the highest TF (6.62 ± 0.36) was found in *A. mangium*, and the lowest (1.12 ± 0.28) was found in *D. costulata*. The TFs of Cu within each level of Cu applied to the growth media were varied among plant species. However, there was no significant difference ($p \leq 0.05$) between plant species exposed to Cu_1 and Cu_2. The TFs among plant species grown in media treated with Cu_5 were highest in *J. curcas* (0.33 ± 0.07), followed by *A. mangium* (0.28 ± 0.01), *D. costulata* (0.19 ± 0.008) and *H. odorata* (0.19 ± 0.009).

Similarly, Bjelková et al. (2011) found that artificially increasing the level of Cd in soil resulted in lower translocation of Cd from root to shoot of flax and linseed cultivars as compared to translocation of this metal in control soil, which may occur due to the toxicity of this metal. Bose et al. (2008) evaluated the uptake and transport of heavy metals (Cu, Pb, Ni, Cr, Zn and Mn) by *Tyaha angustata* L. grown on waste amended soil. They found that the translocation factor (TF) was very small in plants grown in 100% amended soil indicating the toxic effect of these metals. Lorestani et al. (2011) also reported that *Euphorbia macroclada*, which comes from the same family (*Euphorbiaceae*) as *J. curcas*, showed a TF < 1 (0.34) and a BCF > 1 (1.33) compared to the other native plants (*Ziziphora clinopodioides*, *Cousinia* sp. and *Chenopodium botrys*) when grown in heavy-metal-contaminated soil in a Cu mine

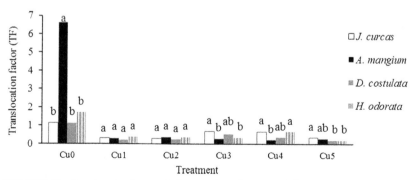

FIGURE 7.13   Translocation factors of plant species at harvest, as influenced by different copper levels.

in Iran. The result obtained by Rosselli et al. (2003) on the potential of five woody species (*Salix viminalis, Betula pendula, Alnusin cana, Fraxinus excelsior* and *Sorbus mougeotii*) to phytoextract Cd, Cu or Zn in polluted soils and translocate those metals to their aerial parts also revealed that none of the woody species translocate Cu efficiently to their above-ground tissues. Similarly, Sinha et al. (2006) studied the translocation and accumulation of metals in vegetables/crops using treated tannery wastewater for irrigation. They reported that the accumulation of metals was less in the above-ground tissues of the most studied plants. Similar results were also attained by Zheljazkov et al. (2008) regarding the metal uptake of medicinal plant species (*Bidens tripartita* L., *Leonurus cardiaca* L., *Melissa officinalis* L. and *Origanum heracleoticum* L.) grown in soil polluted by a smelter. They found that the TFs were < 1, indicating that the species were not suitable to be used for phytoextraction.

The results of our study was also supported by the findings of Mangkoedihardjo and Surahmaida (2008) regarding the phytoremediation potential of *J. curcas* L. for Cd- and Pb-contaminated soil. They found that *J. curcas* L. was not an accumulator. The results of Del Rio Celestino et al. (2006) regarding the uptake of Pb and Zn by wild plants (*Amaranthus blitoides, Cynodon dactylon, Cichorium intybus* and *Silybum marianum*) grown on contaminated soil also showed that the lowest translocation of Pb from root to shoot was recorded in *C. album*, indicating the immobilization of Pb by root and the suitability of this species to be used in phytostabilization. In contrast to our finding, Haque et al. (2008) reported that the TF values for Cu, Zn, Cr and Mo were > 1 for desert broom (*Baccharis sarothroides Gray*) grown on mine tailing, indicating that these metals were transported easily in the plants. The species of plant is one important factor that controls and affects metal translocation (Weis and Weis, 2004).

In general, plant species with a BCF > 1 and a TF < 1 are suitable for use in phytoremediation through phytostabilization, whereas plant species with both TF and BCF > 1 are suitable for use in phytoextraction (Yoon et al., 2006).

SUMMARY, GENERAL CONCLUSION AND RECOMMENDATION FOR FUTURE RESEARCH

Conclusions

Based on our findings, the following conclusions can be drawn:

The control media had a sandy clay texture, 4.62 pH, 14.03 $cmol_c kg^{-1}$ CEC and 2.6 $mg kg^{-1}$ and 9.93 $mg kg^{-1}$ Cd and Cu concentrations, respectively. In both experiments, soil pH was higher at harvest than before planting. Cu concentrations in the growth media at harvest was lower than the concentrations of this metal in the growth media before planting.

1. The growth variables (basal stem diameter, height and number of leaves) were significantly influenced ($p \leq 0.05$) by different Cu levels applied to the growth media at harvest. Higher Cu levels disaffect the growth variable of each tested species.

2. Total dry biomass was significantly different ($p \leq 0.05$) between plant species under different Cu levels in the growth media. Among plant species, *A. mangium* showed the highest total dry biomass at each level of Cu applied to the growth media. *J. curcas* showed the lowest total dry biomass within each level of Cu applied to the growth media except in control media. The total dry biomass of plant species exposed to Cu_5 was varied. *A. mangium* achieved the highest total dry biomass (44.3 g), followed by *H. odorata* (42.93 g), *D. costulata* (36.73 g) and *J. curcas* (34 g).

3. Total Cu concentrations for different plant species were significantly different ($p \leq 0.05$) between different Cu levels added to the growth media. Of all the tested species, *J. curcas* showed the highest total Cu concentration when grown in control media and in the media treated with every Cu level. The total Cu concentrations in plant species exposed to Cu_5 were highest in *J. curcas* (665.04 mg kg^{-1}), followed by *D. costulata* (491.98 mg kg^{-1}), *A. mangium* (395.95 mg kg^{-1}) and *H. odorata* (366.14 mg kg^{-1}).

4. Total dry biomass of plant species and Cu concentration in the growth media were significantly related at $p \leq 0.01$ for *J. curcas*, *D. costulata* and *H. odorata* and at $p \leq 0.05$ for *A. mangium*. The negative correlation indicated that total dry biomass of the plants decreased with increase in the Cu levels applied to the growth media. The Cu concentration in the growth media was significantly ($p \leq 0.01$) related to the total Cu concentration in the four plant species. The positive correlation revealed that the total concentration of Cu in plants increased with an increase in the Cu level in the growth media.

5. Cu removal efficiency (RE), BCF and TF were significantly different ($p \leq 0.05$) between plant species grown under different Cu levels added to the growth media. Of all the tested species, *J. curcas* always attained the highest Cu removals (1.18%, 0.55%, 0.41%, 0.25% and 0.23%) when exposed to Cu_1, Cu_2, Cu_3, Cu_4 and Cu_5, respectively. The Cu removal among plant species grown under Cu_5 was highest in *J. curcas* (0.23%), followed by *D. costulata* (0.18%), *A. mangium* (0.16%) and *H. odorata* (0.13%). *J. curcas* was the most suitable species for removing Cu from Cu-contaminated soils under all Cu levels applied to the growth media, except for control media.

The BCFs of Cu were >1 at each level of Cu and in control media. Among plant species, *J. curcas* always showed the highest BCFs at each level of Cu applied to the growth media and in control media. The BCFs among plant species exposed to the highest Cu level (Cu_5) added to the growth media were highest in *J. curcas* (1.76), followed by *D. costulata* (1.29), *A. mangium* (1.04) and *H. odorata* (1.02). Plant species grown in the control media (Cu_0) showed TFs >1, whereas plant species grown in the media treated with various levels of Cu showed very small TFs (<1), which may imply that root–shoot transfer of Cu was restricted. The TFs among plant species grown in media treated with Cu_5 were highest in *J. curcas* (0.33) followed by *A. mangium* (0.28), *D. costulata* (0.19) and *H. odorata* (0.19).

In general, of the four tested species evaluated for Cu phytoremediation potential, *J. curcas* was the most suitable species at excluding this metal from contaminated soil, especially in its root system. Total Cu concentration for *J. curcas* seedlings grown in media treated separately with Cu_5 was 665.04 mg kg^{-1}. Having a BCF > 1 in all Cu-contaminated media showed the potential of the four plant species for accumulation of this metal; however, because these plants had a lower TF (< 1), they can be used as excluders rather than accumulators. Cu was accumulated mainly in plant roots rather than leaves or stems, indicating that the species tended to restrict root–shoot transfer. Therefore, it can be concluded that all tested species are suitable to be used in phytoremediation through the phytostabilization method to prevent distribution of Cu in contaminated areas.

Recommendations for Future Research

Further study, especially in field conditions, is highly recommended to confirm the efficiency of the selected species for phytoremediation because the current study was performed under greenhouse conditions, which was a controlled environment. Species screening in the future is highly recommended for further verification of the results and the efficiency of the selected species for phytoremediation. Future studies should be extended for longer durations, if possible, to obtain more detailed results, especially for tree species. Moreover, plant physiological processes, such as photosynthesis and respiration rate, should definitely be studied to verify the physiological processes that affect the uptake of heavy metals by the tested species.

REFERENCES

Abdu, A., 2005. Effects of sewage sludge on tree growth, soil properties and groundwater quality. Masters thesis. Faculty of Forestry, Universiti Putra Malaysia.

Abdullah, A.R., Tahir, N.M., Loong, T.S., Hoque, T.M., Sulaiman, A.H., 1999. The GEF/UNDP/IMO Malacca straits demonstration project: Sources of pollution. Mar. Pollut. Bull. 39 (1–12), 229–233.

Abhilash, P., Pandey, V.C., Srivastava, P., Rakesh, P.S., Chandran, S., Singh, N., Thomas, A.P., 2009. Phytofiltration of cadmium from water by *Limnocharis flava* (L.) Buchenau grown in free-floating culture system. J. Hazard. Mater. 170 (2–3), 791–797.

Ait Ali, N., Bernal, M.P., Ater, M., 2002. Tolerance and bioaccumulation of copper in *Phragmites australis* and *Zea mays*. Plant and Soil 239 (1), 103–111.

Akbar, M., Ahmed, O., Jamaluddin, A., Majid, N., Abdul-Hamid, H., Jusop, S., Hassan, A., Yusof, K.H., Abdu, A., 2010. Differences in soil physical and chemical properties of rehabilitated and secondary forests. American J. Appl. Sci. 7 (9), 1200–1209.

Alshaebi, F.Y., Yaacob, W.Z.W., Samsudin, A.R., Alsabahi, E., 2009. Risk assessment at abandoned tin mine in Sungai Lembing, Pahang, Malaysia. Electron. J. Geotechnical Eng. 14, 1–9.

Alvarenga, P., Goncalves, A.P., Fernandes, R.M., de Varennes, A., Vallini, G., Duarte, E., Cunha-Queda, A.C., 2008. Evaluation of composts and liming materials in the phytostabilization of a mine soil using perennial ryegrass. Sci. Total. Environ. 406 (1–2), 43–56.

Alvarenga, P., Gonçalves, A.P., Fernandes, R.M., de Varennes, A., Vallini, G., Duarte, E., Cunha-Queda, A.C., 2009. Organic residues as immobilizing agents in aided phytostabilization: (I) Effects on soil chemical characteristics. Chemosphere 74 (10), 1292–1300.

Amini, M., Khademi, H., Afyuni, M., Abbaspour, K.C., 2005. Variability of available cadmium in relation to soil properties and landuse in an arid region in central Iran. Water, Air, & Soil Pollution 162 (1), 205–218.

Ang, L.H., Tang, L.K., Ho, W.M., Hui, T.F., Theseira, G.W., 2010. Phytoremediation of Cd and Pb by four tropical timber species grown on an Ex-tin Mine in Peninsular Malaysia. World Academy of Sci. Eng. Technol. 2, 244–248.

Ang, L.H., Tang, L.K., Hui, T.F., Ho, W.M., Theisera, G.W., 2003. Bioaccumulation of heavy metals by *Acacia mangium*, *Hopea odorata*, *Intsia palembanica* and *Swietenia macrophylla* grown on slime tailings. Project No. 05-03-10-SF0038. Forest Research Institute Malaysia (FRIM), Malaysia. 22–26.

Ariyakanon, N., Winaipanich, B., 2006. Phytoremediation of copper contaminated soil by *Brassica juncea* (L.) Czern and *Bidens alba* (L.) DC. var. radiata. J. Sci. Res. Chula. Univ. 31, 49–56.

Atienzar, F.A., Cheung, V.V., Jha, A.N., Depledge, M.H., 2001. Fitness parameters and DNA effects are sensitive indicators of copper-induced toxicity in Daphnia magna. Toxicol. Sci. 59 (2), 241.

Babu, T.S., Marder, J.B., Tripuranthakam, S., Dixon, D.G., Greenberg, B.M., 2001. Synergistic effects of a photooxidized polycyclic aromatic hydrocarbon and copper on photosynthesis and plant growth: Evidence that in vivo formation of reactive oxygen species is a mechanism of copper toxicity. Environ. Toxicol. Chem. 20 (6), 1351–1358.

Baccio, D., Tognetti, R., Sebastiani, L., Vitagliano, C., 2003. Responses of *Populus deltoides* × *Populus nigra* (*Populus* × *euramericana*) clone I-214 to high zinc concentrations. New Phytol. 159 (2), 443–452. 410.1046/j.1469-8137.2003.00818.x.

Baker, A.J.M., 1981. Accumulators and excluders—strategies in the response of plants to heavy metals. J. Plant Nutr. (United States) 3, 1–4.

Baker, A.J.M., Brooks, R.R., 1989. Terrestrial higher plants which hyperaccumulate metallic elements. A review of their distribution, ecology and phytochemistry. Biorecovery 1 (2), 81–126.

Baker, A.J.M., McGrath, S.P., Reeves, R.D., Smith, J.A.C., 2000. Metal hyperaccumulator plants: A review of the ecology and physiology of a biological resource for phytoremediation of metal-polluted soils. In: Terry, N., Banuelos, G. (Eds.), Lewis Publishers, Boca Raton, FL, pp. 85–107.

Baker, A.J.M., Whiting, S.N., 2002. In search of the Holy Grail—a further step in understanding metal hyperaccumulation? New Phytol. 155 (1), 1–4.

Banasova, V., Horak, O., Nadubinska, M., Ciamporova, M., 2008. Heavy metal content in *Thlaspi caerulescens* J. et C. Presl growing on metalliferous and non-metalliferous soils in Central Slovakia. Int. J. Environ. Pollut. 33 (2), 133–145.

Bañuelos, G.S., Zambrzuski, S., Mackey, B., 2000. Phytoextraction of selenium from soils irrigated with selenium-laden effluent. Plant and Soil 224 (2), 251–258.

Barceló, J., Poschenrieder, C., 2003. Phytoremediation: Principles and perspectives. Contrib. Sci. 2 (3), 333–344.

Beladi, M., Habibi, D., Kashani, A., Paknejad, F., Nooralvandi, T., 2011. Phytoremediation of lead and copper by Sainfoin (*Onobrychis viciifolia*): Role of antioxidant enzymes and biochemical biomarkers. American–Eurasian J. Agric. Environ. Sci. 10 (3), 440–449.

Benavides, M.P., Gallego, S.M., Tomaro, M.L., 2005. Cadmium toxicity in plants. Braz. J. Plant Physiol. 17, 21–34.

Benimeli, C.S., Medina, A., Navarro, C.M., Medina, R.B., Amoroso, M.J., Gómez, M.I., 2009. Bioaccumulation of copper by *Zea mays*: Impact on root, shoot and leaf growth. Water, Air, & Soil Pollut. 55, 1–6.

Bjelková, M., Gencurová, V., Griga, M., 2011. Accumulation of cadmium by flax and linseed cultivars in field-simulated conditions: A potential for phytoremediation of Cd-contaminated soils. Ind. crops and prod. 33, 761–774.

Blaylock, M.J., Huang, J.W., 2000. Phytoextraction of metals. Phytoremediation of Toxic Metals: Using Plants to Clean-up the Environment. John Wiley & Sons, Inc, New York. 53–70.

Borkert, C.M., Cox, F.R., Tucker, M.R., 1998. Zinc and copper toxicity in peanut, soybean, rice, and corn in soil mixtures. Commun. Soil Sci. Plant Anal. 29 (19), 2991–3005.

Bose, S., Vedamati, J., Rai, V., Ramanathan, A., 2008. Metal uptake and transport by *Tyaha angustata* L. grown on metal contaminated waste amended soil: An implication of phytoremediation. Geoderma 145 (1–2), 136–142.

Brazeau, B.J., Johnson, B.J., Wilmot, C.M., 2004. Copper-containing amine oxidases. Biogenesis and catalysis; a structural perspective. Arch. Biochem. Biophys. 428 (1), 22–31.

Brun, L., Le Corff, J., Maillet, J., 2003. Effects of elevated soil copper on phenology, growth and reproduction of five ruderal plant species. Environ. Pollut. 122 (3), 361–368.

Brunet, J., Repellin, A., Varrault, G., Terryn, N., Zuily-Fodil, Y., 2008. Lead accumulation in the roots of grass pea (*Lathyrus sativus* L.): A novel plant for phytoremediation systems? C. R. Biol. 331 (11), 859–864.

Burke, S.C., Angle, J.S., Chaney, R.L., Cunningham, S.D., 2000. *Arbuscular mycorrhizae* effects on heavy metal uptake by corn. Int. J. Phytoremediation 2, 23–29.

Chehregani, M., Lorestani, B., Yousefi, N., 2011. Introduction of hyperaccumulator plants with phytoremediation potential of lead-zinc mine in Iran. World Acad. Sci. Eng. Technol. 77, 163–168.

Chen, X., Wang, J., Shi, Y., Zhao, M.Q., Chi, G.Y., 2011. Effects of cadmium on growth and photosynthetic activities in pakchoi and mustard. Botanical Studies 52 (1), 41–46.

Chen, Y.X., Lin, Q., Luo, Y.M., He, Y.F., Zhen, S.J., Yu, Y.L., Tian, G.M., Wong, M.H., 2003. The role of citric acid on the phytoremediation of heavy metal contaminated soil. Chemosphere 50 (6), 807–811.

Chiew, H., 2007. Poisonous wasteland. Lifestyle. Retrieved 18 December, 2010, from http://www.sacredland.org/mount-kinabalu/.

Clemens, S., 2001. Molecular mechanisms of plant metal tolerance and homeostasis. Planta 212 (4), 475–486.

Cluis, C., 2004. Junk-greedy Greens: Phytoremediation as a new option for soil decontamination. BioTeach J. 2, 61–67.

Danh, L.T., Truong, P., Mammucari, R., Tran, T., Foster, N., 2009. Vétiver grass, *Vetiveria zizanioides*: A choice plant for phytoremediation of heavy metals and organic wastes. Int. J. Phytoremediation 11 (8), 664–691.

Del Rio Celestino, M., Font, R., Moreno-Rojas, R., De Haro-Bailón, A., 2006. Uptake of lead and zinc by wild plants growing on contaminated soils. Ind. Crops and Prod. 24 (3), 230–237.

Dembitsky, V.M., Rezanka, T., 2003. Natural occurrence of arseno compounds in plants, lichens, fungi, algal species, and microorganisms. Plant Sci. 165 (6), 1177–1192.

Dengyi, L.I.U., Youbao, W., 2002. Effects of Cu and As on germination and seedling growth of crops.[J]. Chin. J. Appl. Ecol. 2 (13), 179–182.

DOE, 1998. Department of environment. Ministry of science, technology and the environment. Retrieved 19 July, 2010, from www.tshe.org/ea/pdf/vol3s%20p50-55.pdf.

Environmetal News Service, 2006. Environmetal News Service. New York.

Ernst, W.H.O., Nelissen, H.J.M., Ten Bookum, W.M., 2000. Combination toxicology of metal-enriched soils: Physiological responses of a Zn- and Cd-resistant ecotype of *Silene vulgaris* on polymetallic soils. Environ. Exp. Bot. 43 (1), 55–71.

Evangelou, M.W.H., 2007. Biochelators as an alternative to EDTA and other synthetic chelators for the phytoextraction of heavy metals (Cu, Cd, Pb) from soil. Universitätsbibliothek, 1–172. PHD Thesis: http://darwin.bth.rwth-aachen.de/opus3/volltexte/2007/1906/pdf/Evangelou_Michael.pdf

Fässler, E., Robinson, B.H., Stauffer, W., Gupta, S.K., Papritz, A., Schulin, R., 2010. Phytomanagement of metal-contaminated agricultural land using sunflower, maize and tobacco. Agriculture. Ecosystems & Environ. 136 (1–2), 49–58.

Gambling, L., Dunford, S., McArdle, H.J., 2004. Iron deficiency in the pregnant rat has differential effects on maternal and fetal copper levels. J. Nutr. Biochem. 15 (6), 366–372.

Garbisu, C., Alkorta, I., 2001. Phytoextraction: A cost-effective plant-based technology for the removal of metals from the environment. Bioresour. Technol. 77 (3), 229–236.

Gardea-Torresdey, J.L., Peralta-Videa, J.R., De La Rosa, G., Parsons, J.G., 2005. Phytoremediation of heavy metals and study of the metal coordination by X-ray absorption spectroscopy. Coordination Chem. Rev. 249 (17–18), 1797–1810.

Ghassemzadeh, F., Yousefzadeh, H., Arbab-Zavar, M.H., 2008. Arsenic phytoremediation by *Phragmites australis*: green technology. Int. J. Environ. Stud. 65 (4), 587–594.

Ghosh, M., Singh, S.P., 2005. A review on phytoremediation of heavy metals and utilization of it's by products. Appl. Ecol. Environ. Res. 3 (1), 1–18.

Grifferty, A., Barrington, S., 2000. Zinc uptake by young wheat plants under two transpiration regimes. J. Environ. Qual. 29 (2), 443–446.

Gupta, A.K., Sinha, S., 2007. Phytoextraction capacity of the plants growing on tannery sludge dumping sites. Bioresour. Technol. 98 (9), 1788–1794.

Haque, N., Peralta-Videa, J.R., Jones, G.L., Gill, T.E., Gardea-Torresdey, J.L., 2008. Screening the phytoremediation potential of desert broom (*Baccharis sarothroides* Gray) growing on mine tailings in Arizona, USA. Environ. Pollut. 153 (2), 362–368.

Hinsinger, P., Courchesne, F., 2007. Mobility and bioavailability of heavy metals and metalloids at soil–root interface. Biophysico-chemical Process. Heavy Metals and Metalloids in Soil Environments 1, 9–15.

Hussain, I., Khan, L., 2010. Comparative study on heavy metal contents in *Taraxacum Officinale*. Int. J. Pharmacognosy and Phytochem. Res. 2 (1), 15–18.

Jadia, C.D., Fulekar, M.H., 2008a. Phytoremediation: The application of vermicompost to remove zinc, cadmium, copper, nickel and lead by sunflower plant. Environ. Eng. Manag. J. 7 (5), 547–558.

Jadia, C.D., Fulekar, M.H., 2008b. Phytotoxicity and remediation of heavy metals by Alfalfa (*Medicago sativa*) in soil–vermicompost media. Adv. Nat. Appl. Sci. 2 (3), 141–151.

Jadia, C.D., Fulekar, M.H., 2009. Phytoremediation of heavy metals: Recent techniques. Afr. J. Biotechnol. 8 (6), 921–928.

Jiang, L.Y., Yang, X.E., He, Z.L., 2004. Growth response and phytoextraction of copper at different levels in soils by *Elsholtzia splendens*. Chemosphere 55 (9), 1179–1187.

Jopony, M., Tongkul, F., 2009. Acid mine drainages at Mamut copper mine, Sabah, Malaysia. Borneo Sci. 24, 83–94.

Justin, V., Majid, N., Islam, M.M., Abdu, A., 2011. Assessment of heavy metal uptake and translocation in *Acacia mangium* for phytoremediation of cadmium contaminated soil. J. Food, Agriculture & Environ. 9 (2), 588–592.

Kabata-Pendias, A., Pendias, H., 1984. Trace Elements in Plants and Soils. Florida: CRC Press, USA. p. 413.

Kabata-Pendias, A., Pendias, H., 2001. Trace Elements in Soils and Plants. Retrieved June, 2010, from http://www.srcosmos.gr.

Karaca, A., 2004. Effect of organic wastes on the extractability of cadmium, copper, nickel, and zinc in soil. Geoderma. 122 (2–4), 297–303.

Karami, A., Shamsuddin, Z.H., 2010. Phytoremediation of heavy metals with several efficiency enhancer methods. Afr. J. Biotechnol. 9 (25), 3689–3698.

Ke, W., Xiong, Z., Jin, Z., Ke, S., 2007a. Differences of Cu uptake and acid phosphatase activities of two *Elsholtzia haichowensis* Sun populations. *Acta Ecologica Sinica* 27 (8), 3172–3181.

Ke, W., Xiong, Z.T., Chen, S., Chen, J., 2007b. Effects of copper and mineral nutrition on growth, copper accumulation and mineral element uptake in two *Rumex japonicus* populations from a copper mine and an uncontaminated field sites. Environ. Exp. Bot. 59 (1), 59–67.

Khatun, S., Ali, M.B., Hahn, E.J., Paek, K.Y., 2008. Copper toxicity in *Withania somnifera*: Growth and antioxidant enzymes responses of in vitro grown plants. Environ. Exp. Bot. 64 (3), 279–285.

Khellaf, N., Zerdaoui, M., 2009. Growth response of the duckweed *Lemna minor* to heavy metal pollution. Iran. J. Environ. Health and Sci. Eng. 6 (3), 161–166.

Khellaf, N., Zerdaoui, M., 2010. Growth response of the duckweed *Lemna gibba* L. to copper and nickel phytoaccumulation. Ecotoxicology 19 (8), 1363–1368.

Kopittke, P.M., Menzies, N.W., 2006. Effect of Cu toxicity on growth of cowpea (*Vigna unguiculata*). Plant and Soil 279 (1), 287–296.

Kos, B., Le tan, D., 2003. Influence of a biodegradable ([S,S]-EDDS) and nondegradable (EDTA) chelate and hydrogel modified soil water sorption capacity on Pb phytoextraction and leaching. Plant and Soil 253 (2), 403–411.

Kuzovkina, Y.A., Knee, M., Quigley, M.F., 2004. Cadmium and copper uptake and translocation in five Willow (*Salix* L.) species. Int. J. Phytoremediation 6 (3), 269–287.

Lasat, M.M., 2000. Phytoextraction of metals from contaminated soil: A review of plant / soil / metal interaction and assessment of pertinent agronomic issues. J. Hazard. Subst. Res. 2 (5), 1–25.

Lasat, M.M., 2002. Phytoextraction of toxic metals: A review of biological mechanisms. J. Environ. Qual. 31, 109–120.

Li, M.S., 2006. Ecological restoration of mineland with particular reference to the metalliferous mine wasteland in China: A review of research and practice. Sci. Total Environ. 357 (1–3), 38–53.

Lim, J.M., Salido, A.L., 2004. Phytoremediation of lead using Indian mustard (*Brassica juncea*) with EDTA and electrodics. Microchemical J. 76 (1–2), 3–9.

Lin, J., Jiang, W., Liu, D., 2003. Accumulation of copper by roots, hypocotyls, cotyledons and leaves of sunflower (*Helianthus annuus* L.). Bioresour. Technol. 86 (2), 151–155.

Liu, W., Zhou, Q., An, J., Sun, Y., Liu, R., 2010. Variations in cadmium accumulation among Chinese cabbage cultivars and screening for Cd-safe cultivars. J. Hazard. Mater. 173 (1–3), 737–743.

Liu, X., Shen, Y., Lou, L., Ding, C., Cai, Q., 2009b. Copper tolerance of the biomass crops Elephant grass (*Pennisetum purpureum Schumach*), Vetiver grass (*Vetiveria zizanioides*) and the upland reed (*Phragmites australis*) in soil culture. Biotechnol. Adv. 27 (5), 633–640.

Liu, Z., He, X., Chen, W., Yuan, F., Yan, K., Tao, D., 2009a. Accumulation and tolerance characteristics of cadmium in a potential hyperaccumulator—*Lonicera japonica* Thunb. J. Hazard. Mater. 169 (1–3), 170–175.

Lombi, E., Zhao, F.J., Dunham, S.J., McGrath, S.P., 2001. Phytoremediation of heavy metal-contaminated soils: Natural hyperaccumulation versus chemically enhanced phytoextraction. J. Environ. Qual. 30 (6), 1919–1926.

Long, X., Yang, X., Ni, W., 2002. Current situation and prospect on the remediation of soils contaminated by heavy metals. J. Appl. Ecol. 13 (6), 757–762.

Lorestani, B., Chehregani, M., Yousefi, N., 2011. Phytoremediation potential of native plants grwoing on a heavy metals contaminated soil of copper mine in Iran. World Academic of Science. Eng. Technol. 77, 377–382.

Luo, C., Shen, Z., Lou, L., Li, X., 2006. EDDS and EDTA-enhanced phytoextraction of metals from artificially contaminated soil and residual effects of chelant compounds. Environmental Pollution 144 (3), 862–871.

Ma, Y., Rajkumar, M., Freitas, H., 2008. Inoculation of plant growth promoting bacterium *Achromobacter xylosoxidans* strain Ax10 for the improvement of copper phytoextraction by *Brassica juncea*. J. Environ. Manag. 90 (2), 831–837.

Macnair, M.R., 1987. Heavy metal tolerance in plants: A model evolutionary system. Trends in Ecol. Evolution 2 (12), 354–359.

MacPherson, I.S., Murphy, M.E.P., 2007. Type-2 copper-containing enzymes. Cell. Mol. Life Sci. 64 (22), 2887–2899.

Malik, R.N., Husain, S.Z., Nazir, I., 2010. Heavy metal contamination and accumulation in soil and wild plant species from industrial area of Islamabad, Pakistan. Pak. J. Bot. 42 (1), 291–301.

Mangkoedihardjo, S., Surahmaida, 2008. *Jatropha curcas* L. for phytoremediation of lead and cadmium polluted soil. World Appl. Sci. J. 4 (4), 519–522.

Manousaki, E., Kadukova, J., Papadantonakis, N., Kalogerakis, N., 2008. Phytoextraction and phytoexcretion of Cd by the leaves of *Tamarix smyrnensis* growing on contaminated non-saline and saline soils. Environ. Res. 106 (3), 326–332.

Mattina, M.J.I., Lannucci-Berger, W., Musante, C., White, J.C., 2003. Concurrent plant uptake of heavy metals and persistent organic pollutants from soil. Environ. Pollut. 124 (3), 375–378.

McGrath, S.P., Zhao, F.J., 2003. Phytoextraction of metals and metalloids from contaminated soils. Curr. Opin. Biotechnol. 14 (3), 277–282.

Meers, E., Lamsal, S., Vervaeke, P., Hopgood, M., Lust, N., Tack, F.M.G., 2005. Availability of heavy metals for uptake by *Salix viminalis* on a moderately contaminated dredged sediment disposal site. Environ. Pollut. 137 (2), 354–364.

Mehta, R., Templeton, D.M., O'Brien, P.J., 2006. Mitochondrial involvement in genetically determined transition metal toxicity: II. Copper toxicity. Chem. Biol. Interact. 163 (1–2), 77–85.

Merian, E., Anke, M., Ihnat, M., Stoeppler, M., 2004. Elements and their compounds in the environment: Occurence, analysis and biological relevance. Vch Verlagsgesellschaft Mbh.1, 1773.

Mohamed, A.F., Wan Yaacob, W.Z., Taha, M.R., Samsudin, A.R., 2009. Groundwater and soil vulnerability in the langat Basin Malaysia. Eur. J. Sci. Res. 27 (4), 628–635.

Monni, S., Salemaa, M., White, C., Tuittila, E., Huopalainen, M., 2000. Copper resistance of *Calluna vulgaris* originating from the pollution gradient of a Cu–Ni smelter, in southwest Finland. Environ. Pollut. 109 (2), 211–219.

Moore, A.D., Alva, A.K., Collins, H.P., Boydston, R.A., 2010. Mineralization of nitrogen from biofuel by-products and animal manures amended to a sandy soil. Commun. Soil Sci. Plant Anal. 41 (11), 1315–1326.

Murakami, M., Ae, N., 2009. Potential for phytoextraction of copper, lead, and zinc by rice (*Oryza sativa* L.), soybean (*Glycine max* [L.] Merr.), and maize (*Zea mays* L.). J. Hazard. Mater. 162 (2–3), 1185–1192.

Nagaraju, A., Karimulla, S., 2002. Accumulation of elements in plants and soils in and around Nellore mica belt, Andhra Pradesh, India: A biogeochemical study. Environ. Geology 41 (7), 852–860.

Nascimento, C.W.A., Xing, B., 2006. Phytoextraction: A review on enhanced metal availability and plant accumulation. Sci. Agricola 63, 299–311.

Niu, Z.X., Sun, L.N., Sun, T.H., Li, Y.S., Wang, H., 2007. Evaluation of phytoextracting cadmium and lead by sunflower, ricinus, alfalfa and mustard in hydroponic culture. J. Environ. Sci. 19 (8), 961–967.

Nwaichi, E.O., Onyeike, E.N., 2010. Cu tolerance and accumulation by *Centrosema Pubescen* Benth and *Mucuna Pruriens* Var Pruriens. Arch. Appl. Sci. Res. 2 (3), 238–247.

Odjegba, V.J., Fasidi, I.O., 2004. Accumulation of trace elements by *Pistia stratiotes*: implications for phytoremediation. Ecotoxicology 13 (7), 637–646.

Perez, D.V., Anjos, L.H.C., Ebeling, A.G., Pereira, M.G., 2009. Comparison of H / Al stoichiometry of mineral and organic soils in Brazil. Rev. Bras. Ciên. do Solo 33 (4), 1071–1076.

Pivetz, B.E., 2001. Ground Water Issue: Phytoremediation of contaminated soil and ground water at hazardous waste sites. 1–36.

Rajakaruna, N., Baker, A.J.M., 2004. Serpentine: A model habitat for botanical research in Sri lanka. Ceylon J. Sci. (Biological Sciences) 32, 1–19.

Rajakaruna, N., Böhm, B.A., 2002. Serpentine and its vegetation: A preliminary study from Sri Lanka. J. Appl. Bot. Angew. Botanik 76, 20–28.

Rajakaruna, N., Tompkins, K.M., Pavicevic, P.G., 2006. Phytoremediation: An affordable green technology for the clean-up of metal-contaminated sites in Sri Lanka. Ceylon J. Sci. (Biological Sciences) 35, 25–39.

Raskin, I., Ensley, B.D., 2000. Phytoremediation of Toxic Metals: Using Plants to Clean up the Environment. John Wiley & Sons, Inc. Publishing, New York. 304.

Raskin, I., Kumar, P.B.A., Dushenkov, S., Salt, D.E., 1994. Bioconcentration of heavy metals by plants. Curr. Opin. Biotechnol. 5 (3), 285–290.

Robinson, B., Schulin, R., Nowack, B., Roulier, S., Menon, M., Clothier, B., Green, S., Mills, T., 2006. Phytoremediation for the management of metal flux in contaminated sites. Forest. Snow Landscape Res. 80 (2), 221–234.

Ross, S.M., 1994. Toxic metals in soil–plant systems. John Wiley & Sons Ltd. 1–484.

Rosselli, W., Keller, C., Boschi, K., 2003. Phytoextraction capacity of trees growing on a metal contaminated soil. Plant and Soil 256, 265–272.

Ruiz, J.M., Blasco, B., Ríos, J.J., Cervilla, L.M., Rosales, M.A., Rubio-Wilhelmi, M.M., Sánchez-Rodríguez, E., Castellano, R., Romero, L., 2009. Distribution and efficiency of the phytoextraction of cadmium by different organic chelates. Terra Latinoam. 27 (4), 296–301.

Saraswat, S., Rai, J.P.N., 2009. Phytoextraction potential of six plant species grown in multimetal contaminated soil. Chem. Ecol. 25 (1), 1–11.

Sarma, H., 2011. Metal hyperaccumulation in plants: A review focusing on phytoremediation technology. J. Environ. Sci. Technol. 4, 118–138.

Schmidt, U., 2003. Enhancing phytoextraction: The effects of chemical soil manipulation on mobility, plant accumulation, and leaching of heavy metals. J. Environ. Qual. 32 (6), 1939–1954.

Sereno, M.L., Almeida, R.S., Nishimura, D.S., Figueira, A., 2007. Response of sugarcane to increasing concentrations of copper and cadmium and expression of metallothionein genes. J. Plant Physiol. 164 (11), 1499–1515.

Sekara, A., Poniedzialeek, M., Ciura, J., Jedrszczyk, E., 2005. Cadmium and lead accumulation and distribution in the organs of nine crops: Implications for phytoremediation. Polish J. Environ. Stud. 14 (4), 509–516.

Sheoran, V., Sheoran, A.S., Poonia, P., 2009. Phytomining: A review. Minerals Eng. 22 (12), 1007–1019.

Shorrocks, V.M., Alloway, B.J., 1988. Copper in plant, animal and human nutrition. Copper in Plant. Animal and Human Nutrition, 1–106.

Shuhe, W., Qixing, Z., Xin, W., Kaisong, Z., Guanlin, G., Qiying, M.A.L., 2005. A newly-discovered Cd-hyperaccumulator *Solanum nigrum* L. Chin. Sci. Bull. 50 (1), 33–38.

Singh, O.V., Labana, S., Pandey, G., Budhiraja, R., Jain, R.K., 2003. Phytoremediation: An over-view of metallic ion decontamination from soil. Appl. Microbiol. Biotechnol. 61 (5), 405–412.

Sinha, S., Gupta, A.K., Bhatt, K., Pandey, K., Rai, U.N., Singh, K.P., 2006. Distribution of metals in the edible plants grown at Jajmau, Kanpur (India) receiving treated tannery wastewater: Relation with physico-chemical properties of the soil. Environ. Monit. Assess. 115 (1), 1–22.

Soleimani, M., Hajabbasi, M.A., Afyuni, M., Mirlohi, A., Borggaard, O.K., Holm, P.E., 2010. Effect of endophytic fungi on cadmium tolerance and bioaccumulation by *Festuca arundinacea* and *Festuca pratensis*. Int. J. Phytoremediation 12 (6), 535–549.

Song, J., Zhao, F.J., Luo, Y.M., McGrath, S.P., Zhang, H., 2004. Copper uptake by *Elsholtzia splendens* and *Silene vulgaris* and assessment of copper phytoavailability in contaminated soils. Environ. Pollut. 128 (3), 307–315.

Sukganah, A., Lee, H.H., Choong, C.Y., Wickneswari, R., 2009. Characterization of lignin genes in *Acacia auriculiformis* × *Acacia mangium* hybrid for wood pulp quality. In: online: Proceedings of Seminar on *Acacia* Research in Malaysia, pp. 1–5. 12 July 2005, Putrajaya.

Sun, Q., Ye, Z.H., Wang, X.R., Wong, M.H., 2007a. Cadmium hyperaccumulation leads to an increase of glutathione rather than phytochelatins in the cadmium hyperaccumulator *Sedum alfredii*. J. Plant Physiol. 164 (11), 1489–1498.

Sun, R.L., Zhou, Q.X., Sun, F.H., Jin, C.X., 2007b. Antioxidative defense and proline / phytochela-tin accumulation in a newly discovered Cd-hyperaccumulator, *Solanum nigrum* L. Environ. Exp. Bot. 60 (3), 468–476.

Sun, Y., Zhou, Q., Diao, C., 2008. Effects of cadmium and arsenic on growth and metal accumula-tion of Cd-hyperaccumulator *Solanum nigrum* L. Bioresource Technol. 99 (5), 1103–1110.

Sun, Y., Zhou, Q., Wang, L., Liu, W., 2009. Cadmium tolerance and accumulation characteris-tics of *Bidens pilosa* L. as a potential Cd-hyperaccumulator. J. Hazard. Mater. 161 (2–3), 808–814.

Susarla, S., Medina, V.F., McCutcheon, S.C., 2002. Phytoremediation: an ecological solution to organic chemical contamination. Ecol. Eng 18 (5), 647–658.

Tan, K.H., 2005. Soil Sampling, Preparation, and Analysis. CRC. 122–127.

Turan, M., Esringu, A., 2007. Phytoremediation based on canola (*Brassica napus* L.) and Indian mustard (*Brassica juncea* L.) planted on spiked soil by aliquot amount of Cd, Cu, Pb, and Zn. Plant Soil and Environ. 53 (1), 7–15.

United States Environmental Protection Agency, 2001. Citizen's Guide to Phytoremediation. Retrieved 4 January, 2011, from http://www.clu-in.org/download/citizens/citphyto.pdf.

United States Protection Agency, 1997. Introduction to Phytoremediation, EPA 600/R-99/107. Environmental Protection Agency (EPA). Office of Research and Development, Cincinnati, OH, US.

Van Nevel, L., Mertens, J., Oorts, K., Verheyen, K., 2007. Phytoextraction of metals from soils: How far from practice? Environ. Pollut. 150 (1), 34–40.

Vidali, M., 2001. Bioremediation. An overview. Pure and Appl. Chem. 73 (7), 1163–1172.

Wei, S., Teixeira da Silva, J.A., Zhou, Q., 2008. Agro-improving method of phytoextracting heavy metal contaminated soil. J. Hazard. Mater. 150 (3), 662–668.

Wei, S.H., Zhou, Q.X., 2004. Discussion on basic principles and strengthening measures for phy-toremediation of soils contaminated by heavy metals. Chin. J. Ecol. 23 (1), 65–72.

Weis, J.S., Weis, P., 2004. Metal uptake, transport and release by wetland plants: Implications for phytoremediation and restoration. Environ. Int. 30 (5), 685–700.

Wu, F., Yang, W., Zhang, J., Zhou, L., 2009. Cadmium accumulation and growth responses of a poplar (*Populus deltoids* × *Populus nigra*) in cadmium contaminated purple soil and alluvial soil. J. Hazard. Mater., 1–6. http://dx.doi.org/10.1016/j.jhazmat.2009.1012.1028.

Xiong, Z.T., Wang, H., 2005. Copper toxicity and bioaccumulation in Chinese Cabbage (*Brassica pekinensis* Rupr.). Environ. Toxicol. 20, 188–194.

Yadav, S.K., Dhote, M., Kumar, P., Sharma, J., Chakrabarti, T., Juwarkar, A.A., 2010. Differential antioxidative enzyme responses of *Jatropha curcas* L. to chromium stress. J. Hazard. Mater. 180 (1–3), 609–615.

Yadav, S.K., Juwarkar, A.A., Kumar, G.P., Thawale, P.R., Singh, S.K., Chakrabarti, T., 2009. Bio-accumulation and phyto-translocation of arsenic, chromium and zinc by *Jatropha curcas* L.: Impact of dairy sludge and biofertilizer. Bioresour. Technol. 100 (20), 4616–4622.

Yan, H., Pan, G., 2002. Toxicity and bioaccumulation of copper in three green microalgal species. Chemosphere 49 (5), 471–476.

Yanai, J., Zhao, F.J., McGrath, S.P., Kosaki, T., 2006. Effect of soil characteristics on Cd uptake by the hyperaccumulator *Thlaspi caerulescens*. Environ. Pollut. 139 (1), 167–175.

Yang, X.E., Long, X.X., Ni, W.Z., Ye, Z.Q., He, Z.L., Stoffella, P.J., Calvert, D.V., 2002. Assessing copper thresholds for phytotoxicity and potential dietary toxicity in selected vegetable crops. J. Environ. Sci. Health. B. 37 (6), 625–635.

Ye, Z.H., Baker, A.J.M., Wong, M.H., Willis, A.J., 2003. Copper tolerance, uptake and accumulation by *Phragmites australis*. Chemosphere 50 (6), 795–800.

Yoon, J., Cao, X., Zhou, Q., Ma, L.Q., 2006. Accumulation of Pb, Cu, and Zn in native plants growing on a contaminated Florida site. Sci. Total Environ. 368 (2–3), 456–464.

Yruela, I., 2005. Copper in plants. Braz. J. Plant Physiol. 17 (1), 145–156.

Yruela, I., 2009. Copper in plants: Acquisition, transport and interactions. Funct. Plant Biol. 36 (5), 409–430.

Zhang, B.Y., Zheng, J.S., Sharp, R.G., 2010. Phytoremediation in engineered wetlands: Mechanisms and applications. Procedia Environ. Scien. 2, 1315–1325.

Zhang, J., Wang, S., Feng, Z., Wang, Q., 2009. Stability of soil organic carbon changes in successive rotations of Chinese fir (*Cunninghamia lanceolata* (Lamb.) Hook) plantations. J. Environ. Sci. 21 (3), 352–359.

Zheljazkov, V.D., Craker, L.E., Xing, B., 2006. Effects of Cd, Pb, and Cu on growth and essential oil contents in dill, peppermint, and basil. Environ. Exp. Bot. 58 (1–3), 9–16.

Zheljazkov, V.D., Jeliazkova, E.A., Kovacheva, N., Dzhurmanski, A., 2008. Metal uptake by medicinal plant species grown in soils contaminated by a smelter. Environ. Exp. Bot. 64 (3), 207–216.

Zhou, Q.X., Song, Y.F., 2004. Principles and methods of contaminated soil remediation. Science Press, Beijing. 568.

Chapter 8

Role of Phytoremediation in Radioactive Waste Treatment

L.F. De Filippis

School of the Environment, University of Technology, Sydney, NSW, Australia

INTRODUCTION

Radiation exists all around us, and radioactive materials provide numerous benefits to humans and society, and play a significant role in daily life. Applications for radiation include scientific, medical, agricultural, industrial and energy generation. Therefore it is inevitable that such diverse activities lead to radioactive waste being generated. The International Atomic Energy Agency (IAEA, 2010) has defined radioactive waste as any material that contains levels of radioactive emitting particles greater than those deemed safe by national and international standards, and for which no use is envisaged. For example, hazardous radioactive waste is generated at every stage of the uranium nuclear fuel cycle. It is unfortunate that at present no country has established permanent repositories and storage facilities for the most dangerous high-level nuclear wastes from nuclear power plants. Radioactive waste can be solid, liquid or gas from a diverse group of operations and activities (uranium mining and nuclear power) and accidents (spills and reactor meltdown), but no matter of its origin, radioactivity poses a risk to human health, the environment and has the potential to disrupt ecosystems. The nuclear accident at Chernobyl, Ukraine alone has been calculated to have increased the risk of cancer to humans by 0.1% (Duschenkov, 2003; IAEA, 2005). Additionally, a recent report from Fukushima, Japan has already detected elevated levels of radioactive caesium in woody plants so soon after the accident (Yoshihara et al. 2013). Radioactive caesium (Cs), strontium (Sr), uranium (U) and plutonium (Pu) are the main radioactive isotopes present in the environment as a consequence of nuclear activities, and are the radionuclides of most concern (for a list of radionuclides of environmental and health concern refer to Table 8.1).

Some of the radioactive waste products have military potential; e.g. depleted uranium is used in munitions, and spent nuclear fuel from reactors contain weapons-usable plutonium. Nuclear waste on the one hand can be short-lived

Soil Remediation and Plants. http://dx.doi.org/10.1016/B978-0-12-799937-1.00008-5

radiation, which is generally of little concern because it disappears quickly by natural radioactive decay, and / or weak radioactive waste with intensities comparable to natural background radiation. However high-level and medium-level long-lived radioactive nuclear waste is more problematic and safe disposal of this waste is necessary (half-life and energy of environmentally dangerous isotopes are listed in Table 8.1). Most of the nuclear waste of concern is produced in nuclear power plants, and much of it is in temporary storage. In particular, the concern is for spent fuel elements recently removed from reactors, not yet reprocessed and held in cooling ponds. In addition, there is an alarming accumulation of separate radioactive material cast in glass or ceramics, encased in stainless steel containers and held in dry storage at deposits all over the world (LeBars and Pescatore, 2004; Comby, 2005). Radioactive waste needs to be managed in a safe and secure manner, and it must be isolated from people and the environment for as long enough a period as the waste remains dangerous.

Phytoremediation (*Phyto* = plant and *remediation* = correct evil) is a relatively new technology (introduced around the 1980s) and is the name given to a set of remediation methods that use plants to clean or partly clean contaminated sites, or render the contaminants less harmful (Barkay and Schaefer, 2001; Duschenkov, 2003; EPA, 2004; Eapen et al., 2007). Many techniques and applications have been called phytoremediation, possibly leading to confusion, especially when very toxic materials are in question (radioactive waste). Phytoremediation has been called green remediation, botano-remediation, agro-remediation and vegetative remediation (Olson and Fletcher, 2000; Pivetz, 2001). Phytoremediation is a general term and this document uses it primarily to refer to a set of plant–contaminant interactions, and not to any specific application method involved in remediation of radioactively contaminated sites. Many of the phytoremediation techniques involve applying information that has been known for decades in agriculture, silviculture, horticulture and ecology to environmental problems. Basic information for phytoremediation comes from a number of research areas; including ecotoxicology, plant ecophysiology, agriculture, soil science, hydrology, constructed wetlands and uptake, toxicity and translocation of radioactive isotopes.

Current engineering practices for remediation of radioactive contaminated soils rely heavily on 'excavation-and-dump' or 'encapsulation', neither of which addresses the issue of removal of the contaminant from the soil. The physical methods mentioned earlier may include removal of topsoil, soil washing, use of chelating agents, sand filtering, co-precipitation, ion exchange, flocculation, reverse osmosis and electrodialysis to aid remediation (Page, 2011; Das, 2012). Immobilization and extraction by physico-chemical techniques can be expensive and is often appropriate only for small areas, where rapid, complete decontamination is necessary, and the contaminant is placed safely elsewhere (Best et al., 2001; Mossman, 2003; Tadevosyan et al., 2011). Some methods, such as soil washing, precipitation and filtration have an adverse effect on biological activity, soil structure and fertility, and most require significant engineering

TABLE 8.1 Isotopes and Radionuclides of Importance in Waste Areas Often Used for Phytoremediation

Element	Radionuclides	Half-Life (h, d, y)	Radiation Emitted	Radiation Energy (MeV)	Notes on Effects and Danger with Reference to Plants and Human Health
Essential Radionuclides to Assess					
Cs	134, 137, 133	30 y–106 y	β	1.2–2.1	Biological properties like K, can substitute for K in cells. Highly toxic when released into the environment
Sr	89, 90, 86, 88	50 d–29 y	βγ	0.6–1.2	Properties like Ca and can substitute for Ca in tissues. Often toxic when released into the environment
U	238, 235, 234, 239	69 y–10⁹ y	αβ	4.2–205	Dangerous and abundant, contains natural fissile elements as reactant. Absorbed by most cells/organisms
Pu	238, 239, 240, 242	88 y–10⁵ y	α	0.1–560	Highly dangerous and accumulates especially in bones. Highly fissile/reactive, sustains chain reaction
Ra	226, 228, 223, 224	11 d–10³ y	αβ	4.9–6.0	All isotopes are highly radioactive/dangerous. A heavy metal which incorporates easily into cells/tissues
Rn	210, 222, 224	2.4 h–3.8 d	αβ	2.8–6.0	Noble gas that has only radioisotopes present. Intensely radioactive and dangerous even from natural sources
I	131, 129, 127, 135	6 h–10⁷ y	γβ	0.2–1.0	Strong mutagenic agent, especially on thyroid gland. Used in radiomedicine for treatment of cancers

Continued

TABLE 8.1 Isotopes and Radionuclides of Importance in Waste Areas Often Used for Phytoremediation—cont'd

Element	Radionuclides	Half-Life (h, d, y)	Radiation Emitted	Radiation Energy (MeV)	Notes on Effects and Danger with Reference to Plants and Human Health
Co	60, 59, 58, 57	71 d–5.3 y	βγ	2.3–2.8	Essential element, except in plants. Mostly man-made isotopes, and used as a strong mutagenic agent
Important Radionuclides to Assess					
H	3, 2	Stable–12.3 y	β	0.0–0.2	Incorporates into water and readily taken-up by vegetation. Easy to shield from and easily volatilized
Po	208, 209, 210	2.8 y–138 y	αβ	1.4–5.2	Highly dangerous, is taken-up with no biological function. Acute radiation toxic, used to treat lung cancers
Th	232, 228, 230, 230	24 d–10^{10} y	αβ	0.3–5.5	Produces radioactive Rn gas. More abundant than U in some ore bodies, alternate nuclear fuel use to U
Pb	204, 206, 207, 208	22 y–10^7 y	αβ	0.5–3.8	Heavy metal widely used in industry, mostly man-made isotopes. Poisonous and accumulates in brain
Hg	194, 203, 199, 201	9.9 h–444 y	β	0.6–2.0	Toxic heavy metal which bioaccumulates and poisonous. Radioactive isotopes can result from U fission
Eu	151, 153, 152	13 y–10^{18} y	αβ	1.8–2.2	Rare, biological role unknown. Lanthanide with unusual properties, similar to Np, a product of U fission
Desirable Radionuclides to Assess					
Ru	96, 102, 103	40 d–380 d	βγ	0.5–3.5	Rare with minor electrical/industrial applications. Natural occurring isotopes in U mines, mine tailings

Am	241, 243, 242	10^3 y–10^4 y	α	5.2–5.6	Product of U fission and often present in high levels in industrial waste. Adheres strongly to soil particles
Pm	147, 145, 146	2.6 y–17.7 y	αβ	0.2–1.5	Product of U fission with high industrial use. Radioisotopes used for luminescence and phosphorescence
Tc	99, 98, 97	4 d–10^6 y	βγ	0.2–0.8	All forms are radioactive. Product of U fission, present in waste and dangerous, used in cancer treatment
Np	237, 236, 235	2.5 d–10^6 y	αβ	1.0–5.2	All forms are radioactive. Product of U fission, present in waste and dangerous, used in cancer treatment
Cm	244, 245, 248, 250	160 d–10^7 y	αβ	5.1–6.2	All forms are radioactive. Product of Pu fission, often present in high levels in waste, more evaluation
Tl	208, 203, 204, 205	73 h–12 d	βγ	0.3–2.6	Element used in radiomedicine and cancer treatment. Dominant as background radiation on nuclear sites
Pd	107, 103, 100,	3.6 d–10^6 y	β	0.1–2.0	Catalytic converter element. Naturally contains stable/unstable isotopes. Widely used in many industries
Bi	209, 213, 208, 209	31 y–10^{19} y	β	0.2–6.2	Dense heavy metal, recently found to be slightly radioactive. Thermal use in industry, needs evaluation

Continued

TABLE 8.1 Isotopes and Radionuclides of Importance in Waste Areas Often Used for Phytoremediation—cont'd

Element	Radionuclides	Half-Life (h, d, y)	Radiation Emitted	Radiation Energy (MeV)	Notes on Effects and Danger with Reference to Plants and Human Health
Ac	227, 225, 226	30h–22y	β	0.6–6.0	Rare element with no apparent applications. Natural occurring isotopes in U mine ores, hard to separate
Ga	67, 69, 71	Stable–3.3 d	β	Very low	Biological properties like Fe, can substitute for Fe. Present in radiomedicine waste, treatment of tumours
Ce	141, 144,140	32d–10^{16}y	αβ	1.2–2.4	Various industrial uses but rare. Product of U fission, often present in high levels in U mine tailing waste
Y	90, 89, 88	2.7d–107d	β	0.5–2.3	Metallic transition element in electronic industry, used widely. Biological effects on lung cancer known
Zr	95, 90, 94, 93	78h–10^{17}y	αβ	0.4–3.3	Metal with high price extracted from U mining. No known biological action but present in U mine waste

Radionuclides are divided into assessment criteria of essential, important, desirable. Half lives of the isotopes are given in hours (h), days (d) or years (y), and notes on the nature of the isotopes, biological effects and health risks are detailed essentially from Chu et al. (1999).

inputs and costs. Consequently, the low-technology in situ approach of phytoremediation is attractive, as it offers partial site restoration, partial decontamination, maintenance of biological activity and physical structure of soils, and is potentially cheap, visually more attractive and there is the possibility of biorecovery of important radioactive nucleotides (Table 8.2 details the advantages and disadvantages of phytoremediation). There are six main subgroups in the technology commonly referred to as phytoremediation:

1. Phytoextraction: plants removing contaminants from the soil/tailings and concentrating the contaminants in the harvestable parts of plants; i.e., in all organs of the plant—leaves, stems and roots (Kumar et al., 1995; Banuelos et al., 1999).
2. Phytodegradation: plants degrading pollutants by using hydrolytic enzymes and metabolites in plants; however, this method may be limited only to degradation of organic contaminants (Burken and Schnoor, 1998; Schroeder et al., 2007).
3. Phytostabilization: plants reducing mobility and bioavailability of pollutants in the soil either by immobilization and precipitation, or by preventing contaminant migration (Smith and Bradshaw, 1979; Vangronsveld et al., 2009).
4. Phytovolatilization: volatilization of pollutants into the air directly or indirectly via plant uptake into tissues and organs, and then transformation of the products into volatile compounds (Banuelos et al., 1997; Burken and Schnoor, 1999).
5. Rhizofiltration: plant roots strongly absorbing, accumulating and/or precipitating contaminants from aqueous waste streams or soil water almost exclusively into the root system (Dushenkov et al., 1995; Prasad and Freitas, 2003).
6. Rhizodegradation: enhancement of naturally occurring biodegradation and destruction of contaminants in the soil through mineralization and transformation of pollutants by plant roots and associated microbes (Rugh, 2001; Sors et al., 2005).

The development of phytoremediation is being driven primarily by the high cost of many other engineering-based tailing and soil-remediation methods, as well as the desire to use a more environmentally friendly and sustainable process. However, with regard to phytoremediation of radioactive waste in general, only some of the phytoremediation technologies/methods above may be possible, and only with some low-level toxic categories of radioactive waste.

This chapter therefore describes the current international best practice in radioactive waste management, as defined primarily by the International Atomic Energy Agency (IAEA), and implemented successfully by many countries that utilize nuclear technology. The chapter draws on previous information provided by the excellent reviews of Pivetz (2001), Duschenkov (2003), Eapen et al. (2007) and Prasad (2007). However, these are now between 6 and 10 years of age and more information and research has become available. The review draws

TABLE 8.2 Advantages and Disadvantages of Phytoremediation Remediation Methods

Advantages	Disadvantages
Low cost in capital investment and equipment (constitutes substantial savings)	Long-term low performance and not capable of complete removal of contaminants
Lower cost in most of the operations used in phytoremediation, including labour costs	Not applicable to all types of wastes, especially high level (very toxic and dangerous) wastes
Can be applied on the site of contaminants, whether in the soil, water or ground water	Applicable mainly to surface or near-surface soils and waste areas and mine tailings
Usually aesthetically pleasing and the landscape is near what is acceptable to the public	A longer time period is required to observe visual remediation site effects
Low environmental impact on soil and water (i.e. it is non-destructive, non-intrusive, highly biologically active)	Plant matter in highly contaminated sites will require proper disposal and risk pathway identified
Very high public acceptance of the use of plants and vegetation to remediate waste areas	Seasonal factors may affect the effectiveness of the phytoremediation process
Use of vegetation will mean reduced erosion of soils, especially thin inorganic soils	In highly toxic and very dangerous wastes, death of plants may occur (not cost effective or pleasing)
Vegetation will also help reduce escape of particulate matter and spread of toxicants and leachate	Introduction and potential spread of inappropriate and invasive species of plants
Phytoremediation is now often required and is sometimes mandated by legislation	Bioconcentration and potential spread of contaminants higher in the food chain
Recovery of contaminant and maybe some cost recovery (i.e. biomedical radionuclides) possible	Wide range of response of plants to toxicants, and careful testing of plants is required
Very effective when low amounts, and low level (low toxicity/danger) contaminated waste is present	Good cultivation and maintenance of plants is required on sites (qualified personnel required)
Possibility of using and remediating wastes on soils that are not productive for agriculture	Cultivation practices and amendments for good plant growth may have adverse effects

Summary data were combined from the reviews of Pivetz (2001), Duschenkov (2003), Pulford and Watson (2003), Eapen et al. (2007) and EPA (2010).

on more recent scientific literature for all classes of radioactive waste, including disused radioactive products and other sources used in the nuclear fuel industry. Examples of existing practices are provided in the review, and conclusions are derived against what could only be described as a 'patchy' and 'inconsistent' international code of 'best practice' for treatment and remediation of radioactive waste.

RADIOACTIVE MATERIAL AND SAFETY

Mankind did not invent radioactivity and radioactive isotopes are found all around the world. The weak doses of natural radiation to which we have been exposed are normally not dangerous. It is worth noting that this natural background radiation is highly variable, changing by a factor of nearly a thousand from one place to another. For example, the city of Guarapari, Brazil contains a natural background radiation that would be considered dangerous for workers at a nuclear reactor or at a nuclear waste storage site. Yet the inhabitants have maintained normal health. This background radiation, mainly due to uranium and thorium diminishes slowly over time as these elements have a half-life comparable to the age of the earth (Brook, 2012; Mitchell et al., 2013). However high levels of radionuclides cause neurological diseases, infertility and birth defects and cancers of various tissues and organs (Benjamin and Wagner, 2006; McCombie, 2009; Das, 2012). It has been argued that the Earth in the past, when life began to develop, was more radioactive due to very active geological (volcanic) processes, and these higher levels of natural radiation have not stood in the way of evolution and adaptation of life. Human activity on our planet has not significantly increased the levels of ionizing radiation except in a few isolated places. But it is true that in 'mining and burning' uranium for energy we have accelerated its disappearance from natural geological deposits deep beneath the earth and into the surface environment.

Radioactive waste includes radioactive material that contains unstable elements (radionuclides), which decay over time. Radioactive decay causes the emission of radiation in the form of charged particles (α, β) and gamma (γ) rays. For any given material, radioactivity diminishes with time as radioactive elements decay into more stable elements. In some classes of radioactive waste, the radioactivity decays relatively quickly to a level low enough for the material to be reused or disposed of as normal industrial waste. The lower level and short-lived intermediate level radioactive waste has already been disposed of in many countries with normal industrial waste (Alvarez, 2007). In this way, radioactive waste differs from some other forms of hazardous environmental waste, such as heavy metals or asbestos which remain hazardous indefinitely. Current total global hazardous waste production is approximately 400 million tonnes per annum (Pandey et al., 2013). Radioactive waste from nuclear power plants and the nuclear fuel cycle comprise approximately 0.4 million tonnes per annum, or 0.1% of global hazardous waste. Over 75% of all radioactive waste (on a volume basis) has already been disposed of or assigned for disposal (calculated from Figure 8.1; ANSTO, 2011). Because of the wide variety of nuclear applications, the amounts, types and even physical forms of radioactive waste varies considerably. Some wastes (e.g. small radioactive sources found in smoke detectors) carry little or no safety or security risk; however, others are highly radioactive and must be managed appropriately to address safety and security issues forever. We can not justify an indefinite delay in finding a final resting place to store the most dangerous radioactive waste, since we already know that effective solutions are available where we can protect ourselves, and the environment from radiation poisoning.

A typical nuclear power reactor produces about 30 tonnes (10 m^3) of spent nuclear fuel annually, including 300 m^3 of low- and intermediate-level waste (ANSTO, 2011). Annually, nuclear power plants around the world produce 12,000–14,000 tonnes of spent fuel, and about 200,000 m^3 of low- and intermediate-level waste. In total, about 340,000 tonnes of spent fuel have been produced in power reactors around the world (IAEA, 2005; ANSTO, 2011). About one-third of this amount has been reprocessed, and the remainder is in storage. These are small amounts of waste compared to the mass or volume of waste generated by coal-fired electricity plants (Mossman, 2003; Ravichandran et al., 2011).

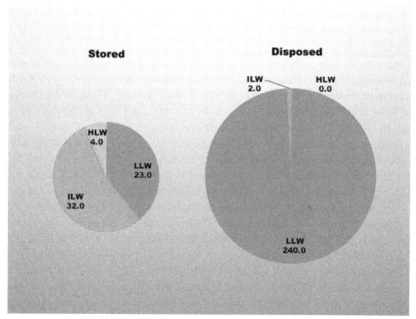

FIGURE 8.1   Worldwide radioactive waste inventory (based on 2008 data). Waste classifications are based on Figure 8.2, and data are for material in storage or stored (left), or disposed or assigned for disposal (right). Values are in (m$^3 \times 10^5$), and data was provided by IAEA (2010).

However, there are very large waste streams generated across the nuclear fuel cycle, not least the many millions of tonnes of uranium mine tailings. Moreover, it is not the volume or mass of spent fuel that is of concern but its extreme toxicity, longevity, heat generation, and the fact that it contains plutonium which can be extracted and diverted to nuclear weapons.

CLASSIFICATION AND CATEGORIES

Radioactive waste is defined as 'radioactive material in gaseous, liquid or solid form for which no further use is foreseen, and which is controlled as radioactive waste by a regulatory organization'. Some countries define used nuclear fuel as waste, whereas others do not. Under international law, used nuclear fuel is not defined as waste unless so designated by the responsible country. Waste categories are determined by the concentration and type of radioactive particles emitted, energy and heat generation. The most recent waste classification scheme for radioactive waste has been adopted in international standards developed by the IAEA (2010). The new and revised classifications are described below and shown schematically in Figure 8.2.

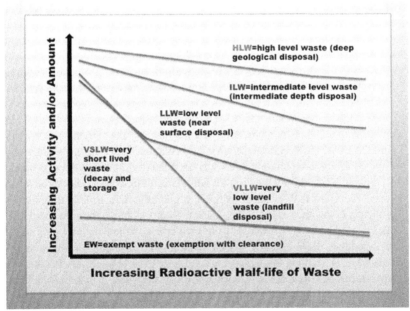

FIGURE 8.2   Conceptual illustration of the latest radioactive waste classification system based on relative radiation amounts or activity, and length of half-life of isotopes present. The figure is based on information provided by the IAEA (2010).

- Exempt Waste (EW)—contains such a low concentration of radionuclides that it can be excluded from nuclear regulatory control because any radiological hazards are considered negligible.
- Very Short Lived Waste (VSLW)—stored for decay over a limited period of up to a few years, subsequently cleared of regulatory control and to be disposed of as regular industrial waste. These types of wastes are often treated to achieve volume reduction and/or conditioned (i.e. waste immobilization) prior to disposal (Page, 2011; Saleh and Eskander, 2012). A variety of other safe and effective treatment options are available, including chemical precipitation and incineration.
- Very Low Level Waste (VLLW)—does not need a high level of containment and isolation, and is therefore suitable for disposal in near-surface landfill (soil) type facilities with limited control. VLLW wastes are nearly always treated to achieve liquidity (volume) reduction and/or the waste is immobilized prior to disposal (Dimitrescu, 2010; Saleh and Eskander, 2012). A number of safe and effective additional treatments are available, such as chemical precipitation and incineration.
- Low Level Waste (LLW)—contains limited amounts of long-lived radionuclides. This classification covers a very wide range of radioactive waste, from waste that does not require shielding for handling or transportation, up to activity levels that require more restricted containment, and isolation periods of up to a few hundred years. LLW includes a wide range of materials that may be slightly contaminated with radiation; for example, paper, glassware, tools and clothing. A range of disposal and storage options are available, from simple near-surface facilities to more complex engineered facilities. Most of this waste is either disposed of in shallow repositories or is stored pending disposal in a more permanent repository. Some low-level waste (especially short-lived radiation waste) is legally disposed of in landfills or incinerated.
- Intermediate Level Waste (ILW)—contains increased quantities of long-lived radionuclides and needs an increase in containment and isolation barriers compared to LLW. ILW needs no provision for heat dissipation during storage and disposal. Long-lived radionuclides such as alpha (α) emitters will not decay to acceptable levels during the time for which institutional controls can be relied upon. Therefore ILW requires disposal at greater depths of tens to hundreds of meters below the surface. ILW includes reactor components, chemical residues, sealed radioactive sources from medicine and industry, and used metal fuel cladding. ILW requires special handling and shielding of radioactivity. This waste is destined for disposal in deep geological repositories, but no such repositories exist (except a military repository, the Waste Isolation Pilot Plant in the USA).
- High Level Waste (HLW)—contains high levels of radiation that generate significant quantities of heat by radioactive decay, which needs to be considered in the design of disposal/storage facilities. Disposal in very deep, stable geological strata, usually several hundreds of metres below the surface is accepted

as the only appropriate option for HLW. The two primary classes of civilian HLW are: (1) used fuel rods from nuclear power plants and (2) waste arising from reprocessing the fuel rods. HLW includes spent nuclear fuel intended for disposal, and the waste stream from reprocessing the spent fuel. The waste contains high concentrations of radioactivity and requires cooling and special shielding, handling and storage. HLW contains both short-lived and long-lived high radiation nucleotides (some with half-lives of many thousands of years). Countries with high-level waste envisage disposing of it in very deep geological repositories, but no such repositories exist.

MANAGEMENT AND DISPOSAL

The greatest component and amount (by volume) of the nuclear fuel cycle waste by far is low level waste (LLW) from the mining and milling of uranium ores. After all, uranium ore to be useful in energy production must contain the naturally occurring isotopes of U and Pu that contain high energy (Table 8.1). Significant waste sources are the mine tailings (finely crushed solid residues from ore crushing), liquid waste from processing the ore and radon gas. After mining ceases, uranium tailings retain about 80% of the radioactivity that was present in the original ore. Tailings emit radioactivity to the environment for tens of thousands of years (Plecas, 2003; WNA, 2010). Before mining, the radioactive elements were generally locked in the impervious rock strata; so little radioactivity reached the surface and the wider environment. After mining, radioactive elements can escape into waterways and the atmosphere. Tailings are finely ground and the radon gas escapes many thousands of times faster than it otherwise would have from the original ore. Wind and water provide a variety of pathways for the spread of this waste. Tailings dams have had a poor track record around the world, with many recorded examples of leaks, spills and dam collapses (WISE, 2010). Serious questions over the long-term management of uranium mine tailings remain mostly unanswered.

The majority of the world's LLW has been safely disposed of (Figure 8.1) in near-surface disposal facilities, which have been operated in numerous countries such as France, UK and the USA for over 35 years. HLW and some ILW contain the most dangerous of radioisotopes that have isolation periods (based on energy and half-life) that are usually long, up to hundreds and thousands of years, thus requiring institutional and administrative control for that period. Many countries do not have central storage or disposal facilities for even ILW waste, and this waste is held at a host of widely dispersed locations. These ground-level repositories for low and short-lived intermediate level waste have experienced problems, and some have been closed. Older repositories of radioactive waste are being decommissioned because of large-scale water infiltration and the possibility of leaching out of radioactivity. Disposal at greater depths has the potential to provide a long period of isolation from the environment, and the likelihood of inadvertent human intrusion is greatly reduced well below the surface. Liquid wastes can be more of a problem and some need to be dried into a solid waste, and unfortunately many of these liquid radioactive compounds are not currently in a form suitable for off-site transport or ultimate disposal (Page, 2011; Veselov, 2011). Finland is well placed to have deep geological repositories for intermediate- and high-level radioactive waste operational by 2025, but most other planned facilities face significant obstacles, for example, poor public acceptance, high cost and lack of suitable sites. In some countries HLW is reprocessed and reprocessing is regarded as a valuable method of recycling because of the high energy uranium and plutonium isotopes that still remain in the spent fuel elements. Countries that recycle radioactive fuel include France, UK, Japan and Russia (IAEA, 2003; Bracke, 2012).

TRANSPORTATION AND RESPONSIBILITY

Dangerous nuclear fuel must be stored to manage higher levels of heat and radiation that may be emitted. Dry cask storage is widely used for storing the most dangerous of the radionuclides. It is interesting to note that tens of millions of shipments of radioactive material are conducted worldwide every year on public roads, railways and ships. In addition to shipment of radioactive waste, there are large distributions of radioactive supplies and isotopes for use in medicine, industry, agriculture, research, test and mineral exploration. Only about 5% of radioactive shipments relate to the nuclear fuel cycle, that is, activities associated with nuclear power (IAEA, 2006, 2010). In several decades of transporting radioactive materials around the world, there has never been an 'in-transit' accident with serious human health, economic or environmental consequences. Since 1971, more than 50,000 tonnes of use nuclear fuel and high-level radioactive waste have been safely transported a total distance of more than 30 million kilometers (IAEA, 2010; Abbott, 2012). Transportation of other hazardous materials, for example, petrol and toxic chemicals can be argued to pose a much greater threat to public health and the environment.

PHYTOREMEDIATION AND NON-PLANT METHODS

Low-level, large-scale contamination presents economic and logistical barriers to effective, timely and economic treatment. A number of remediation technologies have been successfully described and applied to waste sites, and most fall into the two broad categories below.

Engineering-Based Technologies

Engineering methods can be aggressive and are usually applied to clean up more acute pollution point sources. The methods can be uneconomic or even not environmentally justified for marginally (i.e. low-level radioactive waste) affected sites. Methods can be diverse, but usually rely heavily on excavation, movement and entombment or variations of these methods. The methods are not likely to diminish or reduce the hazardous material, and more importantly they cannot reduce total landfill capacity. These engineering-based approaches are usually applied to where more rapid responses are required, but the methods can cause secondary problems in the long term (Pilons-Smits, 2005; Banuelos, 2006). These engineering methods and possible application to radionuclide waste remediation will not be covered in this review.

In Situ Biological Remediation

Biologically based methods could be cost effective and more appropriate corrective options for treatment of low-level radioactive and non-radioactive contamination (De Filippis, 2010; Mohanty et al., 2010; Table 8.2). The methods fall into the two sub-categories:

- bioremediation—microbial-induced remediation
- phytoremediation—plant-based cleanup procedures

Bioremediation through the process of bioabsorption is an interesting intermediate technological innovation that removes (more correctly binds) radionuclides in dead (inactive) biomaterials extracted from living microorganisms or plants. The review of Das (2012) on bioabsorption concludes that bioabsorption is a suitable and cost-effective method for the removal of metals and radionuclides using non-living organisms or their biomass. Phytoremediation on the other hand uses living plants and their living microflora, and can involve a number of different processes, with different techniques operating to differing degrees depending on conditions, media, contaminants and plants. For example, combinations of phytoremediation, bioabsorption, vegetation and crop cover, and forest plantations (sections later) are becoming common for large-scale decontamination and / or degradation at toxic waste sites (EPA, 2000; Bhainsa and D'Souza, 2001).

PHYTOREMEDIATION AND HYPERACCUMULATION

Phytoremediation takes advantage of some natural biology present in plants. Growth of plants depends on photosynthesis, in which water and carbon dioxide are converted into oxygen and carbohydrates using the energy from sunlight. Carbon dioxide uptake from the atmosphere occurs through the stomata (openings) in the leaves and stems, along with the release of oxygen. Respiration of carbohydrates produced during photosynthesis release Adenosine Triphosphate energy necessary for the active transport of nutrients by roots, a process requiring oxygen (Atwell et al., 1999). Roots are effective in extracting water held in soils and upwardly transporting it through the xylem. Transpiration (water vapor lost from the plant to the atmosphere) via the stomata provides most of the driving force necessary for water and mineral transport (Singh et al., 2009a;

Fulekar et al., 2010). Diffusion of oxygen into the soil is necessary for continued plant growth and metabolism, and high or saturated soil water content will
greatly lower oxygen. Plants require macronutrients (N, P, K, Ca, Mg, S, Fe)
and micronutrients (B, Cl, Cu, Mn, Mo, Zn and possibly Si) for growth. For
uptake into a plant, a mineral or chemical (e.g. radionuclides) must be in solution, either in ground water or in the soil water (i.e. water in the unsaturated
soil zone) (Atwell et al., 1999). Water is absorbed from the soil solution into
the outer tissues of the root. Minerals and contaminants dissolved in the soil
water can move easily through the epidermis and endodermis, where they can
be absorbed, bound or metabolized (Ebbs et al., 1998; Zhu and Smolders, 2000;
Gerhardt et al., 2009). Therefore contaminants may be magnified in concentration inside plant cells and tissues (i.e. hyperaccumulate). In some plants there
are strong internal and adaptive mechanisms by which the contaminant may
be tolerated or even excluded by plant roots. Fungi associated with some plant
roots, that is, mycorrhizae and saprophytic fungi can also influence chemical
conditions within the soil and cause greater uptake (Clint and Dighton, 1992;
Westhoff, 1999; Pilon-Smits et al., 2009). Decaying roots and the above-ground
plant material incorporated into the soil will increase organic matter content
of soils, potentially leading to increased sorption of contaminants and humification (binding of compounds onto organic matter and transportation into the
roots) (Delvaux et al., 2000; Pulford and Watson 2003).

   End-user applications for phytoremediation and other related plant based
processes (like hyperaccumulation) have had different aims and objectives in
the past, as outlined below:

- Phytomining—historically metal and radioactive hyperaccumulating plants
  were only important for their ability to identify areas or rock strata useful as
  possible mining sites, most often for seeking deposits of precious metals (phy-
  toprospecting) and uranium-rich ores, and recovery of these valuable metals
  and uranium (Baker et al., 2000).
- Revegetation—more recently, plants that can survive high metal and trans-
  uranium elements (toxic and radioactive) have been used increasingly in
  revegetation projects, some necessary by legislation, and yet others done for
  aesthetic purposes; for example, barren, eroding, mining or industrial impacted
  soils and tailings. Recovery of metals and elements was not a primary objective (as in phytomining) as it was deemed that recovery was too expensive,
  uneconomic and even highly dangerous (Sors et al., 2005). However these
  practices and other technologies have led to the evolution of more refined
  phytoremediation techniques (Ward and Singh, 2004).
- Metal (element) recovery—recovery of soil containing metals taken up by
  plants and their reuse has been described only for nickel (Ni) and thallium
  (Th); which have high economic value. Other metals, for example, mercury
  (Hg), lead (Pb), arsenic (As), cadmium (Cd), caesium (Cs) and most radioactive elements have little added economic value and are extremely toxic
  and dangerous. These contaminants and compounds must be processed as

hazardous waste, and so far have not been identified and used in conjunction with accumulation and recovery (Mohanty et al., 2010).

- Paraphytoremediation—selected minerals like zinc (Zn), iron (Fe) and selenium (Se) have been used in crop fortification to increase levels of these essential minerals in edible crops and animal feed (Finley, 2005; De Filippis, 2010). This is unlikely to happen for radioactive nucleotides, but 'paraphytoremediation' may be a future possibility. Paraphytoremediation is defined as a 'mixed-benefit' strategy that combines and identifies a beneficial (economic) process in the phytoremediation method with their ability to detoxify the environment in which plants are grown (Wu et al., 1999; Pilon-Smits, 2005; Banuelos, 2006). So what may be useful products or by-products that could be obtained from plants loaded with potentially toxic radioactive compounds? One such benefit may be energy production that accompanies incineration, an essential procedure required to process and dispose of hyperaccumulating radioactive toxic plant biomass after phytoremediation. Another possible valued product in the case of radiation waste could be in the purification and manufacture of radio-biopharmaceutical compounds used in cancer treatment (de Mello-Faras et al., 2011). The attraction and benefits of some of these plant-based proposals are that there are economic returns and cost recovery in remediation of nuclear waste. These are only possibilities to think about at this stage!

METHODS IN PHYTOREMEDIATION

There are a number of different forms of phytoremediation for soils and water. Defining physico-chemical parameters in remediation are useful to clarify and understand the different processes that can occur due to vegetation; what happens to a contaminant; where contaminant remediation occurs and what should be done for more effective phytoremediation. Different forms of phytoremediation may apply to specific types of contaminants or contaminated media, and may require different types of plants. (The terms 'plant' and 'vegetation' will be used interchangeably to indicate all plant life, whether trees, shrubs, grasses or other mixed groups of plants.)

Phytoextraction

Phytoextraction is a process in which the radioactive contaminant is taken up by the roots, partly accumulated in roots and subsequently translocated to the aboveground portion of the plant. It is essential that this must be followed by harvest and ultimate disposal of all the plant biomass. Extraction is a slow removal process that carries risks in the case of radionuclides. Phytoextraction has been tested and/or applied successfully to metals (e.g. Ag, Cd, Co, Cr, Cu, Hg, Mn, Mo, Ni, Pb and Zn), metalloids (e.g. As and Se), non-metals (e.g. B) and radionuclides (e.g. ^{90}Sr, ^{137}Cs, ^{234}U,^{238}U) (more examples are detailed in Table 8.3), as these are normally not further degraded or changed in chemical form within the plant. Phytoextraction has generally not been considered for organic radionuclide

TABLE 8.3 List of Plant Species Where Radionuclide Phytoremediation Research Has Been Reported

Radionuclide Phytoremediation	Plant Species	Results and Notes on Phytoremediation	Reference
137Cs Soil Phytoextraction	Melilotus, Sorghum, Trifolium	Melilotus accumulates Cs twice as much compared to the cereals Sorghum and Trifolium	Rogers and Williams, 1986
	Polytrichum, Festuca, Agrostis, Carex	Polytrichum accumulates Cs more than Carex, and other two grasses accumulate much less	Coughtrey et al., 1989
	Field evaluated mixed grasslands	Cs uptake and removal by mixed grasslands at Chernobyl was slow, needed a long time (decades)	Kirk and Staunton, 1989
	Triticum	Cs root uptake kinetics studied in wheat, inhibition by K and ammonium was noted	Shaw and Bell, 1991
	Calluna	Cs uptake by Calluna was increased by mycorrhizal associations, root factors were important	Clint and Dighton, 1992
	Vitis (grapevine)	Foliar uptake of Cs over five seasons was very low, and little was translocated to shoots	Zehnder et al., 1995
	Chenopodiaceae (30 species tested)	Chenopodium, Brassica, Beta, Polygonum were best accumulators, rest had poor uptake	Broadley and Willey, 1997
	Pahlaris, Brassica, Phaseolus	Cs near Chernobyl show Pahlaris best accumulator, followed by Phaseolus and Brassica	Lasat et al., 1997
	Amaranthus, Brassica, Phaseolus	Cs accumulation best in Amaranthus, then Brassica but it was slow, needs over 15 years for lowering levels	Lasat et al., 1998
	Vaccinium, Empetrum, Deschampsia	Cs content in three species decreased continually with distance from nuclear power source	Bunzl et al., 1999

Plants	Description	Reference
Amaranthus, Helianthus, Brassica, Pisum	Screening of plants showed Amaranthus better than Helianthus, and both were better than Brassica	Dushenkov et al., 1999
Sorghum, Panicum, Paspalum	Mycorrhizal association improved Cs uptake in all three grasses, but still low Cs levels detected	Entry et al., 1999
Helianthus, Phaseolus, Hordeum, Triticum	Examined the mechanism of Cs uptake and competition with K in roots, stems, leaves	Zhu and Smolders, 2000
Helianthus, Populus	Helianthus accumulated higher Cs than Populus, and had better translocation to stems and leaves	Vanek et al., 2001
Beta	Beta vulgaris accumulated Cs from soils, but appears low Cs in soils is better for uptake	Watt et al., 2002
Asteraceae (4 sp), Chenopodiaceae (2 sp)	Soils from China with Cs readily taken-up, but concentrations differed in all six species	Tang and Willey, 2003
Brassica, Spinacia, Lactuca, Raphanus	Cs uptake in Brassica best, Lactuca, Spinacia poorest, and affected by K fertilizer application	Chou et al., 2005
Salix (willow)	Short-term greenhouse studies showing Cs accumulation, but remediation takes 50–90 years	Dutton and Humphreys, 2005
Calotropis	Laboratory study of Calotropis Cs accumulation, Cs was reduced to 99% in 15 days	Eapen et al., 2006
Vetiveria	Laboratory study of Vetiveria Cs accumulation, undetectable Cs levels in just 15 days	Singh et al., 2008
Soya, Pisum, Avena	Mycorrhizal association increased Cs in forage species, need 15–20 years for remediation	Goncharova, 2009

Continued

TABLE 8.3 List of Plant Species Where Radionuclide Phytoremediation Research Has Been Reported—cont'd

Radionuclide Phytoremediation	Plant Species	Results and Notes on Phytoremediation	Reference
	Chromolaena	Laboratory study of Chromolaena Cs uptake in shoots, a reduction by 75% in 15 days	Singh et al., 2009b
	Helianthus	Cs uptake was affected positively by mycorrhizae, Cs uptake was up to five times greater	Dubchak et al., 2010
	Catharanthus	Cs was primarily accumulated in shoots, good candidate for low level radioactive waste	Fulekar et al., 2010
	Cannabis	Cs did not accumulate in plant due to chemical forms being complexed, and inactivated in soil	Hoseini et al., 2012
	Evergeen coniferous species	Cs was accumulated in 20 species in less than 6 months, mostly in young foliage of trees	Yoshihara et al., 2013
Water Rhizofiltration	Helianthus, Brassica	Both species hyperaccumulated Cs from aquatic systems, Helianthus better than Brassica	Dushenkov et al., 1997
	Helianthus	Cs uptake and removal by Helianthus was good and could remediate contaminated soil	Salt et al., 1997
	Eichhornia, Catharanthus	Cs was removed up to 73% in 15 days in pots, both plants assessed as suitable for remediation	Prasad and Freitas, 2003
	Helianthus	Cs uptake and removal by Helianthus was good and could remediate contaminated soil	Prasad, 2007

	Catharanthus	Cs was removed by up to 70% in 15 days in pots, plant assessed suitable for remediation	Singh et al., 2009a
^{90}Sr, Soil Phytoextraction	Pinus	Radiata pine and ponderosa pine accumulate more Sr with mycorrhizal associations	Entry et al., 1994
	Vitis (grapevine)	Foliar uptake of Sr over five seasons was very low, and little was translocated to shoots	Zehnder et al., 1995
	Agrostis	Shoot Sr concentrations were affected by Ca in the soils, cation amount and organic matter	Veresoglu et al., 1996
	Sorghum, Panicum, Paspalum	Mycorrhizal association improved Sr uptake in all three grasses, low Sr levels present	Entry et al., 1999
	Salix (willow)	Short-term greenhouse studies showing Sr accumulation, but remediation in 50–90 years	Dutton and Humphreys, 2005
	Calotropis	Detailed study of Calotropis Sr accumulation in pots, Sr was reduced to 99% in 15 days	Eapen et al., 2006
	Vetiveria	Detailed study of Vetiveria Sr accumulation in pots, undetectable Sr levels in 15 days	Singh et al., 2008
	Chromolaena	Laboratory study of Chromolaena Sr uptake in shoots, a reduction by 75–80% in 15 days	Singh et al., 2009b
	Cannabis	Sr was accumulated in roots, stems, leaves; could be suitable plant for phytoremediation	Hoseini et al., 2012
Water Rhizofiltration	Helianthus, Brassica	Both species hyperaccumulated Sr from aquatic systems, Helianthus was better	Dushenkov et al., 1997

Continued

TABLE 8.3 List of Plant Species Where Radionuclide Phytoremediation Research Has Been Reported—cont'd

Radionuclide Phytoremediation	Plant Species	Results and Notes on Phytoremediation	Reference
	Helianthus	Sr uptake and removal by Helianthus was good and could remediate contaminated soil	Salt et al., 1997
	Eichhornia, Catharanthus	Sr was removed by up to 70% in 15-day pot trial, plants assessed suitable for remediation	Prasad and Freitas, 2003
	Helianthus	Sr uptake and removal by Helianthus was good and could remediate contaminated soil	Prasad, 2007
	Catharanthus	Sr was removed by up to 70% in 15-day pot trial, plants assessed suitable for remediation	Singh et al., 2009a
^{60}Co Soil Phytoextraction	Melilotus, Sorghum, Trifolium	Melilotus and Sorghum accumulated Co more than the grass Trifolium; unlike Cs uptake	Rogers and Williams, 1986
^{222}Rn Soil Phytoextraction	Helianthus, Festuca, Zea	Rn taken up by all three species although levels were low and Rn could easily volatilize	Lewis and MacDonnell, 1990
^{226}Ra Soil Phytoextraction	Helianthus, Festuca, Zea	Ra taken up by all three species although levels were low and depended most on leaf area	Lewis and MacDonnell, 1990
^{238,235}U Soil Phytoextraction	Mixed forest trees and shrubs	Field uptake of U into established plants and trees, all plants took-up U in the uranyl form	Cornish et al., 1995
	Helianthus, Brassica, Phaseolus	Helianthus best at accumulating U, next best was Brassica, however most U was in roots	Dushenkov et al., 1997
	Pisum, Beta, Phaseolus	All three plants took up U; however, the chemical form of U was important in uptake	Ebbs et al., 1998

	Brassica, Pisum, Amaranthus	All Brassicas were best at accumulating U, and Pisum and Amaranthus were similar	Huang et al., 1998
Water Rhizofiltration	Brassica, Chenopodium, Eichhornia	Hairy root transformation removed U with a shorter period of time in all three species	Dushenkov et al., 1997
	Apium, Callitriche, Lemna, Fontinalis	All three species accumulated U, the native species show potential in phytoremediation	Eapen et al., 2003
	Aquatic plants (unspecified native species)	Some species were good bioaccumulators, suitability for phytoremediation is discussed	Favas and Pratas, 2013
^{239,240}Pu Soil/Water Rhizofiltration	Sargassum (sea grass in salt water)	Sea grass accumulated Pu significantly, suggesting this salt water species was a good indicator	Noshkin, 1972
	Hordeum, Soya	Both species accumulated Pu, but mixed results, levels were variable from 60% to 90% uptake	Cataldo et al., 1988
^{99}Tc Soil Phytoextraction	Deciduous forest	Tc progressively accumulated seasonally. Wood was the major sink for Tc radioactivity	Garten et al., 1986
	Triticum, Lolium	Triticum accumulated 92–95% Tc, while Lolium accumulated only 62–78% Tc	Garten and Lomax, 1989
	Mixed grass species	Tc uptake was present in all plants, depended on growth rate of plant and soil application	Echevarria et al., 1997
^{40}K Soil Phytoextraction	Vaccinium, Empetrum, Deschampsia	K content in the three species decreased with increase distance from nuclear power source	Bunzl et al., 1999
	Graminaceae species, Taraxcum	High K results in lower Cs levels in all plants, uptake of Cs varied with species and soil	Ciufo and Belli, 2006

Continued

TABLE 8.3 List of Plant Species Where Radionuclide Phytoremediation Research Has Been Reported—cont'd

Radionuclide Phytoremediation	Plant Species	Results and Notes on Phytoremediation	Reference
^{210}Po Soil Phytoextraction	Tabaccum (tobacco)	Po taken up most readily by leaves, lower in stems and roots, but wide range existed	Kovacs et al., 2004
^{210}Pb Soil Phytoextraction	Tabaccum (tobacco)	Pb taken up in low amounts, leaves were higher in Pb than rest of plant (shoot and root)	Kovacs et al., 2004
^{237}Np Soil/Water Rhizofiltration	Triticum, Lolium	Np in water used for plant growth, Np was taken-up by both plants, most Np was present in roots	Garten and Tucker, 1986
^{125}I Soil Phytoextraction	Helianthus, Populus	Helianthus accumulated higher I than Populus, easily translocated to stems and leaves	Vanek et al., 2001
^{65}Ni Soil Phytoextraction	Helianthus, Populus	Helianthus accumulated higher Ni than Populus, easily translocated to stems and leaves	Vanek et al., 2001
^{67}Ga Liquid Waste Phytoextraction	Olea europaea (olive oil pomace waste)	Aqueous solutions of radioactive Ga were adsorbed onto fresh plant extract of the olive oil waste	Eroglu et al., 2009
Eu (III) Soil Phytoextraction	Eichhornia	Eu is surrogate for radionuclide Am, 25% radioactivity removed from water by roots	Rogers and Williams, 1986
	Eichhornia	Eu radionuclides were taken up into the plant, but especially onto the surface of roots	Kelley et al., 1999

Summary results and notes are given to the different phytoremediation methods used.

contaminants, as these can be metabolized, converted or volatilized by the plant itself, thus preventing control over the accumulation process and the risk of spread of radioactivity is high. Nevertheless, some studies have shown the accumulation of unaltered organic contaminants within the above-ground portion of a plant, and this may be the case for some organic radioactive compounds, but biological transformation must be thoroughly examined for any organic radionuclides (Huang et al., 1998; Duschenkov et al., 1999; Eapen et al., 2006). The target media are generally soils and tailings, although contaminants in sediments and sludges may also undergo phytoextraction, perhaps in conjunction with rhizofiltration. The presence of plant roots will often increase the size and variety of microbial populations in the soil surrounding the roots (saprophytes in the rhizosphere) or mycorrhizae (associations of fungi and plant roots), and positive effects have been observed with radionuclides (Entry et al., 1994; Westhoff, 1999; Turnau et al., 2006). Phytoextraction is also known as phytoaccumulation, phytoabsorption and phytosequestration. Phytoextraction has great value and high potential in use with low-level radioactive waste; however, accumulation of radionuclides within roots and in above portions of the plant must be monitored and managed carefully (Zhu and Shaw, 2000; Dubchak et al., 2010). Rhizofiltration (described below) is a type of, or modification of, phytoextraction using aquatic-tolerant plants or even just aquatic vegetation to accumulate radionuclides primarily in the root system. Table 8.3 summarizes the available literature on phytoextraction using various plants and radionuclides.

Phytodegradation

Phytodegradation is the uptake, metabolism and degradation of contaminants mainly within the above-ground parts of plants, leading to a decrease of contaminants in the soil, sediments, sludge, ground water or surface water. The degradation is usually affected by enzymes and released metabolites by the plant itself (Rugh, 2001; Meagher et al., 2000). Phytodegradation is not dependent on microorganisms associated with the rhizosphere. Contaminants suitable for phytodegradation include organic compounds such as munitions, chlorinated solvents, herbicides and insecticides (Gerhardt et al., 2009). Phytodegradation is also known as phytotransformation, and can be a powerful decontamination process. For phytodegradation to occur within the plant, the plant must be able to take up the contaminant and usually translocate it to the stem/leaves, and this is true for most contaminants (Vanek et al., 2001; Singh et al., 2008, 2009b). In phytodegradation, the transformation of a contaminant within the plant to a more toxic form is common, with subsequent release to the atmosphere through transpiration, and this is not desirable with radioactive compounds. The application of phytodegradation has typically focused on the degradation of organic compounds, and there are no reports of research using phytodegradation with radionuclides.

Phytostabilization

Phytostabilization is the use of vegetation to contain radioactive contaminants in situ, through modification of the chemical, biological and physical conditions within the root and/or the soil. Contaminant transport of soils, tailings, sediments or sludges can be reduced through absorption and accumulation onto roots, precipitation and complexation in soils within the root zone, or binding to humic (organic) matter through the process of humification (Lasat et al., 1997). As an additional benefit, vegetation can reduce wind and water erosion of the soil, thus potentially preventing radioactive dispersal of the contaminant in runoff and leachate or in aerial dust particles. Phytostabilization is also known as in-place inactivation or phytoimmobilization. Phytostabilization research to date has generally focused on metal contamination, with Pb, Cr, Zn, Cd and Hg being identified as suitable candidates for phytostabilization (EPA, 1997, 2004). However, there may be potential for phytostabilization of radionuclides as many of the low-level radioactive compounds resemble metals, but this requires further investigation. Therefore phytostabilization has some value in use with low-level radioactive waste; however, accumulation and leaching of radionuclides from the soil must be carefully evaluated (Lasat et al., 1998; Watt et al., 2002).

Phytovolatilization

Phytovolatilization is the uptake of a contaminant by plants, and the subsequent release of a volatile form of the contaminant, a volatile degradation product of the contaminant or a volatile form of an initial non-volatile contaminant (Bizily et al., 1999; De Filippis, 2010). For effective phytoremediation, the degradation product or modified volatile form should be less toxic than the initial contaminant, but this is not always the case. The chemical fate and translocation are most likely concentration-dependent, and this method has only been described for a few contaminants (Meagher, 2000). Phytovolatilization is primarily a contaminant removal process transferring the contaminant from the original medium (pond or ground water or soil water) to the atmosphere; which does not solve the remediation problems with most toxic radioactive compounds.

Tritium (^{3}H) is a special isotope where phytovolatilization has already proved to be a success. Tritium occurs naturally when cosmic radiation reacts with gases in the atmosphere. Tritium combines with oxygen to form water, and reaches the surface as rain, which is easily shielded from the air and skin to produce almost no external radiation exposure. Tritium is also present as a component of nuclear weapons, reactors and nuclear test explosions and contaminates groundwater. Tritium directly incorporates into water, is taken up by plants and trace amounts of tritium are released through the foliage in evapotraspiration (a form of volatilization). For example, tritiated water applied as irrigation to the floor of forests enters the hydrological cycle as it percolates into the soil. Water then evaporates from the soil surface or is absorbed into the transpiration stream by plant roots or drains through the soil to the water table. A small portion of the ^{3}H that was absorbed by plant roots remains in plant tissues in the form of easily exchangeable hydroxyl ions or is incorporated into organic molecules through photosynthesis (Murphy, 2001).

Forest test sites at Brookhaven National Laboratory (BNL), Savannah River Site (SRS), Department of Energy USA and other nuclear facilities (e.g. Mixed Waste Management Facility, Four Mile Branch and Argonne National Laboratory) have led the way in remediation techniques for removal of tritium (Negri et al., 2000). The most cost-effective approach for the minimization of ^{3}H risk was not removal of ^{3}H from the effluent nor isolation, but rather a change in the path of ^{3}H exposure in the environment and its use in forest irrigation water. Calculations based on research at the Savannah River Site indicated that a 40% reduction in dose could be achieved by releasing ^{3}H water in part into the atmosphere, and the rest as irrigation, in contrast to allowing it to flow off-site in surface water (Fulbright et al., 1996). This phytoremediation approach has reduced the risk of spills and exposure to workers, and avoided secondary waste treatments. The presence of trees and vegetation was considered the best fit with planned future land use, and the forest phytoremediation system has provided near full hydraulic control. It has been estimated that there will be minimal risks to workers harvesting trees just 5 years after irrigation has halted (Dushenkov, 2003).

Rhizofiltration

Initial attempts to use plant roots to concentrate and precipitate toxic radionuclides using hydrophytic plants had mixed success, and these results had to be interpreted with some caution and improvements have been introduced. Rhizofiltration has been tested and evaluated on metals (e.g. Ag, Cd, Co, Cr, Cu, Hg, Mn, Mo, Ni, Pb and Zn), metalloids (e.g. As and Se) and radionuclide elements (e.g. ^{90}Sr, ^{137}Cs, ^{234}U, ^{238}U) (more are listed in Table 8.3), but considerably less research has been concluded on rhizofiltration compared to phytoextraction. The appealing features of rhizofiltration include the contaminant accumulating in root tissue below ground (in soil/media systems) or in the pond water (in constructed aquatic pond systems; Dushenkov, 2003) and the potential binding to organic root exudate; therefore translocation of the toxic radioactive compounds from the plant to the atmosphere is less likely than with other phytoremediation methods. Strong binding and any degradation occurs at the source of contamination in the soil/water. Harvesting of plants (especially the root system) is still necessary since there is accumulation of the radionuclides from the soil water or from the filtering pond water (Coughtrey et al., 1989; Kelley et al., 1999). Stimulation of soil/water microbes by plant root exudates can also result in alteration of the biological conditions in the soil and water, such as pH, water content, water and nutrient transport, aeration, structure, temperature or other parameters which may result in changes in the transport of contaminants. Often these changes create more favorable environments for soil microorganisms (Roca et al., 1997; Echevarria et al., 2001; Chou et al., 2005), increasing the size and variety of microbial populations in the soil surrounding the root rhizosphere or mycorrhizae, and positive effects have been observed in radionuclide uptake (Entry et al., 1997; Westhoff, 1999; Turnau et al., 2006). Table 8.3 also summarizes the available literature on rhizofiltration using various plants and radionuclides.

Rhizodegradation

Rhizodegradation is the enhancement of naturally occurring biodegradation in the soil/tailing through the influence of plant roots and microbes and ideally will lead to destruction and detoxification of an organic contaminant in most cases. Rhizosphere biodegradation is another term that has been used for rhizodegradation. Organic contaminants in soils can often be broken down into simpler compounds or completely mineralized into inorganic products and such methods have not been investigated for radioactive compounds. These biologically dynamic and unpredictable degradation reactions have generally been considered undesirable and very low in priority to be used with dangerous radionuclides. Root penetration throughout the soil may allow an insignificant percentage of the soil to be treated, as only a small percentage of the total soil volume is in contact with living roots. It can take a long time for root growth into new areas of the soil in contact with the contaminant, and areas of high contamination are usually inhospitable soil–root conditions (Duschenkov et al., 1997; Huang et al., 1997).

Vegetation Caps and Buffer Strips

A vegetation cap (or cover) is a long-term, self-sustaining plant cover growing in or over contaminated material, designed to minimize exposure pathways and risk. The primary purpose of the vegetation is to provide hydraulic water control and minimize infiltration of water into the contaminated sub-surface, thus preventing or minimizing leachate. This is done by maximizing evapotranspiration in plants and also maximizing the water storage capacity of the soil. A cap designed to incorporate contaminant destruction or removal, in addition to the prevention of infiltration is called a phytoremediation cap. Vegetated caps can be constructed over landfills, or over contaminated soil or groundwater (Bunzl et al., 1999; Dutton and Humphreys, 2005; Kilda et al., 2009). Long-term maintenance of the cap is essential, or the cap vegetation may be designed to allow an appropriate plant succession that will maintain the cap integrity with time. Significant issues remain with the use of vegetative caps on landfills in evapotranspiration control and application to very dangerous contaminants like radionuclides. Buffer strips are an alternate technology where areas of vegetation are placed down-gradient of a contaminant source or a toxic plume, or along a waterway (i.e. riparian corridor). The vegetation contains, extracts and/or destroys contaminants in the soil, surface water and ground water under the buffer zone through hydraulic control. Buffer strips have been proposed for use with any of the conventional phytoremediation methods above, and the same concerns for buffer strips are present as with vegetation caps.

Hydraulic Control

Hydraulic control (or hydraulic plume control) is the use of vegetation to influence the movement of ground water and soil water, through the uptake and consumption of large volumes of water by the vegetation. Hydraulic control may influence and potentially contain movement of a ground-water toxic plume, reduce or prevent infiltration and leaching, and induce upward flow of water from the water table (Hoffman et al., 1982; Olson and Fletcher, 2000). Other conventional phytoremediation processes must be applied as the contaminated water is brought towards the surface and taken up by plants. For this reason, design and operation of phytoremediation hydraulic control will likely require site-specific observations of water levels, flow patterns and water uptake rates. Hydraulic control is still a method with little published data and perhaps a method not to be used with radionuclides because of the many uncertainties. A possible area of use for hydraulic control with radioactive contaminants is when there is radiation leachate or an accidental discharge of radioactivity into ground water, which needs to be controlled quickly.

TOLERANCE AND EXTRACTION

Plant roots generally contain higher radionuclide concentrations than the shoots due to exclusion barriers operating during uptake and translocation. However, radionuclide uptake into roots was not necessarily followed by translocation into shoots or leaves in some plants, and the age of the various plants was important (Cataldo et al., 1988; Shaw and Bell, 1991). Also, plants differed markedly in their ability to tolerate and accumulate radionuclides (Table 8.2). Soil conditions were important and alterations of a number of physical and chemical conditions around plant roots affected the uptake (Veresoglou et al., 1996; Ohnuki et al., 2010). It was also apparent that an upper limit to radionuclide concentration within the roots must be present without toxic effects. Phytoextraction, for example, begins in the root zone of plants, but the root zone typically may be relatively shallow, with the bulk of roots at shallower rather than deeper depths. This can be a limitation of phytoextraction, and remediation of contaminated soils using plants where it has been limited to the top 15–20 cm, with insignificant contaminant removal further down the soil profile (Blaylock et al., 1997, 1999; Goncharova, 2009). Due to the scarcity, small biomass, slow growth rate, uncertain or specialized growing conditions of many suitable (hyperaccumulating) plants for some of the most toxic radioactive contaminants, the effectiveness of some plants for phytoextraction has been questioned, especially if they are able to remove only a relatively small mass of contaminant from soils (Nazina et al., 2010). Table 8.3 cites at least 100 plants that have been used in pilot studies on phytoremediation of various radionuclides, but only a handfull have been seriously considered, and have superior qualities for general adoption; plants with good potential include *Helianthus, Brassica, Amaranthus, Eichhornia,* and trees like *Populus* and *Salix.*

Most of the current available research has been directed towards radioactive Cs and Sr, which are the most important constituents of nuclear fallout as a consequence of spills and accidents. A reasonable amount of data also exists for a number of naturally occurring radioisotopes associated with uranium mining and nuclear energy. Less research has been conducted on undisturbed radionuclide-contaminated sites, such as inactive land treatment or waste sites in nuclear test zones, which will naturally regenerate in time (Wallace and Romney, 1972). The term 'forensic phytoremediation' has been used to describe such research where natural revegetated sites have been used to evaluate ecological effects after new vegetation has established or re-established, and future impacts at the ecosystem level determined (Pivetz, 2001; Ehlken and Kirchner, 2002). Vegetation may become established after the phytotoxic contamination has been reduced through naturally occurring radiation half-life degradation, biotic processes lowering toxicity or through intentional remedial technologies having been applied. In these cases, vegetation analysis would indicate that the contaminants are no longer bioavailable and/or toxic to the established vegetation on site (Kovacs et al., 2004; Schroeder et al., 2007). Alternatively, a plant that can withstand the contaminant might preferentially become established, and perhaps then contribute to additional contaminant loss through other forms of phytoremediation. This has apparently happened at the Nevada nuclear test site, in which native trees and shrubs (Wallace and Romney, 1972; Olson and Fletcher, 2000) have re-established through natural regeneration. It is from such long-term studies and observations that a more refined list of plants (including high-performing native plant species) suitable for phytoremediation of radionuclide waste and mine tailings may prove invaluable (Smith and Bradshaw, 1979; Gerhardt et al., 2009).

UPTAKE AND DISTRIBUTION

Bioassays using plants have been used routinely in the environmental sciences. In many cases, the effectiveness of bioremediation efforts at hazardous waste sites has been assessed and guidelines established using plant bioassays (called 'phytoinvestigation'). Phytotoxicity testing was used to determine the extent of bioremediation required in a number of radionuclide-containing soils (Baud-Grasset et al., 1993; Eapen et al., 2003; Hoseini et al., 2012; Favas and Pratas,

2013). More recent research with radioactive nucleotide hyperaccumulating plants have achieved high levels of radionuclide uptake when using plants grown in control solution cultures in laboratory or greenhouses (Eapen et al., 2006; Singh et al., 2008, 2009a, b). Plant bioassays have been used on radioactive wastes to test for decreased levels of the most important radionuclides in soils (Sandhu et al., 1991; Duschenkov, 2003; Eapen et al., 2007). Geobotany is an alternate approach, as used on the Nevada nuclear test site (Wallace and Romney, 1972) and is beginning to be used around Chernobyl nuclear contamination areas (Ertel and Ziegler, 1991; Kovacs et al., 2004; Ciuffo and Bellini, 2006; Dubchak et al., 2010). The presence of a particular plant species that accumulated radioactivity can be indicative of an underlying amount of radionuclide left in the soil. The change in radiation levels in a particular plant species, over a wide area can also indicate the extent and zone of radioactive contamination. An analysis of previously collected herbarium specimens can also lead to the identification of areas that could contain high radioactivity (Wallace and Romney, 1972; Brooks, 1998). The presence of different species of plants also provides clues as to the presence and depth of ground water contamination (Pivetz, 2001; Goncharova, 2009). Even in a phytoremediation pilot project, failure or part-success might provide information about the nature of the contaminated soil or ground water, as unhealthy or dying vegetation might indicate previously undetected 'hot spots' of high radioactive contamination.

A cautionary note must be delivered at this stage in that despite our good understanding of the mechanisms of phytoremediation and some success from laboratory and greenhouse studies, efforts to transfer these remediation methods to the field have proven more difficult, and have presented numerous challenges (Gerhardt et al., 2009). There are of course many encouraging results, but there have been also numerous inconclusive and unsuccessful attempts at field phytoremediation. There is an urgent need to re-assess the reasons why phytoremediation in the field has been unsuccessful, before negative perceptions of the whole technology sets in and undermines further progress. The most common reasons for lack of success in the field are given below:

- Plant stress factors—not present in the laboratory/greenhouse represent significant challenges in the field, these may include negative features like reduced growth and poor environmental adaptation to field conditions.
- Current assessment methods—for field phytoremediation are not adequate and may show, for example, small amounts of decrease in radioactivity yet active remediation may be occurring more slowly.

WETLANDS AND AQUATIC PHYTOREMEDIATION

Constructed Wetlands

Constructed wetlands or treatment wetlands are artificial wetlands that are used for treating organic, inorganic and excess nutrient contaminants in surface water, municipal wastewater, domestic sewage, refinery effluents, acid uranium mine drainage or landfill leachate. A considerable amount of research and applied work has been conducted using constructed wetlands for some of these applications. Kadlec and Knight (1996), Cole (1998) and Pivetz (2001) provide an overview of constructed wetlands. Wetlands and ground-water treatment have not been examined for treatment of radinucleotide wastes. Except in very few cases, constructed wetlands have not been used in remediation of very hazardous waste. However, in future constructed wetlands might become an option for treatment of radionuclides dissolved in water using rhizofiltration; a phytoremediation method well suited to aquatic environments (Dushenkov, 2003; Eapen, 2007). Integration of radioactive hazardous waste site phytoremediation and constructed wetland technologies requires more research and development.

Combinations of Phytoremediation Methods

At a phytoremediation site, combinations of the phytoremediation processes discussed above may be required simultaneously or in sequence for a particular contaminant, or different processes may act on different contaminants and at different exposure concentrations. For example, trichloroethane, tetrachloroethane, explosives (TNT, nitroglycerine) and oganomercurials in soil can be subject to biodegradation in the root zone (rhizodegradation) and metabolism within the plant (phytodegradation), with loss of some contaminant through volatilization from the plant (phytovolatilization) (Meagher et al., 2000; Pivetz, 2001; Rugh, 2001). Some metals and radionuclides can be accumulated on or within the roots (rhizofiltration) while other metals or radionuclides are simultaneously taken up into the aerial portion of the plant (phytoextraction). These complex systems of different patterns of accumulation in different organs of plants, and soil profiles for many phytoremediation methods can provide considerable difficulties in containing toxic compounds, and introduces many uncertainties and problems. This is most undesirable for phytoremediation of radioactive wastes.

TREATMENT, EVALUATION AND OBJECTIVES

Phytoremediation of radionuclides may be limited by high contaminant concentrations that are very dangerous, as these compounds are likely to be extremely phytotoxic or could cause an unacceptable decrease in plant growth and a danger to the public. Areas of high, phytotoxic and dangerous contaminant concentrations may have to be treated using other technologies of excavation and deep landfill, with phytoremediation being used for the lower-level radioactive contaminants, low radiation and upper soil profile of a waste site. Phytoremediation (such as phytoextraction and rhizofiltration) may be suited for a final step, for example, if active land treatment remediation has ended without having achieved a desired lowering of radiation to acceptable levels.

The EPA (2010) has reported on one such programme, the Hanford 100-N Area (Washington) where phytoremediation was selected as a 'polishing step' for groundwater contaminated with strontium 90. Long-term field/site studies with additional plant species are required, and may indicate that there are fewer limitations in toxicity to plants than currently thought of in phytoremediation of radioactive waste sites (Prasad, 2007; Mitchell et al., 2013). The radioactive contaminant concentrations that are phytotoxic to specific plants are likely to be site-specific, and affected by soil, climate and bioavailability. Aged toxic radioactive compounds in soils can be less bioavailable. This will decrease phytotoxicity, but can also decrease the effectiveness of phytoremediation. Site-specific phytotoxicity data should be used on contaminated soil from natural sites rather than uncontaminated soils spiked with the contaminant. Phytotoxic concentration levels will need to be determined on a site-specific basis, although literature values from previous bioassay studies can provide a first approximation.

Ground Water

Selected radioactive contaminants in ground water may be addressed using phytoextraction and/or rhizofiltration. The primary considerations for ground-water contamination are the depth of the ground water and the depth of the contaminated zone. In situ ground water phytoremediation is essentially limited to unconfined aquifers in which the water table depths are within the reach of plant roots usually in the uppermost portion of the water table (Echevarria et al., 1997; Roca et al., 1997; Ebbs et al., 1998; Zhu and Smolders, 2000). Plant roots will not grow through even clean ground water to great depths to react with the contaminated zone, and this may be a limitation to reaching deep radionuclides. Deep groundwater that is beyond the reach of plant roots could be remediated by rhizofiltration after the water is pumped from the sub-surface using extraction wells. However, this is undesirable for radionuclides because it is bringing a very dangerous contaminant closer to the surface and humans.

Surface Water and Waste Water

Surface water can be treated using rhizofiltration (phytoextraction) in ponds, engineered tanks, natural wetlands or constructed wetlands (Huang et al., 1997; Duschenkov, 2003). In some cases, the contaminated water can be used as irrigation water in which the contaminants then undergo rhizodegradation and phytodegradation. Radionuclides provide a type of contaminant which is dangerous in these surface ponds and surface wetland processes, let alone when such contaminated water is used in irrigation. The exception to this appears to be the phytoremediation of tritium discussed before, but this isotope is in a special category of low-hazard radiation.

Soil, Sediment and Sludge

Phytoremediation of radionuclides in contaminated soils has already been demonstrated on a pilot-scale, but needs to be developed to an industrial and commercial scale. Phytoremediation of soils is still developing to address plant species that will be able to survive and thrive in contaminated waste soil, yet be able to mitigate the toxic radioactive pollutants. Phytoremediation is most appropriate for large areas of a relatively thin near-surface layer of contaminated soil, and most likely sediment or sludge within the root depth of the selected plants. Deeper soil contamination, high contaminant concentrations or small localized civil areas might be more effectively treated using conventional technologies already in use with radionuclides before phytoremediation is used. It is true that through future phytoremediation research, the capabilities of phytoremediation of toxic radioactive sediment, soil and sludge might be better evaluated.

Air

Phytoremediation research and application to radionuclides have focused on contaminated water and soils. There has been little discussion or research into the possible phytoremediation of contaminated air or soil gas, and no such research has been published. However, airborne contaminants can be directly withdrawn from the atmosphere through the uptake of gaseous contaminants by plant leaves or by deposition of contaminated particulate matter on to the leaf surface. Some plants appear to remove volatile compounds from the air, in addition to removing air contaminants through the action of roots and soil microbes (Raloff, 1989; Stutte, 2012). Contaminated air has been bioremediated by drawing the air through soil beds in which microbial activity helps to degrade and bind the contaminants. The process of bioremediation has been enhanced by the presence of the root zone of plants, even slow-growing plants kept indoors (Irga et al., 2013). Phytoremediation of radioactive contaminated air and soil gas may become a subject for future research, especially since some radioactive contaminants are in the gaseous state (e.g. Ra, Rd).

COSTS AND ECONOMICS

When research on phytoremediation began, initial cost estimates predicted that phytoremediation would have significantly lower costs compared to other remedial technologies, and data in general strongly support this (Table 8.4). Actual cost data for phytoremediation technologies are sparse, and currently are from pilot-scale or experimental studies that may not accurately reflect expected costs once the technology matures. Additionally, many of these projected calculations (costs) are in reports that are difficult to obtain and check independently (a list of examples is given in Table 8.4). Phytoremediation costs may or may not include preliminary treatability studies to select the proper plants (vegetation) and to assess their effectiveness. Soil preparation, planting, maintenance such

TABLE 8.4 Cost Estimates of Phytoremediation Compared to Other Remediation Methods Using a Variety of Toxic Contaminants

Phytoremediation	Conditions	Contaminant	Cost Estimates	References
Rhizodegradation	Field-scale	Petroleum hydrocarbons	$240/yd^3 or $184/m^3 or $160/ton	AATDF, 1998
	Full-scale		$20/yd^3 or $15/m^3 or $13/ton	
Phytoextraction	Surface removal	Lead	50%–75% of traditional engineering methods	Blaylock et al., 1999
Phytoremediation	60-cm-deep soil	Lead	$6/m^2 compared with $15/m^2 for soil capping, and $730/m^2 for excavation, removal and stabilization	Berti and Cunningham, 2000
Remediation	20 inch (50 cm) deep sediment in 1.2 acre (0.5 ha) chemical waste disposal pond	Cd, Zn, ^{137}Cs sediment	Phytoextraction costs reduced by one-third of soil washing cost above	Cornish et al., 1997
Remediation	20 ins (50 cm) deep sandy loam in 1 acre (0.4 ha) waste	Toxic metals	$60,000–$100,000 compared to minimum $400,000 for excavation	Salt et al., 1995
Removal of contaminants	Gravel system using aquatic plants	Explosives (TNT, nitroglycerine)	$1.78 per 1,000 gallons (3.8 M litres)	ESTCP, 1999
Rhizofiltration	Sunflower	Radionuclide	$2 to $6 per 1000 gallons (3.8 M litres)	Cooney, 1996
Phytostabilization	Soil assuming one metre root depth	Cropping system	$200–$10,000 per hectare equivalent to $0.02 to $1.00/m^3	Cunningham et al., 1995

Continued

TABLE 8.4 Cost Estimates of Phytoremediation Compared to Other Remediation Methods Using a Variety of Toxic Contaminants—cont'd

Phytoremediation	Conditions	Contaminant	Cost Estimates	References
Remediation hydraulic control	20 foot (6 m) deep aquifer of 1 acre (0.4 ha)	Tree vegetation	$660,000 for conventional pump/treat, and $250,000 using trees	Gatliff, 1994
Conventional and phytovolatilization	Irrigated liquid waste Spray on forest trees	Tritum waste	Cost was greater for conventional decontamination and isolation	Fulbright et al., 1996
Conventional and phytovolatilisation	Irrigated liquid waste Spray on forest trees	Tritum waste	Remediation with trees cost 33% less operations saved additionally 40%	Negri et al., 2000
Phytoremediation	Irrigated forest trees	Salix (willow)	50–80% lower due phytoremediation compared to engineering methods	Pulford and Watson, 2003
Cost estimate presented	Evapotranspiration, compared to mechanical	Hydrocarbons	Large substantial savings	RATDF, 1988

Conditions associated with the method, test area and soil type are outlined. Please note that costs are estimated by different methods and values.

as irrigation and fertilization, monitoring, which may include plant nutrient status, plant contaminant concentrations, as well as soil or water concentrations; and air monitoring in the case of the phytovolatilization needs to be included. Disposal of radioactive contaminated biomass is still a serious and costly issue to be resolved. The one report relevant to radioactive waste is the pilot scale cost estimates made for remediation of 20-inch (40-cm) -thick layer of cadmium-, zinc-, and ^{137}caesium-contaminated sediment from a 1.2 acre (0.5-ha) chemical waste disposal pond, and estimates indicated that phytoextraction would cost about one-third of the amount of soil washing alone (Cornish et al. 1995). Independently, on a site with similar characteristics, costs were estimated to be $60,000–$100,000 using phytoextraction for 1 acre (0.4 ha) of 20-in (40-cm) -thick sandy loam waste soil, compared to a minimum of $400,000 just for excavation and storage of the soil (Salt et al., 1995).

Phytoremediation has very good potential for cleaning-up toxic metals, pesticides, solvents, gasoline, explosives and radionuclides. The US Environment Protection Agency (EPA) has estimated that more than 30,000 sites in the United States alone require hazardous waste treatment (EPA, 1999, 2000). Restoring these areas and their soils, as well as disposing of the waste are costly projects, but costs are significantly reduced if plants provide phytoremediation. Recovery of any remediation costs through energy production during incineration and the sale of recovered radionuclides suitable as radiopharmaceuticals after phytoremediation has some merit; however, it might be difficult to find a company and market the production and use of such contaminated materials. Similarly, recovery of costs by selling a commodity type of vegetation, such as wood products for energy could be difficult or impossible due to potential concerns about radioactive residues in wood. Confirmation that the vegetation is uncontaminated and has no residual radioactivity is essential. Cost recovery, and appropriateness of including such recovery of costs as a possibility is an issue that will likely have to wait until greater experience has been gained in phytoremediation of radionuclides.

A cautionary note on commercial companies using remediation and cleaning-up technologies for contaminated radioactive waste sites: In February 1996, Phytotech Inc., a Princeton, NJ-based remediation company reported that it had developed transgenic strains of sunflowers (*Helianthu sp*) that could remove as much as 95% of toxic contaminants in as little as 24 h. Subsequently, *Helianthus* was planted on styrofoam rafts at one end of a contaminated pond near Chernobyl, and in 12 days the radioactive Cs concentrations within its roots were reported to be 8000-times that of the water, while the radioactive Sr concentrations were 2000-times that of the water. In 1998 Phytotech, along with Consolidated Growers and Processors (CGP), and the Ukraine Institute of Bast Crops, planted industrial hemp, *Cannabis*, for the purpose of removing contaminants near Chernobyl. Hemp was as good as sunflower at extracting radioactive Cs and Sr (Phytotech website: www.hawaii.edu/abrp/Technologies/rhizofil.html). It is interesting to note that, following these reports some 15 years ago, there have been virtually no further reports in the literature or on websites on further developments, improvements, success and costs of field phytoremediation projects in which commercial companies have been involved, not just Phytotech Inc!

TRANSGENIC PHYTOREMEDIATION

Genetic engineering of plants (transgenic plants) has the potential to increase the effectiveness and use of phytoremediation, as plants can be genetically modified using specific bacterial, fungal, animal or plant genes that are known to have useful properties for contaminant uptake, degradation or transformation. Elkens and Kirchner (2002) and Verbruggen and LeDuc (2008) have discussed the potential benefits of genetic engineering for phytoremediation, with some examples of what genetically engineered plants could achieve. Examples of promising research into genetic engineering of plants and possible use in phytoremediation have been listed by Gleba et al. (1999) and De Filippis (2013b). Genetically modified canola and tobacco were able to survive concentrations of Hg (II) that killed non-modified control plants, and the tobacco converted the toxic Hg (II) to the less toxic metallic mercury (Hg^0), and volatilized it (Rugh et al., 1996; Meagher et al., 2000). These results were also confirmed with genetically modified yellow poplar (Rugh, 2001). Genetic engineering of plants has demonstrated that key detoxifying enzymes, such as dehalogenases, peroxidases, catalases (monooxygenases), nitroreductase, phosphatases and cytochrome oxidase/reductase can be up-regulated (Doty et al., 2000; Meagher, 2000; Gerhardt et al., 2009). The use of fast-growing transgenic plants that survive extreme toxic environments could increase the accumulation of toxic metals and radionuclides by having a higher biomass; however, much of the transgenic plants being developed is not necessarily going to contain genes causing high radionuclide accumulation patterns. In a list of transgenic plants at the developmental stage provided by De Filippis (2013b), only a handful would contribute to greater growth, and indirectly lead to better plants for use in phytoremediation of radionuclides. There are now available over 3000 near-complete DNA sequenced organisms in computer databanks, including over 70 plants [a list of sequenced or partially sequenced plants is provided by De Filippis (2013a)]. DNA and RNA sequencing of plants and microbes, molecular markers, protein coding and non-coding transcripts are providing new and novel information (De Filippis, 2013a; Pandey et al., 2013), and future research projects should be directed towards the best combination of transgenic microbes and plants best suited for phytoremediation. The power of genetics and molecular biology is no longer confined to the health sciences and agriculture. New and exciting resources and technologies in bioinformatics, transcriptomics, proteomics and metabolomics (De Filippis, 2013a) are going to contribute to better knowledge of plants and microorganisms essential for a better understanding of phytoremediation. Bioinformatic data could significantly contribute to designing new and more efficient phytoremediation methods. In conjunction with research on genetically engineered plants for phytoremediation, however, regulatory and public concerns will have to be addressed for this relatively new technology. Public opinion and environmental issues surrounding radioactive waste are matters of concern, and it would be inadvisable to add to these issues the concerns from transgenic plant technology.

CONCLUSIONS AND FUTURE DIRECTIONS

Large quantities of radioactive waste and mine tailings are being generated annually worldwide from mining and processing uranium ores for nuclear energy. Radioactive tailings represent a large clean-up challenge to the mining and nuclear energy industries. Therefore in the first few sections of this chapter we detailed the physical, chemical, storage, transportation and disposal of radionuclides, and which ones are more dangerous to the environment and human health. The conclusions arrived at suggest that a common-sense approach to radioactive waste and its safe disposal involves the following four steps:

- Minimize the production of radioactive waste. Before using radioactivity it needs to be demonstrated very clearly that the benefits outweigh the risks.
- Assess all options for the management of radioactive waste. Most radioactive waste management assumes the need for off-site storage, but the option of storing waste where it is produced needs re-evaluation. Even if centralized facilities exist, waste is inevitably stored at the sites of production for too long, and on-site storage facilities must be adequately constructed and regulated.
- Scientific and environmental strict criteria must be used in choosing operational and management options. Since technologically safe solutions are now available for radiation, we have no right to transmit unprocessed nuclear waste to future generations in an open-ended way.
- Invest in phytoremediation of radionuclides, a technology already demonstrated on a pilot-scale; but it needs to be developed at an industrial and commercial level. Phytoremediation of radioactive contaminants is still in development, and more research is required to address plant species that will be most effective in different radioactive waste scenarios. Plants must be able to survive and thrive in contaminated waste, yet be able to mitigate the toxic radioactive pollutants.

To summarize, much of the technology for safe handling and storage of radionuclides is available. There is not one solution but rather several complementary solutions to the problem of managing radioactive waste. We must pursue research on radioactive safety encased in inert substances, continue to reprocess spent fuel and immediately begin work on underground storage. Public involvement in decisions and informed consent to proposals is also essential from a practical point of view, because there is a long history of communities successfully mobilizing to force the abandonment of nuclear projects.

The majority of this chapter, however, is devoted to the growing importance of phytoremediation of radioactive waste sites. A brief description of non-plant based remediation methods is outlined, followed by a detailed evaluation of plant based remediation methods, including hyperaccumulation, radioactive tolerance, uptake and distribution, aquatic phytoremediation and cost estimates. Phytoremediation has become a fast growing field of research and development for application to radionuclide waste. Phytoremediation, although still an emerging technology for radioactive contaminated sites, has become more attractive due to its low cost, high public acceptance and environmental (green) acceptability. It is not a method to be used for all radioactive waste problems, but it is well worth considering as a major supplement to existing technologies. A number of important areas of research and development in phytorermediation still need to be improved, and future research should focus on these. Topics to be better understood include the role of soil chelation and soil acidity on extraction of radionuclides. Increased growth and biomass of selected plants for remediation, and in particular trying to achieve greater rates of transpiration in plants that is the driving force for radionuclide hyperaccumulation is required. There will also need to be better identification of plant species with increased resistance to radiation, and careful selection for better adaptation of plants to radiation poisoning. Clearly, different approaches using different plants are going to be the normal way of dealing with different radionuclides.

Phytoremediation technology has been demonstrated in a number of different situations and for a number of different toxicants (including radionuclides), but has not yet been commercially exploited. More research is required for the development of plants tailored to remediation needs, and the use of transgenic fast-growing woody trees like willow and poplar, where genetic engineering is going to play an important role. The use of trees and transgenic technology is well advanced in some phytoremediation situations and superior seedlings of quick-growing trees are already available. The concept of manipulating plant genes for toxic metal (and perhaps radionuclide) uptake is today a cutting-edge research topic; however, these technologies are not likely to be a total substitute for more basic, site and field testing of various phytoremediation methods. Nevertheless, transgenic plants would provide more suitable plants for increased secretion around roots to aid radionuclide uptake and binding, and better overall growth and biomass to absorb and remediate radioactive sites. The use of trees and forests leads to considerable savings in remediation of radioactive sites, as clearly demonstrated by removal of tritium in wastewater. The likelihood of public acceptance of genetically engineered plants for phytoremediation should be welcomed, since it has the potential to clean up the environment of toxicants; however, as in the whole public debate on radioactive waste, public involvement in decision making and good information is critical. Phytoremediation technology has attracted a great deal of attention in recent years and the expectation is that phytoremediation is likely to capture a significant share of the environmental remediation market. It is expected that phytoremediation of radionuclide wastes will become an integral part of the environmental management and risk reduction strategy for governments, industry and society.

REFERENCES

AATDF (Advanced Applied Technology Demonstration Facility), 1998. AATDF Technology evaluation report on phytoremediation of hydrocarbon-contaminated soil. AATDF Report TR 98–16. Available at: www.ruf.rice.edu/~AATDF.

Abbott, D., 2012. Limits to growth: Can nuclear power supply the world's needs? Bull. Atom. Sci. 68, 23–32.

Alvarez, R., 2007. Radioactive wastes and the global nuclear energy partnership. Bull. Atom. Sci. 63, 57–59.

ANSTO (Australian Nuclear Science and Technology Organisation), 2011. Management of Radioactive Waste in Australia. Australian Government Printing Office. Available at: www.apo.ansto. gov.au/dspace/.

Atwell, B.J., Kriedemann, P.E., Turnbull, C.G.N., 1999. Plants in Action: Adaptation in Nature, Performance in Cultivation. Macmillan Education, Melbourne.

Baker, A.J.M., McGrath, S.P., Reeves, R.D., Smith, J.A.C., 2000. Metal hyperaccumulator plants: A review of the ecology and physiology of a biological resource for phytoremediation of metal polluted soils. In: Terry, N., Banuelos, G.S. (Eds.), Phytoremediation of Contaminated Soil and Water. CRC Press, Boca Raton.

Banuelos, G.S., 2006. Phyto-products may be essential for sustainability and implementation of phytoremediation. Environ. Pollut. 144, 19–23.

Banuelos, G.S., Ajwa, H.A., Mackey, B., Wu, L.L., Cook, C., Akohoue, S., Zambrzuski, S., 1997. Evaluation of different plant species used for phytoremediation of high soil selenium. J. Environ. Qual. 26, 639–646.

Bañuelos, G.S., Shannon, M.C., Ajwa, H., Draper, H.J.H., Jordahl, J., Licht, L., 1999. Phytoextraction and accumulation of boron and selenium by poplar (*Populus*) hybrid clones. Int. J. Phytoremed. 1, 81–96.

Barkay, T., Schaefer, J., 2001. Metal and radionuclide bioremediation: Issues, considerations and potentials. Curr. Opin. Microbiol. 4, 318–323.

Baud-Grasset, F., Baud-Grasset, S., Safferman, S.I., 1993. Evaluation of the bioremediation of a contaminated soil with phytotoxicity tests. Chemosphere 26, 1365–1374.

Benjamin, R., Wagner, J., 2006. Reconsidering the law and economics of low-level radioactive waste management. Environ. Econom. Policy Stud. 8, 33–53.

Berti, W.R., Cunningham, S.D., 2000. Phytostabilization of metals. In: Raskin, I. (Ed.), Phytoremediation of Toxic Metals: Using Plants to Cleanup the Environment. John Wiley and Sons Inc, New York.

Best, E.P.H., Miller, J.L., Larson, S.L., 2001. Tolerance towards explosives, and explosives removal from groundwater in treatment wetland mesocosms. Water Sci. Technol. 44, 515–521.

Bhainsa, K.C., D'Souza, S.F., 2001. Uranium(VI) biosorption by dried roots of *Eichhornia crassipes* (water hyacinth). J. Environm. Sci. Health (A). 36 1621–1631.

Bizily, S.P., Rugh, C.L., Summers, A.O., Meagher, R.B., 1999. Phytoremediation of methylmercury pollution: *merB* expression in *Arabidopsis thaliana* confers resistance to organomercurials. Proc. Natl. Acad. Sci. USA 96, 6808–6813.

Blaylock, M.J., Elless, M.P., Huang, J.W., Dushenkov, S.M., 1999. Phytoremediation of lead-contaminated soil at a New Jersey brownfield site. Remediation 9, 93–101.

Blaylock, M.J., Salt, D.E., Dushenkov, S., Zakharova, O., Gussman, C., Kapulnik, Y., Ensley, B.D., Raskin, I., 1997. Enhanced accumulation of Pb in Indian mustard by soil applied chelating agents. Environ. Sci. Technol. 3, 860–865.

Bracke, G., 2012. Aspects of final disposal of radioactive waste in Germany. Turk. J. Earth Sci. 21, 145–152.

Broadley, M.R., Willey, N.J., 1997. Difference in root uptake of radiocesium by 30 plant taxa. Environ. Pollut. 97, 2–11.

Brook, B.W., 2012. Could nuclear fission energy solve the greenhouse problem? Energy Policy 42 (March). Available at: www.sciencedirect.com/science/article/.

Brooks, R.R., 1998. Plants that Hyperaccumulate Heavy Metals. CAB International, NY.

Bunzl, K., Albersa, B.P., Shimmacka, W., Rissanenb, K., Suomelab M,Puhakainenb, M., Raholab, T., Steinnesc, E., 1999. Soil to plant uptake of fallout 137Cs by plants from boreal areas polluted by industrial emissions from smelters. Sci. Total Environ. 234, 213–221.

Burken, J.G., Schnoor, J.L., 1998. Predictive relationships for uptake of organic contaminants by hybrid poplar trees. Environ. Sci. Technol. 32, 3379–3385.

Burken, J.G., Schnoor, J.L., 1999. Distribution and volatilization of organic compounds following uptake by hybrid poplars. Int. J. Phytoremediation 1, 139–151.

Cataldo, D.A., McFadden, K.M., Garland, T.R., Wildung, R.E., 1988. Organic constituents and complexation of Nickel (II), Iron (III), Cadmium (II), and Plutonium (IV) in soybean xylem exudates. Plant Physiol. 86, 734–739.

Chu, S.Y.F., Ekstrom, L.P., Firestone, R.B., 1999. TORI interactive site, periodic table interface to nuclides. Lunds University Isotope Project. Available at: www.ie.lbl.gov/toi/.

Chou, F.I., Chung, H.P., Teng, S.P., Sheu, S.T., 2005. Screening plant species native to Taiwan for remediation of 137Cs-contaminated soil and the effects of K addition and soil amendment on the transfer of 137Cs from soil to plants. J. Environ. Radioactiv. 80, 175–181.

Ciuffo, L.E.C., Belli, M., 2006. Radioactive trace in semi-natural grassland. Effects of 40K in soil and potential remediation. Electron. J. Biotech. 9, 297–302.

Clint, G.M., Dighton, J., 1992. Uptake and accumulation of radiocaesium by mycorrhizal and non–mycorrhizal heather plants. New Phytol. 121, 555–561.

Cole, S., 1998. The emergence of treatment wetlands. Environ. Sci. Technol. 32, 218A–223A.

Comby, B., 2005. The solutions for nuclear waste. Inter. J. Environ. Stud. 62, 725–736.

Cooney, C.M., 1996. Sunflowers remove radionuclides from water in ongoing phytoremediation field tests. Environ. Sci. Technol. 30, 194A.

Cornish, J.E., Goldberg, R.S., Levine, R.S., Benemann, J.R., 1995. Phytoremediation of soils contaminated with toxic elements and radionuclides. In: Hinchee, R.E., Means JL,Burris, D.R. (Eds.), Bioremediation of Inorganics. Battelle Press, Columbus OH.

Cornish, J.M., Fuhrmann, L., Kochian, L.V., Page, D., 1997. Phytoextraction treatability study: Removal of 137Cs from soils at Brookhaven National Laboratory Hazardous Waste Management Facility Site. Progress Report, US Department of Energy. Available at: www.doe.gov/.

Coughtrey, P.J., Kirton, J.A., Mitchell, N.G., Morris, C., 1989. Transfer of radioactive caesium from soil to vegetation and comparison with potassium in upland grasslands. Environ. Pollut. 62, 281–283.

Cunningham, S.D., Berti, W.R., Huang, J.W.W., 1995. Phytoremediation of contaminated soils. Trends Biotechnol. 13, 393–397.

Das, N., 2012. Remediation of radionuclide pollutants through biosorption – an overview. Clean – Soil, Air, Water 40, 16–23.

De Filippis, L.F., 2010. Biochemical and molecular aspects in phytoremediation of selenium. In: Ashraf, M., Ozturk, M., Ahmad, M.S.A. (Eds.), Plant Adaptation and Phytoremediation. Springer Science & Business Media, Berlin New York.

De Filippis, L.F., 2013a. Bioinformatic tools for crop improvement. In: Hakeem, K.R., Ahmad, P., Ozturk, M. (Eds.), Crop Improvement, New Approaches and Modern Techniques. Springer Science & Business Media, Berlin New York.

De Filippis, L.F., 2013b. Crop improvement through tissue culture. In: Hakeem, K.R., Ahmad, P., Ozturk, M. (Eds.), Crop Improvement, New Approaches and Modern Techniques. Springer Science & Business Media, Berlin New York.

Delvaux, B., Kruyts, N., Cremers, A., 2000. Rhizospheric mobilization of radiocesium in soils. Environ. Sci. Technol. 34, 489–1493.

de Mello-Faras, P.S., Chaves, A.L.S., Lencina, C.L., 2011. Transgenic plants for enhanced phytoremediation – physiological studies. In: Alvarez, M.A. (Ed.), Genetic Transformation. InTech Publications.

Dimitrescu, I., 2010. Technology of underground storage of radioactive waste. Rev. Minerol. Mining 6, 15–19.

Doty, S.L., Shang, T.Q., Wilson, A.M., Tangen, J., Westergreen, A.D., Newman, L.A., Strand, S.E., Gordon, M.P., 2000. Enhanced metabolism of halogenated hydrocarbons in transgenic plants containing mammalian cytochrome P450 2E1. Proc. Natl. Acad. Sci. USA 97, 6287–6291.

Dubchak, S., Ogar, A., Mietelski, J.W., Turnau, K., 2010. Influence of silver and titanium nanoparticles on arbuscular mycorrhiza colonization and accumulation of radiocaesium in *Helianthus annuus*. Span. J. Agric. Res. 8, S103–S108.

Dushenkov, S., 2003. Trends in phytoremediation of radionuclides. Plant Soil 249, 167–175.

Dushenkov, S., Kumar, V., Motto, H., Raskin, I., 1995. Rhizo-filtration: The use of plants to remove heavy metals from aqueous streams. Environ. Sci. Technol. 29, 1239–1245.

Dushenkov, S., Mikheev, A., Prokhnevsky, A., Ruchko, M., Sorochinsky, B., 1999. Phytoremediation of radiocesium-contaminated soil in the vicinity of Chernobyl, Ukraine. Environ. Sci. Technol. 33, 469–475.

Dushenkov, S., Vasudev, D., Kapulnik, Y., Gleba, D., Fleisher, D., Ting, K.C., Ensley, B., 1997. Removal of uranium from water using terrestrial plants. Environ. Sci. Technol. 31, 3468–3474.

Dutton, M.V., Humphreys, P.N., 2005. Assessing the potential of short rotation coppice (Src) for cleanup of radionuclide contaminated sites. Inter. J. Phytoremediation 7, 279–293.

Eapen, S., Singh, S., D'Souza, S.F., 2007. Phytoremediation of metals and radionuclides. In: Singh, S.N., Tripathi RD (Eds.), Environmental Bioremediation Technologies. Springer-Verlag, Berlin Heidelberg.

Eapen, S., Singh, S., Thorat, V., Kaushik, C.P., Raj, K., D'Souza, S.F., 2006. Phytoremediation of radiostrontium (^{90}Sr) and radiocesium (^{137}Cs) using giant milky weed (*Calotropis gigantea* R.Br.) plants. Chemosphere 65, 2071–2073.

Eapen, S., Suseelan, K.N., Tivarekar, S., Kotwal, S.A., Mitra, R., 2003. Potential for rhizofiltration of uranium using hairy root cultures of *Brassica juncea* and *Chenopodium amaranticolor*. Environ. Res. 91, 127–133.

Ebbs, S.D., Brady, D.J., Kochian, L.V., 1998. Role of uranium speciation in the uptake and translocation of uranium by plants. J. Exp. Bot. 49, 1183–1190.

Echevarria, G., Sheppard, N.I., Morel, J., 2001. Effect of pH on sorption of uranium in soils. J. Environ. Radioact. 53, 257–264.

Echevarria, G., Vong, P.C., Morel, J.L., 1997. Bioavailability of Technetium-99 as affected by plant species and growth, application form, and soil incubation. J. Environ. Qual. 26, 947–956.

Ehlken, S., Kirchner, G., 2002. Environmental processes affecting plant root uptake of radioactive tracer elements and variability of transfer factor: A review. J. Environm. Radioact. 58, 97–112.

Entry, J.A., Rygiewicz, P.T., Emmingham, W.H., 1994. Accumulation of cesium 137 and strontium 90 in Ponderosa pine and Monterey pine seedlings. J. Environ. Qual. 22, 742–745.

Entry, J.A., Watrud, L.S., Manasse, R.S., Vance, N.C., 1997. Phytoremediation and reclamation of soils contaminated with radionuclides. In: Kruger, E.L., Anderson, T.A., Coats, J.R. (Eds.), Phytoremediation of Soil and Water Contaminants. Symposium Series No. 664. American Chemical Society, Washington.

Entry, J.A., Watrud, L.S., Reeves, M., 1999. Accumulation of ^{137}Cs and ^{90}Sr from contaminated soil by three grass species inoculated with mycorrhizal fungi. Environ. Pollut. 104, 449–457.

EPA (Environmental Protection Agency USA), 1997. Status of in situ phytoremediation technology. In: Recent Developments for in Situ Treatment of Metal Contaminated Soils. U.S. Environmental Protection Agency PA-542-R-97–004. Available at: www.epa.gov/.

EPA (Environmental Protection Agency USA), 1999. Phytoremediation Resource Guide.U.S. Environmental Protection Agency. Available at: www.epa.gov/.

EPA (Environmental Protection Agency USA), 2000. Introduction to Phytoremediation. Office of Research and Development, Washington EPA/600/R-99/107. Available at: www.epa.gov/.

EPA (Environmental Protection Agency USA), 2004. Radionuclide Biological Remediation Resource Guide Prepared by: Ibeanusi VM, Grab DA. U.S. Environmental Protection Agency, Region 5 Division, Chicago IL. Available at: www.epa.gov/.

EPA (Environmental Protection Agency USA), 2010. Phytotechnology for Site Cleanup, Fact Sheet. EPA U.S. Environmental Protection Agency. Available at: www.epa.gov/.

Eroglu, H., Yapici, S., Nuhoglu, C., Varoglu, E., 2009. Biosorption of Ga-67 radionucleotides from aqueous solutions onto waste pomace of an olive oil factory. J. Hazard. Mater. 172, 729–738.

Ertel, J., Ziegler, H., 1991. $^{134/137}$Cs contamination and root uptake of different forest trees before and after the Chernobyl accident. Rad. Environm. Biophys. 30, 147–157.

ESTCP (Environmental Security Technology Certification Program), 1999. The Use of Constructed Wetlands to Phytoremediate Explosives-Contaminated Groundwater at the Milan Army Ammunition Plant, Milan TN. Cost and Performance Report ESTCP, US Department of Defense. Available at: www.erdc.usace.army.mil/phyto/pubs.

Favas, P.J.C., Pratas, J., 2013. Uptake of uranium by native aquatic plants: Potential for bioindication and phytoremediation. E3S Web of Conferences, owned by the authors; published by EDP Sciences.

Finley, J.W., 2005. Selenium accumulation in plant foods. Nut. Rev. 63, 196–202.

Fulbright, H.H., Schwirian-Spann, A.L., Jerome, K.M., Looney, B.B., Brunt, V.V., 1996. Status and practicality of detritiation and tritium reduction strategies for environmental remediation. Westinghouse Savannah River Company, Aiken, SC. Available at: www.srs.gov/general/srs-home.

Fulekar, M.H., Singh, A., Thorat, V., Kaushik, C.P., Eapen, S., 2010. Phytoremediation of ^{137}Cs from low level nuclear waste using Catharanthus roseus. Indian J. Pure Appl. Phys. 48, 516–519.

Garten, C.T., Lomax, R.D., 1989. Technetium-99 cycle in maple trees: Characterization of changes in chemical form. Health Phys. 57, 299–307.

Garten, C.T., Tucker, C.S., 1986. Plant uptake of neptunium. J. Environ. Radioact. 4, 91–99.

Garten, C.T., Tucker, C.S., Walton, B.T., 1986. Environmental fate and distribution of technetium-99 in a deciduous forest ecosystem. J. Environ. Radioact. 3, 163–188.

Gatliff, E.G., 1994. Vegetative remediation process offers advantages over traditional pump-and-treat technologies. Remediation 4, 343–352.

Gerhardt, K.E., Huang, X.-D., Glick, B.R., Bruce, M., Greenberg, B.M., 2009. Phytoremediation and rhizoremediation of organic soil contaminants: Potential and challenges. Plant Sci. 176, 20–30.

Gleba, D., Borisjuk, N.V., Borisjuk, L.G., Kneer, R., Poulev, A., Skarzhinskaya, M., Dushenkov, S., Logendra, S., Gleba, Y.Y., Raskin, I., 1999. Use of plant roots for phytoremediation and molecular farming. Proc. Natl. Acad. Sci. USA 96, 5973–5977.

Goncharova, N.V., 2009. Availability of radiocesium in plant from soil: Facts, mechanisms and modelling. Global NEST J. 11, 260–266.

Hoffman, F.O., Garten, C.T., Huckabee, J.W., Lucas, D.M., 1982. Interception and retention of technetium by vegetation and soil. J. Environ. Qual. 11, 133–141.

Hoseini, P.S., Poursafa, P., Moattar, F., Amin, M.M., Rezaei, A.H., 2012. Ability of phytoremediation for absorption of strontium and cesium from soils using *Cannabis sativa*. Int. J. Env. Health Eng. 1, 1–5.

Huang, J.W., Blaylock, M.J., Kapulnik, Y., Ensley, B.D., 1998. Phytoremediation of uranium-contaminated soils: Role of organic acids in triggering uranium hyperaccumulation in plants. Environ. Sci. Technol. 32, 2004–2008.

Huang, J.W.W., Chen, J.J., Berti, W.R., Cunningham, S.D., 1997. Phytoremediation of lead contaminated soils: Role of synthetic chelates in lead phytoextraction. Environ. Sci. Technol. 31, 800–805.

IAEA (International Atomic Energy Agency), 2003. The long-term storage of radioactive waste: Safety and sustainability. A Position Paper of International Experts. IAEA, Vienna. Available at: www.iaea.org/books/.

IAEA (International Atomic Energy Agency), 2005. Chernobyl's legacy: Health, environmental and socio-economic impacts, Cherbobyl Forum 2003–2005. IAEA, Vienna. Available at: www.iaea.org/books/.

IAEA (International Atomic Energy Agency), 2006. Nuclear Technology Review 2006. IAEA, Vienna. Available at: www.iaea.org/books/.

IAEA (International Atomic Energy Agency), 2010. Managing Radioactive Waste. International Atomic Energy Agency, Vienna. Available at: www.iaea.org/books/.

Irga, P.J., Torpy, F.R., Burchett, M.D., 2013. Can hydroculture be used to enhance the performance of indoor plants for the removal of air pollutants? Atmosphereric Environ. 77, 267–271.

Kadlec, R.H., Knight, R.L., 1996. Treatment Wetlands. CRC Press, Boca Raton FL.

Kelley, C., Mielke, R.E., Dimaquibo, D., Curtis, A.J., Dewitt, J.G., 1999. Adsorption of Eu(III) onto roots of water hyacinth. Environ. Sci. Technol. 33, 1439–1443.

Kilda, R., Grigaliuniene, D., Bartkus, G., 2009. Potential sites for landfill type radioactive waste repository-analysis of radionuclide migration in the atmosphere. Energetika 55, 144–148.

Kirk, G.J.D., Staunton, S., 1989. On predicting the fate of radioactive caesium in soil beneath grassland. J. Soil Sci. 40, 71–84.

Kovacs, T., Bodrogi, E., Somlai, I., Gorjanacz, Z., 2004. ^{210}Po- and ^{210}Pb-determination in Hungarian grow tobacco. AARMS 3, 165–169.

Kumar, P.B.A.N., Dushenkov, V., Ensley, B.D., Chet, I., Raskin, I., 1995. Phytoremediation: A novel strategy for the removal of toxic metals from environment using plants. Biotechnol 13, 1232–1238.

Lasat, M.M., Norvell, W.A., Kochian, L.V., 1997. Potential for phytoextraction of ^{137}Cs from contaminated soil. Plant Soil 195, 99–106.

Lasat, M.M., Fuhrmann, M., Ebbs, S.D., Cornish, J.E., Kochian, L.V., 1998. Phytoremediation of a radiocesium-contaminated soil: Evaluation of cesium-137 bioaccumulation in the shoots of three plant species. J. Environ. Qual. 27, 165–169.

LeBars, Y., Pescatore, C., 2004. Shifting paradigms in managing radioactive waste. NEA updates, NEA News No. 22.2. pp14–16. Available at: www.oecd-nea.org/nea-news/.

Lewis, B.G., MacDonnell, M.M., 1990. Release of radon-222 by vascular plants: Effect of transpiration and leaf area. J. Environ. Qual. 19, 93–97.

Mc Combie, C., 2009. Evaluating solutions to the nuclear waste problem. Bull. Atomic Sci. 65, 42–48.

Meagher, R.B., 2000. Phytoremediation of toxic elemental and organic pollutants. Curr. Opin. Plant Biol. 3, 153–162.

Meagher, R.B., Rugh, C.L., Kandasamy, M.K., Gragson, G., Wang, N.J., 2000. Engineered phytoremediation of mercury pollution in soil and waters using bacterial genes. In: Terry, N., Banuelos, G. (Eds.). Phytoremediation of Contaminated Soil and Water, Leurs Publ. Boca Raton, FL.

Mitchell, N., Perez-SanchezD, D., Thorne, M.C., 2013. A review of the behaviour of U-238 series radionuclides in soils and plants. J. Radiol. Protec. 33, R17–R48.

Mohanty, M., Dhal, N.K., Patra, P., Das, B., Sita, P., 2010. Phytoremediation: A novel approach for utilization of iron-ore wastes. In: Reddy, R., Whitacre, D.M. (Eds.), Reviews of Environmental Contamination and Toxicology 206. Springer Science & Business Media.

Mossman, K.L., 2003. Restructuring nuclear regulation. Environ. Health Pers. 111, 13–17.

Murphy, C.E.J., 2001. An estimate of the history of tritium inventory in wood following irrigation with tritiated water. Westinghouse Savannah River Company, Aiken SC. Available at: www.srs.gov/general/srs-home.

Nazina, T.N., Safonov, A.V., Kosarev, I.M., Ivoilov, V.S., Poltaraus, A.B., Ershov, B.G., 2010. Microbiological Processes in the Severnyi deep disposal site for liquid radioactive wastes. Microbiol 79, 528–537.

Negri, M.C., Hinchman, R.R., Wozniak, J.B., 2000. Capturing a mixed contaminant plume: Tritium phytoevaporation at Argonne National Laboratory's area 319. Argonne National Laboratory, Argonne, IL. Available at: www.anl.gov/.

Noshkin, V.E., 1972. Ecological aspects of plutonium dissemination in aquatic environments. Health Phys. 22, 537–549.

Ohnuki, T., Kozai, N., Sakamoto, Ozaki T., Nankawa, T., Suzuki, Y., Francis, A.J., 2010. Association of actinides with microorganisms and clay: Implications for radionuclide migration from waste-repository sites. Geomicrobiol. J. 27, 225–230.

Olson, P.E., Fletcher, J.S., 2000. Ecological recovery of vegetation at a former industrial sludge basin and its implications to phytoremediation. Environ. Sci. Pollut. Res. 7, 1–10.

Page, R., 2011. Poisons for orphan wastes: A look at the latest containment techniques for radioactive waste. NUCLEARtcetoday.com. October. Available at: www.tcetcxday.com.

Pandey, A., Kumar, A., Purohit, R., 2013. Sequencing *Closterium moniliferum*: Future prospects in nuclear waste disposal. Egypt J. Med. Human Genet. 14, 113–115.

Pilon-Smits, E.A.H., 2005. Phytoremediation. Annu. Rev. Plant Biol. 56, 15–39.

Pilon-Smits, E.A.H., Quinn, C.F., Tapken, W., Malagoli, M., Schiavon, M., 2009. Physiological functions of beneficial elements. Curr. Opin. Plant Biol. 12, 267–274.

Pivetz, B.E., 2001. Phytoremediation of contaminated soil and ground water at hazardous waste sites. EPA USA ground water issue EPA/540/S-01/500. Available at: www.epa.gov/.

Plecas, I.B., 2003. Comparison of mathematical interpretation in radioactive waste leaching studies. J. Radioanalyt. Nuclear Chem. 258, 435–437.

Prasad, M.N.V., 2007. Sunflower (*Helinathus annuus* L.) – a potential crop for environmental industry. HELIA 30, 167–174.

Prasad, M.N.V., Freitas, H.M.O., 2003. Metal hyperaccumulation in plants – Biodiversity prospecting for phytoremediation technology. Electron. J. Biotech. 6, 285–305.

Pulford, I.D., Watson, C., 2003. Phytoremediation of heavy metal-contaminated land by trees—a review. Environ. Internat. 29, 529–540.

Raloff, J., 1989. Greenery filters out indoor air pollution. Sci. News 136, 212.

Ravichandran, R., Binukumar, J.P., Sreeram, R.L.S., Arunkumar, L.S., 2011. An overview of radioactive waste disposal procedures of a nuclear medicine department. J. Med. Phys. 36, 95–99.

Roca, M.C., Vallejo, V.R., Roig, M., Tent, J., Vidal, M., Rauret, G., 1997. Prediction of cesium-134 and strontium-85 crop uptake based on soil properties. J. Environ. Qual. 26, 1354–1362.

Rogers, R.D., Williams, S.E., 1986. Vesicular-arbuscular mycorrhiza: Influence on plant uptake of cesium and cobalt. Soil Biol. Biochem. 18, 371–376.

RTDF (Remediation Technologies Development Forum), 1998. Summary of the RTDF Alternative Covers Assessment Program Workshop. Las Vegas NV. Available at: www.rtdf.org/public/phyto/minutes/altcov/Alt21798.

Rugh, C.L., 2001. Mercury detoxification with transgenic plants and other biotechnological break-throughs for phytoremediation. In Vitro Cell. Dev. Biol.-Plant, 37, 321–325.

Rugh, C.L., Wildett, H.D., Stack, N.M., Thompson, D.M., Summers, A.O., Meagher, R.B., 1996. Mercuric ion reduction and resistance in transgenic *Arabidopsis thaliana* plants expressing a modified bacterial *merA* gene. Proc. Natl. Acad. Sci. USA 93, 3182–3187.

Saleh, H.M., Eskander, S.B., 2012. Using portland cement for encapsulation of *Epipremnum aureum* generated from phytoremediation process of liquid radioactive wastes. Inter. J. Chem. Environ. Engineer Sys. 3, 1–8.

Salt, D.E., Blaylock, M., Kumar, P.B.A.N., Dushenkov, V., Ensley, B.D., Chet, I., Raskin, I., 1995. Phytoremediation: A novel strategy for the removal of toxic metals from the environment using plants. Biotechnol 13, 468–474.

Salt, D.E., Pickering, I.J., Prince, R.C., Gleba, D., Dushenkov, S., Smith, R.D., Raskin, I., 1997. Metal accumulation by aquacultured seedlings of Indian mustard. Environ. Sci. Technol. 31, 1636–1644.

Sandhu, S.S., Gill, B.S., Casto, B.C., Rice, J.W., 1991. Application of *Tradescantia* micronucleus assay for in situ evaluation of potential genetic hazards from exposure to chemicals at a wood-preserving site. Waste Hazard. Mater. 8, 257–262.

Schroeder, P., Navarro-Avino, J., Azaizeh, H., Golan Goldhirsh, A., DiGregorio, S., Komives, T., Langergraber, G., Lenz, A., Maestri, E., Memon, A., 2007. Using phytoremediation technologies to upgrade waste water treatment in Europe. Environ. Sci. Pollut. Res-Int. 14, 490–497.

Shaw, G., Bell, J.N.B., 1991. Competitive effects of potassium and ammonium on caesium uptake kinetics in wheat. J. Environ. Radioactiv. 13, 283–296.

Singh, A., Eapen, S., Fulekar, M.H., 2009a. Phytoremediation technology for remediation of radio-strontium (^{90}Sr) and radiocaesium (^{137}Cs) by *Catharanthus roseus* (l.) G. Don in aquatic environment. Environ. Engineer Man. J. 8, 527–532.

Singh, S., Eapen, S., Thorat, V., Kaushik, C.P., Raj, K., D'Souza, S.F., 2008. Phytoremediation of ^{137}cesium and ^{90}strontium from solutions and low level nuclear waste by *Vetiveria zizanoides*. Ecotoxicol. Environ. Safe 69, 306–311.

Singh, S., Thorat, V., Kaushik, C.P., Raj, K., Eapen, S., D'Souza, S.F., 2009b. Potential of *Chromolaena odorata* for phytoremediation of 137Cs from solution and low level nuclear waste. J. Hazard. Mater. 162, 743–745.

Smith, R.A.H., Bradshaw, A.D., 1979. The use of metal tolerant plant populations for the reclamation of metalliferous wastes. J. Appl. Ecol. 16, 595–612.

Sors, T.G., Ellis, D.R., Salt, D.E., 2005. Selenium uptake, translocation, assimilation and metabolic fate in plants. Photosynth. Res. 86, 373–389.

Stutte, G.W., , 2012. Phytoremediation of indoor air: Bill Wolverton and development of an industry. NASAPublications ID 20120003454. Available at: www.naca.larc.nasa.gov/search.

Tadevosyan, A.H., Mayrapetyan, S.K., Schellenberg, M.P., Ghalachyan, L.M., Hovsepyan, A.H., Khachatur, Mayrapetyan, S., 2011. Migration and accumulation of artificial radionuclides in the system water-soil-plants depend on polymer. World Acad. Sci. Eng. Technol. 54, 79–82.

Tang, S., Willey, N.J., 2003. Uptake of [134]Cs by four species from Asteraceae and two variants from Chenopodiaceae grown in two types of Chinese soil. Plant Soil 250, 75–81.

Turnau, K., Orlowska, E., Ryszka, P., Szymon Zubek, S., Anielska, T., Gawronski, S., Jurkiewicz, A., 2006. Role of mycorrhizal fungi in phytoremediation and toxicity monitoring of heavy metal rich industrial wastes in southern Poland. In: Twardowska, I. (Ed.), Soil and Water Pollution Monitoring, Protection and Remediation. Springer-Verlag, Berlin.

Venek, T., Soudek, P., Tykva, R., 2001. Study of radiophytoremediation. Minerv. Biotecnol. 13, 117–121.

Vangronsveld, J., Herzig, R., Weyens, N., Boulet, J., Adriaensen, K., Ruttens, A., Thewys, T., Vassilev, A., Meers, E., Nehnevajova, E., van der Lelie, D., Mench, M., 2009. Phytoremediation of contaminated soils and groundwater: Lessons from the field. Environ. Sci. Pollut. Res. 16, 765–794.

Verbruggen, N., LeDuc, D., 2008. Potential of plant genetic engineering for phytoremediation of toxic trace elements. Encyclopedia of Life Science Support Systems (EOLSS).

Veresoglou, D.S., Barbayiannis, N., Matsi, T., 1996. Shoot Sr concentrations in relation to shoot Ca concentrations and to soil properties. Plant Soil 1978, 95–100.

VeselovEIS, 2011. Strategy for the environmental safety of radiologically dangerous objects. Atomic Energy (Atom Energya) 109, 233–237.

Wallace, A., Romney, E.M., 1972. Radioecology and ecophysiology of desert plants at the nevada test site. Environmental Radiation Division. Los Angeles Soil Science and Agricultural Engineering. Univ. California, Riverside.

Ward, O.P., Singh, A., 2004. Soil bioremediation and phytoremediation – An overview. In: Singh, A., Ward, O.P. (Eds.), Applied Bioremediation and Phytoremediation. Springer-Verlag, Berlin Heidelberg.

Watt, N.R., Willey, N.J., Hall, S.C., Cobb, A., 2002. Phytoextraction of [137]Cs: The effect of soil ^{137}Cs concentrationon ^{137}Cs uptake by *Beta vulgaris*. Acta Biotechnol. 22, 183–188.

Westhoff, A., 1999. Mycorrhizal plants for phytoremediation of soils contaminated with radionuclides. Restoration Reclamation Rev. 5, 1–6.

WISE (World Information Service on Energy), 2010. Uranium project, Chronology of uranium tailings dam failures. Available at: www.wise-uranium.org/.

WNA (World Nuclear Association), 2010. Australia's uranium. Available at: www.world-nuclear. org/Information-Library/.

Wu, J., Hsu, F.C., Cunningham, S.D., 1999. Chelate assisted Pb phytoextraction: Pb availability, uptake, and translocation constraints. Environ. Sci. Technol. 3, 1898–1904.

Yoshihara, T., Matsumura, H., Hashida, S.-N., Nagaoka, T., 2013. Radiocesium contamination of 20 wood species and the corresponding gamma-ray dose rates around the canopies 5 months after the Fukushima nuclear power plant accident. J. Environ. Radioactiv. 115, 60–68.

Zehnder, H.J., Kopp, P., Eikenberg, J., Feller, U., Oertl, J.J., 1995. Uptake and transport of radioactive cesium and strontium into grapevines after leaf contamination. Radiat. Phys. Chem. 46, 61–69.

Zhu, Y.G., Shaw, G., 2000. Soil contamination with radionuclides and potential remediation. Chemosphere 41, 121–128.

Zhu, Y.G., Smolders, E., 2000. Plant uptake of radiocesium: A review of mechanisms, regulation and applications. J. Exp. Bot. 51, 635–1645.

Chapter 9

Plant–Microbe Interactions in Phytoremediation

Ibrahim Ilker Ozyigit* and Ilhan Dogan†

*Marmara University, Faculty of Science & Arts, Department of Biology, Goztepe, Istanbul, Turkey, †Izmir Institute of Technology, Faculty of Science, Department of Molecular Biology and Genetics, Urla, Izmir, Turkey

DEFINITION OF PHYTOREMEDIATION

There are various descriptions of phytoremediation. However, when we look at the terminology, we can see that 'φυτο' (phyto) means plant in Ancient Greek, and 'remedium' means restoring balance from Latin. The term was first used in the 1980s to express the usage of plants to recover degraded or polluted areas (Neil, 2007). Today, there are different definitions of phytoremediation, which relate to its cleaning mechanism, pollutant type, used plant species and even the researcher's study area. However, the most accepted ones are:

- the use of plants for environmental remediation, which involves removing organics and metals from soils and water (Raskin et al., 1994);
- the use of plants, including trees and grasses, to remove, destroy or sequester hazardous contaminants from media such as air, water and soil (Prasad and Freitas, 2003);
- an important tool for decontaminating soil, water and air by detoxifying, extracting, hyperaccumulating, and/or sequestering contaminants, especially at low levels where, using current methods, costs exceed the effectiveness (Heinekamp and Willey, 2007);
- the use of plants and their associated soil microorganisms, soil amendments, and agronomic techniques to remove or render harmless environmental contaminants (Wang et al., 2012).

As is clear from the different definitions, some aromatic hydrocarbons (Joner et al., 2006), alkanes (Lin and Mendelssohn, 2009), phenols (Ibanez et al., 2012), polychlorinated solvents (Kassel et al., 2002), pesticides (Knuteson et al., 2002; Merini et al., 2009), chloroacetamides (Hoagland et al., 1997) and some explosives (Medina et al., 2002) could be remediated by using phytoremediation

Soil Remediation and Plants. http://dx.doi.org/10.1016/B978-0-12-799937-1.00009-7

technology (Kvesitadze et al., 2006). Additionally, some trace elements and toxic heavy metals/metalloids such as Ag (Xu et al., 2010), As (Favas et al., 2012), Au (Sasmaz and Obek, 2012), Bi (Wei et al., 2011), Cd (Turgut et al., 2004), Ce (Peili et al., 2011), Co (Saleh, 2012), Cr (Mohanty and Patra, 2012), Cu (Tulod et al., 2012), Fe (Chaturvedi et al., 2012), Hg (Chattopadhyay et al., 2012), Mn (Hua et al., 2012), Ni (Jadia and Fulekar, 2008), Pb (de Souza et al., 2012), Sb (Littera, et al., 2012), Sn (Joseph, 2005), Te (Nolan et al., 1991), Tl (Poscic et al., 2013), U (Pratas et al., 2012), V (Marcano et al., 2006) and Zn (Caraiman et al., 2012) could be accumulated by some plant species and could be removed from any area by using phytoremediation processes.

Nowadays, phytoremediation is becoming an important tool for decontaminating soil, water and air by detoxifying, extracting, hyperaccumulating and/or sequestering contaminants with different types of applications (Heinekamp and Willey, 2007).

Accumulator/Hyperaccumulator Plants

The idea of using plants to remediate metal-polluted soils came from the discovery of hyperaccumulators (Alkorta et al., 2004). These plant species are able to concentrate metals in their above-ground tissues to levels far exceeding those present in the soil or in the non-accumulating species growing nearby (Memon and Schroder, 2009; Ali et al., 2013).

Baker and Brooks (1989) accepted plant species as hyperaccumulators, which accumulate greater than $100\,mg\,kg^{-1}$ dw Cd, or greater than $1000\,mg\,kg^{-1}$ dw Cu, Ni and Pb or greater than $10,000\,mg\,kg^{-1}$ Mn and Zn in their shoots when they grow on metal rich soils.

Recently, van der Ent et al. (2013) recommended a concentration criterion which is given in table form for the different metals/metalloids in dried foliage with plants growing in their natural habitats (Table 9.1). Finally, hyperaccumulator

TABLE 9.1 Relation between Minimum Amount and Metals/Metalloids in Hyperaccumulator Plant Species

Minimun amount	Metals/metalloids
$100\,mg\,kg^{-1}$ dry weight	Cd, Se, Tl
$300\,mg\,kg^{-1}$ dry weight	Co, Cr, Cu
$1000\,mg\,kg^{-1}$ dry weight	As, Ni, Pb
$3000\,mg\,kg^{-1}$ dry weight	Zn
$10,000\,mg\,kg^{-1}$ dry weight	Mn

Source: van der Ent et al. (2013).

plants are accepted as capable of accumulating more than a 100 times more potentially phytotoxic elements than non-accumulators species (Chaney et al., 1997; Raskin and Ensley, 2000; Pulford and Watson, 2003).

Currently, 420 species belonging to 45 plant families are recorded as heavy metal hyperaccumulators and this number is likely to change in the future (Cobbett 2003; Rajakaruna et al., 2006; Mudgal et al., 2010).

Ni is one of the most accumulated metals by many different hyperaccumulator plants (more than 300). Some well known hyperaccumulator plant species, their families and accumulated heavy metal/metalloids are: *Azolla pinnata* for Cd in Azollaceae (Rai, 2008), *Bidens pilosa* for Cd (Sun et al., 2009), *Sonchus asper* for Pb and Zn (Yanqun et al., 2005) and *Helianthus annuus* for Cd, Cr and Ni (Turgut et al., 2004) in Asteraceae, *Alyssum bertolonii* and *Alyssum murale* for Ni (Li et al., 2003; Bani et al., 2010), *Arabidopsis thaliana* for Cu, Mn, Pb and Zn (Lasat, 2002), *Arabidopsis halleri* for Cd and Zn (Reeves and Baker, 2000; Cosio et al., 2004), *Brassica junceae* for Ni and Cr (Saraswat and Rai, 2009) and *Brassica oleracea* for Cd (Salt et al., 1995a), *Cardaminopsis halleri* for Cd and Zn (Sun et al., 2007), *Rorippa globosa* for Cd (Wei et al., 2008), *Thlaspi caerulescens* for Cd, Ni and Zn (Lombi et al., 2001; Cluis 2004) in Brassicaceae, *Sedum alfredii* for Cd, Pb and Zn (Li et al., 2005; Sun et al., 2007) in Crassulaceae, *Euphorbia cheiradenia* for Pb (Chehregani and Malayeri, 2007) in Euphorbiaceae, *Clerodendrum infortunatum* and *Haumaniastrum katangense* for Cu (Rajakaruna and Bohm, 2002; Chipeng et al., 2010) in Lamiaceae, *Astragalus racemosus* and *Astragalus bisulcatus* for Se (Vallini et al., 2005), *Pteris vittata* for As, Cr and Se (Baldwin and Butcher, 2007; Kalve et al., 2011) in Pteridaceae, *Solanum nigrum* and *S. photeinocarpum* for Cd (Sun et al., 2008; Zhang et al., 2011) in Solanaceae and *Viola baoshanensis* for Cd (Wu et al., 2010) in Violaceae.

PHYTOREMEDIATION APPLICATIONS

Phytoremediation can be divided into phytoextraction, rhizofiltration, phytostabilization, phytodegradation, rhizodegradation, phytovolatilization and phytorestauration; and various physiological mechanisms are involved in each of these processes (Figure 9.1).

Phytoextraction deals with the absorption of toxic metals and metalloids by roots and their transportation to and accumulation in above-ground (harvestable) parts of plants resulting in reduced soil metal concentrations (Pulford and Watson, 2003; Kvesitadze et al., 2006; Zhao and McGrath, 2009; Zhang et al., 2010; Ali et al., 2013).

The following are well-known plants used for phytoextraction processes: Indian mustard (*Brassica juncea*) for As, B, Cd, Cr(VI), Cu, Ni, Pb, Se, Sr and Zn (Raskin et al., 1994; Salt et al., 1995a; Salido et al., 2003); Alpine pennycress (*Thlaspi caerulescens*) for Cd and Zn (Zhao et al., 2003; Cosio et al., 2004); alyssum (*Alyssum wulfenianum*) for Ni (Reeves and Brooks, 1983); canola (*Brassica napus*), kenaf (*Hibiscus cannabinus* L. cv. Indian) and tall fescue (*Festuca*

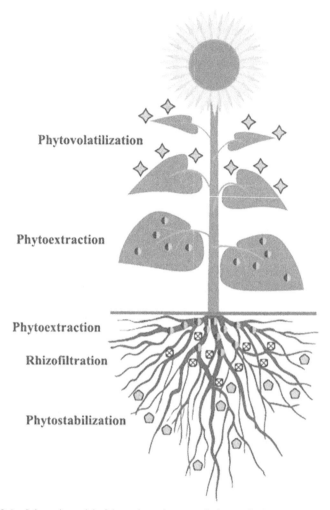

FIGURE 9.1   Schematic model of the various phytoremediation methods.

arundinacea Schreb cv. Alta) for Se (Bañuelos et al., 1997); poplar (*Populus* sp.) for As and Cd (Pierzynski et al., 1994); sunflower (*Helianthus annuus*) for Cs and Sr (Adler, 1996); Sudangrass (*Sorghum vulgare* L.), alfalfa (*Medicago sativa*) and maize (*Zea mays*) for Pb, Zn, Hg and Ni (EPA, 2000).

Rhizofiltration is used for the removal of pollutants, mainly metals from aquatic environments such as damp soil and ground and/or surface waters by adsorption or precipitation onto roots or other submerged organs of metal-tolerant aquatic plants related to their physiological and biochemical characteristics (Young, 1996; Salt et al., 1998; Kvesitadze et al., 2006; Jadia and Fulekar, 2009).

For this aim, the mainly used plants and elements are: *Azolla caroliniana* for As (Favas et al., 2012); *Eichhornia crassipes* for As, Cs, Co, Hg and Mn (Karkhanis et al., 2005; Nateewattana et al., 2010; Chattopadhyay et al., 2012; Saleh, 2012); *Lemna minor* and *Lemna gibba* for Ag, Au and Ni (Khellaf and Zerdaoui, 2010; Sasmaz and Obek, 2012; Favas et al., 2012; Pratas et al., 2012); *Typha angustifolia* for As, Cd, Cr, Cu, Fe, Mn, Ni, Pb and Zn (Chandra and Yadav, 2010); Indian mustard for Cu, Cd, Cr, Ni, Pb and Zn; sunflower for Cu, Cd, Cr, ^{137}Cs, Ni, Pb, ^{90}Sr, U and Zn; *Brassica juncea* for Cd and Pb (Dushenkov et al., 1995; Salt et al., 1995a); and *Medicago sativa* for Cr, Cu, Ni, Pb and Zn (Gardea-Torresdey et al., 1998).

Phytostabilization (phytoimmobilization) is based on plants' ability to immobilize contaminant metals in soils or sediments by sorption, precipitation and complexation. By decreasing metal mobility, these processes prevent leaching and groundwater pollution and minimize soil erosion and migration of sediments (Barceló and Poschenrieder, 2003; Kvesitadze et al., 2006; Ali et al., 2013). This process does not remove the contaminant from the soil, but it reduces the hazard of the contaminant (Arthur et al., 2005).

Sorghum sp. for Cd, Cu Ni, Pb and Zn (Jadia and Fulekar, 2008); *Solanum nigrum* for Ni (Ferraz et al., 2012); *Eucalyptus urophylla* and *Eucalyptus saligna* for Zn (Magalhães et al., 2011); *Vigna unguiculata* for Pb and Zn (Kshirsagar and Aery, 2007) have been used for this process.

Phytodegradation is the elimination of organic pollutants, which are easily entered into plant tissues or the rhizosphere by decomposition through internal or secreted plant enzymes or products (Barceló and Poschenrieder, 2003; Prasad and Freitas, 2003; Peer et al., 2005; Pilon-Smits, 2005).

Hybrid poplars (*Populus trichocarpa* × *Populus deltoides*) (Newman et al., 1997; Gordon et al., 1998), tropical leguminous tree *Leucaena leucocephala* (Doty et al., 2003), hairy root cultures of *Catharanthus roseus* (Bhadra et al., 2001) were studied materials used for phytodegradation processes. *Datura innoxia* and *Lycopersicon peruvianum* containing peroxidase, laccase and nitrilase have been shown to degrade soil pollutants (Schnoor et al., 1995; Lucero et al., 1999). Recently, *Blumea malcolmii* for Malachite Green dye (Kagalkar et al., 2011), *Erythrina crista-galli* for petroleum (de Farias et al., 2009) and *Chlorella pyrenoidosa* for pentachlorophenol (Headley et al., 2008) were used for phytodegradation processes.

Rhizodegradation defines the decomposition of organic pollutants such as xenobiotics in the soil by microorganisms in the rhizosphere (Mukhopadhyay and Maiti, 2010; Ali et al., 2013). Decomposition describes breakdown of a compound into its smaller constituents or its transformation to a metabolite and thus, rhizodegredation is one of the most important phases in the process of remediation of organic pollutants (Arthur et al., 2005).

Some trees, such as orange (*Citrus* sp.), apple (*Pyrus* sp.) and mulberry (*Morus* sp.) are able to excrete flavonoids and coumaric acid into the soil, stimulating the degradation of polychlorinated biphenyls (Donnelly et al., 1994;

Gilbert and Crowley, 1997; Kvesitadze et al., 2006) Additionally, *Salix nigra* was used for rhizodegredation of perchlorate (Yifru and Nzengung, 2008); orchardgrass (*Dactylis glomerata*), smooth bromegrass (*Bromus inermis*), tall fescue (*Festuca arundinacea*), Illinois bundle flower (*Desmanthus illinoensis*), perennial rye-grass (*Lolium perenne*), switchgrass (*Panicum virgatum*) and eastern gamagrass (*Tripsacum dactyloides*) were used for atrazine (Lin et al., 2011); *Kandelia candel* was used for phenanthrene and pyrene (Lu et al., 2011); and *Sesbania cannabina* for petroleum hydrocarbons (Maqbool et al., 2012).

Phytovolatilization is the uptake of pollutants (organics such as tetrachloro-ethane, trichloromethane, tetrachloromethane, etc. and/or certain metals such as As, Hg and Se) and their subsequent release into the atmosphere by tran-spiration, either in their original form or after metabolic modification (Susarla et al., 2002; Ali et al., 2013). Rice, rabbit foot grass, *Azolla* and pickle weed are known as the best Se volatilizers (Hansen et al., 1998; Pilon-Smits et al., 1999; Lin et al., 2000; Zayed et al., 2000; Hooda, 2007). Parrot's feather (*Myriophyllum brasiliense*), iris-leaved rush (*Juncus xiphioides*), cattail (*Typha latifolia*) and club-rush (*Scirpus robustus*) are also potential Se phytoremediator species in wetlands (Pilon-Smits et al., 1999; Arthur et al., 2005). *Brassica juncea* and several species of the genus *Astragalus* were also identified as valuable plants for removing Se from soils (Raskin et al., 1997; Bañuelos et al., 1998).

The hydraulic control technique uses plants that absorb large amounts of water through the plant body by transpiration and prevents the spread of con-taminated wastewater into uncontaminated areas (Quinn et al., 2001; Barceló and Poschenrieder 2003; Kvesitadze et al., 2006). The most convenient plants for this application are poplar, birch, willow, eucalyptus, etc. (Gatliff 1994; Pivetz 2001; Kvesitadze et al., 2006).

Phytorestoration involves the complete remediation of contaminated soils which stabilizes wastes and prevents exposure pathways via wind and water erosion (Bradshaw 1997; Prasad and Freitas, 2003).

INTERACTIONS BETWEEN PLANTS AND MICROBES IN PHYTOREMEDIATION

Soil is unique in having a life-supporting system and provides ecosystem ser-vices essential for planetary functions, including primary production, the reg-ulation of biogenic gases and the earth's climate, biogeochemical and water cycling, and the maintenance of biodiversity (Magdoff and van Es, 2000; Welbaum et al., 2004). A considerable amount of total land is contaminated due to various human activities and a gradual increase of this level is predicted in coming years.

Soils continuously exposed to metals such as As, Cd, Cu, Hg, Ni, Pb and Zn have high levels of accumulation leading to toxicity through various agri-cultural activities including agrochemical utilizations and long-term application of urban sewage sludge in agricultural soils, industrial activities and vehicle exhausts, as well as from anthropogenic sources (Modaihsh et al., 2004; Sabiha-Javied et al., 2009; Akguc et al., 2010). In turn, elevated levels of accumulations could pose potential threats to food safety issues and potential health risks due to transfers of metals from soils to plants (Ahmadpour et al., 2012; Ali et al., 2013). These metals are known to be potentially toxic to plants and animals and are of concern as their abilities to cause DNA damage and their carcinogenic capacity in animals and humans are probably related to their mutagenic abilities (Knasmuller et al., 1998; Baudouin et al., 2002). The potential threat from heavy metals is related to their non-degradable structures and therefore, without disturbance they persist in soils for a very long time (Neil, 2007). In order to minimize the access of heavy metals to the food chain and recover areas contaminated with heavy metals, the employment of cleanup strategies including soil excavation, soil washing or burning or pump and treat systems is obligatory and they are already being used for cleaning soil up. However, the applications of these non-biological remediation technologies for cleaning-up the sites contaminated with heavy metals causes destruction of biotic components of soils and implementation of these technologies are difficult and too expensive (Raskin et al., 1994).

A new practical and cost effective plant-based technology known as phytoremediation that uses plants and their associated microbial flora for environmental cleanup has gained acceptance (Salt et al., 1995a; Salt et al., 1998). Efficient degradation of organic compounds can be accomplished by employing a plant-based remediation technology which acts as a transformation system to metabolize organic compounds while taking advantage of plants' nutrient utilization processes (Kassel et al., 2002; Kvesitadze et al., 2006). Alternatively, toxic elements including heavy metals can be removed from the environment by using this plant-based technology through absorption and bioaccumulation (Zhao and McGrath, 2009; Zhang et al., 2010; Ali et al., 2013).

Phytoremediation efficiency is determined by key factors: the establishment of vital plants with biomass production, active root proliferation and/or root activities with the root symbiosis formation, assisting phytoremediation in the rhizosphere. In turn, assembly of microbial consortia can benefit the plant. Apart from many beneficial interactions, there is resource competition between plants and microbes (Kaye and Hart, 1997). Due to the constraints of limiting resources and resource competition, which commonly occur at polluted sites, microbial growth and biodegradation may become limited (Moorehead et al., 1998; Joner et al., 2006; Unterbrunner et al., 2007). Although excessive supply of nutrients could provide suitable conditions for stimulation of heterotrophic and pollutant adapted bacteria, this may not necessarily lead to achieving enhanced rates of phytoremediation. This was shown for hydrocarbon degradation in a crude-oil-contaminated soil. Pollutant degradation was not affected or even inhibited by the addition of nutrients (Chaineau et al., 2005).

The second aspect to be highlighted relates to rhizospheric control over pollutant toxicity. Plants and microorganisms can become adapted to toxic pollutant concentrations. Phyto-/rhizoremediation are being successfully exploited not only for removing pollutants but also for contributing to alleviation of toxicity by decreasing pollutant concentration in the rhizosphere. Resistance in plants and beneficial rhizospheric soil-borne microbes against (multiple) pollutants is one of the most important features and a prerequisite for their use of any phytoremediation technology (Burd et al., 2000; Belimov et al., 2005).

Soil-borne microbial communities showing different capabilities of genetic and functional activities can vary extensively in soils and have influences on soil functions, due to the fact that they are involved in fundamental metabolic processes (Nannipieri et al., 2003). Interactions between microbes are controlled by specific molecules (Pace 1997) and are shown to be responsible for key environmental processes including the biogeochemical cycling of nutrients and matter and the maintenance of plant health and soil quality (Barea et al., 2004). For an eminence bioremediation, formation of large numbers of metabolically active populations of beneficial soil-borne microbes are a necessity (Metting, 1992). In these soil-borne microbes, the driving factors are high adaptability in a wide variety of environments, fast growth rate, and biochemical versatility to metabolize a variety of natural and xenobiotic chemicals for a well-established ecosystem (Narasimhan et al., 2003).

The mutualistic relationship between plants and their associated rhizospheric microbial communities is complicated. Uptake efficiency of trace elements by plants can be augmented by soil-borne microbes such as rhizospheric bacteria (De Souza et al., 1999; Weyens et al., 2009a, b). In general, for successful rhizoremediation, a plant species must have a dense and highly branched root system to harbor large numbers of bacteria, primary and secondary metabolism, and establishment, survival, and ecological interactions with other organisms (Salt et al., 1998; Kuiper et al., 2004).

The primary contact point between plant tissues and pollutants in the soil or water is roots and therefore, they provide a key point for assessment of the phytoremediation potential of a particular plant species. There is a large volume of traffic between soil-borne microbes, roots and soil constituents at the root–soil interface (Lynch, 1990; Linderman, 1992; Glick, 1995; Kennedy, 1998; Bowen and Rovira, 1999). The high volume of interactions between roots, microbes and microbial communities creates a dynamic environment known as the rhizosphere. The physical, chemical and biological state of the root-associated soil differs from that of the root-free bulk soil, and physical, chemical and biological conditions are shown to influence diversity, numbers and activity of microorganisms in the rhizosphere microenvironment (Kennedy, 1998). Rhizospheric processes resulting from the activities of diverse groups of rhizospheric communities can be expected to affect uptake of heavy metals by plants. It has been shown that plants grown in non-sterile conditions exhibited no iron-deficiency symptoms and had a higher iron level in roots in comparison to plants grown in sterile conditions. This seems to be the result of rhizospheric microbial activity, which facilitates iron acquisition (Masalha et al., 2000).

Some rhizospherial strains have ability to release different substrates such as antibiotics (including antifungals), hydrocyanic acid, indoleacetic acid (IAA), siderophores, 1-aminocyclopropane-1-carboxylic acid (ACC) deaminase which lead to increase in bioavailability and root absorption of heavy metals, including both essential (e.g. Fe and Mn) and non-essential (e.g. Cd) (Barber and Lee, 1974; Crowley et al., 1991; Salt et al., 1995). An investigation was performed to show the correlation between metal resistance and metal mobilization abilities of rhizobacteria under heavy metal stress. The highest degree of the biochemical activity of isolates and metal resistance was recorded for phosphate solubilizers; then, for siderophore producers, and finally for acid producers. The data imply that phosphate solubilization and siderophore-mediated solubilization and acid production are adopted by rhizobacteria and provide mobilizing metals in soil (Abou-Shanab et al., 2005). Uptake and translocation of trace elements may vary considerably and often depend on the species and type of trace elements. To varying degrees, different metals exhibit different mobility rates and the mobilization rate could be higher than others for a particular metal within a plant.

The complex feedback mechanisms allow plant roots and rhizosphere microorganisms to adapt to changing soil conditions as they grow. The vast majority of the soil-borne microbial population found is in contact with the plant roots, and their numbers can reach up to $10^9-10^{12} g^{-1}$ of soil around the roots (Whipps, 1990), leading to a biomass equivalent to $500 kg ha^{-1}$ (Metting, 1992). This enrichment in vegetated soils is due to the availability of nutrients via plant root exudation (Brimecombe et al., 2001).

As growth facilitators, plant hormones released by some rhizospheric microorganisms stimulate root growth and thereby the secretion of root exudates. Plant exudates including carbohydrates, amino acids, lipophilic compounds and chelating agents (citric, acetic, other organic acids, etc.) released from plant roots exert favourable effects on sustaining a broad range of microbial communities in the rhizosphere (Anderson et al., 1993). Plants subjected to phosphorus deficiency attempt to mobilize phosphorous components present in the soil by increasing citric acid level in root exudates (Hooda, 2007). Mucigel, a gelatinous lubricant for root penetration secreted by the root cells contributes to increasing root mass apically through the soil during growth. Soil microorganisms use this supply of compounds for proliferation to generate the plant rhizosphere (Anderson et al., 1993). Plant survival is promoted in toxic and nutrient-limited environments through occurring interactions between plants and rhizosphere- and root-associated microbes. Various metallic ions are found to be immobile; therefore, they have limited bioavailability because of their low solubility (in water) and strong binding (of soil particles) characteristics, restricting their uptake by plants. But microorganisms colonizing plant roots can contribute to increasing the bioavailability of various heavy metal ions for uptake (Hooda, 2007).

For accumulation of metals by the plants from the soil, the metals must be mobilized in the soil solution. The bioavalability of metals is increased in soil through several means. Soil-bounded metal uptakes by roots of metal-starved plants are achieved by secreting phytosidophores (the chelating amino acids) into the rhizosphere for mobilization of metals via chelation and solubilization (Kinnersely, 1993). Also, pH seems to be a major element in controlling of mobilization (Gadd, 2004; Wenzel, 2009). The changes in pH can also affect speciation of metals/metalloids in solution (Anjum et al., 2012). Organic exudates released by soil-borne microorganisms increase bioavailability of metal ions including Fe^{2+} (Crowley et al., 1991; Burd et al., 2000), Mn^{2+} (Barber and Lee, 1974) and Cd^{2+} (Salt et al., 1995) for root absorption. Biosurfactants (e.g. rhamnolipids) are produced by some bacteria to make hydrophobic pollutants more water soluble (Volkering et al., 1998). Rhizosphere chemistry can be modified by two mechanisms, acidification and exudation of carboxylates, assisting in the mobilization of metals (Ma et al., 2001).

Some bacterial strains have detoxification abilities for toxic compounds. *Xanthomonas maltophyla* (Blake et al., 1993), *Escherichia coli* and *Pseudomonas putida* (Lasat, 2002) have been shown to catalyze the reduction and precipitation of highly mobile and environmentally less hazardous compounds. Most bacteria seem to posses multiple metal-resistance systems. Research conducted by Abou-Shanab et al. (2005) showed that all the rhizobacterial strains tested in the study were found to be tolerant to multiple metal ions. Similar findings had been previously reported by Sabry et al. (1997). Moreover, many anaerobic microorganisms may independently remove a number of metals from the environment by reducing them to a lower redox state. Biologically mediated reduction and oxidation of metals/metalloids may affect solubilization. Iron (III) can be used as a terminal electron acceptor by specialized anaerobic bacteria. Various metals including Cr (VI), Fe (III), Hg (II), Mn (IV), Se (VI) and U (VI) can be utilized by specialist dissimilatory metal-reducing bacteria. Microbial reduction is exploited not only for mobilization of Fe and Mn, but also for immobilizing of metals including Cr, U (Gadd, 2004), and Se (Di Gregorio et al., 2006).

Metals can be immobilized by microorganisms in various other ways including accumulation of them in their biomass or on cell walls via intracellular sequestration (Gadd, 2004) or precipitation, or adsorbtion (Leyval and Joner, 2000; Fein et al., 2001). It was shown that *Bacillus subtilis* strain SJ-101 protected *Brassica juncea* against Ni toxicity when inoculated together because of its high capacity for Ni (Zaidi and Musarrat, 2004). Mineralization of dissolved metal–organic complexes may be considered to be another cause of microbially mediated immobilization. The release of oxygen through wetland plants such as cattail (*Typha latifolia*) and common reed (*Phragmites australis*) in anaerobic soils increases the redox potential in the rhizosphere (Flessa and Fischer, 1992; Brix et al., 1996). This leads to induction of the formation of ferric Fe plaque and subsequently causes sorption and immobilization of metals (Doyle and Otte, 1997) and metalloids (Blute et al., 2004).

A number of endophytic and rhizobacteria in plant–bacteria combinations are involved in the degradation of toxic compounds in the rhizosphere. Endophytic bacteria can be defined as bacteria colonizing the internal tissues of plants without causing symptoms of infection or negative effects on their host (Schulz and Boyle, 2006). With the exception of seed endophytes, the root is the primary site for endophytes to gain access into plants. Several microscopic studies confirm this route of colonization (Pan et al., 1997; Germaine et al., 2004). After entry, endophytes either remain localized in specific plant tissues such as the root cortex or the xylem or colonize inner host tissues by transport through the vascular system or the apoplast (Mahaffee et al., 1997; Quadt-Hallmann et al., 1997). Endophyte-bearing plant samples comprise of herbaceous crop plants (Lodewyckx et al., 2002; Malinowski et al., 2004), different grass species (Zinniel et al., 2002; Dalton et al., 2004) and woody tree species (Cankar et al., 2005; Pandey et al., 2005). In general, the most common genera of cultivable endophytic species are Pseudomonaceae, Burkholderiaceae and Enterobacteriaceae (Mastretta et al., 2006).

Plant–endophyte associations rely on beneficial interactions. While plants provide nutrients and a harbour for endophytes, plant growth and health can be improved directly or indirectly via production of metabolites by endophytes (Bacon and White, 2000; Garbeva et al., 2001; Tan and Zou, 2001). Growth promotion effects of endophytic bacteria are related to preventing the growth or activity of plant pathogens through competition for space and nutrients, production of hydrolytic enzymes, antibiosis, and induction of plant defense mechanisms and through inhibition of pathogen-produced enzymes or toxins. Production of plant growth regulators including auxins, cytokinins and gibberellins, suppression of stress ethylene production by ACC-deaminase activity, nitrogen fixation and the mobilization of unavailable nutrients such as phosphorus and other mineral nutrients may be involved in the direct growth promoting mechanisms (Weyens et al., 2009b).

Transpiration rate and concentrations are considered to be critical factors in the phytoremediation of some organic contaminants. Removing trichloroethene (TCE) from soils the plants use a phytovolatilization mechanism (Xingmao and Burken, 2003). Thus, in phytoremediation plants are employed to provide optimum conditions for microbial degradation of pollutants and to accomplish the extraction of pollutants inside the plants (Boominathan et al., 2004; Tamaoki et al., 2008). Therefore, the role of providing optimum conditions for root colonizing bacteria and a simple way of extracting pollutants are two benefits provided by plants in phytoremediation (Suresh and Ravishankar, 2004). Besides symbiotic bacteria, studies concentrated on using fungi in phytoremediation technology. One example is the use of arbuscular mycorrhizae (AM) fungi that is more suitable for establishing symbiotic relationships with 80–90% of land plants (Huang et al., 2004; Khan, 2006). The volume of soil explored by mycorrhizae can be in an efficient manner due to their small diameter. Bioavailablity of pollutants can be modified by mycorrhizae through competition with roots and other microorganisms for water and pollutant uptake, protection of roots from direct interaction with the pollutant via formation of the ectomycorrhizal sheath, and impeded pollutant transport through increased soil hydrophobicity (Meharg and Cairney, 2000). Mycorrhiza associated plants have been reported growing on heavy-metal-contaminated soil (Shetty et al., 1994; Chaudhry et al., 1998, 1999). Root absorption (up to 47-fold) and the acquisition of plant nutrients including metal ions can be increased by the presence of fungal symbiotic associations (Smith and Reed, 1997). However, there are contradictory observations in the results of mycorrhizal effect on metal uptake. There are reports that mycorrhizae may have inhibitory effects on Cu, Zn (Scheupp et al., 1987; Heggo et al., 1990) and Cd accumulation (Weissenhorn and Leyval, 1995; Joner and Leyval, 1997; Schutzendubel et al., 2002). Arbuscular mycorrhizal colonization of *Thlaspi praecox* showed interfering effect on heavy metal (Cd and Pb) uptake (Vogel-Mikus et al., 2005). The inhibition may be related to the mycorrhizal protection of the plants from heavy metal toxicity, although the mechanism of protection is unclear. In addition, mycorrhizae have been reported to both enhance uptake of essential metals in the presence of low levels of metals and decrease uptake of metals in the presence of phytotoxic levels (Frey et al., 2000). The mechanisms in plant–microbe interactions are still broadly unclear; enhanced plant uptake mediated by microbes may be due to a stimulatory effect on root growth, microbial production of metabolites may involve altered plant gene expression of transporter proteins or microbial effect on the bioavailability of the element (DeSouzaet al., 2000).

In addition to effects of microbial associations, a vast number of organic pollutants can be taken up by direct activities of the plants themselves from the soil through their roots. The decomposition of organic residues with the release of plant nutrient elements including C, K, N, S, and phosphate carried out by soil microorganisms is important (Macek et al., 2000).

RHIZOSPHERE MICROBIOME

The microbial inhabitants are an integral natural active part of the biota in soils. Exposure of bacteria to plants in various ecological systems including rhizospere gives rise to stimulation of growth by directly affecting plant metabolism in the absence of a major pathogen. These bacteria belong to diverse genera, including *Acetobacter, Achromobacter, Anabaena, Arthrobacter, Azoarcos, Azospirillum, Azotobacter, Bacillus, Burkholderia, Clostridium, Enterobacter, Flavobacterium, Frankia, Hydrogenophaga, Kluyvera, Microcoleus, Phyllobacterium, Pseudomonas, Serratia, Staphylococcus, Streptomyces, Vibrio* and the well-known legume symbiont *Rhizobium* (Bashan et al., 2008).

Broadly, there are three separate, but interacting, components recognized in the rhizosphere. These are the rhizosphere (soil), the rhizoplane and the root itself. The plants that are colonized by microbial communities have a soil zone termed rhizosphere that is influenced by roots through the release of substrates.

In turn, those substrates stimulate microbial activity in the rhizosphere. The rhizoplane is the root surface, including the strongly adhering soil particles. The root itself is a part of the system. Root tissues are colonized by certain microbes known as the endophytes (Kennedy, 1998; Bowen and Rovira, 1999). Microbial colonization occurs in patches along the root tissues and/or the rhizoplane known as root colonization, whereas the colonization of the adjacent volume of soil exists under the influence of the root known as rhizosphere colonization (Kloepper et al., 1991, 1994). The rhizosphere can comprise up to 10^{11} microbial cells/g of root (Egamberdieva et al., 2008) and more than 30,000 prokaryotic species (Mendes et al., 2011). The rhizospheric microbial diversity is enormously diverse, including tens of thousands of species. This complex plant-associated microbial community is crucial for plant health (Berendsen et al., 2012).

Recent studies showed that the composition of the rhizosphere microbiome can be determined by plants through active secretion of substrates that are known to vary between plant species and be able to shape distinct rhizomacrobial communities. Therefore, different plant species host specific rhizospheric microbial communities when grown on the same soil (Sorensen, 1997; Jaeger et al., 1999). The difference between the chemical conditions of the rhizosphere and the bulk soil is the result of various processes induced by plant roots and/or by the rhizobacteria (Marschner, 1995; Hinsinger, 2001).

Interactions between plants and rhizospheric microbes could stimulate the production of compounds that could alter soil chemical properties in rhziosphere. For example, the increased accumulation of Hg by plants could be the result of decreased pH by rhizobacteria of the plants (De Souza et al., 1999). A study of the influence of hydrogen and aluminum ions on the growth of the associative nitrogen-fixing and growth-promoting bacteria *Azospirillum lipoferum* 137, *Arthrobacter mysorens* 7, *Agrobacterium radiobacter* 10 and *Flavobacterium* sp. L30 showed that the response of plants to the inoculation varied under different values of pH (from positive to negative) (Belimov et al., 1998).

Exudates released by plant roots are used by rhizospheric microbes as nutrients. It is estimated that the proportion of net photosynthetic carbon transferred to the roots is between 30–60% and 10–20% of root needs comes from rhizodeposition (Marschner 1995; Salt et al., 1998). Exudates consist primarily of low molecular weight (LMW) and high molecular weight (HMW) organic acids. The total concentration of organic acids measured in roots usually ranges from 10 to 20 mM, generally comprising of lactate, acetate, oxalate, succinate, fumarate, malate, citrate, isocitrate and aconitate. The remainder of organic solutes in roots consists of sugars (90 mM) and amino acids (10–20 mM) (Jones, 1998). The HMW organic acids are mucilage and ectoenzymes (Knee et al., 2001).

It is essential for bioremediation that using such soils treated with metals and/or acids increases their biological activity (Boon et al., 1998), especially for mineral soils with a low content of organic matter (Priha et al., 1999). The exudates create favourable conditions for rhizospheric microbial populations leading to increase in their mass well beyond those of the bulk soil, attracting motile bacterial and fungal hyphae. Consequently, stimulation of an array of positive, neutral or negative interactions with plants occurs (Gerhardson, 2002).

Microbial cells have the ability to produce and recognize signal molecules. This allows the whole population to form biofilm over large areas of the root surface and a concerted action is taken by the whole population after a critical level is exceeded. This phenomenon is known as quorum sensing. Many microbes control gene expression by employing quorum sensing in response to cell population density. The successful infection and formation of nitrogen fixing nodules by rhizobial bacteria upon legume roots is carried out chemotactically through certain root exudates leading to activation of rhizobial nodulation genes, Nod factors during adhesion and colonization of the legume root surface. Many quorum sensing signal molecules such as N-acyl-homoserine lactones (AHLS) play important roles in the regulation of expression and repression of the symbiotic genes (Daniels et al., 2004).

Plants, as well as their plant-derived chemicals, including those generated from the roots applied to unplanted soil, can nurse degradative microbes when applied to unplanted soil (Shann, 1995). Spoilage of organic matter and many other substrates is found to be two to three times higher in the rhizosphere than in the bulk soil (Jones, 1998). Plants produce a broad range of diverse low molecular mass secondary metabolites. Based on the estimation, the total number of plants exceeds 500,000 (Hadacek, 2002). Secondary metabolites are taken into account as non-essential for the basic metabolic processes of the plant (Dixon, 2001).

The effects of microbial communities in the rhizosphere are apparent in determining plant health and findings also verify the importance of the root microbiome.

STIMULATION OF PLANT GROWTH BY MICROBIAL COMMUNITIES

Many plants are not capable of gaining sufficient biomass for noticeable rates of remediation when elevated levels of pollutants are present (Harvey et al., 2002; Chaudhry et al., 2005). The remediation process of contaminated soils is limited and slowed because of their poor nutrient nature. Soil microbes are thought to exert positive effects on plant health via mutualistic relationships between them. However, microbes are sensitive to pollution, and depletion of microbial populations, both in terms of diversity and biomass, often occur in such contaminated soils (Shi et al., 2002).

Biotic or abiotic stress through a small change in the physico-chemical-biological properties of rhizosphere soils can cause a dramatic effect on plant–microbe interaction. Plant growth promoting microbes as rhizosphere inoculums are receiving attention in profitability of phytoremediation process; this partly depends on the plant's ability to withstand metal toxicity and to yield adequate biomass (Rajkumar and Freitas, 2008; Kuffner et al., 2010; Ma and Wang, 2010; Maria et al., 2011; Aafi et al., 2012). Phytoremediation and bioaugmentation are the terms used in combination to describe rhizoremediation, in conjunction with plant growth promoting rhizobacteria (Kuiper et al., 2004).

The PGPR (plant growth promoting rhizobacteria) are defined by three intrinsic characteristics: the organisms are capable of colonizing the root; the organisms have capability for survival, proliferation and competition in micro-habitats associated with the root surface; and the organisms are able to promote plant growth (Lugtenberg et al., 1999, 2001; Rothballer et al., 2003; Espinosa-Urgel, 2004; Gamalero et al., 2004).

The PGPR have been divided into two groups: those that are found to be involved in nutrient cycling and phytostimulation, including fixing atmospheric nitrogen and supply it to plants, synthesizing siderophores which can sequester iron from the soil and provide it to plant cells, synthesizing phytohormones such as auxins, cytokinins and gibberelins, solubilizing minerals such as phosphorous, making them more readily available for plant growth and synthesizing the enzyme ACC-deaminase, which can lower ethylene levels; and those that are found to be involved in the biocontrol of plant pathogens resulting from any one of a variety of mechanisms including antibiotic production, depletion of Fe from the rhizosphere, induced systemic resistance, production of fungal cell wall lysing enzymes, and competition for binding sites on the root (Bashan and Holguin, 1998; Glick et al., 2007). There are a number of reports stating that some PGPR species are able to be resistant to relatively high concentrations of heavy metals and remain active in moderately acidic soils (Belimov et al., 1998; Ivanov et al., 1999). Such naturally occurring rhizobacteria could assist phytoremediation both indirectly, by increasing the overall fertility of the contaminated soil and enhancing plant growth through nutrient uptake and control of pathogenity, and, also directly, catabolizing certain organics and/or intermediate partly oxidized biodegradation products (Kamnev et al., 1999).

Applied cadmium-resistant, rhizosphere-competent bacterial strains increased root and shoot biomass production of *Brassica napus* grown in cadmium-polluted soil (Sheng and Xia, 2006). Similarly, a PGPR consortium consisting of N_2-fixing *Azotobacter chroococcum* HKN5, P-solubilizing *Bacillus megaterium* HKP-1, and K-solubilizing *Bacillus mucilaginosus* HKK-1 was applied to increase growth and biomass production of *Brassica juncea* grown on Pb–Zn mine tailings and upon inoculation, increased production, in terms of growth and biomass has been observed (Wu et al., 2006).

The growth of *Brassica juncea* on nickel-polluted soil was stimulated by utilization of the PGPR *Bacillus subtilis* strain SJ-101 capable of producing the phytohormone indole acetic acid and solubilizing inorganic phosphates (Zaidi et al., 2006).

The root associated bacteria having ACC-deaminase activity has provided a better root growth and proliferation to the plant in polluted soils (Arshad et al., 2007). Inhibition of root growth and proliferation as a consequence of high concentrations of ethylene produced by plant roots occurs in response to toxicity and other stresses. Bacterial ACC-deaminase can significantly decrease ACC levels by metabolizing its ethylene precursor ACC into a-keto butyric acid and ammonia (Glick, 2005). PGPR with ACC-deaminase activity leading to better metal tolerance were found in the rhizosphere of the Ni tolerant *Thlaspi goesingense* (Idris et al., 2004).

Kluyvera ascorbata SUD165, an interesting example of PGPR was found to be resistant to a range of heavy metals and was reported to protect plants from nickel toxicity without affecting Ni uptake by seedlings or its accumulation in the plant (Burd et al., 1998). The plant growth-promoting effect in the presence of Ni may relate to the ACC-deaminase activity of this bacterium (Shah et al., 1998).

It was shown that *Brassica napus* and *Brassica campestris* were protected against metal toxicity by metal-resistant PGPR containing ACC-deaminase producing bacteria (Burd et al., 1998; Belimov et al., 2001). IAA (auxin) is also produced by many PGPR and is believed to play an important role in plant–bacterial interactions (Lambrecht et al., 2000). Therefore, any direct influence on IAA production by bacteria may in turn affect their phytostimulating efficiency. It has been demonstrated that bacterial excretion of auxins into the soil is beneficial for plants, in conjunction with making the bacterial plant-growth-promoting effect (Steenhoudt and Vanderleyden, 2000; Kamnev, 2003).

It has been found that presence of Cu^{2+} and Cd^{2+} in soils caused significantly reduced level of IAA produced by nonendophytic and facultatively endophytic strains of *Azospirillum brasilense*, leading to alteration of the plant-growth-stimulating efficiency of associative plant–bacterial symbioses in heavy-metal-polluted soils (Kamnev et al., 2005). As a matter of fact, while low levels of bacterial IAA promote root elongation, high levels of bacterial IAA stimulate lateral and adventitious root formation (Glick, 1995) but inhibit root growth (Xie et al., 1996). Hence, plant growth can be improved by altering the hormonal balance within the affected plant via plant growth-promoting bacteria (Glick et al., 1999).

Plants are able to take up microbial Fe complexes with siderophores serving as a Fe source for plants (Reid et al., 1986; Bar-Ness et al., 1991; Wang et al., 1993). Therefore, protection of plants from becoming chlorotic in the presence of high levels of heavy metals can be provided by the utilization of a siderophore-producing bacterium in the rhizosphere of plants. Thus, enhanced plant growth could be accomplished by treating plants with associated plant-growth-promoting bacteria in the presence of heavy metals including Ni, Pb and Zn (Burd et al., 1998, 2000), thereby enabling plants to have longer roots and get better established during the early stages of growth (Glick et al., 1998). Similarly, Cr-resistant *Pseudomonas* were isolated from paint industry effluents and used to stimulate seed germination and growth of *Triticum aestivus* in the presence of potassium bichromate (Hasnain and Sabri, 1996). In this case, the bacterial enhancement of seedling growth was associated with reduced chromium uptake. It was found that axenic saltmarsh bulrush plants inoculated with different rhizospheric bacteria accumulated more Se than plants grown under axenic conditions (De Souza et al., 1999).

Production of organic acids by soil fungi (Gadd, 1999) and bacteria, including rhizobacteria (Goldstein et al., 1999; Nautiyal et al., 2000), may promote solubilization, mobility and bioavailability of metals and accompanying anions by lowering the pH and supplying metal-complexing organic acid ligands (Kamnev and Van der Lelie, 2000). These microbially driven processes are a prerequisite for mineral weathering (Barker et al., 1998; Banfield et al., 1999).

Toxic heavy metal constituents coming from the dissolving minerals increase their bio- and phytoavailability in soils through these microbially driven processes leading to alteration in the fertility of soils. For instance, Hg and Se accumulations are promoted by naturally occurring rhizobacteria in the rhizosphere of wetland plants (De Souza et al., 1999). These plant-growth-promoting rhizobacteria could be deployed to increase efficiency of phytoextraction. Using *Pseudomonas aeruginosa* PAO1 and its three isogenic lipopolysaccharide (LPS) mutants, while the precipitation of essential volumes of Fe and La is controlled on the cell surface, Cu was bound at the cell surface of all the four strains assuming common surface functional groups responsible for Cu binding (Langley and Beveridge, 1999). It was found that the precipitation of Cd compounds was promoted by certain rhizobacteria on the plant root surface leading to a reduction in the bioavailability of Cd uptake by roots and enhancing their growth (Van der Lelie et al., 2001). A novel siderophore, also known as alcaligin E from a metal-tolerant bacterium, *Ralstonia eutropha* CH34 was found to have the capacity for binding Cd resulting in immobilizing and exclusion of Cd from metabolism (Diels et al., 1995; Gilis et al., 1998). The immobilization of heavy metals entrapped within the insoluble crystalline and/or amorphous phases (e.g. phosphate minerals) can be created in natural microbial communities (Douglas and Beveridge, 1998; Lins and Farina, 1999).

The root exudates provide an abundance of energy for the microbial transformation of organic compounds in the resolver zone. Soil microorganisms are also known to produce biosurfactants for facilitating removal of organic pollutants (Volkering et al., 1998). Direct detoxification of metals by utilization of root exudates (through forming chelates with metal ions) can be carried out in such soils contaminated with heavy metals.

The PGPR has important roles in facilitating plant growth on soils contaminated with both heavy metals and organic compounds and detoxification of soils and is exploited for phytoremediation purposes.

ACKNOWLEDGEMENTS

The authors are grateful to Dr Aysen Yumurtaci (Marmara University, Biology Department), Research Assistants Ibrahim Ertugrul Yalcin (Bahcesehir University, Environmental Engineering Department), Onur Zorluer and Zeynep Uzunova (Marmara University, Biology Department) Biologist-Designer Ilke Ertem and Muhammed Emre Ozyigit for their support with designing the figure and table, reviewing and compiling the article.

REFERENCES

Aafi, N.E., Brhada, F., Dary, M., Maltouf, A.F., Pajuelo, E., 2012. Rhizostabilisation of metals in soils using *Lupinus luteus* inoculated with the metal resistant rhizobacterium *Serratia* sp. MSMC541. Int. J. Phy. 14, 261–274.

Abou-Shanab, R.A., Ghozlan, H., Ghanem, K., Moawad, H., 2005. Behaviour of bacterial populations isolated from rhizosphere of *Diplachne fusca* dominant in industrial sites. World J. Microbiol. Biotechnol. 21, 1095–1101.

Adler, T., 1996. Botanical cleanup crews. Sci. News 150, 42–43.

Ahmadpour, P., Ahmadpour, F., Mahmud, T.M.M., et al., 2012. Phytoremediation of heavy metals: A green technology. Afr. J. Biotechnol. 11, 14036–14043.

Akguc, N., Ozyigit II, Yasar, U., et al., 2010. Use of *Pyracantha coccinea* Roem. as a possible bio-monitor for the selected heavy metals. Int. J. Environ. Sci. Te. 7, 427–434.

Ali, H., Khan, E., Sajad, M.A., 2013. Phytoremediation of heavy metals — concepts and applications. Chemosphere 91, 869–881.

Alkorta, I., Hernandez-Allica, J., Becerril, J.M., et al., 2004. Recent findings on the phytoremediation of soils contaminated with environmentally toxic heavy metals and metalloids such as zinc, cadmium, lead and arsenic. Rev. Environ. Sci. Biotechnol. 3, 71–90.

Anderson, T.A., Guthrie, E.A., Walton, B.T., 1993. Bioremediation Environ. Sci. Technol. 27, 2630–2636.

Anjum, N.A., Pereira, M.E., Ahmad, I., et al. (Eds.), 2012. Phytotechnologies: Remediations of environmental contaminants. CRC Press, Boca Raton.

Arshad, M., Saleem, M., Hussain, S., 2007. Perspectives of bacterial ACC-deaminase in phytoremediation. Trends Biotechnol. 25, 356–362.

Arthur, E.L., Rice, P.J., Anderson, T.A., et al., 2005. Phytoremediation—an overview. Crit. Rev. Plant Sci. 24, 109–122.

Bacon, C., White, J., 2000. Microbial endophytes. Marcel Dekker, New York.

Baker, A.J.M., Brooks, R.R., 1989. Terrestrial higher plants which hyperaccumulate metallic elements—a review of their distribution, ecology and phytochemistry. Biorecovery 1, 81–126.

Baldwin, P.R., Butcher, D.J., 2007. Phytoremediation of arsenic by two hyperaccumulators in a hydroponic environment. Microchem. J. 85, 297–300.

Banfield, J.F., Barker, W.W., Welch, S.A., Taunton, A., 1999. Biological impact on mineral dissolution: Application of the lichen model to understanding mineral weathering in the rhizosphere. Proc. Natl. Acad. Sci. USA 96, 3404–3411.

Bani, A., Pavlova, D., Echevarria, G., et al., 2010. Nickel hyperaccumulation by the species of *Alyssum* and *Thlaspi* (Brassicaceae) from the ultramafic soils of the Balkans. Botanica. Serbica. 34, 3–14.

Bañuelos, G.S., Ajwa, H.A., Terry, N., Zayed, A., 1997. Phytoremediation of selenium laden soils: A new technology. J. Soil Water Conserv. 52, 426–430.

Bañuelos, G.S., Ajwa, H.A., Wu, L.L., Zambrzuski, S., 1998. Selenium accumulation by *Brassica napus* grown in Se-laden soil from different depths of Kesterson reservoir. J. Contam. Soil 7, 481–496.

Barber, S.A., Lee, R.B., 1974. The effect of microorganisms on the absorption of manganese by plants. New Phytol. 73, 97–106.

Barceló, J., Poschenrieder, C., 2003. Phytoremediation: Principles and perspectives. Contrib. Sci. 2, 333–344.

Barea, J.M., Azcón, R., Azcón-Aguilar, C., 2004. Mycorrhizal fungi and plant growth promoting rhizobacteria. In: Varma, A., Abbott, L., Werner, D., Hampp, R. (Eds.), Plant surface microbiology. Springer, Heidelberg, pp. 351–371.

Barker, W.W., Welch, S.A., Chu, S., Banfield, J.F., 1998. Experimental observations of the effects of bacteria on aluminosilicate weathering. Am. Mineral 83, 1551–1563.

Bar-Ness, E., Chen, Y., Hadar, Y., et al., 1991. Siderophores of *Pseudomonas putida* as an iron source for dicot and monocot plants. Plant Soil 130, 231–241.

Bashan, Y., Holguin, G., 1998. Proposal for the division of plant growth-promoting rhizobacteria into two classifications: Biocontrol-PGPB (plant growth-promoting bacteria) and PGPB. Soil Biol. Biochem. 30, 1225–1228.

Bashan, Y., Puente, M.E., De-Bashan, L.E., Hernandez, J.-P., 2008. Environmental uses of plant growth-promoting bacteria. In: Barka, E.A., Clément, C. (Eds.), Plant microbe interactions. Research Signpost, Kerala, pp. 69–93.

Baudouin, C., Charveron, M., Tarrouse, R., Gall, Y., 2002. Environmental pollutants and skin cancer. Cell Biol. Toxicol. 18, 341–348.

Belimov, A.A., Kunakova, A.M., Gruzdeva, E.V., 1998. Influence of soil pH on the interaction of associative bacteria with barley. Mikrobiologiya (Moscow) 67, 561–568.

Belimov, A.A., Safronova, V.I., Sergeyeva, T.A., et al., 2001. Characterisation of plant growth promoting rhizobacteria isolated from polluted soils and containing 1-aminocyclopropane-1-carboxylate deaminase. Can. J. Microbiol. 47, 642–652.

Belimov, A.A., Hontzeas, N., Safronova, V.I., et al., 2005. Cadmium-tolerant plant growth-promoting bacteria associated with the roots of Indian mustard (*Brassica juncea* L. Czern.). Soil Biol. Biochem. 37, 241–250.

Berendsen, R.L., Pieterse, C.M.J., Bakker, P.A.H.M., 2012. The rhizosphere microbiome and plant health. Trends Plant Sci. 17, 478–486.

Bhadra, R., Wayment, D.G., Williams, R.K., et al., 2001. Studies on plant-mediated fate of the explosives RDX and HMX. Chemosphere 44, 1259–1264.

Blake, R.C., Choate, D.M., Bardhan, S., et al., 1993. Chemical transformation of toxic metals by a *Pseudomonas* strain from a toxic-waste site. Environ. Toxicol. Chem. 12, 1365–1376.

Blute, N.K., Brabander, D.J., Hemond, H.F., et al., 2004. Arsenic sequestration by ferric iron plaque on cattail roots. Environ. Sci. Technol. 38, 6074–6077.

Boominathan, R.R., Saha-Chaudhury, N.M., Sahajwalla, V., Doran, P.M., 2004. Production of nickel bio-ore from hyperaccumulator plant biomass: Applications in phytomining. Biotechnol. Bioeng. 86, 243–250.

Boon, G.T., Bouwman, L.A., Bloem, J., Romkens, P.F.A.M., 1998. Effects of a copper-tolerant grass (*Agrostis capillaris*) on the ecosystem of a copper-contaminated arable soil. Environ. Toxicol. Chem. 17, 1964–1971.

Bowen, G.D., Rovira, A.D., 1999. The rhizosphere and its management to improve plant growth. Adv. Agron. 66, 1–102.

Bradshaw, A., 1997. Restoration of mined lands—using natural processes. Ecol. Eng. 8, 255–269.

Brimecombe, M.J., De Leij, F.A., Lynch, J.M., 2001. The effect of root exudates on rhizosphere microbial populations. In: Pinton, R., Varanini, Z., Nannipieri, P. (Eds.), The rhizosphere. Marcel Dekker, New York, pp. 95–140.

Brix, H., Sorrell, B.K., Schierup, H.-H., 1996. Gas fluxes achieved by in situ convective flow in *Phragmites australis*. Aquat. Bot. 54, 151–163.

Burd, G.I., Dixon, D.G., Glick, B.R., 1998. A plant-growth-promoting bacterium that decreases nickel toxicity in seedlings. Appl. Environ. Microbiol. 64, 3363–3368.

Burd, G.I., Dixon, D.G., Glick, R.R., 2000. Plant-growth-promoting bacteria that decrease heavy metal toxicity in plants. Can. J. Microbiol. 46, 237–245.

Cankar, K., Kraigher, H., Ravnikar, M., Rupnik, M., 2005. Bacterial endophytes from seeds of Norway spruce (*Picea abies* L. Karts). FEMSMicrobiol. Lett. 244, 341–345.

Caraiman, P., Pohontu, C., Soreanu, G., et al., 2012. Optimisation process of cadmium and zinc removal from soil by phytoremediation using *Brassica napus* and *Triticales* sp. Environ. Eng. Manag. J. 11, 271–278.

Chaineau, C.H., Rougeux, G., Yepremian, C., Oudot, J., 2005. Effects of nutrient concentration on the biodegradation of crude oil and associated microbial populations in the soil. Soil Biol. Biochem. 37, 1490–1497.

Chandra, R., Yadav, S., 2010. Potential of *Typha angustifolia* for phytoremediation of heavy metals from aqueous solution of phenol and melanoidin. Ecol. Eng. 36, 1277–1284.

Chaney, R.L., Malik, K.M., Li, Y.M., et al., 1997. Phytoremediation of soil metals. Curr. Opin. Biotech. 8, 279–284.

Chattopadhyay, S., Fimmen, R.L., Yates, B.J., et al., 2012. Phytoremediation of mercury- and methyl mercury-contaminated sediments by water hyacinth (*Eichhornia crassipes*). Int. J. Phytoremediat. 14, 142–161.

Chaturvedi, N., Dhal, N.K., Reddy, P.S.R., 2012. Phytostabilization of iron pre tailings through *Calophyllum inophyllum* L. Int. J. Phytoremediat. 14, 996–1009.

Chaudhry, Q., Blom-Zandstra, M., Gupta, S.K., Joner, E., 2005. Utilising the synergy between plants and rhizosphere microorganisms to enhance breakdown of organic pollutants in the environment. Environ. Sci. Pollut. Res. 12, 34–48.

Chaudhry, T.M., Hayes, W.J., Khan, A.J., Khoo, C.S., 1998. Phytoremediation focusing on accumulator plants that remediate metal contaminated soils. Aust. J. Ecotoxicol. 4, 37–51.

Chaudhry, T.M., Hill, L., Khan, A.G., Keuk, C., 1999. Colonisation of iron and zinc contaminated dumped filter cake waste by microbes, plants and associated mycorrhizae. In: Wong, M.H., Baker, A.J.M. (Eds.), Remediation and management of degraded land. CRC Press, Boca Raton, pp. 275–283.

Chehregani, A., Malayeri, B.E., 2007. Removal of heavy metals by native accumulator plants. Int. J. Agri. Biol. 9, 462–465.

Chipeng, F.K., Hermans, C., Colinet, G., et al., 2010. Copper tolerance in the cuprophyte *Haumaniastrum katagense* (S. Moore) PA Duvign. & Plancke. Plant Soil 328, 235–244.

Cluis, C., 2004. Junk-greedy Greens: Phytoremediation as a new option for soil decontamination. Biotech. J. 2, 61–67.

Cobbett, C., 2003. Heavy metals and plants-model systems and hyperaccumulators. New Phytol. 159, 289–293.

Cosio, C., Martinoia, E., Keller, C., 2004. Hyperaccumulation of cadmium and zinc in *Thlaspi caerulescens* and *Arabidopsis halleri* at the leaf cellular level. Plant Physiol. 134, 716–725.

Crowley, D.E., Wang, Y.C., Reid, C.P.P., Szaniszlo, P.J., 1991. Mechanisms of iron acquisition from siderophores by microorganisms and plants. Plant Soil 130, 179–198.

Dalton, D., Kramer, S., Azios, N., et al., 2004. Endophytic nitrogen fixation in dune grasses (*Ammophila arenaria* and *Elymus mollis*) from Oregon. FEMS Microbiol. Ecol. 49, 469–479.

Daniels, R., Vanderleyden, J., Michiels, J., 2004. Quorum sensing and swarming migration in bacteria. FEMS Microbiol. Rev. 28, 261–289.

De Farias, V., Maranho, L.T., de Vasconcelos, E.C., et al., 2009. Phytodegradation potential of *Erythrina crista-galli* L., Fabaceae, in petroleum-contaminated soil. Appl. Biochem. Biotechnol. 157, 10–22.

De Souza, M.P., Chu, D., Zhao, M., et al., 1999. Rhizosphere bacteria enhance selenium accumulation and volatilisation by Indian mustard. Plant Physiol. 119, 565–573.

De Souza, M.P., Pilon-Smiths, E.A.H., Terry, N., 2000. The physiology and biochemistry of selenium volatilisation by plants. In: Raskin, I., Ensley, B.D. (Eds.), Phytoremediation of toxic metals: Using plants to clean up the environment. John Wiley & Sons, New York, pp. 171–190.

De Souza, S.C.R., de Andrade, S.A.L., de Souza, L.A., Schiavinato, M.A., 2012. Lead tolerance and phytoremediation potential of Brazilian leguminous tree species at the seedling stage. J. Environ. Manag. 110, 299–307.

Di Gregorio, S., Lampis, S., Malorgio, F., et al., 2006. *Brassica juncea* can improve selenite and selenate abatement in selenium contaminated soils through the aid of its rhizospheric bacterial population. Plant Soil 285, 233–244.

Diels, L., Dong, Q., Van der Lelie, D., et al., 1995. The *czc* operon of *Alcaligenes eutrophus* CH34: From resistance mechanism to the removal of heavy metals. J. Ind. Microbiol. Biotechnol. 14, 142–153.

Dixon, R., 2001. Natural products and plant disease resistance. Nature 411, 843–847.

Donnelly, P.K., Hegde, R.S., Fletcher, J.S., 1994. Growth of PCB-degrading bacteria on compounds from photosynthetic plants. Chemosphere 28, 981–988.

Doty, S.L., Shang, T.Q., Wilson, A.M., et al., 2003. Metabolism of the soil and groundwater contaminants, ethylene dibromide and trichloroethylene, by the tropical leguminous tree, *Leuceana leucocephala*. Water Res. 37, 441–449.

Douglas, S., Beveridge, T.J., 1998. Mineral formation by bacteria in natural microbial communities. FEMS Microbiol. Ecol. 26, 79–88.

Doyle, M.O., Otte, M.L., 1997. Organism-induced accumulation of iron, zinc and arsenic in wetland soils. Environ. Pollut. 96, 1–11.

Dushenkov, V., Nanda Kumar, P.B.A., Motto, H., Raskin, I., 1995. Rhizofiltration: The use of plants to remove heavy metals from aqueous streams. Environ. Sci. Technol. 29, 1239–1245.

Egamberdieva, D., Kamilova, F., Validov, S., et al., 2008. High incidence of plant growth-stimulating bacteria associated with the rhizosphere of wheat grown on salinated soil in Uzbekistan. Environ. Microbiol. 10, 1–9.

EPA, 2000. Introduction to phytoremediation. Environmental Protection Agency (EPA) Report EPA/600/R-99/107. US Environmental Protection Agency, Cincinnati.

Espinosa-Urgel, M., 2004. Plant-associated *Pseudomonas* populations: Molecular biology, DNA dynamics, and gene transfer. Plasmid 52, 139–150.

Favas, P.J.C., Pratas, J., Prasad, M.N.V., 2012. Accumulation of arsenic by aquatic plants in large-scale field conditions: Opportunities for phytoremediation and bioindication. Sci. Total Environ. 433, 390–397.

Fein, J.B., Martin, A.M., Wightman, P.G., 2001. Metal adsorption onto bacterial surfaces: Development of a predictive approach. Geochim. Cosmochim. Acta 65, 4267–4273.

Ferraz, P., Fidalgo, F., Almeida, A., Teixeira, J., 2012. Phytostabilisation of nickel by the zinc and cadmium hyperaccumulator *Solanum nigrum* L. Are metallothioneins involved? Plant Physiol. Bioch. 57, 254–260.

Flessa, H., Fischer, W.R., 1992. Plant-induced changes in the redox potential of rice rhizospheres. Plant Soil 143, 55–60.

Frey, B., Zierold, K., Brunner, I., 2000. Extracellular complexation of Cd in the Hartig net and cytosolic Zn sequestration in the fungal mantle of *Picea abies—Hebeloma crustuliniforme* ectomycorrhizas. Plant Cell Environ. 23, 1257–1265.

Gadd, G.M., 1999. Fungal production of citric and oxalic acid: Importance in metal speciation, physiology and biogeochemical processes. Adv. Microb. Physiol. 41, 47–92.

Gadd, G.M., 2004. Microbial influence on metal mobility and application for bioremediation. Geoderma 122, 109–119.

Gamalero, E., Lingua, G., Capri, F.G., et al., 2004. Colonisation pattern of primary tomato roots by *Pseudomonas fluorescens* A6RI characterised by dilution plating, flow cytometry, fluorescence, confocal and scanning electron mycroscopy. FEMS Microbiol. Ecol. 48, 79–87.

Garbeva, P., Overbeek, L.S., Vuurde, J.W., et al., 2001. Analysis of endophytic bacterial communities of potato by plating and denaturing gradient gel electrophoresis (dgge) of 16s rdna based pcr fragments. Microb. Ecol. 41, 369–383.

Gardea-Torresdey, J.L., Gonzalez, J.H., Tiemann, K.J., et al., 1998. Phytofiltration of hazardous cadmium, chromium, lead and zinc ions by biomass of *Medicago sativa* (Alfalfa). J. Hazard. Mater. 57, 29–39.

Gatliff, E.G., 1994. Vegetative remediation process offers advantages over traditional pump-and-treat technologies. Remediation 4, 343–352.

Gerhardson, B., 2002. Biological substitutes for pesticides. Trends Biotechnol. 20, 338–343.

Germaine, K., Keogh, E., Garcia-Cabellos, G., et al., 2004. Colonisation of poplar trees by *gfp* expressing endophytes. FEMS Microbiol Ecol 48, 109–118.

Gilbert, E.S., Crowley, D.E., 1997. Plant compounds that induce polychlorinated biphenyl biodegradation by *Arthrobacter* sp. strain B1B. Appl. Environ. Microb. 63, 1933–1938.

Gilis, A., Corbisier, P., Baeyens, W., et al., 1998. Effect of the siderophore alcaligin E on the bioavailability of Cd to *Alcaligenes eutrophus* CH34. J. Ind. Microbiol. Biotechnol. 20, 61–68.

Glick, B., 1995. The enhancement of plant growth by free-living bacteria. Can. J. Microbiol. 41, 109–117.

Glick, B.R., 2005. Modulation of plant ethylene levels by the bacterial enzyme ACC deaminase. FEMS Microbiol. Lett. 251, 1–7.

Glick, B.R., Penrose, D.M., Li, J.P., 1998. A model for the lowering of plant ethylene concentrations by plant growth-promoting bacteria. J. Theor. Biol. 190, 63–68.

Glick, B.R., Patten, C.L., Holguin, G., Penrose, D.M., 1999. Biochemical and genetic mechanisms used by plant growth-promoting bacteria. Imperial College Press, London.

Glick, B.R., Cheng, Z., Czarny, J., Duan, J., 2007. Promotion of plant growth by ACC deaminase containing soil bacteria. Eur. J. Plant Pathol. 119, 329–339.

Goldstein, A.H., Braverman, K., Osorio, N., 1999. Evidence for mutualism between a plant growing in a phosphate-limited desert environment and a mineral phosphate solubilising (MPS) rhizobacterium. FEMS Microbiol. Ecol. 30, 295–300.

Gordon, M., Choe, N., Duffy, J., et al., 1998. Phytoremediation of trichloroethylene with hybrid poplars. Environ. Health Persp. 106, 1001–1004.

Hadacek, F., 2002. Secondary metabolites as plant traits: Current assessment and future perspectives. CRC Crit. Rev. Plant Sci. 21, 273–322.

Hansen, D., Duda, P.J., Zayed, A., Terry, N., 1998. Selenium removal by constructed wetlands: Role of biological volatilisation. Environ. Sci. Technol. 32, 591–597.

Harvey, P.J., Campanella, B.F., Castro, P.M., et al., 2002. Phytoremediation of polyaromatic hydrocarbons, anilines and phenols. Environ. Sci. Pollut. Res. Int. 9, 29–47.

Hasnain, S., Sabri, A.N., 1996. Growth stimulation of *Triticum aestivum* seedlings under Cr-stresses by non rhizospheric pseudomonad strains. Abstracts of the 7th International symposium on biological nitrogen fixation with non-Legumes. Kluwer Academic Publishers, Dordrecht. p 36.

Headley, J.V., Peru, K.M., Du, J.L., et al., 2008. Evaluation of the apparent phytodegradation of pentachlorophenol by *Chlorella pyrenoidosa*. J. Environ. Sci. Heal. A. 43, 361–364.

Heggo, A., Angle, J.S., Chaney, R.L., 1990. Effects of vesicular arbuscular mycorrhizal fungi on heavy metal uptake by soybean. Soil. Biol. Biochem. 22, 856–869.

Heinekamp, Y., Willey, N., 2007. Using real-time polymerase chain reaction to quantify gene expression in plants exposed to radioactivity. In: Willey, N. (Ed.), Phytoremediation: Methods and reviews. Humana Press, Totowa, pp. 59–70.

Hinsinger, P., 2001. Bioavailability of soil inorganic P in the rhizospere as affected by root-induced chemical changes: A review. Plant Soil 237, 173–195.

Hoagland, R.E., Zablotowicz, R.M., Locke, M.A., 1997. An integrated phytoremediation strategy for chloroacetamide herbicides in soil. In: Kruger, E.L., Anderson, T.A., Coats, J.R. (Eds.), Phytoremediation of soil and water contaminants, ACS symposium series 664. American Chemical Society, Washington, pp. 92–105.

Hooda, V., 2007. Phytoremediation of toxic metals from soil and waste water. J. Environ. Biol. 28, 367–376.

Hua, J.F., Zhang, C.S., Yin, Y.L., et al., 2012. Phytoremediation potential of three aquatic macrophytes in manganese-contaminated water. Water Environ. J. 26, 335–342.

Huang, X.D., El-Alawi, Y., Penrose, D.M., et al., 2004. A multi process phytoremediation system for removal of polycyclic aromatic hydrocarbons from contaminated soils. Environ. Pollut. 130, 465–476.

Ibanez, S.G., Alderete, L.G.S., Medina, M.I., Agostini, E., 2012. Phytoremediation of phenol using *Vicia sativa* L. plants and its antioxidative response. Environ. Sci. Pollut. R. 19, 1555–1562.

Idris, R., Trifonova, R., Puschenreiter, M., et al., 2004. Bacterial communities associated with flowering plants of the Ni hyperaccumulator *Thlaspi goesingense*. Appl. Environ. Microbiol. 70, 2667–2677.

Ivanov, A., Gavryushkin, A.V., Siunova, T.V., et al., 1999. Investigation of heavy metal resistance of some *Pseudomonas* strains. Mikrobiologiya (Moscow) 68, 366–374.

Jadia, C.D., Fulekar, M.H., 2008. Phytoremediation: The application of vermicompost to remove zinc, cadmium, copper, nickel and lead by sunflower plant. Environ. Eng. Manag. J. 7, 547–558.

Jadia, C.D., Fulekar, M.H., 2009. Phytoremediation of heavy metals: Recent techniques. Afr. J. Biotechnol. 8, 921–928.

Jaeger, C.H., Lindow, S.E.I.I.I., Miller, W., et al., 1999. Mapping of sugar and amino acid availability in soil around roots with bacterial sensors of sucrose and tryptophan. Appl. Environ. Microbiol. 65, 2685–2690.

Joner, E.J., Leyval, C., 1997. Uptake of 109Cd by roots and hyphae of a Glomus mosseae / Tripholium subterraneum mycorrhiza from soil amended with high and low concentration of cadmium. New Phytol. 135, 353–360.

Joner, E.J., Leyval, C., Colpaert, J.V., 2006. Ectomycorrhizas impede phytoremediation of polycyclic aromatic hydrocarbons (PAHs) both within and beyond the rhizosphere. Environ. Poll. 142, 34–38.

Jones, D., 1998. Organic acids in the rhizosphere—a critical review. Plant Soil 205, 24–44.

Joseph, M., 2005. Phytoremediation of tin. Abstracts of Papers of the American Chemical Society. Meeting Abstract, 229: U502.

Kagalkar, A.N., Jadhav, M.U., Bapar, V.A., Govindwar, S.P., 2011. Phytodegradation of the triphenylmethane dye Malachite Green mediated by cell suspension cultures of *Blumea malcolmii* Hook. Bioresour. Technol. 102, 10312–10318.

Kalve, S., Sarangi, B.K., Pandey, R.A., Chakrabarti, T., 2011. Arsenic and chromium hyperaccumulation by an ecotype of *Pteris vittata*—prospective for phytoextraction from contaminated water and soil. Curr. Sci. 100, 888–894.

Kamnev, A.A., 2003. Phytoremediation of heavy metals: An overview. In: Fingerman, M., Nagabhushanam, R. (Eds.), Recent advances in marine biotechnology. Bioremediation, vol. 8. Science Publishers, Enfield, pp. 269–317.

Kamnev, A.A., 2005. Use of spectroscopic methods to study the molecular mechanisms of plant–microbial interactions. In: Ignatov, V.V. (Ed.), Molecular bases of interrelationships between associative microorganisms and plants (in Russian). Nauka, Moscow, pp. 238–260.

Kamnev, A.A., Van der Lelie, D., 2000. Chemical and biological parameters as tools to evaluate and improve heavy metal phytoremediation. Biosci. Rep. 20, 239–258.

Kamnev, A.A., Antonyuk, L.P., Ignatov, V.V., 1999. Biodegradation of organic pollution involving soil iron (III) solubilised by bacterial siderophores as an electron acceptor: Possibilities and perspectives. In: Fass, R., Flashner, Y., Reuveny, S. (Eds.), Novel approaches for bioremediation of organic pollution. Kluwer Academic & Plenum Publishers, New York, pp. 205–217.

Karkhanis, M., Jadia, C.D., Fulekar, M.H., 2005. Rhizofilteration of metals from coal ash leachate. Asian J. Water Environ. Pollut. 3, 91–94.

Kassel, A.G., Ghoshal, D., Goyal, A., 2002. Phytoremediation of trichloroethylene using hybrid poplar. Physiol. Mol. Biol. Plants 8, 1–8.

Kaye, J.P., Hart, S.C., 1997. Competition for nitrogen between plants and soil microorganisms. Trends Ecol. Evol. 12, 139–143.

Kennedy, A.C., 1998. The rhizosphere and spermosphere. In: Sylvia, D.M., Fuhrmann, J.J., Hartel, P.G., Zuberer, D.A. (Eds.), Principles and applications of soil microbiology. Prentice Hall, Upper Saddle River, pp. 389–407.

Khan, A.G., 2006. Mycorrhizoremediation—an enhanced form of phytoremediation. J. Zhejiang Univ. Sci. B. 7, 503–514.

Khellaf, N., Zerdaoui, M., 2010. Growth response of the duckweed Lemna gibba L. to copper and nickel phytoaccumulation. Ecotoxicology 19, 1363–1368.

Kinnersely, A.M., 1993. The role of phytochelates in plant growth and productivity. Plant Growth Regul. 12, 207–217.

Kloepper, J.W., 1994. Plant growth-promoting rhizobacteria (other systems). In: Okon, Y. (Ed.), Azospirillum / plant associations. CRC Press, Boca Raton, pp. 111–118.

Kloepper, J.W., Zablotowick, R.M., Tipping, E.M., Lifshitz, R., 1991. Plant growth promotion mediated by bacterial rhizosphere colonisers. In: Keister, D.L., Cregan, P.B. (Eds.), The rhizosphere and plant growth. Kluwer Academic Publishers, Dordrecht, pp. 315–326.

Knasmuller, S., Gottmann, E., Steinkellner, H., et al., 1998. Detection of genotoxic effects of heavy metal contaminated soils with plant bioassays. Mutat. Res. 420, 37–48.

Knee, E.M., Gong, F.C., Gao, M., et al., 2001. Root mucilage from pea and its utilisation by rhizosphere bacteria as a sole carbon source. Mol. Plant Microbe Interact. 14, 775–784.

Knuteson, S.L., Whitwell, T., Klaine, S.J., 2002. Influence of plant age and size on simazine toxicity and uptake. J. Environ. Qual. 31, 2096–2103.

Kshirsagar, S., Aery, N.C., 2007. Phytostabilisation of mine waste: Growth and physiological responses of Vigna unguiculata (L.) Walp. J. Environ. Biol. 2, 651–654.

Kuffner, M., De Maria, S., Puschenreiter, M., et al., 2010. Culturable bacteria from Zn- and Cd-accumulating Salix caprea with differential effects on plant growth and heavy metal availability. J. Appl. Microbiol. 108, 1471–1484.

Kuiper, I., Lagendijk, E.L., Bloemberg, G.V., Lugtenberg, B.J.J., 2004. Rhizoremediation: A beneficial plant–microbe interaction. Mol. Plant Microbe Interact 17, 6–15.

Kvesitadze, G., Khatisashvili, G., Sadunishvili, T., Ramsden, J.J., 2006. Biochemical mechanisms of detoxification in higher plants: Basis of phytoremediation. Springer, Berlin.

Lambrecht, M., Okon, Y., Vande Broek, A., Vanderleyden, J., 2000. Indole–3-acetic acid: A reciprocal signalling molecule in bacteria–plant interactions. Trends Microbiol. 8, 298–300.

Langley, S., Beveridge, T.J., 1999. Effect of O-side-chain-lipopolysaccharide chemistry on metal binding. Appl. Environ. Microbiol. 65, 489–498.

Lasat, M.M., 2002. Phytoextraction of toxic metals: A review of biological mechanisms. J. Environ. Qual. 31, 109–120.

Leyval, C., Joner, E.J., 2000. Bioavailability of metals in the mycorhizosphere. In: Gobran, G.R., Wenzel, W.W., Lombi, E. (Eds.), Trace elements in the rhizosphere. CRC press, Boca Raton, pp. 165–185.

Li, T.Q., Yang, X.E., Jin, X.F., et al., 2005. Root responses and metal accumulation in two contrasting ecotypes of *Sedum alfredii* Hance under lead and zinc toxic stress. J. Environ. Sci. Heal. A. 40, 1081–1096.

Li, Y.M., Chaney, R., Brewer, E., et al., 2003. Development of a technology for commercial phytoextraction of nickel: Economic and technical considerations. Plant Soil 249, 107–115.

Lin, C.H., Lerch, R.N., Kremer, R.J., Garrett, H.E., 2011. Stimulated rhizodegradation of atrazine by selected plant species. J. Environ. Qual. 40, 1113–1121.

Lin, Q.X., Mendelssohn, I.A., 2009. Potential of restoration and phytoremediation with *Juncus roemerianus* for diesel-contaminated coastal wetlands. Ecol. Eng. 35, 85–91.

Lin, Z.Q., Schemenauer, R.S., Cervinka, V., et al., 2000. Selenium volatilisation from a soil–plant system for the remediation of contaminated water and soil in the San Joaquin valley. J. Environ. Qual. 29, 1048–1056.

Linderman, R.G., 1992. Vesicular-arbuscular mycorrhizae and soil microbial interactions. In: Bethlenfalvay, G.J., Linderman, R.G. (Eds.), Mycorrhizae in sustainable agriculture, ASA Special Publication 54. Agronomy Society of America, Madison, pp. 45–70.

Lins, U., Farina, M., 1999. Phosphorus-rich granules in uncultured magnetotactic bacteria. FEMS Microbiol. Lett. 172, 23–28.

Littera, P., Urik, M., Gardosova, K., et al., 2012. Accumulation of antimony (III) by *Aspergillus niger* and its influence on fungal growth. Fresen. Environ. Bull. 21, 1721–1724.

Lodewyckx, C., Mergeay, M., Vangronsveld, J., et al., 2002. Isolation, characterisation, and identification of bacteria associated with the zinc hyperacccumulator *Thlaspi caerulescens* subsp. *calaminaria*. Int. J. Phytoremediat. 4, 101–115.

Lombi, E., Zhao, F.J., Dunham, S.J., McGrath, S.P., 2001. Phytoremediation of heavy metal-contaminated soils: Natural hyperaccumulation versus chemically enhanced phytoextraction. J. Environ. Qual. 30, 1919–1926.

Lu, H.L., Zhang, Y., Liu, B.B., et al., 2011. Rhizodegradation gradients of phenanthrene and pyrene in sediment of mangrove (*Kandelia candel* (L.) Druce). J. Hazard. Mater. 196, 263–269.

Lucero, M.E., Mueller, W., Hubstenberger, J., et al., 1999. Tolerance to nitrogenous explosives and metabolism of TNT by cell suspensions of Datura innoxia. Vitro Cell Dev. Biol. Plant 35, 480–486.

Lugtenberg, B.J.J., Dekkers, L.C., 1999. What makes *Pseudomonas* bacteria rhizosphere competent? Environ. Microbiol. 1, 9–13.

Lugtenberg, B.J.J., Dekkers, L., Bloemberg, G.V., 2001. Molecular determinants of rhizosphere colonisation by *Pseudomonas*. Annu. Rev. Phytopathol. 39, 461–490.

Lynch, J.M., 1990. The rhizosphere. John Wiley & Sons, New York.

Ma, L.Q., Komar, K.M., Tu, C., et al., 2001. A fern that hyperaccumulates arsenic. Nature 409, 579.

Ma, X., Wang, C., 2010. Fullerene nanoparticles affect the fate and uptake of trichloroethylene in phytoremediation systems. Environ. Eng. Sci. 27, 989–992.

Macek, T., Macková, M., Kás, J., 2000. Exploitation of plants for the removal of organics in environmental remediation. Biotechnol. Adv. 18, 23–34.

Magalhães, M.O.L., Sobrinho, N., dos Santos, F.S., Mazur, N., 2011. Potential of two species of eucalyptus in the phytostabilisation of a soil contaminated with zinc. Rev. Ciênc. Agron. 42, 805–812.

Magdoff, F., van Es, H., 2000. Building soils for better crops. In: Sustainable agriculture network handbook series, book 4, second ed. University of Vermont, Burlington.

Mahaffee, W.F., Kloepper, J.W., Van Vuurde, J.W.L., et al., 1997. Endophytic colonisation of *Phaseolus vulgaris* by *Pseudomonas fluorescens*strain 89B–27 and *Enterobacter asburiae*strain JM22. In: Ryder, M.H.R., Stevens PM,Bowen, G.D. (Eds.), Improving plant productivity in rhizosphere bacteria. CSIRO, Melbourne.

Malinowski, D.P., Zuo, H., Belesky, D.P., Alloush, G.A., 2004. Evidence for copper binding by extracellular root exudates of tall fescue but not perennial ryegrass infected with *Neotyphodium* spp. endophytes. Plant Soil 267, 1–12.

Maqbool, F., Wang, Z.Y., Xu, Y., et al., 2012. Rhizodegradation of petroleum hydrocarbons by *Sesbania cannabina* in bioaugmented soil with free and immobilised consortium. J. Hazard. Mater. 237, 262–269.

Marcano, L., Carruyo, I., Fernandez, Y., et al., 2006. Determination of vanadium accumulation in onion root cells (*Allium cepa* L.) and its correlation with toxicity. Biocell. 30, 259–267.

Maria, S.D., Rivelli, A.R., Kuffner, M., et al., 2011. Interactions between accumulation of trace elements and macronutrients in *Salix caprea* after inoculation with rhizosphere microorganisms. Chemosphere 84, 1256–1261.

Marschner, H., 1995. Mineral nutrition of higher plants, second ed. Academic Press, San Diego.

Masalha, J., Kosegarten, H., Elmaci, O., Mengal, K., 2000. The central role of microbial activity for iron acquisition in maize and sunflower. Biol. Fertil. Soils 30, 433–439.

Mastretta, C., Barac, T., Vangronsveld, J., et al., 2006. Endophytic bacteria and their potential application to improve the phytoremediation of contaminated environments. Biotech. Gen. Eng. Rev. 23, 175–207.

Medina, V.F., Larson, S.L., Agwaramgbo, L., Perez, W., 2002. Treatment of munitions in soils using phytoslurries. Int. J. Phytoremediat. 4, 143–156.

Meharg, A.A., Cairney, J.W.G., 2000. Extomycorrhizas — extending the capabilities of rhizosphere remediation. Soil Biol. Biochem. 32, 1475–1484.

Memon, A.R., Schroder, P., 2009. Implications of metal accumulation mechanisms to phytoremediation. Environ. Sci. Pollut. R. 16, 162–175.

Mendes, R., Kruijt, M., de Bruijn, I., et al., 2011. Deciphering the rhizosphere microbiome for disease-suppressive bacteria. Science 332, 1097–1100.

Merini, L.J., Bobillo, C., Cuadrado, V., et al., 2009. Phytoremediation potential of the novel atrazine tolerant *Lolium multiflorum* and studies on the mechanisms involved. Environ. Pollut. 157, 3059–3063.

Metting Jr, F.B., 1992. Structure and physiological ecology of soil microbial communities. In: Metting Jr, F.B. (Ed.), Soil microbial ecology. Marcel Dekker, New York, pp. 3–25.

Modaihsh, A., Al-Swailem, M., Mahjoub, M., 2004. Heavy metal contents of commercial inorganic fertiliser used in the Kingdom of Saudi Arabia. Agri. Mar. Sci. 9, 21–25.

Mohanty, M., Patra, H.K., 2012. Phytoremediation potential of paragrass — an *in situ* approach for chromium contaminated soil. Int. J. Phytoremediat. 14, 796–805.

Moorehead, D.L., Westerfield, M.M., Zak, J.C., 1998. Plants retard litter decay in a nutrient-limited soil: A case of exploitative competition. Oecologia 113, 530–536.

Mudgal, V., Madaan, N., Mudgal, A., 2010. Heavy metals in plants: Phytoremediation: Plants used to remediate heavy metal pollution. Agr. Biol. J. N. Am. 1, 40–46.

Mukhopadhyay, S., Maiti, S.K., 2010. Phytoremediation of metal enriched mine waste: A review. Global J. Environ. Res. 4, 135–150.

Nannipieri, P., Ascher, J., Ceccherini, M.T., et al., 2003. Microbial diversity and soil functions. Eur. J. Soil Sci. 54, 655–670.

Narasimhan, K., Basheer, C., Bajic, V.B., Swarup, S., 2003. Enhancement of plant–microbe interactions using a rhizosphere metabolomics driven approach and its application in the removal of polychlorinated biphenyls. Plant Physiol. 132, 146–153.

Nateewattana, J., Trichaiyaporn, S., Saouy, M., et al., 2010. Monitoring of arsenic in aquatic plants, water, and sediment of wastewater treatment ponds at the Mae Moh Lignite power plant, Thailand. Environ. Monit. Assess. 165, 585–594.

Nautiyal, C.S., Bhadauria, S., Kumar, P., et al., 2000. Stress induced phosphate solubilisation in bacteria isolated from alkaline soils. FEMS Microbiol. Lett. 182, 291–296.

Neil, W. (Ed.), 2007. Phytoremediation: Methods and reviews. Humana Press, Totowa. pp V-VII.

Newman, L.A., Strand, S.E., Choe, N., et al., 1997. Uptake and biotransformation of trichloroethylene by hybrid poplars. Environ. Sci. Technol. 31, 1062–1067.

Nolan, C., Whitehead, N., Teyssie, J.L., 1991. Tellurium-speciation in seawater and accumulation by marine phytoplankton and crustaceans. J. Environ. Radioactiv. 13, 217–233.

Pace, N.R., 1997. A molecular view of microbial diversity in the biosphere. Science 276, 734–740.

Pan, M.J., Rademan, S., Kuner, K., Hastings, J.W., 1997. Ultrastructural studies on the colonisation of banana tissue and *Fusarium oxysporum* f.sp. cubense race 4 by the endophytic bacterium *Burkholderia cepacia*. J. Phytopathol. 145, 479–486.

Pandey, P., Kang, S.C., Maheshwari, D.K., 2005. Isolation of endophytic plant growth promoting *Burkholderia* sp. MSSP from root nodules of *Mimosa pudica*. Curr. Sci. 89, 177–180.

Peer, W.A., Baxter, I.R., Richards, E.L., et al., 2005. Phytoremediation and hyperaccumulator plants. In: Tamas, M., Martinoia, E. (Eds.), Molecular biology of metal homeostasis and detoxification. Topics in current genetics, vol. 14. Springer, Berlin, pp. 299–340.

Peili, H., Li, J., Zhang, S., Chen, C., et al., 2011. Effects of lanthanum, cerium, and neodymium on the nuclei and mitochondria of hepatocytes: Accumulation and oxidative damage. Environ. Toxicol. Pharmacol. 31, 25–32.

Pierzynski, G.M., Schnoor, J.L., Banks, M.K., et al., 1994. Vegetative remediation at superfund sites. Mining and its environment impact. Royal Soc. Chem. Issues Environ. Sci. Technol. 1, 49–69.

Pilon-Smits, E., 2005. Phytoremediation. Annu. Rev. Plant Biol. 56, 15–39.

Pilon-Smits, E.A.H., de Souza, M.P., Hong, G., et al., 1999. Selenium volatilisation and accumulation by twenty aquatic plant species. J. Qual. 28, 1011–1018.

Pivetz, B.E., 2001. Phytoremediation of contaminated soil and ground water at hazardous waste sites. EPA 540-S-01-500, ground water issue. pp 1–36.

Poscic, F., Marchiol, L., Schat, H., 2013. Hyperaccumulation of thallium is population-specific and uncorrelated with caesium accumulation in the thallium hyperaccumulator, *Biscutella laevigata*. Plant Soil 365, 81–91.

Prasad, M.N.V., Freitas, H.M.D., 2003. Metal hyperaccumulation in plants — biodiversity prospecting for phytoremediation technology. Electron. J. Biotechn. 6, 285–321.

Pratas, J., Favas, P.J.C., Paulo, C., et al., 2012. Uranium accumulation by aquatic plants from uranium-contaminated water in central Portugal. Int. J. Phytoremediat. 14, 221–234.

Priha, O., Grayston, S.J., Pennanen, T., Smolander, A., 1999. Microbial activities related to C and N cycling and microbial community structure in the rhizospheres of *Pinus sylvestris, Picea abies* and *Betula pendula* seedlings in an organic and mineral soil. FEMS Microbiol. Ecol. 30, 187–199.

Pulford, I.D., Watson, C., 2003. Phytoremediation of heavy metal-contaminated land by trees — a review. Environ. Int. 29, 529–540.

Quadt-Hallmann, A., Benhamou, N., Kloepper, J.W., 1997. Bacterial endophytes in cotton: Mechanisms entering the plant. Can. J. Microbiol. 43, 577–582.

Quinn, J.J., Negri, M.C., Hinchman, R.R., et al., 2001. Predicting the effect of deep-rooted hybrid poplars on the groundwater flow system at a large scale phytoremediation site. Int. J. Phytoremediat. 3, 41–60.

Rai, P.K., 2008. Technical note: Phytoremediation of Hg and Cd from industrial effluents using an aquatic free floating macrophyte *Azolla pinnata*. Int. J. Phytoremediat. 10, 430–439.

Rajakaruna, N., Bohm, B.A., 2002. Serpentine and its vegetation: A preliminary study from Sri Lanka. J. Appl. Bot. 76, 20–28.

Rajakaruna, N., Tompkins, K.M., Pavicevic, P.G., 2006. Phytoremediation: An affordable green technology for the clean-up of metal-contaminated sites in Sri Lanka. Cey. J. Sci. (Bio Sci) 35, 25–39.

Rajkumar, M., Freitas, H., 2008. Influence of metal resistant-plant growth-promoting bacteria on the growth of *Ricinus communis* in soil contaminated with heavy metals. Chemosphere 71, 834–842.

Raskin, I., Ensley, B.D. (Eds.), 2000. Phytoremediation of toxic metals: Using plants to clean up the environment. John Wiley & Sons, New York.

Raskin, I.P., Nanda Kumar, B.A., Dushenkov, S., et al., 1994. Phytoremediation—using plants to clean up soils and waters contaminated with toxic metals. Emerging technologies in hazardous waste management VI. ACS Industrial & Engineering Chemistry Division Special Symp, vol 1, Atlanta.

Raskin, I., Smith, R.D., Salt, D.E., 1997. Phytoremediation of metals: Using plants to remove pollutants from the environment. Curr. Opin. Biotechnol. 8, 221–226.

Reeves, R.D., Brooks, R.R., 1983. European species of *Thlaspi* L. (*Cruciferae*) as indicators of nickel and zinc. J. Geochem. Explor. 18, 275–283.

Reeves, R.D., Baker, A.J.M., 2000. Phytoremediation of toxic metals. Wiley, New York. pp. 193–229.

Reid, C.P., Szaniszlo, P.J., Crowley, D.E., 1986. Siderophore involvement in plant iron nutrition. In: Swinburne, T.R. (Ed.), Iron, siderophores and plant diseases. Plenum Press, New York, pp. 29–42.

Rothballer, M., Schmid, M., Hartmann, A., 2003. In situ localisation and PGPR-effect of *Azospirillum brasilense* strains colonising roots of different wheat varieties. Symbiosis 34, 261–279.

Sabiha-Javied, Mehmood T., Chaudhry, M.M., et al., 2009. Heavy metal pollution from phosphate rock used for the production of fertiliser in Pakistan. Microchem. J. 91, 94–99.

Sabry, S.A., Ghozlan, H.A., Abou-Zeid, D.M., 1997. Metal tolerance and antobiotic resistance patterns of a bacterial population isolated from sea water. J. Appl. Microbiol. 82, 245–252.

Saleh, H.M., 2012. Water hyacinth for phytoremediation of radioactive waste simulate contaminated with cesium and cobalt radionuclides. Nucl. Eng. Des. 242, 425–432.

Salido, A.L., Hasty, K.L., Lim, J.M., Butcher, D.J., 2003. Phytoremediation of arsenic and lead in contaminated soil using Chinese Brake Ferns (*Pteris vittata*) and Indian mustard (*Brassica juncea*). Int. J. Phytoremediat. 5, 89–103.

Salt, D.E., Blaylock, M., Kumar, N., et al., 1995a. Phytoremediation — a novel strategy for the removal of toxic metals from the environment using plants. Bio–Technol. 13, 468–474.

Salt, D.E., Prince, R.C., Pickering, I.J., Raskin, I., 1995b. Mechanisms of cadmium mobility and accumulation in Indian mustard. Plant Physiol. 109, 1427–1433.

Salt, D.E., Smith, R.D., Raskin, I., 1998. Phytoremediation. Annu. Rev. Plant Physiol. 49, 643–668.

Saraswat, S., Rai, J.P.N., 2009. Phytoextraction potential of six plant species grown in multimetal contaminated soil. Chem. Ecol. 25, 1–11.

Sasmaz, A., Obek, E., 2012. The accumulation of silver and gold in *Lemna gibba* L. exposed to secondary effluents. Chem. Erde 72, 149–152.

Scheupp, H., Dehn, B., Sticher, H., 1987. Interaktionen zwischen VA-mycorrhizen und schwermetallbelastungen. Agnew Botanik 61, 85–96.

Schnoor, J., Licht, L., Mccutcheon, S., et al., 1995. Phytoremediation of organic and nutrient contaminants. Environ. Sci. Technol. 29, A318–A323.

Schulz, B., Boyle, C., 2006. What are endophytes? In: Schulz, B.J.E., Boyle, C.J.C., Sieber, T.N. (Eds.), Microbial root endophytes. Springer, Berlin, pp. 1–13.

Schutzendubel, A., Polle, A., 2002. Plant responses to abiotic stresses: Heavy metal induced oxidative stress and protection by mycorrhization. J. Exp. Bot. 53, 1351–1365.

Shah, S., Li, J., Moffatt, B.A., Glick, B.R., 1998. Isolation and characterisation of ACC deaminase genes from two different plant growth-promoting rhizobacteria. Can. J. Microbiol. 44, 833–843.

Shann, J.R., 1995. The role of plants and plant–microbial systems in the reduction of exposure. Environ. Health Perspect. 103 (suppl 5), 13–15.

Sheng, X.F., Xia, J.J., 2006. Improvement of rape (*Brassica napus*) plant growth and cadmium uptake by cadmium-resistant bacteria. Chemosphere 64, 1036–1042.

Shetty, K.G., Hetrick, B.A.D., Figge, D.A.H., Schwab, A.P., 1994. Effects of mycorrhizae and other soil microbes on revegetation of heavy metal contaminated mine spoil. Environ. Pollut. 86, 181–188.

Shi, W., Becker, J., Bischoff, M., et al., 2002. Association of microbial community composition and activity with lead, chromium, and hydrocarbon contamination. Appl. Environ. Microbiol. 68, 3859–3866.

Smith, S.E., Reed, D.J., 1997. Mycorrhizal symbiosis. Academic Press, London.

Sorensen, J., 1997. The rhizosphere as a habitat for soil microorganisms. In: Van Elsas, J.D., Trevors, J.T., Wellington, E.M.H. (Eds.), Modern soil microbiology. Marcel Dekker, Inc., New York, pp. 21–45.

Steenhoudt, O., Vanderleyden, J., 2000. *Azospirillum*, a free-living nitrogen-fixing bacterium closely associated with grasses: Genetic, biochemical and ecological aspects. FEMS Microbiol. Rev. 24, 487–506.

Sun, Q., Ye, Z.H., Wang, X.R., Wong, M.H., 2007. Cadmium hyperaccumulation leads to an increase of glutathione rather than phytochelatins in the cadmium hyperaccumulator *Sedum alfredii*. J. Plant Physiol. 164, 1489–1498.

Sun, Y.B., Zhou, Q.X., Diao, C.Y., 2008. Effects of cadmium and arsenic on growth and metal accumulation of Cd-hyperaccumulator *Solanum nigrum* L. Bioresour. Technol. 99, 1103–1110.

Sun, Y.B., Zhou, Q.X., Wang, L., Liu, W.T., 2009. Cadmium tolerance and accumulation characteristics of *Bidens pilosa* L. as a potential Cd-hyperaccumulator. J. Hazard Mater. 161, 808–814.

Suresh, B., Ravishankar, G., 2004. Phytoremediation — a novel and promising approach for environmental clean-up. Crit. Rev. Biotech. 24, 97–124.

Susarla, S., Medina, V.F., McCutcheon, S.C., 2002. Phytoremediation: An ecological solution to organic chemical contamination. Ecol. Eng. 18, 647–658.

Tamaoki, M., Freeman, J.L., Pilon-Smits, E.A.H., 2008. Cooperative ethylene and jasmonic acid signaling regulates selenite resistance in *Arabidopsis*. Plant Physiol. 146, 1219–1230.

Tan, R., Zou, W., 2001. Endophytes: A rich source of functional metabolites. Nat. Prod. Rep. 18, 448–459.

Tulod, A.M., Castillo, A.S., Carandang, W.M., Pampolina, N.M., 2012. Growth performance and phytoremediation potential of *Pongamia pinnata* (L.) Pierre, *Samanea saman* (Jacq.) Merr. and *Vitex parviflora* Juss. in copper-contaminated soil amended with zeolite and VAM. Asia Life Sci. 21, 499–522.

Turgut, C., Pepe, M.K., Cutright, T.J., 2004. The effect of EDTA and citric acid on phytoremediation of Cd, Cr, and Ni from soil using *Helianthus annuus*. Environ. Pollut. 131, 147–154.

Unterbrunner, R., Wieshammer, G., Hollender, U., et al., 2007. Plant and fertiliser effects on rhizodegradation of crude oil in two soils with different nutrient status. Plant Soil 300, 117–126.

Vallini, G., Di Gregorio, S., Lampis, S., 2005. Rhizosphere-induced selenium precipitation for possible applications in phytoremediation of se polluted effluents. Z. Naturforsch C. 60, 349–356.

Van der Ent, A., Baker, A.J.M., Reeves, R.D., et al., 2013. Hyperaccumulators of metal and metalloid trace elements: Facts and fiction. Plant Soil 362, 319–334.

Van der Lelie, D., Schwitzguébel, J.P., Glass, D., Baker, A., 2001. Assessing phytoremediations's progress in the United States and in Europe. Environ. Sci. Technol. 35, 446A–452A.

Vogel-Mikus, K., Drobne, D., Regvar, M., 2005. Zn, Cd and Pb accumulation and arbuscular mycorrhizal colonisation of pennycress *Thlaspi praecox* Wulf. (Brassicaceae) from the vicinity of a lead mine and smelter in Slovenia. Environ. Pollut. 133, 233–242.

Volkering, F., Breure, A.M., Rulkens, W.H., 1998. Microbiological aspects of sulfactant use for biological soil remediation. Biodegradation 8, 401–417.

Wang, H.B., Xie, F., Yao, Y.Z., et al., 2012. The effects of arsenic and induced-phytoextraction methods on photosynthesis in *Pteris* species with different arsenic-accumulating abilities. Environ. Exp. Bot. 75, 298–306.

Wang, Y., Brown, H.N., Crowley, D.E., Szaniszlo, P.J., 1993. Evidence for direct utilisation of a siderophore, ferroxamine B, in axenically grown cucumber. Plant Cell Environ. 16, 579–585.

Wei, C.Y., Deng, Q.J., Wu, F.C., et al., 2011. Arsenic, antimony, and bismuth and accumulation by plants in an old antimony mine, China. Biol. Trace Elem. Res. 144, 1150–1158.

Wei, S.H., Zhou, Q.X., Saha, U.K., 2008. Hyperaccumulative characteristics of weed species to heavy metals. Water Air Soil Pollut. 192, 173–181.

Weissenhorn, I., Leyval, C., 1995. Root colonisation of maize by a Cd-sensitive and a Cd-tolerant *Glomus mossae* and cadmium uptake in sand culture. Plant Soil 175, 223–238.

Welbaum, G., Sturz, A.V., Dong, Z., Nowak, J., 2004. Fertilising soil microorganisms to improve productivity of agroecosystems. Crit. Rev. Plant Sci. 23, 175–193.

Wenzel, W.W., 2009. Rhizosphere processes and management in plant-assisted bioremediation (phytoremediation) of soils. Plant Soil 321, 385–408.

Weyens, N., Van der Lelie, D., Taghavi, S., et al., 2009a. Exploiting plant–microbe partnerships to improve biomass production and remediation. Trends Biotechnol. 27, 591–598.

Weyens, N., Van der Lelie, D., Taghavi, S., Vangronsveld, J., 2009b. Phytoremediation: Plant–endophyte partnerships take the challenge. Curr. Opin. Biotechnol. 20, 248–254.

Whipps, J.M., 1990. Carbon economy. In: Lynch, J.M. (Ed.), The rhizosphere. Wiley, New York, pp. 59–97.

Wu, C.A., Liao, B., Wang, S.L., et al., 2010. Pb and Zn accumulation in a Cd hyperaccumulator (*Viola baoshanensis*). Int. J. Phytoremediat. 12, 574–585.

Wu, S.C., Cheung, K.C., Luo, Y.M., Wong, M.H., 2006. Effects of inoculation of plant growth-promoting rhizobacteria on metal uptake by *Brassica juncea*. Environ. Pollut. 140, 124–135.

Xie, H., Pasternak, J.J., Glick, B.R., 1996. Isolation and characterisation of mutants of the plant growth-promoting rhizobacterium *Pseudomonas putida* GR12-2 that overproduce indoleacetic acid. Curr. Microbiol. 32, 67–71.

Xingmao, M., Burken, J.G., 2003. TCE diffusion to the atmosphere in phytoremediation applications. Environ. Sci. Technol. 37, 2534–2539.

Xu, Q.S., Hu, J.Z., Xie, K.B., et al., 2010. Accumulation and acute toxicity of silver in *Potamogeton crispus* L. J. Hazard. Mater. 173, 186–193.

Yanqun, Z., Yuan, L., Jianjun, C., et al., 2005. Hyperaccumulation of Pb, Zn and Cd in herbaceous grown on lead–zinc mining area in Yunnan. China Environ. Int. 31, 755–762.

Yifru, D.D., Nzengung, V.A., 2008. Organic carbon biostimulates rapid rhizodegradation of perchlorate. Environ. Toxicol. Chem. 27, 2419–2426.

Young, P., 1996. The "new science" of wetland restoration. Environ. Sci. Technol. 30, A292–A296.

Zaidi, S., Musarrat, J., 2004. Characterisation and nickel sorption kinetics of a new metal hyper-accumulator *Bacillus* sp. J. Environ. Sci. Health A. 39, 681–691.

Zaidi, S., Usmani, S., Singh, B.R., Musarrat, J., 2006. Significance of *Bacillus subtilis* strain SJ-101 as a bioinoculant for concurrent plant growth promotion and nickel accumulation in *Brassica juncea*. Chemosphere 64, 991–997.

Zayed, A., Pilon-Smits, E., deSouza, M., et al., 2000. Remediation of selenium-polluted soils and waters by phytovolatilisation. In: Terry, N., Bañuelos, G.S. (Eds.), Phytoremediation of contaminated soil and water. CRC Press, Boca Raton, pp. 61–83.

Zhang, B.Y., Zheng, J.S., Sharp, R.G., 2010. Phytoremediation in engineered wetlands: Mechanisms and applications. In: Yang, Z., Chen, B. (Eds.), International conference on ecological Informatics and ecosystem. Procedia Environmental Sciences, 2, pp. 1315–1325.

Zhang, X.F., Xia, H.P., Li, Z.A., et al., 2011. Identification of a new potential Cd-hyperaccumulator *Solanum photeinocarpum* by soil seed bank–metal concentration gradient method. J. Hazard. Mater. 189, 414–419.

Zhao, F.J., McGrath, S.P., 2009. Biofortification and phytoremediation. Curr. Opin. Plant Biol. 12, 373–380.

Zhao, F.J., Lombi, E., McGrath, S.P., 2003. Assessing the potential for zinc and cadmium phytoremediation with the hyperaccumulator *Thlaspi caerulescens*. Plant Soil 249, 37–43.

Zhao F.J., Wang, JR., Barker, J.H.A., et al., 2003b. The role of phytochelatins in arsenic tolerance in the hyperaccumulator *Pteris vittata*. New Phytol. 159, 403–410.

Zinniel, D.K., Lambrecht, P., Harris, N.B., et al., 2002. Isolation and characterisation of endophytic colonising bacteria from agronomic crops and prairie plants. Appl. Environ. Microbio. l68, 2198–2208.

Chapter 10

Soil Pollution in Turkey and Remediation Methods

Hatice Dağhan* and Münir Öztürk[†,‡,§]

*Eskisehir Osmangazi University, Agricultural Faculty, Department of Soil Science and Plant Nutrition, Eskisehir, Turkey; †Department of Botany, Ege University, Izmir, Turkey; ‡Faculty of Forestry, Universiti Putra Malaysia, Selangor, Malaysia; §ICCBS, Karachi University, Pakistan

INTRODUCTION

Soil covers the earth surface at a depth of anywhere between a few millimetres to a few metres, and is formed by the weathering of various rocks and organic materials allowing the living organisms to live, both inside and on the soil making the soil a dynamic medium. Therefore, soil is one of the most important components of the lifecycle in nature alongside water and air. The mineral composition of soil originates from primary minerals in the rocks, secondary minerals, oxides of Fe, Al, Mn and sometimes carbonates such as $CaCO_3$. The organic matter component of the soil comes from the mesofauna, microorganisms, dead plant debris and their decayed products, and colloidal humus formed by the action of microorganisms. These solid components are usually clustered together in the form of soil aggregates, thus creating a system of interconnected pores (voids) of various sizes filled with either water or air (Pavel and Gavrilescu, 2008). The resulting soil–water–air relations in the environment are very complex. The anthropogenic interferences through an application of chemicals or interferences in the management systems lead to more complex relations in the ongoing reactions among different soil components or phases. Some of these interferences may either cease or accelerate or slow down reactions resulting in future risks to soil systems. The addition of a chemical causes an increase in the soil solution concentration of the chemical in question due to oversaturation of sorption sites. On the other hand, any interference leading to soil acidity or changing redox potential remobilizes both indigenous and human induced pollutants which eventually increase potential risk factor (Pavel and Gavrilescu, 2008).

The formation of a soil layer with a thickness of 1 cm requires at least 250 years, but we can lose this precious life support system and the backbone of our agriculture within minutes to hours by adding a pollutant to it or through

Soil Remediation and Plants. http://dx.doi.org/10.1016/B978-0-12-799937-1.00010-3

erosion if protection measures are not adopted. Soils are continuously contaminated by wastes originating from fertilizers, pesticides, herbicides, plant
and animal wastes, human wastes, population outburst, fossil fuels, mines,
gaseous and liquid wastes as well as nuclear wastes (Tan, 1994; Hooda, 2007;
Ahmadpour et al., 2012; Ismail, 2012). The problem of environmental pollution did not exist or was negligible before the start of industrialization because
of the small population of human beings, as such there was limited addition of
pollutants. However, the industrial revolution and the following demographic
developments after the nineteenth century lead to an increase in the amount
and variety of pollutants in our environment. In particular, contamination of
soil and water bodies by toxic metals produced negative effects on agricultural
practices, agricultural products, life forms and habitats in soil and eventually
human health.

Thousands of hectares of land all over the globe are receiving a variety of
contaminants (Luo et al., 2009). Of these, the heavy metals are irreversibly polluting the soils as one of the most deleterious problem of our times (Singh et al.,
2003). The annual release of heavy metals has reached 22,000 metric tons for
Cd, 939,000 for Cu, 1,350,000 for Zn and 738,000 for Pb during the past few
decades (Ahmadpour et al., 2012). Li and Yang (2008) have reported that at the
end of 2004 the degraded land associated with mining activities, which is one
of the most important causes of the heavy metal pollution, has reached about
3.2 million ha, and the figure is increasing at an alarming rate of 46,000 ha per
year. In some regions of China, soil contamination has reached a serious level.
The crops grown in these regions suffer a yield loss totalling approximately
12 million tons due to contamination from heavy metals, resulting in an economical loss of over 2.5 billion dollars. Moreover, an accummulation in agricultural products causes various diseases and anomalies in humans through the
food chain. The contamination is threatening the health of both natural ecosystems and agricultural ecosystems (Xu, 2007).

The situation in Europe is similar, as there are several million hectares of polluted agricultural land on the continent (Flathman and Lanza, 1998). The number
of contaminated sites in Western Europe is around 400,000 and there are also
many contaminated areas in the central and east European countries (Glass,
1999a). The European Union countries need to spend more than 400 billion dollars to reclaim these areas during the next 20–30 years. Although the spread of soil
pollution across economic sectors differs from country to country, the oil sector
constitutes 14% of this in total and together with other industrial activities these
are responsible for more than 60% of the polluted soils of Europe. The most harmful and common contaminants are heavy metals (37%) and mineral oils (33%).
Estimates from the Union show that currently there are approximately 250,000
sites requiring cleaning up processes and for 80,000 sites data on remediation
is available as these have been cleaned up during the last 30 years (EPA, 2009).

The pollution of soils in Turkey like other parts of the world started from
the beginning of the twentieth century because of the risks taken during the

combination of mechanization in agriculture and rapid industrialization neglected the pollution side (Ceritli, 1997). In this chapter an attempt is made to evaluate the methods used in conservation and remediation, and to present the current situation of contaminated soils in Turkey.

LAND USE OF TURKISH SOILS

Turkey is situated between 26°–45° and 35°51′–42°06′ north latitude in the northern hemisphere. The total surface area is 777.971 km² and 97% of this area is in Asia and the remaining 3% in Europe. The country exhibits a great variety of geological structures, topography, climate, biodiversity and vegetation (ESRT, 2011). We therefore come across several different types of soils in Turkey depending on the combination of soil forming factors. There are seven geographical regions: four of them are interrelated to the coastal zones called the regions of the Black Sea, Marmara, Aegean and the Mediterranean and the remaining three are Central Anatolia, Eastern Anatolia, and South-Eastern Anatolia (Okumus, 2002). In these different geographical regions we find all kinds of soil groups due to varying geological and climatic features; as such Turkey is one of the few countries in the world where we observe this situation. At the end of the first quarter of the twentieth century, I–IV land classes were under cultivation due to low population and lack of agricultural machinery. However, after World War II a rapid increase in population and development together with the availability of agricultural mechanization started heavy deforestation in favour of agriculture and meadows were converted into agricultural lands. The total area in 1934 was only 11,677,000 ha but the arable lands reached 27,699,000 ha (Karaca and Turgay, 2012). The increase was 2.37-fold, 25.85 million ha of agricultural land are suitable for irrigation, but only 8.5 million ha of this land can be classified as economically irrigable (Figure 10.1) (Namli, 2012). The differences in land use have subsequently lead to a degradation of soils through erosion and overgrazing of meadows. At present Turkey is one of the 19 countries in the world which have reached the upper limits of soil sources which can be converted into arable land (Guney, 2004; Karaca and Turgay, 2012). According to Namli (2012) out of the 77.95 million ha, 28.05 million ha (35.98%) of the soil sources can be classified as cultivable soils.

According to land-use classification, the arable lands are the land classes of I–II–III. The area of agricultural soils is decreasing due to poor management strategies related to urbanization, industrialization, transportation and tourism. This can be easily observed around the plains of Cukurova and Amik in the Eastern Mediterranean region, Istanbul, Kocaeli and Bursa in Marmara region, and Izmir in the Aegean region. The area of cultivable soils has reduced as a result of contamination resulting from soil degradation (natural and non-natural) originating from different biotic activities (Guney, 2004). Moreover, an increase in demand for agricultural lands is met through deforestation by intentional fires and logging, which have been commonly practiced in Turkey.

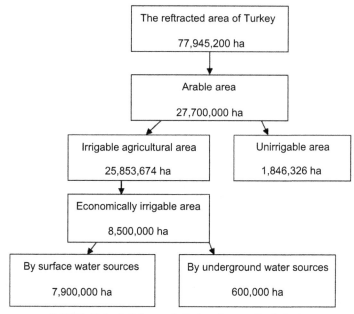

FIGURE 10.1 Soil Sources in Turkey. *Source: Namli (2012).*

The percentage of land suitable for agriculture in the country is around 24%. The lack of environmental measures is threatening 83% of all agricultural products. Conversely, insufficient control along with lack of incentives, penalties and public awareness, education, and farmers participation in decision-making processes are the main handicaps in conserving soil and water resources. Establishing an environmentally friendly and economically feasible policy for the provision of sewerage and appropriate waste treatment for all communities with populations over 20,000 is necessary. The industrial and/or urban wastewaters constitute a very important source of soil pollution in Turkey. The natural ecosystem of the soil is affected by industrial discharges, solid wastes and waste waters. Various hazardous substances resulting from industrial discharges reach the soil (Okumus, 2002).

Data from the Statistical Institute of Turkey (SIT) reveals that the population of the country will reach 100 million by the year 2030. There is need for increased production to meet the housing and food requirements of the increasing population. The degrading agricultural areas need to be reclaimed in order to increase agricultural production and to increase fertility per unit area. However, a detailed study has not been undertaken yet in relation to the importance and dimensions of soil contamination in the country. No doubt approaches regarding protection of the environment in Turkey started in 1980. Especially in the last decades, along with the other issues, within the framework of European Union harmonization efforts, the current situation has been reviewed and environmental regulations have been brought into line with European Union legislation. Therefore, public awareness of environmental issues has increased (ESRT, 2011).

Despite this, studies related to soil contamination in Turkey are scant compared to the countries with more awareness and sensitivity to this aspect. Currently, determination of contaminated sites and severity of pollution and subsequent remediation requires more time and huge expenses (Turkoglu, 2006). The most important effect of soil contamination in terms of environmental health is the mobility of contaminants from the soil to humans via direct consumption of plants growing on polluted sites, or consumption of animals fed with contaminated plants. The transportation of the contaminants through leaching and percolation processes into underground waters and through floods and erosion in to surface waters is the other problem.

SOURCES OF SOIL POLLUTION IN TURKEY

Thousands of hectares of fertile agricultural soil are rapidly getting degraded irreversibly in Turkey by industrialization, urbanization due to demographic explosion and bad land use policies. A majority of Turkish industry (about 60%) is found in the Marmara region resulting in an imbalance in the distribution of the country's industry. In addition to Istanbul and Izmit, another industrial area is fast developing on the Bursa Plain in the same region. Other intensive industrialization has taken place around Izmir in the Aegean region, followed by west Black Sea, Cukurova in the Mediterranean region, and Eskişehir, Kayseri, Sivas, Konya, Kırıkkale, Eregli and Ankara in the Central Anatolia region. In the East and South-East Anatolian regions active local industrial centres are developing at the places where the population is accumulating. The provinces where industry developed in South-East Anatolia are Gaziantep and Diyarbakir. Cement, food, textile, brick and agricultural tools industry are developing at great speed in these provinces. There are a considerable number of cotton, sugar, cement, food and tobacco industries surrounding the big cities in East Anatolia (Ertin, 2012).

The evaluation report on the environmental problems and privileges for the years 2005–2006 published by the Ministry of Forest and Environment reveals that in 29 cities storage of waste is in the open, whereas 22 cities have excessive industrial estates, 17 cities agricultural waste and 12 cities urbanization-induced soil pollution (RTMEF, 2008).

The survey presents preventive measures against soil contamination. These are: liquid, solid and gaseous wastes of industrial establishments are to be incinerated in 71 provinces in accordance with the legislation of the Ministry; storage capacity units should be increased in 51 provinces; a well-planned urbanization and infrastructure should be started in 43 provinces; controlled use of fertilizers, pesticides and irrigation water has been outlined under the legislation of the Ministry of Agriculture and Rural Affairs. Many other complimentary measurements are outlined for six provinces (RTMEF, 2008). In the subsequent two yearly report the problems are outlined as: primitive storage in 31 provinces, facilities for industrial waste in 15 provinces, upgrading of agricultural practices in 23 provinces, and urbanization in 11 provinces out of 80 in the country has been followed. No data is available for Sirnak province in the report (RTMEF, 2010). The report stresses that in order to prevent soil pollution, the following precautions are to be taken: (1) at least 48 incinerators in different provinces are to be built for gas, liquid and solid industrial waste; (2) the capacity of storage areas should be increased in 20 provinces; (3) the urbanization in two provinces should be realized under a theme of recreational projects; (4) agricultural input management such as fertilizer application, use of pesticides and irrigation is to be followed according to the legislation and regulations suggested by the Ministry of Agriculture and Rural Affairs; (5) other different precautions are to be taken in seven provinces.

Non-agricultural use of agricultural land as evaluated in 51 provinces clearly depicts that there is soil loss in an irreversible way. Non-agricultural use is directly related to urbanization and infrastructure build-up in 31 provinces, absence of environmental projects in nine provinces, insufficient precautions in relation to non-agricultural uses along with lack of an environmental project and/or urbanization projects in five provinces, lack of authority and control in four provinces, and a few other problems in two provinces. No data is available for the rest of the 30 provinces (RTMEF, 2010). According to the 2012 report of the Ministry of Environment and Urban Planning, there are no soil pollution problems in Turkey except for in the Mediterranean Region; set against this, water and air pollution are the most prominent environmental problems in Turkey (RTMECP, 2012). The problems causing soil contamination in general in Turkey can be classified as follows: erosion, parent material, agricultural activities, industrial wastes, urbanization and mining.

Erosion

One of the major soil degradation problems in Turkey is erosion, because 81% of the total land surface in the country is affected to differing degrees of severity by erosion. Nearly 1 billion tons of fertile soil is washed away per year into the seas. It is the main problem on cultivated steep lands where necessary measurements are not taken (Ozturk et al., 1996b, 2010b, c, 2011b; Okumus, 2002).

Out of approximately 79% of the land surface of Turkey, as little as 24% of the soils are suitable for agriculture due to shallow soil depth and occurrence of different degrees of erosion. In accordance with the very high erosion risk, approximately 500 million tons of fertile surface soils are transported by erosion annually. Nearly 83% of the cultivated lands of Turkey are subjected to erosion-induced environmental problems due to the lack of required precautions measures (Findik, 2007).

The erosion in the country is 12 and 17 times higher than Europe and Africa, respectively. The rate of erosion risk classes in the Turkish soils are 14, 20 and 63% for slight, medium and severe cases, respectively. Moreover, approximately 57.15 million ha of land in Turkey is subjected to water erosion; wind erosion is not very common and it is effective at different levels on an area of 506,309 ha. Erosion is the main problem for 16.4 million ha out of the total 27.7 million ha of agricultural lands (Erpul and Deviren, 2012).

As a result, soil loss by erosion in Turkey is a very serious problem. In fact it is a natural process, but due to human-induced activities the erosion has currently reached dangerous levels. The expectations are that in the near future, if precautionary measures are not adopted, excessive land degradation will accelerate erosion rates by 48-fold (Erpul and Deviren, 2012), suggesting erosion is a problem which needs high priority to be solved as the rate of soil loss and population increase.

Parent Material

Parent material-induced soil contamination is common in soils formed on rocks of serpentine and other heavy metals containing (asbestos, B, Cr, Cd, etc.) parent materials. High concentrations of Fe and Mg characterizing the ultramafic rocks occurring in many parts of the world are a notable feature of the geology of some countries and regions such as California, New Caledonia, Cuba, central Brazil and parts of Mediterranean Europe. Turkey is one of the countries which embodies these features. According to Reeves and Adiguzel (2004), the significant exposures of ultramafic rocks and soil are found in many parts of Turkey (Figure 10.2), but are not important in the geology of eastern and southeastern provinces. The central part of the north-west (Kutahya and Balikesir provinces), the south-west between Antalya and Marmaris (Antalya and Mugla provinces), the Amanus Mountains (Hatay and Adana provinces), regions of the eastern Taurus (north and north-east of Mersin) and its extensions into the Aladag massif (Nigde and Adana provinces), and numerous areas run in a band generally north-eastwards for several hundred kilometres from near Adana to Erzincan and all these are recorded as the notable areas. Several smaller areas near Ankara and in Canakkale provinces are also included as other significant outcrops (Reeves and Adiguzel, 2004).

High levels of other elements such as Ni, Co and Cr are present in abnormal levels in soils derived from ultramafic rocks; the hydrated Mg silicate mineral known as serpentine is often included in the mineral assemblage, and subsequently these soils are referred to as serpentine soils (Prasad, 2005; Reeves and Adiguzel, 2008). A flora recognizably different from that of adjacent areas of different geology is often supported by these soils owing to the unusual chemistry being generally moderately to extremely deficient in essential plant nutrients such as Ca, K and P. Plant behaviour in response to

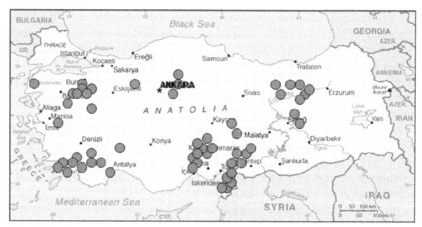

FIGURE 10.2 Map of Turkey showing areas of ultramafic geology (circles in red). *Source: Reeves and Adigüzel (2012).*

the elevated levels of nickel (Ni) on serpentine soils is one of the interesting aspects (Ahmed and Ashraf, 2011). About 700–5000 mg kg^{-1} Ni is available in normal soils; Ni levels of about 0.2–8 mg kg^{-1} in dry leaf tissue is generally present in plants grown on these soils (Reeves and Adiguzel, 2004).

Nearly 70% of the boron (B) reserve of the world is found in Turkey (Ozturk et al., 2010a). The borate deposits of Turkey were deposited in the lacustrine environment during the Miocene, following the volcanic related activities, boron got deposited in the playa-lake sediments (Helvaci, 2004). The boron reserves of Turkey are common in the town of Kirka in Eskisehir, the counties of Emet and Hisarcik in Kutahya, the village of Kestelek in Bursa and the county of Bigadic in Balikesir. The reserves are spread over a wide area and affect plants, animals and people negatively in various ways. It has been reported that the high boron content of the soils in the region of central Anatolia affects the crop yield seriously (Torun et al., 2006).

According to studies carried out in this region by Gezgin et al. (2002) the available boron content of the soils of central Anatolia varies between 0.01 and 63.9 mg kg^{-1} and it has been observed that 26.6% of the soils showed insignificant levels (<0.5 mg kg^{-1}) but 18% of the soils had toxic levels (>3.0 mg kg^{-1}).

Additionally, the important contaminants in terms of human health such as asbestos (As), chrome (Cr), etc. which are naturally present in the soil forming rocks and minerals in some regions (especially central Anatolia) of the country also constitute a risk factor.

Agricultural Activities

Fertilizer Impurities

Cadmium (Cd), fluorine (F), lead (Pb) and mercury (Hg) can be found in the soil amendments including lime and gypsum besides the impurities in fertilizers. The most important of these is cadmium, as it shows high mobility from soil to the edible portions of plants. Due to heavy applications of superphosphate over decades elevated levels of Cd are currently present in large areas of land.

Fertilizers may negatively affect the quality of soil, water, air and plants in cases of improper usage. They can affect soils through soil structure, soil pH, reduction in soil microorganisms and accumulation of toxic elements. The most intensive fertilizer consumption regions in Turkey are the Mediterranean, Marmara and Aegean. While average chemical fertilizer production during 1985–2008 was 3.342 million tons, on the basis of efficient plant nutrient matter the average was around 1.255 million tons.

Excessive application of fertilizers and agrochemicals in the cultivated areas results in direct soil contamination. If the total impact of both locally produced as well as imported fertilizers is considered, they lead to an accumulation of fertilizer bound contaminants due to intensive and ignorant usage of fertilizers (Karaca and Turgay, 2012). In particular, fertilizers containing nitrogen and phosphorus are consumed to a great extent in the country. However, soils are getting contaminated more by phosphorous fertilizers (Gezgin, 2005; Koleli and Kantar, 2006; Karaca and Turgay, 2012). Pure phosphorus of around 200 to $500 \, kg \, ha^{-1}$ has especially been used in central Anatolia. It has been reported that there is a contamination of Cd of around 60% and As, Pb and Cr are also present in some parts of soils of this region. For potato, wheat and other plants grown on these soils, uptake of heavy metals will pose serious health problems for people feeding on plants cultivated on such soils (Gezgin, 2005).

Cadmium is present in the composition of phosphorous fertilizers and accumulates in the soil depending upon the intensive application of the fertilizer. There is an estimated annual increase in Cd concentration of around $1 \, mg \, kg^{-1}$ in soils with $pH < 6$ and $3 \, mg \, kg^{-1}$ in soils with $pH > 6$.

In the production of phosphorous fertilizers raw phosphate rocks of either sedimentary or volcanic origin are used. During the production of these fertilizers 70–80% of Cd present in the raw phosphate rock passes into the fertilizer. In order to decrease the entry of Cd into the soil through phosphorous fertilizers, volcanic-originated raw phosphate rocks with smaller Cd content need to be used in the production of the fertilizer or the Cd content of phosphoric acid used in the production of the fertilizer and sediment-originated raw phosphate materials should be decreased. In order to decrease this accumulation of Cd originating from phosphorous fertilizers in the soil, legislation should be put in place to set the upper limits of Cd in phosphorous fertilizers and/or raw materials, and fertilizers involving low cadmium must be preferred (Namli, 2012).

Pesticides

Three million tons of pesticides are used per year in the world whereas their usage in Turkey lies at around 30,000 tons. However, increase in their usage brings about many environmental problems (Ozturk and Turkan, 1989). These chemicals increase the yield, on the one hand, by killing the pests harming

TABLE 10.1 The Input Amounts of Heavy Metal of Fertilizer into Cultivated Areas (mg kg^{-1})

Elements	Phosphorous Fertilizers	Nitrogenous Fertilizers
Cd	0.1–170	0.05–8.5
Co	1–12	5.4–12
Cr	66–245	3.2–19
Cu	1–300	1–15
Hg	0.01–1.2	0.3–2.9
Mn	40–2000	—
Mo	0.1–60	1–7
Ni	7–38	7–34
Pb	7–225	2–27
Sn	3–19	1.4–16
Te	20–23	—
U	30–300	—
V	2.1–1600	—
Zn	40–1450	1–42

Adapted from Namli (2012).

agricultural products; on the other hand, they pollute the soil and water resources. Soils are getting contaminated with organochlorines such as dichlorodiphenyl-trichloroethane and other pesticides such as arsenic-based compounds. Toxic compounds of persistent pesticides in soils pass into plants in small amounts and subsequently to herbivores and human beings and cause serious health problems upon accumulation (Ozturk,1989; Ozturk and Turkan, 1989; Cepel, 1997; Durmusoglu et al., 2010). The use of pesticides is highly intensive in the regions of the Mediterranean, Aegean and Marmara due to intensive agricultural practices. These region are susceptible to possible pesticide pollution due to two or three crops which involve intensive usage of both fertilizers and pesticides. Thus, non-persistent and non-toxic by-products containing pesticides in the natural ecosystems are to be used (Namli, 2012).

Contaminated Waters and Agricultural Areas

Rapid population growth, development of industry, an increased use of fertilizers and agrochemicals and insufficient awareness of environmental problems are leading towards excessive pollution in some of the aquatic ecosystems.

The degree of contamination reported for the surface waters of some basins is said to have reached level 4. Excessive contaminations have been determined in tstreams, rivers and lakes located in the basins of Meric-Ergene, Marmara, Sakarya, Gediz, Küçük Menderes, Büyük Menderes, Burdur and Akarcay (Afyon) (Guney, 2004; Akin and Akin, 2007; Sakcali et al., 2009). In some water basins, even excessive heavy metal contaminations are met (Akin and Akin, 2007). In particular, some industrial estates are established in these areas and the waste and sewerage without treatment are dumped into water resources such as rivers, lakes and coastal parts. The upper basin of Dicle River is polluted by the by-products of the Etibank Copper Factory (Guney, 2004).

As a result of the use of contaminated water resources in agriculture, the contaminants (microbiological, heavy metals, etc.) are incorporated into the soil (Karaca and Turgay, 2012).

A classification based on the limit (critical) values of the water resources in the river basins encompassing the important agricultural and industrial centres in Turkey on the basis of The Quality Criteria For Continental Water resource Classesî and Water Pollution Control Regulatîons reveals that the quality varies between class II (slightly polluted waters) and class IV (severely polluted waters). According to the data from 2006, approximately 36% of the municipal waste water, 64% of the manufacturing industry water and 25% of the waste water of local industrial regions are not treated before discharge (UCTEA, 2009).

The environmental unawareness and uncontrolled addition of the waste water from industrial estates into Porsuk, Simav, Nilüfer and Ankara streams and Iznik, Eber, Karamuk, Büyükçekmece and Burdur lakes are severely polluting the surface waters (Ozturk et al., 1996a). This dumping of wastes and waste water causes serious problems in the water qualities of rivers such as Büyük Menderes, Kızılırmak, Gediz and the lakes Tuz Gölü, Sapanca, and Mogan (Guney, 2004; Akin and Akin, 2007; Namli, 2012).

As a result of the activities of Soma Lignite and olive oil production foundations located in the basin of the North Aegean, there is some contamination in the Bakir streamlet. Gediz River in the basin of Gediz has highly contaminated surface water. This river has class IV water quality due to the addition of domestic disposals, industrial waste and agricultural waste which enrich the concentration of nitrogen, organic matter and heavy metals in the water. Similarly Buyuk Menderes and Kucuk Menderes rivers have class III and IV water quality.

The basin of Burdur Lake located in the Mediterranean region is facing serious contamination problems as a result of industry, municipal discharges and agriculture based waste. In the lower parts of the Seyhan, Ceyhan and Orontes rivers in the Mediterranean region, contamination levels have reached class III and IV water quality due to the same reasons (Guney, 2004; Akin and Akin, 2007).

Eber Lake placed in the basin of Akarcay of the central Anatolia is severely contaminated (Ozturk et al., 1996). As the concentration of NO_2, O_2, Pb and Cr is considered, the streams of Ankara, Karasu, Göksu, Mudurnu, Seydisu and Kızılırmak, the branches of Sakarya river in the basin of Sakarya located partly in the central Anatolian region have a water quality class of III and IV. The heavy metal contamination in this basin is reported to be at a serious level (Akin and Akin, 2007). Class III and IV levels of contamination are observed in the areas where industries are located alongside the Kizilirmak river in the province of Kirikkale. In the middle and upper parts of the Kizilirmak basin the contamination is not yet at a harmful level. The quality of the Yesilirmak river located in the middle of the Black Sea region sometimes reaches class IV due to the waste water from the food industry and domestic disposals. Conversely, the heavy metal contamination in the districts of Tokat and Amasya is alarming. In the rest of the basins, there is no serious level of contamination depending on relatively lower levels of industrialization, population density and agricultural activity. However, excessive contamination depending upon the presence of industrial foundations such as copper mining in Artvin, Murgul and Ergani are seen (Akin and Akin, 2007).

In the 1991 environmental report prepared by the countries of the Economical Cooperation and Development Organisation it was reported that the Porsuk streamlet is among the most contaminated streams in the world. According to the report, the determined Pb, Cu and Cr concentrations in the Porsuk River were measured as three- to four-fold higher than the ones in the other countries (Yucel et al.,1995; Guney, 2004).

The nitrogen, sugar and magnesite factories in Kutahya; Seyitomer thermal power station; Sumerbank textile industry, sugar factories and locomotive industry in Eskisehir dump their industrial waste directly into the Sakarya River (Guney, 2004; Namli, 2012).

Lakes such as Sapanca, Manya, İznik, Tuz, etc., are getting rapidly contaminated and their qualities are degrading (Guney, 2004; Namli, 2012). The urbanization, industrialization and agricultural activities surrounding Lake Sapanca have changed the quality of the lake waters. The waste water of 34 urbanization areas and 40 industrial estates surrounding Manyas Lake along with fertilizers leaching from agricultural areas are affecting the nitrogen–phosphorus balance and are causing eutrophication in the lake. This has negative effects on the aquatic life and the breeding birds and/or those staying there temporarily (Namli, 2012).

The waste from the caustic soda and chrome production in Kazanli, Mersin is dumped into the Mediterranean Sea without any treatment. Nearly 5 tons of mercury (Hg) are dumped yearly together with 15 tons of tatmium (Ta), 1729 tons of lead (Pb), 1163 tons of zinc (Zn) and 105 tons of Copper (Cu) leading to a severe contamination in the East Mediterranean Sea (Guney, 2004).

Stubble Burning

The plant stalks remaining in the field after the harvest of cereals known as stubble are generally cut high from the ground by the thresher during the crop harvest so as to widen agricultural areas, resulting in an increase of plant remains.

Cereals like barley, rye and wheat are sown on very large areas especially in the regions of Central Anatolia, Thrace and South-East Anatolia. After the harvest of those crops very large quantities of stubble remain on the land. These are generally burnt by farmers to avoid the difficulties faced during recultivation, combat against diseases and pests. Moreover these remnants are regarded as having no economical value for them.

As a result, the stubbles are burnt and the soils become susceptible to erosion with a subsequent decrease in fertility. The increased soil temperature kills the microorganisms, beneficial insects, bacteria and fungi in these soils. They become poorer in terms of organic matter. Subsequently the soils turn into brick-like hard and dry structures, with low porosity, increase in water repellence and difficult tillage (Guney, 2004).

In order to prevent the harmful effects of such primitive practices, stubble burning has been forbidden by Environment Law number 2872 on 9/8/1983 and an earlier Forest Law number 6831.

Industrial Activities

Organized Industrial Zones

Industrial activities are one of the sources of pollution problems in Turkey. The untreated industrial wastes in the form of liquids, gases and solids cause excessive soil, water and air pollution. In order to reduce and control the effects of such pollution impacts from industry, many Organized Industrial Zones (OIZs) have been established in Turkey. Unfortunately, most of the industrial establishments in Turkey are operating close to the or on agricultural lands, in the vicinity of cities, near harbours and coasts, around rivers and lake boundaries (ESRT, 2011). The organized industrial zone edifices have invaded the agricultural land around Cigli-Izmir, the plains in the surroundings of Samsun and Manisa, followed by heavy urbanization on areas covering 570, 300 and 416 ha respectively (Guney, 2004). The reports from the Supreme Council of OIZs reveal that there are 264 established OIZs in the country and out of these 149 are activated on provincial basis (ESRT, 2011).

Industrial Waste

The cumulation of industrial areas in Turkey is as follows: 1st region: Istanbul, Kocaeli, Bursa, Sakarya; 2nd region: Izmir, Balikesir, Aydın, Manisa, Mugla, Antalya; 3rd region: Ankara, Eskisehir, Konya; and 4th region: Adana, Mersin, Hatay, Gaziantep. However, industrial activities rapidly pollute the soil in these regions (Karaca and Turgay, 2012). The most common pollutants coming from these sources are heavy metals, which directly affect the soils. They are taken up by plants and therefore enter the food chain leading to toxic effects on living organisms. At the same time, when heavy metal concentrations increase in the soils, they are washed off by rainfall and transported through infiltration in the

soil profile to the groundwaters. Eventually, these metals decrease the water quality through contamination. The polluted soils also affect the activities and functions of soil organisms and in turn produce adverse affects on ecological cycles.

Industrial waste waters also contain organic and inorganic toxic pollutants
(Seker et al., 2006) . This results in the pollution of agricultural lands via irrigation waters. Approximately 81,312 ha of agricultural lands in Balıkesir, Kepsut,
Susurluk and Karacabey plain have been polluted with contaminated irrigation
water (Ozturk et al. 2010a). The pollution due to industrial waste is also very
high especially in the regions where industry and agriculture are close to each
other (Karaca and Turgay, 2012; Namli, 2012). For instance, because of industrial waste, 40 km^2 were polluted in Murgul and no plants are grown in the
agricultural soils of the wasted area.

The copper factory of Goktas-Artvin is known as one of the oldest indus-
try-based polluters in Turkey. The SO$_2$ emissions have strongly damaged the
vegetation and soils at many places during the last three decades, and the soil
erodibility in these areas has increased up to 6–10 km (Haktanir et al., 2006).

Delibacak et al. (2003) have investigated the pollution level of agricultural
soils by investigating the heavy metal contents of soils in the agroecosystems of
Gediz basin. The Gediz River has been polluted by industrial wastes, and over-
irrigation of agricultural lands by this river water has contaminating the soils
around the river (Ozturk et al., 2010 b, c). Therefore the heavy metal concentrations of soils around the Gediz river have been analysed, by taking 12 soil samples from the beginning to the end of the river. The results show that the levels of
B and S are high in 33% of the soil samples, the total concentrations of Fe, Zn,
Mn, Cd, Ni, Cr and Co are also quite high in the area (Delibacak et al., 2003).

Studies show that the cement and fertilizer factories in different regions of
Turkey and the copper factory in the Samsun province of the Black Sea region
have caused serious pollution in the surrounding agricultural areas (Sheikh
et al.,1976; Haktanir et al., 2006; Uysal et al., 2012) . Soil samples were taken
from different locations around the factories to determine pollution levels. The
results indicated that Fe and Cu metals were accumulated at excessive concentrations and Fluor element accumulated at significant levels in the surface soil
within a distance of 3–6 km from the factories (Haktanir et al., 2006).

Yilmaz et al. (2003) measured the metal concentrations of surface soils in
the urban and rural areas in the intensive industry zone of Izmit Gulf. Cadmium
levels in the soils of three different regions are reported to be very low. However, the concentrations of Co, Cu, Mn, Pb and Zn are high in both the urban
and rural soils. In general, the metal contents in the soils of the gulf region are
higher than the ones in the rural areas. The soils of the Izmit gulf region have
suffered much deterioration because of intensive industrial activities and traffic.

Thermal Power Plants

Chimney gases (SO$_x$, NO$_x$, hydrocarbons, CO, CO$_2$), dust (as particles), and
ash (the remnant) are emitted to the atmosphere directly from thermal power

stations because these establishments use fossil, liquid and gaseous fuels to produce electrical energy (Guney, 2004; ESRT, 2011; Karaca and Turgay, 2012). There are 13 coal-running thermal power plants in Turkey, which use 67,800,457 tons of coal. The amounts of volatile and ground ash formed are approximately 19,664,120 tons (ESRT, 2011). In the city of Yatagan, in Mugla province, the ash and SO_2 polluting the air together with the chimney ash emitted from the thermal power station cause the formation of acid rain. The forests around the power plant have been wiped out and the agricultural crops have suffered seriously (Guney, 2004). The negative effects of the Yatagan thermal power plant have spread out to an area of 37 km diameter. In Afşin-Elbistan, Soma and Kangal thermal power plants we observe severe pollution of the soils around these establishments and natural vegetation (in particular the forests) has been destroyed. The pollution has proved hazardous for vegetable and fruit gardens as well as drinking water in Soma (Haktanir et al., 2006). Four units of Soma thermal power plants have even proved detrimental to plant species 78 km away. In general the air pollution due to industrial activities, exhaust gas emissions or burning-based contaminant gases affects the ecological nature of soils. A variety of pollutants emitted to the atmosphere reach our soils via rainfall. The acidification of the soils due to sulphur gases negatively influences soil biological activity and structure (ESRT, 2011).

Petroleum Pollution

The most important fossil fuel contamination in the soil is caused by petroleum due to leaks, explosions and other accidents in the stages of refining, storing and transportation after drilling. The petroleum primarily fills the pores in soil, preventing air entry–exit which is necessary for soil organisms, it also prevents the entry of the water into the soil as well as movement of water present in soil with the subsequent dysfunction of the transport of the nutrients necessary for the maintenance of the life in the soil (Cepel, 1997; Karaca and Turgay, 2012). Crude oil contains high concentrations of many hazardous compounds such as polycyclic aromatic hydrocarbons, benzene and its derivatives and cyclo-alkane chains. Therefore if petroleum and its side products are incorporated or contaminate natural resources, they end up causing critical environmental and health problems (Karaca and Turgay, 2012).

In addition to this, soil and water resources are contaminated by petroleum wastes like other wastes generally in the regions where petroleum refineries are present. The dumped waste and the toxic items from the petroleum producing establishments in Batman, Izmit, Aliaga (Izmir), Mersin and Kirikkale reach the waters and soils. Studies undertaken by Guney (2004) have revealed that in 12 different places petroleum establishments are contaminating drinking waters. The petroleum refinery in Mersin and two petroleum pipe lines in the Iskenderun Gulf carry a great risk in terms of petroleum-based contamination.

Urbanization

Population

According to an SIT (2013) report, Turkey's population will reach 84.3 million by 2023. This high population growth will lead to higher migration rates to the large cities (Istanbul, Izmit, Izmir, Adana, Mersin, etc.). This accelerated internal migration from rural areas to urban centres has and will cause major environmental changes and problems. In particular, unplanned urbanization can lead to the transfer of rich agricultural land into urban conglomerates and produce severe environmental impacts, including soil erosion and pollution of surface waters (Ozturk, 1999; Okumus, 2002; Nurlu et al., 2008; Ozturk et al., 2011b).

Wastes

Rapid industrialization and urbanization also cause solid and liquid waste problems. Widespread use of packaged products and disposable materials with the change of consumption habits has been the starting point for waste problems which have reached the proportion of crisis currently. A study conducted in New York City has revealed that 100 thousand tons of waste goes to the Fresh Hills waste collection center per week. This amount is equal to 10 times the mass of the Egyptian pyramids. The contents of the constituent materials of waste vary from boxes to plastic bottles, from hospital waste to radioactive waste (Findik, 2007).

Data taken from industry associations and municipal authorities in Turkey in 2008 shows that 500,000 tons/year of sewage sludge is produced from domestic/municipal wastewater treatment institutions and 575,000 tons/year of sludge from industrial manufacturers, bringing the total to 1,075,000 tons of sludge (dry solids) per year (Ozturk et al., 2006a; ESRT, 2011).

In the country, the wastes are controlled by regulations on Control of Solid Wastes, Control of Medical Wastes and Hazardous Waste Control Management concerning clinical waste, but the existing data is incomplete. There are few hazardous waste disposal facilities in Turkey and obviously this amount is inadequate to control the current level of waste produced. Local capacity, including financial resources, equipment and staff, is too low to cope with the problems of rapid urbanization. The level of public awareness and public participation in activities surrounding solid waste management strategies is also limited (Okumus, 2002).

Sewage sludge is an intermediate product obtained from waste water after a series of applications. The least environmentally damaging mode of processing and removal of solid materials is defined as 'sewage sludge' in which heavily polluting elements are transported and collected along with the water. Their treatment is as important as that of wastewater. The last situation on the use and application of waste sludge is its application to agricultural lands in the vicinity of urban areas. This causes soil pollution (Namli, 2012). However, if proper treatments are performed, the sludge and wastewater can be used for this purpose keeping in view future water crises due to water shortages (Ozturk et al., 1991, 1994a, 2005, 2011a).

Highways

The highways generally pass through mixed urban residential / industrial areas for ease of access. Since industry and urbanization areas are established on primary agricultural lands the highway routes close to these areas result in the loss of large areas of land. Establishment of the E-5 highway has caused a loss of 25,000 ha of primary agricultural land in Thrace. Similarly, 3000 ha of soil has been lost in Duzce (Guney, 2004). Similarly, highways such as those between Izmir-Aydin, Adana-Mersin, Adana-Iskenderun, Ankara-Eskisehir and Istanbul-Erbaa-Susehri have consumed substantially large agricultural lands (Namli, 2012).

Mining

Turkey ranks 10th in the amount of mines and minerals among 152 countries with 29 mine products. According to the General Directorate of Mining Affairs, 33,780 mines (gold, silver, platinum, copper, lead, zinc, iron, zeolite, pumice, perlite, sulphur, coal, etc.) were licensed in 2011 and 13,128 of these are operational (ESRT, 2011). All of these can be accepted as sources for the production of solid mineral waste due to mineral mining and processing activities. Coal, especially contaminates soil and pollutes our waters. Boron is produced in the region between the centres of Balıkesir province and Kutahya-Emet. Waste water used during the production of boron pollutes surrounding water resources (Simav stream, Ulubat Lake, etc.). The soils of Balikesir plain, Kepsut plain, Susurluk-Karacabey, Nigde-Aksaray, Burdur, Eskisehir, Igdır, Yuksekova and Hakkari are irrigated by boron-contaminated water and those soils get contaminated by boron (Cepel, 1997; Ozturk et al., 2010a).

The fertile plain soils in Turkey, where kaolinite and illite types of clays are prevalent, are used as raw material by the brick industry with subsequent loss of good soils (Ozturk et. al., 1995 ,1996b; Gokmen et al.,1996). The brick industry uses the most fertile soils by taking 40–50 cm of the land surface as raw material. The texture of the soil left behind decreases with a subsequent loss of the physico–chemical features of soil appropriate for agriculture. In Thrace 1400 da, in Gediz Plain 4400 da, in Corum 2000 da of soil have been lost due to brick manufacturing (Karaca and Turgay, 2012). The 23 brick factories established around the plain of Erbaa remove 300 da of the fertile area per year (Guney, 2004). Generally, the contaminated sites in Turkey are mostly identified after some potential problems become obvious and public, as a result of the efforts of local authorities or concerned citizens, steps are taken to find a solution. The number of contaminated sites is estimated to be in the range of 1000–1500, of which 5–10% require remediation (Gungor, 2008). There are scant studies and not enough data are available on the effects of the environment on human health related to air and soil contamination in Turkey; those published are only from universities besides a few from research institutes (Dogan and Ozturk,1991; Dogan and Ozturk, 1994a, b). It is well known that it takes 100 years for the formation of 1 cm depth of soil and nearly 3000–12,000 years to change the soil into a fertile form; but, we are losing the same soils in a very short time. Sometimes this loss is irreversible. It is extremely important that limited soil sources be conserved and used in a very careful way (Sakcali and Ozturk, 2003).

REMEDIATION METHODS FOR POLLUTED SOILS

Physical, chemical and biological methods are used to clean contaminated soils; however, the magnitude of the contaminated land, topography, soil structure and physico-chemical properties, and type and amount of contaminator should be considered while choosing methods for remediation. Remediation of heavy-metal-contaminated soils is one of the most difficult tasks for environmental engineers. The properties of contaminated soils and behaviour of contaminants together with the relationship between soil properties and contaminants are highly complex. Therefore, the cost of cleaning contaminated soils is extremely high and the applications of waste disposal technologies are limited. Thus, economically feasible soil cleaning and easily applicable soil cleaning technologies are needed (Ozturk et al.,1994b, 2008, 2012a; Kocaer and Başkaya, 2003; Ashraf et al., 2010a, b; Hakeem et al., 2011).

Various engineering-based methods such as soil washing, soil flushing, incineration, solidification, stabilization, soil excavation and land fill, in situ vitrification, electrokinetic systems, soil burning or pump and treat systems are used to remediate metal-contaminated soils (Hooda, 2007; Ismail, 2012).

Physical and chemical cleaning methods used in soil remediation have some advantages such as ease of application and short application period; but they are not preferred due to the higher application cost and difficulties of disposing of by-products which occur during the soil remediation processes.

Physical and chemical remediation methods require high energy input, damage soil structure and decrease soil productivity. Presently, excavation and burial of soil at a hazardous waste site is being adopted as a remedial process for rehabilitation of toxic-metal-contaminated sites at an average cost of $1,000,000 per acre (Ismail, 2012).

One of the alternative methods of chemical remediation is the use of plants for in situ cleaning of organic and heavy-metal-contaminated soils called phytoremediation which is a new low cost remediation approach that does not require any special equipment. Currently, phytoremediation is highly preferred compared with other remediation methods (Ozturk et al., 2005, 2012a; Percin, 2006). Glass (1999b) has reported that the cost of land filling, use of kinetic method and pyhtoextraction come to $20–200 ton^{-1}, $100–500 ton^{-1} and $5–40/ton^{-1}, respectively. According to Kidney (1997) the cost of organic matter separation from ground water costs $2–3 million and removal of heavy metals from the contaminated soil $1–2 million. He has estimated that $20–45 million are to be spent on removing organic matter wastes from the ground water, $40–80 million on removing heavy metals from the contaminated soils and \$40–80 million on the cleaning of radionuclides (EPA, 2000). The phytoremediation market is growing rapidly in the world and comprises approximately \$100–150 million per year. There is not significant use of phytoremediation currently either in developing or developed countries, but this is the cheapest method and may become a technology and patent of choice for remediation projects due to its cost efficiency, ease of implementation except for longer time use (Hooda, 2007).

Compared with other methods, phytoremediation can be applied in large areas and esthetically it improves appearance of the landscape. In addition to this, use of plants in phytoremediation reduces the chances of soil and wind erosion; the movement of pollutants from the contaminated areas to lakes and rivers is also decreased. After application of phytoremediation technique, the heavy metals absorbed by the experimental plants can be converted into controllable form or some like Ni and Ca can be regained by drying, firing, gasification, pyrolysis and anaerobic decaying.

REMEDIATION STUDIES IN TURKEY

Very limited studies have been conducted and the published data available on remediation of polluted soils in Turkey too is scant (Ozturk et al, 2012a). Except for accidents and urgent prevention for pollutants, not many remediation approaches for polluted soils are observed, except for the studies conducted by the universities and research institutes. However, a few remediation case studies are reported which have been carried out by private companies (Izmit Waste Treatment, Incineration and Recyling Co. Inc. (IZAYDAS) and INTERGO)) in Turkey.

The publically known soil pollution incidents are mostly due to illegal disposal of industrial waste, oil leakage resulting from accidental spills around oil storage tanks or pipelines, metal leaching from disposed metal ore processing residues and waste disposal sites. Remedial measures have been carried out for very few of these contaminated sites. Therefore, some information is officially available on the contaminated sites and remediation techniques used. However, no statistical data exists about formerly used remedial technologies and methods.

The important soil pollution problems as reported by Yavuz (2005) and Gungor (2008) in Turkey are outlined here: nearly 640 barrels containing phenols buried illegally around Tuzla Orhanli (Istanbul) were recovered in March 2006. The chemicals leaching from these barrels have largely polluted the soil and this contaminated site was cleaned by IZAYDAS. The IZAYDAS Waste Treatment Plant produces 640 barrels and 2000 sacks of contaminated soil a month. IZAYDAS facilities are the only licensed hazardous waste disposal facility in Turkey. Remediation of contaminated sites is mainly based on the incineration of both toxic barrels and highly contaminated soil. Similarly, five barrels containing asbestos were found in Gebze.

The Environmental Protection Department of the Kocaeli Municipality and IZAYDAS took the necessary precautions and inspection of the barrels. There were also 20,000–25,000 barrels of crude oil leakages reported in the bay of Yiginak village, around the shore of Baglica village and near Sanliurfa. About 500 tons of contaminated soil were removed and transported to a site of 20 acres. The less-contaminated soil was cleaned up technically with a bioremediation method at the arranged site. The highly polluted soil, on the other hand, was transported together with other oily wastes to IZAYDAS incineration plant.

The cleaned water from the water cleaning systems in the Aegean and Mediterranean regions is just used for irrigation purposes. A part is used for irrigating landscapes, parks and gardens of apartments and some of it is used for agricultural purposes after collection in stabilization pools (ESRT, 2011).

In addition to these remediation studies, estimation of contaminated soil acreage and remediation suggestions have been carried out by the universities and research institutes. Duman et al. (2007) determined that the heavy metals in Sapanca lake, and the values of Pb, Cr, Cu, Mn, Ni, Zn and Cd in the lake were $35,67\,\mu g\,l^{-1}$, $61,97\,\mu g\,l^{-1}$, $24,61\,\mu g\,l^{-1}$, $22,57\,\mu g\,l^{-1}$, $46,44\,\mu g\,l^{-1}$, $88,52\,\mu g\,l^{-1}$, $2,97\,\mu g\,l^{-1}$, respectively. Ok (2008) has investigated the inorganic contaminants in 33 soil samples collected at the depth of 0–10 cm from agricultural areas in Sakarya, which showed that in all the samples As, Cr, Cu, Cd, Zn, Pb and Ni concentrations were present with concentrations ranging from $0.98\,mg\,kg^{-1}$ to $18.06\,mg\,kg^{-1}$, 4.17–$173.06\,mg\,kg^{-1}$, 2.98–$108.18\,mg\,kg^{-1}$, 0.02–$0.47\,mg\,kg^{-1}$, 14.10–$201.07\,mg\,kg^{-1}$, 2.10–$27.58\,mg\,kg^{-1}$, 3.01–$219.88\,mg\,kg^{-1}$, respectively.

Daghan et al. (2011) have evaluated the potential of phytoextraction and the effect of metallothionein gene expression isolated from *Saccharomyces cerevisiae* on the ability of tobacco plants for accumulation of heavy metals such as Cd, Cu, Ni and Zn. Transgenic plants were compared to wild type tobacco (*Nicotiana tobaccum* Petit Havana, SR-1) plants with respect to their heavy metal (Cd, Zn, Cu and Ni) tolerance and accumulation both in the soil and under hydroponic culture conditions. Increasing heavy metal doses increased heavy metal uptake in both soil and hydroponic culture. Except for Cu, the transgenic tobacco plants accumulated higher amounts of Cd, Zn and Ni as compared to the non transgenic plants. Transgenic tobacco plants were much more tolerant to increased Zn concentration ($1600\,mg\,kg^{-1}$) and accumulated much more Zn than the non-transgenic tobacco cultivars in both soil and hydroponic cultures. The results also showed that Zn accumulation capacity of transgenic tobacco was not sufficient to use as a phytoextractor to clean up Zn polluted soils. However, Cd uptake levels of 354, 400, 372 and $457\,mg\,kg^{-1}$ in 10-, 20-, 40- and 80-mg Cd kg^{-1} doses, respectively, showed that transgenic tobacco could be used as a good Cd hyperaccumulator. The minimum Cd uptake level was $100\,mg\ kg^{-1}$ in the hyperaccumulator plants. The transgenic tobacco plant accumulated 3.5- and 4.5-fold of Cd compared to the standard level of $100\,mg\ kg^{-1}$ (Daghan et al., 2011).

The pollution data collected and published by the UN from different countries in 2010 includes a 'Human Developed Index'; it includes information on imported environmental problems of Turkey which is ranked as 92. In the same index, the Environmental Performance Index of Turkey is 60.4 (ESRT, 2011).

The first soil pollution control legislation of Turkey was published on May 31, 2005 (official gazette number 25831). After this, soil pollution control and point source polluted areas legislation was published on June 08, 2010 (official gazette number 27605). The rules in the legislation cover technical and administrative aspects dealing with the prevention of soil contamination, estimation of polluted and possible contaminable areas and sectors, keeping records of these areas, remediation of polluted soils and lands and surveillance of these areas.

RADIOACTIVE POLLUTION

Although use of nuclear power provides cheap energy, there are some well-known risks to its use. One of the most affected countries from the accident of Chernobyl nuclear power plant radioactive contamination was Turkey (Ozturk et al., 1987). There was accumulation due to fallout on the soil in the Edirne-Eskikadın, Ismailce, Kapikule and Buyukdoganca areas in Thrace. The coastal area between Hopa and Pazar in the Eastern Black Sea Region became the most affected area due to heavy precipitation during the passage of the radioactive clouds. The coastal area between Hopa and Pazar in the Eastern Black Sea Region were the most affected area due to heavy precipitation during the passage of the radioactive clouds (Kapukaya, 2010).

Still, there is no evidence of a major public health impact attributable to radiation exposure in Turkey. It was indicated that increase in nodule formation, an autoimmune trait, in the Cayeli and Pazar districts of Rize was due to radiation exposure. A study carried out on lichens indicated that the Eastern Black Sea Region was more affected by the disaster than other regions of Turkey and Rize was more affected than many Europian countries (Turkkan, 2006).

CONCLUSION

One of the best approaches to polluted soil remediation is the prevention of soil pollution. However, there have not been enough regulations in this connection and studies on polluted soils are limited. Therefore, there is need to prepare a data bank for polluted areas to be investigated and the areas in need of reclamation and remediation determined (Ozturk et al., 2006b). There is a need for alternative methods for the solution of urban ecological problems facing our coastal zone soils as well (Ozturk et al., 1997).

The best remediation methods must be determined after completing necessary laboratory studies, considering the specialities of pollutants, hydrogeologic properties of the polluted areas and economic feasibilities of the remediation methods (Ozturk, 1989; Kocaer and Başkaya, 2003). It is imperative to increase the studies on prevention and remediation of soil pollution, and prepare administrative rules for its prevention. A national program must be prepared to find out the best remediation methods. Projection, application, evaluation, control, observation and coordination of collaboration mechanisms of natural resources must be improved. The missuse of agricultural land and forest areas must be prevented. Necessary financing must be supplied for reforestation, erosion control and range and meadow improvement (Ozturk et al., 2012b). Inventory studies are necessary for recording, observation and remediation of the polluted areas.

REFERENCES

Ahmad, M.S.A., Ashraf, M., 2011. Essential roles and hazardous effects of nickel in plants. Rev. Environ. Contam. Toxicol. 214, 125–167.

Ahmadpour, P., Ahmadpour, F., Mahmud, T.M.M., Abdul, A., Soleimani, M., Hosseini, T.F., 2012. Phytoremediation of heavy metals: A green technology. Afr. J. Biotechnol. 11 (76), 14036–14043.

Akin, M., Akin, G., 2007. Importance of water, water potential in Turkey, water basins and water Pollution. J. Ankara Univ. Faculty of Lang. Hist. Geography 47 (2), 105–118 (in Turkish).

Ashraf, M., Ozturk, M., Ahmad, M.S.A. (Eds.), 2010a. Plant Adaptation and Phytoremediation. Series: Tasks for Vegetation Science. Springer Verlag. 481 pp.

Ashraf, M., Ozturk, M., Ahmad, M.S.A., 2010b. Toxins and their phytoremediation. In: Ashraf, et al. (Ed.), Plant Adaptation and Phytoremediation. Springer Verlag—Tasks for Vegetation Science, pp. 1–34.

Cepel, N., 1997. Soil pollution, erosion and damage to the environment. TEMA (The Turkish Foundation for Combating Soil Erosion, for Reforestation and the Protection of Natural Habitats) Publications. No. 14. (in Turkish).

Ceritli, I., 1997. Soil problem of Turkey. J. Ecol. 22, 4–8 (in Turkish).

Daghan, H., Arslan, M., Uygur, V., Koleli, N., Onder, D., Agca, N., 2011. Reclamation of heavy metal polluted soils with Phytoextraction Technique. Project Final Report, Project No: 108O161. Supported by The Scientific and Technological Research Council of Turkey (TUBITAK). p.104. (in Turkish).

Delibacak, S., Elamci, O.L., Secer, M., Bodur, A., 2003. Fertility status, trace elements and heavy metal pollution of agricultural land irrigated from the Gediz River. Inter. J. Water 2 (2/3), 184–195.

Dogan, F., Ozturk, M., 1991. Studies on the correlations between air and water pollution with the mortality rates in Turkiye. In: Ozturk, M.A., et al. (Ed.), Int. Urban Ecology Symp., Aydın, pp. 16–23.

Dogan, F., Ozturk, M., 1994a. Ecological evaluation of urban problems in Turkey. Ege Univ. Sci. Fac. Jour. 16, 1–8.

Dogan, F., Ozturk, M., 1994b. Air pollution problems in urban areas of developing countries—A case study from Turkiye. Ege Univ. Sci. Fac. Jour. 16, 9–17.

Duman, F., Sezen, G., Nilhan, T.G., 2007. Seasonal changes of some heavy metal concentrations in Sapanca Lake water, Turkey. Int. J. Nat. Eng. Sci. 1 (3), 25–28.

Durmuşoğlu, E., Tiryaki, O., Canhilal, R., 2010. Pesticide Use, its residue and durability problems in Turkey. VII. Turkey Agricultural Engineering Technical Conference, in proceedings of congress, 2:589–607.

EPA, 2000. Introduction to Phytoremediation. Environmental Protection Agency Office of Solid Waste and Emergency Response Technology. EPA/600/r-99/107, Cincinati, Ohio, U.S.A. p 72 http://www.clu-in.org.

EPA, 2009. Soils Policy: Soil Contamination in Europe. http://www.epa.gov/oswer/international/factsheets/200906_eu_soils_contamination.htm.

Erpul, G., Deviren, S., 2012. Soil Erosion in Our Country: What Should Be Done? J. Soil Sci. Plant Nutr. 1 (1), 26–32 (in Turkish).

Ertin, G., 2012. Industry in Turkey. (in Turkish) www.anadolu.edu.tr/aos/kitap/IOLTP/2291/unite10.pdf.

ESRT, 2011. Environmental Status Report for Turkey. Publish No:11, p: 356, T.R. Ministry of Environment, Ankara (in Turkish) http://www.csb.gov.tr/turkce/dosya/ced/TCDR_2011.pdf.

Findik, M.S., 2007. Green tax within the terms of preventing factors leading to environmental pollution in Turkey. Marmara University, Institute of Social Sciences, Master's Thesis. 135 p. (in Turkish).

Flathman, P.E., Lanza, G.R., 1998. Phytoremediation: Current view on an emerging green technology. Soil Contam. 7 (4), 415–432.

Gezgin, S., 2005. http://cevrekoruma.sitemynet.com/cevre/id1.htm.

Gezgin, S., Dursun, N., Hamurcu, M., Harmankaya, M., Onder, M., Sade, B., Topal, A., .Ciftci, N., Acar, B., Babaoglu, M., 2002. Determination of Boron contents of soils in Central Anatolia Cultivated Lands and its relationship, between soil and water characteristics. In: Goldbah, H.E. (Ed.), Boron in Plant and Animal Nutrition, first ed. Kluwer Academic Pub, New York, pp. 39–400.

Glass, D., 1999a. http://www.dglassassociates.com/INFO/phy99exc.htm.

Glass, D., 1999b. Markets for phytoremediation 1999–2000. Glass Associates, Needham. Mass http://www.channel 1.com/dglassassoc/index.htm.

Gokmen, D., Guvensen, A., Ozturk, M., Sayar, A., 1996. Urban Ecological Problems in Gediz Basin. Gediz Basin II. Agriculture, Erosion and Env. Problems Symp, Manisa.

Guney, E., 2004. Environmental Problems of Turkey. Nobel Publishing. No:705, Ankara. ISBN:975–591-690-3. (in Turkish).

Gungor E. B. O. (2008). Soil pollution and remediation problems in Turkey, Chapter 8, p:111–132. Environmental Technologies. Edited by E. Burcu Ozkaraova Gungor, ISBN 978-3-902613-10-3, Hard cover, 268 pages, Publisher: I-Tech Education and Publishing, Published: January 01, 2008 under CC BY-NC-SA 3.0 license.DOI: 10.5772/5301.

Hakeem, K.R., Ahmad, A., Iqbal, M., Gucel, S., Ozturk, M., 2011. Nitrogen efficient rice genotype can reduce nitrate pollution. Environ. Sci. Pollut. Res. 18, 1184–1193.

Haktanır, K., Cangir, C., Arcak, C., Arcak, S., 2006. Soil resources and its uses. (in Turkish) www.kimyamuhendisi.com.

Helvacı, C., 2004. Borate deposits of Turkey: Geological Position, Economic Importance and Bor Policy. Fifth Industrial Raw Materials Symposium. 13–14 May Izmir, p:11–27. (in Turkish).

Hooda, V., 2007. Phytoremediation of toxic metals from soil and waste water. J. Environ. Biol. 28 (2), 367–376.

Ismail, S., 2012. Phytoremediation: A green technology. Iran. J. Plant Physiol. 3 (1), 567–576.

Kapukaya, C., 2010. Chernobyl nuclear accident and its effects on Turkey. Gazi University, Faculty of Education, Department of Physics Education(in Turkish).

Karaca, A., Turgay, O.C., 2012. Soil Pollution. J. Soil Sci. Plant Nutr. 1 (1), 13–19 (in Turkish).

Kidney, S., 1997. Phytoremediation may take root in Brownfields. The Brownfield Rep. 2 (No:14), 167.

Kocaer, F.O., Başkaya, H.S., 2003. Remediation technologies for metal-contaminated soils. Uludag Univ. J. Fac. Eng. Archit. 8 (1), 121–131 (in Turkish).

Koleli, N., Kantar, C., 2006. Heavy metal hazard in phosphorus fertiliser. Journal of Ecology Magazine. Issue: 9 http://www.ekolojimagazin.com/?id=44&s=magazin (in Turkish).

Li, M.S., Yang, S.X., 2008. Heavy metal contamination in soils and phytoaccumulation in a manga-
    nese mine wasteland, South China. Air, Soil and Water Res. 1, 31–41.
Luo, Y., Wu, L., Liu, L., Han, C., Li, Z., 2009. Heavy metal contamination in Asian agricultural
    land. Marco Symposium, Challenges for Agro-Environmental Research in Monsoon Asia, 5–7
    October. National Institute for Agro-Environmental Science (NIAES), Tsukuba Japan. http://
    www.niaes.affrc.go.jp/marco/marco2009/english/program/S-1_LuoYM.pdf.
Namli, A., 2012. Environmental problems in the soil. Text book. Ankara University, Institute of Sci-
    ence, Ankara. p.137 http://www.agri.ankara.edu.tr/soil_sciences/1250__ayten_CTS_dersnotu.
    pdf (in Turkish).
Nurlu, E., Erdem, U., Ozturk, M., Guvensen, A., Turk, T., 2008. Landscape, Demographic Develop-
    ments, Biodiversity and Sustainable Land Use Strategy: A Case Study on Karaburun Peninsula,
    Izmir, Turkey. In: Petrosillo, I., et al. (Ed.), Use of Landscape Sciences for The Assessment of
    Environmental Security. Springer, The Netherlands, pp. 357–368.
Ok, G., 2008. A survey of inorganic contaminants in soil samples collected from agricultural areas
    in Sakarya. Sakarya University, Institute of Science, Department of Environmental Engineer-
    ing, MSc Thesis. p. 49 (in Turkish).
Okumus, K., 2002. Turkey's Environment—A review and evaluation of Turkey's environment and
    its stakeholders. The Regional Environmental Center for Central and Eastern Europe. ISBN:
    963 9424 09 9.
Ozturk, M., Turkan, I., Selvi, S., 1987. Radioactive pollution and plants. Turk. J. Bot. 11/3,
    322–329.
Ozturk, M. (Ed.), 1989. Plants and Pollutants in Developed and Developing Countries. Ege Univ.
    Press, Izmir,Turkey. 759 pp.
Ozturk, M., Turkan, I., 1989. Role of biocides in Turkish agriculture. In: Ozturk, M. (Ed.), Intern.
    Symp. on Plants and Pollutants in Developed and Developing Countries, pp. 475–483. Izmir.
Ozturk, M., Akgun, S., Turkan, I., Pirdal, M., 1991. The possible use of sewage sludge for
    the cultivation of plants. In: Ozturk, M., et al. (Eds.), Int. Urban Ecology Symp. Aydın,
    pp. 266–270.
Ozturk, M., Uysal, T., Guvensen, A., 1994b. Lemna minor L. as a Water Cleaner. XII. National Biol-
    ogy Congress, Edirne. 68–70.
Ozturk, M., Erdem, U., Butuner, H., Dalgic, R., 1995. Non-agricultural use of land in Turgutlu-
    Gediz Basin. I Gediz Basin Erosion and Env. Symp. 354–369.
Ozturk, M., Secmen, O., Leblebici, E., 1996a. Plants and pollutants in the Eber lake. Ekoloji 20, 14–16.
Ozturk, M., Celik, A., Nurlu, E., Erdem, U., 1996b. Land degradation in relation to urbanisation
    and industrialisation in the West Anatolian region, Turkey. Int. Conf. on Land Degradation,
    Adana. 305–317.
Ozturk, M., Guvensen, A., Gokmen, D., 1997. Alternative methods for the solution of urban eco-
    logical problems facing our coastal zones. Tourism and Environment Competition. OLEYİS
    Foundation, Ankara. 119–140.
Ozturk, M., 1999. Urban ecology and land degradation. In: Farina, Almo (Ed.), Perspectives in
    Ecology. Backhuys Publishers, Leiden, NL, pp. 115–120.
Ozturk, M., Alyanak, I., Sakcali, S., Guvensen, A., 2005. Multipurpose plant systems for renovation
    of waste waters. The Arabian J. Sci. Eng. 30/2C, 17–28.
Ozturk, M., Ergin, M., Kucuk, M., 2006a. Sustainable use of biomass energy in Turkey. Proc.of
    the 13th IAS Science Conference on "Energy for Sustainable Development" and "Science for
    the Future of the Islamic World and Humanity," Kuching/Sarawak, Malaysia (2003). In: Ergin,
    M., Zou'bi, M.R. (Eds.), Islamic World Academy of Sciences (IAS). National Printing Press,
    Amman, Jordan, pp. 231–242.

Ozturk, M., Guvensen, A., Aksoy, A., Beyazgul, M., 2006b. An overview of the soils and sustainable land use in Turkiye. Proceedings of the Fifth Int. GAP Engineering Congress, Şanlıurfa, Turkey. 1548–1555.

Ozturk, M., Yucel, E., Gucel, S., Sakcali, S., Aksoy, A., 2008. Plants as biomonitors of trace elements pollution in soil. Trace Elements: Environmental Contamination, Nutritional Benefits and Health Implications. In: Prasad, M.N.V. (Ed.), Chapter 28. John Wiley & Sons, USA, pp. 723–744.

Ozturk, M., Sakcali, S., Gucel, S., Tombuloğlu, H., et al., 2010a. Boron and Plants. Plant Adaptation and Phytoremediation. In: Ashraf (Ed.). Springer Verlag, pp. 275–311. Part 2.

Ozturk, M., Mermut, A., Celik, A., 2010b. Land Degradation, Urbanisation, Land Use and Environment, NAM S. & T. (Delhi India). 445 pp.

Ozturk, M., Okmen, M., Guvensen, A., Celik, A., Gucel, S., 2010c. Land degradation, urbanisation and biodiversity in the Gediz Basin Turkiye. Urbanisation, land use, land degradation and environment. In: Ozturk, et al. (Ed.), NAM Proceedings. Daya Publishing House, Delhi, India, pp. 74–93.

Ozturk, M., Gucel, S., Sakcali, S., Guvensen, A., 2011a. An overview of the possiblities for waste water utilisation for agriculture in Turkey. Isr. J. Plant Sci. 59, 223–234.

Ozturk, M., Okmen, M., Guvensen, A., Celik, A., Gucel, S., 2011b. Land degradation, urbanisation and biodiversity in the Gediz Basin Turkiye. Urbanisation, land use, land degradation and environment. In: Ozturk, et al. (Ed.), NAM Proceedings. Daya Publishing House, Delhi, India, pp. 74–93.

Ozturk, M., Memon, A.R., Gucel, S., Sakcali, M.S., 2012a. *Brassicas* in Turkey and their possible role in the phytoremediation of degraded habitats. In: Anjum, N.A., et al. (Ed.), The Plant Family Brassicaceae: Contribution Towards PhytoremediationSpringer Verlag, Environmental Pollution Book Series, 21. 265–288.

Ozturk, M., Altay, V., Gucel, S., Aksoy, A., 2012b. Aegean grasslands as endangered ecosystems in Turkey. Pak. J. Bot. 44, 7–17.

Pavel, L.V., Gavrilescu, M., 2008. Overview of ex situ decontamination techniques for soil cleanup. Environ. Eng. Manag. J. 7 (6), 815–834.

Percin, B., 2006. Cleaning of heavy metal contaminated soils. Akkuş Chamber of Agriculture. (in Turkish) http://akkus.ziroda.com/forum/viewtopic.php?pid=102.

Prasad, M.N.V., 2005. Nickelophilous plants and their significance in phytotechnologies. Braz. J. Plant Physiol. 17 (1), 113–128.

Reeves, R.D., Adiguzel, N., 2004. Rare plants and nickel accumulators from Turkish serpentine soils, with special reference to *Centaurea* species. Turk. J. Bot. 28, 147–153.

Reeves, R.D., Adiguzel, N., 2008. The nickel hyperaccumulating plants of the serpentines of Turkey and adjacent areas: A Review with New Data. Turk. J. Bot. 32, 143–153.

Reeves, R.D., Adigüzel, N., 2012. Important serpentine areas of Turkey and distribution patterns of serpentine endemics and nickel accumulators. Bocconea 24, 7–17. ISSN: 1120–4060.

RTMEF, 2008. Republic of Turkey Ministry of Environment and Forestry, Inventory of Environmental Problems and Priorities Assessment Report of Turkey (2005–2006). Issue Number: 6 Ankara, Turkey. (in Turkish).

RTMEF, 2010. Republic of Turkey, Ministry of Environment and Forestry, Inventory of Environmental Problems and Priorities Assessment Report of Turkey (2007–2008). Issue Number: 9 Ankara, Turkey. (in Turkish).

RTMECP, 2012. Republic of Turkey, Ministry of Environment and City Planning. Problems of the Mediterranean Region. Ankara, Turkey www.csb.gov.tr/gm/ced/ (in Turkish).

Sakcali, M.S., Ozturk, M., 2003. Eco-physiological behaviour of some Mediterranean plants as suitable candidates for reclamation of degraded areas. J. Arid Environ. 57, 141–153.

Sakcali, S., Yilmaz, R., Gucel, S., Yarci, C., Ozturk, M., 2009. Water pollution studies in the rivers of Edirne State Turkey. Aquat. Ecosystem Health & Manag. 12, 3,313–319.

Seker, S., Ileri, R., Ozturk, M., 2006. Evaluation of activated sludge by white rot fungi for decolorisation of textile wastewaters. J. World Assoc. Soil and Water Conserv. J1–7, 81–87.

Sheikh, K.H., Ozturk, M., Secmen, O., Vardar, Y., 1976. Field studies of the effects of cement dust on the growth and yield of olive trees in Turkey. Environ. Conserv. 3,117–121.

Singh, O.V., Labana, S., Pandey, G., Budhiraja, R., Jain, R.K., 2003. Phytoremediation: An overview of metallic ion decontamination from soil. Appl. Microbiol. Biotechnol. 61 (5), 405–412.

SIT, 2013. Statistical Institute of Turkey. Report of Population Projections, 2013–2075. Issue: 15844, 14 February http://www.tuik.gov.tr/PreHaberBultenleri.do?id=15844.

Tan, H.K., 1994. Environmental Soil Science. Marcel Dekker, New York.

Torun, A.T., Yazici, A., Erdem, H., Cakmak, I., 2006. Genotypic variation in tolerance to boron toxicity in 70 durum wheat genotypes. Turk. J. Agriculture and Forestry 30, 49–58.

Turkkan, A., 2006. The effects of Chernobyl nuclear power plant accident on Turkey. Cancer in Turkey after the Chernobyl Nuclear Accident. Publications of the Turkish Medical Association. 17p, ISBN 975–6984-80-5 (in Turkish).

Turkoglu, B., 2006. Soil Pollution and remediation of polluted soils. Cukurova Uni., Natural and applied Science Institute, Soil Sci. Dept., MSc thesis. 134 p.

UCTEA, 2009. The Union of Chambers of Turkish Engineers and Architects. Water report: Global Water Policy and Turkey. ISBN: 978-9944-89-682-5. (in Turkish).

Uysal, I., Ozdilek, H.G., Ozturk, M., 2012. Effect of kiln dust from a cement factory on growth of *Vicia faba* L. J. Environ. Biol. 33 (Suppl. 02), 525–530.

Yavuz, S., 2005. Enhanced bioremediation of contaminants in soil. The Graduate School of Natural and Applied Sciences of Dokuz Eylül University, Environmental Engineering, Environmental Technology Program. 115 p.

Yilmaz, F., Yilmaz, Y.Z., Ergin, M., Erkol, A.Y., Muftuoglu, A.E., Karakelle, B., 2003. Heavy metal concentrations in surface soils of Izmit gulf region, Turkey. J. Trace and Microprobe Tech. 21 (39), 523.

Yucel, E., Dogan, F., Ozturk, M., 1995. Heavy metal status of Porsuk stream in relation to public health. Ekoloji 17, 29–32.

Xu, Q., 2007. Facing up to 'invisible pollution'. China Dialogue. 29 January 2007. Available at http://www.chinadialogue.net/article/show/single/en724-Facing-up-to-invisible-pollution.

Chapter 11

Soil Pollution Status and Its Remediation in Nepal

Anup K.C.* and Subin Kalu[†]

Department of Environmental Science, Amrit Campus, Tribhuvan University, Thamel, Kathmandu, Nepal, [†]Central Department of Environmental Science, Tribhuvan University, Kirtipur, Kathmandu, Nepal

INTRODUCTION

Nepal is located in Southeast Asia between $80°04'–88°12'E$ longitude and $26°22'–30°27'N$ latitude, having borders with China in the north and India in the east, west and south. The total area is $147,181 km^2$, extending 800 km from east to west and 144 km to 240 km north to south. The country is blessed with tremendous geographical diversity ranging from an altitude of 60 m to 8848 m from south to north. In this roughly rectangular outlined country, 83% of the land is mountainous and 17% is formed by the alluvial plains of the Gangetic basin (Paudyal, 2002).

Physiographically, Nepal can be divided into eight distinct divisions from south to north (Hagen, 1969):

1. The Terai,
2. The Siwalik (Churia) Range,
3. The Dun Valleys,
4. The Mahabharat Range,
5. The Midlands,
6. The Fore Himalayas,
7. The Higher Himalayas, and
8. The Inner and Trans Himalayan Valleys.

The provinces run from east to west and are therefore incorporated into the Indian Himalayan belt. Each of these divisons has distinct altitudinal, topographical, climatic and vegetational characteristics. A detailed description of physiographic provinces is presented in Table 11.1 and Figures 11.1 and 11.2.

Soil Remediation and Plants. http://dx.doi.org/10.1016/B978-0-12-799937-1.00011-5

TABLE 11.1 Physiographical Division of the Nepal Himalaya (Upreti, 1999)

SN	Geomorphic unit	Width (km)	Altitudes (m)	Main rock types	Main processes for Landform development
1	Terai (Northern edge of the Gangetic Plain)	20–50	100–200	Alluvium: coarse gravels in the north near the foot of the mountains, gradually becoming finer southward	River deposition, erosion and tectonic upliftment
2	Churia Range (Siwaliks)	10–50	200–1300	Sandstone, mudstone, shale and conglomerate.	Tectonic upliftment, erosion and slope failure
3	Dun Valleys	5–30	200–300	Valleys within the Churia Hills filled up by coarse to fine alluvial sediments	River deposition, erosion and tectonic upliftment
4	Mahabharat Range	10–35	1000–3000	Schist, phyllite, gneiss, quartzite, granite and limestone belonging to the Lesser Himalayan Zone	Tectonic upliftment, weathering, erosion and slope failure
5	Midlands	40–60	300–2000	Schist, phyllite, gneiss, quartzite, granite, limestone geologically belonging to the Lesser Himalayan Zone	Tectonic upliftment, weathering, erosion and slope failure
6	Fore Himalaya	20–70	2000–5000	Gneisses, schists, phyllites and marbles mostly belonging to the northern edge of the Lesser Himalayan Zone	Tectonic upliftment, weathering, erosion and slope failure
7	Higher Himalaya	10–60	>5000	Gneisses, schists, migmatites and marbles belonging to the Higher Himalayan Zone	Tectonic upliftment, weathering, erosion (rivers and glaciers) and slope failure
8	Inner and Trans Himalaya	5–50	2500–4500	Gneisses, schists and marbles of the Higher Himalayan Zone and Tethyan sediments (limestones, shale, sandstone, etc.) belonging to the Tibetan-Tethys Zone	Tectonic upliftment, wind and glacial erosion and slope degradation by rock disintegrations

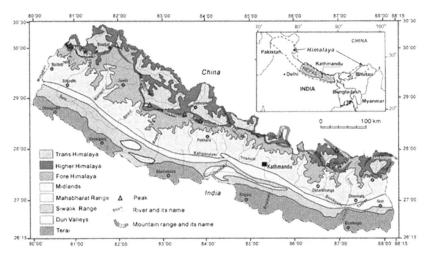

FIGURE 11.1   Physiography of the Nepal Himalaya. For a colour version of this figure, please see the section at the end of this book. *Source: Dahal, 2006.*

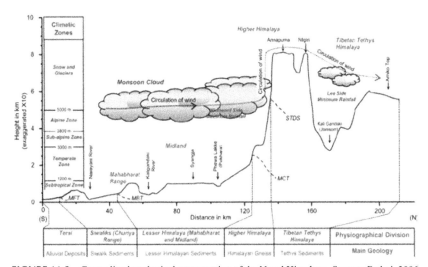

FIGURE 11.2   Generalized geological cross section of the Nepal Himalaya. *Source: Dahal, 2006.*

SOIL CHARACTERISTICS

Soil, except for open water surfaces and rock outcrops, is a thin layer covering the entire surface of the earth and is a complex mixture of mineral nutrients, organic matter, water, air and living organisms (Kang and Tripathi, 1992). There is a large number of different kinds of soils whose characteristics are determined by various environmental parameters such as climate, parent material, relief, organisms and time factor, reflecting different kinds and degrees of soil formation factors and their combinations (Kang and Tripathi, 1992).

The major factor related to the agricultural output is soil fertility – one of the key factors in determining soil productivity. The most important problem currently is the land degradation which is a primary constraint to improve food security in industrializing countries (Drechsel et al., 2004). Physiography and the water resources help to make the soil fertile to some extent, but the total nutrient content varies from soil to soil depending on the nature of parent material and other soil-forming processes (Anup et al., 2013). Only the plant-available form of the nutrients in the soil is relevant for the crops and is chemically determined through appropriate testing methods (Reddy et al., 2012). The primary nutrients for plant growth are nitrogen, phosphorus and potassium (known collectively as NPK). When they are insufficient, they will be responsible for limiting crop growth (Gruhn et al., 2000). Poor soil fertility, low levels of mineral nutrients in soil, improper nutrient management and lack of plant genotypes are major constraints contributing to food insecurity, malnutrition and ecosystem degradation in developing countries (Cakmak, 2002).

SOILS OF NEPAL

The soils of Nepal are highly variable and are derived mainly from young parent material (Manandhar, 1989). Various factors such as geology, climate and vegetation types have resulted in the variations in soil characteristics. The classification given for the soils in Nepal on the basis of soil texture, mode of transportation, and color, can broadly be outlined as: alluvial, sandy and alluvial, gravelly, residual and glacial soils (Figure 11.3).

Alluvial Soil

Alluvial soil is found in the valleys of the Terai region and in the middle hill valleys around Kathmandu and Pokhara. The valleys lie between the Siwalik and Mahabharat hills which widen out in places to form flat fertile valleys called Dun valleys. New alluvial soil with more sand and silt than clay is being deposited in the flood plain areas along the river courses. Alluvial soil is also found in the higher areas above the flood plain covering a greater part of the Terai. The nutrient content of new alluvial soil is fair to medium depending on how long it has been cultivated. Conversely, the nutrient content of old alluvial soils is very low.

Sandy and Alluvial Soil

Valleys in the mid-hills of Kathmandu and Pokhara are composed of sandy and silty alluvial soils, which are fairly fertile. In the Kathmandu valley, some deposits of peat mare (Kumero) have been found. This is diatomaceous clay which is used for painting house walls during festivals in rural areas.

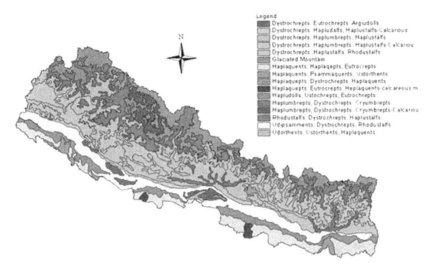

FIGURE 11.3   Soils of Nepal. For a colour version of this figure, please see the section at the end of this book. *Source: Pariyar, 2008.*

In addition, the Kathmandu valley is a source of dark clay or silty clay (Kalimati) soil which is obtained from deep underground pits and is used as manure for potato and other vegetable crops. This soil is rich in humus, potash and calcium.

Gravelly Soil

The foot of the Churia hills has soils of mixed gravel and pebbles. This soil is not useful for agriculture as it has a very coarse texture and cannot hold sufficient moisture for plant growth. Such soils were deposited by rivers originating in the Churia hills and have high lime content. Some soils in high mountain areas are also coarse-textured gravels.

Residual Soil

Residual soil is found mostly on the ridges and slopes of the mountains. Soils of the Churia hills are very young and coarse, and are dry for most of the year. Soils on the slopes of the mid-hills are medium to low in plant nutrients but less productive due to moisture and climatic limitations.

Glacial Soil

Glacial soil is found in high Himalayan regions having rocky terrain with ice blocks. They are covered with snow for most of the year. The soil is much less exposed to the air due to snow cover.

LRMP (1986) reported that 14 dominant soil groups covering four soil orders are encountered in Nepal. Major soil orders of Nepal according to

US Department of Agriculture (USDA) taxonomy are Entisols, Inceptisols, Mollisols and Alfisols. Soil orders like Spodosols, Histosols, Utisols and Aridisols are occasionally found in Nepal. The country resource profile of Nepal prepared by Food and Agriculture Organization (FAO) has also described these soil groups and orders (Pariyar, 2008).

Entisols

These are the youngest and least-developed soils, generally found on hillsides and adjacent to river courses. They are formed through deposition of colluvium and alluvium and are present throughout the country.

Three great groups of this order are recorded, namely, ustifluvents, ustorthents and fluvaquents.

Ustifluvents

These are commonly found in the depositional stage of rivers. Horizons of deposition are identifiable but soil does not show any pedogenetic development. They are mostly coarse textured, highly permeable and well drained. Depending upon the type of materials transported by rivers, they can be calcareous or non-calcareous.

Ustorthents

These develop through colluvial deposition and are found in landslide scars and on slopes of more than 35 degrees. As the soil develops it is constantly removed by erosion. They are shallow, near the bedrock, coarse-textured and poorly vegetated, but used for grazing, fodder and firewood collection.

Fluvaquents

These entisols are also found adjacent to rivers but are poorly to imperfectly drained, vary in texture and occasionally flooded. If they are suited to cultivation, rice can be grown.

Inceptisols

These cover the largest area in Nepal and are the most important soils. They occur on more stable slopes and show distinct weathering in the subsoil. The vast difference in landscape, climate, geology and parent materials have helped to develop a variety of Inceptisols.

Haplaquents

This soil is dominant in the lower piedmont plain of the Terai where drainage is restricted. It is also found in duns (broad flat valleys) valleys and limited areas of the Middle hills. The B-horizon is well developed. As water remains for more than 3 months, the subsoil shows gleying and mottling. Due to the moisture regime rice grows well on the soils, whereas crops requiring aeration do not thrive. These soils are common in the low-relief areas adjacent to major river systems.

Dystrochrepts

These are the common soils in the Terai as well as in the Middle hills, mostly below 1500 m. They have developed on the acidic or neutral bedrock including lacustrine deposits, with a well-developed B-horizon and base saturation below 60%. They develop under forest and are found on steeper slopes, can be stony, well drained and deeper with ample stones and gravel. The pH is below 5.5 and they have low base saturation. Organic matter plays an important role in retaining soil plant nutrients suppressing the possibility of aluminium toxicity. These soils are used cautiously by maintaining high organic matter content. Prolonged use of nitrogenous fertilizers alone may increase acidity of these soils and need to be amended with high rates of lime. Erosion control on the hill slopes is a must to maintain the productivity of Dystrochrepts.

Ustochrepts

These are commonly found on alluvial plains of the Terai and Siwalik regions and develop on phyllite, schists, quartzite and limestone on the Middle and High hills. They are common on the Western and Middle hills and are diagnosed by a well-developed B-horizon, pale surface soils, high base saturation, variable soil texture and structure. The soils developing on colluvial deposits are stonier, those on calcareous parent materials are non-calcareous at the surface. As depth increases, calcium carbonate increases due to the leaching and precipitation of the calcium carbonate in the lower horizons. On hilly areas these soils are prone to heavy erosion. Ustochrepts in the Terai are deep, well drained, with loamy texture, non-stony and non-calcareous with high base saturation. When they are irrigated they have wide productive potential. Ustochrepts in the Siwalik, Middle and Mountain regions are deep to shallow, stony, coarse to loamy texture, well-drained calcareous or non-calcareous forms but have high base saturation.

Cryumbrepts

These are the soils of the High Himalayan and High Hill regions, generally found above 3000 m. Annual mean temperature is below 8°C. Soils of this great group have dark A-horizon, high organic matter with wide C/N ratio, low base saturation and contain no free carbonate. They are rubbly and silty in texture. Near the settlements trees are cleared for fodder and firewood so bare areas are prone to soil erosion. Pathways of gullies caused by melting snow are common. Areas under these soils are extensively used for seasonal grazing.

Haplumbrepts

These are the soils of the High and Middle hill regions developed in cool temperatures on the acidic bedrocks in mixed forest. They have low base saturation and moisture regimes. Soils under forest and on steep slopes are shallow and stony but the cultivated ones are fertile due to a high organic matter content, which inactivates the toxic effect of aluminium by its chelating action. Frequency of stones on the surface hinders cultivation. Soil fertility is regularly maintained by grazing animals, leaving fallow for 2–3 year periods. Barley, millet and potato are the main crops grown on these soils.

Cryochrepts

These soils are similar to Ustochrepts and are found above 3000 m. They are of no importance for agricultural production.

Eutrochrepts

These soils are similar to Ustochrepts but develop on calcium-rich parent materials under a definite moisture regime.

Spodosols

These are soils with high organic matter. The active amorphous materials contain Al with or without Fe. They develop between 3000- and 4000-m altitudes in a humid, cool climate, mainly occurring in the higher parts of the High Hills and the lower parts of the high Himalayan region, but occupy a very small area. Agriculturally they are of very little importance, have low pH, which restricts growth of agricultural crops.

Mollisols

Soils with high organic matter content, usually under thick grass or forest, dark colour and high base saturation. They develop on basic parent materials at higher elevations.

Haplustolls

These are common in the sub-tropical mixed forest of the Terai and inner valleys. They develop on alluvial materials and are distinguished by a soft and dark-coloured mollic A horizon with high base saturation and a well-developed B-horizon under a moisture regime. Haplustolls develop under forest but not under grassland. Land with old alluvial deposition and forest litter which, on decomposition, contributes high base saturation helps in the development of mollisols. They are usually very fertile and produce high crop yields for the first few years after clearing. The yields decrease as organic matter content decreases.

Cryoborolls

These differ from Haplustolls mainly in their development on base-rich parent materials under thick grassland of the high mountains in high Himalayan regions. They are found in cooler climates and moisture regimes.

Alfisols

These soils are found on the higher river terraces with accumulation of a leached layer of lattice of silicate clays in their B horizon and high base saturation. They are available on stable slopes of the Middle and High hill regions where climate helps the development of mature pedogenetic argillic horizons. The great groups of Alfisols found in Nepal are as follows:

Rhodustalfs

These soils are found in Upper River terraces especially in the Siwaliks and Middle hills and mostly develop on green phyllite. They are not present in the Terai and High hills. Base saturation is more than 35%. Fertility is maintained with the application of ample organic matter. Decrease in the organic matter content from loss of fertile soil decreases crop productivity. These occur on ancient river terraces (tars) and the upper alluvial terraces where water for irrigation is scarce. Rainfed cultivation is practiced with maize/millet being the major crops.

Eutroboralfs

These Alfisols develop on calcium-rich material under cold temperatures in the high Himalayan region.

Haplustalfs

These are Alfisols similar to the Rhodustalf but do not meet the criteria of the Rhodustalfs.

Ultisols

Ultisols are not very common in Nepal. Only one great group, Rhodudults, is found in small pockets of upper terraces formed by rivers. They are similar to the Rhodustalfs but soil pH is low in these soils. Phosphorous management is a problem to maintain productivity.

Aridosols

These soils are rare in Nepal and occur in the north of Jhomsom in Mustang district where rainfall is less than 250 mm a year. Soils have calcium and other salts accumulated on the surface. Depending on the local microclimate, these soils can be fertile and produce good crops.

Soils originating from weathered soft rocks (Phyllite, Quartzite, Sandstone, Granites, Gneiss and Schists) are characterized by a high degree of porosity, poor slope stability, shallow soil depth, course texture and acidic reaction.

Nepal has complex topography and diversified agricultural practices. Soil fertility loss is the main problem in the hills due to uncontrollable soil erosion and improper soil management. The problem of soil fertility deterioration is increasing with the increase in cropping intensity, use of high-yielding crop varieties, low and unbalanced application of chemical fertilizers and decreasing use of organic manure are major problems (Mandal, 2002). Agricultural expansion has converted steep shrub and grasslands to rainfed agriculture, leading to soil erosion. The forests expanded during 1980s due to an intensive afforestation program and these are now playing a key role in sustaining agriculture. Both expansion of agricultural land into marginal grass and shrub land and agricultural intensification are taking place simultaneously, and nutrient inputs appear to be insufficient to sustain long-term productivity. Nutrients present in the forest biomass are continuously removed which is expected to lead to a remarkable decline on a long-term basis in the forest soil fertility. There is clear evidence of soil acidification, leading to phosphorus deficiencies and impairing decomposition processes. Organic matter and associated nitrogen, soil pH, exchangeable Ca, and available phosphorus are all key indicators of soil fertility degradation. The forest soils have the worst fertility status followed by grassland and rainfed agricultural land (Schreier et al., 2007).

Subsistence agriculture and lack of technical knowledge and inputs lead to improper soil nutrient management in Nepal. Cropping intensity has resulted in depletion of nutrients, while socioeconomic and cultural factors are leading to poor nutrient concentration. Pedologically, most of the soil in the hilly areas is derived from phyllite and schist (leading to modestly inherent soil fertility) or sandstones, quartzite and granite (leading to infertile sandy soil (Andersen, 2007)).

NUTRIENT AND HEAVY METAL STATUS IN THE SOILS OF NEPAL

High concentration of heavy metals in the soils has negative impacts on both the environment and human health. Population growth, industrialization and unsustainable urbanization in developing countries like Nepal leads to an accumulation of heavy metals in the soil system. These heavy metals can be bio-accumulated and bio-magnified to have serious ecological and health problems. Heavy metal pollution is threatening atmosphere, water and soil systems in the country. The indigenous technology of Nepal in activities like gold-plating techniques heavily depends on the use of mercury and the use of leaded gasoline adds to it. The latter is used in greater amounts in vehicles which release more and more lead in the form of exhaust gas into the atmosphere (Shrestha, 2003). It gets deposited in the soil due to gravity polluting, the soil. The soil in the river bank of Nepal is polluted by cadmium,salts of lead and ferrous etc. The effluents of battery industries, leather factories and dye factories are directly dumped into the river system of urban areas in Nepal (Shrestha, 2003).

Several soil scientists have studied the nutrient status of the soils throughout the country. Some of the outcomes of these studies are presented below.

A report by Sippola and Lindstedt (1994) in central Mid Hills of Nepal was based on 150 soil samples collected from cultivated fields. Paddy crop rotations had the lowest concentrations of all nutrients, especially phosphate (P), potassium (K), sulphur (S), manganese (Mn) and zinc (Zn). Heavy metals such as cadmium (Cd) and lead (Pb) were not high at the sites investigated. Sulphur (S), calcium (Ca) and magnesium (Mg) content was generally low with acidic pH (average 5.8, minimum 4.4). The main deficiency was of boron (B), where 58% of the samples had very low, and 36% had low boron content. Zinc was very low in 14% of the samples, and low in 42% of the samples. Only a few samples were very low in Mo, but half the samples were in the lower category.

Gupta et al. (1989) analyzed the soils of citrus orchards in Dhankuta District in the Eastern Hills of Nepal for their N, P, K, Mg, Mn, Cu, Zn and B contents. The soil samples were tested for total element concentration. Leaf samples were analyzed and cross-correlated with soil values. The results show widespread deficiency of Zn, B, N, Mg and Cu.

Reports from the Terai plains show similar situations. One report from the Chitwan District in central Terai by Khatri-Chhetri and Ghimire (1992) mentions that 100% of 70 soil samples were 'very low to low' in B, 83% were 'very low to low' in Zn, and 23% were 'very low to low' in Mn. The trends were confirmed by plant tissue analysis and yield responses in trials.

The study of Turton et al. (1997) shows that at low altitudes, the majority of the farmers reported unchanged or increasing soil fertility. However, soil analysis in these villages highlighted critically low levels of organic matter, as well as nitrogen and acidification problems. Management of soil fertility increasingly relies on chemical fertilizers. The soil fertility is declining at higher altitudes. The measured soil nutrient levels are higher than for villages. From group discussions, several key factors were identified that may account for the perceived decline in soil fertility. These include the deterioration in forest resources and a decline in livestock numbers. These villages do not have access to alternative nutrient sources. In the midlands, there was a pattern of low nutrient status and declining soil fertility. The soil samples analyzed in the laboratory of Department of Agriculture and in NARC indicate that 48% of soils are high in available P and 39% in available K but low in nitrogen. Available B, Mo, Zn and N content in the samples is low. Soil erosion is one of the major causes that threatens soil sustainability in Nepal. Due to mountainous physiography, poorly managed slopy terraces and degraded rangelands, erosion on these lands is highest (Turton et al., 1997). In the Western Hills of Nepal, Tripathi (1999) found that 87% of samples are deficient ($< 1 \, \mathrm{mg \, kg^{-1}}$) in B, and 10–20% of samples are low in Zn ($< 0.5 \, \mathrm{mg \, kg^{-1}}$), Mn ($< 10 \, \mathrm{mg \, kg^{-1} kg}$) and Cu ($< 0.5 \, \mathrm{mg \, kg^{-1}}$).

Micronutrients often vary in relation to soil types. Zinc is less available in sandy and alkaline soils, whereas Mo deficiency is only a problem in acidic soils.

Andersen and Sandvold (2000) studied 102 samples from altitudes of 300–2200 m in the Arun Valley in Eastern Nepal. Boron was deficient in 86 samples and zinc was deficient in 34 samples, with deficiency limits of 0.5 and 0.6 mg kg^{-1}, respectively. Zinc values were larger under maize–potato and horticulture at high altitudes, whereas B was deficient everywhere. The Zn values were largely deficient in khet fields. Tripathi (2003) presented data from the Western hills by correlating nutrient content, altitude and land type. The main division of land types in Nepal is between khet (paddy fields) and bari (dryland terraces). On altitudes ranging from 600 to 2200 m the mean values of available nutrients are Zn 0.92 mg kg^{-1}, Fe 180.6 mg kg^{-1}, Mn 58 mg kg^{-1}, Cu 1.50 mg kg^{-1} and B 0.59 mg kg^{-1}. In this study, altitude did not affect micronutrient concentrations except for B, which increased with altitude. Comparing khet with bari, there were larger Zn values on khet, whereas all other micronutrients were lower in khet soil than in bari. Bhatta et al. (2005) gave a map of the areas in Nepal affected by wheat sterility in Terai and in the districts surrounding the Kathmandu Valley. Trials with and without boron supply at a rate of 2 kg ha^{-1} proved that sterility was caused by boron, as reported by Ozturk et al. (2010). But the susceptibility of wheat varieties varied greatly from 0 to almost 100% sterility. A paper by Karki et al. (2005) included maps of the district-wide distribution of B, Zn, Cu, Fe and Mn shown in classes of low, medium, and high concentrations. Of the 75 Nepalese districts, 21 are represented in the survey. It was concluded that Fe and Mn are sufficient in most soils, but B, Zn and Mo are commonly deficient in the soils. Rai et al. (2005) has presented a statistical analysis of Zn in the soils of Rupandehi district in Terai with a mean value of 0.29 mg kg^{-1} (0.002–1.641 mg kg^{-1}). Sapkota and Andersen (2005) presented a case study of intensive horticulture from the Kathmandu Valley. About 75 soil samples had sufficient macro- and micronutrients, except for B.

Geology and soil type affect supply of micronutrients such as selenium (Se) and zinc (Zn). Zinc, Mo and B deficiencies affect the yields of pulse crops and reduce the availability of protein, iron, folate and other nutrients. Most micronutrient problems in the Nepal region are due to soil deficiencies, but excess can also be a problem. Calcium (Ca), boron (B), zinc (Zn) and magnesium (Mg) content in soils are low. In addition, Mo deficiency is likely to occur in Nepal, most probably in the soils of the Bagmati area. Among the other elements studied, the status of copper (Cu), iron (Fe) and manganese (Mn) are not considered to be a problem (Andersen, 2007). Soil acidification, associated with high inputs of acid-causing fertilizers (urea and ammonium based fertilizers) and acid bedrock geology, is becoming a major problem in the double and triple crop rotation systems in the Jhikhu Khola area. This acidification has serious implications as low soil pH (<5.0) slows down the rate of organic matter decomposition, and leads to the leaching of base cations (calcium and magnesium) and the fixing of available phosphorous in the soil, making it unavailable to plants and causing aluminium toxicity and micronutrient deficiencies.

REMEDIATION OF TOXICITY FROM SOIL

Remediation deals with the removal of contaminants of pollutants from the resource. In case of soils, once polluted, it needs to be cleaned and purified for betterment of its quality and fertility. The process of purifying and revitalizing the soil is known as soil remediation. Remedy technologies can be categorized as ex situ and in situ methods. Ex situ methods involve excavation of affected soils and subsequent treatment at the surface, whereas in situ methods seek to treat the contamination without removing the soils. Among the processes used in soil remediation, excavation and dredging are the most common. This process involves extracting soil that is contaminated and deemed to be unrecoverable using current technology, and transporting it to a landfill set aside for this purpose. Often, purified soil is used to fill in the area where the extraction took place. Soil remediation is also accomplished by using a process known as pump and treat. Essentially, this approach involves the removal of contaminated ground water, using various methods to purify the extracted liquid. While the water is purified, the soil is extracted and filtered to remove various contaminants and returned back to its original position. The purified water is pumped back into the purified soil, effectively restoring the ecological balance of the area. As technology advances, newer methods of reclaiming contaminated soil are also in thedevelopment phase. This will make it possible to purify land and use the area for growing food. It will also help in creating wildlife preserves, allowing humans to safely construct dwellings and commercial buildings in the area.

Heavy metal pollution is the severe toxic substance pollution of the soils. Of these, arsenic is a toxic metalloid of global concern which can be intensified by human activities such as applications of pesticides and wood preservatives, mining and smelting operations and coal combustion. Bioremediation of arsenic-contaminated soils and groundwaters shows a great potential for future development due to its environmental compatibility and possible cost-effectiveness. It relies on microbial activity to reduce, mobilize, or immobilize arsenic through sorption, biomethylation, complexation, and oxidation–reduction processes. Microbially mediated redox reactions involving organic carbon, Fe, Mn and S are the basic underlying mechanisms affecting arsenic mobility. Microorganisms have evolved biochemical mechanisms to exploit arsenic oxyanions, either as an electron acceptor for anaerobic respiration, or as an electron donor to support chemoautotrophic fixation of carbon dioxide (CO_2) into cell carbon. A number of investigations have been performed to remediate arsenic-contaminated soils and groundwater using biologically based method. Plant growth promoting *Rhizobacteria* combats heavy-metal stress through the processes of: exclusion – the metal ions are kept away from the target sites; extrusion – the metals are pushed out of the cell through chromosomal/plasmid-mediated; accommodation – metals form complexes with the metal binding proteins (e.g. metallothienins, a low molecular weight proteins) or other cell components; and bio-transformation – toxic metal is reduced to less toxic forms and methylation and demethylation (Jaiswal, 2011).

REMEDIATION STUDIES ON REMOVAL OF TOXICITY IN SOIL OF NEPAL

In Nepal very few studies have been carried out on the remediation of toxic substances from the soil. As water resources are of great priority for human beings in Nepal, as in other industrializing countries, most of the toxicity remediation research and projects are focused on the drinking-water sector. Some investigations have been carried out on the effect of pine litter in compost to soil acidification on red soils originating from phyllitic parent materials and brown (non-red) soils from quartzitic materials. The soil was analyzed for pH, exchangeable cations, carbon, and available phosphorous (Bray-1) using standard procedures. No acidification was detected after the first year, but in second year soil acidification was taking place. Initially the rate of acidification was higher in the non-red soils than the red soils but acidification was significant in both soils in the second year. While the carbon and calcium content improved with pine litter addition, the pH decreased. In the non-red soils, the available phosphorous content increased. This suggests that pine litter is acidifying the soils and the phosphorous availability in the red soils. These results suggest that the addition of other types of litter is needed to have a positive impact on nutrient management.

A test has also been conducted on the effects of applying lime to the acidic soils of the Jhikhu Khola watershed. Eight sites with low soil pH were selected. A recommended dose of lime was applied to five ropanis of land (1 ropani = 508 m²) by each of three farmers (three sites) in Lamdihi to test its effects on maize. For example: 120, 230 and 294 kg of lime per ropani on clay loam were used with soil pH of 6.0, 5.5 and 5.2, respectively. Likewise, lime was applied to five vegetable farming sites (cauliflower, potato, tomato and brinjal). At each site, control plots were established, and soil pH and production before and after the lime application was studied. The results show that there is a slight increase in soil pH (by 0.1–0.3) after one crop season following lime application. Interestingly, the production of potato increased by about 50% in plots where lime was applied. A few research farmers pointed out that it was easier to till the land after lime application. The effect of lime on soil pH and production demands much more intensive scientific study including cost–benefit analysis, proper design, and accuracy of measurement.

CONCLUSIONS

Nepal is a mountainous country with distinct altitudinal, topographical, climatic and vegetational characteristics. The deficiency of nutrients such as phosphate, potassium, sulphur, manganese, zinc, cadmium, lead, calcium, magnesium, boron and molybdenum in the soils is widespread in different regions of the country. Population growth, industrialization and unsustainable urbanization are leading to an accumulation of heavy metals in the soils. Bioremediation of arsenic-contaminated soils and groundwater shows a great potential for future developments in Nepal. Plant growth promoting *Rhizobacteria* seem to be fruitful for combating heavy-metal stress through the process of exclusion, extrusion, accommodation, bio-transformation, methylation and demethylation. Since Nepal is a rice-consuming country, heavy rice cultivation may also lead to arsenic and nitrate pollution; therefore, at least nitrogen-efficient genotypes can be used to reduce nitrate pollution in the country (Hakeem et al., 2011).

There is a need for further research on assessment of soil nutrients and other toxic pollutants present in soil throughout the country. Very few remediation studies of soil toxicity and remedial practices have been carried out in Nepal. Degradation of soil fertility and toxicity is causing nutritional, health and sanitation problems in addition to food adequacy; therefore, there is an urgent need for studies on the phytoremediation technology (Ashraf et al., 2010a, b; Ozturk et al., 2008, 2012).

REFERENCES

Andersen, P., 2007. A review of micronutrient problems in the cultivated soil of Nepal. An Issue with Implications for Agriculture and Human Health. Mountain Research. Dev. 27 (4), 331–335. http://dx.doi.org/10.1659/mrd.0915.

Andersen, P., Sandvold, S., 2000. Nutrient deficiencies in cultivated Soils, a study of macro and micro nutrients from Koshi Hills, Eastern Nepal. In: Lag, J. (Ed.), Geomedical Problems in Developing Countries. The Norwegian Academy of Science and Letters, Oslo, Norway, pp. 235–240.

Anup, K.C., Bhandari, G., Wagle, S.P., Banjade, Y., 2013. Status of soil fertility in a community forest of Nepal. Int. J. Environ. 1 (1).

Ashraf, M., Ozturk, M., Ahmad, M.S.A. (Eds.), 2010a. Plant Adaptation and Phytoremediation. Series: Tasks for Vegetation Science. Springer Verlag. 481 pp.

Ashraf, M., Ozturk, M., Ahmad, M.S.A., 2010b. Toxins and their phytoremediation. In: Ashraf, et al. (Ed.), Plant Adaptation and Phytoremediation. Springer Verlag, pp. 1–34.

Bhatta, M.R., Ferrera, G.O., Duveiller, E., Justice, S., 2005. Wheat sterility induced by boron Deficiency in Nepal. In: Andersen, P., Tuladhar, J.K., Karki, K.B., Maskey, S.L. (Eds.), Micronutrients in South and South East Asia. International Centre for Integrated Mountain Development (ICIMOD), Kathmandu, Nepal, pp. 221–229.

Cakmak, I., 2002. Plant Nutrition Research: Priorities to Meet Human Needs for Food in Sustainable Ways. Plant and Soil 247, 22.

Dahal, R.K., 2006. Geology for Technical Students. Bhrikuti Academic Publications, Kathmandu, Nepal.

Drechsel, P., Giordano, M., Gyiele, L., 2004. Valuing Nutrients in Soil and Water: Concepts and Techniques with Examples from IWMI Studies in the Developing World. International Water Management Institute, Colombo, Sri Lanka.

Gruhn, P., Goletti, F., Yudelman, M., 2000. Integrated Nutrient Management, Soil Fertility, and Sustainable Agriculture: Current Issues and Future Challenges. International Food Policy Research Institute, Washington, U.S.A.

Gupta, R.P., Pandey, S.P., Tripathi, B.P., 1989. Soil Properties and Availability of Nutrient Elements in Mandarin Growing Areas of Dhankuta District Pakhribas Agricultural Research Centre Technical Paper. Pakhribas Agricultural Research Centre, Pakhribas, Nepal.

Hagen, T., 1969. Report on the Geological Survey of Nepal Preliminary Reconnaissance. Memoires De La Soc, Helvetique Des Sci. Naturelles, Zurich.

Hakeem, K.R., Ahmad, A., Iqbal, M., Gucel, S., Ozturk, M., 2011. Nitrogen efficient rice genotype can reduce nitrate pollution. Environ. Sci. Pollut. Res. 18, 1184–1193.

Jaiswal, S., 2011. Role of rhizobacteria in reduction of arsenic uptake by Plants: A Review. J. Bioremed. Biodegrad. 2 (126). http://dx.doi.org/10.4172/2155-6199.1000126.

Kang, B.T., Tripathi, B., 1992. Technical Paper 1: Soil Classification and Characterization The AFNETA Alley Farming Training Manual, Source Book for Alley Farming Research, 2.

Karki, K.B., Tuladhar, J.K., Uprety, R., Maskey, S.L., 2005. Distribution of Micronutrients Available to Plants in Different Ecological Regions of Nepal. In: Andersen, P., Tuladhar, J.K., Karki, K.B., Maskey, S.L. (Eds.), Micronutrients in South and South East Asia. International Centre for Integrated Mountain Development (ICIMOD), Kathmandu, Nepal, pp. 17–29.

Khatri-Chhetri, T.B., Ghimire, S.K., 1992. A Review of the Boron Deficiency Problem in Nepal. Unpublished paper presented at International Workshop on Boron Deficiency Problems in Cereals. Chiang Mai University, Thailand.

LRMP, 1986. Land Utilization Report. Kenting Earth Sciences Limited.

Manandhar, D.N., 1989. Climate and Crops of Nepal. Nepal Agricultural Research Council and Swiss Agency for Development and Cooperation, Nepal.

Mandal, S.N., 2002. An Inventory of Current Soil Fertility Status of Mahottary District, Nepal. Soil Testing and Service Section, Department of Agriculture, Kathmandu, Nepal.

Ozturk, M., Sakcali, S., Gucel, S., Tombuloğlu, H., 2010. Boron and Plants. Plant Adaptation & Phytoremediation. In: Ashraf, et al. (Ed.). Springer Verlag. Part 2, pp: 275–311.

Ozturk, M., Yucel, E., Gucel, S., Sakcali, S., Aksoy, A., 2008. Plants as biomonitors of trace elements pollution in soil. Trace Elements: Environmental Contamination, Nutritional Benefits and Health Implications. In: Prasad, M.N.V. (Ed.), Chapter 28. John Wiley & Sons, USA, pp. 723–744.

Ozturk, M., Memon, A.R., Gucel, S., Sakcali, M.S., 2012. Brassicas in Turkey and their possible role in the phytoremediation of degraded habitats. In: Anjum, N.A., et al. (Ed.), The Plant Family Brassicaceae: Contribution Towards Phytoremediation Springer Verlag, Environmental Pollution Book Series 21, pp. 265–288. http://dx.doi.org/10.1007/978-94-007-3913-0-10.

Pariyar, D., 2008. (Cartographer). Country Pasture/Forage Resource Profiles of Nepal.

Paudyal, K., 2002. Geology for Civil Engineers. Oxford International Publication.

Rai, S.K., Pandey, S.P., Khadka, Y.G., Karki, K.B., Uprety, R., 2005. Mapping Spatial Zinc Distribution in Rupandehi District, Nepal. In: Andersen, P., Tuladhar, J.K., Karki, K.B., Maskey, S.L. (Eds.), Micronutrients in South and South East Asia. International Centre for Integrated Mountain Development (ICIMOD), Kathmandu, Nepal, pp. 49–56.

Reddy, E.N.V., Devakumar, A.S., Charan Kumar, M.E., Madhusudana, M.K., 2012. Assessment of Nutrient Turnover and Soil Fertility of Natural Forests Of Central Western Ghats. Int. J. Sci. Nat. 3 (1), 5.

Sapkota, K., Andersen, P., 2005. Commercial Horticulture Farming and its Effect on Soil Fertility: A Case Study from Peri Urban Agriculture in the Kathmandu Valley. In: Andersen, P., Tuladhar, J.K., Karki, K.B., Maskey, S.L. (Eds.), Micronutrients in South and South East Asia. International Centre for Integrated Mountain Development (ICIMOD), Kathmandu, Nepal, pp. 153–165.

Schreier, H., Brown, S., Shah, P.B., 2007. The Jhikhu Khola Watershed Project in the Nepalese Himalayas: Approaches Used and Lessons Learned. International Development Research Centre, Ottawa, Canada.

Shrestha, H.D., 2003. Heavy metals pollution in the environment of Kathmandu. J. de Physique IV (Proceedings) 107 (1), 1239–1246.

Sippola, J., Lindstedt, L., 1994. Report on the Soil Test Results for Samples from Nepal. Jokioinen, Finland: Unpublished report by the Agricultural Research Centre of Finand, Institute of Soils and Environment.

Tripathi, B.P., 1999. Soil Fertility Status in the Farmers' Fields of the Western Hills of Nepal. Paper presented at the Lumle Seminar Paper. Lumle, Nepal.

Tripathi, B.P., 2003. Soil Fertility Issues and Approaches in the Hills of Nepal. Unpublished Paper Presented at a Seminar/ Workshop on Land and Forest Degradation in Marginalised Agriculture of Nepal. Kathmandu, Nepal.

Turton, C., Vaidya, A., Tuladhar, J., Joshi, K., 1997. The Use of Complementary Methods to Understand the Dimensions of Soil Fertility in the Hills of Nepal PLA Notes. pp. 37–41. IIED London.

Upreti, B.N., 1999. An overview of the stratigraphy and tectonics of the Nepal Himalaya. J. Asian Earth Sci. 17, 741–753.

Chapter 12

Transfer of Heavy Metals and Radionuclides from Soil to Vegetables and Plants in Bangladesh

Mahfuza S. Sultana,* Y.N. Jolly,[†] S. Yeasmin,[†] A. Islam,* S. Satter* and Safi M. Tareq*

*Department of Environmental Sciences, Jahangirnagar University, Savar, Dhaka, Bangladesh; [†]Chemistry and Health Physics Division, Atomic Energy Centre, Dhaka, Bangladesh

INTRODUCTION

Humans have continuously been exposed to natural ionizing radiation from both terrestrial and extra-terrestrial origins. Terrestrial radiation is emitted from natural radionuclides present in varying amounts in all types of soils, rocks, air, water, food, in the human body itself and other environmental materials around us. The mobility of the natural radionuclides which have long half-lives largely depends on their chemical properties and, usually, their chemical toxicity exceeds their radiological toxicity.

Environmental pollution by heavy metals and radionuclides is a worldwide phenomenon and has adverse effects on human health, plants and animals. It is severe in Bangladesh due to rapid and unplanned urbanization and industrialization specifically surrounding Dhaka, where industries, nuclear facilities and residential areas are located in the same regions such as the Dhaka Export Processing Zone, Savar. A recent survey showed that more than 1200 industries in Bangladesh discharge about 35,000 m^3 of waste products that pollute air, water and land, ultimately leading to environmental degradation; around 49% of the industries are located in the areas surrounding Dhaka (DoE, 2009). Although most industries are required to have effluent treatment plants, so far only a few have installed these plants; and even then, most of the installed treatment plants operate only occasionally. This is one of the major causes of environmental pollution in Bangladesh. Further, major industrial hot-spots are located close to or in the major cities and are normally sited adjacent to rivers or water bodies that facilitate the disposal of effluents. Discharge of untreated and semi-treated

Soil Remediation and Plants. http://dx.doi.org/10.1016/B978-0-12-799937-1.00012-7

wastewater from most of these industries into the river, canal or wetland causes serious water pollution. Therefore, river pollution and fresh water depletion are viewed as the major forms of environmental pollution in Bangladesh. The main sources of industrial pollution in Bangladesh are from textiles, tanneries, fertilizer industries, steel mills, sugar industries, paper industries, etc. (Mahfuza et al., 2011a, 2012; Sultana et al., 2003; Islam et al., 2012; Kamal et al., 2007).

Industrial or municipal wastewater irrigation is a common reality in three-quarters of the cities in Asia, Africa and Latin America (Gupta et al., 2008; Rattan et al., 2005). Long-term use of industrial wastewater in irrigation is known to make a significant contribution to heavy metals such as Cd, Cu, Zn, Cr, Ni, Pb and Mn in surface soil (Mapanda et al., 2005). Agricultural soil contamination with heavy metals through the repeated use of untreated or partially treated wastewater from various industries, polluted river water and unlimited use of fertilizers and pesticides is one of the most severe ecological problems in Bangladesh (Ahmad and Goni, 2010). In the Bangladeshi context, some of the major known toxic bio-accumulative metal pollutants from industrial sectors which are particularly dangerous are Hg, Pb, As, Cr, Ni, Cu, Zn and Cd (Faisal et al., 2004; Ahmad and Goni, 2010; Chowduhry et al., 2010; Mahfuza et al., 2011b; Naser et al., 2011). Heavy metal contamination of soil resulting from wastewater irrigation is a cause of serious concern due to the potential health impacts of consuming contaminated products. Studies have shown that heavy metals are potentially toxic to crops, animals and humans when contaminated soils are used for crop production, because heavy metals easily accumulate in vital organs to threaten growing crops and human health (Sharma et al., 2009). Intake of heavy metals through the food chain by the human population has been widely reported throughout the world (Muchuweti et al., 2006). Due to their non-biodegradable and persistent natures, heavy metals are accumulated in vital organs in the human body such as the kidneys, bones and liver and are associated with numerous serious health disorders (Duruibe et al., 2007).

Moreover, Bangladesh has planned to set up the country's first nuclear power plant at Rooppur and then progressively other plants in the *Southern Part of Bangladesh* to alleviate the energy crisis. As Bangladesh is a small and densely populated country, any kind of radioactivity release to the environment may have serious impacts on human health. Moreover, analogs to heavy metals, radionuclides can reach the human body through several food chains in the environment. Once the radionuclide is taken into the human body by ingestion of food and water or inhalation, it can distribute into the bone where it has long biological half-life. Exposure to radioactivity can cause cancers and other body disorders. Therefore, it is important to measure radioactivity in both terrestrial and aquatic environments to determine the background level and also the extent of change of the natural background activity with time resulting from any radioactive release from anthropogenic sources.

The uptake of radionuclides and heavy metals by plants from the soil in most cases is described by soil to plant transfer factor (PTF). In this case, the phenomenon of radionuclides and heavy metals from soil to biota can also be described in ecological models by transferring these from soil to various compartments of the environment. Knowledge of the level of these toxic substances in human diet or animal feed is of particular concern for the assessment of the possible human health risk.

However, some research has been carried out on the heavy metal contamination in soil by irrigation water in the vicinity of industrial areas of Bangladesh (Ahmad and Goni, 2010; Chowdhury et al., 2010; Naser et al., 2009; Rahman et al., 2012). But, the uptake of heavy metal by plants or food crops from the soil to biota and its impacts on the human health in Bangladesh has been rarely reported up to now. Further, the consumption of rice crop and vegetables are very high in Bangladesh, so it is necessary to obtain data of the PTF for the crops and vegetables.

In view of the scarcity of information on the level of radionuclides and heavy metals in various environmental components in Bangladesh, an extensive research programme has been conducted on the measurement of radionuclides and heavy metals to determine their degree of pollution, transfer mechanism, health risk assessment and bioremediation in terrestrial environments. Thus, this study was conducted to calculate the PTF of heavy metals Fe, V, Cr, Zn, As, Mn, Co, Cu, Ni, Sr, Mo, Cd, Pb, Hg and natural radionuclides ^{226}Ra, ^{232}Th, ^{238}U, ^{137}Cs and ^{40}K in various species of commonly consumable vegetables and plants from the soils of various regions of Bangladesh and to assess the potential health risk for inhabitants through the consumption of food crops contaminated with heavy metals.

MATERIALS AND METHODS

Geology and Geomorphology of the Study Area

Bangladesh is a country in Asia, located between latitude 20°34′ and 26°38′N and longitude 88°3′ and 92°42′E. It has a tropical climate characterized by three main seasons: (1) a dry season from November to March; (2) a rainy monsoon season from June to October; and in addition to the summer monsoon, winter depressions originating from the Mediterranean and the so-called (3) 'Nor Westers' in April and May contribute to annual precipitation. The climatic feature of Bangladesh varies region to region and season to season.

The soils of Bangladesh were formed from different kinds of parent materials occurring in various topographic and drainage conditions. They are spread over three physiographic units: (1) northern and eastern hills of tertiary formations, covering 12% of the total area; (2) Pleistocene terraces of the Madhupur and Barind tracts, covering 8% of the total area; (3) recent flood plains, covering 80% of the total area. Sandstone, siltstone, shale and clay stones are the main ocean sedimentary rock types existing all over Bangladesh. Details of the geology and geomorphology of Bangladesh are given elsewhere (Islam et al., 2000).

Sampling

Sampling was carried out from January 2012 to August 2013 around the industrial areas of Dhaka, the capital city, and other rural areas of Bangladesh. Different kinds of plants, leafy and non-leafy vegetables, which are grown in these areas throughout the year for the consumption of surrounding inhabitants and associated rooting zone soils (5–15 cm) were collected randomly in triplicate for the analysis of heavy metals.

The study areas were located in industrial and non-industrial areas of Bangladesh surrounded by various rivers which exposed different degrees of environmental pollution. These are:

Area I – the agricultural lands of the southwestern end of Dhaka and irrigated with Buriganga river water which receives pollution load from the clusters of hazaribug tannery industries and also from Naraynganj knitwear industries, sewerage wastes and urban pollution. The food crops and associated rooting zone soil were collected from this area.

Area II – the agricultural lands on the south side of Dhaka irrigated with Dhaleswari river water which receives pollution load from the southern end of Naraynganj industrial clusters and agricultural practices. The food crops and associated soils were collected from the sampling area.

Area III – the agricultural lands irrigated with Bangshi river water bordering the western edge of Dhaka which receives pollution load from Dhaka Export Processing Zone (DEPZ) and Savar industrial clusters. The food crops and associated soils were collected from the sampling area.

Area IV – The vegetable garden of the rural areas of Pabna irrigated with Padma river surface water. The food crops and associated soils were collected from the sampling area.

Area V – The vegetable garden of the urban areas of Dhaka which is supposedly polluted mainly by agricultural practices. The food crops and associated soils were collected from the sampling area.

Further, the samples were collected from the agricultural lands and vegetable gardens (Area VI) of the sandy beach zones of southeastern part at Teknaf, Cox's Bazar of Bangladesh for the analysis of radionuclides. The Teknaf beach formation is a recent geological feature. Heavy minerals are found as placer deposits in this area in huge quantities due to high beach sediment deposition. A recent radiometric study in this area has revealed the presence of a very high radiometric anomalous zone (Mollah and Chakraborty, 2009). Tulatoli village of Teknaf is an old beach and a well-populated region. Therefore, there is great potential for the incorporation of radionuclides in the food chain from the soil of this area. Papaya (*Carica papaya*) was selected as a food crop in this study. About 2–3 kg (fresh weight) of food crop sample was collected from each point. Soil samples were collected from 5–15 cm depth of associated root zone.

The details of the plant and vegetable samples collected from these areas are given in Table 12.1 and the map of the sampling areas is shown in Figure 12.1.

Analysis of Heavy Metals and Radioactivity

Sample Preparation for Heavy Metals

The collected soil samples were dried at 60°C in an oven until constant weight was achieved. The dried samples were ground to fine powder in an Agate mortar with a pestle and preserved in polyethylene bags in a desiccators for further analysis. The whole plants and the edible part of the vegetable samples were

TABLE 12.1 Description of Vegetables and Plants for Heavy Metal Analysis of the Different Studied Areas

Sampling Area	Common Name	Sample ID	Scientific Name	Vegetable/ Plant
Area I	Helencha	H-I	Enhydra fluctuans	Leafy
	Fern	F-I	Nephrolepis exaltata	Plant
	Rice	R-I	Oryza sativa	Plant
Area II	Helencha	H-I	Enhydra fluctuans	Leafy
	Pui shak	Pui-II	Basella alba	Leafy
	Ladies' fingers	Lf-II	Abelmoschus esculentus	Non-leafy
	Amaranthus	Am-II	Amaranthus caudatus L	Leafy
	Pop corn	Pc-II	Zea Mays	Plant
Area III	Pui shak	Pui-III	Basella alba	Leafy
	Helencha	H-III	Enhydra fluctuans	Leafy
	Kuchu	Kc-III	Colocasia	Leafy
	Egg plant	Ep-III	Solanum melongena L	Leafy
Area IV	Spinach	SP-IV	Beta vulgaris L	Leafy
	Amaranthus	AM-IV	Amaranthus caudatus(red) L	Leafy
	Egg plant	Ep-IV	Solanum melongena L	Non-leafy
Area V	Pui shak	Pui-V	Basella alba	Leafy
	Ladies' fingers	Lf-V	Abelmoschus esculentus	Non-leafy

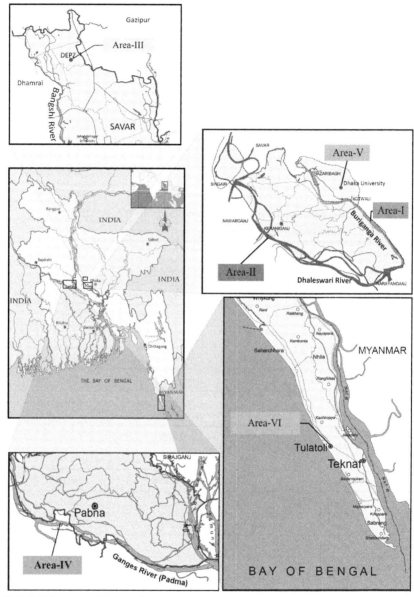

FIGURE 12.1   Map showing different sampling areas of Bangladesh. For a colour version of this figure, please see the section at the end of this book.

first thoroughly washed with tap water and finally with deionized water. The samples were then dried in an oven at 60°C until constant weight was achieved. The dried samples were finally ground in a carbide mortar with a pestle and preserved in polyethylene bags in desiccators until subsequent analysis.

Heavy Metal Measurement

The heavy metal (Fe, V, Cr, Zn, As, Mn, Co, Cu, Ni, Sr, Mo, Cd, Pb and Hg) analysis of the soil and plant samples was carried out using isotopic Energy Dispersive X-ray Fluorescence in Trace Metal Analysis Laboratories of Atomic Energy Centre, Dhaka.

For irradiation of the sample with an X-ray beam, 2 g of each powdered material was pressed into a pellet of 25 mm diameter with a pellet maker (Specac) and loaded into the X-ray excitation chamber with the help of an automatic sample changer system. The irradiation of all samples was performed by assigning a time-based programme controlled by a software package provided with the system. The standard materials were also irradiated under similar experimental conditions for construction of the calibration curves for quantitative elemental determination in the respective samples. The generated X-ray spectra of the materials were stored into the computer. The X-ray intensities of the elements in sample spectrum were calculated using the system software by integration of area of the respective X-ray peak areas using peak fitting deconvolution software. All the analysis was replicated three times. The precision and accuracy were checked by analysis of IAEA certified Standard Reference Material, Soil (Montana-1) and Plant (Spinach) for soil and plant, respectively. Good agreements between the certified and measured values were observed in the reference sample analyzed in this study.

Sample Preparation for Radioactivity

All the samples were pretreated according to the recommendations given by IAEA, 1989. The soil samples were dried at 105–110°C. Then these samples were smashed with mortar and pestle, homogenized, screened with a laboratory test sieve of aperture 425 μm (Mesh No: 40 and Serial No: 238523) and weighted with the weighing balance (Table balance, capacity 1000 g) supplied by shanghai, Japan. The weight of the soil samples were in between 350 and 380 g. All samples were then packed individually into cylindrical plastic containers (7.3 cm diameter × 8.9 cm height) and tightly sealed.

The vegetable (papaya) samples were first washed with distilled water. Then they were kept in a tray and were sundried for 48 h. After that they were dried in the oven at 105–110°C for 48 h and the dry weights were recorded. The weights of the papaya samples were between 30 and 95 g. Papaya samples were packed in same type of containers in same way.

All of the samples were then kept at room temperature for about 30 days to ensure that ^{238}U and its daughter products were in secular equilibrium.

Radioactivity Measurement

The activity concentrations of gamma ray emitting radioisotopes in the samples were measured using a gamma ray spectrometer with a high-resolution high-purity germanium (HPGe) coaxial detector coupled with a Silena Emcaplus multichannel analyzer. The effective volume of the detector was 83.469 cm^3 and energy resolution of the 1.33-MeV energy peak for ^{60}Co was found to be

1.69 keV at full width half maximum with a relative efficiency of 19.6%. All samples were counted for 5000 s. The activity concentration of ^{226}Ra was calculated through 609.3 keV of ^{214}Bi. ^{238}U was calculated through 351.92 keV of ^{214}Pb and 1120.3 keV, 1764.5 keV of ^{214}Bi. ^{232}Th activity was calculated through and 238.63 keV of ^{212}Pb, 583.14 keV ^{208}Tl, 911.07 keV, 969.11 keV of ^{228}Ac, respectively. The ^{40}K and ^{137}Cs activity were calculated through 1460.75 keV and 661.66 keV, respectively. The radioactivity concentration of each radionuclide was calculated using the following equation (IAEA, 1989):

$$A_i \left(Bq\ kg^{-1} \right) = \frac{N}{W \times \varepsilon \times P\gamma}$$

where A_i is the activity concentration of each radionuclide in the sample, N is the net count of each radionuclide which is found by subtracting the sample radionuclide activity counting from background activity counting, ε is the detector efficiency of the specific γ ray, P_γ is the transition probability of the specific γ ray and W is the mass of the sample (kg).

Plant Transfer Factor (PTF) Determination

PTF is the ratio of the concentration of heavy metals/radionuclides in a crop (in Bq/kg or mg/kg dry weight) to the concentration of the heavy metals/radionuclides in the soil (in Bq/kg or mg/kg dry weight).

$$PTF\ for\ heavy\ metal = \frac{Heavy\ Metal\ (mg/kg\ dry\ weight)\ in\ plant\ or\ vegetable\ tissues}{Metal\ concentration\ in\ soil\ (mg/kg\ dry\ weight)}$$

$$PTF\ for\ radionuclides = \frac{Radionuclides\ (Bq/kg\ dry\ weight)\ in\ vegetable\ tissues}{Radionuclides\ concentration\ in\ soil\ (Bq/kg\ dry\ weight)}$$

Data Analysis

Single Pollution Load Index (P_{SLI}) of Soil

The degree of soil pollution for each metal was measured using the single pollution load index (P_{SLI}) evaluation method. As there is no established permissible limit available for heavy metals in soil of Bangladesh, evaluation criteria standards for determining the level of heavy metal content in soil were compared with the permissible limit set by NEPA, China (2001) and Kabata-Pendias and Pendias (1992). The following equation was used to assess the P_{SLI} level in soils,

$$P_{SLI} = \frac{C_S}{C_P}$$

where P_{SLI} is the single pollution index, Cs and Cp represents the heavy metal concentrations in the studied soils and permissible limit set for evaluation criteria, respectively (Table 12.2). This evaluation method assesses the degree of individual heavy metal pollution in terms of four levels (Hakanson, 1980).

TABLE 12.2 Mean Concentrations (n = 10) of Heavy Metals in the Various Study Areas of Bangladesh with Permissible Limits

Elements	Area I	Area II	Area III	Area IV	Area V	Permissible Limit of Soil [a,b]
Fe	39864.75 ± 567	38749 ± 465	38398.04 ± 460	34318 ± 432	34906 ± 430	40000[b]
Mn	623.24 ± 52	595.57 ± 49	673 ± 57	691 ± 56	765 ± 66	270[a]
Sr	153.68 ± 43	171.26 ± 52	171.24 ± 53	145 ± 36	109 ± 22	87[a]
Mo	3.58 ± 0.52	6.13 ± 0.63	5.84 ± 0.79	<0.53	2.3 ± 0.44	1.3[a]
Cd	1.12 ± 0.01	1.58 ± 0.02	2.02 ± 0.02	0.84 ± 0.01	0.715 ± 0.01	1[b]
V	89.33 ± 8.89	<3.53	94.44 ± 10.01	83.63 ± 8.95	79.885 ± 8.83	67[a]
Pb	47 ± 5.06	<1.35	<1.35	<1.35	<1.35	50[b]
Cr	74.67 ± 5.69	109.23 ± 8.98	79.21 ± 3.33	58.17 ± 2.21	52.425 ± 2.58	100[b]
Co	16.8 ± 1.11	17.28 ± 1.12	16.20 ± 1.02	9.10 ± 0.5	8.01 ± 0.9	47[a]
Ni	30 ± 0.48	<0.19	30.69 ± 0.44	20.63 ± 0.22	10.77 ± 0.16	30[a]
Cu	87.66 ± 2.23	38.22 ± 1.50	36.83 ± 1.2	52.76 ± 1.3	56 ± 1.9	30[a]
Zn	444.75 ± 20	116.21 ± 15	174.32 ± 16	97.77 ± 15	181 ± 15	100[b]
As	40.66 ± 5	38.03 ± 4	37.52 ± 3	40.98 ± 5	42.37 ± 5	30[a]
Hg	<0.68	<0.68	<0.68	<0.68	<0.68	0.5[a]

[a]Source: Kabata-Pendias and Pendias (1992).
[b]Source: National Environmental Protection Agency of China (GB 15618, 2001).

Composite Pollution Load Index (P$_{CLI}$) for Soil

Quality of soil environment classification of the studied soil was measured using the composite Load Index (Newmerow) evaluation method (Liang et al., 2011). It represents the total pollution for all heavy metals in the studied soil. Further, the Newmerow Composite Index Method does not only take account of all the individual evaluation factors, but also highlights the importance of the most polluted heavy metals. The Concrete calculation formula for Composite Metal Pollution Index (P$_{CLI}$) is as follows:

$$P_{CLI} = \frac{\sqrt{P_{ave}^2 + P_{max}^2}}{2}$$

where P$_{ave}$ was the average pollution index, and P$_{max}$ was the maximum value of the pollution index.

Heavy Metal Pollution Index (MPI) in Food Crops

To compare the overall heavy metal concentrations in different food crops analyzed in the studied soils, the MPI of food crops was computed. This index was obtained by calculating the average concentrations of all the heavy metals in the plants, leafy and non-leafy vegetables (Usero et al., 1997).

$$MPI\,(mg/kg) = (Cf_1 \times Cf_2 \times ... \times Cf_n)^{1/n}.$$

Cf$_n$ = concentration of heavy metal n in the sample.

The Daily Oral Intake of Heavy Metals (DIM) in Vegetables

The daily oral intake of metals from soil through vegetables was calculated by the following equation:

$$DIM = \frac{C_{metal} \times C_{factor} \times D_{food\,intake}}{B_{average\,weight}}$$

where C$_{metal}$, C$_{factor}$, D$_{food\,intake}$ and B$_{average\,weight}$ represent the heavy metal concentrations in plants (mg/kg), conversion factor (0.085), daily intake of vegetables and average body weight, respectively. The conversion factor, daily intake of vegetables and average weight to dry weight, as described by Rattan et al., 2005. The average daily vegetable intakes for adults and children were considered to be 0.345 and 0.232 kg/person day^{-1}, respectively, while the average child and adult body weights for Bangladesh were considered to be 49.5 and 16.5 kg, respectively (GMA News, 2012).

Health Risk Index (HRI) of Heavy Metal Pollution from Vegetables

The health risk index in this study was defined as a quotient between the estimated exposure to daily metal intake (DIM) from soil through the food chain and reference oral dose (R$_f$D) for each metal (USEPA 2002, RAIS 2007). Therefore,

$$\text{Health Risk Index (HRI)} = \frac{\text{DIM}}{R_f D}$$

Here, $R_f D$ represents the safe levels of oral exposure for a lifetime. A HRI < 1 means that the exposed population is assumed to be safe.

Statistical Analysis

Pearson correlation between the concentrations of heavy metals in soil and vegetables was statistically analyzed using statistical package, SPSS 16, with a significance level $p < 0.01$ (two-tailed).

RESULTS AND DISCUSSION

Concentrations of Heavy Metals in the Top Soil

Heavy metal contents in the agricultural soils of various industrial areas of Bangladesh were compared with the agricultural soils of non-industrial areas which are summarized in Table 12.2. A wide variation in soil heavy metal concentrations were observed in the studied soils where Fe was the most abundant metal with average concentrations of 39864.75 and 34612.23 mg/kg and Cd was the least abundant metal with average concentrations of 1.57 and 0.78 mg/kg in soils of both industrial and non-industrial areas of Bangladesh, respectively. However, the highest deposition of iron in the studied soil was probably due to both industrial exposure of this area and from natural sources, where the reddish brown surfacial clay layer of the Madhupur formation containing Fe minerals release much iron into the studied soil. Mn is the second most abundant heavy metal, followed by Zn, Sr, V, Cr, Cu, As, Ni, Co, Mo and Cd in the soils of both industrial and non-industrial areas. The content of Hg was found to be below the detection limit (0.68 mg/kg) in all the studied soils.

The trend of the heavy metal contents according to average concentration was observed to be: Fe > Mn > Zn > Sr > V > Cr > Cu > As > Ni > Co > Mo > Cd in the soils of both industrial and non-industrial areas, though the average values of the heavy metal contents in the soils of the industrial areas are higher than those in the non-industrial areas. The variation of results is probably due to the variation of heavy metal sources in the studied soils. Moreover, elevated contents of Pb, Zn and Mn were found only in the soils of an industrial area near a steel mill and along the highway, which might be from the atmospheric deposition from automobile exhausts due to heavy traffic. The results were in good agreement with the previous findings near another highway in Bangladesh by Huq and Islam (1999).

Previously, the background concentration of the soil of Bangladesh was found to be 0.01–0.2 mg/kg for Cd, 12–20 mg/kg for Pb and 68 mg/kg for Zn (Kashem and Singh, 1999). The total heavy metal contents of the studied soils contain higher levels than these previous data except Cd, reflecting severe pollution. The extent of pollution of heavy metals in the agricultural soil of Bangladesh has been increasing gradually through the repeated use of polluted wastewater directly from various industries or polluted river water used for irrigation, and also due to the excess use of chemical fertilizers and pesticides. Similar results were observed for the heavy metal contents in different industrial sites of Bangladesh (Alam et al., 2003; Naser et al., 2009; Ahmad and Goni, 2010; Rahman et al., 2012).

Pollution Index Assessment of Soil

Single Pollution Load Index (P_{SLI})

Single Pollution Load Index (P_{SLI}) and Composite Pollution Load Index (P_{CLI}) methods are used for the quantitative assessment of the degree of heavy metal pollution in the agricultural soil of the various regions near industrial and non-industrial areas of Bangladesh. As there is no established safe limit available for heavy metal in the soil of Bangladesh, evaluation criteria standards for determining the level of heavy metal content in soil were compared with the permissible limit set by NEPA of China (2001) and Kabata-Pendias and Pendias (1992) (Table 12.2). The single pollution index factor assesses quantitatively the individual heavy metal pollution of the area which directly indicates the pollution of the specific environmental indicators.

Single Pollution Load Index (P_{SLI}) and pollution level of the analyzed heavy metals in the soils of the both industrial and non-industrial areas are shown in Table 12.3. The table shows that Area I is seriously contaminated by Zn and moderately contaminated by Pb, Fe, Mn, Sr,Mo, Cd, V, Ni, As and Cu. Area II is seriously contaminated by Mo and moderately contaminated by Fe, Mn, Sr, Cd, Cr, Cu, Zn, As, whereas Area III is seriously contaminated by Mo and moderately contaminated by Fe, Mn, Sr, Cu, Zn, As, V. However, P_{SLI} values for heavy metals in the soils of the non-industrial agricultural soils show moderate pollution for Mn, Sr, Mo, V, Cu, Zn and As and low pollution for Fe, Ni, Cr, Cd, Co. According to the average value of P_{SLI} of the 12 different heavy metals, the ranking pattern in the industrial and non-industrial areas are in the following order: Mo > Zn > Mn > Cu > Sr > Cd > As > Fe > V > Cr > Ni > Co and Mn > Sr > Cu > As > Zn > V > Mo > Cd > Ni > Cr > Co, respectively. Therefore, the single pollution load index P_{SLI} value of the studied soil varied with the different heavy metals and also with locations suggesting pollution from different sources.

Therefore, in the present study Fe, Mn, Sr, Mo, Cd, V, Ni, As, Cr, Mo and Cu were identified as potential heavy metal toxicants in the soils of the different areas of Bangladesh which are related to human health. The heavy metal pollution in the studied soil is mainly due to polluted river water irrigation, urban pollution, application of fertilizers and pesticides and possible atmospheric deposition.

TABLE 12.3 Single Pollution Load Index and Pollution Level of the Heavy Metals in the Studied Soil

Element	Area I		Area II		Area III		Area IV		Area V	
	P_{SLI} Value	Pollution Level	P_{SLI} Value	Pollution Level	P_{SLI} Value	Pollution Level	P_{SLI} Value	Pollution Level	P_{SLI} Value	Pollution Level
Fe	1	Moderate	1	Moderate	1	Moderate	0.86	Low	0.86	Low
As	1.35	Moderate	1.26	Moderate	1.42	Moderate	1.37	Moderate	1.39	Moderate
Mn	2.3	Moderate	2.22	Moderate	2.49	Moderate	3	Moderate	2.92	Moderate
Cu	2.92	Moderate	1.27	Moderate	1.40	Moderate	1.23	Moderate	1.40	Moderate
Zn	4.45	Serious	1.16	Moderate	2.66	Moderate	0.97	Low	1.36	Moderate
Cr	0.75	Low	1.1	Moderate	0.79	Low	0.58	Low	0.55	Low
Pb	1	Moderate	—	—	—	—	—	—	—	—
Ni	1	Moderate	—	—	1.02	Low	0.79	Low	0.57	Low
Cd	1.12	Moderate	1.58	Moderate	0.02	Low	0.84	Low	0.78	Low
Sr	1.77	Moderate	1.97	Moderate	1.83	Moderate	1.66	Moderate	1.45	Moderate
Mo	2.75	Moderate	4.72	Serious	4.10	Serious	—	—	0.88	Low
V	1.33	Moderate	—	—	1.35	Moderate	1.24	Low	1.2	Moderate
Co	0.36	Low	0.37	Moderate	0.36	Low	0.22	Low	0.2	Low

Composite Pollution Load Index (P_{CLS})

The quantitative assessment of the overall pollution of the studied agricultural soil was based on the Composite Pollution Load Index (P_{CLS}). It provides some understanding to the public of the area about the quantity of the environmental pollution and classifies the area depending on the total heavy metal pollution load in the soil. Composite Pollution Load Index (P_{CLS}) of the studied soils showed that the pollution trend in Area I > Area III > Area II > Area V > Area IV. Further, the mean P_{CLS} value of the soil of industrial areas was observed to be > 3, which is classified as a progressive deterioration level and suggests that crops grown in the soil have been affected by the toxic metal (Sridhara et al., 2008). The higher pollution level was also observed in the steel mill area (Area I) than the other industrial areas (II and III) which might be due to an excessive amount of heavy metals in the discharged effluents and also from the atmospheric deposition from the surrounding metal industries. Moreover, this area is near Mawa highway which connects central Dhaka to the western part of Bangladesh so it might also be polluted by automobile exhausts from heavy traffic. Further, Composite Pollution Load Index (P_{CLS}) value of the agricultural and garden soils of the non-industrial area showed $P_{CIS} > 1$ which indicates mild levels of pollution due to the application of excess fertilizers and pesticides.

Therefore, it is necessary to protect the soil of these areas from heavy metal pollution by controlling the source of pollution through proper treatment of waste water or by preventing further discharge of waste water into the river. Moreover, optimization of the application of agricultural practices such as fertilizers and pesticides uses in the agricultural fields of Bangladesh are necessary to protect the agricultural products from heavy metal pollution.

Levels of Heavy Metals in Plants and Vegetables

Heavy metal content of the most commonly consumable leafy and non-leafy vegetables and food crop plants grown in various industrial areas and also various rural and urban vegetable gardens of Bangladesh were investigated, as shown in Figure 12.2a and b.

Heavy metal concentration showed wide variations among different plants and vegetables grown in the studied soils of industrial and non-industrial areas which may be attributed to differential absorption capacity of vegetables for different heavy metals. Mean concentrations of all the heavy metals were observed to be higher in plants as compared with leafy and non-leafy vegetables, except Hg which was detected below the detection limit (0.04 mg/kg) in all the vegetables. Among all the heavy metals Fe, Zn and Mn showed the highest accumulation, Sr, Cr, Cu, Ni, Co, Pb showed moderate accumulation and V, Co, Cd and As showed the lowest accumulation in the all kinds of vegetables. Sharma et al. (2009) have also found the highest concentration of Zn as compared to Cu, Cd and Pb in the vegetables of Varansasi city, India.

Further, highest deposition of Fe within the 13 heavy metals was observed in all of the green vegetables and plants which varied in leafy vegetables from 109 to 2758 mg/kg, in non-leafy vegetables 33 to 545 mg/kg and in green crop plants 369 to 3569 mg/kg. The average Fe content shows the following trend: Green plant>Leafy vegetable>Non-leafy vegetable. Usually, Fe participates in chlorophyll synthesis and photosynthesis, so the green of the plant contains higher Fe than the fruit.

(a)

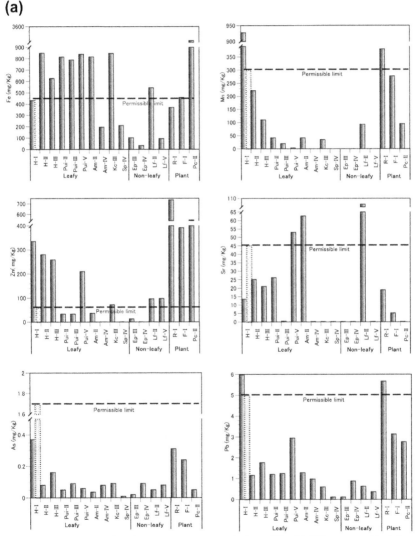

FIGURE 12.2   (a) Variation of heavy metals (Fe, Mn, Zn, Sr, As, Pb) with different vegetables/plants grown in the studied soil. The broken line represents the permissible limit. (b) Variation of heavy metals (V, Cr, Co, Cd, Ni, Cu) with different vegetables and plants grown in the studied soil. The broken line represents the permissible limit.

(b)

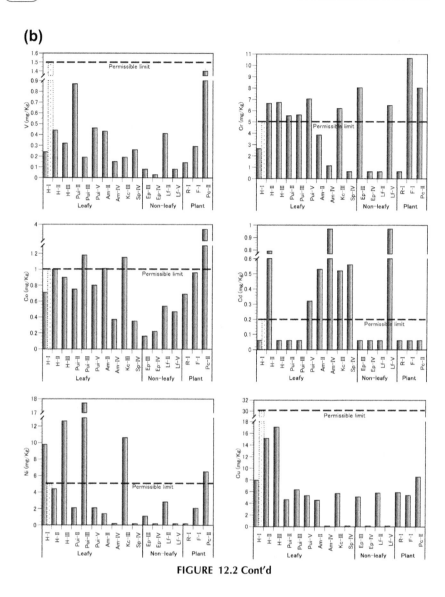

FIGURE 12.2 Cont'd

Besides this, higher content of Zn and Mn was also observed in plants and leafy vegetables than non-leafy vegetables. Zn and Mn are airborne pollutants so might accumulate from the atmospheric deposition of the surrounding metal industries as well as vehicle exhausts. High content of Pb (5.98 mg/kg) was found in the Helencha grown in an industrial area near a steel mill and a highway. High concentrations of Cd (0.97 mg/kg) were found in the plants grown in agricultural soil where fertilizer was applied. Cd was found in vegetables of Bangladesh grown in soil irrigated by wastewater and using phosphate fertilizer which contains Cd as an impurity (Naser et al., 2009). Elevated levels of heavy metals Pb, Cd and Zn in plants near metal smelters have previously been demonstrated in other investigations (Farago and O'Connell, 1983; Cui et al., 2004). The results for heavy metals in vegetables in this work were compared with similar work reported by Ahmad and Goni (2010) and Naser et al. (2009). This comparison revealed that heavy metals Zn and Fe were higher, Cr, Ni and Cu were similar, and Pb and Cd were lower in this work.

The trends of the heavy metal contents according to average concentrations found in plants, leafy vegetables and non-leafy vegetables were as follows: $Fe > Zn > Mn > Sr > Cr > Cu > Ni > Pb > Co > V > As > Cd$; $Fe > Zn > Mn > Sr > Cu > Ni > Cr > Pb > Co > Cd > V > As$; and $Fe > Mn > Zn > Sr > Cr > Cu > Ni > Cd > Pb > Co > V > As$, respectively.

Moreover, the degree of pollution in the plants, leafy and non-leafy vegetables for each heavy metal was evaluated compared with the permissible limit recommended by WHO / FAO (1984) and Kabata-Pendias and Pendias (1992) (Figure 12.2a and b). The sequence of heavy metals according to pollution for leafy, non-leafy vegetables and plants are $Zn > Cd > Ni > Fe > Cr > Mn > Co > Pb > Cu > V > As$, $Cr > Zn > Fe > Ni > Co > Pb > Mn > Cu > V > As$ and $Zn > Fe > Cr > Co > Ni > Pb > Mn > Cu > V > As$, respectively. The Fe, Zn and Cr contents exceeded the permissible limits in all vegetables grown in the industrial areas while Ni exceeded the permissible limit for few leafy vegetables grown in industrial areas. Further, Mn and Pb exceeded the permissible limits only for the vegetables grown in industrial Area I. Mn and Pb are airborne pollutants which might be deposited from surrounding metal industries or from vehicle emissions and readily fix to the hairy or waxy cuticles of leaves, which are then taken in to the foliar cells of the studied vegetables of Area I. The content of Cd exceeded the standard level for a few vegetables grown in both industrial and non-industrial areas where excess fertilizer was used for agricultural production. However, As, V and Cu content showed lower levels than the permissible limits in all the vegetables and plants grown in the studied soils. Generally, the variations of the degrees of pollution in heavy metal concentrations in vegetables may be ascribed to the variability in the absorption of metals in plants and their further translocation within the plants (Vousta et al., 1996).

The higher levels of heavy metals Cd, Cr, Pb, Zn, Mn and Fe in the studied vegetables cause adverse health effects in the surrounding inhabitants. In general, Fe and Mn are interrelated in their metabolic functions and their appropriate proportion ($Fe / Mn = 2.5$) is necessary for healthy plant. Most of the leafy and non-leafy vegetables exceeded this range, indicating toxicity of Fe and consequently deficiency of Mn in the studied vegetables. However, Helencha grown in Area I has an Fe / Mn ratio of 0.46, which might cause Mn toxicity to the surrounding residents through the ingestion of these vegetables. Manganese can bring a variety of serious toxic effects upon prolonged exposure to elevated concentrations. The common symptom is called 'manganism' which is similar to symptoms found in Parkinsonism (Chu et al., 1995).

Further, Mn–Fe , Mn–Cr antagonistic interactions have been observed in the studied vegetables in which uptake of one element is competitively inhibited by the other. This may indicate the same carrier sites in absorption mechanisms of both metals. It is also supported by the observed negative Pearson correlation values of Mn–Fe (−0.03) and Mn–Cr (−0.11).

Therefore, the results of the present study demonstrate that the plants and vegetables grown in the studied soils are greatly polluted with heavy metals and pose a major health concern.

Pollution Index Assessment of Different Crops

Metal pollution index (MPI) is a reliable and precise method for the quantitative assessment of heavy metal pollution of the crops grown in the polluted soil and used for regular monitoring of pollution. MPI evaluated the overall heavy metal pollution in the different types of crops analyzed in the studied soil. Figure 12.3 represents the MPI value of the different plants grown in both industrial and non-industrial areas. The MPI value varied with sampling areas as well as plant species.

In Area I the metal accumulation was observed to be highest in the rice plant, followed by Helencha and Fern. In Area II, the metal accumulation was observed to be highest in *Zea mays*, followed by Helencha, Ladies' fingers and Pui shak. In Area III, the metal accumulation was observed to be highest in Helencha, followed by Pui shak, Kochu shak and Eggplant. In Area IV, the metal accumulation was observed to be highest in Amaranthsus, followed by spinach and Eggplant. In Area V, the metal accumulation was observed to be highest in Puishak, followed by Ladies' fingers. The variations of MPI values in vegetables and plants of the

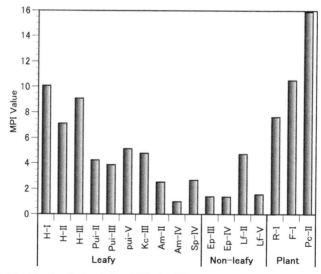

FIGURE 12.3   Metal pollution index (MPI) in different vegetables and plants grown in the studied area.

same area of Bangladesh may be attributed to the differences in their morphology and physiology for heavy metal uptake, exclusion, accumulation and retention.

Moreover, among different species of crops, plants showed the highest MPI value (11.5) followed by leafy vegetables (6.27), and the lowest MPI value (2.21) was observed for non-leafy vegetables. Further, among the various non-leafy vegetables, the highest MPI value was observed in helencha which also varied with locations. According to MPI value, the order of Helencha is Helencha–I (10.07) > Helencha–III (9.08) > Helencha–II (7.13). These plants had higher concentrations of almost all metals evaluated in this study, because they can uptake elements from soil easily. Among the non-leafy vegetables, Ladies' fingers showed the highest value of MPI and the order was Ladies' fingers–II > Ladies' fingers–V. The observed MPI trend in the studied vegetables and plants is consistent with the C_{PLI} among the studied soils of industrial and non-industrial areas of Bangladesh. Therefore, the high MPI values of the vegetables grown in the industrial areas might be due to high contents of heavy metals in the soil as caused by irrigation with metal-contaminated water released from industries. The study area Keranigonj (Areas I and II) is an indirectly polluted area where the river Buriganga and its tributary Dhaleswari are highly polluted by industrial effluents, sewage sludge, municipal waste water and urban pollution. During the rainy season, the river flooded, and submerged the banks and polluted the cultivated land. Further, the higher MPI values in the vegetables also found in industrially polluted area near DEPZ (Area III) where the river Bangshi is highly polluted by industrial effluents of textile, paint, battery, rice milling and chemical industries which also flooded during the rainy season and polluted the surrounding agricultural lands.

Higher MPI values were observed in all the plants and vegetables grown in industrial areas when compared to non-industrial areas. Therefore, the use of polluted irrigation water in the industrial areas increased the uptake and accumulation of the heavy metals in the plants and vegetables from the studied soil. This is consistent with the reports of higher concentrations of all the heavy metals in vegetables from wastewater-irrigated areas of Bangladesh (Ahmad and Goni, 2010).

Higher MPI values of Helencha, Ladies' fingers, Puishak and Kuchu shak of these industrial areas suggest that these vegetables may pose a severe human health risk due to the higher accumulation of heavy metals in the edible portions.

Many researches have shown that leafy vegetables accumulate the highest concentration of heavy metals and frequently exceed the phytotoxic levels found in non-leafy vegetables (Kashem and Singh, 1999; Ahmad and Goni, 2010).

Comparative Study of the Heavy Metal Contents in Soil and Plants

In order to ascertain whether the uptake of heavy metals in the studied vegetables is from their growth media, soil or other sources, a Pearson Correlation was calculated between vegetables and soil of the respective areas of Bangladesh. The results showed (Table 12.4) significant correlation between heavy metal

TABLE 12.4 Pearson Correlation between Heavy Metal Contents in Various Vegetable/Plant Species with Soil of the Respective Areas of Bangladesh

Vegetable / Plant Species	Area	Pearson Correlation	Significance Level
H-I	A-I	0.31	Non-significant
H-II	A-II	0.93	Significant at 0.01 level
H-III	A-III	0.92	Significant at 0.01 level
Pui-II	A-II	0.99	Significant at 0.01 level
Pui- III	A-III	0.99	Significant at 0.01 level
Pui-V	A-V	0.97	Significant at 0.01 level
Am-IV	A-IV	1.00	Significant at 0.01 level
Sp-IV	A-IV	1.00	Significant at 0.01 level
Kc- III	A-III	0.99	Significant at 0.01 level
Ss-II	A-II	0.99	Significant at 0.01 level
Ep-III	A-III	0.99	Significant at 0.01 level
Ep-IV	A-IV	1.00	Significant at 0.01 level
Lf-II	A-II	0.96	Significant at 0.01 level
Lf-V	A-V	0.65	Non-significant
R-I	A-I	0.29	Non-significant
F-I	A-I	0.66	Significant at 0.05 level
Pc-II	A-II	0.99	Significant at 0.01 level

contents in vegetables and plants grown in the studied respective soils at 0.01 level except in Area I. It revealed that the heavy metal contents in the vegetables and plants were translocated efficiently from the soil through the root systems in Areas II, III and IV. However, the adsorption of heavy metal in the vegetables and plant by atmospheric deposition might be more dominant than intake from soil in Area I.

Heavy Metal Transfer from Soil to Food Crops

Plant Transfer Factor of different heavy metals from soil to crops was calculated as the ratio between the concentrations of heavy metals in vegetables and their respective concentrations in the soil. The PTF is one of the key components of human exposure to metal through the food chain. Higher values of PTF suggest poor retention of metals in soil and more translocation in plants. The ability of a metal species in its different forms to migrate from the soil through the plant parts and make itself available for consumption was represented by PTF. The PTF values are given in Table 12.5.

The results indicated that the PTF was at a higher level for Zn, Mn, Cd, Cu, Ni , Sr (0.12–4.4), at a moderate level for Pb, Fe, Co, Cr (0.02–0.14) and at a lower level for V and As (0.009–0.001). Variation in PTF among different vegetables may be attributed to differences in the concentrations of metals in soils and different capacities of heavy metal uptake by different vegetables (Cui et al., 2004).

Generally, a PTF of 0.1 indicates that the plant is excluding the element from its tissues. The greater the transfer factor value than 0.2 the greater the chances of metal contamination of vegetables by anthropogenic activities (Khan et al., 2009). In this study, V, Co and As are excluded from the plant tissues which could be justified by comparing these contents in soil and plants which significantly dropped from 91.89, 87.70, 49.24 mg/kg to 0.44, 0.69, 0.07 mg/kg, respectively. However, Zn, Mn, Cd, Cu, Ni and Sr have PTF values >0.2 in various vegetables grown in the studied soil indicating pollution of these by various anthropogenic activities.

The PTF value for Zn (4.5) in the PC-II was highest among all considered heavy metals in this study which could be attributed to the low retention rate of the metal in soil and therefore it is more mobile in the soil. The lowest PTF was found to be 9.4×10^{-4} for Fe which is the most abundant element in the soil of all areas, but is possibly retained in the soil by strongly binding with the soil matrix.

Moreover, a higher PTF value (>0.2) for Zn was observed in all the vegetables and plants except egg plants grown in agricultural and garden soils. It suggests that Zn is the most phytoavailable heavy metal in the studied agricultural and garden soils which is in good agreement with other studies (Chowdhury et al., 2010). Higher PTF values of 1.15, 0.78, 0.74, 0.5 were found for Cd in amaranth, Pui shak, Ladies' fingers and Helecha, respectively, possibly due to higher mobility of cadmium and, subsequently, lower retention in the soil than other toxic cations (Alam et al., 2003; Lokeshwari and Chandrappa, 2006). The result also supports the finding that accumulation of Pb and Ni is comparatively less than that of Cd in plants because activity of Pb and Ni are low and not easy to migrate. The mobility of metals from soil to plant is a function of the physical and chemical properties of the soil and of vegetable species and is altered by innumerable environmental and human factors (Zurera et al., 1987).

According to PTF, it can be concluded that *Zea mays* is a high Zn accumulator, *Oryza sativa* and *Asplenium platyneuro* are high Mn and Zn accumulators, *Enhydra fluctuans* is a high Mn and Zn accumulator, *Basella alba* is a high Sr, Zn and Cd accumulator, *Amaranthus* is a high Zn and Cd accumulator, Ladies' fingers is a high Sr, Zn and Cd accumulator and *Colocasia* is a high Ni and Zn accumulator in the studied areas.

Soil Remediation and Plants

TABLE 12.5 Heavy Metal Transfer Factors (on Dry Weight Basis) for Vegetables and Plants Grown in the Studied Soil

Plants	Fe	Mn	Sr	V	Cr	Co	Ni	Cu	Zn	As	Pb	Cd
						PTF_{value} (mean)						
Leafy												
H-I	0.01	1.49	0.09	2.69E-03	8.44E-03	0.04	0.32	0.09	0.75	9.09E-03	0.13	
H-II	0.02	0.37	—	—	0.06	0.06	—	0.40	2.4	2.1E-03	—	0.5
H-III	0.016	0.16	0.12	3.4E-03	0.09	0.06	0.4	0.46	1.48	4.32E-03	—	
Pui-II	0.02	0.07	0.16	—	0.05	0.04	—	0.12	0.28	2.2E-03	—	0.21
Pui-III	0.02	0.03	0.3	2.02E-03	0.07	0.07	0.58	0.18	0.19	2.4E-03	—	
Pui-V	0.02	0.04	0.49	5.8E-03	0.10	0.1	0.2	0.095	0.60	9.09E-03	—	0.78
Am-II	0.02	0.07	0.37	—	0.04	0.06	—	0.12	0.30	9.01E-04	—	—
Am-IV	2.95E-03	—	—	1.8E-03	0.02	0.04	—	—	—	2.0E-03	—	1.15
Sp-IV	2.91E-03	—	—	2.1E-03	—	0.04	—	—	—	—	—	
Kc-III	0.02	0.05	—	1.7E-03	0.08	0.07	0.35	0.16	0.4	2.4E-03	—	
Non-Leafy												
LF-II	0.01	0.15	0.69	—	—	0.03	—	0.15	0.83	9.01E-03	—	0.33
LF-V	2.7E-03	—	—	1.0E-03	0.12	0.06	—	—	0.54	1.9E-03	—	0.74

TABLE 12.5 Heavy Metal Transfer Factors (on Dry Weight Basis) for Vegetables and Plants Grown in the Studied Soil—cont'd

PTF_{Value} (mean)

Plants	Fe	Mn	Sr	V	Cr	Co	Ni	Cu	Zn	As	Pb	Cd
EP-III	2.65E-03	—	—	8.5E-03	0.1	0.01	0.04	0.14	0.07	4.76E-03	—	
EP-IV	9.49E-04	—	—	—	—	—	—	—	—	2.2E-03	—	
Plant												
R-I	0.019	0.61	0.12	2.5E-03	—	0.04	—	0.06	1.66	7.7E-03	0.12	
F-I	0.02	0.44	0.034	2.0E-03	0.14	0.06	0.07	0.06	0.88	6.0E-03	0.016	
Pc-II	0.09	0.16	—	—	0.07	0.23	—	0.22	4.5	2.33E-03	—	

Further, PTF for heavy metals Mn, Cr, Ni, Cu, Zn and Cd was observed to be higher in leafy vegetables than in non-leafy vegetables. Similar findings were also reported by Sridhara et al. (2008) and Khan et al. (2008). The higher uptake of heavy metals in leafy vegetables may be due to higher transpiration rates to maintain the growth and moisture content of these plants (Tani and Barrington, 2005).

Conversely, higher PTF was found in *Zea mays* and Fern for Zn and Helencha for Zn and Mn which could be used for phytoremediation from the contaminated soil of Bangladesh to eliminate risk to humans or the environment from these toxic metals.

Health Risk Assessment of Heavy Metals in Vegetables

The transfer factor does not present the risk associated with the metals in any form. The degree of toxicity of heavy metals to human being depends upon their daily intake (Singh et al., 2010). Therefore, in order to assess the health risk of any pollutant, it is essential to estimate the level of exposure by quantifying the routes of exposure of a pollutant to the target organisms. There are various exposure pathways of pollutants to humans but the food chain is one of the most important pathways. Consumption of the heavy-metal-contaminated food can cause human health risks. Thus, to assess the health risks associated with heavy metal contamination of the edible parts of leafy and non-leafy vegetables grown in studied soils, estimated Daily Metal Intake (DIM) and Health Risk Index (HRI) were calculated. Estimated exposure (DIM) and risk index (HRI) of heavy metal contamination to the inhabitants for various vegetables grown in studied soils are shown in Table 12.6 and Table 12.7.

The health risk for heavy metal of the studied vegetables varied indistinctly irrespective of species or location. The daily heavy metal intake (DIM) for both adults and children through the consumption of vegetables in studied soil was higher than $R_f D$ limit set by US-EPA and RAIS for Helencha, Puishak, Kochushak, Eggplant and Ladies' fingers. However, the DIM value of Fe, Mn, Pb, Cr Zn, Ni, As and Cd for the ingestion of these vegetables by children were greater in comparison with those in adults. In addition, inhabitants from industrial areas had the highest intake of Fe, Cr, Mn, Pb, Zn and As through consumption of vegetables whereas only Cd intake through vegetables was higher in non-industrial areas than in industrial areas. Intake of Co, Cu and V through vegetables was similar both in industrial and non-industrial areas.

This simply means that the inhabitants (both adults and children) are highly exposed to health risks (HRI > 1) associated with these heavy metals in the order Mn > Fe > Cr > Pb > Zn > Cd > As in the industrial areas of Bangladesh. In general, Pb, Cr, Ni, Zn, Fe, Mn and Cd are the well-known toxic elements which induce gastrointestinal disorder, diarrhoea, haemoglobinuria, pneumonia, kidney damage, cancer and paralysis in the inhabitants (McCluggage, 1991).

TABLE 12.6 DIM for Individual Heavy Metals in Different Vegetables Grown in Studied Soil

Plants	Individual		Fe	Mn	Cd	V	Cr	Co	Ni	Cu	Zn	As	Pb
Leafy													
H-I	Adult	DIM	0.26	0.55	—	1.42E-04	3.7E-04	4.21E-04	5.79E-03	4.7 E-03	0.20	2.19 E-04	3.54 E-03
	Children	DIM	0.51	1.11	—	4.26E-04	7.53E-04	8.48E-04	0.011	9.49 E-03	0.4	4.42 E-04	7.14 E-03
H-II	Adult	DIM	0.50	0.13	4.68 E-04	2.6E-04	3.93E-03	5.92E-04	2.6E-03	8.96 E-03	0.16	4.74 E-05	6.81 E-04
	Children	DIM	1.01	0.26	9.36 E-04	5.2E-04	7.87E-03	1.18E-03	5.2E-03	0.02	0.33	9.48 E-05	1.37 E-03
H-III	Adult	DIM	0.37	0.06	—	1.89E-04	3.96E-03	5.33E-04	7.46E-03	0.01	0.15	9.48 E-05	1.04 E-03
	Children	DIM	0.74	0.13	—	3.79E-04	7.92E-03	1.06E-03	0.015	0.02	0.3	1.90 E-05	2.08 E-03
Pui-II	Adult	DIM	0.48	0.02	1.89 E-04	5.15E-04	3.29E-03	4.44E-04	1.25E-03	2.74 E-03	0.02	2.99 E-05	7.11 E-04
	Children	DIM	0.96	0.04	3.79 E-04	1.04E-03	6.58E-03	8.88E-04	2.5E-03	5.47 E-03	0.04	5.98 E-05	1.42 E-03
Pui-III	Adult	DIM	0.47	0.01	—	1.12E-04	3.34E-03	6.99E-04	0.01	3.74 E-03	0.02	5.33 E-05	7.20 E-04
	Children	DIM	0.93	0.02	—	2.25E-04	6.69E-03	1.40E-04	0.02	7.48 E-03	0.04	1.06 E-04	1.45 E-03
Pui-V	Adult	DIM	0.50	0.02	3.31 E-04	2.72E-04	2.99E-03	4.86E-04	1.25E-03	3.16 E-03	0.07	3.55 E-05	1.75 E-03
	Children	DIM	1.02	0.04	6.64 E-04	5.45E-04	5.98E-03	9.73E-04	2.5E-03	6.32 E-03	0.13	7.10 E-05	3.40 E-03

Continued

TABLE 12.6 DIM for Individual Heavy Metals in Different Vegetables Grown in Studied Soil—cont'd

Plants	Individual		Fe	Mn	Cd	V	Cr	Co	Ni	Cu	Zn	As	Pb
Am-II	Adult	DIM	0.454	0.02	—	2.37E-04	2.14E-03	5.64E-04	5.79E-04	2.52E-03	0.020	2.0E-05	7.13E-04
	Children	DIM	0.917	0.05	—	4.79E-04	4.33E-03	1.13E-03	1.57E-03	5.08E-03	0.040	4.04E-05	1.43E-03
Am-IV	Adult	DIM	5.8 E-04	—	5.75 E-04	8.89 E-05	6.81 E-04	2.19 E-04	—	—	—	4.74 E-04	3.43E-03
	Children	DIM	1.16 E-03	—	1.15 E-03	1.78 E-04	1.36 E-03	4.38 E-04	—	—	—	9.48 E-04	—
Sp-IV	Adult	DIM	0.12	—	—	1.5 E-04	-	2.15 E-04	—	—	—	—	—
	Children	DIM	0.25	—	—	3.0 E-04	-	4.30 E-04	—	—	—	—	—
Kc-III	Adult	DIM	0.50	0.02	—	1.12 E-04	3.68 E-03	6.81 E-04	6.3 E-03	3.36 E-03	0.04	5.33 E-05	3.55 E-04
	Children	DIM	1.01	0.04	—	2.25 E-04	7.36 E-03	1.36 E-03	0.03	6.73 E-03	0.08	1.06 E-04	7.11 E-04
Non-Leafy													
Lf-II	Adult	DIM	0.32	0.05	3.08 E-04	2.35E-04	—	3.19E-04	1.68E-03	3.42 E-03	0.06	2.99 E-05	3.73 E-04
	Children	DIM	0.65	0.11	6.16 E-04	4.50E-04	—	6.40E-04	3.35E-03	6.84 E-03	0.11	5.98 E-05	7.47 E-04

TABLE 12.6 DIM for Individual Heavy Metals in Different Vegetables Grown in Studied Soil—cont'd

Plants	Individual		Fe	Mn	Cd	V	Cr	Co	Ni	Cu	Zn	As	Pb
Lf-V	Adult	DIM	0.06	—	3.10 E-04	4.74E-05	3.84E-03	2.74E-04	—	—	0.06	4.74 E-05	2.19 E-04
	Children	DIM	0.11	—	6.22 E-04	9.48E-05	7.7E-03	5.49E-04	—	—	0.11	9.48 E-05	4.38 E-04
Ep-III	Adult	DIM	0.06	—	—	4.74 E-04	4.76 E-03	9.48 E-05	6.51 E-04	3.03 E-03	7.61 E-03	1.18 E-05	—
	Children	DIM	0.12	—	—	9.48 E-04	9.53 E-03	1.90 E-04	1.30 E-03	6.05 E-03	0.015	2.4 E-05	—
EP-IV	Adult	DIM	5.21 E-04	—	—	—	—	—	—	—	—	3.5 E-05	3.08E-03
	Children	DIM	1.17 E-03	—	—	—	—	—	—	—	—	7.17 E-05	—

TABLE 12.7 HRI for Individual Heavy Metals in Different Vegetables Grown in Studied Soil

Plants	Individual		Fe	Mn	Cd	V	Cr	Co	Ni	Cu	Zn	As	Pb
Leafy													
H-I	Adult	HRI	0.87	3.93	—	0.03	0.12	0.02	0.30	0.12	0.67	0.73	1.01
	Children	HRI	1.7	7.93	—	0.08	0.25	0.04	0.55	0.24	1.33	1.47	2.04
H-II	Adult	HRI	1.67	0.93	0.47	0.05	1.31	0.03	0.13	0.22	0.53	0.16	0.19
	Children	HRI	3.37	1.86	0.94	0.1	2.62	0.05	0.26	0.5	1.1	0.32	0.39
H-III	Adult	HRI	1.23	0.2	—	0.04	1.32	0.03	0.37	0.25	0.5	0.32	0.3
	Children	HRI	2.47	0.43	—	0.07	2.64	0.05	0.75	0.5	1	0.64	0.59
Pui-II	Adult	HRI	1.6	0.83	0.20	0.1	1.1	0.02	0.06	0.07	0.07	0.1	0.2
	Children	HRI	3.2	1.66	0.4	0.19	2.19	0.04	0.12	0.14	0.13	0.2	0.4
Pui-III	Adult	HRI	1.56	0.42	—	0.02	1.11	0.03	0.5	0.09	0.07	0.18	0.21
	Children	HRI	3.1	0.83	—	0.04	2.23	0.07	1	0.19	0.13	0.35	0.41
Pui-V	Adult	HRI	1.66	0.83	0.33	0.05	0.99	0.01	0.06	0.08	0.23	0.12	0.05
	Children	HRI	3.4	1.66	0.62	0.1	1.99	0.03	0.12	0.16	0.43	0.24	0.21
Am-II	Adult	HRI	1.513	0.164	—	0.047	0.713	0.028	0.028	0.063	0.067	0.067	0.203
	Children	HRI	3.057	0.332	—	0.095	1.443	0.056	0.079	0.127	0.133	0.135	0.409
Am-IV	Adult	HRI	0.12	—	0.56	1.76E-02	2.27 E-01	1.09E-02	—	—	—	1.58E-01	—
	Children	HRI	0.24	—	1.15	3.53E-02	4.53 E-01	2.19E-02	—	—	—	3.16E-01	—

TABLE 12.7 HRI for Individual Heavy Metals in Different Vegetables Grown in Studied Soil—cont'd

Plants	Individual		Fe	Mn	Cd	V	Cr	Co	Ni	Cu	Zn	As	Pb
Sp-IV	Adult	HRI	0.4	—	—	2.97E-02	—	1.07E-02	—	—	—	—	—
	Children	HRI	8.3E-01	—	—	5.95E-02	—	2.15E-02	—	—	—	—	—
Kc-III	Adult	HRI	1.66	0.43	—	2.23E-02	1.22	3.4E-02	3.15E-01	8.4E-02	1.3E-01	1.77E-01	1.01E-01
	Children	HRI	3.36	0.87	—	4.46E-02	2.45	6.8E-02	1.5	1.68E-01	2.66E-01	3.53E-01	2.03E-01
Non-Leafy													
Lf-II	Adult	HRI	1.06	0.39	0.31	0.04	—	1.59E-02	0.08	0.09	0.2	0.1	0.1
	Children	HRI	2.16	0.79	0.62	0.08	—	3.2E-02	0.17	0.17	0.37	0.2	0.21
LF-V	Adult	HRI	0.2	—	0.31	0.09	1.28	0.02	—	—	0.2	1.58E-01	0.06
	Children	HRI	0.37	—	0.62	0.12	2.57	0.04	—	—	0.37	3.16E-01	0.12
Ep-III	Adult	HRI	0.2	—	—	9.4E-02	1.59	4.74E-03	3.25E-02	7.57E-02	2.53E-02	3.93E-02	—
	Children	HRI	0.4	—	—	1.88E-01	3.18	9.5E-03	6.5E-02	1.51E-01	5.0E-02	8.0E-02	—
Ep-IV	Adult	HRI	0.02	—	—	—	—	—	—	—	—	1.16E-01	—
	Children	HRI	0.04	—	—	—	—	—	—	—	—	2.39E-01	—

Further, health risks for leafy vegetables are found to be higher than those for non-leafy vegetables, irrespective of their locations. Similar findings were obtained for heavy metal contents in vegetables grown from waste water irrigated areas of India (Singh et al., 2010). The leafy vegetables grown in industrial areas have HRI values > 1 for Fe, Pb, Zn, Cr, Mn, Cd, Ni, whereas non-leafy vegetables, such as Ladies' fingers have HRI values > 1 for Fe, and Eggplant has a HRI value > 1 for Cr. On the contrary, most of the leafy and non-leafy vegetables grown in the non-industrial area have HRI values < 1 except Puishak–V.

Moreover, Helencha which grown near a steel mill and a highway has HRI values > 1 for Pb in both adult and children. Among different vegetables, Helencha showed the highest values of DIM and HRI, whereas Spinach and Egg plant grown in the non-industrial area showed the lowest values of DIM and HRI. The trend of the vegetables according to HRI value is as follows: Helencha > Puishak > Kochu shak > Amerenthus > Ladies' fingers > Egg plant > Spinach.

However, HRI values < 1 were found for heavy metals Cu, V, Hg and Co in the vegetables analyzed in the studied soil. Hence the heavy metals As, Cu, V, Hg and Co through the consumption of studied vegetables generate risk neither for children nor for adults of the surrounding population. On the other hand, the vegetables particularly, Helecha, Puishak, Kochu shak are contaminated with toxic heavy metals and the consumption of such vegetables can cause human health risks related to carcinogenic and non-carcinogenic effects.

Radioactivity in Soil

The radioactivity concentrations of ^{226}Ra, ^{238}U, ^{232}Th and ^{40}K and ^{137}Cs in different soil samples are shown in the Figure 12.4. Results show that the concentrations of ^{226}Ra, ^{238}U, ^{232}Th and ^{40}K in soil samples varied from 28.64 to 480.42 Bq kg^{-1} with a mean of 254.53 ± 6.66 Bq kg^{-1}, from 20.94 to 366.96 Bq kg^{-1} with a mean of 193.95 ± 7.23 Bq kg^{-1}, from 38.62 to 685.74 Bq kg^{-1} with a mean of 362.18 ± 6.45 Bq kg^{-1}, from 87.20 to 184.37 Bq kg^{-1} with a mean of 135.79 ± 24.63 Bq kg^{-1}, respectively. The mean activity of ^{226}Ra, ^{238}U and ^{232}Th were 9, 5.5 and 8 times higher than the world averages, respectively (Figure 12.5) (UNSCEAR, 2000, 1993). The high radiation level found in this area may be due to deposition of excessive heavy minerals as placer deposits. The concentration of ^{232}Th was found to be greater in soil than that for ^{226}Ra, ^{238}U and ^{40}K because high amounts of monazite are present in the beach sands of the studied area which contain thorium (Chowdhury, 2003). The activity levels of ^{137}Cs in some of the locations was below the detection limit of 0.2 Bq kg^{-1} and in other regions, range from 0.39 to 4.02 Bq kg^{-1} with a mean of 2.21 ± 0.49 Bq kg^{-1} which might be due to fallout of previous worldwide nuclear explosion and reactor accidents. The ^{40}K found in soil lower than the average, and might be due to the fact that potassium is much more readily lost by leaching in sandy soils.

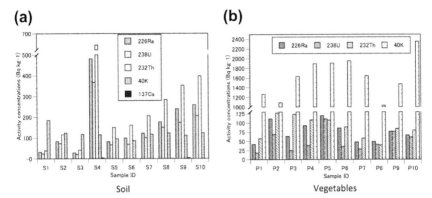

FIGURE 12.4   Activity concentrations ($Bq\,kg^{-1}$) of radionuclides in the studied area. (a) Soil; (b) vegetables.

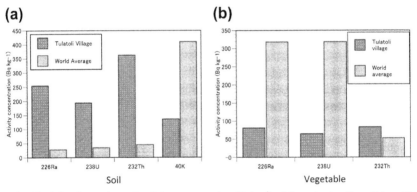

FIGURE 12.5   Comparison of activity concentration ($Bq\,kg^{-1}$) of the natural radionuclides with world average values. (a) Soil; (b) vegetables.

Radioactivity in Vegetables

The radioactivity concentration of Papaya at the studied village are presented in Figure 12.4. Results show that the concentrations of ^{226}Ra, ^{238}U, ^{232}Th and ^{40}K in papaya samples varied from 41.82–120.08 $Bq\,kg^{-1}$ with a mean of 80.95 ± 13.61 $Bq\,kg^{-1}$, from 18.57–110.98 $Bq\,kg^{-1}$ with a mean of 64.77 ± 38.47 $Bq\,kg^{-1}$, from 39.58–127.48 $Bq\,kg^{-1}$ with a mean of 83.53 ± 20.50 $Bq\,kg^{-1}$, from 1030 to 2352 $Bq\,kg^{-1}$ with a mean of 1691 ± 244.98 $Bq\,kg^{-1}$, respectively. ^{137}Cs was not found in these samples. The concentration of ^{40}K was found to be very high. Potassium is a macronutrient, so the concentration may be high. Beside this, beach soil contains a high amount of salts from sea water. The concentrations of ^{40}K in papaya were found to be higher than in the associated soils; this may be due to soil characteristics which favour the potassium to mobilize into the subsurface of the soil and migrate into

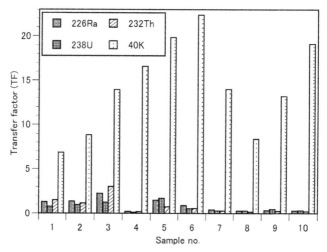

FIGURE 12.6   Transfer factors of radionuclides in food crop of the study area.

plants by root uptake. The concentration of ^{226}R and ^{238}U in the papaya samples are lower than the world values suggested by the UNSCEAR for root vegetables and fruits while that of ^{232}Th is comparable (UNSCEAR, 2000). A comparison of average activity concentration (Bq kg^{-1}) of the natural radionuclides in root vegetables / fruits with the results of the present study is given in Figure 12.5.

Transfer Factor

The calculated PTF of the radionuclides in the papaya samples from the studied village are tabulated in Figure 12.6. Among the studied samples the TF of ^{226}Ra ranges from 0.19 to 2.22 with a mean of 1.21, the TF of ^{238}U ranges from 0.11 to 1.69 with a mean of 0.90, the TF of ^{232}Th ranges from 0.14 to 3.00 with a mean of 1.57 and the TF of ^{40}K ranges from 6.83 to 22.38 with a mean of 14.61. The ^{137}Cs was not transferred from the soil to the food crop. The soil–plant transfer of naturally occurring radionuclides is mainly affected by numerous physical, chemical and biological conditions of the soil. The uptake of these radionuclides can also be affected by their chemical properties (Chen et al., 2005). The average values of TF found in the study for all the natural radionuclides are generally higher in respect to those provided by UNSCEAR (2010).

CONCLUSIONS

Industrial wastewater and urban pollution associated with inappropriate agricultural practices lead to increases in the accumulation of heavy metals such as Cr, Zn, As, Mn, Cu, Ni, Sr, Cd and Pb in the vegetables and soils of the studied areas of Bangladesh. Various indices (MPI, DIM and

HRI) revealed that leafy vegetables had higher heavy metal accumulation than non-leafy vegetables, irrespective of their locations. Among different vegetables, Enhydra fluctuans, Basella alba, Colocasia and Ladies' fingers showed higher values of MPI, DIM and HRI, whereas Spinach and Eggplant showed lower values. Further, although few individual heavy metals occur at particularly harmful levels in the studied vegetables, their cumulative effects may cause health risks for the inhabitants, especially for children. The elevated level of heavy metal accumulation in the edible portion of the vegetables is significantly correlated with their sources. In addition, higher PTF values of the vegetables grown in the studied soil also indicate accumulation of these heavy metals from anthropogenic sources. Long-term consumption of the heavy-metal-contaminated vegetables can cause different diseases like manganism, arteriosclerosis, schizophrenia, thalassaemia, dermatitis and cancer in the human body.

Therefore, legislative measures and regular monitoring must be necessary to control the industrial discharge into the surrounding environment as well as management of agricultural practices which induce heavy metal related diseases in Bangladesh. Further, it is also possible to reduce health risk in the studied areas by selecting the particular vegetable cultivation based on the level of heavy metal uptake from the polluted soil. It is suggested that the vegetables Enhydra fluctuans, Basella alba, Colocasia and Ladies' fingers of the studied areas are unsafe and should be avoided for human consumption. On the contrary, *Zea mays*, *Oryza sativa*, *Nephrolepis exaltata* and *Enhydra fluctuans* could be used for phytoremediation of the heavy metals such as Zn and Mn, Cu from the studied soils which is cost effective and environmentally friendly.

Additionally, the elevated radioactivity levels of ^{226}Ra, ^{238}U and ^{232}Th were found in the soil of Teknaf in Bangladesh. The high radiation level found in this area may be due to deposition of excessive heavy minerals as placer deposits. The higher transfer factor of ^{226}Ra(> 1), ^{238}U (~ 1) and ^{232}Th(> 1) in the food crops in this area may pose potential health risks to the surrounding inhabitants, because chronic exposure to low radiation doses can cause cancer, kidney damage, mutation, birth defects, etc., to human and non-human biota.

Finally it can be concluded that the findings of the research could be useful in supporting ecological and radiological assessment as well as in providing basic information in the design of national nuclear and food safety policies.

ACKNOWLEDGEMENTS

The authors would like to thank Shirin Akter of Bangladesh Atomic Energy Commission for her technical support during this research work. Further, our appreciation goes to Toufick Imam, Shiblur Rahman, Sohel Rana and Sharmin Yousuf Rikta, graduate students of the Department of Environmental Sciences for their kind cooperation during the preparation of the manuscript.

REFERENCES

Ahmad, J.U., Goni, M.A., 2010. Heavy metal contamination in water, soil, and vegetables of the industrial areas in Dhaka, Bangladesh. Environ. Monit. Assess. 166, 347–357.

Alam, M.G.M., Snow, E.T., Tanaka, A., 2003. Arsenic and heavy metal concentration of vegetables grown in Samta village, Bangladesh. Sci. Total Environ. Vol. 111, 811–815.

Chen, S.B., Zhu, Y.G., Hu, Q.H., 2005. Soil to plant transfer of ^{238}U, ^{226}Ra and ^{232}Th on a uranium mining-impacted soil from south-eastern China. J. Environ. Radioact. 82, 223–236.

Chowdhury, M.I., 2003. Status of radioactive materials occurring in beach sand of Cox's Bazar. Bangladesh J. Geol. Vol. 22, 109–119.

Chowdhury, M.T.A., Nesa, L., Kashem, M.A., Hoq, S.M.I., 2010. Assessment of the Phytoavailability of Cd, Pb and Zn using Various Extraction Procedures. Pedilogist 53 (3), 80–95.

Chu, N.S., Hochberg, F.H., Calne, Olanow, C., 1995. Handbook of neorotoxicology of manganese. In: Chang, L.L., Dyyer, R. (Eds.). Marcel Dekker Inc, New York, NY. 91.

Cui, Y.J., Zhu, Y.G., Zhai, R.H., Chen, D.Y., Huang, Y.Z., Qiu, Y., Liang, J.Z., 2004. Transfer of metals from soil to vegetables in an area near a smelter in Nanning, China. Environ. Int. 30, 785–791.

DoE (Department of Environment), 2009. Environmental Quality Standard (EQS) for Bangladesh.

Duruibe, J.O., Ogwuegbu, M.D.C., Egwurugwu, J.N., 2007. Heavy metal pollution and human bio toxic effects. Int. J. Phys. Sci. 2 (5), 112–118.

Faisal, I., Shammin, R., Junaid, J., 2004. Industrial pollution in Bangladesh. World Bank Report.

FAO / WHO, 1984. first ed. Food standards Program, vol XVII, Codex Alimentarious, Geneva.

Farago, M.E., O'Connell, J.T., 1983. Soil and plant concentrations of cadmium and zinc in the vicinity of asmelter. Miner. Environ. 5, 71–78.

GMA, Online news, (accessed 5.07.12). http://www.gmanetwork.com/news/.

Gupta, N., Khan, D.K., Santra, S.C., 2008. An assessment of heavy metal contamination in vegetables grown in waste water-irrigated areas of Titagarh, West Bengal. India. Bull. Environ. Contam. Toxicol. 80, 115–118. http://dx.doi.org/10.1007/00128-007-3327.

Hakanson, L., 1980. An ecological risk index for aquatic pollution control, a sedimentilogical approach. Water Research 14, 975–1001.

Huq, I.S.M., Islam, M.N., 1999. Contamination of soil and plant by lead, zinc and other heavy metals from motor vehicle exhausts along a highway in Bangladesh. In Contaminants and the Soil Environment. Proceedings of the Second International Conference on Contaminants and Soil Environment in the Australasia–Pacific Region, New Delhi. 12–17 December 1999, SIC-14. pp. 487–489.

IAEA (International Atomic Energy Agency). Measurement of radio-nuclides in food and the environment, 1989. Technical Report Series No. 295.

Islam, M.R., Lahermo, P., Salminen, R., Rojstaczer, S., Peuraniemi, V., 2000. Lake and reservoir water quality affected my metal leaching from tropical soils. Bangladesh Environ. Geology 39 (10), 1083–1089.

Islam, S., Jolly, Y.N., Akter, S., Islam, A., Mahfuza, S.S., 2012. Background chemical study of relocated hazaribagh tannery complex environment. Savar, J. Bangladesh Acad. Sci. 36 (1), 45–51.

Kabata-Pendias, A., Pendias, H., 1992. Trace elements in soil and plants, second ed. CRC, Boca Raton. 365.

Kamal, F.M., Ahmed, N., Mohee, F.M., Abedin, M.J., Shariff, A., Sadat, A.H.M., Fazlul Hoque, A.K.M., 2007. Trace element analysis by PIXE in soil samples of Hazaribagh Tannery Area. J. Environ. Sci. (Dhaka) 5, 37–44.

Kashem, M.A., Singh, B.R., 1999. Heavy metal contamination of soil and vegetation in the vicinity of industries in Bangladesh. Water Air Soil Pollt. 115, 347–361.

Khan, S., Cao, Q., Zheng, Y.M., Huang, Y.Z., Zhu, Y.G., 2008. Health risks of heavy metals in contaminated soils and food crops irrigated with wastewater in Beijing, China. Environ. Pollut. 152, 686–692.

Khan, S., Farooq, R., Sahabaz, S., Khan, M.A., Sadique, M., 2009. Health risk assessment of heavy metals for population via consumption of vegetables. World Appl. Sci. J. 6 (12), 1602–1606.

Liang, J., Chen, C., Song, X., Han, Y., Liang, Z., 2011. Assessment of heavy metal in soil and plant from Dunhua sewage irrigation area. Int. J. Electrochemical Sci. 6, 5314–5324.

Lokeshwari, H., Chandrappa, G.T., 2006. Impact of heavy metal concentration of Bellandur Lake on soil and cultivated vegetation. Curr. Sci. 91 (5), 1–6.

Mahfuza, S.S., Nushrat, J., Shakila, A., Mala, K., 2011a. Impact of wastewater and soil of fertilizer production plant on the water of the Meghna River Bangladesh. J. Environ. Res. 9, 17–26.

Mahfuza, S.S., Shahidul, I.M., Shahrin, R., Al-Mansur, M.A., 2011b. Study of Surface Water and Soil Quality affected by Heavy Metals of Pabna Sadar, Bangladesh J. Sci. Ind. Res. 46 (1), 133–140.

Mapanda, F., Mangwayana, E.N., Nyamangara, J., Giller, K.E., 2005. The effect of long-term irrigation using wastewater on heavy metal contents of soils under vegetables in Harare, Zimbabwe Agr. Eco. Environ. 107, 151–165.

McCluggage, D., 1991. Heavy metal poisoning, NCS Magazine. The Bird Hospital, CO. USA. www.cockatiels.org/articles/Diseases/metals.html.

Mollah, A.S., Chakraborty, S.R., 2009. Radioactivity and radiation levels in and around the proposed nuclear power plant site at Rooppur. Jpn. J. Phys. 44 (4), 408–413.

Muchuweti, M., Birkett, J.W., Chinyanga, E., Zvauya, R., Scrimshaw, M.D., Lister, J.N., 2006. Heavy metal content of vegetables irrigated with mixtures of wastewater and sewage sludge in Zimbabwe: Implication for human health. Agri. Ecosys. Environ. 112, 41–48.

Naser, H.M., Shil, N.C., Mahmud, N.U., Rashid, M.H., Hossain, K.M., 2009. Lead, cadmium and nickel contents of vegetables grown in industrially polluted and non polluted areas of Bangladesh. Bangladesh J. Agr. 34 (4), 545–554.

Naser, H.M., Sultana, S., Mahamud, N.U., Gomes, R., Noor, S., 2011. Heavy metal levels in vegetables with growth stage and plant species variations. Bangladesh J. Agril. Res. 36 (4), 563–574.

Rahman, S.H., Khanam, D., Adyel, M.T., Islam, S.M., Ahsan, A.M., Akbor, M.A., 2012. Assessment of heavy metal contamination of agricultural soil around Dhaka export processing zone (DEPZ), Bnagladesh: Implication of Seasonal Variation and indices. Appl. Sci. 2, 584–601.

RAIS, 2007. The risk information system. http//rais.oml.govt/tox/rap_toxp.shtml.

Rattan, R.K., Datta, S.P., Chonkar, P.K., Suribahu, K., Singh, A.K., 2005. Long term impact of irrigation with sewage effluents on heavy metal contents in soil, crops and ground water—A case study. Agriculture Ecosystems & Environ. 109, 310–322.

Sharma, R.K., Agrawal, M., Marshall, F.M., 2009. Heavy metals in vegetables collected from production and market sites of tropical urban area of India. Food Chem. Toxicol. 47, 583–591.

Singh, A., Sharma, R.K., Agrawal, M., Marshall, F.M., 2010. Health risk assessment of heavy metals via dietary intake of foodstuffs from the wastewater irrigated site of a dry tropical area of India. Food. Chem. Toxicol. 48, 611–619.

Sridhara Chary, N., Kamala, C.T., Samuel Suman Raj, D., 2008. Ecotoxicol. Environ. Saf. 69, 513–524.

Sultana, M.S., Muramatsu, M.S., Majibur, Y., Rahman, M.R., Rashidul Alam, A.K.M., Saadat, M., 2003. Analysis of Some Trace Elements in the Uncultivated Soils of Savar by ICP-MS. Bangladesh J. Environ. Sci. 1, 57–62.

Tani, F.H., Barrington, S., 2005. Zinc and copper uptake by plant under two transpiration ratios Part 1. Wheat (TriticumaestivumL. Environ. Pollut. 138, 538–547.

UNSCEAR, 1993. Sources, Effects and Risk of Ionization Radiation, United Nations Scientific Committee on the Effects of Atomic Radiation. Report to General Assembly, UN, New York.

UNSCEAR, 2000. Sources, Effects and Risk of Ionization Radiation, United Nations Scientific Committee on the Effects of Atomic Radiation. Report to General Assembly, UN, New York.

UNSCEAR, 2010. Sources, Effects and Risk of Ionization Radiation, United Nations Scientific Committee on the Effects of Atomic Radiation, Annex A: Exposure from Natural Sources. Report to General Assembly, UN, New York.

USEPA, 2002. United States Environmental Protection Agency. OSWER, 9355.4–24.

Usero, j., Gonzalez-Regalado, E., Gracia, I., 1997. Trace metals in the bivalve mollusks *Ruditapes decussates* and *Ruditapes philliippinarum* from the Atlantic Coast of Southern Spain. Environ. Int. 23 (3), 291–298.

Vousta, D., Grimanis, A., Samara, C., 1996. Trace elements in vegetables grown in an industrial area in relation to soil and air particulate matter. Environ. Pollut. 94, 325–335.

Zurera, G., Estrada, B., Rincon, F., Pozo, R., 1987. Lead and cadmium concentration levels in edible vegetables. Bull. Environ. Contamin. Toxicol. 38, 805–812.

Chapter 13

Remediating Cadmium-Contaminated Soils by Growing Grain Crops Using Inorganic Amendments

Muhammad Zia-ur-Rehman,* Muhammad Sabir,*,†
Muhammad Rizwan,‡ Saifullah,* Hamaad Raza Ahmed*
and Muhammad Nadeem§

*Institute of Soil and Environmental Sciences, University of Agriculture, Faisalabad,
Pakistan; †School of Plant Biology, University of Western Australia, Crawley, WA,
Australia; ‡Department of Environmental Sciences, Government College University,
Faisalabad, Pakistan; §Department of Environmental Sciences, COMSATS Institute of
Information Technology (CIIT), Vehari, Pakistan

INTRODUCTION

Cadmium (Cd) is a transition metal and is located in group 2B and period 5 of the periodic table. The atomic number of Cd is 48 and its atomic weight is 112.4 with specific density of $8.65 \, g \, cm^{-3}$. The melting point of Cd is $320.9°C$ and the boiling point is $765°C$ (Wuana and Okieimen, 2011). Cd is widely used in Ni/Cd batteries, as a pigment and stabilizer for plastics, in alloys and electronic compounds (di Toppi and Gabbrielli, 1999). Moreover, Cd is also produced as a by-product of Zn and lead refining. However, Cd is non-essential and is toxic to plant and animal existence. The oral uptake of cadmium in humans via food and drinking water contributes the largest part of the total Cd uptake (EC, 2000). For example in Japan, higher Cd concentration in rice was considered a major source of Cd entry in a human's body. This larger Cd concentration was considered for itai-itai disease in Japan in the mid 1950s and 1960s (Yamagata and Shigematsu, 1970). Currently in Japan, it is still known that rice is the main source of Cd entry to humans (Watanabe et al., 2000) which is a threat to human health (Ueno et al., 2010). Mean tolerable weekly intake (TWI) of Cd for the European population was established at $2.5 \, \mu g \, kg^{-1}$ of body weight. However, this TWI may exceed about two-fold in people living in highly contaminated areas (EFSA, 2009).

Soil Remediation and Plants. http://dx.doi.org/10.1016/B978-0-12-799937-1.00013-9

It has been widely reported that accumulation of Cd in plants may cause many biochemical, structural and physiological changes (Das et al., 1997; di Toppi and Gabrielli, 1999; Benavides et al., 2005). Plant growth and photosynthesis are negatively affected by Cd (Sandalio et al., 2001). In plants, the most visible symptom of Cd is the reduction in root length (Guo and Marschner, 1995). However, the mechanisms behind these toxicities are not yet fully explored. In addition, Cd also affects the absorption and translocation of essential micro- (Zn, Cu and Mn) and macronutrients (N, P, K, Ca and Mg) by plants (Jalil et al., 1994; Sarwar et al., 2010). One of the main pathways for heavy metals to enter the plants is through the roots because root is a primary organ that participates primarily in the heavy metal uptake due to the direct contact with the soil solution, containing these metals (Lux et al., 2011).

Currently a number of technologies exist for the remediation of Cd-contaminated soils. These techniques include in situ and ex situ remediation, isolation, immobilization, phytoextraction, phytostabilization, physical separation and extraction, etc. (Wuana and Okieimen, 2011). Among these techniques, phytoremediation is mostly used for clean-up of metal-contaminated soils. In this technique, green plants are used to remove metal and other toxic compounds from contaminated sites. There are a number of different phytoremediation techniques, which include phytovolatilization, phytostabilization and phytoextraction. Phytostabilization, also called in-place inactivation, is primarily used to immobilize toxic materials in soils and sediments with the help of plants and amendments (Adriano, 2001; Wuana and Okieimen, 2011). Contaminants are absorbed and accumulated in roots or precipitated in the rhizosphere, so in this way their mobility and bioavailability is limited (Bes and Mench, 2008). This technique is useful for quick stabilization of heavy metals including Cd and is advantageous because disposal of hazardous material is not required (Mench et al., 2006). In the present chapter, we will review concentration of Cd in soils and crops, its toxic effects, as well as stabilization of Cd and reduction of its bioavailability and toxicity with the help of inorganic amendments.

NATURAL CADMIUM LEVELS IN SOIL

Cadmium occurs naturally in Earth's crust and is widely distributed but it is a relatively rare element ($0.1–0.2 \, mg \, kg^{-1}$). Natural sources of Cd are bedrocks located below the soil and/or nature of the parent material. In nature, Cd is mostly present in complex forms in ores containing zinc, lead and copper (UNEP, 2010). In Earth's crust, natural Cd varies from 0.1 to 0.5 ppm but different factors may increase or decrease Cd concentrations. Cd in sedimentary rocks varies from 0.1 to 25 ppm whereas igneous and metamorphic rocks have lower values generally ranging from 0.02 to 0.2 ppm (Cook and Morrow, 1995). Higher concentrations are found in association with Zn, Pb and Cu ores. The Cd content of phosphate fertilizers varies from 2 to 200 $mg \, kg^{-1}$. Sedimentary rocks and marine phosphates contain about $15 \, mg \, Cd \, kg^{-1}$ (EC, 2000).

Phosphate rocks (PRs) of igneous origin generally contain less than 15 mg Cd per kilogram P_2O_5 (phosphate fertilizer) compared with 20–245 mg Cd kg^{-1} in sedimentary counterparts (Çotuk, et al., 2010). Earth's normal Cd concentrations in the soils are reported to range between 0.02 to 6.2 mg Cd kg^{-1} and the soils containing 5–20 mg Cd kg^{-1} are likely required remedial actions as being toxic to the surrounding environment (Adriano, 2001). In France, the Cd concentration increased to over 100 mg Cd kg^{-1} of soil (Baize et al., 1999). In agricultural or horticultural soil, Cd concentrations vary from 0.2 to 1.0 mg kg^{-1} in rural and 0.5–1.5 mg kg^{-1} in urban areas (EC, 2000). However, anthropogenic or natural activities can increase Cd concentrations in the soil (He et al., 2005).

SOURCES OF CADMIUM CONTAMINATION OF AGRICULTURAL SOILS

Cd level in agricultural soils depends on parent material, aerial deposition, pedogenic processes, use of phosphate fertilizers, pesticides, sewage sludge and disposal of industrial wastes containing Cd (Figure 13.1) (Grant et al., 1998; Wuana and Okieimen, 2011). According to the French ASPITET program, Cd in agricultural soils can vary from 0.02 to 6.9 mg kg^{-1} (Baize, 1997; Mench et al., 1997). In soil solution, Cd concentration is low and ranges from 0.2 to 6.0 µg l^{-1}. However, larger values up to 300 µg l^{-1} have also been reported and this soil is considered to be contaminated (Itoh and Yumura, 1979). Moreover, domestic sewage sludge contains a higher concentration of Cd when compared with other heavy metals; for example, cigarette butts containing Cd flushed down toilets and Cd from car tyres. In addition, composted sludge may contain higher levels of Cd. For example, the composted sludge from Topeka, Kansas, which is applied to crop land, contains up to 4.2 mg Cd kg^{-1} compost (Liphadzi

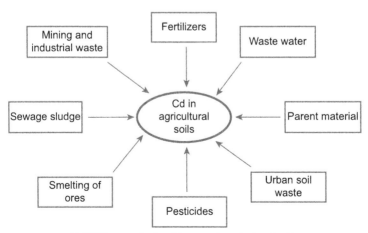

FIGURE 13.1   Major Cd sources in agricultural soils.

and Kirkham, 2006). Phosphate fertilizers, manures, disposal of industrial and urban wastes and pesticides also contain Cd as an impurity (Nagajyoti et al., 2010; Wuana and Okieimen, 2011). The application of these agricultural inputs increases the total Cd in soils. However, Cd concentrations of less than $3\,mg\,kg^{-1}$ of dry soil are recommended for agricultural soils in order to reduce its concentrations in edible parts (Lux et al., 2011).

BIOASSESSMENT OF CADMIUM IN SOILS

In order to assess the metal phytoavailability it is very important to extract and measure the portion of the metals that plants can take up from the soil. For this purpose, several methods have been developed in an attempt to predict phytoavailability of heavy metals. The choice of extractant is an essential step to evaluate the availability of metals to the plant. The extractant should reflect the same phenomenon as the absorption of heavy metal by the plant. It should only remove bioavailable metals. Bioavailability varies with the soil and plant studied. However, there is no universal extractant but many chemical extractants are often used to approach the bioavailability of metals to plants (Lakanen and Ervio, 1971; Gupta and Aten, 1993). Ideally, the chosen extractant should not affect the solid solution equilibrium of the soil, so it does not change the pH, the form of complexes or participate in exchange reactions of ions (Kennedy et al., 1997). DTPA-TEA (DiethylenetriaminePentaacetic Acid Triethanolamine) buffered at pH 7.3 presents the best results for the extractable fraction of heavy metals in soils (Borges, 2002). Similarly, Keller et al., (2005) showed that the DTPA-TEA can be relied on as extraction method for estimating the bioavailability of metals for the tobacco plant on the soils. Based on the above discussion, we selected DTPA-TEA for the extraction of bioavailable metals from the soil.

FACTORS INFLUENCING THE ACCUMULATION OF CADMIUM IN CROPS

Uptake of Cd in crop plants depends mainly on soil and plant factors and microorganisms.

Soil Factors

The major pollution problem involving Cd can be correlated with the behaviour of Cd in soil. Cd in the soil exists in different forms such as a free metal ion, exchangeable (free ion Cd^{2+}), adsorbed on the surfaces of organic matter, clay particles and Fe-oxides. Cd also exists in the form of complexes with amino acids and carboxylic acids and SO_4^{2-} and Cl^- complexes (Sammut et al., 2010; Vega et al., 2010). Cd behaviour in soil depends on these different forms and is mainly controlled by complex interactions (Adriano et al., 2004).

The bioavailable Cd is more important for plants rather than total Cd concentration in the soil because plants can uptake bioavailable Cd (Ok et al., 2004; Kirkham, 2006).

The bioavailability of Cd to plants depends on a number of soil factors and plant species. Among the soil factors are available Cd and total Cd concentrations, pH and organic matter, (Sauve et al., 2000a, b; Kirkham, 2006; Jung, 2008), speciation (Sammut et al., 2010), cation exchange capacity (CEC) (Lehoczky et al., 2000; Vega et al., 2010) and chloride (Grant et al., 1998; Degryse et al., 2004; Weggler et al., 2004; Kirkham, 2006). Cd solubility and bioavailability are also affected by dissolved organic carbon present in the soil solution which is known to complex metals such as Cd, Zn and Cu and affect their solubility as well as plant uptake (Antoniadis and Alloway, 2002; Zhao et al., 2007). Cd bioavailability also decreased with time (Kirkham, 2006). Nitrogen fertilizers, soil types, genotypes and management practices also affect Cd bioavailability (Zhang et al., 2009; Perilli et al., 2010; Gao et al., 2010, 2011). Plant nutrient levels in the soil also affect Cd bioavailability (Kirkham, 2006; Sarwar et al., 2010). The amount of Cd in soil solution is also affected by adsorption, complexation and sorption capacity of the soil. Metal adsorption on the soil solid phase is the major process that control metals present in the soil solution. With the increase in adsorption and sorption capacity of soil, the seed in the soil solution decreases (Grant et al., 1998).

Among all soil factors, pH is a major factor influencing Cd bioavailability (McBride et al., 1997; Grant et al., 1998; Jung, 2008). Under acidic conditions, Cd is mobile between pH 4.5 and 5.5, but in the alkaline soils Cd is relatively immobile (Kirkham, 2006). The pH can also affect the bioavailability of Cd by altering different processes in soil such as adsorption, complexation, sorption or desorption processes (Naidu et al., 1994, 1997; Bolan et al., 1999b). In general, all soil processes controlling Cd behaviour in soil are of special importance in the root developing zone of soil.

Plant Factors

Cd bioavailability also depends on plant species (Mench et al., 1989). The dominant form of CD in the soil solution is Cd^{2+} but also exists in the form of Cd-chelates (Tudoreanu and Phillips, 2004). Specific chelating compounds are also released by some plants called phytosiderophores and/or root exudates (Mench and Martin, 1991). In general, phytosiderophores are produced by graminaceous plants such as barley, wheat and rice under Fe deficiency which mobilizes Fe from sparingly soluble forms (Marschner et al., 1986; Reichard et al., 2005). These phytosiderophores can complex metals such as Cd and Cu and affect their bioavailability. Root exudates strongly affect Cd bioavailability by affecting the characteristics of rhizosphere (Hill et al., 2002; Dong et al., 2007). Root exudates may influence the Cd bioavailability and toxicity by modifying the rhizosphere pH and redox potential (Eh), chelating/complexing and depositing with Cd ions. In addition, many plant species can exude organic acids that can form complexes with metal salter metal speciation in the rhizosphere. Organic acids also increased the uptake of Cd by solubilizing particulate bound Cd into the soil solution (Cieslinski et al., 1998). Similarly in maize plants, presence of organic acids increased Cd mobilization, plant availability and accumulation (Nigam et al., 2001; Han et al., 2006). Such factors may alter Cd behaviour in the soil and affect the Cd uptake by the plants, and also alter the community and activities of microbes present in rhizosphere (Shenker et al., 2001; Dong et al., 2007).

Microorganisms

Microorganisms also release chelating compounds called siderophores (Neubauer et al., 2000 and the references therein) which may solubilize Cd (Dimkpa et al., 2009). A bacterium can produce many types of siderophores, including hydroxomates and carboxylic acids (Klumpp et al., 2005). These siderophores can desorb Cd at a moderate pH range (Hepinstall et al., 2005). Microorganisms may also decrease Cd solubility by the formation of insoluble metal sulphides and also sequestration of the toxic metal via the cell walls or by proteins and extracellular polymers, etc. (Francis, 1990; Dong et al., 2007).

Climatic Factors

Climate strongly influences metal mobility and bioavailability. Climate strongly influences organic matter concentration in the soil that ultimately affects Cd availability to plants. Arid climates generally result in the smaller soil organic matter and humid climates contain large amounts of organic matter. Organic matter can bind Cd on the exchange complex. In tropical climate conditions, presence of iron, manganese and aluminum oxide minerals in soil profiles may limit the mobility and bioavailability of Cd. Temperature exerts an important effect on metal speciation, because most chemical reaction rates are highly sensitive to temperature changes (Elder, 1989). An increase of $10°C$ can double biochemical reaction rates and enhance the tendency of a system to reach equilibrium. Temperature may also affect quantities of metal uptake by an organism (Prosi, 1989). Acid rain also affects Cd bioavailablity by inducing the release of metals due to cation exchange with Mg^{2+}, Ca^{2+} and H^+, etc. (Probst et al., 2000; Hernandez et al., 2003).

CADMIUM UPTAKE AND ACCUMULATION IN PLANTS

Plants mainly take up Cd by the roots from the soil solution. Cd uptake by roots seems to occur via different transporters such as Mn^{2+} and Zn^{2+}, Ca^{2+}, Fe^{2+} (Clemens, 2006). However, Cd uptake and accumulation widely differ among different crop species (Liu et al., 2003; Grant et al., 1998, 2008). In general, Cd accumulation in roots, shoots and grains depends on three transport processes: (1) root uptake of Cd; (2) xylem loading; and (3) re-translocation to seeds. Root is the main pathway by which water, nutrients and pollutants including heavy metals enter the plant body. Cd enters the plant through root uptake from the soil solution and is a key process in the accumulation of Cd by plants. A part of Cd present in the soil solution is also adsorbed on the surface of plant roots. Cd uptake by roots increased with increasing exposure periods and Cd concentrations (Hentz et al., 2012).

Cadmium uptake by roots is also affected by root morphology, root apics and root surface area (Kubo et al., 2011). Root endodermis and exodermis play an important role in Cd uptake in maize plants and act as barriers to the solute flow (Redjala et al., 2011) and the presence of high Cd accelerated the maturation of the maize root endodermis (Lux et al., 2011). During plant uptake, Cd ions can compete for the same trans-membrane carriers as those which take up plant nutrients (Benavides et al., 2005). The presence of other elements can inhibit the Cd uptake, such as Zn^{2+} in the nutrient solution inhibited the uptake of Cd in wheat plants (Welch et al., 1999; Hart, et al., 2002). Molecular mechanisms of Cd uptake by roots are still poorly understood. After adsorption on the root surface, Cd enters in the root cells as Cd^{2+} through ZIP transporters like Zn-regulated transporter- or Fe-regulated transporter-like-protein (Lux et al., 2011).

Once Cd has been taken up by root system, part of the Cd accumulates in the roots and another part is translocated to the shoots (Kabata-Pendias and Pendias, 2001; Gill et al., 2011). In general, Cd ions are retained in the roots and a small portion is transferred to the shoots but it depends on plant species (Abe et al., 2008). After absorption by roots, Cd can move towards xylem through apoplastic and/or symplastic pathways (Salt et al., 1995; Benavides et al., 2005). Cd in roots can also be complexed with several ligands including organic acids and/or phytochelatins (PCs) and is mainly concentrated in vacuoles and nuclei (Hart et al., 2006; Lux et al., 2011). Xylem loading is an important process for long-distance transport of Cd (Clemens et al., 2002). Cd transport into the central cylinder is also regulated by casparian strip and plasmalemma of the endodermis (Seregin et al., 2004).

After absorption by the roots, Cd is transported by xylem and phloem to the aerial parts of plants (Tudoreanu and Phillips, 2004). However, Cd translocation to shoots depends on plant species and genotypes within species (Dunbar et al., 2003). Translocation of Cd from roots to shoots is a passive process driven by transpiration (Salt et al., 1995; Hart et al., 2006) or it is translocated actively through different transporters such as Fe transporters (Nakanishi et al., 2006). Recently, it has been suggested that citrate might play a significant role in Cd transportation in the xylem vessels (Zorrig et al., 2010). Similarly, Van der Vliet et al. (2007) reported that most of the Cd translocation from the root to the shoot in durum wheat plants takes place via the symplastic pathway.

Studies on Cd translocation showed that Cd translocation in the edible parts of plants depends on genotypes (Meyer et al., 1982; Cakmak et al., 2000). Cd accumulation in grains also depends on Cd concentration in shoots and flag leaf

(Greger and Löfstedt, 2004). In grains, Cd accumulation may occur through phloem-mediated Cd transport from the leaves and stalks to maturing grains (Hart et al., 1998; Harris and Taylor, 2001; Liu et al., 2007; Yoneyama et al., 2010). Presence of other ions may inhibit or increase phloem loading of Cd (Welch et al., 1999; Cakmak et al., 2000).

Cadmium does not appear to be an essential nutrient for plant growth and development because it has no known biological function (Marschner, 1995). Among the toxic heavy metals, Cd is of more concern than others due to its high toxicity at very low concentrations and high solubility in water (Benavides et al., 2005). The higher Cd accumulation in plants induces a series of stress factors in plants. Major toxic effects (direct or indirect) of this metal on plant growth and physiological processes are listed in Figure 13.2.

Effects on Seed Germination

Cd toxicity adversely affects seed germination and plant growth. Seed germination is inhibited by Cd in many plants such as *Arabidopsis thaliana* (Li et al., 2005), mustard (*Sinapis arvensis* L.) (Heidari and Sarani, 2011),

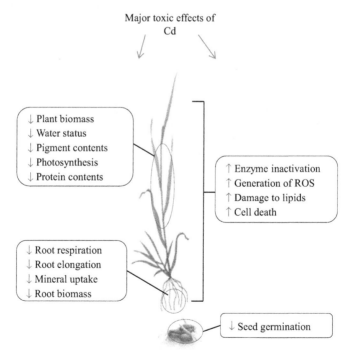

FIGURE 13.2   Major toxic effects of Cd in plants. Upward arrows heads indicate increase in value and downward arrow heads indicate decrease in value.

soybean (*Glycine max* L.) (Liu et al., 2011), *Pisum sativum* (Smiri, 2011), cowpea (*Vigna unguiculata* L.) (Vijayaragavan, et al., 2011) and milk thistle (*Silybum marianum* L.) (Khatamipour, et al., 2011). This inhibition of seed germination is probably due to selective uptake of different ions by the intact seeds (Li et al., 2005). Inhibition of seed germination is mainly due to reduced water uptake by seeds that consequently reduces water availability to the embryo (Poschenreider et al., 1989; Vijayaragavan et al., 2011).

Effects on Plant Growth and Biomass

Cd toxic effects on plant growth and biomass are largely studied. Reduction in root length is among the most visible symptoms of Cd toxicity in plants (Guo and Marschner, 1995; Lux et al., 2011; Haouari et al., 2012). Higher Cd concentration in the roots accelerated the development of endodermis of the roots and also changed the relative size and proportion of root tissues (Seregin et al., 2004; Lux et al., 2011).

It is well known that Cd negatively affects the plant growth and biomass. For example, pea plants grown hydroponically with $50\,\mu M$ Cd for 28 days significantly decreased plant growth and biomass (Sandalio et al., 2001). More recently Haouari et al., (2012) reported that the root and shoot fresh and dry biomass in tomato plants grown hydroponically with increasing Cd concentrations were negatively affected. In *Thlaspi caerulescens* plants, accumulation of $319\,mg\,Cd\,kg^{-1}$ (± 12) dry weight of shoots resulted in visual toxicity symptoms in plants (Wójcik and Tukiendorf, 2005). Similarly, high Cd ($100\,mg\,Cd\,kg^{-1}$ of soil) perturbed the growth of garden cress (*Lepidium sativum* L.) plants (Gill et al., 2012). In cowpea plants, increasing Cd (10, 30 and $50\,mg\,Cd\,kg^{-1}$ soil) concentrations also decreased the growth and development (Vijayaragavan et al., 2011).

A number of previous studies have reported that Cd supply in durum wheat cultivars reduced the shoot and root dry matter, root length and leaf area (Jalil et al., 1994). Higher Cd concentration in plants resulted in leaf chlorosis, wilting and leaf abscission in plants (Bavi et al., 2011). Similarly, Shi et al. (2010a) showed that Cd exposure depressed peanut plant growth. This decrease in growth and biomass under Cd stress is related to a decrease in photosynthesis, inhibition of the metabolic enzymes production and a decrease in the uptake and translocation of macro- and micronutrient contents in plants (Sandalio et al., 2001; Gonçalves et al., 2009; Feng et al., 2010). However, the toxic effects of Cd on plant growth and biomass depend on dose and time of exposure to Cd stress (Das et al., 1997; Di Toppi and Gabrielli, 1999). Similarly, chlorosis symptoms and necrotic spots appeared in leaves of tomato (*Lycopersicon esculentum*) plants at 10 and $100\,\mu M$ Cd in the nutrient solution, respectively (López-Millán et al., 2009). Moreover, the toxic effects of Cd differ with plant species (Das et al., 1997; Di Toppi and Gabrielli 1999).

Effects on Mineral Nutrients

Cd interactions with mineral nutrients are of public concern because decrease in mineral nutrition by Cd is directly related to growth and yield of plants. Numerous studies regarding Cd effects on mineral nutrients have provided contradicting results. Toxic effects of Cd on mineral nutrients uptake and accumulation in plants have been widely reported (Ouzounidou et al., 1997; Sandalio et al., 2001; Wang et al., 2008; Gonçalves et al., 2009). Toxic effects of Cd on mineral nutrients depend on time and intensity of Cd stress imposed and plant size (Hernandez et al., 1998; Street et al., 2010). Toxic effects of Cd on mineral nutrients uptake and translocation vary with plant and nutrient types (Hernandez et al., 1998; Dong et al., 2006). The effects of Cd on Zn uptake and accumulation in plants are not consistent. For example, $0.2\,\mu M$ Cd concentrations in the nutrient solution decreased Zn uptake in durum wheat when Zn concentration was lower $(1.0\,\mu M)$ in the nutrient solution while at higher Zn (10 and $19\,\mu M$) concentrations this effect was synergistic (Welch et al., 1999). Conversely, in spring wheat with $20\,\mu M$ Cd in the nutrient solution decreased Zn concentration and increased Cd concentrations in roots and shoots when Zn was lower $(1\sim200\,\mu M)$ in the nutrient medium while at higher $(>200\,\mu M)$ Zn levels Cd concentration significantly reduced in seedlings while Zn concentration increased indicating antagonistic effects on each other (Zhao et al., 2005). In tomato plants, excess Cd (50 and $100\,\mu M$) in the nutrient solution decreased the uptake of K^+, Ca^{2+} and Mg^{2+} ions by roots and consequently decreased in shoots (Haouari et al., 2012).

Excess Cd in the medium not only changed the uptake of nutrients but also changed the deposition and translocation. However, again, this trend varies according to plant species and Cd stress imposed (Yang et al., 1998; Zhang et al., 2002). For example, in soybean seedlings Cd concentration decreased Cu, Zn and Mn concentrations in roots but was not affected in shoots (Drazic et al., 2004). In contrast, addition of Cd in the growth medium increased P, K and Mn concentrations in wheat roots and inhibited their translocation to shoots (Zhang et al., 2002). Similarly, another expression of Cd interactions with essential elements is the effect of Cd when there is element deficiency in plants. For example, Ca deficiency increased Cd toxicity in rice seedlings (Cho et al., 2012).

Effects on Photosynthetic Pigments

Cd is highly toxic to photosynthetic machinery of plants. Reduction in photosynthesis is a well-known symptom of Cd toxicity in plants. This reduction in photosynthesis takes place by many processes including decrease in net photosynthetic rate, chlorophyll and carotenoid contents (Sandalio et al., 2001; Mobin and Khan, 2007; Vijayaragavan et al., 2011; Haouari et al., 2012). In addition, in cucumber (*Cucumis sativus* L.) plant, Cd treatment decreased photosynthesis which was related to damage or functional loss of the photosynthetic machinery and enzyme inhibition of nitrate metabolism (Feng et al., 2010). Similarly, Cd application in wheat seedlings also damaged the leaf photosystem II (PS II) and structure of chloroplast which resulted in the reduction of chlorophyll contents and consequently inhibition of photosynthesis (Ouzounidou et al., 1997; Ci et al., 2009). In peanut (*Arachis hypogaea*) plants, Cd treatments inhibited the net photosynthetic rate which may be due to reduction of stomatal conductance and photosynthetic pigments and alteration in leaf structure (Shi and Cai, 2008). More recently, Gill et al. (2012) reported that Cd at a higher dose reduced the photosynthesis and nitrogen metabolism in garden cress (*Lepidium sativum* L.). Similarly, in tomato (*Lycopersicon esculentum*) plants, low Cd (10 µM) concentration did not affect the photosynthesis but higher Cd (100 µM) in the nutrient solution decreased the photosynthetic rates and photosynthetic pigment concentrations (López-Millán et al., 2009).

Cd toxic effects on photosynthetic machinery depend not only on the quantity of Cd but also depend on the exposure period. For example, Chugh and Sawhney (1999) reported that in 1-month-old pea seedlings, different Cd treatments for 6 days had a more pronounced effect on the activity of PS II but on prolongation of the Cd exposure time for 12 days, the functioning of photosystem I (PS I) was also equally affected.

Cd-Induced Oxidative Stress in Plants

Excess Cd generates free radicals and reactive oxygen species (ROS) those causes oxidative stress in plants (Khan et al., 2007; Mobin and Khan, 2007). ROS react with proteins and pigments and cause lipid peroxidation, inactivation of enzymes and membrane damage (Khan et al., 2007). Excess Cd in plants also resulted in the modification of antioxidant enzymes and sulphur assimilation activity (Shi et al., 2010b; Gill et al., 2012). Ascorbic acid concentration decreased in the shoots and roots of durum wheat cultivars by supplying 75 and 150 µM Cd in the nutrient solution (Ozturk et al., 2003). Excess Cd also decreased the activity of glutamine synthetase (GS) enzyme in several plant species (Balestrasse et al., 2006). Excess Cd also increased the production of H_2O_2 in pea plants and consequently enhanced activity of SOD and low activity of CAT, POD and APX (Pandey and Singh, 2012). However, physiological response of plant to Cd stress depends on species and cultivars (Shi et al., 2010a).

Miscellaneous Toxic Effects

Cd toxicity affected normal plant growth and development in a number of ways, as discussed above. However, there are still large numbers of toxic effects of Cd on plants. Excess Cd decreased the percentage of water content of shoot and root in pea plants (Lozano-Rodriguez et al., 1997). Influence of Cd on plant–water relationships is directly related to the reduction in the absorption surfaces by inhibiting the formation of root hairs (Pál et al., 2006). Cd application in wheat seedlings also decreased the total soluble sugar concentration and increased the free amino acid concentration in both shoots and roots (Ci et al., 2009). The excess Cd also limited the stomatal conductance in safflower (*Carthamus tinctorius* L.) plants (Shi et al., 2010b).

PLANT RESPONSE TO CD CONCENTRATIONS

Plants developed a number of defence strategies to cope with Cd stress and can be divided into avoidance strategies and tolerance strategies. Avoidance strategies lead to a reduction in Cd uptake. Tolerance strategies include accumulation and storing of Cd into metabolically inactive parts of plants (Pál et al., 2006). The first defence mechanism is the reduced uptake by plants then immobilization in the cell wall, synthesis of PCs and sequestration into the vacuole (Pál et al., 2006). In plant cells, PCs are the best-characterized metal-binding ligands and are frequently cited as metal protective proteins in plants (Maestri et al., 2010). In addition, activation of various antioxidants to cope with Cd-induced ROS production constitutes a secondary defence system (Shi et al., 2010a, b; Heidari and Sarani, 2011; Sharma et al., 2012). However, plant responses to excess Cd vary depending on plant species, tissues, stages of development and Cd concentration (Sharma et al., 2012). Use of agronomic management practices also helps the plant to minimize the uptake and toxic effects of Cd in the plant (Gao et al., 2010; Perilli et al., 2010; Rizwan et al., 2012). However, information is still lacking in this respect and further research is needed.

THRESHOLD BIO-AVAILABLE CONCENTRATION OF CD

The phytotoxicity threshold concentration of Cd in the plant tissue, corresponding to 50% growth reduction (PT50) is considered as 110.6 mg kg^{-1} DM. The PT50 value depends on plant and metal species. The PT50 values of >10 mg kg^{-1} for soybean (Miller et al., 1976), >500 mg kg^{-1} for radish tops and >300 mg kg^{-1} for radish roots (John et al., 1972) grown in Cd(NO$_3$)$_2$-treated soils where Cd was applied at 10–190 mg kg^{-1} soil have been proposed. While 2.5, 2.0, 150 and 158 mg kg^{-1} DM for beetroot, carrot, Swiss chard and tomato, respectively, grown in nutrient solution are reported (Turner, 1973). Street et al. (1978) and Sauve et al. (2000b) noticed that more than 75% Cd remain unavailable to plants because of its organic complexation in soils containing high levels of organic matter. However, there are possibilities for the formation of inorganic complexes such as CdCO$_3^0$ and CdOH$^+$ in limed soils (Street et al., 1978).

Animal and human health are at serious threat due the bioaccumulation of Cd in wheat and rice crops. Cd bioaccumulation is also negatively affecting local and international markets of cereal grains (Nogawa and Kido, 1996). Such problems have been identified in southern China due to Cd uptake by rice grown on Cd-contaminated soils (Li et al., 2005). Rice grown on Cd-contaminated soils can contain as much as $3\,mg\,kg^{-1}$ DM of Cd, which is higher than the limits of the World Health Organization recommendations of $<0.2\,mg\,kg^{-1}$ DM (Schnoor, 2004). In a long-term (40 years) wastewater disposal on a sandy soil site in Germany, a positive linear relationship was recorded between Cd concentration in many plants including sugar beet (*Beta vulgaris*), potato (*Solanum tuberosum*) or winter wheat (*Triticum aestivum*) and the Cd concentration in soil solution (Ingwersen and Streck, 2005).

REMEDIATION OF CD-CONTAMINATED SOILS

Cadmium-contaminated soils are a threat to plants and animals and should be remediated in order to decrease Cd entry to the food chain. Highly Cd-contaminated soils should be covered by crops which will help to reduce the spread of contaminant through wind or water erosion (Vangronsveld and Cunningham, 1998). However, crop growth is reduced due to the toxic effects of Cd in crop plants. A range of soil amendments, has been found effective in immobilizing seed in the soil, thereby reducing Cd availability to plants. These amendments include P-compounds like apatite, hydroxyapatite (Knox et al., 2001, 2003), rock phosphate (Basta et al., 2001), DAP (McGowen et al., 2001) and KH_2PO_4, phosphogypsum and dolomitic residue (Illera et al., 2004), liming agents like $CaCO_3$ and CaO (Tsadilas et al., 2005) and S-compound-like elemental sulphur, $(NH_4)_2SO_4$ and H_2SO_4 (Seidel et al., 2006). Calcium (Ca) containing amendments are widely used to decrease Cd bioavailability and toxicity in soils. Precipitation as metal hydroxides or carbonates is one of the mechanisms to immobilize metals in soils, such as Pb, Zn and Cd by liming materials (Pierzynski and Schwab, 1993). Immobilization of metals plays a major role in highly contaminated soils as compared to normal soils, especially under conditions of alkaline pH. However, the addition of Ca compounds (lime and gypsum) could increase plant availability of metals (John et al., 1972; Williams and David, 1976). Other materials such as phosphorus, iron, zeolites and clays can also be used to remediate these soils by binding Cd in different forms and also affecting soil pH.

Metals Response to Calcium Containing Inorganic Amendments

Initially primarily liming was done to ameliorate soil acidity; but now it is widely accepted as an important tool for managing heavy metals toxicity in soils (Brown et al., 1997; Bolan et al., 2003a). Liming materials include burnt lime (CaO), calcite ($CaCO_3$), dolomite $CaMg(CO_3)_2$, slaked lime $Ca(OH)_2$ and slag ($CaSiO_3$). Lime application is a part of normal cultural practices in acid soils but it has also proved effective to decrease Cd and other metals concentration in edible parts of many crops. It is also a good source of Ca^{2+} for plants in soils. It increased the Ca^{2+} concentration in soil solution on dissolution which is a well-known component of signal transduction pathways in plants (Hepler and Wayne, 1985). Other alkaline waste materials such as coal fly ash can also be used to decrease Cd uptake by plants. In such cases, effect of liming materials decreased Cd mobility in soils and increased competition between Ca^{2+} and Cd^{2+} ions for absorption at root surfaces. In soils treated with lime materials, Cd can be precipitated as $CdCO_3$ and $Cd(OH)_2$ (Street et al., 1978).

Lime can increase the negative charge (CEC) in variable charge soils, form strongly bound hydroxyl metal species, precipitate metals as hydroxides. Calcium released from lime followed by plant absorption could inhibit the translocation of metals including Cd from roots to shoots. However, there are evidences from some case studies that addition of Ca-containing compounds (lime, gypsum) increased the plant availability of metals in soils (John et al., 1972; Williams and David, 1976). This was attributed to the exchange of Ca^{2+} with the Cd^{2+} on exchange sites and the subsequent increase in Cd^{2+} concentration in soil solution which was subjected to plant absorption. Precipitation of Cd as cadmium hydroxides or carbonates is usually considered one of the major mechanisms for immobilization of Cd by liming materials both in acid and neutral soils (Street et al., 1978; Pierzynski and Schwab, 1993).

The effect of liming on Cd immobilization and subsequent phytoavailability depends on the changes in pH and Ca^{2+} concentration in soil solution. Several studies showed that the adsorption of Cd^{2+} increased with increasing soil pH (Naidu et al., 1994; Bolan et al., 1999a) resulting in low phytoavailability of Cd in alkaline soils. The equilibrium solution concentration at which zero Cd^{2+} sorption–desorption occurred (PZC) decreased with increasing pH, which indicated that adsorption continued to occur at high pH even at low solution Cd^{2+} concentration (Filius et al., 1998). A significant relationship generally exists between the increase in pH-induced surface charge and Cd^{2+} adsorption. However, only a small fraction of the pH-induced surface charge (7–11%) was occupied by Cd^{2+}, and the ratio of pH-induced increase in Cd^{2+} adsorption. Calcium often competes strongly with Cd^{2+} for adsorption (Boekhold et al., 1983), that results in decreased Cd^{2+} adsorption in the $Ca(OH)_2$ treated soil and limited plant absorption.

Soil pH is one of the most important factors which control the heavy metal uptake by plants (Seuntjens et al., 2004; Amini et al., 2005; Basta et al., 2005). Decrease in soil pH results in the increased Cd uptake by plants especially in alkaline soils (Kuo et al., 2004; Tudoreanu and Phillips, 2004; Tsadilas et al., 2005).

Adams et al. (2004) reported that grain Cd concentrations influenced by the total Cd and pH of soils in 162 wheat and 215 barley samples. An increase in soil pH caused an increase in the solublilization of organic matter, resulting in turn in an increase in the concentration of dissolved organic carbon (Temminghoff, 1998). Bolan et al. (2003a, b) showed a significant relationship between increase in pH-induced surface charge and Cd^{2+} adsorption. In alkaline soils, solubility and uptake of Cd was facilitated due to Cd complexation with humic and organic acids (Harter and Naidu, 1995). Ionic composition of soil solution could greatly affect the metals, including Cd^{2+} and its concentration in soils and plants (Harter, 1992). Ionic radius of Cd^{2+} is 0.109 nm which is closer to that of Ca^{2+} (0.114 nm) as compared to Zn^{2+} (0.088 nm). Calcium suppressed the Cd ion exchange on smectite soils (Zachara et al., 1993).

It is obvious from the above literature that Ca-containing amendments can be used for the remediation of soils with toxic levels of Cd. These amendments include liming material such as $CaCO_3$, CaO, $Ca(OH)_2$ and gypsum ($CaSO_4.2H_2O$) which cause insolubilization of Cd in the forms of $CdSO_4$, $CdCO_3$, $Cd(OH)_2$, etc. Ultimately, these amendments help to decrease the available concentration of Cd in soil thus its uptake by grain crops like rice and wheat is limited, and as a result decreased the expected hazards associated with elevated concentrations of Cd. So due importance should be given to testing the effects of different rates of such amendments in Cd-contaminated soils for a safe environment.

Metals Response to Phosphorus Containing Amendments

Phosphorus (P) is an essential nutrient and is required for normal plant growth and development. It affects the bioavailability of other elements. It is the structural component of cells and many plant metabolites and is involved in energy storage and transfer of genetic code being a part of DNA (Hopkins and Ellsworth, 2003). Phosphorus enters the soils from parent material, pedogenic and anthropogenic sources. On solubility bases, phosphatic fertilizers are divided into water soluble and insoluble fertilizers (Bolan et al., 1993). Water soluble include mainly triple super phosphate (TSP), single super phosphate (SSP), monoammonium phosphate (MAP), diammonium phosphates (DAP) and nitrophos (NP). Water-insoluble fertilizers include PRs and basic slag. Soluble phosphate sources could rapidly increase phosphorus concentration in soil solution and facilitate the formation of metal–phosphate complexes (Ma et al., 1993; Berti and Cunningham, 1997; Hettiarachchi et al., 1997; Cooper et al., 1998). This complex formation will reduce metal solubility and mobility that will reduce transport of metal to plants and ground water (Cotter-Howells, 1996). Complexation of P with Cd [$Cd_3(PO_4)_2$, Ksp ¼ 2.53×10^{-33}] form a stable complex so the solubility of Cd could be minimized in soil solution (Hettiarachchi and Pierzynski, 1999). Thus application of P fertilizers is highly cost-effective as compared to other remedial techniques and may be a long-term solution to remediate metal-contaminated soils (Vangronsveld and Cunningham, 1998).

Recently, the application of phosphatic compounds to reduce metal availability has shown very good promise and is accepted as an alternative method to remediate metal-contaminated soils (Berti and Cunningham, 1997). Solubility of metals is reduced by the formation of metal–phosphate precipitates which are sparingly soluble and more stable which leads to a reduction in the solubility of metals (Mench et al., 1998). The mechanism of co-precipitation appeared more significant for Cd sorption as compared to Zn and other heavy metals (Xu and Schwartz, 1994) owing to competition between their ionic species having similar ionic size.

On the one hand, it has also been reported that the use of Cd-containing fertilizers like P-fertilizers increased Cd uptake by carrot (*Daucus carota L.*), oat (*Avena sativa* L.), ryegrass (*Lolium multiflorum* L.), and spinach (*Spinacia oleracea* L.) (He and Singh, 1994; Guttormsen et al., 1995). On the other hand, P fertilizer application did not increase Cd uptake in many plants including pasture plants, grass and clover (Smilde and Van Luit, 1983; Mortvedt, 1987; Escrig and Morel, 1998). This could be related to the quality of fertilizer applied and genetic architecture of plants. Some low-quality fertilizers applied could increase the Cd entrapment by P fertilizers.

Although P application has proved the most effective treatment in reducing Cd concentration in both plant and soils, the most striking question is to have a look at its potential effect on the eutrophication especially of surface water (Djodjic et al., 2004). Thus application rates of P amendments to immobilize Cd must be given due consideration. In case of co-contamination with As, P application solubilized As in soil system (Peryea, 1991).

For in situ remediation of metals, water soluble P compounds including super phosphate fertilizers, added to soils, the dissolution of MCP could result in the formation of slowly soluble dicalcium phosphates with a release of phosphoric acid close to the fertilizer granules. The dissociation of phosphoric acid subsequently released phosphate and hydrogen ions (H^+) which reduced soil pH around fertilizer granules. When ammonium phosphate fertilizers are added to soils, these dissociate into ammonium and phosphate ions, and subsequent oxidation of NH_4^+ to NO_3^- results in the release of protons causing a decrease in pH of soils, at least around the fertilizer granules.

Brown et al. (2004) reported that phosphates decreased Cd availability at a soil pH value of 7.15. It was found that PRs had no effect on plant Cd uptake while TSP remained effective in binding Cd in soils rendering it unavailable to tall fescue plants. Seven different types of soils were used by Bolan et al. (2003a) to evaluate the effect of different phosphate fertilizers on Cd immobilization. All the soils had both positive and negative charges, but were net negatively charged because their point of zero charge (PZC) was less than the corresponding soil pH. Negative charge, as indicated by Na^+ adsorption, increased with increasing phosphate adsorption and pH in response to KH_2PO_4 addition. The phosphate-induced increase in negative charge was higher in soils dominated by allophanes than in non-allophanic soils while the increase in negative charge per unit amount of phosphate adsorption decreased with increasing absorption. Adsorption of Cd (II) increased with increasing phosphate adsorption and pH of soils, the effect being more pronounced in allophones than in non-allophanic soils. The inhibitory effect of Cd on plant growth decreased with increasing phosphate application.

Phosphate amendments can reduce mobility of metals by means of two possible reasons: (1) phosphate induced Cd adsorption, and (2) precipitation of Cd as $Cd(OH)_2$ and $Cd_3(PO_4)_2$. Anion–cation interactions in soils have been reported, including anion-induced adsorption for a number of cations (Kuo, 1986). Several mechanisms are involved in phosphate-induced Cd^{2+} adsorption, including: (i) increase in pH; (ii) increase in surface charge; (iii) co-adsorption of phosphate and Cd as an ion pair; and (iv) surface complex formation of Cd on the phosphate compounds. Cd adsorption to metal binding sites in soils is increased by increased soil pH and decreased Cd in the soil solution that help to reduce phytoavailable Cd (McBride et al., 1997; Sauve et al., 2000a). Plant tissue Cd concentration is predicted through soil pH, total soil Cd, organic matter and Fe/Mn oxides (McBride, 2002).

Several studies have shown that treatment of contaminated water and soil with hydroxiapatite, phosphate rock or phosphoric acid effectively decreased available Cd and Zn by making compounds of low solubility (Ma et al. 1994; Seaman et al., 2001). When insoluble P fertilizers like PRs are added into soils, phosphate mineral apatite needs to be dissolved to release available P for plants. Dissolution of PRs is necessary not only for the availability of P to plants (Rajan et al., 1996), but also for the insolubilization of Cd in the form of Cd phosphates (Laperche and Traina, 1998). In acid soils, PRs are very effective as a nutrient source because it dissolves more in the acidic than alkaline conditions. After dissolution of PRs, P released undergoes same reactions of adsorption and precipitation as in the case of soluble fertilizers. Rock phosphates are not only effective in decreasing Cd solubility but it could also decrease the Cd bioavailability very efficiently, thus help in reducing plant toxicity and restricting its entry into the food chain (Laperche et al., 1997; Basta and Gradwohl, 1998; Zhang et al., 1998).

The main source of P in soils is due to increasing amounts of PRs added to soils. PRs have a liming value in addition to supplying P and Ca^{2+}. This liming action of PRs may be due to some free $CaCO_3$ present in PRs, which itself can act as a liming agent. Similarly, soil acidity may be decreased through the dissolution of P mineral component (apatite) in soils that consumes H^+ and as a result soil pH is increased. According to an estimate, every 1 kg of P dissolved from PRs generated a liming value equivalent to 3.2 kg $CaCO_3$ (Bolan et al. 2003b). Free $CaCO_3$ in PRs dissolved reasonably fast that provide a small amount of immediate liming value, the apatite dissolved at a variable but slow rate that helped to provide liming value over a longer period of time.

Among chemical immobilization techniques, use of P-containing amendments is a good alternative to soil excavation being cost-effective, and could provide a long-term solution for Cd-contaminated soils. This is because of the formation of stable compounds with Cd on solubilization of P. Further investigations are needed to test different rates of P-containing amendments in Cd-contaminated soils.

Metals Behaviour in Response to Ammonium Containing Amendments

Nitrogen is known to be one of the most important elements in plant nutrition for better growth and development. Influence of N fertilizers on heavy metal uptake has been investigated by several authors for a long period. The application of ammonium-containing fertilizers can increase Cd availability in soils due to soil acidification (Eriksson, 1990). Soil pH is decreased due to the application of high NH_4^+-containing fertilizers, leading to an increase in Cd uptake by plants (Willaert and Verloo, 1992). NH_4^+ ion inhibited Cd absorption by grain crops like wheat and barley in water culture experiments (Chaudhry and Loneragan, 1972; Cox and Reisenauer, 1977; Tills and Alloway, 1981). However, few experiments on the effects of NH_4^+ salts have been reported in soils, and few studies have been undertaken on the effects of complimentary anions to NH_4^+ salts in N fertilizers. The effect of NH_4^+ salts on heavy metal uptake by plants seems to be controlled by a number of factors such as plant species, plant growth stages and soil types.

Cadmium transport into the cells from the roots is decreased due to binding of Cd cation exchange sites in mucilage excretions of maize root tips and to sites in root cell walls (Morel et al., 1986; Rauser, 1987). This kind of CD sequestration may decrease due to low pH (6.5) in the rhizosphere and competition with other cations like Ca^{2+} originating from nitrate of lime. It is reported that there is no evidence for a regulation mechanism in plants themselves for Cd absorption and translocation to grains of durum wheat (*Triticum durum*) (Mitchell et al., 2000). However, the effects of N fertilizer on the Cd concentration in plants are still ambiguous. For instance, it is reported that the Cd concentration in wheat grains decreased with increasing N application rate and N concentration in both soils and nutrient solution (Landberg and Greger, 2003). In long-term experiments, there were no effects of N fertilizers on the grain-Cd concentration (Gavi et al., 1997).

Metals Behaviour in Response to Sulphur-Containing Amendments

Sulphur (S) is used as a fertilizer source to alleviate S deficiency in soils. Elemental S is a cheap fertilizer source and its efficiency on crops could be governed by its oxidation rate in soils. Soil moisture, temperature, added phosphates and organic matter are common factors that govern the oxidation of S in the soil (Watkinson and Bolan, 1998; Li et al., 2000).

It is reported that elemental sulphur is a good soil acidification material because sulphur-oxidizing bacteria (*Thiobucillus*) can oxidize sulphur to sulphate and protons (Moser and Olson, 1953). This acidification process increased the solubility and uptake of Cd and Zn by sunflower and maize crops (Kayser et al., 2000). However, application of acids or acid formers have certain limitations like adverse effects on soil fertility and soil structure, or they may lead to ground water contamination with metals. To avoid such constraints, application of elemental sulphur in several splits has been suggested to decrease soil pH and increase solubility of heavy metals in soils (Kayser et al., 2000; Tichy et al., 1997) owing to its slow bio-oxidation reactions in soils.

The above literature shows some controversial effects of S-containing amendments in Cd-contaminated soils. As Pakistani soils are calcareous in nature and have an alkaline soil pH, there is a need to test different rates of S-amendments to evaluate their effects in calcareous Cd-contaminated soils.

CONCLUSIONS

Cadmium concentration in agricultural soil is increasing continuously due to natural sources and anthropogenic activities. Cd contamination originates from natural sources by means of mineral dissociation, weathering of parent material and atmospheric deposition as well as anthropogenic sources related to mining, industrial emissions, disposal or leakage of industrial wastes, application of sewage sludge to agricultural soils, fertilizer and pesticide use. Cd uptake and accumulation in plants is a function of the complex interaction of soil, plant and environmental factors which influence Cd phytoavailability. Cadmium behaviour in soil depends on soil and biological factors. Soil factors are: parent material, organic matter content, pH, CEC, redox conditions, competing ions and the amount of organic and inorganic ligands, etc. Biological factors are: plants species, root morphology and microbial conditions. The main pathway of Cd uptake by plants is through the roots. Once in roots it sequesters there or translocates to the aerial parts by different pathways depending on the plant species. Higher Cd accumulation in plants causes toxicity in plants. CD decreased the seed germination, reduced root elongation and shoots and root biomass. Higher Cd also resulted in leaf chlorosis, inhibition of photosynthesis and reduced the essential element concentration in different plant parts. Cd induced the oxidative stress in plants by producing the reactive oxygen species and modifying the production of antioxidant enzymes. However, the intensity of these effects varies and depends on the concentration of metal, duration of exposure, plant species and stage of plant. Plants have different mechanisms to cope with Cd stress

Different technologies exist for in situ remediation of Cd toxicity in soil and sediments and to reduce metal uptake by plants. These technologies include immobilization, phytoextraction, phytostabilization, physical separation and extraction, etc. Chemical immobilization is an in situ remediation strategy where inexpensive and easily available chemicals are used to decrease plant availability of metals in contaminated soils. These chemical amendments include Ca and P-compounds and some other alkaline-stabilizing solids which are quite effective at immobilizing metals, thereby decreasing their bioavailability to plants. Use of Ca, P, NH_4 and S-containing amendments is a good alternative to soil excavation, being cost-effective, and could provide a long-term solution for Cd-contaminated soils under grain crops cultivation. Therefore, due importance should be given to test the effects of different rates of such amendments in Cd-contaminated soils for a safe environment. However, there is still a need not only to understand the effects of inorganic amendments on Cd immobilization/insolubilization in soils and its phytoavailability, but also to investigate the effectiveness of different amendments.

REFERENCES

Abe, T., Fukami, M., Ogasawara, M., 2008. Cadmium accumulation in the shoots and roots of 93 weed species. Soil Sci. Plant Nutr. 54, 566–573.

Adams, M.L., Zhao, F.J., McGrath, S.P., Nicholson, F.A., Chambers, B.J., 2004. Predicting cadmium concentrations in wheat and barley grain using soil properties. J. Environ. Qual. 33, 532–541.

Adriano, D.C., 2001. Trace elements in terrestrial environments: Biogeochemistry, Bioavailability, and Risks of Metals, second ed. Springer, New York.

Adriano, D.C., Wenzel, W.W., Vangronsvel, J., Bolan, N.S., 2004. Role of assisted natural remediation in environmental cleanup. Geoderma 122, 121–142.

Amini, M., Khademi, H., Afyuni, M., Abbaspour, K.C., 2005. Variability of available cadmium in relation to soil properties and land use in an arid region in central Iran. Water, Air, Soil Pollut 162, 205–218.

Antoniadis, V., Alloway, B.J., 2002. The role of dissolved organic carbon in the mobility of Cd, Ni and Zn in sewage sludge-amended soils. Environ. Pollut. 117, 515–521.

Baize, D., 1997. Teneurs Totales en Eléments Traces Métalliques dans les Sols Français. Références et Stratégies d'Interprétation. INRA Editions, Paris, France.

Baize, D., Deslais, W., Gaiffe, M., 1999. Anomalies naturelles en cadmium dans les sols de France. Étude et Gestion des Sols 6, 85–104.

Balestrasse, K.B., Gallego, S.M., Tomaro, M.L., 2006. Oxidation of the enzymes involved in nitrogen assimilation plays an important role in the cadmium induced toxicity in soybean plants. Plant Soil 284, 187–194.

Basta, N.T., Gradwohl, R., 1998. Remediation of heavy metal-contaminated soil using rock phosphate. Better Crops Plant Food 82, 29–31.

Basta, N.T., Gradwohl, R., Snethen, K.L., Schroder, J.L., 2001. Chemical immobilization of lead, zinc and cadmium in smelter-contaminated soils using biosolids and rock phosphate. J. Environ. Qual. 30, 1222–1230.

Basta, N.T., Rayan, J.A., Chaney, R.L., 2005. Trace elements chemistry in residual treated soil: Key concepts and metal availability. J. Environ. Qual. 34, 49–63.

Bavi, K., Kholdebarin, B., Moradshahi, A., 2011. Effect of cadmium on growth, protein content and peroxidase activity in pea plants. Pak. J. Bot. 43, 1467–1470.

Benavides, M.P., Gallego, S.M., Tomaro, M.L., 2005. Cadmium toxicity in plants. Braz. J. Plant Physiol. 17, 21–34.

Berti, W.R., Cunningham, S.D., 1997. In-place inactivation of Pb in Pb-contaminated soils. Environ. Sci. Technol. 31, 1359–1364.

Bes, C., Mench, M., 2008. Remediation of copper-contaminated topsoils from a wood treatment facility using in situ stabilisation. Environ. Pollut. 156 (3), 1128–1138.

Boekhold, A.E., Temminghoff, E.J.M., van der Zee, S.E.A.T.M., 1983. Influence of electrolyte composition and pH on cadmium adsorption by an acid sandy soil. J. Soil Sci. 44, 85–96.

Bolan, N.S., Adriano, D.C., Mani, A., Duraisamy, P., Arulmozhiselvan, K., 2003a. Immobilization and phytoavailability of cadmium in variable charge soils: I. Effect of phosphate addition. Plant Soil 250, 83–94.

Bolan, N.S., Adriano, D.C., Mani, A., Duraisamy, P., 2003b. Immobilization and phytoavailability of cadmium in variable charge soils. II. Effect of lime addition. Plant Soil 251, 187–198.

Bolan, N.S., Hedley, M.J., Loganathan, P., 1993. Preparation, forms and properties of slow-release phosphate fertilizers. Fert. Res. 35, 13–24.

Bolan, N.S., Naidu, R., Syers, J.K., Tillman, R.W., 1999a. Surface charge and solute interactions in soils. Adv. Agron 67, 88–141.

Bolan, N.S., Khan, M.A.R., Tillman, R.W., Naidu, R., Syers, J.K., 1999b. The effects of anion sorption on sorption and leaching of cadmium. Aus. J. Soil Res. 37, 445–460.

Borges, M., 2002. Extratabilidade do cádmio: influência de atributos de solos muito intemperizados em extratores convencionais e potencialidade de ácidos orgânicos de baixo peso molecular. 76 f. (Dissertação em Agronomia) – Escola superior de Agricultura Luiz de Queiroz, USP. Piracicaba.

Brown, S., Channey, R., Angle, J.S., 1997. Subsurface liming and metal movement in soils amended with lime-stabilized biosolids. J. Environ. Qual. 26, 724–733.

Brown, S., Chaney, R., Hallfrisch, J., Ryan, A.J., Berti, W.R., 2004. In situ soil treatments to reduce the phyto- and bio-availability of lead, zinc and cadmium. J. Environ. Qual. 33, 522–531.

Cakmak, I., Welch, R.M., Hart, J., Norvell, W.A., Ozturk, L., Kochian, L.V., 2000. Uptake and retranslocation of leaf-applied cadmium (Cd^{109}) in diploid, tetraploid and hexaploid wheat. J. Exp. Bot. 51, 221–226.

Chaudhry, F.M., Loneragan, J.F., 1972. Zinc absorption by wheat seedlings: II. Inhibition by hydrogen ions and by micronutrient cations. Soil Sci. Am. Proc. 36, 327–331.

Cho, S.C., Chao, Y.Y., Kao, C.H., 2012. Calcium deficiency increases Cd toxicity and Ca is required for heat-shock induced Cd tolerance in rice seedlings. J. Plant Physiol. 169, 892–898.

Chugh, L.K., Sawhney, S.K., 1999. Photosynthetic activities of Pisum sativum seedlings grown in presence of cadmium. Plant Physiol. Biochem 37, 297–303.

Ci, D., Jiang, D., Dai, T., Jing, Q., Cao, W., 2009. Effects of cadmium on plant growth and physiological traits in contrast wheat recombinant inbred lines differing in cadmium tolerance. Chemosphere 77, 1620–1625.

Cieslinski, G., Van Rees, K.C.J., Szmigielska, A.M., Krishnamurti, G.S.R., Huang, P.M., 1998. Low-molecular-weight organic acids in rhizosphere soils of durum wheat and their effect on cadmium bioaccumulation. Plant and Soil 203, 109–117.

Clemens, S., 2006. Toxic metal accumulation, responses to exposure and mechanisms of tolerance in plants. Biochem 88, 1707–1719.

Clemens, S., Palmgren, M.G., Kramer, U., 2002. A long way ahead: understanding and engineering plant metal accumulation. Trends in Plant Sci. 7, 309–315.

Cook, M.E., Morrow, H., 1995. Anthropogenic Sources of Cadmium in Canada, National Workshop on Cadmium Transport into Plants. Canadian Network of Toxicology Centres, Ottawa, Ontario, Canada. June 20–21, 1995.

Cooper, E.M., Strawn, D.G., Sims, J.T., Sparks, D.L., Onken, B.M., 1998. Effect of chemical stabilization by phosphate amendment on the desorption of P and Pb from a contaminated soil. In: Agron.abstr. ASA, Madison, WI, p. 343.

Cotter-Howells, J., 1996. Remediation of contaminated land by formation of heavy metal phosphates. Appl. Geochem. 11, 335–342.

Çotuk, Y., Belivermis, M., Kiliç, O., 2010. Environmental biology and pathophysiology of cadmium. IUFS J. Biol. 69, 1–5.

Cox, W.J., Reisenauer, H.M., 1977. Ammonium effects on nutrient cation absorption by wheat. J. Agron 69, 868–871.

Das, P., Samantaray, S., Rout, G.R., 1997. Studies on cadmium toxicity in plants: a review. Environ. Pollut. 98, 29–36.

Degryse, F., Buekers, J., Smolders, E., 2004. Radiolabile cadmium and zinc in soil as affected by pH and source of contamination. Eur. J. Soil Sci. 55, 113–121.

Di Toppi, S.L., Gabrielli, R., 1999. Response to cadmium in higher plants. Environ. Exp. Bot. 41, 105–130.

Dimkpa, C.O., Merten, D., Svatos, A., Buchel, G., Kothe, E., 2009. Siderophores mediate reduced and increased uptake of cadmium by Streptomyces tendae F4 and sunflower (Helianthus annuus), respectively J. Appl. Microbiol. 107, 1687–1696.

Djodjic, F., Borling, K., Bergstrom, L., 2004. Phosphorus leaching in relation to soil type and soil phosphorus content. J. Environ. Qual. 33, 678–684.

Dong, J., Wu, F., Zhang, G., 2006. Influence of cadmium on antioxidant capacity and four micro-element concentrations in tomato seedlings (Lycopersicon esculentum). Chemosphere 64, 1659–1666.

Dong, J., Mao, W.H., Zhang, G.P., Wu, F.B., Cai, Y., 2007. Root excretion and plant tolerance to cadmium toxicity—a review. Plant Soil Environ. 53, 193–200.

Drazic, G., Mihailovic, N., Stojanovic, Z., 2004. Cadmium toxicity: the effect on macro- and micro-nutrient contents in soybean seedlings. Biol. Plant 4, 605–607.

Dunbar, K.R., McLaughlin, M.J., Reid, R.J., 2003. The uptake and partitioning of cadmium in two cultivars of potato (Solanum tuberosum L.). J. Exp. Bot. 54, 349–354.

EC, 2000. Position Paper: Ambient air pollution by As, Cd and Ni compounds. Position Paper. European Commission, Directorate—General Environment, October 2000. Available at http://ec.europa.eu/environment/air/pdf/pp_as_cd_ni.pdf.

EFSA (European Food Safety Authority), 2009. Cadmium in Food: Scientific Opinion of the Panel on Contaminants in the Food Chain. ESFA J. 980, 1–139.

Elder, J.F., 1989. Metal biogeochemistry in surface-water systems-A review of principles and concepts: U.S. Geological Survey Circular, 1013, 43 p.

Eriksson, J.E., 1990. Factors influencing Cd levels in soil and grain of oats and winter wheat: A field study on Swedish soils. In: Swedish Univ., Agric. Sci. Uppsala, Sweden, Report Dissertations. 4.

Escrig, I., Morel, I., 1998. Effect of calcium on the soil adsorption of cadmium and zinc in some Spanish sandy soils. Water, Air, Soil Pollut. 105, 507–520.

Feng, J., Shi, Q., Wang, X., Wei, M., Yang, F., Xu, H., 2010. Silicon supplementation ameliorated the inhibition of photosynthesis and nitrate metabolism by cadmium (Cd) toxicity in Cucumis sativus L. Scient. Horti 123, 521–530.

Filius, A., Streck, T., Richter, J., 1998. Cadmium sorption and desorption in limed top soils as influenced by pH: Isotherms and simulate leaching. J. Environ. Qual. 27, 12–18.

Francis, A.J., 1990. Microbial dissolution and stabilization of toxic metals and radionuclides in mixed wastes. Experientia 46, 840–851.

Gao, X., Brown, K.R., Racz, G.J., Grant, C.A., 2010. Concentration of cadmium in durum wheat as affected by time, source and placement of nitrogen fertilization under reduced and conventional-tillage management. Plant Soil 337, 341–354.

Gao, X., Mohr, R.M., McLaren, D.L., Grant, C.A., 2011. Grain cadmium and zinc concentrations in wheat as affected by genotypic variation and potassium chloride fertilization. Field Crops Res. 122, 95–103.

Gavi, F., Basta, N.T., Raun, W.R., 1997. Wheat grain cadmium as affected by long-term fertilization and soil acidity. J. Environ. Qual. 26, 265–271.

Gill, S.S., Khan, N.A., Tuteja, N., 2011. Differential cadmium stress tolerance in five indian mustard (*Brassica juncea* L.) cultivars. An evaluation of the role of antioxidant machinery. Plant Signal. Behav. 6, 293–300.

Gill, S.S., Khan, N.A., Tuteja, N., 2012. Cadmium at high dose perturbs growth, photosynthesis and nitrogen metabolism while at low dose it up regulates sulfur assimilation and antioxidant machinery in garden cress (*Lepidium sativum* L.). Plant Sci. 182, 112–120.

Gonçalves, J.F., Antes, F.G., Maldaner, J., Pereira, L.B., Tabaldi, L.A., Rauber, R., Rossato, L.V., Bisognin, D.A., Dressler, V.L., de Moraes Flores, E.M., Nicoloso, F.T., 2009. Cadmium and mineral nutrient accumulation in potato plantlets grown under cadmium stress in two different experimental culture conditions. J. Plant Physiol. Biochem. 47, 814–821.

Grant, C.A., Buckley, W.T., Bailey, L.D., Selles, F., 1998. Cadmium accumulation in crops. Can. J. Plant Sci. 78, 1–17.

Grant, C.A., Clarke, J.M., Duguid, S., Chaney, R.L., 2008. Selection and breeding of plant cultivars to minimize cadmium accumulation. Sci. Total Environ. 390, 301–310.

Greger, M., Lofstedt, M., 2004. Comparision of uptake and distribution of cadmium in different cultivars of bread and durum wheat. Crop Sci. 44, 501–507.

Guo, Y., Marschner, H., 1995. Uptake, distribution and binding of cadmium and nickel in different plant species. J. Plant Nutr. 18, 2691–2706.

Gupta, S.K., Aten, C., 1993. Comparison and evaluation of extraction media and their suitability in a simple model to predict the biological relevance of heavy metal concentrations in contaminated soils.Int. J. Environ. Anal. Chem. 51, 25–46.

Guttormsen, G., Singh, B.R., Jeng, A.S., 1995. Cadmium concentrations in vegetable crops grown in a sandy soil as affected by Cd levels in fertilizer and soil pH. Fert. Res. 41, 27–32.

Han, F., Shan, X., Zhang, S., Wen, B., Owens, G., 2006. Enhanced cadmium accumulation in maize roots—the impact of organic acids. Plant Soil 289, 355–368.

Haouari, C.C., Nasraoui, A.H., Bouthour, D., Houda, M.D., Daieb, C.B., Mnai, J., Gouia, H., 2012. Response of tomato (*Solanum lycopersicon*) to cadmium toxicity: Growth, element uptake, chlorophyll content and photosynthesis rate. Afr. J. Plant Sci. 6, 1–7.

Harris, N.S., Taylor, G.J., 2001. Remobilization of cadmium in maturing shoots of near isogenic lines of durum wheat that differ in grain cadmium accumulation. J. Exp. Bot 52, 1473–1481.

Hart, J.J., Welch, R.M., Norvell, W.A., Sullivan, L.A., Kochian, L.V., 1998. Characterization of cadmium binding, uptake and translocation in intact seedlings of bread and durum wheat cultivars. Plant Physiol 116, 1413–1420.

Hart, J.J., Welch, R.M., Norvell, W.A., Kochian, L.V., 2002. Transport interactions between cadmium and zinc in roots of bread and durum wheat seedlings. Physiol.Plant 116, 73–78.

Hart, J.J., Welch, R.M., Norvell, W.A., Kochian, L.V., 2006. Characterization of cadmium uptake, translocation and storage in near-isogenic lines of durum wheat that differ in grain cadmium concentration. New Phytol. 172, 261–271.

Harter, R.D., 1992. Competitive sorption of cobalt, copper and nickel ions by a calcium saturated soil. Soil Sci. Soc. Am. J. 56, 444–449.

Harter, R.D.R., Naidu, R., 1995. Role of metal–organic complexation in metal sorption by soils. Adv. Agron 55, 219–264.

He, Q.B., Singh, B.R., 1994. Plant availability of cadmium in soils. 2. Factors related to the extractability and plant uptake of cadmium in cultivated soils. Acta Agric. Scand 43, 142–150.

He, Z.L., Yang, X.E., Stoffella, P.J., 2005. Trace elements in agroecosystems and impacts on the environment. J. Trace Elem. Med. Biol. 19, 125–140.

Heidari, M., Sarani, S., 2011. Effects of lead and cadmium on seed germination, seedling growth and antioxidant enzymes activities of mustard (*Sinapis arvensis* L.). ARPN. J. Agri. Biol. Sci. 6, 44–47.

Hentz, S., McComb, J., Miller, G., Begonia, M., Begonia, G., 2012. Cadmium uptake, growth and phytochelatin contents of *Triticum Aestivum* in response to various concentrations of cadmium. World Environ. 2, 44–50.

Hepinstall, S.E., Turner, B.F., Maurice, P.A., 2005. Effects of siderophores on Pb and Cd adsorption to kaolinite. Clays and Clay Minerals 53, 557–563.

Hepler, P.K., Wayne, R.O., 1985. Calcium and plant development. Ann. Rev. Plant Physiol. 36, 391–439.

Hernandez, L.E., Lozano, E., Garate, A., Carpena, R., 1998. Influence of cadmium on the uptake, tissue accumulation and subcellular distribution of manganese in pea seedlings. Plant Sci. 132, 139–151.

Hernandez, L., Probst, A., Probst, J.L., Ulrich, E., 2003. Heavy metal distribution in some French forest soils: Evidence for atmospheric contamination. The Science of the Total Environment. 312, 195–219.

Hettiarachchi, G.M., Pierzynski, G.M., 1999. Effect of phosphorus and other soil amendments on soil lead, cadmium and zinc bioavailability. In: Proc. Extended Abstr. 5th Int. Conf. on the Biogeochem. Trace Elements (ICOBTE), Vienna, Austria, pp. 514–515. 11–15 July, 1999.

Hettiarachchi, G.M., Pierzynski, G.M., Zwonitzer, J., Lambert, M., 1997. Phosphorus source and rate effects on cadmium, lead, and zinc bioavailabilities in a metal-contaminated soil. In: Extended Abstr. 4th Int. Conf. on the Biogeochem. Trace Elements (ICOBTE), Berkeley, CA, pp. 463–464. 23–26 June, 1997.

Hill, O.A., Lion, L.W., Ahner, B.A., 2002. Reduced Cd accumulation in *Zea mays*: a protective role for phytosiderophores? Environ. Sci. Technol. 36, 5363–5368.

Hopkins, B., Ellsworth, J., 2003. Phosphorus nutrition in potato production. Jason Ellsworth at the Idaho Potato Conference on January 22–23.

Illera, V., Garrido, F., Serrano, S., Lez, M.T.G.A., 2004. Immobilization of the heavy metals Cd, Cu and Pb in an acid soil amended with gypsum and lime rich industrial by products. Europ. J. Soil Sci. 55, 135–145.

Ingwersen, J., Streck, T., 2005. A regional-scale study on the crop uptake of cadmium from sandy soils: Measurement and modeling. J. Environ. Qual 34, 1026–1035.

Itoh, S., Yumura, Y., 1979. Studies on the contamination of vegetables crops by excessive absorption of heavy metals. Bull, Veg. Ornamental crops Res. Stn. 6a 123 (Ja).

Jalil, A., Selles, F., Clark, J.M., 1994. Effect of cadmium on growth and uptake of cadmium and other elements by durum wheat. J. Plant Nutr. 17, 1839–1858.

John, M.K., van Laerhoven, C.J., Chuah, H., 1972. Factors affecting plant uptake and phytoavailability of cadmium added to soils. Environ. Sci. Technol. 6, 1005–1009.

Jung, M.C., 2008. Heavy metal concentrations in soils and factors affecting metal uptake by plants in the vicinity of a Korean cu-w mine. Sensors 8, 2413–2423.

Kabata-Pendias, A., Pendias, H., 2001. Trace Elements in Soils and Plants, third ed. CRC Press, Boca Raton, FL, USA.

Kayser, A., Wenger, K., Keller, A., Attinger, W., Felix, H.R., Gupta, S.K., Schulin, R., 2000. Enhancement of phytoextraction of Zn, Cd and Cu from calcareous soil: The use of NTA and sulfur amendments. Environ. Sci. Technol. 34, 1178–1783.

Keller, C., Marchetti, M., Rossi, L., Lugon-Moulin, N., 2005. Reduction of cadmium availability to tobacco (*Nicotiana tabacum*) plants using soil amendments in low cadmium-contaminated agricultural soils: A pot experiment. Plant and Soil 276, 69–84.

Kennedy, V.H., Sanchez, A.L., Oughton, D.H., Rowland, A.P., 1997. Use of single and sequential chemical extractants to assess radionuclide and heavy metal availability from soils for root uptake. Analyst 122, 89–100.

Khan, N.A., Samiullah, A., Singh, S., Nazar, R., 2007. Activities of antioxidative enzymes, sulphur assimilation, photosynthetic activity and growth of wheat (*Triticum aestivum*) cultivars differing in yield potential under cadmium stress. J. Agron. Crop Sci. 193, 435–444.

Khatamipour, M., Piri, E., Esmaeilian, Y., Tavassoli, A., 2011. Toxic effect of cadmium on germination, seedling growth and proline content of Milk thistle (*Silybum marianum*). An. Biol. Res. 2, 527–532.

Kirkham, M.B., 2006. Cadmium in plants on polluted soils: Effects of soil factors, hyperaccumulation, and amendments. Geoderma 137, 19–32.

Klumpp, C., Burger, A., Mislin, G.L., Abdallah, M.A., 2005. From a total synthesis of cepabactin and its 3:1 ferric complex to the isolation of a 1:1:1 mixed complex between iron (III), cepabactin and pyochelin. Bioorg. Med. Chem. Lett. 15, 1721–1724.

Knox, A.S., Seaman, J.C., Mench, M.J., Vangronsveld, J., 2001. Remediation of metal- and radionuclides- contaminated soils by in situ stabilization techniques. In: Iskandar, I.K. (Ed.), Environmental Restoration of Metals-Contaminated Soils. CRC Press, Inc., Boca Raton, FL, USA, pp. 21–60.

Knox, A.S., Kaplan, D.I., Adriano, C.I., Hinton, T.Z., Wilson, M.D., 2003. Apatite and phillipsite as sequestering agents for metals and radionuclides. J. Environ. Qual. 32, 515–525.

Kubo, K., Watanabe, Y., Matsunaka, H., Seki, M., Fujita, M., Kawada, N., Hatta, K., Nakajima, T., 2011. Differences in cadmium accumulation and root morphology in seedlings of Japanese wheat varieties with distinctive grain cadmium concentration. Plant Prod. Sci. 14, 148–155.

Kuo, S., 1986. Concurrent adsorption of phosphate and zinc, cadmium, or calcium by a hydrous ferric oxide. Soil Sci. Soc. Am. J 50, 1412–1419.

Kuo, S., Huang, B., Bembenek, R., 2004. The availability to lettuce of zinc and cadmium in a zinc fertilizer. Soil Sci. 169, 363–373.

Lakanen, E., Ervio, R., 1971. A comparison of eight extractants for the determination of plant available micronutrients in soils. Acta Agr.Fenn 123, 223–232.

Landberg, T., M. Greger, M., 2003. Influence of N and N supplementation on Cd accumulation in wheat grain, 7th International Conference on the Biogeochemistry of Trace Elements. Uppsala 03, Conference Proceedings. 1, 90–91.

Laperche, V. Traina, S.J., 1998. Immobilization of Pb by hydroxyapatite. pp. 225–276. In: Everett, J.A., (ed.), Adsorption of Metals by Geomedia: Variables, Mechanisms and Model Applications. Academic Press, Orlando, FL.

Laperche, V., Logan, T.J. Gaddam, P., Traina, S.J., 1997. Effect of apatite amendments on plant uptake of lead from contaminated soil. Environ. Sci. Technol. 31, 2745–2753.

Lehoczky, E., Marth, P., Szabados, I., Lukacs, P., 2000. Influence of soil factors on the accumulation of cadmium by lettuce. Comm. Soil Sci. Plant Anal. 31, 2425–2431.

Li, S., Lin, B., Zhou, W., 2005. Effects of previous elemental sulfur applications on oxidation of additional applied elemental sulfur in soils. Biol. Fertil. Soils 42, 146–152.

Li, Y., Wan, J., Gu, Z., 2000. The formation of cadmium sulfide nanowires in different liquid crystal systems. Mater. Sci. Eng. 286, 106–109.

Liphadzi, M.S., Kirkham, M.B., 2006. Availability and plant uptake of heavy metals in EDTA-assisted phytoremediation of soil and composted biosolids. S. Afr. J. Bot 72, 391–397.

Liu, J.G., Li, K., Xu, J.K., Liang, J.S., Lu, X.L., Yang, J.C., Zhu, Q.S., 2003. Interaction of Cd and five mineral nutrients for uptake and accumulation in differentrice cultivars and genotypes. Field Crop Res. 83, 271–281.

Liu, J.G., Qian, M., Cai, G.L., Yang, J.C., Zhu, Q.S., 2007. Uptake and translocation of Cd in different rice cultivars and the relation with Cd accumulation in rice grain. J. Hazard. Mater. 143, 443–447.

Liu, T.T., Wu, P., Wang, L.H., Zhou, Q., 2011. Response of soybean seed germination to cadmium and acid rain. Biol. Trace Elem. Res. 144, 1186–1196.

López-Millán, A.F., Sagardoy, R., Solanas, M., Abadia, A., Abadia, J., 2009. Cadmium toxicity in tomato (*Lycopersicon esculentum*) plants grown in hydroponics. Environ. Exp. Bot. 65, 376–385.

Lozano-Rodriguez, E., Hernandez, L.E., Bonay, P., Carpena–Ruiz, R.O., 1997. Distribution of cadmium in shoot and root tissues of maize and pea plants: Physiological disturbances. J. Exp. Bot., 48,123–128.

Lux, A., Martinka, M., Vaculik, M., White, P.J., 2011. Root responses to cadmium in the rhizosphere: A review. J. Exp. Bot 62, 21–37.

Ma, L.Q., Traina, S.J., Logan, T.J., Ryan, J.A., 1993. In situ lead immobilization by apatite. Environ. Sci. Technol. 27, 1803–1810.

Ma., Q.Y., Traina, S.J., Logan, T.J., Ryan, J.A., 1994. Effects of aqueous Al, Cd, Cu, Fe(II), Ni, and Zn on Pb immobilization by hydroxyapatite. Environ. Sci. Technol. 28, 1219–1228.

Maestri, E., Marmiroli, M., Visioli, G., Marmiroli, N., 2010. Metal tolerance and hyperaccumulation: Costs and trade-offs between traits and environment. Environ. Exp.Bot. 68, 1–13.

Marschner, H., 1995. Mineral nutrition of higher plants, second ed. Academic Press, San Diego, New York.

Marschner, H., Romheld, V., Kissel, M., 1986. Different strategies in higher-plants in mobilization and uptake of iron. J. Plant Nutr. 9, 695–713.

McBride, M., Sauve, S., Hendershot, W., 1997. Solubility control of Cu, Zn, Cd and Pb in contaminated soils. Eur. J. Soil Sci. 48, 337–346.

McBride, M.B., 2002. Cadmium uptake by crops estimated from soil total Cd and pH. Soil. Sci. 167, 62–67.

McGowen, S.L., Basta, N.T., Brown, G.O., 2001. Use of diammonium phosphate to reduce heavy metal solubility and transport in smelter-contaminated soil. J. Environ. Qual. 30, 493–500.

Mench, M., Martin, E., 1991. Mobilization of cadmium and other metals from two soils by root exudates of *Zea mays* L., *Nicotiana tabacum* L. and *Nicotiana rustica* L. Plant and Soil 132, 187–196.

Mench, M., Baize, D., Mocquot, B., 1997. Cadmium availability to wheat in five soil series from the Yonne district, Burgundy, France. Environ. Pollut. 95, 93–103.

Mench, M., Tancogne, J., Gomez, A., Juste, C., 1989. Cadmium bioavailability to *Nicotiana tabacum* L., *Nicotiana rustica* L., and *Zea mays* L. grown in soil amended or not amended withcadmium nitrate. Biol. Fertil. Soils 8, 48–53.

Mench, M., Vangronsveld, J., Lepp, N.W., Edwards, R., 1998. Physico-chemical aspects and efficiency of trace element immobilization by soil amendments. In: Vangronsveld, J., Cunningham, S.D. (Eds.), Metal-contaminated soils: In situ inactivation and phytorestoration. Springer-Verlag, Berlin.

Mench, M., Renella, G., Gelsomino, A., Landi, L., Nannipieri, P., 2006. Biochemical parameters and bacterial species richness in soils contaminated by sludge-borne metals and remediated with inorganic soil amendments. Environ. Pollut. 144 (1), 24–31.

Meyer, M.W., Fricke, F.L., Holmgren, G.G.S., Kubota, J., Chaney, R.L., 1982. Cadmium and lead in wheat grain and associated surface soils of major wheat production areas of the United States. p.34. In: Agronomy Abstracts. The Am. Soci.Agron, Madison, WI.

Miller, J.E., Hassett, J.J., Koeppe, D.E., 1976. Uptake of cadmium by soybeans as influenced by soil cation exchange capacity, pH, and available phosphorus. J. Environ. Qual. 6, 157–160.

Mitchell, L., Grant, C., Racz, G., 2000. Effect of nitrogen application on concentration of cadmium and nutrient ions in soil solution and in durum wheat. Can. J. Soil Sci. 80, 107–115.

Mobin, M., Khan, N.A., 2007. Photosynthetic activity, pigment composition and antioxidative response of two mustard (*Brassica juncea*) cultivarsdiffering in photosynthetic capacity subjected to cadmium stress. J. Plant Physiol. 164, 601–610.

Morel, J.L., Mench., M., Guckert, A., 1986. Measurement of Pb^{2+}, Cu^{2+} and Cd^{2+} binding with mucilage exudates from maize (*Zea Mays* L.) roots. Biol. Fertil. Soil 2, 29–34.

Mortvedt, J.J., 1987. Cadmium levels in soils and plants from long term soil fertility experiments in the United States of America. J. Environ. Qual. 16, 137–142.

Moser, U.S., Olson, R.V., 1953. Sulfur oxidation in 4 soils as influenced by soil moisture tension and sulfur bacteria. Soil Sci. 76, 251–257.

Nagajyoti, P.C., Lee, K.D., Sreekanth, T.V.M., 2010. Heavy metals, occurrence and toxicity for plants: A review. Environ. Chem. Lett. 8, 199–216.

Naidu, R., Bolan, N.S., Kookana, R.S., Tiller, K.G., 1994. Ionic strength and pH effects on the sorption of cadmium and the surface charge of soils. Eur.. J. Soil Sci. 45, 419–429.

Naidu, R., Kookana, R.S., Sumner, M.E., Harter, R.D., Tiller, K.G., 1997. Cadmium sorption and transport in variable charge soils: A review. J. Environ. Qual. 26, 602–617.

Nakanishi, H., Ogawa, I., Ishimaru, Y., Mori, S., Nishizawa, N.K., 2006. Iron deficiency enhances cadmium uptake and translocation mediated by the Fe^{2+} transporters OsIRT1 and OsIRT2 in rice. Soil Sci. Plant Nutr. 52, 464–469.

Neubauer, U., Nowack, B., Furrer, G., Schulin, R., 2000. Heavy metal sorption on clay minerals affected by the siderophore Desferrioxamine B. Environ. Sci. Technol. 34, 2749–2755.

Nigam, R., Srivastava, S., Prakash, S., Srivastava1, M.M., 2001. Cadmium mobilisation and plant availability – the impact of organic acidscommonly exuded from roots. Plant and Soil 230, 107–113.

Nogawa, K., Kido, M., 1996. Itai-Itai disease and health effects of cadmium. In: Chang, L.W. (Ed.), Toxicology of Metals. CRC Lewis Publishers, New York, NY. USA, pp. 353–370.

Ok, Y.S., Lee, H., Jung, J., Song, H., Chung, N., Lim, S., Kim, J.G., 2004. Chemical Characterization and bioavailability of cadmium in artificially and naturally contaminated soils. Agric. Chem. Biotechnol. 47, 143–146.

Ouzounidou, G., Moustakas, M., Eleftheriou, E.P., 1997. Physiological and ultrastructural effects of cadmium on wheat (*Triticum aestivum* L.) leaves. Arch. Environ. Contam.Toxicol. 32, 154–160.

Ozturk, L., Eker, S., Ozkutlu, F., 2003. Effect of cadmium on growth and concentrations of cadmium. Ascorbic acid and sulphydryl groups in durum wheat cultivars. Turk. J. Agric For. 27, 161–168.

Pál, M., Horvath, E., Janda, T., Paldi, E., Szalai, G., 2006. Physiological changes and defense mechanisms induced by cadmium stress in maize. J. Plant Nutr. Soil Sci. 169, 239–246.

Pandey, N., Singh, G.K., 2012. Studies on antioxidative enzymes induced by cadmium in pea plants (*Pisum sativum* L.) J. Environ. Biol. 33, 201–206.

Perilli, P., Mitchell, L.G., Grant, C.A., Pisante, M., 2010. Cadmium concentration in durum wheat grain (*Triticum turgidum*) as influenced by nitrogen rate, seeding date and soil type. J. Sci. Food Agric. 90, 813–822.

Peryea, F.J., 1991. Phosphate-induced release of arsenic from soils contaminated with lead arsenate. Soil Sci. Soc. Am. J. 55, 1301–1306.

Pierzynsky, G.M., Schwab, A.P., 1993. Bioavailability of Zinc, cadmium and lead in a metal contaminated alluvial soil. J. Environ. Qual. 22, 247–254.

Poschenreider, C.R., Gunse, B., Barcelo, L., 1989. Influence of cadmium on water relations, stomatal resistance and abscisic acid content in expanding bean leaves. Plant Physiol. 90, 1365–1371.

Probst, A., El Ghmari, A., Aubert, D., Fritz, B., McNutt, R., 2000. Strontium as a tracer of weathering processes in a silicate catchment polluted by acid atmospheric inputs, Strengbach, France, Chemical Geology 170, 203–219.

Prosi, F., 1989. Factors controlling biological availability and toxic effects of lead in aquatic organisms: The Science of the Total Environment 79, 157–169.

Rajan, S.S.S., Watkinson, J.H., Sinclair, A.G., 1996. Phosphate rock for direct application to soils. Adv. Agron 57, 77–159.

Rauser, W.E., 1987. Compartmental efflux analysis and removal of extracellular cadmium from roots. Plant Physiol. 85, 62–65.

Redjala, T., Zelkoa, I., Sterckemana, T., Leguec, V., Lux, A., 2011. Relationship between root structure and root cadmium uptake in maize. Environ. Exp. Bot. 71, 241–248.

Reichard, P.U., Kraemer, S.M., Frazier, S.W., Kretzschmar, R., 2005. Goethite dissolution in the presence of phytosiderophores: Rates, mechanisms, and the synergistic effect of oxalate. Plant and Soil 276, 115–132.

Rizwan, M., Meunier, J.D., Miche, H., Keller, C., 2012. Effect of silicon on reducing cadmium toxicity in durum wheat (*Triticum turgidum* L. cv. Claudio W.) grown in a soil with aged contamination. J. Hazard. Mater. 209–210, 326–334.

Salt, D.E., Prince, R.C., Pickering, I.J., Raskin, I., 1995. Mechanisms of cadmium mobility and accumulation in Indian mustard. Plant Physiol. 109, 1427–1433.

Sammut, M.L., Noack, Y., Rose, J., Hazemann, J.L., Proux, O., Depoux, M., Ziebel, A., Fiani, E., 2010. Speciation of Cd and Pb in dust emitted from sinter plant. Chemosphere 78, 445–450.

Sandalio, L.M., Dalurzo, H.C., Gomez, M., Romero-Puertas, M.C., Del Rio, L.A., 2001. Cadmium-induced changes in the growth and oxidative metabolism of pea plants. J. Exp. Bot. 52, 2115–2126.

Sarwar, N., Saifullah, Malhi, S.S., Zia, M.H., Naeem, A., Bibi, S., Farid, G., 2010. Role of mineral nutrition in minimizing cadmium accumulation by plants. J. Sci. Food Agri. 90, 925–937.

Sauve, S., Norvell, W.A., McBride, M., Hendershot, W., 2000a. Speciation and complexation of cadmium in extracted soil solutions. Environ. Sci. Technol. 34, 291–296.

Sauve, S., Hendershot, W., Allen, H.E., 2000b. Solid–solution partitioning of metals in contaminated soils: Dependence on pH, total metal burden, and organic matter. Environ. Sci. Technol. 34 (7), 1125–1131.

Schnoor, J., 2004. Australasian soil contamination gets attention. Environ. Sci. Technol 38, 53A.

Seaman, J.C., Arey, J.S., Bertsch, P.M., 2001. Immobilization of nickel and other metals in contaminated sediments by hydroxyapatite addition. J. Environ. Qual. 30, 460–469.

Seidel, H., Wennrich, R., Hoffmann, P., Loser, C., 2006. Effect of different types of elemental sulfur on bioleaching of heavy metals from contaminated sediments. Chemosphere 62, 1444–1453.

Seregin, I.V., Shpigun, L.K., Ivanov, V.B., 2004. Distribution and toxic effects of cadmium and lead on maize roots. Russ. J. Plant Physiol. 51, 525–533.

Seuntjens, P., Nowack, B., Schulin, R., 2004. Root-zone modeling of heavy metal uptake and leaching in the presence of organic ligands. Plant Soil 265, 61–73.

Sharma, P., Jha, A.B., Dubey, R.S., Pessarakli, M., 2012. Reactive oxygen species, oxidative damage, and antioxidative defense mechanism in plants under stressful conditions. J. Bot., 1–26.

Shenker, M., Fan, T.W.M., Crowley, D.E., 2001. Phytosiderophores influence on cadmium mobilization and uptake by wheat and barley plants. J. Environ. Qual. 30, 2091–2098.

Shi, G., Liu, C., Cai, Q., Liu, Q.Q., Hou, C., 2010b. Cadmium accumulation and tolerance of two safflower cultivars in relation to photosynthesis and antioxidantive enzymes. Bull. Environ. Contam.Toxicol. 85, 256–263.

Shi, G., Cai, Q., Liu, C., 2010a. Silicon alleviates cadmium toxicity in peanut plants in relation to cadmium distribution and stimulation of antioxidative enzymes. Plant Growth Regul. 61, 45–52.

Shi, G.R., Cai, Q.S., 2008. Photosynthetic and anatomic responses of peanut leaves to cadmium stress. Photosynthetica 46, 627–630.

Smilde, K.W., Van Luit, B., 1983. The effect of phosphate fertilizer on cadmium in soils and crops. Coden:IBBRAH 6, 1–17.

Smiri, M., 2011. Effect of cadmium on germination, growth, redox and oxidative properties in *Pisum sativum* seeds. J. Environ. Chem. Ecotoxicol 3, 52–59.

Street, J.J., Sabey, B.R., Lindsay, W.L., 1978. Influence of pH, phosphorus, cadmium, sewage sludge, and incubation time on the solubility and plant uptake of cadmium. J. Environ. Qual. 7, 286–290.

Street, R.A., Kulkarni, M.G., Stirka, W.A., Southway, C., Staden, J.V., 2010. Effect of cadmium on growth and micronutrient distribution in wild garlic (*Tulbaghia violacea*). S. Afr. J. Bot. 76, 332–336.

Temminghoff, E., 1998. Chemical speciation of heavy metals in sandy soils in relation to availability and mobility. Ph.D. Thesis. Wageningen Agric. Uni., Wageningen, The Netherlands.

Tichy, R., Nydl, V., Kuzel, S., Kolar, L., 1997. Increased cadmium availability to crops on a sewage-sludge amended soil. Water, Air, Soil Pollut. 94, 361–372.

Tills, A.R., Alloway, B.J., 1981. The effect of ammonium and nitrate nitrogen sources on copper uptake and amino acid status of cereals. Plant Soil 62, 279–290.

Tsadilas, C.D., Karaivazoglou, N.A., Tsotsolis, N.C., Stamatiadis, S., Samaras, V., 2005. Cadmium uptake by tobacco as affected by liming, N form, and year of cultivation. Environ. Pollut. 134, 239–246.

Tudoreanu, L., Phillips, C.J.C., 2004. Modeling cadmium uptake and accumulation in plants. Adv. Agron, 84, 121–157.

Turner, M.A., 1973. Effects of cadmium treatment on cadmium and zinc uptake by selected vegetable species. J. Environ. Qual. 2, 118–119.

Ueno, D., Yamaji, N., Kono, I., Huang, C.F., Ando, T., Yano, M., Ma, J.F., 2010. Gene limiting cadmium accumulation in rice. PNAS 107, 16500–16505.

UNEP (united nations environment programme), 2010. Final review of scientific information on cadmium –Version of December 2010.

Van der Vliet, L., Peterson, C., Hale, B., 2007. Cd accumulation in roots and shoots of durum wheat: The roles of transpiration rate and apoplastic bypass. J. Exp. Bot. 58, 2939–2947.

Vangronsveld, J., Cunningham, S.D., 1998. Introduction to the concepts. In: Vangronsveld, J., Cunningham, S.D. (Eds.), Metal-Contaminated Soils: In situ Inactivation and Phytorestoration. Springer-Verlag, Berlin, pp. 1–15.

Vega, F., Andrade, M., Covelo, E., 2010. Influence of soil properties on the sorption and retention of cadmium, copper and Pb, separately and together, by 20 soil horizons: Comparison of linear regression and tree regression analyses. J. Hazard. Mater. 174 (1–3), 522–533.

Vijayaragavan, M., Prabhahar, C., Sureshkumar, J., Natarajan, A., Vijayarengan, P., Sharavanan, S., 2011. Toxic effect of cadmium on seed germination, growth and biochemical contents of cowpea (*Vigna unguiculata* L.) plants. Int. Multidis. Res. J. 1 (5), 01–06.

Wang, L., Zhou, Q., Ding, L., Sun, Y., 2008. Effect of cadmium toxicity on nitrogen metabolism in leaves of *Solanum nigrum* L. as a newly found cadmium hyperaccumulator. J. Hazard. Mater. 154, 818–825.

Watanabe, T., Moon, C.S., Zhang, Z.W., Shimbo, S., Nakatsuka, H., Matsuda-Inoguchi, N., Higashikawa, K., Ikeda, M., 2000. Cadmium exposure of women in general populations in Japan during 1991–1997 compared with 1977–1991. Int. Arch. Occup. Environ. Health 73, 26–34.

Watkinson, J.H., Bolan, N.S., 1998. Modeling the rate of elemental sulfur oxidation in soils. In: Maynard, D.G. (Ed.), Sulfur in the environment. Canadian Forest Service Pacific Forestry Center, Victoria, British Columbia, pp. 135–171.

Weggler, K., McLaughlin, M.J., Graham, R.D., 2004. Effect of chloride in soil solution on the plant availability of biosolid-borne cadmium. J. Environ. Qual. 33, 496–504.

Welch, R.M., Hart, J.J., Norvell, W.A., Sullivan, L.A., Kochian, L.V., 1999. Effects of nutrient solution zinc activity on net uptake, translocation, and root export of cadmium and zinc by separated sections of intact durumwheat (*Triticum turgidum* L. var *durum*) seedling roots. Plant and Soil 208, 243–250.

Willaert, G., Verloo, M., 1992. Effects of various nitrogen fertilizers on the chemical and biological activity of major and trace elements in a cadmium contaminated soil. Pedologie 43, 83–91.

Williams, C.H., David, D.J., 1976. The accumulation in soil of cadmium residues from phosphate fertilizers and their effect on the cadmium content of plants. Soil Sci. 121, 86–93.

Wójcik, M., Tukiendorf, A., 2005. Cadmium uptake, localization and detoxification in Zea mays. Biol. Plant 49, 237–245.

Wuana, R.A., Okieimen, F.E., 2011. Heavy metals in contaminated soils: a review of sources, chemistry, risks and best available strategies for remediation. ISRN Ecol. http://dx.doi.org/10.5402/2011/402647. Article ID 402647.

Xu, Y., Schwartz, F.W., 1994. Lead immobilization by hydroxyapatite in aqueous solutions. J. Contain. Hydrol. 15, 187–195.

Yamagata, N., Shigematsu, 1970. Cadmium pollution in perspective. Bull. Inst. Publ. Health 19, 1–27.

Yang, M.G., Lin, X.Y., Yang, X.E., 1998. Impact of Cd on growth and nutrient accumulation of different plant species. Chin. J. Appl. Ecol. 9, 89–94.

Yoneyama, T., Gosho, T., Kato, M., Goto, S., Hayashi, H., 2010. Xylem and phloem transport of Cd, Zn and Fe into the grains of rice plants (*Oryza sativa* L.) grown in continuously flooded Cd-contaminated soil. Soil Sci. Plant Nutr. 56, 445–453.

Zachara, J.M., Smith, S.C., MicKinley, J.P., Resch, C.T., 1993. Cadmium sorption on soil smectites in sodium and calcium electrolytes. Soil Sci. Soc. Am. J. 57, 1491–1501.

Zhang, H., Davison, W., Knight, B., McGrath, S., 1998. In situ measurements of solution concentrations and fluxes of trace metals in soils using DGT. Environ. Sci. Technol. 32, 704–710.

Zhang, G.P., Fukami, M., Sekimoto, H., 2002. Influence of cadmium on mineral concentration and yield components in wheat genotypes differing in cadmium tolerance at seedling stage. Field Crop Res. 79, 1–7.

Zhang, H., Dang, Z., Zheng, L.C., Yi, X.Y., 2009. Remediation of soil co-contaminated with pyrene and cadmium by growing maize (*Zea mays* L.). Int. J. Environ. Sci. Tech. 6, 249–258.

Zhao, M.T., Wang, J., Lu, B., Lu, H., 2005. Certification of the cadmium content in certified reference materials for Cd rice flour. Rapid Commun. Mass Spectrom. 19, 910–914.

Zhao, L.Y.L., Schulin, R., Weng, L., Nowack, B., 2007. Coupled mobilization of dissolved organic matter and metals (Cu and Zn) in soil columns. Geochimica et Cosmochimica Acta 71, 3407–3418.

Zorrig, W., Rouached, A., Shahzad, Z., Abdelly, C., Davidian, J.C., Berthomieu, P., 2010. Identification of three relationships linking cadmium accumulation to cadmium tolerance and zinc and citrate accumulation in lettuce. J. Plant Physiol. 167, 1239–1247.

Chapter 14

Phytoremediation of Pb-Contaminated Soils Using Synthetic Chelates

Saifullah,* Muhammad Shahid,† Muhammad Zia-Ur-Rehman,*
Muhammad Sabir*,‡ and Hamaad Raza Ahmad*

*Institute of Soil and Environmental Sciences, University of Agriculture, Faisalabad, Pakistan,
†Department of Environmental Sciences, COMSATS Institute of Information Technology, Vehari,
Pakistan, ‡School of Plant Biology, University of Western Australia, Crawley, WA, Australia

INTRODUCTION

Rapid industrial development, urbanization, population explosion, lack of pollution control measures, and intensive agricultural activities have resulted in an enormous increase in soil contamination with heavy metals (Sarwar et al., 2010). Heavy metals are the chemicals that cannot be degraded biologically or chemically and upon release into the environment could pose a serious threat to environmental sustainability. Most commonly occurring heavy metals in soils due to anthropogenic activities include: arsenic (As), lead (Pb), nickel (Ni), cadmium (Cd), copper (Cu), cobalt (Co), zinc (Zn) and mercury (Hg). Among many heavy metals, Pb can impose serious risks to ecosystem health because of its potential toxicity to living organisms, and persistence in the environment. It is one of the most common heavy metal contaminants in soils because of its widespread use in industries. The Pb in soils may arise from various anthropogenic activities including mining, smelting of metals, lead–acid batteries, bullets, gasoline, phosphate fertilizers and pesticides. Presence of Pb in soils at high levels can lead to impaired crop productivity, nutrient status and quality of agricultural products. It is therefore imperative to clean up the Pb-contaminated soils in order to produce pollution-safe food from such soils (Saifullah et al., 2009a; Shahid et al., 2013a).

With increasing demands for uncontaminated soils to produce safe food, a large number of techniques are being used worldwide to clean up contaminated soils. Conventional engineering-based methods used for the remediation of polluted soil are highly expensive and destroy the soil fertility, structure and could

Soil Remediation and Plants. http://dx.doi.org/10.1016/B978-0-12-799937-1.00014-0

even permanently change soil texture. Moreover, these techniques are feasible only for small but heavily polluted sites under commercial use. Plants are known as solar driven pumps which extract water and nutrients from soils along with other non-essential elements. The ability of plants to extract and accumulate non-essential elements into their harvestable parts (phytoextraction) has been successfully exploited to remediate polluted soils. Plant-based remediation methods have many advantages compared to other conventional engineering/chemical-based technologies. Plants are solar driven pumps that generally remove the toxic metals from soil without any ill effects on soil health (Mench et al., 2009), and this technique requires relatively lesser inputs and expertise.

Heavy metal phytoextraction can be natural or chemically enhanced/assisted (Saifullah et al., 2009a). Natural phytoextraction relies on plant species with the ability to extract and accumulate high levels of heavy metals in harvestable biomass (Shahid et al., 2012a). Such plants, also known as metal hyperaccumulators, should show: (1) fast growth, high tolerance to metals and high biomass, and (2) high metal accumulation in harvestable biomass. However, most Pb-hyperaccumulator plant species have generally slow growth rates with low biomass production (McGrath and Zhao, 2003) and are able to take up specifically one or a few metals. One of the most important factors behind successful heavy metal remediation of polluted soils is the phytoavailable portion of a metal in the soil solution phase (Evangelou et al., 2007). Although the natural levels of Pb in Earth's crust may be high, a major fraction of soil Pb is found in the non-available form due to strong binding with different soil organic and inorganic compartments (Saifullah et al., 2009a; Shahid et al., 2012b). The tissue concentration of plants growing in Pb-contaminated soils is generally below $50\,mg\,g^{-1}$ Pb. Moreover, a major fraction of root-absorbed Pb does not transport to above-ground harvestable parts of plants. Retention of Pb by soil and root adsorption is greater than Zn, Ni and Cd (Sheoran and Sheoran, 2006), and that could be the possible reason for fewer Pb-hyperaccumulating plants. Natural hyperaccumulator plants lack the ability to extract high amounts of soil bound Pb. These traits associated with natural hyperaccumulators limits the effectiveness of natural phytoextraction in the case of Pb-contaminated soils because of limited availability of Pb for plant uptake.

In the second phytoextraction approach, chemicals especially natural or synthetic chelants are used to increase the availability, uptake by roots and translocation from roots to shoots to concentrate the metal in harvestable biomass (Saifullah et al., 2009a; Shahid et al., 2014). This technology is particularly useful in phytoremediation of metals having a very low solubility and availability like Pb. During the past two decades, several synthetic chelants have been experimented with to enhance Pb uptake by plants from contaminated growth media. These include: EDTA (ethylenediaminetetraacetic acid), NTA (nitrilotriacetate), EDDS (ethylenediaminedisuccinic acid), EGTA [ethyleneglycol-bis-tetraacetic acid], EDDHA [etylenediamine-di (o-hydroxyphenylacetic acid)] and CDTA (trans-1, 2-diaminocyclohexane-N, -tetraacetic acid) (Meers et al., 2008).

The synthetic chelants applied to soil form mobile and thermodynamically stable soluble complexes within pore water (Saifullah et al., 2008, 2010a; Shahid et al., 2012b). Synthetic chelants show very high affinity for Pb present in soil solution as well as that adsorbed on to soil particles. This metal-chelator formation helps release Pb from soil solids into the soil solution. From soil solution, metal–chelator complex moves to roots through diffusion and/or mass flow processes and are taken up along the apoplastic pathway (Tanton and Crowdy, 1971). This complex is then entered into stele through a leaky/underdeveloped Casparian strip. From stele, metals are translocated to above-ground harvestable parts via xylem. Therefore, chelating agents applied to soil not only enhance the Pb solubilization in the soil but also increase the uptake and its translocation from roots to shoots (Tandy et al., 2006).

During the recent past a large number of studies have reported success stories about chelating-agent-assisted Pb phytoextraction. However, many of the scientists have advocated avoiding application of chelating agents for plant-based remediation of heavy metals including Pb due to a number of risks associated with this technology (Saifullah et al., 2009b). The potential risks associated with this technology include: (1) enhanced mobility and thus leaching of heavy metals and other essential macro- and micronutrients from soils; (2) ground water pollution due to lixiviation of heavy metals and chelating agents; (3) toxicity of mobilized metals and chelating agents to soil flora and fauna; (4) high cost of synthetic coolants; (5) adverse effects over soil conditions due to high amounts of amendments (Saifullah et al., 2009).

THE PROBLEM OF PB

Pb pollution is considered a serious threat to ecosystem health and has been reported since a thousand years ago (Pourrut et al., 2011). The average Pb content in Earth's crust is estimated as $15\,mg\,kg^{-1}$ soil. Lead is released into the environment through many natural and anthropogenic sources. The amount of Pb released through anthropogenic activities is much higher than sourced from natural processes (Pain, 1995). Upon release into the soil environment, Pb forms thermodynamically stable complexes and precipitates of low solubility. The very long half-life (740–5900 years) of metallic Pb in soil indicates that the solubility of metallic lead is very low (Rooney et al., 1999). Translocation of Pb from roots to shoots is generally very low due to its strong binding at root surfaces and cell walls (Jarvis and Leung, 2002; Pourrut et al., 2013).

CHELATING AGENTS

Chelating agents are macromolecular compounds with the ability to form several coordinate bonds to a single metal ion. A chelating agent therefore has the ability to form a complex with metal. The complex form of metal behaves differently in a water–soil–plant system than in its free state. Complex formation can free the metals bound to soil components/cation exchange sites or it can decrease the adsorption/precipitation of metals present in soil solution. After its removal from cation exchange sites, metal is available for plant uptake in the immediate vicinity of the roots. There are a number of factors that affect the efficiency of chelating agents in the mobility of metals in soil–plant systems. Among others stability constant (Ks) of the metal-chelating agent's is the most important. Among many of the chelating agents, EDTA has a very strong affinity for Pb due to its higher value of logK and thus forms complex with highly stable over a wide range of pH (Martell et al., 2001).

Ethylene Diamine Tetraacetic Acid (EDTA)

EDTA is a highly stable molecule, and has high industrial and household uses. In agriculture, EDTA is used to increase the uptake of essential nutrients. EDTA is capable of changing metal speciation and thereby its mobility and bioavailability (Nowack et al., 2006; Shahid et al., 2011). In the late twentieth century, EDTA was proposed as a chelating substance for metal phytoextraction. EDTA is one of the most widely used chelating agents for remediation of contaminated soils especially those with metals having low bioavailability (Saifullah et al., 2009b; Shahid et al., 2014). A large number of studies confirmed EDTA as one of the most efficient synthetic chelating agents to increase Pb mobilization in a soil–plant system (Blaylock et al., 1997; Huang et al., 1997; Saifullah et al., 2009a; Bareen and Tahira, 2011; Shahid et al., 2011).

Effect of EDTA on Pb Solubilization in Soil

Lead is a toxic pollutant that can be found in several different forms in soils. These forms include: Pb as the free ion in the pore water, Pb complexed with different inorganic and organic ligands, Pb adsorbed to soil mineral and organic fractions or precipitated as different compounds. Among all these, Pb as the free ion in soil solution is the most bioavilable from soil (Saifullah et al., 2009a). Like all other elements, Pb bound to cation exchange sites is in direct equilibrium with that found in soil solution. That is why Pb present at exchange sites is also considered a bioavailable fraction. Lead associated with other soil components like organic matter, oxide and silicate clay minerals, or precipitated as the carbonate, sulphate, or phosphate is the least available form. Complexation with EDTA in soil solution results in decreasing the concentration of free Pb in soil solution, thus releasing more Pb from exchange sites. This shift in equilibrium enhances the release of Pb from other solid-phase fractions of soil such as organic or sulphide substances (Ramos-Miras et al., 2011). EDTA is capable of extracting Pb from all non-silicate-bound phases in the soil.

A number of factors can affect the efficiency of EDTA in enhancing solubilization of Pb. These factors are: (1) soil physical (texture, structure) and chemical (pH, redox potential, cation exchange capacity, soil buffering capacity, calcareousness, soil organic matter content) properties; (2) concentration of Pb and EDTA; (3) time and mode of EDTA application; (4) presence of competing cations; (5) Pb species and their distribution among soil fractions; (6) adsorption of free and complexed Pb on to charged soil particles; and (6) the formation constant of metal–ligand complexes (Saifullah et al., 2008, 2009a, b, 2010a, b). Upon the addition of EDTA to Pb-contaminated soils, 600- and 217-fold increases in Pb concentration in soil solution have been reported by Wu et al. (2003) and Nascimento (2006), respectively, in separate studies. Although a large number of studies have confirmed the linear relationship between EDTA concentration and Pb solubility (Hong and Jiang, 2005; Kim et al., 2003), other factors like metal species, concentration in the soil, soil properties, as well as the amount of EDTA applied, also play their specific role in affecting the efficacy of EDTA to solubilize soil-bound Pb.

Effect of EDTA on Pb Uptake by Plants

EDTA application greatly increases Pb mobility in soil and plants. The first and most important mechanism behind EDTA-enhanced uptake is the increased mobility of Pb in the soil–plant system with applied EDTA. The literature reports two different theories of EDTA-enhanced Pb uptake by plants. The first theory believes in the free metal ion concept, which states that EDTA–Pb complexes cannot be taken up by plants. According to this theory, the EDTA makes a complex with Pb and facilitates the Pb movement to the root surface where it is broken down into its components. Pb is absorbed by the roots and EDTA remains in soil for further metal complexation. With the advancement in analytical techniques and availability of equipments to study the metal speciation and behaviour within a plant body, recently, many scientists (Sarret et al., 2001; Vassil et al., 1998; Cooper et al., 1999; Epstein et al., 1999) were able to find Pb as a Pb–EDTA complex within the plant body. The size and polarity of the Pb–EDTA complex does not permit it to enter directly through cell membranes or the symplastic pathway. However, from soil solution Pb–EDTA complex may be taken up along the apoplastic pathway and is then entered into stele through a leaky/underdeveloped Casparian strip. From stele it is translocated to above-ground harvestable parts via xylem under the force of plant transpiration (Tanton and Crowdy, 1971).

Effect of EDTA on Pb Translocation to Shoot Tissues

Lead is a unique element which shows very limited movement both outside (soil) and inside the plants. Most of the Pb absorbed by plants is retained by roots with very limited translocation to above-ground plant parts. The limited translocation of Pb from roots to shoots could be attributed to many factors including: (1) high affinity of free ionic Pb to plasma membrane (Jiang and Liu, 2010); (2) its precipitation at the surface of roots as insoluble salts (Pourrut et al., 2011); (3) hindrance offered by the Casparian strip (Mingorance et al., 2012); (4) precipitation in intercellular space (Meyers et al., 2008); or (5) accumulation in the vacuoles

(Kopittke et al., 2007). However, Pb chelation by EDTA results in the formation of a complex that behaves differently in the soil–plant system. Chelated lead is less retained by plant roots and is rapidly translocated to shoot tissues (Crist et al., 2004; Zhivotovsky et al., 2011). The translocation of Pb–EDTA complex across the root cortex into the xylem may occur via the symplastic or the apoplastic pathway and varies with plant type and the concentration of EDTA (Saifullah et al., 2009a; Shahid et al., 2012b). A number of studies reported several-fold enhanced translocation of Pb from roots to aerial parts by EDTA (Epstein et al., 1999; López et al., 2005; Barrutia et al., 2010). Blaylock et al. (1997) reported about 1000–10,000 times greater Pb accumulation by *Brassica juncea* (L.) Czern with the application of EDTA compared to untreated controls. In another study, Jarvis and Leung (2001) using transmission electron microscopy found free ionic Pb species in root tissue plants while Pb complexed with chelating agents was mainly found in the shoots of *Chamaectisus palmensis*. Likewise, other studies (Huang et al., 1997; Vassil et al., 1998) also confirmed the transportation of Pb as Pb–EDTA complex in *B. Juncea*.

Although concentration of EDTA considerably affects the Pb uptake, both time and method of application also play significant roles in the effectiveness of EDTA-enhanced Pb accumulation. Greman et al. (2001) reported about 105 times higher concentration of Pb in the shoots of *Brassica rapa* grown in contaminated soil and treated with a single dose addition of $10\,mmol\,EDTA\,kg^{-1}$, while the Pb concentration in root was 1.7 times lower compared to control treatments. Application of EDTA in many small doses shows more effectiveness compared to the single-dose application method. Saifullah et al. (2009b) found that the split application of EDTA significantly enhanced the shoot Pb contents of *Triticum aestivum* L. compared to single-dose application.

Plant species do respond differently to EDTA-enhanced uptake of heavy metals including Pb. Shen et al. (2002) reported that cabbage mung bean and wheat crops accumulated Pb differently when exposed to EDTA at $3.0\,mmol\,EDTA\,kg^{-1}$ of soil. Although all the plants showed increased concentration of Pb in shoots and roots, cabbage plants accumulated higher Pb than other plant species. It has been reported by Saifullah et al., (2010c) that crop cultivars also differed significantly in accumulating EDTA-mobilized Pb. The authors investigated the effects of EDTA on lead accumulation by two spring wheat varieties (Auqab-2000 and Inqalab-91). The results indicated that the addition of EDTA to the soil significantly increased Pb solubility and Inqalab-91 was superior in tolerance to Pb than Auqab-2000.

Risks Associated with EDTA-Assisted Pb Phytoextraction

Although EDTA has been the most efficient synthetic chelating agent for assisted phytoremediation of Pb, many concerns have been raised about the use of EDTA in remediation technology (Alkorta et al., 2004; Evangelou et al., 2008; Saifullah et al., 2009a). In spite of many efforts to decrease the adverse effects of EDTA-assisted Pb phytoextraction (changing mode, time and rate of application), its large-scale use is still controversial. EDTA is not an ideal candidate for Pb phytoextraction for the following reasons:

1. Soil-applied EDTA can adversely impact soil enzymatic and microbial activities.
2. At high concentrations EDTA can negatively affect soil fungi and plants.
3. Pb–EDTA is a highly soluble complex and thus can easily leach to groundwater.
4. EDTA and Pb–EDTA complexes are not easily biodegradable and may persist in soil for several months.
5. The complex is not ideal in terms of plant uptake and translocation.
6. EDTA salt can destroy the physical and chemical properties of soil.
7. EDTA can enhance the stimulation of Pb as well as other elements leaching into groundwater.
8. Competition of major cations like Ca and Fe with Pb for EDTA can lower the efficiency the EDTA to solubilize Pb.
9. High costs associated with EDTA-assisted Pb phytoextraction.
10. At higher rates, soil-applied EDTA can result in eutrophication because of excessive release of nitrogen from EDTA.
11. EDTA can affect soil nutrients status due to unspecific co-mobilization of macro- and micronutrients.

Ethylene Diamine Disuccinic Acid

Ethylenediamine disuccinic acid (EDDS), $C_{10}H_{16}N_2O_8$, is an isomer of EDTA that is produced either artificially or naturally by various microorganisms (Takahashi et al., 1999; Ullmann et al., 2013). EDDS was first separated from the filtrate of *Amycolatopsis orientalis* filtrate (Nishikiori et al., 1984; Alkorta et al., 2004). EDDS posses two chiral centres, and therefore four isomers: two enantiomers (–S,S and –R,R) and two mesoisomers (–S,R and –R,S) (Ullmann et al., 2013). The –R,R isomer of EDDS is non-biodegradable, whereas the –R,S, and –S,R are partially degradable; therefore, the SS-isomer is commonly used in remediation studies (Schowanek et al., 1997; Fabbricino et al., 2013). Recently EDDS has appeared as a potential substitute to EDTA in heavy metals remediation studies (Kos and Leštan, 2003; Meers et al., 2005; Quartacci et al., 2007; Wang et al., 2012; Chang et al., 2013; Lan et al., 2013). It has the ability to improve solubilization/mobilization and the accumulation of heavy metals such as Cd, Pb, Cu, Ni, Hg and Zn by different plant varieties. It is highly biodegradable and more than 90% of [S,S]-EDDS is reported to mineralize within four to eight weeks of its application (Schowanek et al., 1997; Meers et al., 2005; Yang et al., 2013). In some reports it has been shown to be more efficient for both ex situ washing and phytoextraction treatments than EDTA or NTA (Wang et al., 2007; Xu and Thomson, 2007).

Effect of EDDS on Pb Solubilization in Soil

Owing to less disruptive effects on the environment, EDDS is gaining attention as an amendment to enhance Pb phytoextraction (Yan et al., 2010; Suzuki et al., 2013). EDDS is capable of metal ions complexation. Like EDTA, EDDS can form complexes with Pb and other metals through its oxygen atom of carboxylate groups or nitrogen item of imine groups. Cao et al. (2013) reported up to 99.8% desorption of Pb by EDDS (10 mM) in soil. Mohtadi et al. (2013) observed that 95% of Pb was in the chelated form in a solution containing 1 μM Pb and 3.1 μM EDDS. Presence of EDDS is reported to enhance several times the solubilization/mobilization of Pb and other heavy metals (Fässler et al., 2010; Lozano et al., 2011; Fabbricino et al., 2013; Hseu et al., 2013; Prieto et al., 2013; Ullmann et al., 2013). The ability of EDDS to desorb/solubilize metals including Pb is dependent on many factors including amount of EDDS, concentration of Pb and its distribution among different soil fractions, ratio of EDDS to Pb (stoichiometric effects), soil pH and soil organic matter content (Begum et al., 2012, 2013; Fabbricino et al., 2013). It has been reported to desorb Pb from soil solids very rapidly and making stable Pb–EDDS complex in soil solution (Yan et al., 2010; Cao et al., 2013; Suzuki et al., 2013; Chang et al., 2013; Yang et al., 2013). Generally EDDS-induced extraction of Pb and other metals is a quite fast process. The process of metal extraction by EDDS is greater and faster for metal fractions bound to the carbonate and exchangeable phases (Yip et al., 2009; Ardwidsson et al., 2010). Once EDDS–Pb complex is formed, it prevents the re-adsorption of metals on to solid phases, thus increasing the solubility of Pb in soil. However, the degradation of EDDS in soil results in Pb release to soil, and subsequent re-adsorption or re-precipitation in the soil (Yan et al., 2010; Lo et al., 2011; Yang et al., 2013). It is reported that EDDS-induced solubilization/mobilization of Pb in soil first increases and then gradually decreases with time due to the degradation of EDDS (Wang et al., 2009; Yang et al., 2013).

Effect of EDDS on Pb Accumulation by Plants

Like many other chelating agents, EDDS has the ability to enhance the Pb uptake by plant roots and translocation towards plant shoots. Several previous studies have reported the use of EDDS in chelant assisted phytoextraction of Pb and other metals in polluted soil (Kulli et al., 1999; Kayser et al., 2000; Kos and Leštan, 2003; Quartacci et al., 2007; Wang et al., 2009; Mohtadi et al., 2013; Fabbricino et al., 2013). Metal–EDDS complexes are more mobile inside the plants than free metal ions including Pb (Prieto et al., 2013). The mechanism of EDDS-induced Pb uptake is still unidentifiable. The absorption pathway varies with concentration of EDDS. At low concentrations, diffusion across the root apoplastic spaces into root xylem is considered as the main absorption mechanism. However, free EDDS at high concentrations, whether in hydroponics or in soil, can result in destruction of physiological barriers in roots through removal of Fe^{2+}, Ca^{2+} and other cations from the plasma membrane (Gunawardana et al., 2010; Niu et al., 2011). Mohtadi et al. (2013) reported increased Pb uptake and translocation in hydroponically grown *Matthiola flavida* only at higher levels of EDDS and Pb application. Tandy et al. (2006) proposed passive uptake mechanism for EDDS-induced Pb uptake by *Helianthus annuus*.

Lead is reported to be sequestered in plant roots for the majority of the plants (Pourrut et al., 2011). Unlike EDTA, EDDS is not always reported to facilitate root to shoot Pb translocation (Mohtadi et al., 2013). This may be due to weak Pb–EDDS (log K 12.7) complexes compared to Pb–EDTA complexes (log K 18) (Martell et al., 2001). Pb–EDDS complex may dissociate before plant uptake or inside the plant, thus giving rise to free Pb and free EDDS inside plants. Free-EDDS has been detected in shoots and xylem sap of *Helianthus annuus* (Tandy et al., 2006).

Degradation of EDDS

One of the main advantages of EDDS utilization for soil phytoremediation is certainly its biodegradability (Jones and Williams 2001; Vandevivere et al., 2001; Finžgar et al., 2006; Prieto et al., 2013; Yang et al., 2013). The biodegradability of EDDS depends on its two l-aspartic amino acids, which are the main structural components of EDDS (Ullmann et al., 2013). The EDDS has a range of degradation rates, and the majority of the studies reported its half-life to be from 2 to 9 days in soils (Schowanek et al., 1997; Tandy et al., 2006). Hauser et al. (2005) reported that approximately 18–42% of EDDS in soil was degraded over 7 weeks. According to Meers et al. (2005), EDDS may degrade fully within 54 days. Prieto et al. (2013) showed a significant decrease in metal leaching in the presence of EDDS which was due to the degradation of EDDS. Cao et al. (2007) also reported that solubilization/mobilization of metals decreased to control values within 50 days in the presence of EDDS.

Another problem associated with chelate-assisted phytoremediation is the possible residual effects of chelate on the following crop (Leštan et al., 2008; Yang et al., 2013), which may affect the successful phytoextraction process of a polluted site (Luo et al., 2005). Because of its rapid biodegradability in natural mediums, very little residual effects of EDDS on the environment has been reported (Evangelou et al., 2007; Quartacci et al., 2007; Lan et al., 2013; Mohtadi et al., 2013; Yang et al., 2013). Luo et al. (2006) reported no significant residual effects on Pb concentrations in corn plants six months after EDDS application to soil. Komárek et al. (2010) also reported no marked toxic effect on poplar growth 1 year after EDDS application. It has been reported that Pb solubilization by EDDS sharply decreases within 30–50 days (Wang et al., 2009). Different factors affect the rate of EDDS degradation in soil which include the applied rate and soil physico-chemical properties (Meers et al., 2008; Lo et al., 2011; Yang et al., 2013). Moderately contaminated soils display almost the same lag phase despite their different textures (Fabbricino et al., 2013). The lag phase increases from 3 to 4 weeks with increasing level of contamination for soils with the same texture (Meers et al., 2008). The stability constant of the metal–chelate complex has no effect on the degradation of EDDS complexes (Kołodyńska et al., 2013). These findings suggest that EDDS might be a better substitute of EDTA for the remediation of Pb-polluted soil due to its rapid degradation and limiting potential of metal leaching.

Drawbacks of EDDS-Assisted Phytoremediation

Although EDDS is highly biodegradable, and is capable of increasing Pb solubilization, the threat of Pb leaching to deep soil profiles or groundwater is not fully addressed (Tandy et al., 2006; Leštan et al., 2008; Fedje et al., 2013). Therefore, it is highly necessary to predict the possible consequences of EDDS application on the environment before its real field-scale application. Kos and Leštan (2003) reported that EDDS application caused Pb leaching of 20–25% from 27-cm soil column. Similarly, EDDS-induced metal leaching from a soil column of 60 cm depth was reported by Wang et al. (2012). Likewise, Hauser et al. (2005) and Hu et al. (2007) observed Pb and/or Cu leaching from surface to subsurface soil with the use of EDDS as metal solubilizer. Metal leaching with EDDS application is suggested by the two possible processes. Firstly, EDDS has shown rapid degradation in soil, that might result in short-term complexation and then release of Pb into soil atmosphere (Yan et al., 2010; Lo et al., 2011). Secondly, EDDS-induced metal leaching may occur via its exchange by Fe. Fe–EDDS complex has a high stability constant compared to Pb–EDDS or other metal–EDDS complexes. Therefore, EDDS is generally applied to soil in continuous or multiple additions of EDDS solution (Yan et al., 2010).

Nitrilotriacetic Acid (NTA)

Nitrilotriacetic acid (NTA), $C_6H_9NO_6$, is a tertiary amino-polycarboxylic acid which has been used widely as a chelating agent in the recent past. NTA forms coordination compounds with metal. NTA can chelate Pb and other metal ions to form soluble complexes. NTA is proposed as an alternative to EDTA, for chelate-assisted phytoextraction of Pb and other heavy metals (Quartacci et al., 2007). NTA has high biodegradability and degrades in soil as fast as oxalate or citric acid (Quartacci et al., 2005; Ruley et al., 2006). NTA has been used as a chelating agent in several metal-remediation studies owing to its positive properties (Quartacci et al., 2006; Ruley et al., 2006; Chang et al., 2013 Reinoso-Maset et al., 2013).

Use of NTA for Pb Phytoremediation

NTA is a strong potential synthetic chelate for Pb phytoremediation owing to its relatively fast biodegrability and chelating ability. NTA application to soil reduces the adsorption of Pb and other metals on pure minerals and soils (Howard and Shu, 1996). The addition of NTA increased the exchangeable fraction of Pb with a simultaneous decrease of Pb bound to organic matter. However,

Pb–NTA complex is highly degradable and the dissociation of this complex can reduce the effect of aqueous complexation on Pb sorption (Nowack and Sigg, 1997). Several studies used NTA for the solubilization/extraction of Pb and other metals from polluted soils (Hseu et al., 2013; Lan etal., 2013). Zhao et al. (2013) reported that application of NTA increased Pb uptake 3.8 times by *Festuca arundinacea* and at the same resulted in minimal leaching of Pb. Quartacci et al. (2006) also found that application of NTA enhanced metal concentrations in Indian mustard shoots.

Application of NTA as Pb remediation chelant has the advantage of being non-toxic to plants and soil microorganisms. Zhao et al. (2011) reported that application of NTA has no toxic effect on plant growth of *Festuca arundinacea*. The low toxicity of NTA to plants is probably because of fast degradation of metal–NTA complex (Evangelou et al., 2007; Quartacci et al., 2007). Compared to EDTA, several other studies also reported reduced toxicity of Pb in the presence of NTA (de Souza Freitas and do Nascimento, 2009). NTA is even reported to have positive effects on plant growth (Duo et al., 2010; Zhao et al., 2011).

COMPARISON OF SYNTHETIC CHELATING AGENTS

Several studies compared synthetic chelates with respect to their efficiency to improve metal uptake by plants (Saifullah et al., 2010a, 2010b). In most cases, the effect of EDTA was many times higher than that of other synthetic chelates (Hadi et al., 2010; Jean-Soro et al., 2012). For example, Meers et al. (2004) observed that EDTA was more effective than NTA and DTPA at increasing Pb uptake by *Zea mays*. Lesage et al. (2005) reported that EDTA was more efficient than citric acid in enhancing heavy metals (Pb, Cd, Cu and Zn) uptake by *Helianthus annuus*. Ruley et al. (2006) showed that chelates increased Pb accumulation in *Sesbania drummondii* cultivated on polluted soil in the order EDTA > HEDTA > DTPA > NTA. Huang et al. (1997) reported that the effect of synthetic chelates on Pb uptake by *Pisum sativum* and *Zea mays* was: EDTA > H EDTA > DTPA > EGTA > EDDHA. Similarly, Sekhar et al. (2005) revealed that EDTA was the most effective in enhancing Pb accumulation by *Hemidesmus indicus*. EDDS is reported to desorb more Pb than NTA because of its highest affinity constants with Pb (Meers et al., 2005). Theoretically, the metal remediation efficiency of a chelate depends on the binding constant (log K) of the metal complex. EDTA has higher affinity for Pb than other synthetic chelates and is therefore more efficient for the remediation of Pb-polluted sites.

CONCLUSIONS

Heavy metal contamination has become a serious threat to biological systems. Lead contamination can be decreased to acceptable levels through high biomass plants aided with the applied EDTA as an amendment which forms highly soluble and stable metal–EDTA complexes with Pb. A large number of factors affect the efficiency of EDTA-induced metal phytoremediation. A better understanding of soil, plant and climatic factors can help improve the effectiveness of chelant-assisted phytoremediation of Pb.

Many concerns have been raised about the adverse impacts of EDTA-assisted phytoextraction. Therefore, it has been advocated to avoid EDTA-assisted phytoremediation. Recently many biodegradable synthetic chelants such as EDDS and NTA have been employed to assist high biomass crops in removing metals from contaminated soils. Although chemically assisted phytoextraction has always shown more disruptive effects on the ecosystem than natural phytoextraction, more research on the role of non-persistent chelants with minimal effects on the environment should be continued under real-field conditions to improve understanding of the mechanisms involved in assisted phytoextraction.

REFERENCES

Alkorta, I., Hernández-Allica, J., Becerril, J.M., Amezaga, I., Onaindia, M., Garbisu, C., 2004. Chelate-enhanced phytoremediation of soils polluted with heavy metals. Rev. Environ. Sci. Biot. 3, 55–70.

Ardwidsson, Z., Elgh-Dalgren, K., von Kronhelm, T., Sjoberg, R., Allrd, B., van Hees, P., 2010. Remediation of heavy metal contaminated soil washing residues with amino polycarboxylic acids. J. Hazard. Mater. 173, 697–704.

Bareen, F.E., Tahira, S.A., 2011. Metal accumulation potential of wild plants in tannery effluent contaminated soil of Kasur, Pakistan: Field trials for toxic metal cleanup using *Suaeda fruticosa*. J. Hazard. Mater. 186, 443–450.

Barrutia, O., Garbisu, C., Hernández-Allica, J., García-Plazaola, J.I., Becerril, J.M., 2010. Differences in EDTA-assisted metal phytoextraction between metallicolous and non-metallicolous accessions of *Rumex acetosa* L. Environ. Pollut. 158, 1710–1715.

Begum, Z.A., Rahman, I.M., Tate, Y., Sawai, H., Maki, T., Hasegawa, H., 2012. Remediation of toxic metal contaminated soil by washing with biodegradable aminopolycarboxylate chelants. Chemosphere 87, 1161–1170.

Begum, Z.A., Rahman, I.M., Sawai, H., Mizutani, S., Maki, T., Hasegawa, H., 2013. Effect of extraction variables on the biodegradable chelant-assisted removal of toxic metals from artificially contaminated European reference soils. Water Air Soil Pollut. 224, 1–21.

Blaylock, M.J., Salt, D.E., Dushenkov, S., Zakharova, O., Gussman, C., Kapulnik, Y., Ensley, B.D., Raskin, I., 1997. Enhanced accumulation of Pb in Indian mustard by soil-applied chelating agents. Environ. Sci. Technol. 31, 860–865.

Cao, A., Carucci, A., Lai, T., Colla, P.L., Tamburini, E., 2007. Effect of biodegradable chelating agents on heavy metals phytoextraction with *Mirabilis jalapa* and on its associated bacteria. Eur. J. Soil Biol. 43, 200–206.

Cao, M., Hu, Y., Sun, Q., Wang, L., Chen, J., Lu, X., 2013. Enhanced desorption of PCB and trace metal elements (Pb and Cu) from contaminated soils by saponin and EDDS mixed solution. Environ. Pollut. 174, 93–99.

Chang, F.C., Wang, Y.N., Chen, P.J., Koc, C.H., 2013. Factors affecting chelating extraction of Cr, Cu, and As from CCA-treated wood. J. Environ. Manage. 122, 42–46.

Cooper, E.M., Sims, J.T., Cunningham, S.D., Huang, J.W., Berti, W.R., 1999. Chelate-assisted phytoextraction of lead from contaminated soils. J. Environ.Qual. 28, 1709–1719.

Crist, R.H., Martin, J.R., Crist, D.R., 2004. Ion-exchange aspects of toxic metal uptake by Indian mustard. Int. J. Phytoremed 6, 85–94.

de Souza Freitas, E.V., do Nascimento, C.W.A., 2009. The use of NTA for lead phytoextraction from soil from a battery recycling site. J. Hazard. Mater. 171, 833–837.

Duo, L.A., Lian, F., Zhao, S.L., 2010. Enhanced uptake of heavy metals in municipal solid waste compost by turfgrass following the application of EDTA. Environ. Monit. Assess. 165, 377–387.

Epstein, A.L., Gussman, C.D., Blaylock, M.J., Yermiyahu, U., Huang, J.W., Kapulnik, Y., Orser, C.S., 1999. EDTA and Pb–EDTA accumulation in *Brassica juncea* grown in Pb-amended soil. Plant Soil 208, 87–94.

Evangelou, M.W.H., Ebel, M., Schaeffer, A., 2007. Chelate assisted phytoextraction of heavy metals from soil. Effect, mechanism, toxicity, and fate of chelating agents. Chemosphere 68, 989–1003.

Evangelou, M.W.H., Ebel, M., Koerner, A., Schaeffer, A., 2008. Hydrolysed wool: A novel chelating agent for metal chelant-assisted phytoextraction from soil. Chemosphere 72, 525–531.

Fabbricino, M., Ferraro, A., Del Giudice, G., d'Antonio, L., 2013. Current views on EDDS use for ex situ washing of potentially toxic metal contaminated soils. Rev. Environ. Sci. Bio/Technol. http://dx.doi.org/10.1007/s11157-013-9309-z.

Fässler, E., Evangelou, M.W., Robinson, B.H., Schulin, R., 2010. Effects of indole-3-acetic acid (IAA) on sunflower growth and heavy metals uptake in combination with ethylene diamine disuccinic acid (EDDS). Chemosphere 80, 901–907.

Fedje, K.K., Yillin, L., Stromvall, A.M., 2013. Remediation of metal polluted hotspot areas through enhanced soil washing – evaluation of leaching methods. J. Environ. Manage. 128, 489–496.

Finžgar, N., Žumer, A., Leštan, D., 2006. Heap leaching of Cu contaminated soil with [S, S]-EDDS in a closed process loop. J. Hazard. Mater 135, 418–422.

Greman, H., Velikonja-Bolta, Š., Vodnik, D., Kos, B., Leštan, D., 2001. EDTA enhanced heavy metal phytoextraction: Metal accumulation, leaching and toxicity. Plant Soil 235, 105–114.

Gunawardana, B., Singhal, N., Johnson, A., 2010. Amendments and their combined application for enhanced copper, cadmium, lead uptake by *Lolium perenne*. Plant Soil 329, 283–294.

Hadi, F., Bano, A., Fuller, M.P., 2010. The improved phytoextraction of lead (Pb) and the growth of maize (*Zea mays* L.): The role of plant growth regulators (GA3 and IAA) and EDTA alone and in combinations. Chemosphere 80, 457–462.

Hauser, L., Tandy, S., Schulin, R., Nowack, B., 2005. Column extraction of heavy metals from soils using the biodegradable chelating agent EDDS. Environ. Sci. Technol. 39, 6819–6824.

Hong, P.K.A., Jiang, W., 2005. Factors in the selection of chelating agents for extraction of lead from contaminated soil: effectiveness, selectivity and recoverability. In: Nowack, B., van Briesen, J. (Eds.), Biogeochemistry of Chelating Agents, ACS Symposium Series, vol. 910. Am. Chem. Soc., pp. 421–431.

Howard, J.L., Shu, J., 1996. Sequential extraction analysis of heavy metals using a chelating agent (NTA) to counteract resorption. Environ. Pollut. 91, 89–96.

Hseu, Z.Y., Jien, S.H., Wang, S.H., Deng, H.W., 2013. Using EDDS and NTA for enhanced phytoextraction of Cd by water spinach. J. Environ. Manage. 117, 58–64.

Hu, N., Luo, Y., Wu, L., Song, J., 2007. A field lysimeter study of heavy metal movement down the profile of soils with multiple metal pollution during chelate-enhanced phytoremediation. Int. J. Phytorem 9, 257–268.

Huang, J.W., Chen, J., Berti, W.R., Cunningham, S.D., 1997. Phytoremediation of lead contaminated soils: Role of synthetic chelates in lead phytoextraction. Environ. Sci. Technol. 31, 800–805.

Jarvis, M.D., Leung, D.W.M., 2001. Chelated lead transport in *Chamaecytisus proliferus* (L. f.) ling ssp. *proliferus* var. *palmensis* (H. Chris): An ultrastructural study. Plant Sci. 161, 433–441.

Jarvis, M.D., Leung, D.W.M., 2002. Chelated lead transport in *Pinus radiata*: an ultrastructural study. Environ. Exp. Bot. 48, 21–32.

Jean-Soro, L., Bordas, F., Bollinger, J.C., 2012. Column leaching of chromium and nickel from a contaminated soil using EDTA and citric acid. Environ. Pollut. 164, 175–181.

Jiang, W., Liu, D., 2010. Pb-induced cellular defense system in the root meristematic cells of *Allium sativum* L. BMC. Plant Biol. 10, 1–40.

Jones, P.W., Williams, D.R., 2001. Chemical speciation used to assess [S,S']-ethylenediaminedisuccinic acid (EDDS) as a ready-biodegradable replacement for EDTA in radiochemical decontamination formulations. Appl. Radiat. Isot. 54, 587–593.

Kayser, A., Wenger, K., Keller, A., Attinger, W., Felix, R., Gupta, S.K., Schulin, R., 2000. Enhancement of phytoextraction of Zn, Cd and Cu from calcareous soil: The use of NTA and sulfur amendments. Environ. Sci. Technol. 34, 1778–1783.

Kim, C., Lee, Y., Ong, S.K., 2003. Factors affecting EDTA extraction of lead from lead contaminated soils. Chemosphere 51, 845–853.

Kołodyńska, D., 2013. Application of a new generation of complexing agents in removal of heavy metal ions from different wastes. Environ. Sci. Pollut. Res. 20, 5939–5949.

Komárek, M., Vaněk, A., Mrnka, L., Sudová, R., Száková, J., Tejnecký, V., 2010. Potential and drawbacks of EDDS-enhanced phytoextraction of copper from contaminated soils. Environ. Pollut. 158, 2428–2438.

Kopittke, P.M., Asher, C.J., Kopittke, R.A., Menzies, N.W., 2007. Toxic effects of Pb^{2+} on growth of cowpea (*Vigna unguiculata*). Environ. pollut. 150, 280–287.

Kos, B., Leštan, D., 2003. Induced phytoextraction/soil washing of lead using biodegradable chelate and permeable barriers. Environ. Sci. Technol. 37, 624–629.

Kulli, B., Balmer, M., Krebs, R., Lothenbach, B., Geiger, G., Schulin, R., 1999. The influence of nitrilotriacetate on heavy metal uptake of lettuce and ryegrass. J. Environ. Qual. 28, 1699–1705.

Lan, J., Zhang, S., Lin, H., Li, T., Xu, X., Li, Y., Jia, Y., Gong, J., 2013. Efficiency of biodegradable EDDS, NTA and APAM on enhancing the phytoextraction of cadmium by *Siegesbeckia orientalis* L. grown in Cd-contaminated soils. Chemosphere 91, 1362–1367.

Lesage, E., Meers, E., Vervaeke, P., Lamsal, S., Hopgood, M., Tack, F.M.G., Verloo, M.G., 2005. Enhanced phytoextraction: II. Effect of EDTA and citric acid on heavy metal uptake by *Helianthus annuus* from a calcareous soil. Int. J. Phytoremt 7, 143–152.

Leštan, D., Luo, C.L., Li, X.D., 2008. The use of chelating agents in the remediation of metal contaminated soils: A review. Environ. Pollut. 153, 3–13.

Lo, I.M.C., Tsang, D.C.W., Yip, T.C.M., Wang, F., Zhang, W., 2011. Influence of injection conditions on EDDS-flushing of metal-contaminated soil. J. Hazard. Mater. 192, 667–675.

López, M.L., Peralta-Videa, J.R., Benitez, T., Gardea-Torresdey, J.L., 2005. Enhancement of lead uptake by alfalfa (*Medicago sativa*) using EDTA and a plant growth promoter. Chemosphere 61, 595–598.

Lozano, J.C., Blanco Rodríguez, P., Vera Tomé, F., Calvo, C.P., 2011. Enhancing uranium solubilization in soils by citrate, EDTA, and EDDS chelating amendments. J. Hazard. Mater. 198, 224–231.

Luo, C.L., Shen, Z.G., Li, X.D., 2005. Enhanced phytoextraction of Cu, Pb, Zn and Cd with EDTA and EDDS. Chemosphere 59, 1–11.

Luo, C.L., Shen, Z.G., Lou, L.Q., Li, X.D., 2006. EDDS and EDTA-enhanced phytoextraction of metals from artificially contaminated soil and residual effects of chelant compounds. Environ. Pollut. 144, 862–871.

Martell, A.E., Smith, R.M., Motekaitis, R.J., 2001. NIST critically selected stability constants of metal complexes. Version 6.0. NIST, Gaithersburg, MD.

McGrath, S.P., Zhao, F.G., 2003. Phytoextraction of metals and metalloids from contaminated soils. Curr. Opin. Biotechnol. 14, 277–282.

Meers, E., Hopgood, M., Lesage, E., Vervaeke, P., Tack, F.M.G., Verloo, M.G., 2004. Enhanced phytoextraction: In search of EDTA alternatives. Int. J. Phytorem. 6, 95–109.

Meers, E., Ruttens, A., Hopgood, M.J., Samson, D., Tack, F.M.G., 2005. Comparison of EDTA and EDDS as potential soil amendments for enhanced phytoextraction of heavy metals. Chemosphere 58, 1011–1022.

Meers, E., Tack, F.M.G., Van Slycken, S., Ruttens, A., Du Laing, G., Vangronsveld, J., Verloo, M.G., 2008. Chemically Assisted Phytoextraction: A review of potential soil amendments for increasing plant uptake of heavy metals. Int. J. Phytorem 10, 390–414.

Mench, M., Schwitzguébel, J.P., Schroeder, P., Bert, V., Gawronski, S., Gupta, S., 2009. Assessment of successful experiments and limitations of phytotechnologies: Contaminant uptake, detoxification and sequestration, and consequences for food safety. Environ. Sci. Pollut. Res. 16, 876–900.

Meyers, D.E., Auchterlonie, G.J., Webb, R.I., Wood, B., 2008. Uptake and localisation of lead in the root system of *Brassica juncea*. Environ. Pollut. 153, 323–332.

Mingorance, M.D., Leidi, E.O., Valdés, B., Rossini, Oliva, S., 2012. Evaluation of lead toxicity in *Erica andevalensis* as an alternative species for revegetation of contaminated soils. Int. J. Phytorem 14, 174–185.

Mohtadi, A., Ghaderian, S.M., Schat, H., 2013. The effect of EDDS and citrate on the uptake of lead in hydroponically grown *Matthiola flavida*. Chemosphere 93, 986–989.

Nascimento, C.W.A.D., 2006. Organic acids effects on desorption of heavy metals from a contaminated soil. Sci. Agricola 63, 276–280.

Nishikiori, A., Naganawa, Okuyama, Takita, T., Hamada, M., Takeuchi, T., Aoyagi, T., Umezawa, H., 1984. Production by actinomycetes of (S,S)-N,N′-ethylenediamine-disuccinic acid, an inhibitor of phospholipidase. C. J. Antibiot. 37, 426–427.

Niu, L., Shen, Z., Wang, C., 2011. Sites, pathways, and mechanism of absorption of Cu-EDDS complex in primary roots of maize (*Zea Mays* L.): Anatomical, chemical and histochemical analysis. Plant Soil 343, 303–312.

Nowack, B., Sigg, L., 1997. Dissolution of Fe(III)-(hydr)oxides by metal-EDTA complexes. Geochim. Cosmochim. Acta 61, 951–963.

Nowack, B., Schulin, R., Robinson, B.H., 2006. Critical assessment of chelant-enhanced metal phytoextraction. Environ. Sci. Technol. 40, 5225–5232.

Pain, D.J., 1995. Lead in the environment. In: Hoffman, D.J., Rattner, B.A., Allen Burton Jr., G., Cairns Jr., J. (Eds.), Handbook of Ecotoxicology. CRC Press, Lewis Publishers, USA, pp. 356–391.

Pourrut, B., Shahid, M., Dumat, C., Winterton, P., Pinelli, E., 2011. Lead uptake, toxicity, and detoxification in plants. Rev. Environ. Contam. Toxicol 213, 113–136.

Pourrut, B., Shahid, M., Douay, F., Dumat, C., Pinelli, E., 2013. Molecular mechanisms involved in lead uptake, toxicity and detoxification in higher plants. In: Gupta, D.K., Corpas, F.J., Palma, J.M. (Eds.), Heavy Metal Stress in Plants. Springer, Berlin Heidelberg, pp. 121–147.

Prieto, C., Lozano, J.C., Blanco Rodriguez, P., Vera Tome, F., 2013. Enhancing radium solubilization in soils by citrate, EDTA, and EDDS chelating Amendments. J. Hazard. Mater. 250–251, 439–446.

Quartacci, M.F., Baker, A.J.M., Navari-Izzo, F., 2005. Nitriloacetate- and citric acid-assisted phytoextraction of cadmium by Indian mustard (*Brassica juncea* (L.) Czernj, *Brassicaceae*). Chemosphere 59, 1249–1255.

Quartacci, M.F., Argilla, A., Baker, A.J.M., Navari-Izzo, F., 2006. Phytoextraction of metals from a multiply contaminated soil by Indian mustard. Chemosphere 63, 918–925.

Quartacci, M.F., Irtelli, B., Baker, A.J.M., Navari-Izzo, F., 2007. The use of NTA and EDDS for enhanced phytoextraction of metals from a multiply contaminated soil by *Brassica carinata*. Chemosphere 68, 1920–1928.

Ramos-Miras, J., Roca-Perez, J., Guzmán-Palomino, L., Boluda, R.M., Gil, C., 2011. Background levels and baseline values of available heavy metals in Mediterranean greenhouse soils (Spain). J. Geochem. Expl. 110, 186–192.

Reinoso-Maset, E., Worsfold, P.J., Keith-Roach, M.J., 2013. Effect of organic complexing agents on the interactions of Cs^+, Sr^{2+} and with silica and natural sand. Chemosphere 91, 948–954.

Rooney, C.P., McLaren, R.G., Cresswell, R.J., 1999. Distribution and phytoavailability of lead in a soil contaminated with lead shot. Water Air Soil Pollut. 116, 535–548.

Ruley, A.T., Sharma, N.C., Sahi, S.V., Singh, S.R., Sajwan, K.S., 2006. Effects of lead and chelators on growth, photosynthetic activity and Pb uptake in Sesbania drummondii grown in soil. Environ. Pollut. 144, 11–18.

Saifullah, Ghfoor, A., Sabir, M., Rehman, M.Z., Yaseen, M., 2008. Removal of lead from contaminated soils by organic acids. Int. J. Agri. Biol. 10, 173–178.

Saifullah, Meers, E., Qadir, M., de Caritat, P., Tack, F.M.G., Du Laing, G., Zia, M.H., 2009a. EDTA-assisted Pb phytoextraction. Chemosphere 74, 1279–1291.

Saifullah, Ghafoor, A., Qadir, M., 2009b. Lead phytoextraction by wheat in response to EDTA application method. Int. J. Phytorem 11, 268–282.

Saifullah, Ghafoor, A., Zia, M.H., Murtaza, G., Waraich, E.A., Bibi, S., Srivastava, P., 2010a. Comparison of organic and inorganic amendments for enhancing soil lead phytoextraction by wheat (*Triticum aestivum* L.). Int. J. Phytorem 12, 633–649.

Saifullah, Zia, M.H., Meers, E., Ghafoor, A., Murtaza, G., Sabir, M., Rehman, M.Z., Tack, F.M.G., 2010b. Chemically enhanced phytoextraction of Pb by wheat in texturally different soils. Chemsophere 79, 652–658.

Saifullah, Ghafoor, A., Murtaza, G., Waraich, E.A., Zia, M.H., 2010c. Effect of EDTA on growth and phytoremediative ability of two wheat varieties. Commun. Soil Sci. Plant Anal. 41, 1478–1492.

Sarret, G., Vangronsveld, J., Manceau, A., Musso, M., D'Haen, J., Menthonnex, J.J., Hazemann, J.L., 2001. Accumulation forms of Zn and Pb in Phaseolus vulgaris in the presence and absence of EDTA. Environ. Sci. Technol. 35, 2854–2859.

Sarwar, N., Saifullah, Malhi, S.S., Zia, M.H., Naeem, A., a Bibi, S., Farid, G., 2010. Role of Mineral Nutrition in Minimizing Cadmium Accumulation by Plants. J. Sci. Food Agri. 90, 925–937.

Schowanek, D., Feijtel, T.C.J., Perkins, C.M., Hartman, F.A., Federle, T.W., Larson, R.J., 1997. Biodegradation of S, S, R, R and mixed stereoisomers of ethylene diamine disuccinic acid (EDDS), a transition metal chelator. Chemosphere 34, 2375–2391.

Sekhar, K.C., Kamala, C.T., Chary, N.S., Balaram, V., Garcia, G., 2005. Potential of *Hemidesmus indicus* for the phytoextraction of lead from industrially contaminated soils. Chemosphere 58, 507–514.

Shahid, M., Pinelli, E., Pourrut, B., Silvestre, J., Dumat, C., 2011. Lead-induced genotoxicity to *Vicia faba* L. roots in relation with metal cell uptake and initial speciation. Ecotoxicol. Environ. Saf. 74, 78–84.

Shahid, M., Arshad, M., Kaemmerer, M., Pinelli, E., Probst, A., Baque, D., Pradere, P., Dumat, C., 2012a. Long-term field metal extraction by Pelargonium: Phytoextraction efficiency in relation to plant maturity. Int. J. Phytorem. 14, 493–505.

Shahid, M., Pinelli, E., Dumat, C., 2012b. Review of Pb availability and toxicity to plants in relation with metal speciation; role of synthetic and natural organic ligands. J. Hazard. Mater., 219–220. 1–12.

Shahid, M., Xiong, T., Masood, N., Leveque, T., Quenea, K., Austruy, A., Foucault, Y., Dumat, C., 2013a. Influence of plant species and phosphorus amendments on metal speciation and bioavailability in a smelter impacted soil: A case study of food-chain contamination. J. Soils Sediment. http://dx.doi.org/10.1007/s11368-013-0745-8.

Shahid, M., Austruy, A., Echevarria, G., Arshad, M., Sanaullah, M., Aslam, M., Nadeem, M., Nasim, W., Dumat, C., 2014. EDTA-Enhanced Phytoremediation of Heavy Metals: A Review. Soil Sediment. Contam. http://dx.doi.org/10.1080/15320383.2014.831029.

Shen, Z.-G., Li, X.-D., Wang, C.-C., Chen, H.-M., Chua, H., 2002. Lead phytoextraction from contaminated soil with high-biomass plant species. J. Environ. Qual. 31, 1893–1900.

Sheoran, A.S., Sheoran, V., 2006. Heavy metal removal mechanism of acid mine drainage in wetlands: A critical review. Miner. Eng., 19, 105–116.

Suzuki, T., Nakamura, A., Niinae, M., Nakata, H., Fujii, H., Tasaka, Y., 2013. Lead immobilization in artificially contaminated kaolinite using magnesium oxide-based materials: Immobilization mechanisms and long-term evaluation. Chem. Eng. J. 232, 380–387.

Takahashi, R., Fujimoto, N., Suzuki, M., Endo, T., 1997. Biodegradabilities of ethylenediamine-N,N′-disuccinic acid (EDDS) and other chelating agents. Biosci. Biotechnol. Biochem. 61, 1957–1959.

Tandy, S., Schulin, R., Nowack, B., 2006. The influence of EDDS on the uptake of heavymetals in hydroponically grown sunflowers. Chemosphere 62, 1454–1463.

Tanton, T.W., Crowdy, S.H., 1971. The distribution of lead chelate in the transpiration stream of higher plants. Pestic. Sci. 2, 211.

Ullmann, A., Brauner, N., Vazana, S., Katz, Z., Goikhman, R., Seemann, B., Marom, H., Gozin, M., 2013. New biodegradable organic-soluble chelating agents for simultaneous removal of heavy metals and organic pollutants from contaminated media. J. Hazard. Mater. 260, 676–688.

Vandevivere, P., Hammes, F., Verstraete, W., Feijtel, T., Schowanek, D., 2001. Metal decontamination of soil, sediment, and sewage sludge by means of transition metal chelant [S,S]-EDDS. J. Environ. Eng. 127, 802–811.

Vassil, A.D., Kapulnik, Y., Raskin, I., Salt, D.E., 1998. The role of EDTA in lead transport and accumulation by Indian mustard. Plant Physiol. 117, 447–453.

Wang, A.G., Luo, C.L., Yang, R.X., Chen, Y.H., Shen, Z.G., Li, X.D., 2012. Metal leaching along soil profiles after the EDDS application – a field study. Environ. Pollut. 164, 204–210.

Wang, G., Koopmans, G.F., Song, J., Temminghoff, E.J., Luo, Y., Zhao, Q., Japenga, J., 2007. Mobilization of heavy metals from contaminated paddy soil by EDDS, EDTA, and elemental sulfur. Environ. Geochem. Health 29, 221–235.

Wang, X., Wang, Y., Mahmood, Q., Islam, E., Jin, X., Li, T., Yang, X., Liu, D., 2009. The effect of EDDS addition on the phytoextraction efficiency from Pb contaminated soil by *Sedum alfredii* Hance. J. Hazard. Mater. 168, 530–535.

Wu, C.H., Lin, C.F., Ma, H.W., Hsi, T.Q., 2003. Effect of fulvic acid on the sorption of Cu and Pb onto γ-Al$_2$O$_3$. Water Res. 37, 743–752.

Xu, X., Thomson, N.R., 2007. An evaluation of the green chelant EDDS to enhance the stability of hydrogen peroxide in the presence of aquifer solids. Chemosphere 69, 755–762.

Yan, D.Y.S., Yui, M.M.T., Yip, T.C.M., Tsang, D.C.W., Lo, I.M.C., 2010. Influence of EDDS-to-metal molar ratio, solution pH, and soil-to-solution ratio on metal extraction under EDDS deficiency. J. Hazard. Mater. 178, 890–894.

Yang, L., Luo, C.L., Liu, Y., Quan, L.T., Chen, Y.H., Shen, Z.G., 2013. Residual effects of EDDS leachates during EDDS-assisted phytoremediation oKf copper contaminated soil. Sci. Total Environ. 444, 263–270.

Yip, T.C.M., Tsang, D.C.W., Ng, K.T.W., Lo, I.M.C., 2009. Empirical modelling of heavy metal extraction by EDDS from single-metal and multi-metal contaminated soils. Chemosphere 74, 301–307.

Zhao, S., Lian, F., Duo, L., 2011. EDTA-assisted phytoextraction of heavy metals by turfgrass from municipal solid waste compost using permeable barriers and associated potential leaching risk. Bioresour. Technol. 102, 621–626.

Zhao, S., Jia, L., Duo, L., 2013. The use of a biodegradable chelator for enhanced phytoextraction of heavy metals by *Festuca arundinacea* from municipal solid waste compost and associated heavy metal leaching. Bioresour. Technol. 129, 249–255.

Zhivotovsky, O.P., Kuzovkina, Y.A., Schulthess, C.P., Morris, T., Pettinelli, D., 2011. Lead uptake and translocation by Willows in pot and field experiments. Int. J. Phytorem. 13, 731–749.

Chapter 15

Spatial Mapping of Metal-Contaminated Soils

H.R. Ahmad, T. Aziz, Z.R. Rehman and Saifullah

Institute of Soil and Environmental Sciences, University of Agriculture, Faisalabad, Pakistan

INTRODUCTION

Industrialization has caused huge changes in the heavy metal concentration at the earth's surface. Soils are the continued recipients of heavy metals and other pollutants. Excessive inputs of heavy metals and synthetic chemicals into urban soils may lead to the deterioration of the soil biology and function, changes in the soil physicochemical properties and other environmental problems (Papa et al., 2010; Xia et al., 2011; Zhang et al., 2012, 2013). Metals added into soils remain there forming insoluble complexes with soil constituents; heavy metal present in soils usually interact with the soil inorganic and organic colloids (Adriano et al., 2004). The fate of heavy metals in contaminated soils is governed by soil characteristics and climatic factors. Metals mobility and reactivation may be controlled by a complex set of chemical, physical and biological processes occurring within soils and with soil components (Kabata-Pendias, 2004; Acosta et al., 2011). Availability of heavy metals in soils and their uptake by plants not only depend on the labile portion of metals in soils but also on a variety of interacting soil and plant factors like type of clay minerals, presence of organic matter, soil reaction, type of genotype, plant species and their architecture (Ghafoor et al., 2004; Sparks, 2005; Li et al., 2009; Zhao et al., 2014).

Urban agricultural soils are often irrigated with raw effluent mainly because of a shortage of canal irrigation. Secondly, city sewerage/effluent often contains certain plant nutrients and is available free of cost throughout the year, so farmers apply this untreated effluent to reduce input cost of commercial fertilizers. Although this sewage disposal is an economic method for resource-poor countries like Pakistan, it has certain environmental implications due to the presence of soluble salts and harmful substances including heavy metals like iron (Fe), lead (Pb), nickel (Ni), chromium (Cr), zinc (Zn), cobalt (Co), cadmium(Cd) and lead (Pb) (Ghafoor et al., 1994; Qadir et al., 1999; Hussain, 2000; Murtaza et al., 2003; Ahmad et al., 2011). Continuous application of this water onto

Soil Remediation and Plants. http://dx.doi.org/10.1016/B978-0-12-799937-1.00015-2

agricultural soils without any treatment causes accumulation of metal ion in biosphere (Ahmad et al., 2011; Sabir et al., 2011; Ghafoor et al., 2004). Heavy metals emitted from mobile and immobile sources can be transported in soil, water, air and can even be taken up by the plants, entering the food chain (Pruvot et al., 2006). Heavy metals can enter the soil in three primary ways: by continuous use of wastewater for irrigation, leakage of tailings and deposition of aerosol particulates (Ahmad et al., 2011; Weldegebriel et al., 2012). Several factors, both natural and anthropogenic, control the movement of heavy metals and metalloids between the soil, water, plants and even atmosphere and it is part of a complex and intricately organized biogeochemical cycling processes in nature (Singh et al., 2010).

Bioavailability of heavy metals is mainly governed by physical and chemical characteristics of soils, and it is also influenced by plant genetic characteristics. Metal uptake by plants is variable over the years without a clear relationship to weather conditions, and due to such variations the behavior of metals is difficult to predict (Romic and Romic, 2003). Heavy metal uptake depends on the type of plant species and their genetic make-up (Chen et al., 2006; Peris et al., 2007); therefore, their effects vary significantly with plant species. Still, vast areas of arable soils affected by heavy metals have not been reclaimed; for example, about 2×10^7 ha arable soils in China are polluted with heavy metals (Gu et al., 2005).

The major concern of increasing heavy metal contents of soils results from anthropogenic activities. This elevated concentration of heavy metals may lead to human and animal exposure and the associated health risks. In order to prevent potential health hazards of heavy metals in agricultural lands, it is a prerequisite to monitor the sources, reasons and surveying of soil, water and plants, together with the prevention of heavy metals' entry into the food chain via plant uptake (Henning et al., 2001; QuSheng et al., 2010). Heavy metals enter the food chain through plants and animals. Humans can be chronically exposed to heavy metals from drinking water and food ingestion (NRCS, 2000), or through breathing in contaminated dust (Abrahams, 2002; Amato et al., 2010). Children are more susceptible to exposure because of their play-habits (USEPA, 2010). These metal ions exhibit various effects including toxicity to aquatic organisms and man if present beyond respective safe limits in soils, waters and plants (Nriagu, 1990; Shukry, 2001). Nowadays, computerized techniques are available to study the geochemistry of heavy metals. The most popular techniques include Remote Sensing, Global Positioning System (GPS) and Geographic Information System (GIS) (Lee et al., 2006; Wong et al., 2006). Geo-statistic mapping techniques are commonly used to demonstrate the hotspots of heavy-metal-contaminated sites (Chaoyang et al., 2009). These techniques add a spatial component to the heavy metal concentration that will help the farmers and environmentalists to make more informed management decisions, by geospatially defining 'hot spot areas' of metal contamination. This chapter illustrates how geophysical techniques such as GPS, GIS and remote sensing can be integrated to support the management of metal-contaminated soils.

GEOPHYSICAL TECHNIQUES TO ASSESS SPATIAL VARIABILITY

Global Positioning System (GPS)

The GPS is a satellite-based triangulation system which is used to mark the exact location of a point (in latitudes and longitudes) on earth. The GPS gives the accurate position from where the soil sample is taken. The GPS is a setup of 27 satellites revolving the earth (http://www.howstuffworks.com/gps.htm). The American Heritage Dictionary defines GPS as a system for finding the position on the Earth's surface by using the radio signals of several satellites. The GPS has three fragments, namely, space, control and a receiver. The GPS satellites revolving around the Earth are spreading continuous navigation signals. A receiver fragment of GPS captures data from at least six satellites to calculate the time it takes for each satellite signal to reach the GPS receiver, and the difference in reception time determines your location (Anonymous, 2000; Theiss et al., 2005). The coordinates given by the GPS receiver are then transferred to GIS software to develop the geochemical maps of heavy metals. Furthermore, given the ease of manipulating and transferring the data, the data can readily be incorporated with other existing GIS information. Now GPS is a necessary part of the geographic information system, with applications ranging from digital soil mapping, forest management, spatial and temporal surveying to road traffic, hospital and water management, etc.

Remote Sensing

Normally soil scientists make holes using soil-augers to take samples from the contaminated areas followed by laboratory analysis and interpolation of results by developing digital maps (Kemper and Sommer, 2002). However, such an approach is time-consuming and costly. Remote sensing has been used in investigations of the contents of heavy metals as a rapid method of preliminary analysis, and high-quality imaging spectrometer (hyperspectral) data have also been used. Hyperspectral remote sensing observations may act as an alternative to traditional ground-based methods to detect a variety of plants facing environmental stress by providing both spatial and temporal key information for precision farming because of its ability to measure biophysical indicators and spatial variations. Existing studies show that the metal stress spectral characteristics of data from the field can be distinguished by a high spectral resolution of the instrument and analysis techniques (Collins et al., 1983; Clevers et al., 2004). Remote sensing technology can be applied to determine the extent of heavy metal contamination (Wu et al., 2005; Guan and Cheng, 2008; Wu et al., 2013). Liu et al. (2011) monitored the extent of Cu and Cd in rice using red edge position and fractal dimension of reflectance with wavelet transform methods. They reported that wavelet transform methods, along with fractal analysis, are capable of determining heavy metal stress in rice. Similarly Liu et al. (2011) reported that the heavy metal stress in rice fields can be visualized if the reflectance is obtained with a typical spectrum with higher reflectance around 550 nm and a weak absorption peak around 680 nm. Sridhara et al. (2011) monitored the bio-assimilation of heavy metals in soils amended with biosolids and their uptake in soybean. They applied remote sensing to monitor plants facing heavy metal stress. They used the LandSat TM images and a handheld spectro-radiometer. The normalized difference vegetation index (NDVI) of both Spectral and Landsat TM images can be used to identify metal stress plants from normal plants. However, NDVI exhibits a negative correlation with the soil copper concentration. This study suggests that remote sensing can be used to identify plants suffering from metal stress.

GEOGRAPHIC INFORMATION SYSTEM

Geographic Information System (GIS) manipulates and manages both spatial and attribute data and display information on a geographical location. The GIS technology is a combination of computer-based tools, data, people and methods that work together to perform tasks related to spatial data. Although GIS is mostly associated with mapping, it is just one aspect of the functions of GIS. It can serve as a database and can also create geographic models by analyzing different sets of data in the GIS (Lerner and Lerner, 2008).

Planimetric and topographic maps are developed based on field data; these maps are used to identify the exact location of a geographic feature on the earth's surface. For example, accurate location of a sampling or monitoring site can be readily observed from the measured grid present on large-scale maps (Artiola et al., 2004). The position of any geographic feature expressed in longitude and latitude is usually described in terms of degrees, minutes, seconds and fractions of seconds. The exact location of the Washington Monument, for example, is 38°54′21″ north latitude, 78°02′07.55″ west longitude. The application of GIS is increasing and it has become an essential tool for planning of conserving forests, grasslands, ecosystems and nature preserves. GIS has revolutionized the science of planning and conservation-examining problems using quantitative data, just as it may have revolutionized the way you plan a driving trip. The National Research Council, Canada recommended using geophysical techniques like remote sensing, GPS and GIS in environment-related problems to assist in data collection, their processing and analysis steps. Digital maps developed through these techniques are very useful to resource managers, policy makers and researchers. The GIS displays the analytical data with additional information like road network, canals, towns, river, etc., offering several benefits in assessing the heavy-metal-contaminated soil, water, air and plants (Lee, 2005). This approach of mapping the affected areas with the heavy metals or for micronutrients is utilized by scientists. Geostatistics can be used to analyze the data and to map the spatial patterns of soil heavy metals by using GIS. GIS can be very helpful in monitoring the pollution-hit areas and reports are available showing that GIS contour maps are very effective in identifing the pollution-hit areas (Imperato et al., 2003; Li et al., 2004). Spatial distribution maps are being created using the geostatistical analyst tool of ARCGIS v 10.1 software. Heavy-metal-contaminated hotspot maps associated with city area, cultural features and the regions close to industrial zones can be developed. The GIS maps are helpful in identifying the source of heavy metals (Cd, Ni, Pb, Zn and Cu) mainly originating from industrial, municipal wastes, traffic load and human activities or from a single point of discharge (Shi et al., 2008; Ahmad et al., 2011). Zhang (2006) used the Inverse Distance Weighted (IDW) technique to investigate the distribution of Pb, Cu and Zn in surface soils and concluded that spatial distribution of Pb, Cu and Zn were related to the traffic pollution in Galway city in Ireland, which has relatively heavy traffic densities and has been polluted with heavy metals for many years. The copper originated from brake discs and engine wear, lead from the combustion of gasoline and batteries and zinc came from metal oxidation and tyre wear. A similar type of survey was conducted by Li et al. (2004) using the GIS technique. They estimated the Cd, Co, Cr, Cu, Ni and Zn variations in soil samples drawn from the soil surface (0–15 cm) of urbanized Kowloon area of Hong Kong following a systematic soil-sampling method. Geochemical maps developed from GIS marked many hot spots of heavy metal contamination in the industrial and residential areas. They concluded that traffic load and industrial units are the main sources of heavy metal contamination in these areas.

It is very difficult to collect data about heavy metal concentrations from a city level as it is laborious, time consuming and expensive. Although grid sampling describes the information from the sampling points, you need another method like interpolation to address the non-sampling points while surveying large areas to estimate or predict values of unsampled sites. There are various interpolation methods available to predict the heavy metal concentration from non-sampled points. Some of the methods are briefly described in the following sections.

Histogram

One of the frequently used methods in the GIS environment to explore data obtained from metal-contaminated soils is displaying the data as a histogram. It indicates whether the data obtained is normally distributed or skewed. The GIS software ARCGIS version 10.1 has the ability to transform the data to make it normal. Zarrar (2013) studied the spatial distribution of lead in the top 15 cm of soil of Iqbal Town in Faisalabad. The Pb concentration in soil samples was used to construct a histogram. The software used for the construction of the histogram was ARCGIS version 10. Figure 15.1 shows that the spatial distribution pattern of lead concentration was not uniform as the Histogram of lead appears in a skewed form. According to Zhang and Selinus (1998), heavy metal

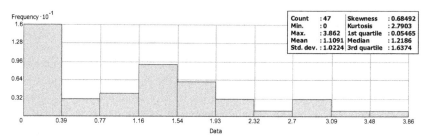

Count	: 47	Skewness	: 0.68492
Min.	: 0	Kurtosis	: 2.7903
Max.	: 3.862	1st quartile	: 0.05465
Mean	: 1.1091	Median	: 1.2186
Std. dev.	: 1.0224	3rd quartile	: 1.6374

FIGURE 15.1   Spatial distribution of lead in Iqbal Town, Faisalabad. *Data source: Zarrar (2013).*

concentration was greater where the untreated effluent was used for irrigation purposes, which provided high values in the data sets. This gives the probability a bell-shaped distribution. Using the GIS technique, zinc spatial dynamic distribution in the Multan district of Pakistan was monitored by Maqsood and Aziz (2012). They determined the distribution pattern of zinc concentration in 15-cm soils by interpolating sampling points using a histogram of zinc. Figure 15.2 shows that the spatial distribution of zinc concentration in surface soils of Multan was uniform as the histogram of zinc is a bell form.

The QQPlot

The QQplot technique is used to monitor the distribution pattern of heavy metals. In a QQ (Quantile–Quantile) plot we determined whether the data obtained is normally distributed or not. If the plot is straight along the line $Y = X$, then the data follow a normal distribution. Abbas (2013) collected 59 soil-surface samples from Lyallpur Town following an approximate grid of 4×4 km. Analytical data generated were used in ARC GIS 10 software for spatial analysis. He used the lead concentration of 59 sites to describe the trend of data distribution using QQ plot. The Figure 15.3 shows that the spatial distribution of lead concentration in surface soil of Lyallpur town was uniform.

Semivariogram

The semivariogram measures the variation in heavy metal contents as the difference in values between all pairs of sampling points plotted as a function of the distances that separate them. It indicates how changes in one variable are associated with changes in a second variable. It assumes that closer things are more similar than things that are farther. It determines the correlation as a function of distance. The semivariogram value where it flattens out is called a 'sill'. The distance range for which there is a slope is called the 'neighborhood'; this is where there is positive spatial structure. The intercept is called the 'nugget' and represents the random noise that is spatially independent. Figure 15.4 shows the graphical picture of a semivariogram. Khalid (2013) constructed the semivariogram for the copper concentration of 47 soil samples collected from Iqbal

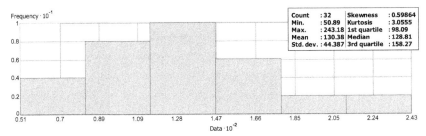

FIGURE 15.2   Spatial distribution of zinc in Multan district. *Data source: Maqsood and Aziz (2012).*

Town of Faisalabad following an approximate grid of 4×4 km. He observed that the semivariogram and covariance of Copper changes not only with distance but also with direction. For example, Figure 15.5 is a semvariogram / covariance graph showing significant correlation in the north-west direction.

INVERSE WEIGHTED DISTANCE

The main focus of using the inverse distance weighted (IDW) method is to understand and describe the spatial patterns of heavy metals distribution. The inverse distance weighting interpolator is based on the assumption that the influence of the sampling point decreased with distance (ESRI). The Inverse Distance Weighted method assumes that the value of heavy metal (numeric attribute to be interpolated) at a non-sampling point is a distance-weighted average of sampled points lying within a defined neighbourhood around that unsampled point. Maqsood and Aziz (2012) monitored the distribution of zinc concentration in surface soil of Lodhran. They used the geostatistic tool of IDW to determine the spatial pattern of Zn concentration on the surface soil of Lodhran using a geostatistical extention of the ARCGIS 10 software. Zinc concentration was obtained using atomic absorption spectrophotometer and GIS map was developed which indicate that zinc concentration increases in a south-east direction (Figure 15.6). They concluded that hot-spot areas of zinc in the south-east direction might be due to the release of textile industrial waste.

KRIGGING

Krigging is also used nowadays to locate hot-spot areas of metal-contaminated soils. This technique is different from the IDW interpolation, though krigging uses the values of nearby measured sampling sites to forecast the values at non-sampling sites. Inverse Distance Weighted interpolation technique gives more weightage to the values of nearby sampling points However, in Krigging the weightage of a close-measured value is more compared to IDW. The IDW is a simple process depending on distance, while krigging uses a semivariogram that was established from the spatial data structure. To develop a spatial map having

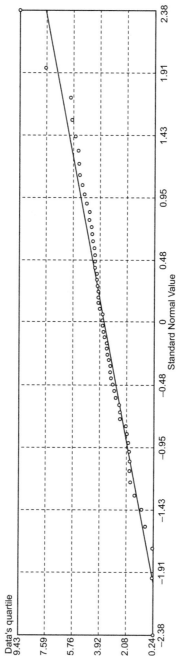

FIGURE 15.3 Normal QQ Plot transformation of lead concentration in soils Lyallpur Town. *Data source: Qamber (2013).*

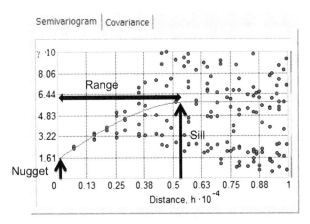

FIGURE 15.4   Graphical presentation of a variogram.

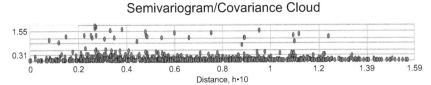

FIGURE 15.5   Semivaroigram of copper concentration in soils of Iqbal Town. *Source: Zarrar (2013).*

a continuous surface of heavy metals or any other attribute, predictions are made usually for sampling sites using a semivariogram and the spatial arrangement of sampled values that are nearby. Tahir (2012) conducted a survey study to find out the spatial variation in Cd concentration in soils of Jinnah Town using a geostatistical analysis extension of the ARCGIS version 10 software following the krigging method to create a surface of Cd concentration in soils of Jinnah Town (Figure 15.7).

Creating a Database File for GIS Environment

Heavy-metal concentration and spatial data collected using GPS are used in Microsoft Excel as a database file (dbf) in Microsoft Office 2006, while in Microsoft Office 2007 and above the data is saved as a comma delimited file (CVs) that will be then opened into ArcMap version 10.1 software ESRI software. The Microsoft Excel file consists of columns: X-coordinates, Y-coordinates and heavy-metal concentration. The titles of columns should be entered in the first row. Figure 15.8 shows how a GIS database file is created for spatial analysis in Microsoft Excel.

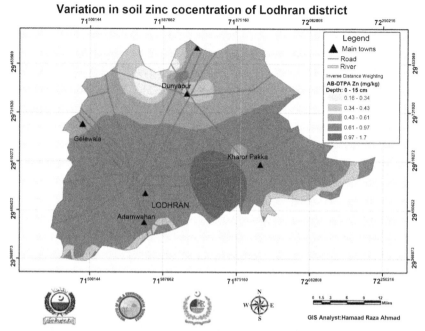

FIGURE 15.6   Spatial variation in zinc concentration in soils of Ladhran district.

1. Importing database file
   - Open Arc GIS component Arc Map
   - Click file. In the Add data click add XY Data. window, choose
   - Choose database file having extension xls, cvs or dbf.

   The titles of first two columns of database file should be mentioned in X-Field and Y-Field as X- and Y-coordinates.
   - For the selection of coordinate system use Edit icon and select the appropriate coordinate system for the database file, usually GIS analyst, select the GCS-WGS-1984.Coordinate system.
   - Then click OK (Figure 15.9).

2. Spatial analyses using Arc GIS geostatistical analyst extension
   - Open the ARC GIS component Arc map
   - On Arc Map toolbar click on customize then select Geostatistical extenstion.
   - Using this extension you will able to create a raster surface of a particular heavy metal in the ArcGIS environment (Figure 15.10)

3. Spatial Interpolation

   Geostatistical extension is commonly applied to create raster surfaces of heavy metal concentration. The raster surfaces consist of rows and columns called cells, and each cell has a particular concentration of heavy metal. Geostatistical extetion uses statistical methods to predict the

Spatial variation in cadmium concentration in soils of Jinnah Town

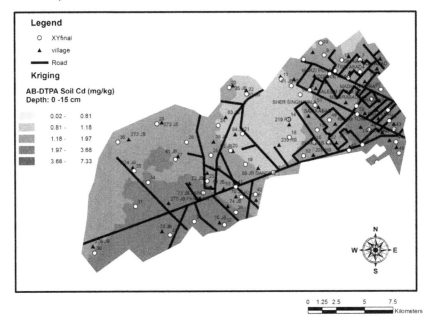

FIGURE 15.7   Spatial variation in cadmium concentration in soils of Jinnah Town.

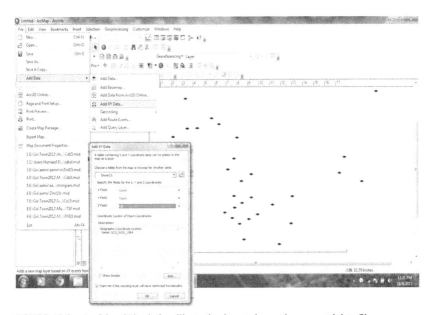

FIGURE 15.8   Arc Map GIS window illustrating how to import heavy metal data file.

SrNo	Xcord	Ycord		Cd	Pb	Ni	Zn	Cu	Fe	Pbpt	Fept	Znpt	Cdpt	Nipt	Cupt
1	31.2528	73.0359	Gulberg Chowk	0.04	3.00	0.19	9.83	3.83	16.04	3.24	10.11	7.27	0.05	1.61	14.13
2	31.2535	73.0413	Jinah colony Chowk	0.05	2.02	0.46	6.62	1.32	10.80	3.10	9.67	6.95	0.04	0.56	9.52
3	31.2513	73.0444	Nurwala Chowk	0.02	2.88	0.17	9.44	1.48	15.40	4.21	13.14	9.44	0.15	0.63	13.57
4	31.2525	73.0457	Aminpur Bazar	0.02	1.31	0.20	4.29	9.30	7.01	4.63	14.45	10.39	0.22	3.96	6.17
7	31.2524	73.0515	Hilal-e-Ahmar Chowk	0.25	3.47	0.32	11.38	0.75	18.56	1.78	5.55	3.99	0.01	0.82	16.35
8	31.2502	73.0546	GTS Chowk	0.52	1.11	0.40	3.64	2.87	5.94	5.43	16.94	12.18	0.09	1.22	5.23
9	31.2512	73.0575	Railway Station Chowk	0.01	2.18	0.08	7.15	2.40	11.66	4.32	13.46	9.69	0.20	1.02	16.27
10	31.2517	73.0621	Abdullah Pur Chowk	0.13	6.18	1.07	20.26	8.06	33.05	9.14	28.52	20.50	0.88	3.43	29.11
11	31.2539	73.0533	Chenab Chowk	0.03	1.10	0.29	3.61	3.08	5.68	3.47	10.83	7.78	0.05	1.31	5.18
12	31.2572	73.0452	University Chowk	0.09	2.32	0.13	7.61	5.15	12.41	5.68	17.72	12.74	0.35	2.21	16.93
13	31.2438	73.5401	Jhaal Wala Chowk	0.06	8.97	0.91	29.41	7.92	47.97	7.63	23.81	17.12	0.97	1.37	42.26
14	31.2417	73.0633	McDonald Chowk	0.05	2.62	0.45	8.59	3.97	14.01	3.22	10.05	7.22	0.53	1.69	12.34
15	31.2455	73.0657	D-Ground chowk	0.03	0.97	0.28	2.85	2.64	4.65	2.01	6.27	4.51	0.03	0.87	4.10
16	31.2439	73.0656	Pak-China Chowk	0.04	0.74	0.20	2.43	2.34	3.96	1.68	5.24	3.77	0.06	0.42	3.49
18	31.2498	73.0431	Imam Bargah Chowk	0.02	2.83	0.19	9.28	0.78	15.13	3.64	11.36	8.17	0.14	0.33	13.33
19	31.2509	73.0416	PC/GC Tower Chowk	0.03	2.75	0.31	9.02	6.96	14.71	5.31	16.57	11.91	0.43	2.98	12.96
20	31.2521	73.0401	LAL Dispensary Chowk	0.02	2.97	0.48	9.74	2.61	15.88	6.38	19.91	14.31	0.23	1.11	13.99
21	31.2583	73.0346	Kabrastan Chowk	0.04	2.60	0.46	8.52	4.16	13.90	4.95	15.44	11.10	0.47	1.77	12.25
22	31.2592	73.0275	Chornoran Chowk	0.02	1.18	0.32	3.87	1.53	6.31	1.99	5.90	4.24	0.06	0.65	5.56
23	31.2720	73.4315	Panj-Pulisa Chowk	0.02	3.63	0.23	11.90	1.67	19.41	1.94	6.05	4.35	0.04	0.71	17.10
25	31.2683	73.4813	Akbara-Abad Mor	0.04	2.74	0.64	8.98	0.75	14.65	2.76	8.61	6.19	0.13	0.32	12.91
26	31.2711	73.0514	Allied Mor	0.08	3.00	0.44	9.83	3.81	16.04	4.53	14.13	10.16	0.19	1.62	14.13
27	31.2603	73.0474	Ambe Wali Puli	0.03	3.51	0.37	11.51	1.08	18.77	2.13	6.65	4.78	0.03	0.46	16.54
28	31.2549	73.0426	Eid-Gah Road	0.02	2.38	0.26	7.80	1.86	12.73	3.42	10.67	7.67	0.01	0.79	11.21

FIGURE 15.9 Screen shot of heavy metal concentration file created in Microsoft Excel for spatial analysis.

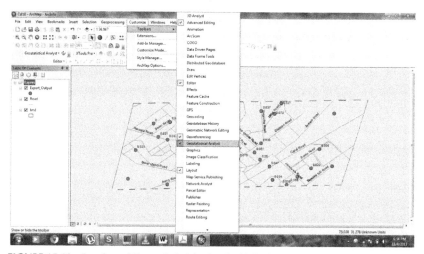

FIGURE 15.10 Interface of Geostatistical Analyst in ARC Map version 10 software.

heavy-metal concentration at non-sampled locations and the procedure is called interpolation. Geostatistical extension consists of many tools used to create raster surface like krigging, IDW (inverse distance weighting) explained earlier. Figures 15.10 and 15.11 show how to use the geostatistical analyst following the IDW method.

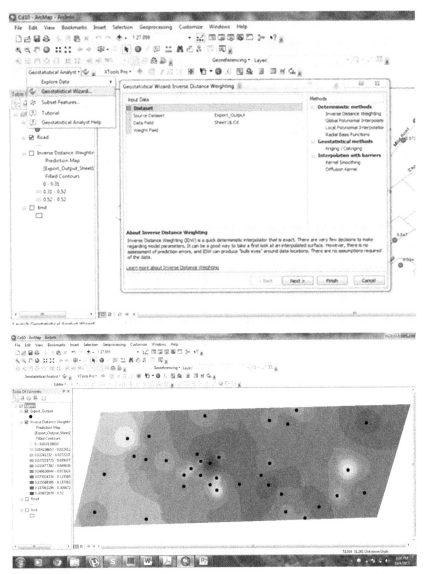

FIGURE 15.11   Procedure for using Geostatistical Analyst in ARC Map version 10 software.

CONCLUSIONS

GIS is an important tool for understanding spatial variability of heavy metals in soil and providing an indication of their non-point sources. One of the main advantages of using GIS-based mapping techniques is that the visualization of geospatial relationships between heavy metal concentrations and

other earth features like roads, rails, canals, rivers, industries, etc., can be
achieved which, in turn, spatially highlights the influence of human activities on heavy metals contents. These techniques are very useful in specula-
tion of the contamination of areas in future and their possible sources, and
are being used by policy-making institutes and agencies to make necessary
action plans.

REFERENCES

Abbas, Q., 2013. GIS based mapping on spatial variation of Pb, Ni and Cu in soil, water and plants
　in Jinnah and Iqbal towns of Faisalabad. M.Sc. Thesis, Inst. Soil Environ. Sci. Univ. Agri,
　Faisalabad, Pakistan.
Abrahams, P.W., 2002. Soils: Their implications to human health. Sci. Total Environ. 291,
　1–32.
Acosta, J.A., Faz, A., Martínez-Martínez, S., Zornoza, R., Carmona, D.M., Kabas, S., 2011. Multi-
　variate statistical and GIS-based approach to evaluate heavy metals behavior in mine sites for
　future reclamation. J. Geochem. Expl. 109, 8–17.
Adriano, D.C., Wenzel, W.W., Vangronsveld, J., Bolan, N.S., 2004. Role of assisted natural reme-
　diation in environmental cleanup. Geoderma 122, 121–142.
Ahmad, H.R., Ghafoor, A., Corwin, D.L., Aziz, M.A., Saifullah, 2011. Amendments affect soil
　concentration and accumulation of cadmium and lead in wheat in calcareous alkaline soils.
　Commun. Soil Sci. Plant Anal., 42,111–122.
Amato, F., Nava, S., Lucarelli, F., Querol, X., Alastuey, A., Baldasano, J.M., Pandolfi, M., 2010.
　A comprehensive assessment of PM emissions from paved roads: Real world emission factors
　and intense street cleaning trials. Sci. Total Environ. 408, 4309–4318.
Anonymous, 2000. Definition of global positioning systems. In: The American Heritage Dictionary,
　Fourth edn. Houghton Mifflin, Houghton Mifflin Pub. Co, Boston, MA.
Artiola, J., Pepper, I.L., Brusseau, M.L., 2004. Environmental Monitoring and Characterization.
　Publisher: Elsevier Science & Technology Books, USA.
Chaoyang, W., Cheng, W., Linsheng, Y., 2009. Characterizing spatial distribution and sources of
　heavy metals in the soils from mining-smelting activities in Shuikoushan, Hunan Province,
　China. J. Environ. Sci. 21, 1230–1236.
Chen, T.B., Song, B., Zheng, Y.M., Huang, Z.C., Zheng, G.D., Lei, M., Chen, H., 2006. A survey of
　nickel concentrations in vegetables and vegetable soils of Beijing and their health risk. J. Nat.
　Resour. 21, 349–361.
Clevers, J., Kooistra, L., Salas, E.A.L., 2004. Study of heavy metal contamination in river
　floodplains using the red-edge position in spectroscopic data. Int. J. Remote Sens. 25,
　3883–3895.
Collins, W., Chang, Raines, S.H., Canney, G.F., Ashley, R., 1983. Airborne biogeophysical mapping
　of hidden mineral-deposits. Econ. Geo. 78, 737–749.
Ghafoor, A., Rauf, A., Arif, M., Muzaffar, W., 1994. Chemical composition of effluents from differ-
　ent industries of the Faisalabad city. Pak. J. Agri. Sci. 33, 367–370.
Ghafoor, A., Qadir, M., Sadiq, M., Murtaza, G., Brar, M.S., 2004. Lead, copper, zinc and iron
　concentrations in soils and vegetables irrigated with city effluent on urban agricultural lands.
　J. Indian Soc. Soil Sci. 52, 114–117.
Gu, J.G., Lin, Q.Q., Hu, R., Zhuge, Y.P., Zhou, Q.X., 2005. Heavy metals pollution in soil–plant
　system and its research prospect. Chin. J. Soil Sci. 36, 128–133.

Guan, Li., Cheng, C., 2008. Remote sensing mechanism model for heavy metal Cd pollution in rice farm land based on hyperspectral data.Proc. SPIE 7145, Geoinformatics 2008 and Joint Conference on GIS and Built Environment: Monitoring and Assessment of Natural Resources and Environments. http://dx.doi.org/10.1117/12.813004. 71450R (November 03, 2008).

Henning, B.J., Snyman, H.G., Aveling, T.A.S., 2001. Plant soil interactions of sludge - borne heavy metals and the effect on maize (*Zea mays* L.) seedling growth. Water Soil Air 27, 71–78.

Hussain, S.I., 2000. Irrigation of crops with sewage effluent: Implication and movement of Pb and Cr as affected by soil texture, lime, gypsum and organic matter. Ph.D. Thesis, Dept. Soil Sci., Univ. Agric., Faisalabad, Pakistan.

Imperato, M., Adamo, P., Naimo, D., Arienzo, M., Stanzione, P., Violante, P., 2003. Spatial distribution of heavy metals in urban soils of Naples city (Italy). Environ Pollut 124, 247–256.

Kabata-Pendias, A., 2004. Soil plant transfer of trace elements – an environmental issue. Geoderma 122, 143–149.

Kemper, T., Sommer, S., 2002. Estimate of heavy metal contamination in soils after a mining accident using reflectance spectroscopy. Environ.Sci. Technol. 36, 2742–2747.

Khalid, M., 2013. Spatial variability of aerosols and metals concentration in soils, water and plants. M.Sc. Thesis, Inst. Soil. Environ. Sci., Univ. Agric., Faisalabad, Pakistan.

Lee, C.S.L., Li, X.D., Shi, W.Z., Cheung, S.C.N., Thornton, I., 2006. Metal contamination in urban, suburban and country park soils of Hong Kong: a study based on GIS and multivariate statistics. Sci. Total Environ 356, 45–61.

Lerner, K.L., Lerner, B.W., 2008. Environmental Science. Part 1 and 2. Gale. Cengage Learning, China. p. 341.

Li, J., Lu, Y., Yin, W., Gan, H., Zhang, C., Deng, X., Lian, J., 2009. Distribution of heavy metals in agricultural soils near petrochemical complex in Guangzhou, China. Environ. Monit. Assess. 153, 365–375.

Li, X., Lee, S.L., Wong, S.C., Shi, W., Thornton.I, 2004. The study of metal contamination in urban soils of Hong Kong using a GIS-based approach. Environ. Pollut. 129, 113–118.

Liu, M., Liu, X., Ding, W., Wu, L., 2011. Monitoring stress levels on rice with heavy metal pollution from hyperspectral reflectance data using wavelet–fractal analysis. Int. J. Appl. Earth Obs. Geoinfor. 13, 246–255.

Maqsood, M.A., Aziz, T., 2012. Zinc biofortification in grain of wheat grown on Zn deficient soils of Punjab through framer friendly approaches. Ist. Proj. Report, HEC, Islamabad, Pakistan.

Murtaza, G., Ghafoor, A., Qadir, M., 2003. Accumulation and bioavailability of Cd, Co and Mn in soils and vegetables irrigated with city effluent. Pak. J. Agri. Sci. 40, 18–24.

NRCS, 2000. Heavy metal soil contamination. Urban Technical Note 3. SoilQuality Inst., Auburn, AL.

Nriagu, J.O., 1990. Trace metal pollution of lakes: a global perspective. Proceeding of 2nd International Conference on Trace Metals in Aquatic Environment, Sydney, Australia.

Papa, S., Bartoli, G., Pellegrino, A., Fioretto, A., 2010. Microbial activities and trace element contents in an urban soil. Environ. Monit. Assess. 165, 193–203.

Peris, M., Mic, O., Recatal, C., Anchez, L.S., Anchez, J., 2007. Heavy metal contents in horticultural crops of a representative area of the European Mediterranean region. Sci. Total Environ. 378, 42–48.

Pruvot, C., Douay, F., Herve, F., Waterlot, C., 2006. Heavy metals in soil, crops and grass as a source of human exposure in the former mining areas. J. Soil Sediment 6, 215–220.

Qadir, M., Ghafoor, A., Murtaza.G, 1999. Irrigation with city effluent for growing vegetables: A silent epidemic of metal poisoning. Proc. Pak. Acad. Sci. 36, 217–222.

QuSheng, Li., Cai, S.C.M., Chu, B., Peng, L., Bing, F., Yang, 2010. Toxic effects of heavy metals and their accumulation in vegetables grown in a saline soil. Ecotox. Environ. Safe. 73, 84–88.

Romic, M., Romic, D., 2003. Heavy metals distribution in agricultural topsoils in urban area. Environ. Geol. 43, 795–805.

Sabir, M., Ghafoor, A., Saifullah, Rehman, M.Z., Ahmad, H.R., Aziz, T., 2011. Growth and Metal Ionic Composition of Zea mays as Affected by nickel Supplementation in the Nutrient Solution. Int. J. Agro. Bio. 2, 186–190.

Shi, G., Chen, Z., Xu, S., Zhang, J., Wang, Li., Bi, C., Teng, J., 2008. Potentially toxic metal contamination of urban soils and roadside dust in Shanghai, China. Environ. Pollut. 156, 251–260.

Shukry, W.M., 2001. Effect of industrial effluents polluting the river Nile on growth, metabolism and productivity of Triticum aestivum and Vicia faba plants. Pak. J. Biol. Sci. 4, 1153–1159.

Singh, R., Singh, D.P., Kumar, N., Bhargava, S.K., Barman, S.C., 2010. Accumulation and translocation of heavy metals in soil and plants from fly ash contaminated area. J. Environ. Biol. 31, 421–430.

Sparks, D.L., 2005. Toxic metals in the environment: The role of surfaces. Element 1, 193–197.

Sridhara, B.B.M., Vincenta, R., Robertsa, S.J., Czajkowsk, K., 2011. Remote sensing of soybean stress as an indicator of chemical concentration of biosolid amended surface soils. Int. J. Appl. Earth Obs. Geoinfor. 13, 676–681.

Tahir, M.U., 2012. Spatial variations of particulate matter in air and metals in plant, soil and water of Lyallpur and Jinnah towns of Faisalabad and preparation of GIS based maps. M.Sc. Thesis, Inst. Soil Environ. Sci. Univ. Agri, Faisalabad, Pakistan.

Theiss, A., Yen, D.C., Ku.C.Y, 2005. Global Positioning Systems: An analysis of applications, current development and future implementations. Comput. Stand. Inter. 27, 89–100.

USEPA, 2010. Exposure pathways. Available at http://www.epa.gov/emergencies/content/hazsubs/pathways.htm (verified 30 Oct. 2013).

Weldegebriel, Y., Chandravanshi, B.S., Wondimu, T., 2012. Concentration levels of metals in vegetables grown in soils irrigated with river water in Addis Ababa, Ethiopia. Ecot. Environ. Saf. 77, 57–63.

Wong, C.S.C., Li, X., Thornton, I., 2006. Review: Urban environmental geochemistry of trace metals. Environ. Pollut. 142, 1–16.

Wu, L., Liu, X., Wang, P., Zhou, B., Liu, M., Li, X., 2013. The assimilation of spectral sensing and WOFOST model for the dynamic simulation of cadmium accumulation in rice tissue. Int. J. Appl. Earth Obs. Geoinform. 25, 66–75.

Wu, Y.Z., Chen, J., Ji, J.F., Tian, Q.J., Wu, X.M., 2005. Feasibility of reflectance spectroscopy for the assessment of soil mercury contamination. Environ. Sci. Technol. 39, 873–878.

Xia, X., Chen, X., Liu, R., Liu, H., 2011. Heavy metals in urban soils with various types of land use in Beijing, China. J. Hazard. Mater. 186, 2043–2050.

Zarrar, M.J., 2013. Bioassessment of heavy metal contamination in soil, plants and water of IQbal towns of Faisalabad and mapping through GIS. M.Sc. Thesis, Inst. Soil Environ. Sci. Univ. Agri, Faisalabad, Pakistan.

Zhang, C., 2006. Using multivariate analyses and GIS to identify pollutants and their spatial patterns in urban soils in Galway. Ireland. Environ. Pollut 142, 501–511.

Zhang, C., Selinus, O., 1998. Statistics and GIS in environmental geochemistry—some problems and solutions. J. Geochem. Expl 64, 339–354.

Zhang, J., Dai, J., Du, X., Li, F., Wang, W., Wang, R., 2012. Distribution and sources of petroleum-hydrocarbon in soil profiles of the Hunpu wastewater-irrigated area, China's northeast. Geoderma 173, 215–223.

Zhang, J., H. L. Li., Chen, J., Wamg, M., Tao, R., Liu, D., 2013. Assessment of heavy metal contamination status in sediments and identification of pollution source in Daye Lake, Central China. Environ. Earth. Sci. http://dx.doi.org/10.1007/s12665-014-3047-6.

Zhao, L., Xu, Y., Hou, H., Shangguan, Y., Li, F., 2014. Source identification and health risk assessment of metals in urban soils around the Tanggu chemical industrial district, Tianjin, China. Sci. Total Environ. 469, 654–662.

Chapter 16

Arsenic Toxicity in Plants and Possible Remediation

Mirza Hasanuzzaman,* Kamrun Nahar,[†,‡] Khalid Rehman Hakeem,[§] Münir Öztürk[¶] and Masayuki Fujita[†]

*Department of Agronomy, Faculty of Agriculture, Sher-e-Bangla Agricultural University, Sher-e-Bangla Nagar, Dhaka, Bangladesh; [†]Laboratory of Plant Stress Responses, Department of Applied Biological Science, Faculty of Agriculture, Kagawa University, Miki-cho, Kita-gun, Kagawa, Japan; [‡]Department of Agricultural Botany, Faculty of Agriculture, Sher-e-Bangla Agricultural University, Sher-e-Bangla Nagar, Dhaka, Bangladesh; [§]Faculty of Forestry, Universiti Putra Malaysia, Serdang, Selangor, Malaysia; [¶]Department of Botany, Ege University, Bornova, Izmir, Turkey

INTRODUCTION

Arsenic (As) toxicity has been recognized for centuries. In the last few decades a large number of articles have been published with regard to As (Figure 16.1). Arsenic has become a great concern because of its chronic and epidemic effects on human, plant and animal health (Hughes et al., 2011). Presenting in the terrestrial, marine and freshwater environments in various chemical forms, As may cause substantial damages to the plant and animal kingdoms (Meharg and Hartley-Whitaker, 2002). Arsenic contamination in the groundwater is becoming more of a threat day by day due to its hazardous effects. Both human and plant species are affected by groundwater As contamination. Worldwide, nearly 150 million people spanning over 70 countries are affected by As contamination (Brammer and Ravenscroft, 2009). Countries of South and South-East Asia are mainly affected by As and over 110 million people of these areas are affected by As. Environmental exposure of As through contaminated drinking water severely affects human health. In recent decades, large-scale groundwater pollution by geogenic As in Bangladesh and India (West Bengal) has been detected, which has largely promoted this element into the environment and largely contaminated water, soils and crops. Evoked by the risk of As entering the food chain, the detection of As-contaminated agricultural soils has renewed interest in studying the dynamics of As in the soil environment.

Entrance of As into the biological food chain through crops or fodder has also been a danger for humans. Groundwater irrigation is chiefly responsible for As contamination and its entrance into the food chain through crops (Polizzotto

Soil Remediation and Plants. http://dx.doi.org/10.1016/B978-0-12-799937-1.00016-4

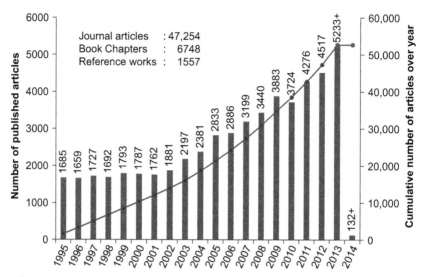

FIGURE 16.1 Number of articles published on As during the last 20 years as recorded in Scopus®. Retrieved from www.scopus.com on October 15, 2013. Note: 2013 records are partial and 2014 records are based on online-first articles. Horizontal bars represent the number of published articles in each year and trend line indicates the cumulative number of articles.

et al., 2008; Brammer, 2009). After entering the human body, As creates complex problems: it may cause severe chronic diseases that damage the liver, lungs, kidneys, bladder, prostate and so on; even As-induced cancer is not uncommon (Benbrahim-Tallaa and Waalkes, 2008). Arsenic is non-essential and toxic for plants, especially in higher concentrations. Roots are usually the first tissue to be exposed to As; root extension and proliferation are inhibited by As. Translocating through shoots, As can severely inhibit plant growth by slowing or arresting cell expansion or biomass accumulation. Arsenic can reduce plants' reproductive capacity through losses in fertility, reducing development of reproductive organs. Thus, the yield components of plants are also hampered and the result is reduced yield or fruit production (Garg and Singla, 2011). Arsenic can actively react with enzymes and proteins, and thus disrupts biochemical functions of cells; and thus photosynthesis, respiration, transpiration and plant metabolism processes are seriously hampered in other aspects by As toxicity (Meharg and Hartley-Whitaker, 2002). Arsenic is responsible for reactive oxygen species (ROS) generation within the cell that may cause lipid peroxidation and protein oxidation, thus severely affecting cellular and subcellular organelles, and it can even damage the DNA. Sufficiently high concentrations of As may cause plant death by interfering with metabolic processes (Finnegan and Chen, 2012).

Considering its damaging effects, remediation of As toxicity has become a great concern and is highly complex. Various approaches have been developed for the remediation of As toxicity. Plants have genetic potential to remove, degrade and metabolize heavy metals including As (Hasanuzzaman and Fujita, 2012). Soil microbes also play a large role in reducing As toxicity. Moreover, remediations of As by employing different agronomic practices are also important. This chapter reviews sources, status and hazards of As; it also reviews some As-detoxifying mechanisms within plants. This chapter sheds light on the effects of As toxicity on rice and various aproaches for remediation of As from the soil or the environment.

Environmental Chemistry of Arsenic

Arsenic is a naturally occurring element widely distributed in the Earth's crust. Arsenic is a metalloid and it occurs in many minerals, either in conjunction with other elements or as a pure elemental crystal. It is the most common cause of acute toxic metal poisoning (Hasanuzzaman and Fujita, 2012). It is one of the toxic metals/metalloids in the environment having diverse chemical behaviour. Arsenic, with atomic number 33, and situated in Group 15 of the periodic table, exists in four different oxidation states, namely, $(-III)$, (0), (III) and (V); however, oxidized As(III) and As(V) are the most widespread forms in nature (Goessler and Kuehnelt, 2002). Since As is a metalloid and it is seated under nitrogen (N) and phosphorus (P) in the periodic table, it has an excess of electrons and unfilled orbitals that stabilize formal oxidation states $(+5$ to $-3)$. Thus, As has a greater oxidation potential (i.e. the ability to lose electrons) than N and P, which increase its cationic character (O'Day 2006). There are many forms, or species, of As in the environment, including: (1) inorganic arsenate (AsO_4^{3-}) and arsenite (AsO_3^{3-}), (2) methylated anionic species, (3) volatile As hydrides and (4) organoarsenic species in food. Under moderately reducing conditions, arsenite $(+3)$ may be the dominant form (Figure 16.2). Under normal environmental conditions, the most stable forms, and thus most readily detected forms, are in oxidation state +5. They are the most common mobile species in water, soil and sediment. Arsenic also has the capacity

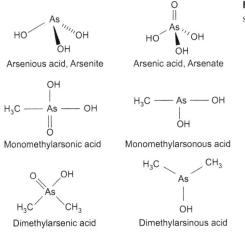

Arsenious acid, Arsenite

Arsenic acid, Arsenate

Monomethylarsonic acid

Monomethylarsonous acid

Dimethylarsenic acid

Dimethylarsinous acid

FIGURE 16.2 Chemical structure of some major arsenic compounds.

to bind to a variety of organic ligands with different coordination geometries and its greater redox potential than P helps to explain why mechanisms of As toxicity in living organisms still remain unclear. Important factors affecting As chemistry, and its mobility in soils, are soil solution chemistry (pH and redox conditions), solid composition, As-bearing phases, adsorption and desorption and biological transformations, volatilization and cycling of As in soil (Sadiq, 1997). In general, inorganic As species are more mobile and toxic than organic forms to living organisms, including plants, animals and humans (Meharg and Hartley-Whitaker, 2002). In general, arsenite is known to be more toxic and mobile than arsenate (Panstar-Kallio and Manninen, 1997; Stronach, et al., 1997).

SOURCES OF ARSENIC CONTAMINATION IN SOIL AND ENVIRONMENT

Arsenic ranks 52nd in crustal abundance and it is a major constituent in more than 245 minerals and that is why it is widely distributed worldwide (O'Neill, 1995). However, the concentration is variable depending on the parent materials. For instance, the mean concentrations of As in igneous rocks range from 1.5 to 3.0 mg kg^{-1}, whereas in sedimentary rocks they range from 1.7 to 400 mg kg^{-1} (Smith et al., 1998). The alluvial and deltaic sediments containing pyrite have favoured the As contamination of groundwater in Bangladesh. Most regions of Bangladesh are composed of a vast thickness of alluvial and deltaic sediments, which can be divided into two major parts—the recent flood plain and the terrace areas.

Arsenic can be introduced into the environment from various sources; mainly categorized as geological (geogenic) and anthropogenic (human activities) sources (Figure 16.3). In addition, small amounts of As also enter the soil and water through various biological sources (biogenic) that are rich in As (Mahimairaja et al., 2005; Figure 16.3). Anthropogenic sources of As contamination are the predominant sources of As contamination in industrialized countries. However, unlike other toxic metals, geological reasons for As contamination are more prominent than anthropogenic sources in some areas of the world. In Bangladesh, for instance, the recent episode of extensive As contamination in groundwaters is mostly of geological origin, transported by rivers from sedimentary rocks over many years (Mahimairaja et al., 2005). Although many are anthropogenic, their capacity to release As greatly depends on their bioavailability in nature as well as the chemical nature (speciation). The major anthropogenic sources of As include: mining, ore dressing and smelting of non-ferrous metals; production of As and As compounds; petroleum and chemical industries; pesticides, beer, table salt, tap water, paints, pigments, cosmetics, glass and mirror manufacture; fungicides, insecticides, treated wood and contaminated food; dyestuff and the tanning industry (Hasanuzzaman and Fujita, 2012; Figures 16.3 and 16.4). For centuries, humans have utilized As compounds [especially, realgar (As_4S_4), orpiment (As_2S_3) and arsenolite (As_2O_3)] in a wide variety of products,

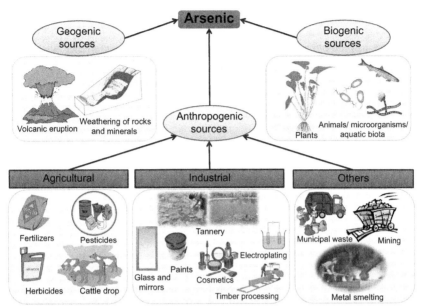

FIGURE 16.3 Major sources and routes of arsenic in soil and aquatic ecosystems. *Source: Mahimairaja et al. (2005).*

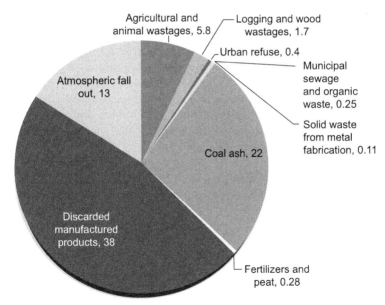

FIGURE 16.4 Worldwide input of arsenic into soils (1000 tons yr^{-1}). *Sources: Nriagu and Pacyna (1988), Purakayastha and Chhonkar (2010).*

which include: pigments, medicines, alloys, pesticides, herbicides, glassware, embalming fluids and as a depilatory in leather manufacturing (Eisler, 2004).

The involvement of As with agricultural practices began after the green revolution when the use of pesticides and fertilizers increased rapidly. Some pesticides, herbicides and fertilizers contain As in their formulation, and thus the industries that manufacture As-containing pesticides and herbicides release As into soil and water. The chemical compounds which are released in soil through pesticides are mainly calcium arsenate $Ca_3(AsO_4)_2$, magnesium arsenate $(Mg_3(As_2O_4)_2)$, zinc arsenate $Zn_3(As_2O_4)_2$, zinc arsenite $Zn_3(AsO_3)_2$, lead arsenate $PbHAsO_4$ and Paris Green $Cu(C_2H_3O_2)_2 \cdot 3Cu(AsO_2)_2$ (Mahimairaja et al., 2005). Chatterjee and Mukherjee (1999) reported that indiscriminate discharge of industrial effluents from the manufacturing of Paris Green (an arsenical pesticide) caused As contamination in soil and groundwater in the residential area of Kolkata, India. The two other herbicides, monosodium methanearsonate and disodium methanearsonate, were reported to release significant amounts of As into the soil (Smith et al., 1998). In the United States, Adriano (2001) reported that the use of sodium arsenite ($NaAsO_2$) to control aquatic weeds has contaminated small fish ponds and lakes in many regions. As reported by Bolan and Thiyagarajan (2001) timber treatment effluent is considered to be the major source of As contamination in aquatic and terrestrial environments in New Zealand. A large portion of As is released during mining and smelting processes of different metals like Au, Cu, Pb and Zn because As is extensively distributed in the sulphide ores of these metals (Mahimairaja et al., 2005). In addition, the flue gases and particulate from smelters can contaminate nearby ecosystems downwind from the operation with a range of toxic metals, including As (Adriano, 2001). Coal combustion also emits As to the atmosphere in a gaseous form.

Some plants and micro- and macroorganisms affect the redistribution of As through their bioaccumulation (e.g. biosorption), biotransformation (e.g. biomethylation) and transfer (e.g. volatilization). However, the biological sources contribute only small amounts of As into soil and water ecosystems. Sometimes, As can be transferred from soil to plants and then to animals and humans, involving terrestrial and aquatic food chains. For example, poultry manure addition is considered to be one of the major sources of As input to soils. Both inorganic and organic species of As undergo various biological and chemical transformations in soils, including adsorption, desorption, precipitation, complexation, volatilization and methylation (Figure 16.5).

STATUS OF ARSENIC TOXICITY IN THE WORLD

Arsenic is not a new hazard in the world. It is one of the oldest poisons in the world and an important global environmental contaminant (Vaughan, 2006). In the Middle Ages, As was used as an efficient suicidal agent and also used in many high-profile murders, and thus it was often known as the 'king of poisons' and the 'poison of kings' (Hughes et al., 2011). In the last century, As was found to be

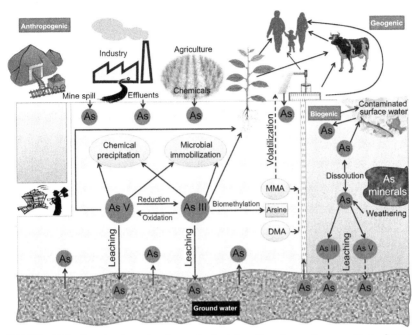

FIGURE 16.5 Arsenic dynamics in contaminated soil and aquatic ecosystem. *Source: Mahimairaja et al. (2005).*

highly abundant in soils, sediments and waters because of intensive mining and metallurgical activities or the alteration of geothermal systems (Vaughan 2006). In nature, As ranks 20th, 14th and 12th in order of abundance in the Earth's crust, seawater and the human body, respectively (Mandal and Suzuki, 2002). Considerable increase in metal mining and combustion of fossil fuel in the last century resulted in the release of significant amounts of toxic metals including As (Callender, 2003). Different major sources which are responsible for emitting As in various parts of the world are depicted in Figure 16.6. In the 1990s, the high levels of As in well water came to light – outbreaks of skin disease and cancers in many parts of Bangladesh and part of West Bengal, India showed the 'worst mass poisoning in history'. Later, As contamination in drinking water and agricultural soil has been studied and their adverse effects are observed in crop and human health. In several reports, As was assumed to be an ubiquitous element and to be the twentieth most abundant element in the biosphere (Woolson, 1977; Mandal and Suzuki, 2002) because of its presence in soil, water, air and all living matter in any of the forms of solid, liquid and gas. Although the average concentration of As (in the inorganic form) in the earth's crust is 2–5 mg kg^{-1} (Tamaki and Frankenberger, 1992), it has become a widespread problem in many parts of the world. Toxic levels of As have been reported in many countries such as Argentina, Bangladesh, Cambodia, Chile, China, Ghana, Hungary,

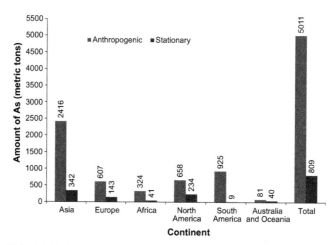

FIGURE 16.6 Worldwide emission of heavy metals from major anthropogenic and stationary sources to the atmosphere. *Source: Pacyna and Pacyna (2001).*

FIGURE 16.7 Distribution of documented world problems with As in groundwater in major aquifers as well as water and environmental problems related to mining and geothermal sources. Areas in blue are lakes. *Adapted from Smedley and Kinniburgh (2002), with permission from Elsevier.*

India, Mexico, Nepal, New Zealand, the Philippines, Taiwan, the United States and Vietnam (Das et al., 2004; Hasanuzzaman and Fujita 2012; Figure 16.7).

In a recent report published in *Toxicological Science*, Hughes et al. (2011) stated that, 'Due to improved understanding of As measurement, one cannot readily "get away with murder" by using As anymore. Nonetheless, incidents do still occur'. Their statement was based on a report where As poisoning made headlines in 2003 when As was detected in the coffee served at a church meeting in Maine (Zernike, 2003; Health, 2013). In agricultural practices As-based pesticides started to be used in 1867 with the introduction of Paris Green to control Colorado potato beetles and mosquitoes (Cullen, 2008). Later, another As-based pesticide, lead arsenate, came to be extensively used in late 1800s as a pesticide for apple and cherry orchards which was less toxic to plants and more effective than Paris Green (Peryea, 1998a). However, the use of these pesticides became limited as they have toxic effects for human and plant health.

The National Academy of Sciences, Medical and Biological Effects of Environmental Pollutants (NAS, 2000), established that the average As content of the Earth's crust is 2.5 mg kg^{-1}; however, it may exceed the limit, ranging from 45 to 3275 mg kg^{-1} (Nagy et al., 2005). In soil, As is present in different amounts ranging from 0.1 to 50 mg kg^{-1} (mean 5–6 mg kg^{-1}) which varies considerably between various regions in the world (Mandal and Suzuki, 2002). Arsenic is present in soils in higher concentrations than those in rocks (Mandal and Suzuki, 2002). When rocks weather, As may mobilize as salts of arsenous and As acid (H_3AsO_4) (Irgolic et al., 1995). The variation of As concentration in soil also depends on several factors, where soil characteristics are important. Several reports indicated that As content is 10.3–8.5 mg kg^{-1} in Japanese paddy soils, 4.6–2 mg kg^{-1} in Korean soils and 9.1–6.7 mg kg^{-1} in soils from Thailand (Mandal and Suzuki, 2002). The contents of As in the soils of different regions are shown in Table 16.1 and Figure 16.7 where the maximum amounts of As are found in Bangladesh. Based on these differences, the drinking water limits and soil threshold values for As are also given different ranges by different authorities (Table 16.2).

Arsenic contamination in agricultural crops has gained more attention in recent years. However, the presence of As greatly varies with soil types and crop species, and even cultivars. The As concentrations in agricultural plants varied from 0.007 to about 7.50 mg kg^{-1} as reported in different studies (Mandal and Suzuki, 2002; Roychowdhury et al., 2002; Liao et al., 2005; Dahal et al., 2008). Mandal and Suzuki (2002) found that As concentration in plants varied from less than 0.01 to about 5.0 mg kg^{-1}. However, the exact information on the current status of As in soil, water and plants is still a matter of survey or research.

ARSENIC HAZARD: A BANGLADESH PERSPECTIVE

The problem of As contamination in the groundwater of Bangladesh is the worst in the world. In Bangladesh, arsenopyrite has been identified as the prime source of As pollution (Fazal et al., 2001a) and in the groundwater As is found to be prsent in all four important forms, namely, $H_2AsO_3^-$, $H_2AsO_4^-$, methyl As acid [$CH_3AsO(OH)_2$] and dimethyl As acid [$(CH_3)_2As(OH)$] (Fazal et al., 2001b). In 1990s, the As problems came to the attention of researchers. The problem of As is currently a national concern with vital consequences on human lives and crop productivity.

TABLE 16.1 Arsenic Contents in the Soils and Groundwater of Various Countries

Region	Types of Soil/Sediment	Concentration of As in soil (mg kg^{-1}) Range	Mean	Concentration in Groundwater (µg l^{-1})
Bangladesh	Sediments	9.0–28.0	22.1	<10–>1000
India	Sediments	10.0–196.0	—	0.003–3700
Japan	All types	0.4–70.0	11.0	0.001–0.293
Japan	Paddy	1.2–38.2	9.0	—
China	All types	0.01–626.0	11.2	0.05–850
France	All types	0.1–5.0	2.0	—
Germany	All types	2.5–4.6	3.5	—
Argentina	All types	0.8–22.0	5.0	100–3810
USA	All types	1.0–20.0	7.5	34–490

Source: Mandal and Suzuki (2002).

TABLE 16.2 Soil Threshold Values and Drinking Water Limits for As

Soil Threshold Value Approving Authority	Values (mg kg^{-1})	Water Limit Approving Authority	Values (µg l^{-1})
Californian Assessment Manual (Cal. Ass.)	500	World Health Organization	10
European Union (EU)	—	European Union (EU)	50
Dutch standards (NL)	29–55	Dutch standards (NL)	10–60
German threshold values for maximum permissible soil concentrations (KSVO-D)	20	German drinking water standards (TVO-D)	10
German threshold values for different soil uses (D-Test)	20–130	German surface water (DVGW)	10–30

Source: Matschullat et al. (2000).

Arsenic in tube well water was first identified in 1993 by Department of Public Health and Engineering (DPHE) (BGS, 1999). The prevalence of As in drinking water has been identified in 61 out of 64 districts of the country. However, the degree of contamination varies from 1% to over 90% with an average contamination of 29% (DPHE, 2013). In some parts of Bangladesh where the concentration of As is high, As in drinking water is a major cause of chronic health effects and its reduction is considered a high-priority. In a survey jointly conducted by DPHE-UNICEF, it was revealed that 271 of 463 upazilas of the country suffered (more or less) from the As problem (DPHE, 2013). According to BGS/DPHE (2001), more than 35 million people in Bangladesh are exposed to As contamination in drinking water exceeding the national standard of $50\,\mu g\,l^{-1}$ (Figures 16.8 and 16.9), and an estimated 57 million people are at the risk of exposure to As contamination where underground water is used mainly for drinking, cooking and other household activities (Rabbani et al., 2002; Das et al., 2004; Huq, 2008). As a result, at least 1 million people are likely to be affected by arsenicosis, which has been termed as the 'greatest mass poisoning in human history'. Among 64 districts of Bangladesh, the worst districts affected with As contamination are Chandpur (90%), Munsiganj (83%), Gopalganj (79%), Noakhali (69%), Madaripur (69%), Satkhira (67%), Comilla (65%), Shriatpur (65%), Faridpur

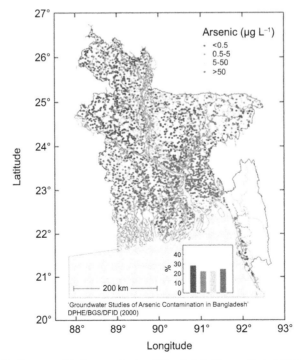

FIGURE 16.8   Presence of As in groundwater of Bangladesh. For a colour version of this figure, please see the section at the end of this book. *Source: BGS / BGS (2001).*

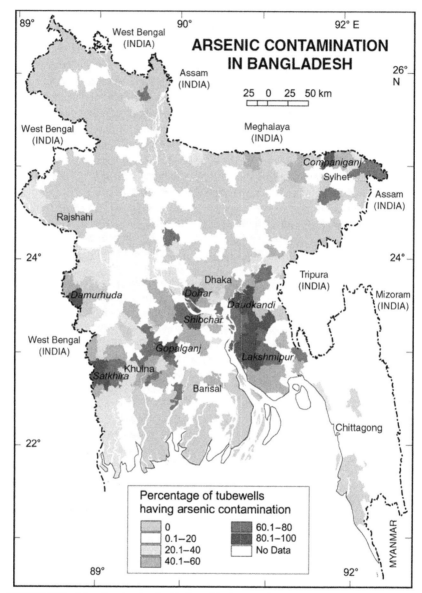

FIGURE 16.9   Map showing the distribution of As in groundwater from shallow (<150 m) tube-wells in Bangladesh. For a colour version of this figure, please see the section at the end of this book. *Source: Banglapedia (2013).*

(65%), Meherpur (60%), Bagerhat (60%) and Laxmipur (56%) (DPHE, 2013; Figure 16.8). In Bangladesh, shallow groundwater is contaminated with As which clearly exceeds the maximum level recommended by WHO, that is, $10\,\mu g\,l^{-1}$. In most cases, As-contaminated water exists at depths of 80–100 m or less below land surface as found by BGS / DPHE (2001). The number of polluted wells is insignificant at depths greater than 100–150 m, while the maximum contamination has been suggested to be between 15 and 30 m which could be associated with a zone of mixing between shallower and deeper groundwaters induced by pumping for drinking water and irrigation (Harvey et al., 2005; Smedley, 2005; Klump et al., 2006). A UNICEF (2008) report reveals that of the total 4.7 million tube wells in Bangladesh, as many as 1.4 million had traces of As more that delineated by the Bangladesh government. In case of Bangladesh, deep tube wells (> 150 m depth) are generally safe. Some studies reported As concentration in uncontaminated land of Bangladesh, which varies from 3–9 mg As kg^{-1} (Ullah, 1998; Alam and Sattar, 2000). Conversely, elevated As concentrations were observed in many studies in agricultural land irrigated with As-contaminated water, which is in some cases found to be about 10- to 20-times higher than As concentrations in non-irrigated land. Ullah (1998) reported As concentrations in top agricultural land soil (up to 0–30 cm depth) up to 83 mg As kg^{-1}. However, this finding is not identical in agreement with other studies, such as Islam et al. (2005) (up to 80.9 mg As kg^{-1}), Alam and Sattar (2000) (up to 57 mg As kg^{-1}) of soil samples collected from different districts of Bangladesh.

Arsenic contamination in crop land has been identified recently and came to the attention of researchers in the last decade (Ali et al., 2003). The problem of As is very prominent in rice cultivation as many rice fields are dependent on groundwater irrigation, which is often contaminated by As because the area irrigated is increasing day by day (Figure 16.10). In practice, ordinary crop plants do not accumulate enough As to be toxic to humans; growth reductions

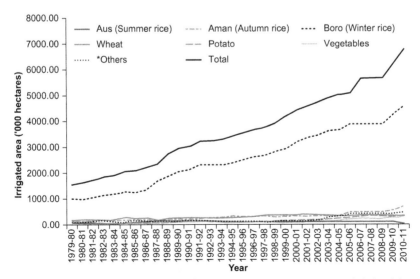

FIGURE 16.10   Irrigated Area under Different Crops, 1978–79 to 2010–11, Bangladesh. *Others means others cereals, pulses, oilseeds, sugarcane, cotton and other minor crops. For a colour version of this figure, please see the section at the end of this book. *Source: BRRI (2013).*

and crop failure are likely to be the main consequences. It is likely that because of the soil/plant barrier effect, increased As concentrations in soils will reduce crop production substantially before enhanced food-chain accumulation occurs. Long-term use of As-contaminated underground water in irrigation has resulted in an increase of its concentration in agricultural soil and eventually in crop plants (Ullah, 1998; Huq et al., 2003; Rahman et al., 2007a, c). Surveys on paddy soil throughout Bangladesh showed that As concentrations were higher in agricultural soils of those areas where shallow tube wells (STWs) have been in operation for longer periods of time, and As-contaminated underground water from those STWs have been irrigated to the crop fields (Meharg and Rahman, 2003).

Most of the research on the As toxicity in agriculture is carried out in rice-farming systems because it is the main food crop of Bangladesh and most of the rice is grown with irrigation water. In their survey, Ross et al. (2005) calculated that 76% of the boro rice is grown in areas where shallow tube wells have less than $0.050\,mg\,l^{-1}$, 17% in areas with 0.050–$0.100\,mg\,l^{-1}$ and 7% in areas with more than $0.100\,mg\,l^{-1}$. In a report, Ali et al. (2003) observed clear differences in As accumulation in rice soil grown with or without irrigation. High levels of As were detected in the irrigated field (3.2–$27.5\,mg\,kg^{-1}$), whereas in areas where irrigation water does not contain As, the soil As level was very low (0.10–$2.75\,mg\,kg^{-1}$). However, the higher levels of As were found in the top 7.5–$15\,cm$ (i.e. the plough zone) and they decreased with soil depth. Based on the results of 71 soil samples in rice fields, Meharg and Rahman (2003) found that the highest soil concentration ($46\,mg\,kg^{-1}$) was obtained in soils irrigated with high As-content water, while the minimum concentration ($<10\,mg\,kg^{-1}$) was observed in soils irrigated with low-As water.

Huq and Naidu (2003) concluded that the As loading in crop field depends on the amount of irrigation water per season. For instance, in irrigated boro (winter) rice irrigated with $1000\,mm$ of water per season, the As loading was calculated to be 1.36–$5.5\,kg\,ha^{-1}\,yr^{-1}$. Conversely, spring wheat (*Triticum aestivum*) irrigated with $150\,mm$ of irrigation water per season, the loading of As was only 0.12–$0.82\,kg\,ha^{-1}\,yr^{-1}$. More importantly, Meharg and Rahman (2003) reported that the soil irrigated with water having only $0.1\,mg\,l^{-1}$ may increase top soil ($10\,cm$ depth) As levels as much as $1\,mg\,kg^{-1}$ As in a year. Similarly, Duxbury and Zavala (2005) estimated that rice fields irrigated with As-contaminated water for 10 years may add 5–$10\,mg$ of As per kg of soil. Arsenic accumulation in plants due to irrigation water application also differs significantly in different cultivars. For instance, between two boro rice cultivars, BRRI dhan28 was found to accumulate more As than BRRI dhan29 grown in a pot with the same soil (Huq et al., 2006). They also noted that in either variety, As(III) was more prominent than As(V). In another study, Rabbi et al. (2007) found that the high-yielding variety (HYV) of rice accumulated more As than the local variety. In a study, Huq and Naidu (2003) found almost similar amounts of As ($<1.0\,mg\,kg^{-1}$ dry weight) in wheat grain compared to rice grain, but the amount of As was much lower in wheat roots and other parts compared to rice.

An extensive study with 330 rice samples grown in aman and boro seasons were studied by Williams et al. (2006). The study area was throughout Bangladesh. They found a positive correlation between As in the groundwater and in the rice plant which was much stronger in boro rice than aman. The As concentration in rice also differed significantly in different locations with southwestern parts contaminated with higher concentration of As following the pattern: Faridpur > Satkhira > Chuadanga > Meherpur. This was due to the higher levels of As in irrigated soils compared to nearby non-irrigated soils. Conversely, other parts of the country did not show differences in As concentrations either in irrigated or non-irrigated soils. Several other reports indicated that the soils in the western and southwestern parts of the country have the highest concentration of As (FAO, 2006). Apart from Williams et al. (2006), Islam et al. (2005) also conducted an extensive study on the levels of As in water, soil and crops at 456 locations in five upazilas. They found that the classified the upazilas according to soil concentrations were: Faridpur > Tala > Brahmanbaria > Paba > Senbag. Most of the studied soil samples (53%) contained less than $10 \, mg \, As \, kg^{-1}$, 26% soil contained $10.1–20 \, mg \, As \, kg^{-1}$ and the rest soil samples (18%) contained more than $20 \, mg \, As \, kg^{-1}$. There was also a seasonal variation of As concentration in water where boro season allowed more As in boro than in aman season. Yamazaki et al. (2003) showed that As concentration in Bangladesh soil also varied depending on the types of sediment. They studied five soil profiles collected from 15 depths from southwest Bangladesh. They reported that clay sediment contained the high concentration of As $(20–111 \, mg \, kg^{-1})$ followed by clay $(4–18 \, mg \, kg^{-1})$ and sand $(3–7 \, mg \, kg^{-1})$.

In last few decades, rice has extensively been studied to determine its level of contamination in relation to health risk. A group of studies analyzed a total of 788 rice samples (USAID, 2003; Duxbury et al., 2003; Hironaka and Ahmad, 2003; Meharg and Rahman, 2003; Shah et al., 2004) that indicated that the daily As intakes from 84% of the Bangladeshi rice samples exceed daily As intake from water at a standard of $10 \, mg \, l^{-1}$. This provides an indication that rice alone may be an important source of exposure in the Bangladeshi food system. Rahman et al. (2007a) investigated the accumulation and distribution of As in different fractions of rice cv. BRRI dhan28 and BRRI hybrid dhan1 collected from an As-affected area of Bangladesh (Sathkhira, Bangladesh) which was irrigated with As-contaminated groundwater $(0.07 \, mg \, l^{-1})$. They found that As content was about 28- and 75-fold higher in root than in shoots and raw rice grain, respectively. The order of As concentrations in different parts was: rice hull > bran-polish > brown rice > raw rice > polished rice. Laizu (2007) also analyzed a total of 400 vegetable samples under 20 varieties and found a large range of As concentrations, indicating the risk of potential hazard of the affected crops to the health of the people in Bangladesh. Ullah et al. (2009) analyzed 330 soil samples taken from 11 districts and they observed the As level to be above permissible limits in 10 districts. They found the highest contamination of $54.74 \, mg \, kg^{-1}$ dry soil in Jessore district which was more than double than the permissible limit $(20 \, mg \, kg^{-1}$ dry soil) indicating the severity of the problem.

FIGURE 16.11 Awareness build-up poster by the non-government organizations towards drinking As-free safe water. Green ($<50\,\mu g\,l^{-1}$) and Red ($>50\,\mu g\,l^{-1}$). Text below the left oval: Tube well with Green colour is 'Arsenic free'; text below the right oval: Tube well with Green colour is 'Arsenic contaminated'; text in the upper balloon: use 'geen' coloured tube well water for cooking, drinking and ther works; text in the lower balloon: use 'red' coloured tube well water for other works except cooking and drinking. These posters created huge awareness among rural communities in Bangladesh in the 2000s.

However, in Bangladesh huge attempts have been made to raise awareness of As problems among the people. Many Government and NGOs are working in this area. The DPHE has measured As levels of shallow tubules of the country and identified the As-contaminated and the safe ones (Figure 16.11). Although lots of effort have been made to find safe drinking water there, no suitable measure has been established yet. In addition to the drinking water problem, continued irrigation with As-contaminated water increases the extent of As contamination in agricultural land soil in Bangladesh (Ullah, 1998; Alam and Sattar, 2000; Ali et al., 2003; Islam et al., 2004; Figure 16.10).

ARSENIC UPTAKE AND TRANSPORTATION IN PLANTS

Whether from anthropogenic or geogenic sources, As has accumulated in the soils and groundwater, and even the atmosphere. Soil is the primary medium for plant growth and water is a basic need for plants; so As accumulated in those is easily uptaken by plants. However, the rate of uptake or accumulation of As greatly depends on several factors such as soil type, plant species and mechanisms of uptake. The capacity of As uptake by plants also varied with the speciation of As. Arsenite, is known to be more toxic and mobile than arsenate (Stronach, et al., 1997).

Roots are the first point of entry of As from where it moves to different plant parts of terrestrial and emergent plants. However, submerged plants may also take up As through leaves from the water column (Wolterbeek and van der Meer, 2002; Figure 16.12). In general, As is primarily found in the roots rather than in the shoots of plants. In rice, Rahman et al. (2007c) observed 28- to 75-times-higher As concentration in roots than in shoots. Later, Rofkar and Dwyer (2011) obtained a more than 15-times higher As

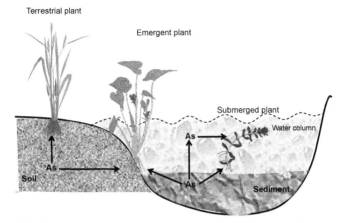

FIGURE 16.12   Uptake routes of As in terrestrial, emergent and submerged plants.

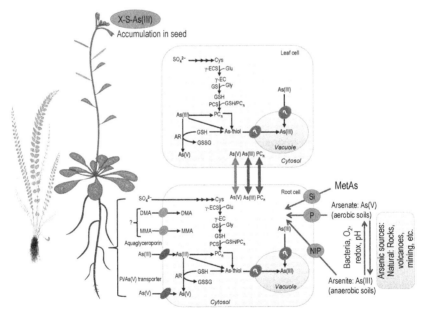

FIGURE 16.13   Mechanism of Arsenic uptake, translocation and detoxification in plants. For a colour version of this figure, please see the section at the end of this book. *Sources: Tripathi et al. (2007), Briat (2010).*

concentration in *Spartina pectinata* roots than in shoots. However, some species, for example radish, have a higher concentration of As in the shoots compared with the roots (Smith et al., 2008). Apoplast is the first entry point of As to the plant body (Figure 16.13). Localization of As in the apoplast may make a significant contribution to the total amount of As in a plant. In their report Bravin et al. (2008) observed that about 60% of the total As (173 mg As kg^{-1}) was accumulated in rice root which was grown under reducing conditions and separated into an apoplastic fraction, while only 39% (112 mg As kg^{-1}) of the total As was retained in the non-apoplastic fraction. Compared to roots, the shoot tissue contained only 1% (2.3 mg As kg^{-1}) of total As. Chen et al. (2005) found 17% of the total As in the apoplast of As-hyperaccumulating fern, *Pteris vittata*. From the apoplast As can enter into the plant cytosol (Figure 16.13). This movement is governed by sevaral factors. Uptake of AsO$_4^{3-}$ into the cell cytoplasm has been suggested to occur by high-affinity phosphate transporters, while AsO$_3^{3-}$ has been suggested to be taken up by aquaglyceroporins (Meharg and Macnair, 1992; Meharg and Jardine, 2003), which are also responsible for the uptake of other As compounds such as momethylarsonic acid (MMA) and dimethylarsinic acid (DMA) (Rahman et al., 2011).

In some cases, large amounts of As were locked up in plant roots and only relatively small amounts of As (as AsO$_4^{3-}$ and AsO$_3^{3-}$) were exported via the xylem to above-ground tissues, as observed by Pickering et al. (2000). Arsenic in soil is found mainly in the oxidized form (AsO$_4^{3-}$). Later, a portion of the AsO$_4^{3-}$ taken up into plant roots may be held in cell walls (possibly as FeAsO$_4$). Some portion of As passes through root membranes, most likely via transporters that facilitate the uptake of phosphate (Pickering et al., 2000), which is chemically similar to AsO$_4^{3-}$. Inside the root cells, some AsO$_4^{3-}$ is reduced to AsO$_3^{3-}$. A small fraction of the AsO$_4^{3-}$ and AsO$_3^{3-}$ move from root to leaf via the xylem (Pickering et al., 2000). Arsenate is then reduced to AsO$_3^{3-}$ by the activity of the enzyme *arsenate* reductase (ArsC), which combines with thiol (–SH) compounds to form As–thiolates. In this form, As is probably shuttled irreversibly into the vacuoles, where it can no longer damage the plant. The pathway of As uptake and transportation is presented in Figure 16.13. For the long-distance movement of As, little or no information is available on the nature of the transporter involved. As described by Tripathi et al. (2007), plants take up AsO$_4^{3-}$ and AsO$_3^{3-}$ through phosphate transporters and aquaglyceroporins (NIP), respectively. Small amounts of organic As MMA and DMA can also be taken up through unknown transporters (Tripathi et al., 2007). Long-distance transport of As from root-to-shoot takes place in the form As(V) and As(III). Plants assimilate sulphate (SO$_4^{2-}$) to form cysteine (Cys) for the synthesis of glutathione (GSH) in two adenosine triphosphate (ATP)-dependent steps: in the first, rate-limiting step, γ-glutamylcysteine (γEC) is synthesized by γ-glutamylcysteine synthetase (γ-ECS) using Cys and γ-glutamic acid (γ-Glu) as substrates; and in the second step, GSH is synthesized by glutathione synthetase (GS), using glycine (Gly) as a substrate (Tripathi et al., 2007; Figure 16.13). In response to As, plants induce synthesis of phytochelatins (PC), through the action of phytochelatin synthase (PCS) which can be transported through the vascular tissues. As a process of detoxification, AsO$_4^{3-}$ is reduced to AsO$_3^{3-}$ by arsenate reductase (AR) using GSH as a reducing agent. Phytochelatins and GSH coordinate with As(III) to form a variety of complexes. These complexes can be sequestered in the vacuole by ABC-type transporters (Figure 16.13). However, there is no clear evidence so far in this regard. Apart from this, large amounts of unbound AsO_3^{3-} are found in vacuoles, but whether these are transported as free As(III) is still unknown (Tripathi et al., 2007; Verbruggen et al., 2009). The role of PC in mitigating As toxicity is described in the later part of this chapter.

The uptake of As from root to shoot and its accumulation and sequestration broadly depends on the transporters involved in As uptake which are also mediated by the constitutive expression of genes that encode the various transporters. Generally, the expression of those genes is greater in hyperaccumulator plants compared with non-accumulators (Krämer, 2005). For instance, Lei et al. (2012) reported that P transporter e has a large role in As translocation in As hyperaccumulator *P. vittata*. Geng et al. (2006) found different capacities of As uptake and distribution by two cultivars of winter wheat (*T. aestivum*) (Jing 411 and Lovrin 10) grown with or without P supply. Between the two cultivars, Lovrin 10 invested more biomass production to roots than Jing 411, which might be partly responsible for higher shoot P and As concentrations and higher shoot-to-root ratios of As concentrations. In addition, Lovrin 10 allocated less As to roots than Jing 411 and the difference disappeared with decreasing P supply. Sharma and Travlos (2012) reported that a concentration of 100 μM phosphate was able to protect the test plant from As (100μM)-mediated oxidative stress in pearl millet (*Pennisetum typhoides*).

PLANT RESPONSES TO ARSENIC TOXICITY

Recently As toxicity in plants has come into light as a number of studies indicated its negative impact on plants. Arsenic has different forms which can be frequently taken up by roots, transported from the root to shoots via the xylem, further accumulated to the aerial parts of plants, including the seeds (Pickering et al., 2006). Several research findings indicated the adverse effects of As on the growth physiology and the ultimate yield of crops (Tables 16.3 and 16.4).

Seed Germination

Marked decreases in germination percentage and shoot and root elongation were recorded with As treatments in rice seedlings. The germination decreased significantly with the increase in As concentration. Root length, shoot length, root and shoot fresh weight of rice seedlings were greatly inhibited at 50 mM As (III) and 500 mM As (V) (Shri et al., 2009). In *Vigna mungo*, control treatment resulted in germination percentage over 97%, while it was only 83 and 70% when exposed to 100 and 200 μM As, respectively. Shoot and root length, and fresh weight of seedling also decreased (Srivastava and Sharma, 2013).

TABLE 16.3 Growth, Physiology and Productivity of Plants as Affected by Arsenic

Plant Species	Dose and Duration of As Exposure	Major Effects	References
Triticum aestivum	20 mg kg^{-1} As, 7 days	Decreased germination energy, germination percentage, germination index, vitality index, Chl content	Chun-xi et al., 2007
Phaseolus aureus	50 μM As, 7 days	Decreased root length, shoot length, root/shoot ratio	Singh et al., 2007
Oryza sativa	40 mg kg^{-1} As, 30 days	Decreased root and shoot dry weight	Sun et al., 2008
Oryza sativa	100 mM As(III), 10 days	Decreased root and shoot length, fresh weight, germination percentage	Shri et al., 2009
Brassica juncea	25 μM As, 96 h	Decreased root and shoot length, root and shoot dry weight	Khan et al., 2009
Brassica rapa	67 μM NaAsO$_2$, 14 days	Decreased dry weight in shoot (72.1%) and root (61.1%)	Shaibur and Kawai, 2009
Cicer arietinum	60 mg kg^{-1} As, 90 days	Decreased fresh weight and dry weight	Gunes et al., 2009
Brassica juncea	30 mg kg^{-1} As As(III), 30 days	Decreased total Chl, fresh weight, carotenoid	Sinha et al., 2010
Festuca arundinacea	25 μM As, 4 days	Decreased dry weight, increased relative ion leakage	Jin et al., 2010
Zea mays	668 μM As(V), 8 days	Decreased root and shoot length	Duquesnoy et al., 2010
Pteris vittata	50 mg kg^{-1} As, 45 days	Reduced germination percentage and biomass production	Raj et al., 2011

TABLE 16.3 Growth, Physiology and Productivity of Plants as Affected by Arsenic—cont'd

Plant Species	Dose and Duration of As Exposure	Major Effects	References
Atriplex atacamensis	1000 μM As (V), 14 days	Increase in the free soluble polyamine, NPT	Vromman et al., 2011
Arabidopsis thaliana	500 μM As(V), 7 days	Decreased fresh weight	Leterrier et al., 2012
Phaseolus aureus	10 μM As, 10 days	Decreased root and shoot growth, Chl content, increased electrolyte leakage	Malik et al., 2012
Trigonella foenum-graecum	30 mg kg^{-1} As, 9 weeks	Decreased plant height, root length, shoot dry weight, root dry weight, leaf photosynthetic rate, Chl content, pods/plant, seed yield/plant	Talukdar, 2013a
Leucaena esculenta	100 mg kg^{-1} As, 60 days	Decreased shoot length, leaf area, shoot dry weight, seeds pod^{-1}, seed yield, photosynthesis rate, Chl stability index. Increased root diameter, root dry weight	Talukdar, 2013b
Luffa acutangula	50 μM As, 7 days	Decreased fresh weight, root length, shoot length, Chl a, Chl b, carotenoid	Singh et al., 2013
Arabidopsis thaliana	50 μM As, 5 days	Decreased root growth, leaves growth	Gupta et al., 2013
Oryza sativa	25 μM As, 10 days	Decreased shoot length and dry weight	Tripathi et al., 2013
Triticum aestivum	0.25 and 0.5 mM Na$_2$HAsO$_4$, 72 h	Decreased in RWC and Chl content	Hasanuzzaman and Fujita, 2013

TABLE 16.4 Yield Reduction of Crop Plants as Affected By Arsenic

Plant Species	Dose of As	Yield Reduction	References
Solanum tuberosum	120 mg kg^{-1}	21.25%	Juzl and Stefl, 2002
Oryza sativa cv. BRRI hybrid dhan 1	90 mg kg^{-1}	About 57%	Rahman et al., 2007b
Fagopyrum esculentum	100 mg kg^{-1}	12.5%	Mahmud et al., 2007
Triticum aestivum cv Jimai	100 mg kg^{-1}	10.72% (grains ear^{-1})	Zhang et al., 2009
Amaranthus spp.	50 mg l^{-1}	89.20%	Choudhury et al., 2009
Oryza sativa cv. BRRI dhan29	30 mg kg^{-1}	Grain: 67.89% Straw: 65.19%	Islam and Jahiruddin, 2010
Zea mays cv. 31H50	50 mg kg^{-1}	97% (dry matter yield)	Namgay et al., 2010
Vigna radiata	10 mg kg^{-1}	87.22%	Sultana et al., 2012
Brassica napus	100 mg kg^{-1}	8.9%	Liu et al., 2012
Triticum aestivum	100 mg kg^{-1}	35.33%	Liu et al., 2012
Oryza sativa cv. BR-11	4 mg l^{-1}	Grain: 69.03% Straw: 14.25%	Azad et al., 2012
Zea mays L	100 mg kg^{-1}	Grain: 39.12%	Xiao-Ke et al., 2012

In wheat seedlings, the effects of As were determined on seed germination, root and shoot length and biomass production under different concentrations. However, low concentrations of As (0–1 mg kg^{-1}) stimulated seed germination and the growth of root and shoot. Treatment of *Phaseolus aureus Roxb* with different concentrations of As decreased the percentage of germination, germination index and vigour index. More inhibitory effects were observed at higher doses of As. Compared to untreated control, the As treatments of 1, 25, 50 and 100 mg l^{-1} reduced the germination percentage by 13, 43, 52, 74; 14, 44, 56 and 74%; the seedling vigour reduced by 14, 44, 56 and 74%, respectively (Patel et al., 2013). In *Vigna mungo* seed placed in As-free conditions resulted in 100% seed germination, whereas in 100 µM As treatment germination percentage was as low as 67%. Apart from a reduction in germination percentage, the root and shoot length of germinated seedlings were also reduced by As toxicity (Srivastava and Sharma, 2013). While studying the effects of As (0, 0.5, 1, 5, 15, 20 mg kg^{-1}) on *T. aestivum* cv. Zhengzhou-9023, Li et al. (2007) observed that

FIGURE 16.14   Toxicity symptoms in rice plants exposed to different levels (mg l⁻¹) of arsenic. For a colour version of this figure, please see the section at the end of this book. *Photo Courtesy of Prof. Dr Md. Asaduzzaman Khan, SAU, Dhaka.*

germination performance of the plant was better in lower concentrations and with the increase of concentration germination was adversely affected. At an As concentration of $0.5\,mg\,kg^{-1}$ the germination energy and germination index increased by 104.33% and 101.97%, respectively. The vitality index increased by 6.56% at an As concentration of $1\,mg\,kg^{-1}$. Germination energy, germination percentage, germination index and vitality index reduced significantly, by 9.60%, 4.04%, 10.73% and 46.61%, respectively, with higher As concentration of $20\,mg\,kg^{-1}$ (Li et al., 2007).

Growth

Higher As concentration can inhibit growth because of interference with plant metabolic processes that often leads to death. Several reports described the morphological changes, reduced fresh and dry biomass of roots and shoots under As-rich growing medium (Mokgalaka-Matlala et al., 2008; Shaibur et al., 2008; Srivastava et al., 2009; Figure 16.14). Plant height of *Lens culinaris* decreased with increasing concentrations of As in irrigation water. As also decreased leaf number, root length and biomass production (Ahmed et al., 2006). Accumulation of As in in *Ceratophyllum demersum* L. increased linearly (13–15% each day) depending on both As concentration (As^V, 0, 10, 50, 250 µM Na$_2$HAsO$_4$) and duration (1, 2, 4 and 7 days) of exposure. At 50 µM, plants accumulated about 76 µg As g⁻¹ dry weight after 4 days but no visible symptoms of toxicity were found. After 7 days, the same treatment resulted in an As level of 201 µg g⁻¹ dry weight and biomass production decreased significantly. Plant exposure to 250 µM As(V) resulted in a sharp increase in As accumulation (525 µg As g⁻¹ dry weight) after 7 days of exposure with a great decrease in biomass (Mishra et al., 2008). In rice plants, height and shoot biomass production were reduced to a great extent with the increase of soil As concentrations. The highest number of tiller (15.51 pot⁻¹) was recorded for control As treatment in BRRI hybrid dhan 1. The same treatment produced 12.34, 10.24, 9.14 and 8.43 tillers in BRRI dhan29, BRRI dhan35, BRRI dhan28 and BRRI dhan36, respectively. At $30\,mg\,kg^{-1}$ As in soil produced the lowest number of tiller (4.01) in BRRI dhan36 (Rahman et al., 2007b). In *Lactuca sativa* L. cv. Hanson, As accumulation in leaves and roots increased with the increase of As(V) and As(III) concentration (6.6, 13.2, 26.4 and 52.8 µM As). The highest As concentration has been found in roots. Highest As transported to the leaves occurred when plants were exposed to As(III). Relative growth rate (RGR) of leaves and roots were decreased with the accumulation of As. When exposed to 52.8 µM As(V) and As(III), reduction of leaf growth was recorded as 39 and 29%, respectively. With exposure to As(V), root growth reduced by 50%; it reduced to 33% by As(III). At 52.8 µM As(V) the roots' relative growth reduced by 50% (Gusman et al., 2013).

Arsenic accumulation increased in *Ceratophyllum demersum* L. with increased concentrations and durations of As, which was correlated with PC content. The percentage of As chelated by PCs dropped by 19% at 50 µM and by 14% at 100 µM, after 4 days that increased As accumulation (Mishra et al., 2008). Root length of rice seedlings decreased by 23 and 37%, respectively, after exposure at 25 and 50 µM As for 24h. The reduction in coleoptiles length was only in the range from 8 to 15% with the same As dose. Without decreasing the root length significantly, increase of As accumulation was evident due to As exposure. Roots accumulated 21.2- and 35.0-fold higher As at 25 and 50 µM As exposure compared to control. Coleoptiles accumulated only 8.5- and 15.5-fold greater As at 25 and 50 µM As-exposure compared to control (Singh et al., 2009). Arsenic treatments of 5 and 10 mg kg^{-1} irrigation water reduced *Vigna radiata* plant growth. Different growth parameters like root length, fresh and dry weight and shoot length, fresh and dry weight were significantly reduced compared to those in plants treated with irrigation water without As treatment. Moreover, As irrigation water reduced different nutrient accumulation in the shoot including P, K, S, Ca and Mg. With the increase of As concentration this trend was more severe. In case of iron accumulation in the shoot, this trend was inverted and with the increase of As concentration, iron accumulation increased (Sultana et al., 2012). Number of plants per pot, stem length, number of leaves per plant, fresh yield *Amaranthus retroflexus* L. decreased drastically and gradually from lower to higher concentrations of As (5, 10, 15, 20, 25, 30, 35, 40, 45, 50, 55, 60 mg l^{-1}). Arsenic application also increases the As uptake of this plant and thus resulted in increased As concentration in root, stem and leaf (Choudhury et al., 2009). The biomass and height of wheat plant increased by 24.1 and 7.8% separately at 60 mg As kg^{-1} soil. Under exposure to high levels of As, the wheat height and biomass decreased significantly (Liu et al., 2012).

Water Relations

Under toxic metal stress, plant organs show a wide-range of secondary stresses; water or osmotic stresses are common among them (Requejo and Tena, 2005; Zhao et al., 2009). Plants respond in various ways to maintain the water status to cope with these kinds of stresses. Osmotic adjustment, regulation of water content or water potential are adapted by plants in response to As exposure

(Grasielle et al., 2013). In response to As stress, accumulation of proline, glycine betaine or osmotic adjustment by Na^+ within plants have been reported to maintain water balance (Ben Hassine et al., 2008; Grasielle et al., 2013). Among different doses of As (6.6, 13.2, 26.4 and 52.8 μM As) water use efficiency of *Lactuca sativa* L. cv. Hanson did not change up to 26.4 μM concentration. Significant reduction of water use efficiency was recorded at the highest concentration of (52.8 μM) As(V) and As(III) application. As(V) and As(III) reduced plant transpiration. But compared As(V), As(III) exposure affected more plant transpiration and reduced by 81% at 52.8 μM As(III). Stomatal conductance in lettuce was reduced by 61% at 52.8 μM As(V). The same concentration of As(III) caused a higher reduction (78%) of stomatal conductance (Grasielle et al., 2013).

Exposure of As for 96 h at 500 μM concentration reduced water use efficiency, increased transpiration rate and did not affect stomatal conductance in *Hydrilla verticillata* (L.f.) Royle (Srivastava et al., 2013). As (0.25 and 0.5 mM $Na_2HAsO_4 \cdot 7H_2O$) reduced the leaf relative water content (RWC) of wheat seedlings. The leaf RWC reduced by 15 and 17% with 0.25 and 0.5 mM As, respectively, as compared to control plants (Hasanuzzaman and Fujita, 2013). A decrease in RWC due to As treatment was also reported in bean plants (Stoeva et al., 2005). In xerohalophyte plant species *Atriplex atacamensis* after 14 days of As treatment (100 μM), root or leaf water contents did not change, but stem water content increased. Root water potential was decreased when exposed to 1000 μM As after 14 days. Root water potential decreased more after 28 days with both the As doses. Leaf water potential remain unchanged with the As doses. Shoot water potential decreased in As in a dose- and day-dependent manner. Stomatal conductance remained unaltered at 100 μM As and decreased at 1000 μM As (Vromman et al., 2011). In pea leaf compared to the control treatment, the RWC declined with increasing concentration of As(V) (60–90 mg kg^{-1}) (Garg and Singla, 2012). Osmolytes were also increased significantly. Exposure of plants to varying As (V) doses improved the protein and soluble sugar synthesis and accumulation. Protein increased by 48% and soluble sugar increased by 172%; increase in As also caused significant increase in the accumulation of proline and glycine betaine (Garg and Singla, 2012). In *Spartina densiflora* Brongn, as compared to the control plant, the water use efficiency, leaf water content and stomatal conductance decreased with increasing As concentrations (0.7, 2.7, 6.7 and 13.4 mM As) (Mateos-Naranjo et al., 2012).

Photosynthesis

Different forms and concentrations of As application were recorded to slow down net photosynthesis (Stoeva and Bineva, 2003). Light-harvesting apparatus can be affected by As stress. Decreases in chlorophyll (Chl) content and / or decreased photosystem II activity are some common effects of As toxicity (Singh etal., 2006; Rahman et al., 2007c; Duman et al., 2010). Photosynthetic electron flow through thylakoid membranes may be reduced under As stress. In the carbon fixation reactions centre, As reduces the potential to produce ATP and nicotinamide adenine dinucleotide phosphate (NADPH). Replacement of P by As(V) in photophosphorylation may cause uncoupling of thylakoid electron transport from ATP synthesis (Avron and Jagendorf, 1959; Watling- Payne and Selwyn, 1974). Decreased activity or availability of enzyme of Chl biosyntehsis [δ-aminolevulinic acid (ALA) dehydratase] or Chl degradation by chlorophyllase may reduce Chl content (Jain and Gadre, 2004). As exposure in many plant species also reduced photosynthetic pigment (Rahman et al., 2007c; Duman et al., 2010). Under As stress, limitation of photosynthesis might be due to chlorophyll degradation. As is also supposed to limit the supply of photosynthetic carbon that may trigger Chl degradation (Finnegan and Chen, 2012). Arsenic negatively affected the photosynthetic process of *Lactuca sativa* L. cv. Hanson by impairing the photochemical and biochemical steps of photosynthesis. Chl *a* fluorescence results indicate that the photosynthetic apparatus and chloroplastic pigments were not damaged by the As(V) and As(III) concentrations used (6.6, 13.2, 26.4 and 52.8 μM As). But liquid-phase photosynthesis of lettuce plants were reduced by different As(V) and As(III) concentrations. Stomatal conductance (gs) showed a similar trend like liquid-phase photosynthesis (Grasielle et al., 2013). A decrease of the ribulose-1,5-bisphosphate carboxylase/oxygenase (RuBisCO) large subunit was investigated under As stress in rice leaves (arsenate, 50 or 100 μM) that resulted in downregulation of RuBisCO and chloroplast 29 kDa. These were responsible for decreased levels of photosynthesis. Protein expression profile investigation of rice leaves revealed that As increased expression of several proteins associated with energy production and metabolism (Ahsan et al., 2010).

Arsenic interferes with carbon fixation capacity. The RuBisCO large subunit content in the carbon fixation side of photosynthesis of rice leaves decreased with As(V) treatment (Ahsan et al., 2010). As RuBisCO large subunit is encoded by the plastid DNA, it is supposed that As also interferes with chloroplast DNA gene expression that may decrease this protein (Bock, 2007). In As(V)-treated *Arabidopsis* RuBisCO small subunit transcripts increased (Abercrombie et al., 2008). In isolated pea chloroplasts, As(III) inhibits the light activation of photosynthetic CO_2 fixation (Marques and Anderson, 1986). As has been attributed to an impairment of electron transport activity that inhibits photosynthesis (Nwugo and Huerta, 2008). As at 500 μM concentration and 96h exposure led to a significant decline in Chl *a* content, Chl *b* and carotenoids after 96h in *Hydrilla verticillata* (L.f.) Royle. Net photosynthetic rate, electron transport rate and PS II efficiency also declined (Srivastava et al., 2013). In lettuce the photosynthetic electron transport values were reduced by 62 and 47% at 52.8 μM for As(V) and As(III), respectively, compared to control values. But non-photochemical quenching values increased significantly with As(V) and As(III) exposure, especially at the highest concentrations. The electron transport rate values reduced gradually with the increase of As concentration, The highest reduction values are 47 and 37% compared to control for As(V) and As(III), respectively (at 52.8 µM) (Grasielle et al., 2013). Under As stress when carbohydrate availability is low, Chl can be used as a source of metabolic carbon that may reduce the Chl content (Araújo et al., 2011). The Chl contents in the leaves of wheat plants showed the opposite trend, with lower and higher dose of As where As concentration ranged from 0–20 mg kg^{-1}. An As concentration up to 0.5 mg kg^{-1} increased Chl a, Chl b and total Chl contents by 18.46, 20.98 and 19.20%, respectively. At the highest concentration of As (20 mg kg^{-1}) they reduced by 32, 35 and 33%, respectively (Li et al., 2007). Different As treatments (10, 20 and 30 mg kg^{-1}) reduced Chl a content up to 59% and decreased Chl b content up to 75% in different rice varieties (BRRI dhan28, BRRI dhan29 , BRRI dhan35, BRRI dhan36, BRRI hybrid dhan 1) (Rahman et al., 2007b).

Yield

Several reports indicated a huge decline in crop yield under As toxicity (Table 16.4). By reducing plant reproductive capacity through losses in fertility As resulted in reduced fruit production and yield (Garg and Singla, 2011). Irrigation water containing 1, 2, 5 and 10 mg l^{-1} As concentration were applied on *Vigna radiata* plant. With negative effects on different growth parameters, As also affected its number of pods and seed numbers per pod. Reduction of these yield components resulted in significant reduction of yield (Sultana et al., 2012). Arsenic treatment resulted in a marked decrease in effective tillers per pot, filled grains per panicle and 1000-grain weight; these together contributed reduced grain yield. Rice grain yield of rice was reduced by 20.6% for 15 mg kg^{-1} As treatment and 63.8% due to 30 mg kg^{-1} As. Conversely, straw yield was reduced by 21.0 and 65.2 % with the same dose of As treatments (Islam and Jahiruddin 2010). Different As levels 5, 10, 15, 20, 25, 30, 35, 40, 45, 50, 55, 60 mg l^{-1} were applied on *Amaranthus retroflexus* L. and their effects were studied as compared to control. Compared to control only 9% yield reduction was recorded for 5 mg l^{-1}. This reduction increased with the increase of As concentration, 30 mg l^{-1} As resulted 63% yield reduction, whereas 55 mg l^{-1} or higher concentration resulted in 100% yield reduction (Choudhury et al., 2009).

Impairment of reproductive development was suggested to reduce yield under severe As stress. Anomalous reproductive structures morphogenesis, fertilization, abortion of female gametophyte, abnormal anther, disruption of sporo- and gametogenesis processes induced by As might reduce the yield (Ghanem et al., 2009; Vromman et al., 2013). Even if after successful completion of the above-mentioned processes, As may interfere with grain filling or assimilate transfer process together with other physiological processes that might lead to reduced grain yield (Vromman et al., 2013). Arsenic application in rice significantly reduced yield and different yield-contributing parameters including the number of panicles per plant, panicle dry weight, the number of spikelets and full grains per plant and 1000-grain weight (Vromman et al., 2013). Varietal difference of rice plant performance was evident under As stress. Yield was measured considering panicle number, filled grain production and weight of total grain. Number of panicles and filled grain in BRRI hybrid dhan 1 were 12 and 101, respectively, under As stress. Conversely, in BRRI dhan36 they were 7 and 53; in BRRI dhan35: 9 and 77; in BRRI dhan29: 9 and 83; in BRRI dhan28: 7 and 54, respectively. Grain yield was 42, 11, 14, 16 and 10g per pot for BRRI hybrid dhan 1, BRRI dhan36, BRRI dhan35, BRRI dhan29 and BRRI dhan28, respectively (Rahman et al., 2007a). Lower levels of As ($<60\,mg\,kg^{-1}$) resulted in slight increase of wheat grain yield. The inverse result was found for higher As doses. Length of spike and 1000-grain weight also showed similar trends. The yield reduction by $80\,mg\,kg^{-1}$ and $100\,mg\,kg^{-1}$ were 20 and 36%, respectively (Liu et al., 2012).

Oxidative Stress

Plant exposure to As leads to production of ROS: these may include superoxide ($O_2^{\cdot-}$), hydroxylradical ($^{\cdot}OH$) and hydrogen peroxide (H_2O_2) (Ahsan et al., 2008; Mallick et al., 2011). ROS are capable of peroxidating membrane lipids, readily damaging the proteins, amino acids, nucleotides and nucleic acids (Møller et al., 2007). Membrane damage resulted in a product of lipid peroxidation, malondialdehyde (MDA) that is considered as an indicator of oxidative stress (Shri et al., 2009). Lipid peroxidation, besides impairing the cellular function, generates lipid-derived radicals which also have detrimental effects (Van Breusegem and Dat, 2006; Møller et al., 2007). Arsenic treatment resulted in increases of As accumulation, H_2O_2 accumulation and lipid peroxidation in rice roots (Ahsan et al., 2008). Another study with rice roots reveled that As exposure at 25 and $50\,\mu M$ for 4h, $O_2^{\cdot-}$ content increased significantly by 20 and 53%, respectively, over untreated controls. Further, at 8 and 24h, $50\,\mu M$ As-exposure, $O_2^{\cdot-}$ content increased by 72 and 79%, respectively. In response to $25\,\mu M$ As, H_2O_2 content increased by 7, 6 and 5.5 times after 4, 8 and 24h. As at $50\,\mu M$ showed similar trends in increase of H_2O_2 but its content increased many fold higher than corresponding lower doses (Singh et al., 2009). At $25\,\mu M$ As-treatment, root oxidizability (RO) of rice roots increased by 1.2- to 5.3-fold after 4–24h and at $50\,\mu M$ As the RO increased 2.6- and 8.1-fold at the same time duration (Singh et al., 2009). Shri et al. (2009) also found that As treatment accumulated As within the rice seedlings and increased malondialdehyde (MDA) content. The wheat seedlings exposed to 0.25 and 0.5mM As increased the MDA level by 58 and 180%, respectively, compared to the control. Treatment with 0.25 and 0.5mM As increased the H_2O_2 content by 41 and 95% compared to control (Hasanuzzaman and Fujita, 2013). Compared to control, the wheat seedlings treated with 0.25 and 0.5mM As treatment increased the oxidized glutathione and GSSG content (increased by 50 and 101%) (Hasanuzzaman and Fujita, 2013). From further study, Hasanuzzaman and Fujita (2013) observed that significant reduction of the antioxidant-scavenging system and the glyoxalase system were downregulated which might increase the oxidative stress.

As (100 and $500\,\mu M$) exposure for 24 and 96h in *Hydrilla verticillata* (L.f.) Royle increased the rate of $O_2^{\cdot-}$ production. The increase in rate was higher in higher and longer periods of As concentration. The higher increase in $O_2^{\cdot-}$ production is supposed to be associated with decreased content of ASC at 500 μM As (Srivastava et al., 2013). Higher and longer duration of As exposure increased oxidative stress by decreasing activities of antioxidant enzymes. Activity of superoxide dismutase (SOD) increased up to 50 μM As(V) until 4 days, then its activity declined. Ascorbate peroxidase (APX) activity decreased at 250 μM As exposure of 2 days. Glutathione peroxidase (GPX) decreased after 4 days exposure to 50 μM As(V). Catalase (CAT) activity decreased at the lower dose of 10 μM As(V) concentration after 2 days in *Ceratophyllum demersum* L. While studying the oxidative stress level the opposite trend was observed. That is, with increasing dose and duration of As exposure, the oxidative stress was reduced. Seven days' exposure to 250 μM As resulted in significant increases in the levels of $O_2^{\cdot-}$, H_2O_2, MDA and ion leakage by 40, 23, 64 and 80%, respectively (Mishra et al., 2008). With increasing concentrations of As, $O_2^{\cdot-}$ content in leaves of wheat seedlings increased. It was amplified by 3, 28, 42, 50 and 57% at 0.5, 1, 5, 15, 20 mg kg^{-1} of As, respectively. Slight increase of MDA level was observed when the concentration was lower than 1 mg kg^{-1}. Conversely, at higher As concentrations (5–20 mg kg^{-1}) the extent of increase was higher. MDA content increased by 22.68% under As concentration of 20 mg kg^{-1} (Li et al., 2007). Under stronger As concentrations, ROS generation increases so high that the antioxidant defence mechanisms may be overwhelmed, leading to cellular damage. This damage can lead to cell death (Van Breusegem and Dat, 2006). Thus, the mechanism of ROS production under As should be known. The mechanism of the As-induced production of ROS is not well understood. One of the proposed ideas is that ROS might be generated from As detoxification processes like reduction of As(V) to As(III), the induction of PC synthesis and so on (Meharg and Hartley-Whitaker, 2002). Besides generation of ROS under As stress (Ahsan et al., 2008; Mishra et al., 2008; Shri et al., 2009), As can upregulate those genes which regulate antioxidant enzymes (Abercrombie et al., 2008; Norton et al., 2008; Chakrabarty et al., 2009), proteins (Requejo and Tena, 2005; Ahsan et al., 2008) and enzymatic activities as well (Srivastaba et al., 2005; Shri et al., 2009); these effects might be considered very important for reducing the oxidative stress within the plant.

ANTIOXIDANT DEFENCE IN PLANTS IN RESPONSE TO ARSENIC STRESS

Plants are equipped with various enzymes and other components which are distributed in chloroplast, cytoplasm, apoplast, mitochondria, peroxisome, membranes and those components composed an effectual antioxidant system to fight against different ROS. Chief non-enzymatic compounds are ascorbate (AsA), glutathione (GSH), carotenoids, tocopherols and some other phenolic compounds. The common enzymes of antioxidant systems are SOD, catalase (CAT), ascorbate peroxidase (APX), monodehydroascorbate reductase (MDHAR), dehydroascorbate reductase (DHAR), glutathione reductase (GR), glutathione *S*-transferase (GST), glutathione peroxidase (GPX) and peroxidases (POX) (Mittler et al., 2004; Ashraf, 2009; Gill and Tuteja, 2010; Hasanuzzaman and Fujita, 2013; Figure 16.15). Content of non-enzymatic components and activities of antioxidant enzymes have been documented to modulate under different stress conditions including As stress (Table 16.5).

In *O. sativa* seedlings, As treatment accumulated As within the plant and MDA in seedlings were increased significantly with increasing As concentration. At the same time compared to control, upregulation of antioxidant enzyme activities and the isozymes of SOD, APX, POD and GR substantiated under As stress (Shri et al., 2009). As treatment resulted in increases of As accumulation, H_2O_2 accumulation and lipid peroxidation in rice roots. GSH has been suggested to play a central role in protecting rice roots from these damages by upregulating SAMS (*S*-adenosylmethionine synthetase), CS (cysteine synthase), GST and GR. However, roles of GSH or activation of these enzymes largely depends on the dose of As together with many other factors (Ahsan et al., 2008). Accumulation or generation of thiol compounds has been correlated with a reduction of oxidative stress. Thiol compounds increased up to a certain dose and duration of As exposure and then they declined. Arsenate exposure significantly increased the synthesis of NP-SH (nonprotein thiol), cysteine and GSH in a concentration- and duration-dependent manner. Maximum increase in the levels of NP-SH (107%), cysteine (41%) and GSH (60%) was noticed when plants were exposed to $50\,\mu M$ As(V) for 2 days. Beyond these exposure their levels started declining and oxidative stress started to increase in *Ceratophyllum demersum* L. The activities of SOD, APX, GPX and CAT played important roles in keeping ROS lower in lower doses and duration of As exposure (Mishra et al., 2008).

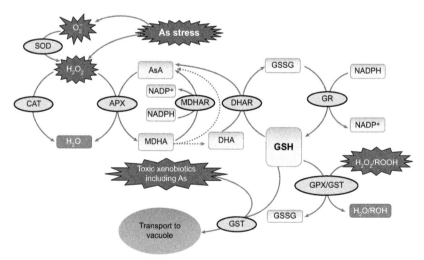

FIGURE 16.15 Antioxidant defence in plants in response to As toxicity. *Source: Hasanuzzaman et al. (2012).*

TABLE 16.5 Regulation of Antioxidant Defence in Plants in Response to Arsenic Toxicity

Plant Species	Dose and Duration of As Exposure	Regulation of Antioxidant Defence	References
Triticum aestivum	20 mg kg⁻¹ As, 7 days	Decreased the activity of CAT (74.92%) Increased the activities of SOD (16.25%) and APX (36.36%)	Chun-xi et al., 2007
Phaseolus aureus	50 μM As, 7 days	Increased the activities of SOD, GPX and GR Decreased the activity of APX	Singh et al., 2007
Oryza sativa	40 mg kg⁻¹ As, 30 days	Decreased GSH (16%) and AsA (24%) content Decreased the activities of SOD (15%) and POD (11%)	Sun et al., 2008
Pteris vittata	10 mg l⁻¹ As(III), 3 days	Increased CAT activity	Kertulis-Tartar et al., 2009
Oryza sativa	100 μM As(III), 10 days	Increased the activities of SOD, APX,POD and GR	Shri et al., 2009
Cicer arietinum	60 mg kg⁻¹ As, 90 days	Decreased the activity of SOD (44%) and total antioxidant activities (14%). Increased activities of CAT (180%) and APX (54%)	Gunes et al., 2009
Oryza sativa	50 μM As, 24 h	Increased the activities of SOD, APX, CAT and GPX	Singh et al., 2009
Brassica juncea	25 mM As, 96 h	Increased the activities of SOD, CAT, APX and GR Increased the levels of Asc and GSH	Khan et al., 2009
Hydrilla verticillata	5 μM As(III), 1 days	Increased the activities of SOD, GPX, APX and MDHAR Increased Asc, GSH, Asc/DHA ratio and GSH/GSSG ratio	Srivastava and D'Souza, 2010
Zea mays	668 μM As(V), 8 days	Decreased the activities of SOD, CAT. Increased the activity of APX	Duquesnoy et al., 2010
Festuca arundinacea	25 μM As, 4 days	Increased the activities of SOD and CAT	Jin et al., 2010
Pteris vittata	50 mg kg⁻¹ As, 45 days	Increased the activity of SOD, CAT, APX, GR and GST Increased the levels of, AsA, DHA, GSH and GSSG	Raj et al., 2011

Continued

TABLE 16.5 Regulation of Antioxidant Defence in Plants in Response to Arsenic Toxicity—cont'd

Plant Species	Dose and Duration of As Exposure	Regulation of Antioxidant Defence	References
Oryza sativa	10 μM As(V), 7 days	Increased the activities of APX, GPX and SOD	Dave et al., 2012
Phaseolus aureus	10 μM As, 10 days	Increased the activities of SOD, CAT, APX and GR	Malik et al., 2012
Trigonella foenum-graecum	30 mg kg⁻¹ As, 9 weeks	Increased the acticvity of SOD (449%). Decreased AsA content Decreased the activities of APX (20%) and CAT (46%)	Talukdar, 2013a
Leucaena esculenta	100 mg kg⁻¹ As, 60 days	Increased DHA content Decreased AsA content, AsA and GSH redox Decreased the activities of APX, GR, CAT, DHAR and GST.	Talukdar, 2013b
Lactuca sativa	52.8 mM As(V), 3 days	Increased the activities of SOD, CAT, GR and GPX	Gusman et al., 2013
Zantedeschia aethiopica	34 ± 11 μg l⁻¹ As, 6 months	Increased the activities of APX, GR and CAT	Del-Toro-Sánchez et al., 2013
Luffa acutangula	50 μM As, 7 days	Increased the activities of SOD and CAT. Increased the level of DHA. Decreased the activity of APX and GST Decreased the level of AsA	Singh et al., 2013
Arabidopsis thaliana	50 μM As, 5 days	Increase the level of AsA, DHA, GSH and GSSG Decrease the activity of CAT	Gupta et al., 2013
Oryza sativa	25 μM As, 10 days	Increased the activities of SOD, APX, DHAR and GST Decreased the activity of CAT	Tripathi et al., 2013
Triticum aestivum	0.25 and 0.5 mM Na₂HAsO₄, 72 h	14 and 34% reductions in the AsA content 46 and 34% increases in GSH content 50 and 101% increases in GSSG content Increase in the activities of APX, GR, GST Decrease in the activities of DHAR and GPX	Hasanuzzaman and Fujita, 2013

Activation of antioxidant enzymes has also been reported in gene expression levels. *Arabidopsis*, encoding genes for different SOD (FeSOD, MnSOD, Cu/ZnSOD) showed different responses under As(V) stress. Cu/ZnSOD were induced twofold by As(V) exposure in the transcripts; these encoded a chloroplastic and a cytosolic gene. The transcripts for an FeSOD showed fivefold downregulation (Abercrombie et al., 2008). In response to As exposure, catalase, GST, GRX and peroxidase transcript and enzymatic activity often increase (Ahsan et al., 2008; Norton et al., 2008; Chakrabarty et al., 2009). In rice As(V) exposure at least 10 GST genes were upregulated and two GST genes were downregulated (Norton et al., 2008; Chakrabarty et al., 2009).

Responses of low-As-accumulating (LARG) and high-As-accumulating (HARG) rice (*Oryza sativa*) genotypes were studied under different forms and doses of As [As(III) (NaAsO$_2$; 0, 10 and 25 µM) and As(V) (Na$_2$HAsO$_4$; 0, 10 and 50 µM) for 1, 4 and 7 days]. As compared to LARG, the HARG genotypes accumulated up to 23 times higher As in the root and 15 times higher As in the shoot. Antioxidant enzymes APX, GPX and SOD played major roles in reducing the oxidative stress induced by As. In roots, compared to control, the highest APX activity of 227% was recorded on 7 days exposure to 50 µM As(V) in LARG, while HARG displayed the highest induction of 309% at the same exposure. Shoots of LARG at 50 µM As(V) exposure led to a significant enhancement of SOD by 512%, and in HARG its activity increased by 208% (as compared to control). For GPX activity, in all durations HARG genotypes showed higher activity as compared to LARG shoots. Higher induction of antioxidant enzymes in HARG genotypes (compared to LARG) was correlated to MDA content (Dave et al. 2012). Application of sodium arsenite (50 and 150 µM and 300 µM; 2 and 4 days) in two varieties (Varuna and Pusa Bold) of *Brassica juncea* showed differential accumulation of As. Increased As tolerance in Pusa Bold was due to the defensive roles of antioxidant enzymes including SOD, CAT and GPX, induction of mitogen-activated protein kinase (MAPK) and upregulation of PCS transcript that produces metal-binding peptides (Gupta et al., 2009).

Modulation of non-enzymatic and enzymatic components was documented in many other plant species. As resulted in elevation of GSH and phytochelatin contents in *Hydrilla verticillata* (Srivastava et al., 2007). Exogenous GSH and cystine supplementation under As stress reduced oxidative stress and restored the growth of rice seedlings (Shri et al., 2009). Exposure of cucumber plant to As increased AsA concentration in hypocotyls and decreased its content in roots (Czech et al., 2008). A marked increase in AsA level and AsA/DHA (ratio of reduced ascorbate, AsA and oxidized ascorbate, DHA) was recorded during As exposure in the fronds of As-hyperaccumulator *Pteris vittata* compared to As-sensitive *P. ensiformis* (Singh et al., 2006). Light harvesting and membrane-associated antioxidant activity are important functions of carotenoid. Carotenoids are able to protect from harmful free radicals and ROS produced naturally during photosynthesis (Collins 2001). As (133 µM , 10 days) increased carotenoid content in hyperaccumulator *P. Vitteta* and decreased in *P. ensiformis* (Singh et al., 2006).

Carotenoid levels reduced in oat, bean and red clover plants grown in As-contaminated soil (Stoeva et al., 2005). Increase of APX activity has been reported in many crops by other researchers including in mung bean (Singh et al., 2007), bean (Stoeva et al., 2005) and maize (Miteva and Peycheva 1999) exposed to As. As(V) stress caused induction of chloroplast Cu/Zn SOD, Cu/Zn SOD and SOD copper chaperone and resulted in reduction of Fe SOD (Abercrombie et al., 2008). Stimulation of GR activity in rice seedlings was correlated with higher GSH content in the As-induced oxidatively stressed rice seedlings (Shri et al., 2009). Enhanced GR activity has also been observed in roots of *Pteris vittata*, *Ptries ensiformis*, *Nephrolepis exaltata* (Srivastava et al., 2005). Guaiacol peroxidase activity increased As-induced oxidative stress in mung bean plant (Singh et al., 2007). The GST activity in mesquite and maize plant were also increased by As stress (Mylona et al., 1998; Mokgalaka-Matlala et al., 2009).

In our recent study, we observed that As at different doses (0.25 and 0.5 mM) created oxidative stress which is revealed by the As-induced elevated H_2O_2 and MDA levels (Table 16.6). Moreover application of SNP or sodium nitroprusside as a source of NO improved non-enzymatic antioxidants (Table 16.6) and the enzymatic components (Table 16.7) of antioxidant systems. Enhanced activities of antioxidant enzymes and improved levels of non-enzymatic components

TABLE 16.6 Malondialdehyde (MDA), H_2O_2 Content, Reduced Ascorbate (AsA), Reduced Glutathione (GSH) and Oxidized Glutathione (GSSG) Contents in *Triticum aestivum* L. cv. Pradip Seedlings Induced by NO (0.25 mM SNP or Sodium Nitroprusside, $Na_2[Fe(CN)_5NO] \cdot 2H_2O$ — A NO donor) under As Stress (0.25 and 0.5 mM $Na_2HAsO_4 \cdot 7H_2O$, 72 h)

Treatment	MDA Content (nmol g⁻¹ fresh weight)	H₂O₂ Content (nmol g⁻¹ fresh weight)	AsA Content (nmol g⁻¹ fresh weight)	GSH Content (nmol g⁻¹ fresh weight)	GSSG Content (nmol g⁻¹ fresh weight)
Control	22.81d	7.19c	3889.8a	251.2e	15.2c
As0.25	36.09c	10.16b	3349.7b	367.5bc	22.6b
As0.5	63.66a	14.03a	2581.5c	336.4cd	30.7a
SNP	22.13d	7.20c	3671.5ab	290.0de	16.1c
SNP+As0.25	27.99d	7.95c	3916.4a	433.1a	16.0c
SNP+As0.5	43.29b	10.43b	3316.8b	394.9ab	19.1bc

Mean (± SE) was calculated from three replicates for each treatment. Values in a column with different letters are significantly different at $P < 0.05$ applying DMRT.
Source: Hasanuzzaman and Fujita (2013).

TABLE 16.7 Activities of Antioxidant Enzymes in *Triticum aestivum* L. cv. Pradip Seedlings Induced by NO (0.25 mM SNP or Sodium Nitroprusside, $Na_2[Fe(CN)_5NO]•2H_2O$ — a NO donor) under As Stress (0.25 and 0.5 mM $Na_2HAsO_4•7H_2O$, 72 h)

Treatments	CAT Activity (μmol min^{-1} mg^{-1} protein)	APX Activity (μmol min^{-1} mg^{-1} protein)	MDHAR Activity (nmol min^{-1} mg^{-1} protein)	DHAR Activity (nmol min^{-1} mg^{-1} protein)	GR Activity (nmol min^{-1} mg^{-1} protein)	GST Activity (nmol min^{-1} mg^{-1} protein)	GPX Activity (nmol min^{-1} mg^{-1} protein)
Control	41.64c	0.574c	36.66abc	150.78ab	11.43c	86.81c	0.102b
As$_{0.25}$	42.07c	0.710ab	30.58c	101.72c	14.98b	112.80ab	0.087bc
As$_{0.5}$	38.01c	0.767a	29.53c	105.31c	12.71bc	125.20a	0.073c
SNP	48.47bc	0.636bc	35.42bc	121.73bc	11.35c	94.87bc	0.102b
SNP + As0.25	54.60ab	0.756a	41.90ab	157.84ab	19.33a	127.21a	0.142a
SNP + As0.5	62.54a	0.780a	45.18a	172.38a	18.90a	125.41a	0.133a

Mean (±SE) was calculated from three replicates for each treatment. Values in a column with different letters are significantly different at P<0.05 applying DMRT.
Source: Hasanuzzaman and Fujita (2013).

together acted to reduce the oxidative damage induced by As stress (i.e. H_2O_2 and MDA levels) (Table 16.6).

REMEDIATION OF ARSENIC HAZARDS

Since As pollution in soil and water is becoming a serious threat for crop production worldwide, it is necessary to find suitable mechanisms to cope with As problems. Like other toxic metals, remediation of As toxicity is complex. However, the possible ways to mitigate As problems may include genetic improvement, antioxidant defence, alteration of metabolism and synthesis of PC. Phytoremediation would be an excellent way to remediate As from soils which may limit the As toxicity as plants get a safe environment in which to grow. Different agronomic management, especially water and nutrient management are also found to be somehow effective in mitigating As toxicity. Genetic variation in the response of plants to As in terms of uptake and metabolism provides the potential to devise agronomic cultivars better suited to As-enriched soils.

Agronomic Management

Since irrigation water is one of the most important sources of As contamination in crop fields, the proper management of irrigation may be a strategy to mitigate As toxicity. Farmers often use more irrigation water than needed. Optimizing water input would be a sound option to reduce As input while saving water. Furthermore, aerobic growth conditions in paddy fields may reduce the bioavailability and uptake of As in rice. Breeding of crops plants tolerant to As or low accumulation of As in harvestable parts, cultivation of crops other than rice when water is scarce, etc., may be other options (Shahariar and Huq 2012). Based on a study conducted by Shahariar and Huq (2012), it was reported that growing rice at reduced moisture levels could reduce As toxicity without compromising the yield. Reduced moisture may also minimize As uptake due to the reduction in phytoavailability, thereby preventing the exposure of As to food chain as well. Huq et al. (2006) investigated the role of water regime management in mitigation of As in Bangladeshi rice (cv. BRRI dhan28 and BRRI dhan29 grown with added As (0, 10, 20 and 40 mg As kg^{-1} soil). They examined two different water regimes (100 and 75% of field capacity). In roots of BRRRI dhan28, the maximum As accumulation from As(III)-treated soil was 17.6 mg kg^{-1} dry weight at saturated soil (100% field capacity), whereas at 75% field capacity it was 15.4 mg kg^{-1}. In the case of BRRI dhan29, As accumulation was a bit higher which was determined as 31.04 and 22.65 mg kg^{-1} at irrigation regimes 100 and 75% of field capacity, respectively. Thus it was concluded that As accumulation can be minimized even when rice is grown at 75% field capacity without hampering the yield (Huq et al., 2006). However, it is not suitable where irrigation cut-off has adverse effects on crops' productivity. For instance, in the United States, growing rice in the field with completely dry conditions for

10–14 days just before panicle initiation, significantly reduced As accumulation but the yield was reduced drastically (Brammer, 2007). In such a case, proper timing of irrigation cut-off is an important concern.

Both irrigation and tillage practices can be managed towards the mitigation of As from soil. Talukder et al. (2011) reported that As content of rice straw and grain grown under permanent raised bed (PRB) was six- and three-times, respectively, lower over the As content of rice grown in flat bed conditions. Later, Sarkar et al. (2012) recommended deficit irrigation as an option to mitigate As load of rice grain in West Bengal, India. They observed that addition of As through irrigation water had a direct impact on total As load of the soil. For instance, application of 690 mm of water up to 45 days after transplanting (DAT) under continuous ponding (CP) resulted in the highest concentration (18.18 mg kg^{-1}) of total As in surface soil (0–150 mm). A reduction in the amount of irrigation to 200 and 390 mm, respectively, under intermittent ponding (IP) and aerobic (AER) conditions during 16–45 DAT decreased the As load by 2.98 and 11.22%. Arsenic content in soil, rice root, rice shoot, rice leaf and rice grain was 18.74, 22.14, 2.48, 4.20, 0.56 mg kg^{-1}, respectively, with CP, whereas the As content was 16.22, 20.54, 1.86, 3.51, 0.46 mg kg^{-1}, respectively, with adoption of aerobic conditions. At harvest stage (80 DAT) the differences were 3.06 and 13.43%, respectively. However, they concluded that imposition of IP only during the vegetative stage was found to be optimum in terms of reduction of As content in straw and grain, respectively, by 23 and 33%, over farmers' irrigation practices with insignificant decrease in grain yields (Sarkar et al. 2012).

Fertilizer management was also found to be beneficial in mitigation of As toxicity. Since, phosphorus (P) is a chemical analogue of As (Adriano, 2001) and competes with As in plant uptake (Meharg and Macnair, 1992), the effect of P on the sorption/desorption of As in soil environments has received great attention, especially when P is used as a crop fertilizer (Peryea, 1998b). In a study, shown in this work, the role of P fertilization in preventing As uptake and translocation in wheat plants is highlighted. According to Li et al. (2009) S played important roles in As detoxification and accumulation in As hyperaccumulator *Pteris vittata* and the As-hypertolerant plant *Adiantum capillus-veneris* (A. capillus-veneris). Under As exposure, sulphydryl groups (–SH) increased in both plants, indicating that arsenate enhanced sulphur assimilation and S then played different roles in the two plants (Li et al., 2009). While studying with *Vigna mungo*, Srivastava and Sharma (2013) found that As-treated plants showed reduction in their growth and pigment content. Arsenic significantly enhanced lipid peroxidation, electrolyte leakage and level of proline and also significantly altered the antioxidant metabolism. However, application of phosphate with As resulted in significant alterations in most of the traits and led to better growth in black gram (*Vigna mungo*). In a recent study, Tripathi et al. (2013) found a beneficial effect of Si on the mitigation of As toxicity in rice. They found that rice plants treated with 1 mM Si reduced As content in rice exposed to 25 µM As. However, in tolerant rice cultivar (Triguna), the effect of

Si was more prominent in mitigating As accumulation as compared to sensitive cultivar (IET-4786). Reduction of As content in tolerant and sensitive cultivar was 24 and 18% on addition of 1 mM Si. This might be due to the reduction in As(III) uptake by rice because of the direct competition between Si and As(III) during uptake (Ma et al., 2008; Guo et al., 2009).

Alternative farming may be a possible way to mitigate As problems. Substituting dry-land crops such as wheat or maize for rice, or growing rice as a dry-land crop, could reduce the problem of As contamination of soils and food crops. In some countries, research is in progress to breed As-tolerant rice varieties, but the use of such varieties would not reduce the accumulation of As in soils, so it might provide only a short-term, interim measure until an alternative safe irrigation supply could be provided.

Phytoremediation of Arsenic-Contaminated Soils

Phytoremediation or plant-based remediation technology has recently come to the fore as a strategy to clean up contaminated soils and water. Suthersan (2002) defined it as 'the engineered use of plants in situ and ex situ for environmental remediation'. It refers to the use of higher plants and their associated microbiota for the clarification of soil, sediment and groundwater; selected plant species acquire the genetic potential to remove, degrade, metabolize or immobilize a number of contaminants including metals and metalloids. Phytoremediation is a cost-effective 'green' technology which is non-intrusive, aesthetically pleasing, socially accepted technology to remediate polluted soils (Garbisu et al., 2002; Kachout et al., 2010; Hasanuzzaman and Fujita, 2012). There are different processes of phytoremediation such as phytoextraction/phytoaccumulation, phytostabilization, phytodegradation/phytotransformation, rhizofiltration, rhizodegradation, phytovolatilization and phytorestauration (Hasanuzzaman and Fujita, 2012; Table 16.8).

Although there are several processes of phytoremediation, phytoextraction is the most important and has been extensively studied. This remediation technique is employed to recover metals from contaminated soils using non-food crops where metals are absorbed by roots and subsequently translocated within the above-ground parts of plants. Genetic and physiological potential of some plants species making them effective to accumulate, translocate and tolerate high concentration of metals and are used for phytoextraction. According to Chaney et al. (1997), naturally occurring plants called 'metal hyperaccumulators' can accumulate 10- to 500-times higher levels of toxic metals than cultivated crops. The efficiency of phytoextraction is determined by the accumulation factor, indicating the ratio of toxic metal concentrations in the plant organs (shoots, roots) and in the soil and the seasonally harvestable plant biomass (Kvesitadze et al., 2006). There are three processes that regulate the movement of toxic metals from root tips to xylem (root symplasm). The sequestration of toxic metals occurs inside cells; they are then subjected to symplastic transport to the stele,

TABLE 16.8 Various Processes of Phytoremediation Processes

Phytoextraction/ Phytoaccumulation	It refers to the use of plants to remove contaminants from the environment and accumulate them in above-ground plant tissue. In the case of phytoextraction, the metals and metalloids are absorbed by roots and subsequently translocated within the above-ground parts of plants. In this case the plants which have genetic potential to tolerate high concentration of contaminants are used.
Phytostabilization	This is an indirect way of phytoremediation where soil metals are removed through absorption and accumulation by roots, either being adsorption onto roots or precipitation within the root zone. The objective of phytostabiliation is to stabilize the metals to prevent the transportation of these into the environment rather than direct removal from a site.
Phytodegradation/ Phytotransformation	This process includes the uptake of contaminants by plants followed by breakdown into simple or non-toxic/less toxic products by the help of agent (e.g. enzymes). These simple products further synthesized and metabolized in plants' vascular systems.
Rhizifiltration	This process is especially applicable to remediate contaminants from aquatic environments. In this process, plants uptake, concentrate and precipitate metals from contaminated sources exclusively through their root systems. It involves two major strategies, namely, sorption of metals by the root system and their removal; these are mostly dependent on physiological and biochemical characteristics of the plant species.
Rhizodegradation	In this process, soil contaminants are broken down through the microbial activity of the rhizosphere. It is a slower process compared to others but works on a large variety of contaminants.
Phytovolatilization	In this process, plants accumulate toxic compounds and then transform these into a volatile form to release into the atmosphere through stomata. This process is based on the facts that some of the toxic contaminants can pass through the plants to the leaves and volatilize into the atmosphere at comparatively low concentrations.
Phytorestauration	This is a special type of phytoremediation which denotes the complete remediation of metal-contaminated soils using native plants making it as suitable for normal activities as non-contaminated soils.

Source: Hasanuzzaman and Fujita (2012).

and subsequent release into the xylem (Figure 16.16). Generally, the toxic metal such as As content in a plant body decreases in the following sequence: root > 1 eaves > stems > inflorescence > seed (Lal, 2010). However, the metal-accumulating capacity greatly varies with plant species (Table 16.9).

To date, more than 400 hyperaccumulating species have been reported, but only a few species belong to As hyperaccumulators such as *Pteris vittata, P. cretica, P. longifolia, P. umbrosa, Pityrogramma calomelanos, Dryopteris juxtaposita* and *Nephrodium molle* are able to hyperaccumulate As when grown in soils added with As (Ma et al., 2001; Visoottiviseth et al., 2002; Zhao et al., 2002; Meharg, 2003; Chen et al., 2003; Hossain et al. 2006; Figure 16.17). Some other higher classes of terrestrial and aquatic plant species have been reported for their potential towards As removal and remediation from soils and water (Figure 16.18). Mediterranean aquatic plants such as *Apium nodiflorum* have widespread applications in removing As(V) from water, sediments and soils (Vlyssides et al., 2005). In fact, separating As from soil is difficult because As is strongly retained by the soil, especially in oxidizing environments and at low pH. Even after its entry into the roots, As is not readily transported to the leaves of most plants (Pickering et al., 2000). Therefore, to make the phytoextraction of As more effective, it is essential to find the means to enhance the transport of As from the root system to stems and leaves, where the metal can be easily harvested.

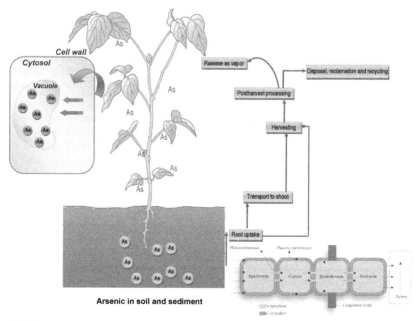

FIGURE 16.16 Phytoextraction mechanism in plants to remediate arsenic.

TABLE 16.9 Identified Naturally Grown Arsenic Hyperaccumulating Plants with Arsenic Concentrations in Different Plant Parts

Plant Species	Organs	Quantity of As (mg kg^{-1})	References
Eichhornia crassipes	Shoot	909.58	Delgado et al., 1993
Eichhornia crassipes	Shoot	805.20	Low et al., 1994
Typha latifolia	Shoot	1120	Ye et al., 1997
Spartina patens	Shoot	250	Carbonell et al., 1998
Phragmites australis	Shoot	119.55	Windham et al., 2001
Pteris vittata	Frond	27,000	Wang et al., 2002
Asplenium nidus	Frond	100	Zhao et al., 2002
Pteris cretica	Frond	2000–2800	Zhao et al., 2002
Pteris longifolia	Frond	5000	Zhao et al., 2002
Pteris umbrosa	Frond	5000	Zhao et al., 2002
Pityrogramma calomelanos	Frond	5000–8350	Francesconi et al., 2002
Ceratophyllum demersum	Shoot	963	Saygideger et al., 2004
Ceratopteris richardii	Frond	<500	Gumaelius et al., 2004
Pteris vittata	Shoot	22,000	Huang et al., 2004
Pteris vittata	Frond	3100	Bondada et al., 2004
Lemna gibba	Shoot	1021	Mkandawire and Dudel, 2005
Pteris vittata	Frond	1748	Luongo and Ma, 2005
Pteris vittata	Root	503	Luongo and Ma, 2005

Continued

Soil Remediation and Plants

TABLE 16.9 Identified Naturally Grown Arsenic Hyperaccumulating Plants with Arsenic Concentrations in Different Plant Parts—cont'd

Plant Species	Organs	Quantity of As $(mg\,kg^{-1})$	References
Nephrolepis exaltata	Frond	200	Srivastava et al., 2006
Pteris cretica, P. biaurita, P. quadriaurita and P. ryukyuensis	Root	182–507	Srivastava et al., 2006
Pteris cretica, P. biaurita, P. quadriaurita and P. ryukyuensis	Frond	1770–3650	Srivastava et al., 2006
Pteris vittata	Frond	57–9677	Wang et al., 2006
Pteris multifida	Frond	624–4056	Wang et al., 2006
Pteris cretica var. nervosa	Frond	1162–2363	Wang et al., 2006
Pteris oshimensis	Frond	301–2142	Wang et al., 2006
Talinum cuneifolium	Leaf	2000	Sekhar et al., 2007
Pteris calomelanos	Frond	16,415	Kachenko et al., 2007
Brassica rapa L. var. pervirdis	Shoot	90.3	Shaibur and Kawai, 2009
Brassica rapa L. var. pervirdis	Root	4840	Shaibur and Kawai, 2009
Brassica juncea	Shoot	322.1	Khan et al., 2009
Brassica juncea	Root	210.7	Khan et al., 2009
Jussieua repens	Shoot	46.50	Gani Molla et al., 2010
Echinochloa cruss-galli	Root	61.25	Gani Molla et al., 2010
Echinochloa cruss-galli	Shoot	67.82	Gani Molla et al., 2010
Xanthium italicum	Root	40.69	Gani Molla et al., 2010

TABLE 16.9 Identified Naturally Grown Arsenic Hyperaccumulating Plants with Arsenic Concentrations in Different Plant Parts—cont'd

Plant Species	Organs	Quantity of As $(mg\,kg^{-1})$	References
Xanthium italicum	Shoot	23.16	Gani Molla et al., 2010
Cynodon dactylon	Shoot	67.82	Gani Molla et al., 2010
Azolla sp	Shoot	27.65	Gani Molla et al., 2010
Oryza sativa	Root	12.10	Gani Molla et al., 2010
Oryza sativa	Shoot	9.43	Gani Molla et al., 2010
Pistia stratiotes	Root	12.00	Gani Molla et al., 2010
Pistia stratiotes	Shoot	33.70	Gani Molla et al., 2010
Monochoria hastata	Root	40.78	Gani Molla et al., 2010
Colocasia esculenta	Root	42.08	Gani Molla et al., 2010
Pteris longifolia	Root	34.47	Gani Molla et al., 2010
Pteris longifolia	Shoot	32.14	Gani Molla et al., 2010
Cyperus rotundus	Root	11.14	Gani Molla et al., 2010
Cyperus rotundus	Shoot	32.49	Gani Molla et al., 2010
Eichhornia crassipes	Root	67.82	Gani Molla et al., 2010
Alternanthera philoxeroides	Shoot	67.30	Gani Molla et al., 2010
Nasturtium officinale	Leaf	1012	Ozturk et al., 2010
Echinochloa crus-galli	Root	56.93	Islam et al., 2013

Continued

TABLE 16.9 Identified Naturally Grown Arsenic Hyperaccumulating Plants with Arsenic Concentrations in Different Plant Parts—cont'd

Plant Species	Organs	Quantity of As (mg kg^{-1})	References
Echinochloa crus-galli	Shoot	26.50	Islam et al., 2013
Eichhornia crassipes	Root	67.9	Islam et al., 2013
Eichhornia crassipes	Shoot	46.83	Islam et al., 2013
Scirpus maritimus	Shoot	65.25	Afrous et al., 2011
Hydrilla verticillata	Shoot	121–231	Srivastava et al., 2011
Azolla caroliniana	Shoot	284	Rahman and Hasegawa, 2011
Spirodela polyrhiza	Shoot	400–900	Zhang et al., 2011
Atriplex atacamensis	Shoot	400	Vromman et al., 2011
Atriplex atacamensis	Root	3500	Vromman et al., 2011
Ceratophyllum demersum	Shoot	862	Xue et al., 2012
Leucaena leucocephala	Root	1.88	Talukdar, 2013a
Leucaena leucocephala	Stem	67.8	Talukdar, 2013a
Leucaena leucocephala	Leaf	3.51	Talukdar, 2013a
Leucaena leucocephala	Pod	45.39	Talukdar, 2013a
Leucaena leucocephala	Seed	2.17	Talukdar, 2013a
Leucaena esculenta	Root	36.47	Talukdar, 2013a
Leucaena esculenta	Stem	0.63	Talukdar, 2013a
Leucaena esculenta	Leaf	9.37	Talukdar, 2013a
Leucaena esculenta	Pod	2.37	Talukdar, 2013a
Leucaena esculenta	Seed	1.00	Talukdar, 2013a

FIGURE 16.17   Some As hyperaccumulator ferns widely grown in As-contaminated soils of Bangladesh. (a) fronds of *Pteris vittata;* (b) whole plant of *Pteris cretica*; (c) frond of *Pteris cretica*; (d) whole plant; (e) frond and (f) root mass of *Nephrodium molle* (Dryopteridaceae). For a colour version of this figure, please see the section at the end of this book. *Photo credit: Nafia Jahan Rashmi.*

In As-hyperaccumulating species, such as Chinese brake fern (*Pteris vittata*), As is rapidly transported as AsO_3^{3-} through the xylem to the fronds rather than being immobilized in the rhizoids (Su et al., 2008). There are many reports on the potential use of *Pteris* spp. in remediating As-contaminated sites. Ma et al. (2001) first reported the As-hyperaccumulator *P. vittata* that accumulated As in shoots up to 22,000 mg kg^{-1} (Huang et al., 2004). Later, Salido et al. (2003) observed that that *P. vittata* could accumulate a huge amount of As in the above-ground plant tissue as much as 200-fold higher than most other plant species examined in As-contaminated soil. Hossain et al. (2006) compared the ability of two Bangladeshi indigenous ferns, namely, *P. vittata* and *Nephrodium molle*, for their hyperaccumulation of As when grown in 0–100 mg As kg^{-1} soil. It was observed that that *P. vittata* could remove more than 95% of the soil As, whereas the removal by *N. molle* was about 68% (Hossain et al., 2006).

Some ferns such as *Salvinia natans* (Mukherjee and Kumar, 2005) and *P. vittata* (Fitz et al., 2003; Embrick et al., 2005; Pickering et al., 2006; Baldwin and Butcher, 2007; Wei et al., 2007) showed enhanced capacity to accumulate As. Huang et al. (2004) reported that *P. vittata* suspended in water with 20–500 µg l^{-1} of A(V) could remove 98.6% of 200 µg/l^{-1} of As within 24 h. Luongo

FIGURE 16.18 Some As-hyperaccumulators in terrestrial and aquatic higher plants. (a) *Colocasia esculenta*; (b). *Azolla* sp., (c) *Pistia stratiotes*; (d) *Hydrilla verticillata*; (e) *Monochoria hastata*; (f) *Eichhornia crassipes*; (g) *Echinochloa cruss-galli*; (h) *Cyperus rotundus*; (i) *Lemna minor*; (j) *Xanthium italicum*; (k) *Typha latifolia*; (l) *Leucaena leucocephala*. For a colour version of this figure, please see the section at the end of this book.

and Ma (2005) attempted to understand the underlying mechanisms of As hyperaccumulation in *P. vittata* comparing with other non-*Pteris* ferns in terms of As hyperaccumulation properties. Seven *Pteris* ferns (*P. vittata*, *P. cretica rowerii*, *P. cretica parkeri*, *P. cretica albo lineata*, *P. quadriavrita*, *P. ensiformis* and *P. dentata*) and six non-*Pteris* (*Arachniodes simplicior*, *Didymochlaena truncatula*, *Dryopteris atrata*, *Dryopteris erythrosora*, *Cyrtomium falcatum* and *Adiantum hispidulum*) ferns were analyzed after exposing them hydroponically to various levels of As ($10 \, mg \, l^{-1}$ $NaH_2AsO_4.H_2O$, 14 days). They concluded that *Pteris* ferns were more efficient in As accumulation than the non-*Pteris* ferns, where *P. vittata* performed the best by accumulating 1748 and 503 $mg \, kg^{-1}$, As in fronds and roots, respectively. Bondada et al. (2004) investigated the proficiency of *P. vittata* L. in absorbing As in the form of sprays. They observed a variability of As absorption by *P. vittata* depending on the frond age. The young fronds with immature sori absorbed more As ($3100 \, mg \, kg^{-1}$) than the fertile mature fronds ($890 \, mg \, kg^{-1}$). Baldwin and Butcher (2007) observed the phytofiltration capacity of two As hyperaccumulators, *P. cretica* cv Mayii (Moonlight fern) and *P. vittata*. Aresnic was shown to preferentially accumulate in the leaves and stems of *P. cretica* cv. Mayii compared to roots. In another field study by Salido et al.

(2003) on an As-contaminated (orchard) site, *P. vittata* was reported to accumulate As in the fronds efficiently, with As concentrations of up to 2740 mg As kg^{-1} dry weight. Conversely, in a field study in Australia, Reichmann et al. (2004) indicated that the ability of *P. vittata* to accumulate As from an As-contaminated soil decreased after subsequent frond harvests; however, fern species could not regrow well in the field which indicates their inefficiency for practical applications for the phytoremediation of As in Australian conditions. In a copper-chromium arsenate (CCA)-contaminated site of the United States, Kertulis–Tartar et al. (2006) reported that *P. vittata* reduced the mean total As content in the surface (0–15 cm) soil from 190 to 140 mg kg^{-1} over 2 years and it was estimated that *P. vittata* would take 8 years to reduce the total As content in soil below 3.7 mg kg^{-1}. In China, a field survey was conducted by Wang et al. (2006) investigating 24 fern species where fronds of four species (*Pteris vittata, P. multifida, P. cretica* var. *nervosa* and *P. oshimensis*) accumulated up to 9677 (range 57–9677), 4056 (624–4056), 2363 (1162–2363) and 2142 (301–2142) mg As kg^{-1}, respectively. In their report, Sridokchan et al. (2005) reported that As-tolerant ferns (*Adiantum capillus-veneris, Pteris cretica* var. albolineata, *Pteris cretica* var. wimsetti and *Pteris umbrosa*) were from the Pteridaceae family. *P. cretica* and *P. umbrosa* accumulated the majority of As in their fronds (up to 3090 mg As kg^{-1} dry weight) compared to the roots (up to 760 mg As kg^{-1}) and these ferns may be useful for phytoremediating As-contaminated sites because of their ability to hyperaccumulate and tolerate high As levels. Srivastava et al. (2006) showed that *P. cretica, Pteris biaurita* L., *P. quadriaurita* Retz and *P. ryukyuensis Tagawa* could be used as new hyperaccumulators of As because the average As concentration ranged from 1770 to 3650 mg kg^{-1} dry weight in the fronds and 182–507 mg kg^{-1} dry weight in the roots of *P. cretica, P. biaurita, P. quadriaurita* and *P. ryukyuensis* when grown in soil with 100 mg As kg^{-1}. It is not always true that a species must be As-tolerant to be a hyperaccumulator. They first reported that ferns that are sensitive to As are As hyperaccumulators. For instance, *Asplenium australasicum* and *A. bulbiferum* could hyperaccumulate As up to 1240 and 2630 mg As kg^{-1} dry weight, respectively, in their fronds after 7 days at 100 mg As l^{-1} (Sridokchan et al. 2005). However, not all members of the *Pteris* genus are able to hyperaccumulate As. For example, Meharg (2003) found that *Pteris tremula* and *Pteris stramina* do not hyperaccumulate As. To date, the only non-*Pteris* fern to exhibit this ability is *P. calomelanos* (Francesconi et al., 2002).

In Bangladesh, marigold (*Tagetes patula*) and ornamental arum (*Syngonuma* sp.) were also found to have the capacity to accumulate large amount of As which made them good candidates for phytoremediation of As-contaminated soil (Huq et al., 2005). Japanese mustard spinach (*Brassica rapa* L. var. *pervirdis*) exposed to 67 µM As in a hydroponic system could accumulate 90.3 mg kg^{-1} As in shoots; however, in roots, the As concentration was much higher, that is, 4840 mg kg^{-1} without showing visible toxicity symptoms (Shaibur and Kawai, 2009). Khan et al. (2009) exposed Indian mustard to 25 µM of As for 96 h and observed that root tissues accumulated $211\,mg\,As\,kg^{-1}$ dry weight while As accumulation in the shoot was $322\,mg\,As\,kg^{-1}$ dry weight. Alvarado et al. (2008) monitored the removal of As by water hyacinth (*Eichhornia crassipes*) and lesser duckweed (*Lemna minor*) grown in a site having $0.15\,mg\,l^{-1}$ As and with plant densities of $1\,kg\,m^{-2}$ for lesser duckweed and $4\,kg\,m^{-2}$ for water hyacinth (wet basis). Although no significant difference was observed in the bioaccumulation capability of these species, the removal rate of As was significant which accounted for $600\,mg\,As\,ha^{-1}$ per day and $140\,mg\,As\,ha^{-1}$ per day by *Eichhornia crassipes and L. minor*, respectively, with a removal recovery of 18 and 5%. Ozturk et al. (2010) grew watercress (*Nasturtium officinale*) in 1, 3, 5, 10 and $50\,\mu M$ AsO_3^{3-} and found that the leaves accumulated $1012\,mg\,kg^{-1}$ dry weight of AsO_3^{3-} after 7 days of exposure. However, growth was negatively affected by higher doses of As. In another experiment conducted in a desert area, Vromman et al. (2011) found that *Atriplex atacamensis* exposed to elevated As, could accumulate up to $400\,mg\,kg^{-1}$ dry weight in the shoots and $3500\,mg\,kg^{-1}$ dry weight in the roots which may act as a promising plant species that can be tested in field trials for its phytoremediation of As contamination. Islam et al. (2013) investigated the phytoaccumulation potential of *E. crassipes* (water hyacinth), *Echinochloa crus-galli* (barnyard grass) and *Monochoria hastata* (water taro) in crop land soils contaminated by As naturally and artificially. In artificially As-contaminated soils, the highest As concentration was recorded in water hyacinth (68 and $47\,mg\,kg^{-1}$ root and shoot, respectively) followed by water taro and barnyard grass at $100\,mg\,As\,kg^{-1}$ treated soil. In naturally As-contaminated soils, the highest accumulation of As in barnyard grass (57 and $27\,mg\,kg^{-1}$ root and shoot, respectively) followed by water taro and water hyacinth in *Paranpur* soils ($116\,mg\,As\,kg^{-1}$ soil). Talukdar (2013a) investigated two *Leucaena* species for their As phytoaccumulation capacity in soil. They observed that between the two species, *Leucaena leucocephala* shoots accumulated significantly higher amounts of soil As compared to *L. esculenta*. *Leucaena leucocephala* grown in $100\,mg\,kg^{-1}$ As accumulated 1.88, 67.88, 3.51, 45.99 and $2.17\,mg\,kg^{-1}$ of As in their root, stem, leaf, pod and seed, respectively; while *L. esculenta* trapped most of the added As in their roots and could accumulate it in the above-ground parts. This high bioaccumulation was due to efficient translocation of As from roots to above-ground parts (Talukdar, 2013b).

   The bioconcentration factors and translocation factors (TF) of plants are important for them to be potential candidates in phytoremediation, and these vary among the plant species (Table 16.10; Dabrowska et al., 2012). Bioconcentration is the ratio of the As concentration in plants to the concentration in soil, while the TF, also termed as shoot : root quotient is the ratio of the As concentration in roots to the concentration in shoots (Baker and Whiting, 2002; Mudhoo et al., 2011) expressed as As concentration in the shoot in relation to As concentration in the root. Arsenic-tolerant plants translocate less than 2.8% As to the parts above-ground leading to low accumulation. For example, hyperaccumulators had values about two magnitudes higher, for example, 23.86 for *P. vittata* (Ma et al., 2001), and 90.91 for *Pityrogramma calomelanos* (Franseconi et al., 2002).

TABLE 16.10 Some As-Tolerant Plant Species Having Different Values of Bioconcentration Factor (BCF) and Translocation Factor (TF) in Soil and Sediments

Plant Species	As Treatment in Soil (mg kg^{-1})	As Treatment in Water (mg l^{-1})	BCF (roots)	BCF (shoots)	TF	References
P. vittata	500	—	—	210	4.5	Ma et al., 2001
P. vittata	500	—	—	2.3	—	Ma et al., 2001
P. vittata	38.9	—	0.8	0.7	—	Tu and Ma, 2002
Leymus cinereus	50	—	24.07	2.23	0.083	Knudson et al., 2003
P. vittata	—	10	66	309	4.82	Fayiga et al., 2005
P. vittata	—	10	71	211	2.97	Faisal and Hasnain, 2005
Pteris cretica	—	10	54.3	98	1.81	Faisal and Hasnain, 2005
Pteris cretica	—	10	107	402	4.80	Faisal and Hasnain, 2005
Talinum cuneifolium	2000	—	0.13	0.82	6.3	Sekhar et al., 2007
Salvinia natans	—	0.132	—	60	—	Rahman et al., 2008
Najas marina	10.6	—	1.5	0.045	0.0301	Mazej and Germ, 2009

The use of algae in bioremediation of As has been reported in a number of studies (Megharaja et al., 2003; Shamsuddoha et al., 2006). It has been observed that algae can hyperaccumulate As from water (Huq, et al., 2005; Shamsuddoha et al., 2006). Huq et al. (2007) reported that the presence of algae in the growth medium reduces As accumulation in rice and the algae *Gloeotrichia, Oscillatoria, Chlamydomona, Lyngbya, Chlorella. Nitzschi* and *Navicula* were found to accumulate considerabe amounts of As in rice fields. Several research results suggested that some As-resistant bacteria may be applied to the bioremediation of As-contaminated sites. Interest in using soil bacteria in bioremediation of As has increased over the years, and the continuous advancements in molecular as well as biochemical techniques have helped to identify and characterize these microbes, to play a major role in understanding the underlying mechanisms of these processes (Bachate et al., 2009; Hossain, 2012). Bachate et al. (2009) isolated a total of 21 As-resistant bacteria in agricultural soils of Bangladesh and most of them were found to be capable of reducing As(V) to As(III) leading to the detoxification of As under anaerobic conditions.

In recent years, the use of arbuscular mycorrhizal (AM) fungi to alleviate metal toxicity has come into focus (Long et al., 2010). The role of mycorrhiza in the growth of plants and in reducing As toxicity in plants has been demonstrated by Ahmed et al. (2006). They examined the effects of As and inoculation with an AM fungus, *Glomus mosseae,* on *Lens culinaris* and it was observed that, besides improvements in growth and yield of the crop, *mycorrhizal inoculation* significantly reduced As concentration in roots and shoots. Jankong and Visoottiviseth (2008) reported that the effects of AM on reducing As toxicity vary with the host plant species. While studying the effects of AM fungi (*Glomus mosseae, G. intraradices* and *G. etunicatum*) on biomass production and As accumulation in *Pityrogramma calomelanos, Tagetes erecta* and *Melastomama labathricum*, they observed a great variability in the effects of AM fungi on phytoremediation of As by different plant species. For *P. calomelanos* and *T. erecta*, AM fungi reduced only As accumulation in plants, but had no significant effect on plant growth. In contrast, AM fungi improved growth and As accumulation in *M. malabathricum*. According to Cozzolino et al. (2010), with phosphorus addition, inoculation with commercial inoculum of AM fungi reduced As toxicity which was due to the improvement of phosphorus nutrition and inhibiting As accumulation in plants. Christophersen et al. (2009) revealed a strong interaction between As exposure and colonization of AM fungi retarded the As uptake which is lacking in non-mycorrhizal plants.

An agricultural field is a complex system where a number of factors such as crop plants, weeds, microbes, management activities and environmental factors including climate interact with each other (Hossain, 2012). Success of the application of a biological agent for bioremediation will depend on the interactions of all these factors. Therefore, it is important to have a detailed knowledge of the effects of all these factors before starting a bioremediation program (Hossain, 2012).

Presumed Phytochelatin-Mediated Detoxification of As Toxicity

To avoid the toxic effects of metals and metalloids plants have developed mechanisms to deactivate and scavenge metal ions that penetrate into the cytosol (Lyubenova and Schröder, 2010). Synthesis of PCs is one of them. Phytochelatins are a family of small enzymatically synthesized peptides having a general structure related to GSH which is (γ-Glu-Cys)n-Gly, where n ranges between 2 and 11. Thus, PCs constitute a number of structural species with increasing repetitions of γ-Glu-Cys units. These peptides are rapidly synthesized in response to toxic levels of heavy metals in all tested plants (Hasanuzzaman and Fujita, 2012). PCs are considered to be part of the detoxifying mechanism of higher plants because the immobilized metals are less toxic than the free ions (Zenk, 1996). Numerous plant studies showed that GSH (γ-GluCys-Gly, GSH) acts as the substrate for synthesis of PCs. Here, an enzyme that catalyzes the formation of PCs from GSH is PCS whose activity has dependency on metal ions. Chelation of toxic metal ions, for example, by newly synthesized PCs, inactivates PC synthase, thereby providing an easy method to regulate the synthesis of PCs and providing a constant protective mechanism against metal toxicity (Hasanuzzaman and Fujita, 2012; Figure 16.19).

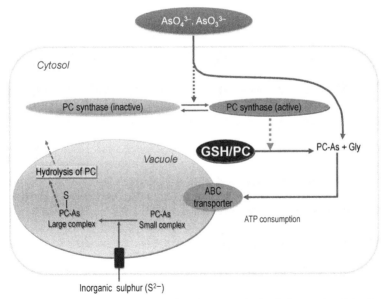

FIGURE 16.19   Control of PC biosynthesis and presumed subcellular transfer of As. Arsenic ions penetrate into the cell and activate the synthesis of PCs. Phytochelatins are synthesized from GSH where inorganic S may play an important role. The As–PC complex is actively transported into the vacuole. The metal is stored there in a different form (i.e. complexed with organic acids), while the PCs are degraded and recycled to the cytosol. *Modified from Hasanuzzaman and Fujita (2012).*

Although As is not a true metal, many studies revealed the vital role of PCs in both constitutive and adaptive tolerances to As. Schulz et al. (2008) investigated six non-hyperaccumulating plant species (*Agropyron repens, Glechoma hederacea, Leonurus marrubiatrum, Lolium perenne, Urtica dioica* and *Zea mays*) and observed the role of PCs in the detoxification of As. They found that long-term (5-week) PC synthesis was positively, but non-linearly correlated with As, which proved that probably not all As is bound by PC. In the more tolerant grasses, *A. repens* and *L. perenne*, it was chiefly the dithiol PCs which were observed. Conversely, the dominant PC species found in the less tolerant plants *V. dicica, G. herderaceae, L. marrubiastrum* and *Z. mays* was PCs, while PC2 and PC3 were detected as well (Schulz et al., 2008). Tolerance to As is enhanced by increased thiol synthesis in transgenic plants overexpressing the genes of thiol (cysteine, GSH or PC) biosynthesis pathway (Tripathi et al., 2007; Gasic and Korban 2007; Wojas et al., 2008, 2010). An As-accumulating, hypertolerant *Brassica, Isatis cappadocica* (Karimi et al., 2009), collected from Iranian As-contaminated mine had PC-based tolerance (>50% As complexed with PCs) rather than through suppression of high-affinity PO_4^{3-}/AsO_4^{3-} root transport. These findings provide conclusive evidence that thiols, particularly PCs, play a crucial role in As detoxification. Although the essentiality of PCs in both constitutive and adaptive tolerance to As is reported (Schat et al., 2002), PCs do not contribute significantly towards As tolerance phenotype in hypertolerant *(H. lanatus* and *Silene paradoxa)* and hyperaccumulator plants (*P. vittata* and *P. cretica*) because of lower synthesis of PCs in these cases (Raab et al., 2004; Arnetoli et al., 2008).

Schmöger et al. (2000) reported that both As(III) and As(V) anions efficiently induce the biosynthesis of PC in vivo and in vitro. The rapid induction of the metal-binding PCs has been observed in cell suspension cultures of *Rauvolfia serpentina*, in seedlings of *Arabidopsis* and in enzyme preparations of *Silene vulgaris* upon challenge to arsenicals. Wojasa et al. (2010) determined overexpression of either of two phytochelatin synthase genes, *AtPCS1* and *CePCS* in *Nicotiana tabacum* with respect to As tolerance and accumulation, and observed how responses relate to non-protein thiol metabolism. The expression of both genes resulted in an increase in As tolerance which was higher at *CePCS*. At the less toxic 50 µM As(V), *AtPCS1* and *CePCS* transformants accumulated more As in roots and leaves than wild type. An increase in PC production and the level of PC2 species were detected in leaves of *AtPCS1* and *CePCS* plants, which might explain their enhanced As accumulation and tolerance. Caonversely, at the highly toxic 200 µM As(V), several disturbances in −SH metabolism of PCS-overexpressing plants were found, unexpectedly, including a decrease in PC levels both in roots and leaves of transgenic plants compared to wild type. Importantly, the plants overexpressing *AtPCS1* and *CePCS* showed increased tolerance to As, making these genes potential candidates for plant engineering for phytoremediation.

Role of Biotechnology in Remediation of Arsenic Toxicity

Biotechnology has changed our perception in almost every field of life. From food security to remediating hazardous elements from the environment, it has proved its mettle. Bioremediation, a cost effective and nature-friendly biotechnology that is powered by microbial enzymes, is effectively being used to remove hazardous materials from soil and water. Since heavy metals are ubiquitous in the environment, microorganisms have developed mechanisms to resist the toxic effects of these metals (White and Gadd, 1986). A large number of microorganisms are capable of growing in the presence of high concentrations of heavy metals including As (Anderson and Cook, 2004; Rehman et al., 2007). Many strains of *Aeromonas, Exiguobacterium, Acinetobacter, Bacillus* and *Pseudomonas* can tolerate high concentrations of As species (up to 100 mM arsenate or up to 20 mM arsenit). Several Bacteria belonging to the genera *Acidithiobacillus, Bacillus, Deinococcus, Desulfitobacterium* and *Pseudomonas* (Suresh et al., 2004; Ahmed and Rehman, 2009) have also been reported to be resistant to As. The information on the mechanisms of bioremediation-related enzymes such as oxido-reductases and hydrolases has been extensively studied.

Arsenic is also phytoremediated by genetically modified plants such as *Arabidopsis thaliana* which expresses two bacterial genes. One of these genes allows the plant to modify arsenate into arsenite, and the second one binds the modified arsenite and stores it in the vacuoles (Leung, 2004).

Anawar (2006) reported that the plant species *Spergula arvensis, Phalasis caerulescens* and *Junsus effuses* can be used to partly remove the bioavailable toxic As, or transgenic plants may be transplanted via *Escherichia coli* genes for the phytoremediation of As-contaminated soils. These hyperaccumulating fern species available in the tropics may be commercially used in south-east Asia as a form of bioremediation technology.

CONCLUSION AND FUTURE PERSPECTIVES

There is growing apprehension regarding As toxicity because of its severe adverse effects on both plants and animals, including humans. Because of the entrance of As into the food chain from different routes, it has become a great concern (Benbrahim-Tallaa and Waalkes, 2008). In recent years, considerable research has been carried out into understanding the physiological and molecular mechanisms of As uptake and transport in plants. However, the exact molecular basis of As hyperaccumulation by plants is still unknown. Therefore, future research should focus on the molecular approaches and biotechnological tools and, perhaps, transgenic plants with increased resistance and uptake. Adaption of multiple approaches is important for managing As toxicity. Phytoremediation can be considered as an effective technique to clean up soil by growing As-hyperaccumulating plants in As-toxic soil. Phytochelatin formation within the plants' cells is, thus, one of the basic mechanisms and is very important for the accumulation of As within the cell. Development of plant species having higher phytochelatin synthase activity or capacity to accumulate higher amounts of As by phytochelain formation, or development of plant species having thiol metabolism through transgenic or biotechnological approaches, might be good tools for reducing As toxicity (Wojasa et al., 2010). Polyomic approaches like proteomics, genomics, micromics, transcriptomics, metablomics, inomics, metallomics, etc. have emerged as powerful tools for understanding the mechanisms of plant responses under different stresses (Hasanuzzaman et al., 2013), and similar approaches might be considered in As stress. In relation to this, research on As transportation and safe sequestration in the cell are vital. Identification and characterization of microbes degrading As or its compounds in the soil is important (Hasanuzzaman and Fujita, 2012). In fact, the aptitude of existing plant species or microbes, especially those grown in As-toxic environments, should be studied in detail; their metabolic profiles should be explored. Suitability of different agronomic practices should be studied, selected and applied considering soil, water and environmental status of a certain area where presence and types of biota including human, animal, plant and microbial, are important considerations. Planning on combined utilization of plants, microbes and agronomic practices with a view to remediating the dangerous As toxicity is vital, and cooperation among plant physiologists, microbiologists, agronomists and environmentalists is necessary, not to mention the importance of the involvement of government or non-government organizations in this combined project.

ACKNOWLEDGEMENTS

We wish to thank Mr Md. Mahabub Alam and Mr Anisur Rahman, Laboratory of Plant Stress Responses, Faculty of Agriculture, Kagawa University, Japan for their vital assistance during the preparation of the manuscript. We also express our sincere thanks to Shamim Hasan Mondal, Nafia Jahan Rashmi, Md. Moniruzzaman Majumder, Rizwana Khondoker Aurin, Mr Md. Shashwata Mahboob and Mr Ashraful Islam Siam of Sher-e-Bangla Agricultural University, Dhaka and Mr S.M. Farhad Sazib, Mymensingh Medical College, Mymensingh, Bangladesh for their editing of some images and collecting information on arsenic. We thankfully acknowledge Prof. Dr Kamal Uddin Ahamed, Department of Agricultural Botany, Sher-e-Bangla Agricultural University, Dhaka, Bangladesh for his continuous encouragement and constructive suggestion during the manuscript preparation. Thanks are also due to Mr Md. Iqbal Hosen, Key Laboratory of Biodiversity and Biogeography, Kunming Institute of Botany, Chinese Academy of Sciences, Kunming 650201, Yunnan, China for providing several supporting articles and books regarding arsenic. We apologize to all researchers for those parts of their work that were not cited in the manuscripts because of the page limitation.

REFERENCES

Abercrombie, J.M., Halfhill, M.D., Ranjan, P., Rao, M.R., Saxton, A.M., Yuan, J.S., Stewart Jr, C.N., 2008. Transcriptional responses of *Arabidopsis thaliana* plants to As(V) stress. BMC Plant Biol. 8, 87.

Adriano, D.C., 2001. Trace Elements in Terrestrial Environments: Biogeochemistry, Bioavailability and Risks of Metals, second edn. Springer, New York.

Afrous, A., Manshouri, M., Liaghat, A., Pazira, E., Sedghi, H., 2011. Mercury and arsenic accumulation by three species of aquatic plants in Dezful, Iran. Afr. J. Agric. Res. 6, 5391–5397.

Ahmed, F.R.S., Killham, K., Alexander, I., 2006. Influences of arbuscular mycorrhizal fungus *Glomus mosseae* on growth and nutrition of lentil irrigated with arsenic contaminated water. Plant Soil 258, 33–41.

Ahmed, A., Rehman, A., 2009. Isolation of Cr(VI) reducing bacteria from industrial effluents and their potential use in bioremediation of chromium containing wastewater. J. Environ. Sci. 21, 814–820.

Ahsan, N., Lee, D.G., Alam, I., Kim, P.J., Lee, J.J., Ahn, Y.O., Kwak, S.S., Lee, I.J., Bahk, J.D., Kang, K.Y., Renaut, J., Komatsu, S., Lee, B.H., 2008. Comparative proteomic study of arsenic-induced differentially expressed proteins in rice roots reveals glutathione plays a central role during As stress. Proteomics 8, 3561–3576.

Ahsan, N., Lee, D.G., Kim, K.H., Alam, I., Lee, S.H., Lee, K.W., Lee, H., Lee, B.H., 2010. Analysis of arsenic stress-induced differentially expressed proteins in rice leaves by two-dimensional gel electrophoresis coupled with mass spectrometry. Chemosphere 78, 224–231.

Alam, M.B., Sattar, M.A., 2000. Assessment of arsenic contamination in soils and waters in some areas of Bangladesh. Water Sci. Technol. 42, 185–192.

Ali, M.A., Badruzzaman, A.B.M., Jalil, M.A., Hossain, M.D., Ahmed, M.F., Masud, A.A., Kamruzzaman, M., Rahman, M.A., 2003. Arsenic in plant and soil environment of Bangladesh. In: Feroze, A.M., Ashraf, A.M., Adeel, Z. (Eds.), Fate of Arsenic in the Environment. Bangladesh University of Engineering and Technology, Dhaka, Bangladesh, and United Nations University, Tokyo, pp. 85–112.

Alvarado, S., Guedez, M., Lue-Meru, M.P., Nelson, G., Alvaro, A., Jesus, A.C., Gyula, Z., 2008. Arsenic removal from waters by bioremediation with the aquatic plants water hyacinth (*Eichhornia crassipes*) and lesser duckweed (*Lemna minor*). Bioresour. Technol. 99, 8436–8440.

Anderson, C.R., Cook, G.M., 2004. Isolation and characterization of arsenate-reducing bacteria from arsenic contaminated sites in New Zealand. Curr. Microbiol. 48, 341–347.

Anawar, H.M., 2006. Exposure and bioavailability of arsenic in contaminated soils from the La Parilla Mine, Spain. Environ. Geol. 50, 170–179.

Araújo, W.L., Tohge, T., Ishizaki, K., Leaver, C.J., Fernie, A.R., 2011. Protein degradation—an alternative respiratory substrate for stressed plants. Trends Plant Sci. 16, 489–498.

Arnetoli, M., Vooijs, R., ten Bookum, W., Galardi, F., Gonnelli, C., Gabbrielli, R., Schat, H., Verkleij, J.A.C., 2008. Arsenate tolerance in *Silene paradoxa* does not rely on phytochelatin-dependent sequestration. Environ. Pollut. 152, 585–591.

Ashraf, M., 2009. Biotechnological approach of improving plant salt tolerance using antioxidants as markers. Biotechnol. Adv. 27, 84–93.

Avron, M., Jagendorf, A., 1959. Evidence concerning the mechanism of adenosine triphosphate formation by spinach chloroplasts. J. Biol. Chem. 234, 967–972.

Azad, M.A.K., Mondal, A.H.M.F.K., Hossain, M.I., Moniruzzaman, M., 2012. Effect of Arsenic Amended Irrigation Water on Growth and Yield of BR-11 Rice (*Oryza sativa* L.) Grown in Open Field Gangetic Soil Condition in Rajshahi. J. Environ. Sci. Nat. Resour. 5, 55–59.

Bachate, S.P., Cavalca, L., Andreoni, V., 2009. Arsenic resistant bacteria isolated from agricultural soils of Bangladesh and characterization of arsenate-reducing strains. J. Appl. Microbiol. 107, 145–156.

Baker, A.J.M., Whiting, S.N., 2002. In search of the holy grail—a further step in understanding metal hyperaccumulation? New Phytol. 155, 1–7.

Baldwin, P.R., Butcher, D.J., 2007. Phytoremediation of arsenic by two hyperaccumulators in a hydroponic environment. Microchem. J. 85, 297–300.

Banglapedia, 2013. Arsenic. In: Banglapedia – The National Encyclopedia of Bangladesh. Asiatic Society of Bangladesh, Dhaka.

Ben Hassine, A., Ghanem, M.E., Bouzid, S., Lutts, S., 2008. An inland and a coastal population of the Mediterranean xero-halophyte species *Atriplex halimus* L. differ in their ability to accumulate proline and glycinebetaine in response to salinity and water stress. J. Exp. Bot. 59, 1315–1326.

Benbrahim-Tallaa, L., Waalkes, M.P., 2008. Inorganic Arsenic and Human Prostate Cancer. Environ. Health Perspect. 116, 158–164.

BGS, 1999. British Geological Survey. Groundwater studies for arsenic contamination in Bangladesh: Final report. V. I. Summary. Department of Public Health Engineering, Government of Bangladesh, Dhaka. p. 21.

BGS / DPHE (British Geological Survey / Department of Public Health Engineering [Bangladesh]), 2001. In: Kinniburgh, D.G., Smedley, P.L. (Eds.), Arsenic Contamination of Groundwater in Bangladesh. Report WC/00/19, British Geological Survey, Keyworth.

Bock, R., 2007. Structure, function, and inheritance of plastid genomes. In: Bock, R. (Ed.), Cell and Molecular Biology of Plastids, Topics in Current Genetics, vol. 19. Springer, Berlin, pp. 29–62.

Bolan, N.S., Thiyagarajan, S., 2001. Retention and plant availability of chromium in soils as affected by lime and organic amendments. Aust. J. Soil Res. 39, 1091–1103.

Bondada, B.R., Tu, S., Ma, L.Q., 2004. Absorption of foliar-applied arsenic by the arsenic hyperaccumulating fern (*Pteris vittata* L.). Sci. Total Environ. 332, 61–70.

Brammer, H., 2007. Arsenic Accumulation in Irrigated Paddy Soils and Possible Mitigation Methods. Proceedings of a Symposium: Arsenic—The Geography of a Global problem, Royal Geographical Society (RGS) in London.

Brammer, H., 2009. Mitigation of arsenic contamination in irrigated paddy soils in South and South-east Asia. Environ. Int. 35, 856–863.

Brammer, H., Ravenscroft, P., 2009. Arsenic in groundwater: A threat to sustainable agriculture in South and South-east Asia. Environ. Int. 35, 647–654.

Bravin, M.N., Travassac, F., Le Floch, M., Hinsinger, P., Garnier, J.M., 2008. Oxygen input controls the spatial and temporal dynamics of arsenic at the surface of a flooded paddy soil and in the rhizosphere of lowland rice (*Oryza sativa* L.): A microcosm study. Plant Soil 312, 207–218.

Briat, J.F., 2010. Arsenic tolerance in plants: "Pas de deux" between phytochelatin synthesis and ABCC vacuolar transporters. Proc. Natl. Acad. Sci. USA. 107, 20853–20854.

BRRI, 2013. Bangladesh Rice Research Institute. www.brri.gov.bd.

Callender, E., 2003. Heavy metals in the environment—Historical tends. Treat. Geochem. 9, 67–105.

Carbonell, A.A., Aarabi, M.A., Delaune, R.D., Gambrell, R.P., Patrick Jr., W.H., 1998. Arsenic in wetland vegetation: Availability, phytotoxicity, uptake and effects on plant growth and nutrition. Sci. Total. Environ. 217, 189.

Chakrabarty, D., Trivedi, P.K., Misra, P., Tiwari, M., Shri, M., Shukla, D., Kumar, S., Rai, A., Pandey, A., Nigam, D., Tripathi, R.D., Tuli, R., 2009. Comparative transcriptome analysis of arsenate and arsenite stresses in rice seedlings. Chemosphere 74, 688–702.

Chaney, R.L., Malik, M., Li, Y.M., Brown, S.L., Brewer, E.P., Angle, J.S., Baker, A.J., 1997. Phytoremediation of metals. Curr. Opin. Biotechnol. 8, 279–284.

Chatterjee, A., Mukherjee, A., 1999. Hydrogeological investigation of ground water arsenic contamination in South Calcutta. Sci. Total Environ. 225, 249–262.

Chen, T., Yan, X., Liao, X., Xiao, X., Huang, Z., Xie, H., Zhai, L., 2005. Subcellular distribution and compartmentalization of arsenic in *Pteris vittata* L. Chin. Sci. Bull. 50, 2843–2849.

Chen, T.B., Huang, Z.C., Huang, Y.Y., Xie, H., Liao, X.Y., 2003. Cellular distribution of arsenic and other elements in hyperaccumulator *Pteris nervosa* and their relations to arsenic accumulation. Chin. Sci. Bull. 48, 1586–1591.

Choudhury, Q.R.M.D., Islam, T.S., Alam, R., Sen, R., Hasan, J.G.M., Chowdhury, I.A.M.D., 2009. Effect of arsenic contaminated irrigation water on the cultivation of red amaranth. Am-Eur J. Sci. Res. 4, 14–19.

Christophersen, H.M., Smith, F.A., Smith, S.E., 2009. Arbuscularmycorrhizal colonization reduces arsenate uptake in barley via downregulation of transporters in the direct epidermal phosphate uptake pathway. New Phytol. 184, 962–974.

Chun-xi, L., Shu-li, F., Yun, S., Li-na, J., Xu-yang, L., Xiao-li, H., 2007. Effects of arsenic on seed germination and physiological activities of wheat seedlings. J. Environ. Sci. 19, 725–732.

Collins, A., 2001. Carotenoids and genomic stability. Mutat. Res. 475, 1–28.

Cozzolino, V., Pigna, M., Di Meo, V., Caporale, A.G., Violante, A., 2010. Effects of arbuscular mycorrhizal inoculation and phosphorus supply on the growth of Lactuca sativa L. and arsenic and phosphorus availability in an arsenic polluted soil under nonsterile conditions. Appl. Soil Ecol. 45, 262–268.

Cullen, W.R., 2008. Is arsenic an aphrodisiac? The Sociochemistry of an Element. Royal Society of Chemistry, Cambridge, U.K.

Czech, V., Czovek, P., Fodor, J., Boka, K., Fodor, F., Cseh, E., 2008. Investigation of arsenate phytotoxicity in cucumber plants. Acta Biol. Szeged 52, 79–80.

Dabrowska, B.B., Vithanage, M., Gunaratna, K.R., Mukherjee, A.B., Bhattacharya, P., 2012. Bioremediation of arsenic in contaminated terrestrial and aquatic environments. In: Lichtfouse, E., Schwarzbauer, J., Robert, D. (Eds.). Environmental Chemistry for a Sustainable World: Remediation of Air and Water Pollution, vol. 2. Springer, New York, pp. 475–509.

Dahal, B.M., Fuerhacker, M., Mentler, A., Karki, K.B., Shrestha, R.R., Blum, W.E.H., 2008. Arsenic contamination of soils and agricultural plants through irrigation water in Nepal. Environ. Pollut. 155, 157–163.

Das, H.K., Mitra, A.K., Sengupta, P.K., Hossain, A., Islam, F., Rabbani, G.H., 2004. Arsenic concentrations in rice, vegetables, and fish in Bangladesh: A preliminary study. Environ. Int. 30, 383–387.

Dave, R., Tripathi, R.D., Dwivedi, S., Tripathi, P., Dixit, G., Sharma, Y.K., Trivedi, P.K., Corpas, F.J., Barroso, J.B., Chakrabarty, D., 2012. Arsenate and arsenite exposure modulate antioxidants and amino acids in contrasting arsenic accumulating rice (Oryza sativa L.) genotypes. J. Hazard. Mater. http://dx.doi.org/10.1016/j.jhazmat.

Delgado, M., Bigeriego, M., Guardiola, E., 1993. Uptake of Zn, Cr and Cd by water hyacinth. Water Res. 27, 269–272.

Del-Toro-Sánchez, C.L., Zurita, F., Gutiérrez-Lomelí, M., Solis-Sánchez, B., Wence-Chávez, L., Rodríguez-Sahagún, A., Castellanos-Hernández, O.A., Vázquez-Armenta, G., Siller-López, F., 2013. Modulation of antioxidant defense system after long term arsenic exposure in Zantedeschia aethiopica and Anemopsis californica. Ecotoxicol. Environ. Saf. 94, 67–72.

DPHE, 2013. Department of Public Health and Engineering, Government of the People's Republic of Bangladesh. www.dphe.gov.bd.

Duman, F., Ozturk, F., Aydin, Z., 2010. Biological responses of duckweed (Lemna minor L.) exposed to the inorganic arsenic species As(III) and As(V): Effects of concentration and duration of exposure. Ecotoxicol. 19, 983–993.

Duquesnoy, I., Champeau, G.M., Evray, G., Ledoigt, G., Piquet-Pissaloux, A., 2010. Enzymatic adaptations to arsenic-induced oxidative stress in Zea mays and genotoxic effect of arsenic in root tips of Vicia faba and Zea mays. C. R. Biol. 333, 814–824.

Duxbury, J.M., Zavala, Y.Z., 2005. What are safe levels of arsenic in food and soils? In Behavior of Arsenic in Aquifers, Soils and Plants: Implications for Management. International symposium held in Dhaka, Bangladesh. Jan. 16–18, 2005. Organized by CIMMYT, CU, TAMU, USGS, and GSB.

Duxbury, J.M., Mayer, A.B., Lauren, J.G., Hassan, N., 2003. Food chain aspects of arsenic con-tamination in Bangladesh: Effects on quality and productivity of rice. J. Environ. Sci. Health 38, 61–69.

Eisler, R., 2004. Arsenic hazards to humans, plants, and animals from gold mining. Rev. Environ. Contam. Toxicol. 180, 133–165.

Embrick, L.L., Porter, K.M., Pendergrass, A., Butcher, D.J., 2005. Characterization of lead and arsenic contamination at Barber Orchard, Haywood County, NC. Microchem. J. 81, 117–121.

Faisal, M., Hasnain, 2005. Chromate resistant *Bacillus cereus* auguments sunflower growth by reduc-ing toxicity of Cr(VI). J Plant Biol. 48, 187–194.

FAO (Food and Agriculture Organization), 2006. Arsenic Contamination of Irrigation Water, Soil and Crops in Bangladesh: Risk Implications for Sustainable Agriculture and Food Safety in Asia. RAP Publication 2006/20. FAO Regional Office for Asia and the Pacific, Bangkok, Thailand. 38 pp.

Fayiga, A.O., Ma, L.Q., Santoa, J., Rathinasabapathi, B., Stamps, B., Littell, R.C., 2005. Effects of arsenic species and concentrations on arsenic accumulation by different fern species in a hydroponic system. Int. J. Phytoremed. 7, 231–240.

Fazal, M.A., Kawachi, T., Ichion, E., 2001a. Validity of the latest research fi ndings on causes of groundwater arsenic contamination in Bangladesh. Water Int. 26, 380–389.

Fazal, M.A., Kawachi, T., Ichion, E., 2001b. Extent and severity of groundwater arsenic contamina-tion in Bangladesh. Water Int. 26, 370–379.

Finnegan, P.M., Chen, W., 2012. Arsenic Toxicity: The Effects on Plant Metabolism. Front Physiol. 182, 1–18.

Fitz, W.J., Wenzel, W.W., Zhang, H., et al., 2003. Rhizosphere characteristics of the arsenic hyper-accumulator *Pteris vittata* L. and monitoring of phytoremoval efficiency. Environ. Sci. Technol. 37 (21), 5008–5014.

Franseconi, K., Visootiviseth, P., Sridokchan, W., Goessler, W., 2002. Arsenic species in an arsenic hyperaccumulating fern, *Pityrogramma calomonanos*: A potential phytoremediator of arsenic-contaminated soils. Sci. Total Environ. 284, 27–35.

Gani Molla, M.O., Islam, M.A., Hasanuzzaman, M., 2010. Identification of arsenic hyperaccumu-lating plants for the development of phytomitigation technology. J. Phytol. 2, 41–48.

Garbisu, C., Hernandez-Allica, J., Barrutia, O., Alkorta, I., Becerril, J.M., 2002. Phytoremediation: A technology using green plants to remove contaminants from polluted areas. Rev. Environ. Health 17, 75–90.

Garg, N., Singla, P., 2011. Arsenic toxicity in crop plants: Physiological effects and tolerance mech-anisms. Environ. Chem. Lett., 9, 303–321.

Garg, N., Singla, P., 2012. The role of *Glomus mosseae* on key physiological and biochemical parameters of pea plants grown in arsenic contaminated soil. Sci. Hortic. 143, 92–101.

Gasic, K., Korban, S.S., 2007. Transgenic Indian mustard (*Brassica juncea*) plants expressing an *Arabidopsis* phytochelatin synthase (*AtPCS1*) exhibit enhanced As and Cd tolerance. Plant Mol. Biol. 64, 361–369.

Geng, C.N., Zhu, Y.G., Tong, Y.P., Smith, S.E., Smith, F.A., 2006. Arsenate (As) uptake by and distribution in two cultivars of winter wheat (*Triticum aestivum* L.). Chemosphere, 62, 608–615.

Ghanem, M.E., van Elteren, J., Albacete, A., Quinet, M., Martinéz-Andújar, C., Kinet, J.M., Perez-Alfocea, F., Lutts, S., 2009. Impact of salinity on early reproductive physiology of tomato (*Solanum lycopersicum*) in relation to a heterogeneous distribution of toxic ions in flowers organs. Funct. Plant Biol. 36, 125–136.

Gill, S.S., Tuteja, N., 2010. Reactive oxygen species and antioxidant machinery in abiotic stress tolerance in crop plants. Plant Physiol. Biochem. 48, 909–930.

Goessler, W., Kuehnelt, D., 2002. Analytical methods for the determination of arsenic and arsenic compounds in the environment. In: Frankenberger Jr., W.T. (Ed.), "Environmental chemistry of arsenic". Marcel Dekker, New York, USA, pp. 27–50.

Gumaelius, L., Lahner, B., Salt, D.E., Banks, J.A., 2004. Arsenic hyperaccumulation in gametophytes of *Pteris vittata*. A new model system for analysis of arsenic hyperaccumulation. Plant Physiol. 136, 3198–3208.

Gunes, A., Pilbeam, D.J., Inal, A., 2009. Effect of arsenic–phosphorus interaction on arsenic-induced oxidative stress in chickpea plants. Plant Soil 314, 211–220.

Guo, W., Zhang, J., Teng, M., Li, H., Wang, J., 2009. Arsenic uptake is suppressed in a rice mutant defective in silicon uptake. J. Plant Nutr. Soil Sci. 172, 867–874.

Gupta, D.K., Inouhe, M., Rodríguez-Serrano, M., Romero-Puertas, M.C., Sandalio, L.M., 2013. Oxidative stress and arsenic toxicity: Role of NADPH oxidases. Chemosphere 90, 1987–1996.

Gupta, M., Sharma, P., Sarin, N.B., Sinha, A.K., 2009. Differential response of arsenic stress in two varieties of *Brassica juncea* L. Chemosphere 74, 1201–1208.

Gusman, G.S., Oliveira, J.A., Farnese, F.S., Cambraia, J., 2013. Arsenate and arsenite: The toxic effects on photosynthesis and growth of lettuce plants. Acta Physiol. Plant 35, 1201–1209.

Harvey, C.F., Swartz, C.H., Badruzzaman, A.B.M., et al., 2005. Groundwater arsenic contamination on the Ganges delta: Biogeochemistry, hydrology, human perturbations, and human suffering on a large scale. C.R. Geosci. 337, 285–296.

Hasanuzzaman, M., Fujita, M., 2012. Heavy metals in the environment: Current status, toxic effects on plants and possible phytoremediation. In: Anjum, N.A., Pereira, M.A., Ahmad, I., Duarte, A.C., Umar, S., Khan, N.A. (Eds.), Phytotechnologies: Remediation of environmental contaminants. Taylor and Francis/CRC Press, USA, pp. 7–73.

Hasanuzzaman, M., Fujita, M., 2013. Exogenous sodium nitroprusside alleviates arsenic-induced oxidative stress in wheat (*Triticum aestivum* L.) seedlings by enhancing antioxidant defense and glyoxalase system. Ecotoxicol. 22, 584–596.

Hasanuzzaman, M., Nahar, K., Fujita, M., Ahmad, P., Chandna, R., Prasad, M.N.V., Ozturk, M., 2013. Enhancing plant productivity under salt stress: Relevance of poly-omics. In: Ahmad, P., Azooz, M., Prasad, M.N.V. (Eds.), Salt Stress in Plants. Signalling, Omics and Adaptations, Springer, New York, pp. 113–156.

Health, M.R., 2013. Maine Rural Health Association 2013 Outstanding Service Award. Available at http://www.maineruralhealth.org/award.htm (accessed 10.22.13).

Hironaka, H., Ahmad, S.A., 2003. Arsenic concentration of rice in Bangladesh. In: Ahmed, M.F. (Ed.), International Symposium on Fate of Arsenic in the Environment. BUET, Dhaka, Bangladesh, pp. 123–130.

Hossain, A., Joardar, J.C., Imamul Huq, S.M., 2006. Comparison of arsenic accumulation by two ferns: *Pteris vittata* and *Nephrodium molle*. Dhaka Univ. J. Biol. Sci. 15, 95–103.

Hossain, M.Z., 2012. Bioremediation of arsenic: Prospects and limitations in the agriculture of Bangladesh. Int. J. Econ. Env. Geol. 3, 6–14.

Huang, J.W., Poynton, C.Y., Kochian, L.V., Elless, M.P., 2004. Phytofiltration of arsenic from drinking water using arsenic-hyperaccumulating ferns. Environ. Sci. Technol. 38, 3412–3417.

Hughes, M.F., Beck, B.D., Chen, Y., Lewis, A.S., Thomas, D.J., 2011. Arsenic Exposure and Toxicology: A Historical Perspective. Toxicol. Sci. 123, 305–332.

Huq, S.M.I., 2008. Fate of arsenic in irrigation water and its potential impact on the food chain. In: Ahuja, S. (Ed.), Arsenic Contamination of Groundwater: Mechanism, Analysis, and Remediation. Wiley, New Jersey, pp. 23–49.

Huq, S.M.I., Naidu, R., 2003. Arsenic in groundwater of Bangladesh: Contamination in the food chain. In: Ahmed, M.F. (Ed.), Arsenic Contamination: Bangladesh Perspective. ITN-Bangladesh, Bangladesh University of Engineering and Technology, Dhaka, Bangladesh, pp. 203–226.

Huq, S.M.I., Abdullah, M.B., Joardar, J.C., 2007. Bioremediation of arsenic toxicity by algae in rice culture. Land Contam. Reclam. 15, 327–333.

Huq, S.M.I., Shila, U.K., Joardar, J.C., 2006. Arsenic mitigation strategy for rice, using water regime management. Land Contam. Reclam. 14, 805–814.

Huq, S.M.I., Rahman, A., Sultana, S., Naidu, R., 2003. Extent and severity of arsenic contamination in soils of Bangladesh. In: Ahmed, F., Ali, M.A., Adeal, Z. (Eds.), Fate of Arsenic in the Environment. Preprints of BUET-UNU International Symposium, Dhaka, Bangladesh, pp. 69–84.

Huq, S.M.I., Bulbul, A., Choudhury, M.S., Alam, S., Kawai, S., 2005. Arsenic bioaccumulation in a green alga and its subsequent recycling in soil of Bangladesh. In: Bundschuh, J., Bhattacharya, P. (Eds.), Proceeding of Natural Arsenic in Groundwater: Occurrence, Remediation and Management. Balkema, New York, USA, pp. 119–124.

Irgolic, K.J., Greschonig, H., Howard, A.G., 1995. Arsenic. In: Townshend, A. (Ed.), The Encyclopedia of analytical Science, 1995. Academic Press, pp. 168–184.

Islam, M.R., Jahiruddin, M., 2010. Effects of arsenic and its interaction with phosphorus on yield and arsenic accumulation in rice. 19th World Congress of Soil Science, Soil Solutions for a Changing World. 1–6 August 2010. Brisbane, Australia. Published on DVD.

Islam, M.R., Jahiruddin, M., Rahman, G.K.M.M., Miah, M.A.M., Farid, A.T.M., Panaullah, G.M., Loeppert, R.H., Duxbury, J.M., Meisner, C.A., 2005. Arsenic in paddy soils of Bangladesh: Levels, distribution and contribution of irrigation and sediments. In: Behavior of Arsenic in Aquifers, Soils and Plants (Conference Proceedings). Dhaka.

Islam, M.S., Wahid-Uz-Zaman, M., Rahman, M., 2013. Phytoaccumulation of arsenic from arsenic contaminated soils by *Eichhornia crassipes* L., *Echinochloa crus-galli* L. and *Monochoria hastata* L. Bangladesh. Int. J. Environ. Protec. 3, 17–27.

Islam, S.M.A., Fukushi, K., Yamamoto, K., 2004. Severity of arsenic concentration in soil and arsenic-rich sludge of Bangladesh and potential of their biological removal: A novel approach for tropical region. Proc. Second Intl. Symp. Southeast Asian Water Environ. Hanoi, Vietnam.

Jain, M., Gadre, R.P., 2004. Inhibition of 5-amino levulinic acid dehydratase activity by arsenic in excised etiolated maize leaf segments during greening. J. Plant Physiol. 161, 251–255.

Jankong, P., Visoottiviseth, P., 2008. Effects of arbuscular mycorrhizal inoculation on plants growing on arsenic contaminated soil. Chemosphere 72, 1092–1097.

Jin, J.W., Xu, Y.F., Huang, Y.F., 2010. Protective effect of nitric oxide against arsenic-induced oxidative damage in tall fescue leaves. Afr. J. Biotechnol. 9, 1619–1627.

Juzl, M., Stefl, M., 2002. The effect of leaf area index on potatoes in soils contaminated by some heavy metals. Rostlinna Vyroba 48, 298–306.

Kachenko, A.G., Bhatia, N.P., Singh, B., Siegele, R., 2007. Arsenic hyperaccumulation and localization in the pinnule and stipe tissues of the gold-dust fern (*Pityrogramma calomelanos*) (L.) Link var. austroamericana (Domin) Farw. using quantitative micro-PIXE spectroscopy. Plant and Soil 300, 207–219.

Kachout, S.S., Mansoura, A.B., Leclerc, J.C., Mechergui, R., Rejeb, M.N., Ouerghi, Z., 2010. Effects of heavy metals on antioxidant activities of: *Atriplex hortensis* and *A. rosea*. EJEAFChe 9, 444–457.

Karimi, N., Ghaderian, S.M., Raab, A., Feldmann, J., Meharg, A.A., 2009. An arsenic accumulating, hyper tolerant brassica, Isatis cappadocica. New Phytol. 184, 41–47.

Kertulis-Tartar, G.M., Rathinasabapathi, B., Maa, L.Q., 2009. Characterization of glutathione reductase and catalase in the fronds of two *Pteris ferns* upon arsenic exposure. Plant Physiol. Biochem. 47, 960–965.

Kertulis-Tartar, G.M., Ma, L.Q., Tu, C., Chirenje, T., 2006. Phytoremediation of an arsenic contaminated site using *Pteris vittata* L.: A two-year study. Int. J. Phytorem. 8, 311–322.

Khan, I., Ahmad, A., Iqbal, M., 2009. Modulation of antioxidant defence system for arsenic detoxification in Indian mustard. Ecotoxicol. Environ. Saf. 72, 626–634.

Klump, S., Kipfer, R., Cirpka, O.A., et al., 2006. Groundwater dynamics and arsenic mobilization in Bangladesh assessed using noble gases and tritium. Environ. Sci. Technol. 40, 243–250.

Knudson, J.A., Meikle, T., DeLuca, T.H., 2003. Role of mycorrhizal fungi and phosphorus in the arsenic tolerance of basin wildrye. J. Environ. Qual. 32, 2001–2006.

Krämer, U., 2005. Phytoremediation: Novel approaches to cleaning up polluted soils. Curr. Opin. Biotechnol. 16, 133–141.

Kvesitadze, G., Khatisashvili, G., Sadunishvili, T., Ramsden, J.J. (Eds.), 2006. Biochemical Mechanisms of Detoxification in Higher Plants. Springer, Berlin.

Laizu, J., 2007. Speciation of arsenic in vegetables and their correlation with inorganic phosphate level. Bangladesh. J. Pharmacol. 2, 88–94.

Lal, N., 2010. Molecular mechanisms and genetic basis of heavy metal toxicity and tolerance in plants. In: Ashraf, M., Ozturk, M., Ahmad, M.S.A. (Eds.), Plant Adaptation and Phytoremediation. Springer, Dordrecht, pp. 35–58.

Lei, M., Wan, X.M., Huang, Z.C., Chen, T.B., Li, X.W., Liu, Y.R., 2012. First evidence on different transportation modes of arsenic and phosphorus in arsenic hyperaccumulator. *Pteris vittata*. Environ. Pollut. 161, 1–7.

Leterrier, M., Airaki, M., Palma, J.M., Chaki, M., Barroso, J.B., Corpas, F.J., 2012. Arsenic triggers the nitric oxide (NO) and S-nitrosoglutathione (GSNO) metabolism in *Arabidopsis*. Environ. Pollut. 166, 136–143.

Li, C., Feng, S., Shao, Y., Jiang, L., Lu, X., Hou, X., 2007. Effect of arsenic on seed germination and physiological activities of wheat seedlings. J. Environ. Sci. 19, 725–732.

Li, X.W., Lei, M., Chen, T.B., Wan, X.M., 2009. Roles of Sulfur in the Arsenic Tolerant Plant *Adiantum capillus-veneris* and the Hyperaccumulator *Pteris vittata*. J. Korean Soc. Appl. Biol. Chem. 52, 498–502.

Liao, X.Y., Chen, T.B., Xie, H., Liu, Y.R., 2005. Soil As contamination and its risk assessment in areas near the industrial districts of Chenzhou city, southern China. Environ. Intl. 31, 791–798.

Liu, Q.J., Zheng, C.N., Hu, C.X., Tan, Q.L., Sun, X.C., Su, J.J., 2012. Effects of high concentrations of soil arsenic on the growth of winter wheat (*Triticum aestivum* L) and rape (*Brassica napus*). Plant Soil Environ. 58, 22–27.

Long, L.K., Yao, Q., Guo, J., Yang, R.H., Huang, Y.H., Zhu, H.H., 2010. Molecular community analysis of arbuscular mycorrhizal fungi associated with five selectged plant species form heavy metal polluted soils. Eur. J. Soil Biol. 46, 288–294.

Low, K.S., Lee, C.K., Tai, C.H., 1994. Biosorption of copper by water hyacinth roots. J. Environ. Sci. Health 29, 171.

Leung, M., 2004. Bioremediation: Techniques for cleaning up a mess. J. Biotech. 2, 18–22.

Luongo, T., Ma, L.Q., 2005. Characteristics of arsenic accumulation by *Pteris* and non-*Pteris ferns*. Plant Soil 277, 117–126.

Lyubenova, L., Schröder, P., 2010. Uptake and effect of heavy metals on the plant detoxification cascade in the presence and absence of organic pollutants. In: Sherameti, I., Varma, A. (Eds.), Soil Heavy Metals. Soil Biol, 9. Springer, Berlin Heidelberg, pp. 65–85.

Ma, L.Q., Komar, K.M.M., Tu, C., Zhang, W., Cai, Y., Kenence, E.D., 2001. A fern that hyperaccumulates As. Nature Lond 409, 579.

Ma, J.F., Yamaji, N., Mitani, N., Xu, X.Y., Su, Y.H., McGrath, S.P., Zhao, F.J., 2008. Transporters of arsenite in rice and their role in arsenic accumulation in rice grain. Proc. Natl. Acad. Sci. USA 105, 9931–9935.

Mahimairaja, S., Bolan, N.S., Adriano, D.C., Robinson, B., 2005. Arsenic contamination and its risk management in complex environmental settings. Adv. Agron. 86, 1–82.

Mahmud, R., Inoue, N., Kasajima, S., Shaheen, R., 2007. Effect of Soil Arsenic on Yield and As and P Distribution Pattern among Plant Organs of Buckwheat and Castor Oil Plant. Proceedings of the 10th International Symposium on Buckwheat. Section D Physiology and Cultivation.

Malik, J.A., Goel, S., Kaur, N., Sharma, S., Singh, I., Nayyar, H., 2012. Selenium antagonises the toxic effects of arsenic on mungbean (*Phaseolus aureus* Roxb.) plants by restricting its uptake and enhancing the antioxidative and detoxification mechanisms. Environ. Exp. Bot. 77, 242–248.

Mallick, S., Sinam, G., Sinha, S., 2011. Study on arsenate tolerant and sensitive cultivars of *Zea mays* L. Differential detoxification mechanism and effect on nutrients status. Ecotoxicol. Environ. Saf. 74, 1316–1324.

Mandal, B.K., Suzuki, K.T., 2002. Arsenic round the world: A review. Talanta 58, 201–235.

Marques, I.A., Anderson, L.E., 1986. Effects of arsenite, sulfite, and sulfate on photosynthetic carbon metabolism in isolated pea (*Pisum sativum* L., cv Little Marvel) chloroplasts. Plant Physiol. 82, 488–493.

Mateos-Naranjo, E., Andrades-Moreno, L., Redondo-Gómez, S., 2012. Tolerance to and accumulation of arsenic in the cordgrass *Spartina densiflora* Brongn. Bioresour. Technol. 104, 187–194.

Matschullat, J., Perobelli Borba, R., Deschamps, E., Figueiredo, B.R., Gabrio, T., Schwenk, M., 2000. Human and environmental contamination in the Iron Quadrangle, Brazil. Appl. Geochem. 15, 181–190.

Mazej, Z., Germ, M., 2009. Trace element accumulation and distribution in four aquatic macrophytes. Chemosphere 74, 642–647.

Megharaja, M., Ragusa, S.R., Naidu, R., 2003. Metal–algae interactions: Implication of bioavailability. In: Naidu, R., Gupta, V.V.V.S.R., Rogers, S., Kookana, R.S., Bolan, N.S., Adriano, D.C. (Eds.), Bioavailability, Toxicity and Risk Relationships in Ecosystems. Science Publishers, Enfield, New Hampshire, UK, pp. 109–144.

Meharg, A., Jardine, L., 2003. Arsenite transport into paddy rice (*Oryza sativa*) roots. New Phytol. 157, 39–44.

Meharg, A.A., Hartley-Whitaker, J., 2002. Arsenic uptake and metabolism in arsenic resistant and non-resistant plant species. Tansley Rev. New Phytol. 154, 29–43.

Meharg, A.A., Macnair, M.R., 1992. Suppression of the high affinity phosphate uptake system a mechanism of arsenate tolerance in *Holcus lanatus* L. J. Exp. Bot. 43, 519–524.

Meharg, A.A., Rahman, M.M., 2003. Arsenic contamination of Bangladesh paddy field soils: Implications for rice contribution to arsenic consumption. Environ. Sci. Technol. 37, 229–234.

Mishra, S., Srivastava, S., Tripathi, R.D., Dwivedi, S., Trivedi, P.K., 2008. Thiol metabolism and antioxidant systems complement each other during arsenate detoxification in *Ceratophyllum demersum* L. Aquat. Toxicol. 86, 205–215.

Miteva, E., Peycheva, S., 1999. Arsenic accumulation and effect on peroxidase activity in green bean and tomatoes. Bulg. J. Agric. Sci. 5, 737–740.

Mittler, R., Vanderauwera, S., Gollery, M., Van Breusegem, F., 2004. Reactive oxygen gene network of plants. Trends Plant Sci. 9, 490–498.

Mkandawire, M., Dudel, G., 2005. Accumulation of arsenic in *Lemna gibba* (duckweed) in tailing waters of two abandoned uranium mining sites in Saxony, Germany. Sci. Total Environ. 336, 81–89.

Mokgalaka-Matlala, N.S., Flores-Tavizön, E., Castillo-Michel, H., Peralta-Videa, J.R., Gardea-Torresdey, J.L., 2008. Toxicity of arsenic (III) and (V) on plant growth, element uptake, and total amylolytic activity of mesquite (*Prosopis juliflora × P. velutina*). Int. J. Phytoremed 10, 47–60.

Mokgalaka-Matlala, N.S., Flores-Tavizo'n, E., Castillo-Michel, H., Peralta-Videa, J.R., Gardea-Torresdey, J.L., 2009. Arsenic tolerance in mesquite (*Prosopis* sp.): Low molecular weight thiols synthesis and glutathione activity in response to arsenic. Plant Physiol. Biochem. 47, 822–826.

Møller, I.M., Jensen, P.E., Hansson, A., 2007. Oxidative modifications to cellular components in plants. Annu. Rev. Plant Biol. 58, 459–481.

Mudhoo, A., Sharma, S.K., Lin, Z.Q., Dhankher, O.P., 2011. Phytoremediation of Arsenic-Contaminated Environment: An Overview. In: Sharma, S.K., Mudhoo, A. (Eds.), Green Chemistry for Environmental Sustainability. CRC Press, Boca Raton, pp. 127–149.

Mukherjee, S., Kumar, S., 2005. Adsorptive uptake of arsenic (V) from water by aquatic fern *Salvinia natans*. J. Water Supply: Res. Technol. – AQUA 54, 47–53.

Mylona, P.V., Polidoros, A.N., Scandalios, J.G., 1998. Modulation of antioxidant responses by arsenic in maize. Free Radic. Biol. Med. 25, 576–585.

Nagy, M.L., Johansen, J.R., Clair St, L.L., Webb, B.L., 2005. Recovery patterns of microbiotic soil crusts, 70 years after arsenic contamination. J. Arid Environ. 63, 304–323.

Namgay, T., Singh, B., Singh, B.P., 2010. Plant availability of arsenic and cadmium as influenced by biochar application to soil. World Congress of Soil Science, Soil Solutions for a Changing World. 1 – 6 August 2010. Brisbane, Australia. Published on DVD.

NAS, 2000. National Academy of Sciences, Medical and Biological Effects of Environmental Pollutants. 2000. Arsenic. Natl. Acad. Sci, Washington, DC. USA. 24p.

Norton, G.J., Lou-Hing, D.E., Meharg, A.A., Price, A.H., 2008. Rice– arsenate interactions in hydroponics: Whole genome transcriptional analysis. J. Exp. Bot. 59, 2267–2276.

Nriagu, J.O., Pacyna, J.M., 1988. Quantitative assessment of worldwide contamination of air, water, and soils by trace metals. Nature 333, 134–139.

Nwugo, C.C., Huerta, A.J., 2008. Silicon-induced cadmium resistance in rice (*Oryza sativa*). J. Plant Nutr. Soil Sci. 171, 841–848.

O'Day, P.A., 2006. Chemistry and mineralogy of arsenic. Elements 2, 77–83.

O'Neill, P., 1995. Arsenic. In: Alloway, B.J. (Ed.), Heavy Metals in Soil. Blackie Academic and Professional, Glasgow.

Ozturk, F., Duman, F., Leblebici, Z., Temizgul, R., 2010. Arsenic accumulation and biological responses of watercress (*Nasturtium officinale* R. Br.) exposed to arsenite. Environ. Exp. Bot. 69, 167–174.

Pacyna, J.M., Pacyna, E.G., 2001. An assessment of global and regional emissions of trace metals to the atmosphere from anthropogenic sources worldwide. Environ. Rev. 9, 269–298.

Panstar-Kallio, M., Manninen, P.K.G., 1997. Speciation of mobile arsenic in soil samples as a function of pH. Sci. Total Environ. 204, 193–200.

Patel, H.V., Parmar, S.R., Chudasama, C.J., Mangrola, A.V., 2013. Interactive studies of zinc with cadmium and arsenic on seed germination and antioxidant properties of *Phaseolus aureus* Roxb. Int. J. Plant Anim. Environ. Sci. 3, 166–174.

Peryea, F.J., 1998a. Historical use of lead arsenate insecticides, resulting soil contamination and implications for soil remediation. In: Proceedings, 16th World Congress of Soil Science, 20–26 August 1998. Montpellier, France, p. 7. Available at http://soils.tfrec.wsu.edu/leadhistory.htm.

Peryea, F.J., 1998b. Phosphate starter fertilizer temporarily enhances soil arsenic uptake by apple trees grown under field conditions. Hort. Sci. 33, 826–829.

Pickering, I.J., Prince, R.C., George, M.J., Smith, R.D., George, G.N., Salt, D.E., 2000. Reduction and Coordination of Arsenic in Indian Mustard. Plant Physiol. 122, 1171–1178.

Pickering, I.J., Gumaelius, L., Harris, H.H., Prince, R.C., Hirsch, G., Banks, J.A., Salt, D.E., George, G.N., 2006. Localizing the biochemical transformations of arsenate in a hyperaccumulating fern. Environ. Sci. Technol. 40, 5010–5014.

Polizzotto, M.L., Kocar, B.D., Benner, S.G., Sampson, M., Fendorf, S., 2008. Near-surface wetland sediments as a source of arsenic release to ground water in Asia. Nature 454, 505–508.

Purakayastha, T.J., Chhonkar, P.K., 2010. Phytoremediation of heavy metal contaminated soils. In: Sherameti, I., Varma, A. (Eds.), Soil heavy metals. Soil Biol, 19. Springer, Berlin/Heidelberg, pp. 389–429.

Raab, A., Feldmann, J., Meharg, A.A., 2004. The nature of arsenic–phytochelatin complexes in *Holcus lanatus* and *Pteris cretica*. Plant Physiol. 134, 1113–1122.

Rabbani, G.H., Saha, S.K., Marni, F., Akhtar, M., Alauddin, M., Mitra, A.K., Nasir, M., Chowdhury, A.K.A., 2002. Antioxidants in detoxification of arsenic-induced oxidative injury in rabbits. Bangladesh Arsenic Control Society, Dhaka. pp. 69–77.

Rabbi, S.M.F., Rahman, A., Islam, M.S., Kibria, Q.K., Imamul Huq, S.M., 2007. Arsenic uptake by rice (*Oryza sativa* L.) in relation to salinity and calcareousness in some soils of Bangladesh. Dhaka Univ. J. Biol. Sci. 16 (1), 29–39.

Rahman, M.A., Hasegawa, H., Rahman, M.M., Miah, M.A.M., Tasmin, A., 2007a. Arsenic accumulation in rice (*Oryza sativa* L.): Human exposure through food chain. Ecotoxicol. Environ. Saf. 69, 317–324.

Rahman, M.A., Hasegawa, H., Rahman, M.M., Islam, M.N., Miah, M.A.M., Tasmen, A., 2007b. Effect of arsenic on photosynthesis, growth and yield of five widely cultivated rice (*Oryza sativa* L.) varieties in Bangladesh. Chemosphere 67, 1072–1079.

Rahman, M.A., Hasegawa, H., Rahman, M.M., Rahman, M.A., Miah, M.A.M., 2007c. Accumulation of arsenic in tissues of rice plant (Oryza sativa L.) and its distribution in fractions of rice grain. Chemosphere 69, 942–948.

Rahman, M.A., Hasegawaa, H., Ueda, K., Makia, T., Rahman, M.M., 2008. Influence of phosphate and iron ions in selective uptake of arsenic species by water fern (*Salvinia natans* L). Chem. Eng. J. 145, 179–184.

Rahman, M.A., Hasegawa, H., 2011. Aquatic arsenic: Phytoremediation using floating macrophytes. Chemosphere 83, 633–646.

Rahman, M.A., Kadohashi, K., Maki, T., Hasegawa, H., 2011. Transport of DMAA and MMAA into rice (*Oryza sativa* L.) roots. Environ. Exp. Bot. 72, 41–46.

Raj, A., Pandey, A.K., Sharma, Y.K., Khare, P.B., Srivastava, P.K., Singh, N., 2011. Metabolic adaptation of *Pteris vittata* L. gametophyte to arsenic induced oxidative stress. Bioresour. Technol. 102, 9827–9832.

Rehman, A., Shakoori, F.R., Shakoori, A.R., 2007. Heavy metal resistant *Distigma proteus* (Euglenophyta) isolated from industrial effluents and its possible role in bioremediation of contaminated wastewaters. World J. Microbiol. Biotechnol. 23, 753–758.

Reichmann, K.G., Gravel, M.R., Burren, B.G., Mayer, D.G., Wright, C.L., 2004. Bioremediation of soil arsenic at a contaminated dip site using *Pteris vittata*. Paper presented at the Proceedings of the 25th Biennial Conference of the Australian Society of Animal Production. University of Melbourne, Victoria, Australia.

Requejo, R., Tena, M., 2005. Proteome analysis of maize roots reveals that oxidative stress is a main contributing factor to plant arsenic toxicity. Phytochemistry 66, 1519–1528.

Rofkar, R., Dwyer, D.F., 2011. Effects of light regime, temperature, and plant age on uptake of arsenic by *Spartina pectinata* and *Carex stricta*. Int. J. Phytoremed. 13, 528–537.

Ross, Z., Duxbury, J.M., Paul, D.N.R., DeGloria, S.D., 2005. Potential for arsenic contamination of rice in Bangladesh: Spatial analysis and mapping of high risk areas. In: Behavior of Arsenic in Aquifers, Soils and Plants (Conference Proceedings). Dhaka.

Roychowdhury, T., Uchino, T., Tokunaga, H., Ando, M., 2002. Survey of arsenic in food composites from arsenic affected area of West Bengal, India. Food Chem. Toxicol. 40, 1611–1621.

Sadiq, M., 1997. Arsenic chemistry in soils: An overview of thermodynamic predictions and field observations. Water Air Soil Pollut. 93, 117–136.

Salido, A., Hasty, K.L., Lim, J., Butcher, D.J., 2003. Phytoremediation of arsenic and lead in contaminated soil using Chinese brake fern (*Pteris vittata*) and Indian mustard (*Brassica juncea*). Int. J. Phytorem 5, 89–103.

Sarkar, S., Basu, B., Kundu, C.K., Patra, P.K., 2012. Deficit irrigation: An option to mitigate arsenic load of rice grain in West Bengal, India. Agric. Ecosys. Environ. 146, 47–152.

Saygideger, S., Dogan, M., Keser, G., 2004. Effect of lead and pH on lead uptake, chlorophyll and nitrogen content of *Typha latifolia* L. and *Ceratophyllum demersum* L. Int. J. Agric. Biol. 6, 168–172.

Schat, H., Llugany, M., Vooijs, R., Hartley-Whitaker, J., Bleeker, P.M., 2002. The role of phytochelatins in constitutive and adaptive heavy metal tolerances in hyperaccumulator and nonhyperaccumulator metallophytes. J. Exp. Bot. 53, 2381–2392.

Schmöger, M.E.V., Oven, M., Grill, E., 2000. Detoxification of arsenic by phytochelatins in plants. Plant Physiol. 122, 793–801.

Schulz, H., Hartling, S., Tanneberg, H., 2008. The identification and quantification of arsenic-induced phytochelatins—comparison between plants with varying As sensitivities. Plant Soil 303, 275–287.

Sekhar, K.C., Kamala, C.T., Chary, N.S., Mukherjee, A.B., 2007. Arsenic accumulation by *Talinum cuneifolium*– application for phytoremediation of arsenic-contaminated soils of Patancheru, Hyderband, India. In: Bhattacharya, P., Mukherjee, A.B., Bundschuh, J., Zevenhoven, R., Loeppert, R.H. (Eds.), Arsenic in Soil and Groundwater Environment: Biogeochemical Interactions, Health Effects and Remediation. Trace metals and other contaminants in environment, vol. 9. Elsevier, Amsterdam, pp. 315–338.

Shah, A.L., Jahiruddin, M., Rahman, M.S., Rashid, M.A., Rashid, M.H., Gani, M.A., 2004. Arsenic accumulation in rice and vegetables grown under arsenic contaminated soil and water. In: Shah, M.A.L. (Ed.), Proceedings of Workshop on Arsenic in the Food Chain: Assessment of Arsenic in the Water–Soil–Crop Systems. Bangladesh Rice Research Institute, Gazipur, Bangladesh, pp. 23–37.

Shahariar, S., Huq, S.M.I., 2012. Experiments on alleviating arsenic accumulation in rice through irrigation management. In: Lee, T.S. (Ed.), Irrigation systems and practices in challenging environments. InTech, Rijeka, pp. 193–206.

Shaibur, M.R., Kawai, S., 2009. Effect of arsenic on visible symptom and arsenic concentration in hydroponic Japanese mustard spinach. Environ. Exp. Bot. 67, 65–70.

Shaibur, M.R., Kitajima, N., Sugewara, R., Kondo, T., Alam, S., Huq, S.M.I., Kawai, S., 2008. Critical toxicity of arsenic and elemental composition of arsenic-induced chlorosis in hydroponic sorghum. Water Air Soil Pollut. 191, 279–292.

Shamsuddoha, A.S.M., Bulbul, A., Huq, S.M.I., 2006. Accumulation of arsenic in green algae and its subsequent transfer to the soil–plant system. Bangladesh J. Microbiol. 22, 48–151.

Sharma, I., Travlos, I.S., 2012. Phosphate supply as a promoter of tolerance to arsenic in pearl millet. Int. J. Plant Prod 6, 443–456.

Shri, M., Kumar, S., Chakrabarty, D., Trivedi, P.K., Mallick, S., Misra, P., Shukla, D., Mishra, S., Srivastava, S., Tripathi, R.D., Tuli, R., 2009. Effect of arsenic on growth, oxidative stress, and antioxidant system in rice seedlings. Ecotoxicol. Environ. Saf. 72, 1102–1110.

Singh, N., Ma, L.Q., Shrivastava, M., Rathinasapathi, B., 2006. Metabolic adaptation to arsenic-induced oxidative stress in *Pteris vittata* L. and *Pteris ensiformis* L. Plant Sci. 170, 274–282.

Singh, H.P., Batish, D.R., Kohali, R.K., Arora, K., 2007. Arsenic induced root growth inhibition in mung bean (*Phaseolus aureus* Roxb.) is due to oxidative stress resulting from enhanced lipid peroxidation. Plant Growth Regul. 53, 65–73.

Singh, H.P., Kaur, S., Batish, D.R., Sharma, V.P., Sharma, N., Kohli, R.K., 2009. Nitric oxide alleviates arsenic toxicity by reducing oxidative damage in the roots of *Oryza sativa* (rice). Nitric Oxide 20, 289–297.

Singh, V.P., Srivastava, P.K., Prasad, S.M., 2013. Nitric oxide alleviates arsenic-induced toxic effects in ridged Luffa seedlings. Plant Physiol. Biochem. 71, 155–163.

Sinha, S., Sinam, G., Mishra, R.K., Mallick, S., 2010. Metal accumulation, growth, antioxidants and oil yield of *Brassica juncea* L. exposed to different metals. Ecotoxicol. Environ. Saf. 73, 1352–1361.

Smedley, P.L., Kinniburgh, D.G., 2002. A review of the source, behaviour and distribution of arsenic in natural waters. Appl. Geochem. 17, 517–568.

Smedley, P., 2005. Arsenic Occurrence in Groundwater in South and East Asia: Scale, Causes and Mitigation. Towards a More Effective Operational Response: Arsenic Contamination of Groundwater in South and East Asian Countries II. Technical Report, World Bank Report No. 31303, World Bank, Washington, DC.

Smedley, P.L., Kinniburgh, D.G., Macdonald, D.M.J., Nicollib, H.B., Barrosb, A.J., Tullioc, J.O., Pearced, J.M., Alonso, M.S., 2005. Arsenic associations in sediments from the loess aquifer of La Pampa, Argentina. Appl. Geochem. 20, 989–1016.

Smith, A.H., Goycolea, M., Haque, R., Biggs, M.L., 1998. Marked increase in bladder and lung cancer mortality in a region of northern Chile due to arsenic in drinking water. Am. J. Epidemiol. 147, 660–669.

Smith, P.G., Koch, I., Reimer, K.J., 2008. Uptake, transport and transformation of arsenate in radishes (*Raphanus sativus*). Sci. Total Environ. 390, 188–197.

Sridokchan, W., Markich, S., Visoottiviseth, P., 2005. Arsenic tolerance, accumulation and elemental distribution in twelve ferns: A screening study. Aust. J. Ecotoxicol. 11, 101–110.

Srivastava, M., Ma, L.Q., Santos, J.A.G., 2006. Three new arsenic hyperaccumulating ferns. Sci. Total Environ. 364, 24–31.

Srivastava, M., Ma, L.Q., Singh, N., 2005. Antioxidant responses of hyper- accumulator and sensitive fern species to arsenic. J. Exp. Bot. 56, 335–1342.

Srivastava, S., D'Souza, S.F., 2010. Effect of variable sulfur supply on arsenic tolerance and antioxidant responses in *Hydrilla verticillata* (L.f.). Royle. Ecotoxicol. Environ. Saf. 73, 1314–1322.

Srivastava, S., Sharma, Y.K., 2013. Impact of arsenic toxicity on black gram and its amelioration using phosphate. ISRN Toxicol. http://dx.doi.org/10.1155/2013/340925.

Srivastava, S., Suprasanna, P., D'Souza, S.F., 2011. Redox state and energetic equilibrium determine the magnitude of stress in *Hydrilla verticillata* upon exposure to arsenate. Protoplasma 248, 805–815.

Srivastava, S., Srivastava, A.K., Suprasanna, P., D'Souza, S.F., 2009. Comparative biochemical and transcriptional profiling of two contrasting varieties of *Brassica juncea* L. in response to arsenic exposure reveals mechanisms of stress perception and tolerance. J. Exp.Bot. 181, 1–13.

Srivastava, S., Srivastava, A.K., Singh, B., Suprasanna, P., D'Souza, S.F., 2013. The effect of arsenic on pigment composition and photosynthesis in *Hydrilla verticillata*. Biol. Plant. 57, 385–389.

Stoeva, N., Bineva, T., 2003. Oxidative changes and photosynthesis in Oat plants grown in As-contaminated soil. Bulg. J. Plant. Physiol. 29, 87–95.

Stoeva, N., Berova, M., Zlatev, Z.L., 2005. Effect of arsenic on some physiological parameters in bean plants. Biol. Plant 49, 293–296.

Stronach, S.A., Walker, N.L., Macphee, D.E., Glasser, F.P., 1997. Reactions between cement and As (III) oxide: The system CaO–SiO$_2$–As$_2$O$_3$–H$_2$O at 25°C. Waste Manage. 17, 9–13.

Su, Y.H., McGrath, S.P., Zhu, Y.G., Zhao, F.J., 2008. Highly efficient xylem transport of arsenite in the arsenic hyperaccumulator *Pteris vittata*. New Phytol. 180, 434–441.

Sultana, R., Rahman, A., Kibria, K.Q., Islam, M.S., Haque, M.M., 2012. Effect of arsenic contaminated irrigation water on growth, yield and nutrient accumulation of *Vigna radiata*. Indian J. Innov. Dev. 1, 682–686.

Sun, Y., Li, Z., Guo, B., Chu, G., Wei, C., Liang, Y., 2008. Arsenic mitigates cadmium toxicity in rice seedlings. Environ. Exp. Bot. 64, 264–270.

Suresh, K., Reddy, G.S.N., Sengupta, S., Shivaji, S., 2004. *Deinococcus indicus* sp. nov., an arsenic-resistant bacterium from an aquifer in West Bengal, India. Int. J. Syst. Evol. Microbiol. 54, 457–461.

Suthersan, S.S., 2002. Natural and enhanced remediation system. Lewis Publishers, Boca Raton. pp. 240–268.

Talukder, A.S.M.H.M., Meisner, C.A., Sarkar, M.A.R., Islam, M.S., 2011. Effect of water management, tillage options and phosphorus status on arsenic uptake in rice. Ecotoxicol. Environ. Saf. 74, 834–839.

Talukdar, D., 2013a. Arsenic-Induced Changes in Growth and Antioxidant Metabolism of Fenugreek. Russ. J. Plant Physiol. 60, 652–660.

Talukdar, D., 2013b. Bioaccumulation, growth and antioxidant defense responses of *Leucaena* species differing in arsenic tolerance. Int. J. Bot. Res. 3, 1–18.

Tamaki, S., Frankenberger, W.T., 1992. Environmental biochemistry of arsenic. Rev. Environ. Contam. Toxicol. 124, 79–110.

Tripathi, P., Tripathi, R.D., Singh, R.P., Dwivedi, S., Goutam, D., Shri, M., Trivedi, P.K., Chakrabarty, D., 2013. Silicon mediates arsenic tolerance in rice (*Oryza sativa* L.) through lowering of arsenic uptake and improved antioxidant defence system. Ecol. Engin. 52, 96–103.

Tripathi, R.D., Srivastava, S., Mishra, S., Singh, N., Tuli, R., Gupta, D.K., Maathuis, F.J.M., 2007. Arsenic hazards: Strategies for tolerance and remediation by plants. Trends Biotechnol. 25, 158–165.

Tu, C., Ma, L.Q., 2002. Effects of arsenic concentrations and forms on arsenic uptake by the hyperaccumulator ladder brake. J. Environ. Qual. 31, 641–647.

Ullah, S.M., 1998. Arsenic contamination of groundwater and irrigated soils in Bangladesh. Proc. International Conference on Arsenic Pollution on Groundwater in Bangladesh: Causes, Effects and Remedies, Dhaka, Bangladesh.

Ullah, S.M., Hossain, M.Z., Islam, M., Jahan, S., Bashirullah, M., 2009. Extent of arsenic poisoning in the food chain of arsenic-affected areas. Dhaka Univ. J. Bio. Sci. 18, 159–171.

UNICEF, 2008. Arsenic Mitigation in Bangladesh. United Nations Children's Fund, Dhaka, Bangladesh.

USAID Bangladesh, 2003. Arsenic contamination on agricultural sustainability and food quality. Project Annual Report. CIMMYT-Bangladesh, Dhaka, Bangladesh.

Van Breusegem, F., Dat, J.F., 2006. Reactive oxygen species in plant cell death. Plant Physiol. 141, 384–390.

Vaughan, 2006. Arsenic. Elements 2, 71–75.

Verbruggen, N., Hermans, C., Schat, H., 2009. Mechanisms to cope with arsenic or cadmium excess in plants. Curr. Opin. Plant Biol. 12, 364–372.

Visoottiviseth, P., Francesconi, K., Sridokchan, W., 2002. The potential of Thai indigenous plant species for the phytoremediation of arsenic contaminated land. Environ. Pollut. 118, 453–461.

Soil Remediation and Plants

Vlyssides, A., Barampouti, E.M., Mai, S., 2005. Heavy metal removal from water resources using the aquatic plant *Apium nodiflorum*. Commun. Soil Sci. Plant Anal. 36, 1075–1081.

Vromman, D., Flores-Bavestrello, A., Šlejkovec, Z., Lapaille, S., Teixeira-Cardoso, C., Briceño, M., Kumar, M., Martínez, J.P., Lutts, S., 2011. Arsenic accumulation and distribution in relation to young seedling growth in *Atriplex atacamensis* Phil. Sci. Total Environ., 412–413. 286–295.

Vromman, D., Lutts, S., Lefèvre, I., Somer, L., De Vreese, O., Šlejkovec, Z., Quinet, M., 2013. Effects of simultaneous arsenic and iron toxicities on rice (*Oryza sativa* L.) development, yield-related parameters and As and Fe accumulation in relation to As speciation in the grains. Plant Soil 371, 199–221.

Wang, H., Ye, Z.H., Shu, W.S., Li, W.C., Wong, M.H., Lan, C.Y., 2006. Arsenic uptake and accumulation in fern species growing at arsenic-contaminated sites of Southern China: Field surveys. Int. J. Phytoremed 8, 1–11.

Wang, J., Zhao, F.J., Meharg, A.A., Raab, A., Feldmann, J., McGrath, S.P., 2002. Mechanisms of arsenic hyperaccumulation in *Pteris vittata*. Uptake kinetics, interactions with phosphate, and arsenic speciation. Plant Physiol. 130, 1552–1561.

Watling-Payne, A.S., Selwyn, M.J., 1974. Inhibition and uncoupling of photophosphorylation in isolated chloroplasts by organotin, organomercury and diphenyleneiodonium compounds. Biochem. J. 142, 65–74.

Wei, C.Y., Wang, C., Sun, X., Wang, W.Y., 2007. Arsenic accumulation by ferns: A field survey in southern China. Environ. Geochem. Health 29, 169–177.

White, C., Gadd, G., 1986. Uptake and cellular distribution of copper, cobalt and cadmium in strains of *Saccharomyces cerevisiae* cultured on elevated concentration of these metals. FEMS Microbiol. Rev. 38, 227–283.

Williams, P.N., Islam, M.R., Adomako, E.E., Raab, A., Hossain, S.A., Zhu, Y.G., Meharg, A.A., 2006. Increase in rice grain arsenic for regions of Bangladesh irrigating paddies with elevated arsenc in groundwater. Environ. Sci. Technol. 40, 4903–4908.

Windham, L., Weis, J.S., Weis, P., 2001. Lead uptake, distribution and effects in two dominant salt marsh macrophytes *Spartina alterniflora* (cordgrass) and *Phragmites australis* (common reed). Mar. Pollut. Bull. 42, 811.

Wojas, S., Clemens, S., Hennig, J., Sklodowska, A., Kopera, E., Schat, H., Bal, W., Antosiewicz, D.M., 2008. Overexpression of phytochelatin synthase in tobacco: Distinctive effects of AtPCS1 and CePCS genes on plant response to cadmium. J. Exp. Bot. 59, 2205–2219.

Wojas, S., Clemens, S., Sklodowska, A., Antosiewicz, D.M., 2010. Arsenic response of AtPCS1- and CePCS-expressing plants-effects of external As(V) concentration on As-accumulation pattern and NPT metabolism. J. Plant Physiol. 167, 169–175.

Wolterbeek, H.T., van der Meer, A.J.G.M., 2002. Transport rate of arsenic, cadmium, copper and zinc in *Potamogeton pectinatus* L.: Radiotracer experiments with ^{76}As, ^{109,115}Cd, ^{64}Cu and ^{65,69m}Zn. Sci. Total Environ. 287, 13–30.

Woolson, E.A., 1977. Generation of alkylarsines from soil. Weed Sci. 25, 412–416.

Xiao-ke, C., Hua-lin, L., Yu-bo, H., Ji-wang, Z., Peng, L., Shu-ting, D., 2012. Arsenic Distribution, Species, and Its Effect on Maize Growth Treated with Arsenate. J. Integr. Agric. 11, 416–423.

Xue, P., Yan, C., Sun, G., Luo, Z., 2012. Arsenic accumulation and speciation in the submerged macrophyte *Ceratophyllum demersum* L. Environ. Sci. Pollut. Res. Int. 19, 3969–3976.

Yamazaki, C., Ishiga, H., Ahmed, F., Itoh, K., Suyama, K., Yamamoto, H., 2003. Vertical distribution of arsenic in ganges delta sediments in Deuli Village, Bangladesh. Soil Sci. Pant. Nutr. 49, 567–574.

Ye, Z.H., Baker, A.J.M., Wong, M.H., Willis, A.J., 1997. Zinc, lead and cadmium tolerance, uptake and accumulation by *Typha latifolia*. New Phytol. 136, 469.

Zenk, M.H., 1996. Heavy metal detoxification in higher plants: A review. Gene 179, 21–30.

Zernike, K., 2003. Arsenic Case is Considered Homicide, Maine Police Say. The New York Times, New York, NY.

Zhang, W.D., Liu, D.S., Tian, J.C., He, F.L., 2009. Toxicity and accumulation of arsenic in wheat (*Triticum aestivum* L.) varieties of China. Phyton 78, 147–154.

Zhang, X., Hu, Y., Liu, Y., Chen, B., 2011. Arsenic uptake, accumulation and phytofi ltration by duckweed (*Spirodela polyrhiza* L.). J. Environ. Sci. (China) 23 (4), 601–606.

Zhao, F.J., Dunham, S.J., McGrath, S.P., 2002. Arsenic hyperaccumulation by different fern species. New Phytol. 156, 27–31.

Zhao, F.J., Ma, J.F., Meharg, A.A., McGrath, S.P., 2009. Arsenic uptake and metabolism in plants. New Phytol. 181, 777–794.

Chapter 17

Phytoremediation of Metal-Contaminated Soils Using Organic Amendments: Prospects and Challenges

Muhammad Sabir,* Muhammad Zia-ur-Rehman,*
Khalid Rehman Hakeem† and Saifullah*

*Institute of Soil and Environmental Sciences, University of Agriculture, Faisalabad, Pakistan;
†Faculty of Forestry, Universiti Putra Malaysia, Serdang, Selangor, Malaysia

BACKGROUND

Soil serves as a sink for all the pollutants present in the environment. In the era of industrial development and urbanization, the contamination of raw city effluent with toxic substances is on the rise concurrently with its increasing volume being generated on a daily basis. Among these toxic substances, heavy metals are important due to their persistence in the environment for longer periods due to their non-biodegradable nature (Wilson and Maliszewska-Kordybach, 2000). Heavy metals like iron (Fe), molybdenum (Mo), manganese (Mn), zinc (Zn), copper (Cu) and nickel (Ni) are essential for various physiological/metabolic functions in living cells while others like lead (Pb), cadmium (Cd), mercury (Hg), arsenic (As) and uranium (U) are toxic even at very low concentrations (Kadem et al., 2004; Meharg, 2005; Schützendübel and Polle, 2002; Sun et al., 2001).

Naturally, concentration of heavy metals is low in soils but different anthropogenic activities lead to increased concentrations above critical limits (Facchinelli et al., 2001; Manta et al., 2002). Heavy metals enter into the soil through pedogenic and anthropogenic sources (Park et al., 2011). Metals in the soil parent material occur in forms which are not phytoavailable due to poor solubility and thus have very little impact on biota. However, metals added to soils anthropogenically are highly mobile and thus exert adverse effects on biota (Lamb et al., 2009; Naidu, 1996). Industrial effluents, agricultural and domestic waste waters and industrial waste materials are the major anthropogenic sources of metals in soils (Adriano, 2001; Kabata-Pendias, 2001). Consumption of Pb-based petrol

Soil Remediation and Plants. http://dx.doi.org/10.1016/B978-0-12-799937-1.00017-6

and paints are other important sources of Pb in soils in urban/peri-urban areas (Mielke et al., 2006). Generally, raw city effluent is a rich source of plant nutrients owing to which, farmers preferably use it for irrigation in peri-urban areas in developing countries. Concurrently, raw city effluent contains heavy metals and hence its use creates opportunities and problems for agriculture in peri-urban areas (Murtaza et al., 2010). Presence of high contents of heavy metals in soils results in bioaccumulation of heavy metals due to enhanced absorption of heavy metals by crops (Muchuweti et al., 2006). Heavy metal accumulation in soils and plants is a serious concern as the metals enter the human body through this pathway. Plant uptake of metals from soils is affected by different physico-chemical and biological reactions occurring in soils and various plant factors (Sun et al., 2001). Phytoavailability of metals in the soil is affected by cation exchange capacity (CEC), pH, redox potential, soil texture and organic matter (OM) contents (Mellis et al., 2004). Organic matter is the most important factor that influences phytoavailability of metals due to its content and nature (Karaca, 2004b).

Various physical, chemical and biological processes are already being used for remediation of contaminated soils. These processes either remove the metals from the soil or immobilize the metals within the soil by altering the soil chemistry that ultimately sequesters or absorbs the pollutant into the matrix (Cunningham et al., 1995). The remediation technology being used at specific sites depends upon the nature of the contaminant, level of contamination and soil type. Depending upon the situation, metal-contaminated soils may be subjected to acid leaching, excavation and landfilling, physical separation of contaminant, electro-chemical processes and in situ immobilization of contaminants in the soils (Cunningham et al., 1995). Recently, development of such technologies which critically consider human health and ecological risks have been encouraged. Traditionally used remediation technologies are disruptive and costly, which may lead to further ecological disturbances and health problems in humans (Işikli et al., 2006; Zheng et al., 2010).

In situ immobilization of metals with the application of soil amendments may decrease their availability and modify their adverse effects on living organisms (van Herwijnen et al., 2007). Lime, gypsum, phosphatic fertilizers, sulphate (SO_4^{2-}) carriers and OM containing amendments are conventionally used for in situ immobilization of metals in soils. Application of organic amendments is beneficial for soil health due to improvement in soil physico-chemical properties. Phytoavailability and transport of metals in aquatic and terrestrial environments is generally controlled by adsorption reactions. Adsorption reactions are affected by soil reaction (pH), cation exchange capacity, organic matter, lime and mineral nutrients in soils, specific surface area, clay contents and back ground electrolyte (Appel and Ma, 2002; Sauve et al., 2000; Seregin and Kozhevnikova, 2006). Among these factors, OM is the most important owing to its effect on other physico-chemical and biological properties of soils. Amount and nature of OM greatly influences mobility of metals as metals form complexes and chelates of varying stabilities with OM (Zhou and Wong, 2001) thus organic amendments may enhance plant growth on metal-contaminated soil (Chamon et al., 2005b). Organic amendments also affect distribution and chemical portioning of metals (Narwal and Singh, 1998) which may influence the phytoavailability of metals. Metal–organic matter complexation can take place in solution and solid phases of soils (Silveira et al., 2003). Generally, metal-contaminated soils are amended with OM to improve fertility, physico-chemical properties and to immobilize the metals in soils (Walker et al., 2004). OM could influence metal availability in the soils through various mechanisms (Huang and Lin, 1981). Organic matter could alter the availability of metals by converting available metals into the forms bound with OM, oxides or carbonates which are immobile in soils (Walker et al., 2004) . A variety of organic materials can be used to immobilize metals in soils, including manures (farmyard, pig and poultry manure), composts of different origins, wood ash, saw dust and biosolids, sewage sludge, bark chips and wood chips (Karaca, 2004b; Narwal and Singh, 1998; Sabir et al., 2013). These amendments are effective for metal immobilization in contaminated soils (Adesodun and Mbagwu, 2008; Clemente et al., 2003; Walker et al., 2004); however, source of OM, amount and time since application could affect availability of metals in soils (Arnesen and Singh, 1998). Interaction of fresh OM with metals is different compared to decomposed OM and metals bound with fresh OM may be released upon decomposition of OM (Karaca, 2004b). Mobility of metals increased with decomposition of OM due to less OM in soils as a result of OM decomposition (Singh and Oste, 2001). Metal mobility in soil is affected by the nature and type of organic matter which may change due to long-term transformations of OM (Martínez et al., 2003a). Transformation of OM may be due to hydrolysis, oxidation or depolymerization which may result in the release of dissolved OM and increased metal availability (Martínez et al., 2003a). Conversely, during decomposition of OM, low molecular weight organic compounds are converted into high molecular weight compounds thus increasing the metal retention in the soil (Clemente et al., 2005). Decomposition of OM produces humic substances, organic acids and amino acids that could alter metal availability by forming complexes with metals (Misra and Pande, 1974b). Use of organic amendments for remediation of metal-contaminated soils is an economic approach ensuring better soil quality and fertility with an added benefit of safer environment through C sequestration. However, this approach is still not practiced on a large scale. We have reviewed the prospects of this approach by focusing on key challenges in the approach that scientists still have to resolve in the near future for adoption of this technology on a commercial basis.

SOURCES OF METALS

Metal contamination of soils is ubiquitous, thus posing a threat to all types of living organisms. Naturally, metals are low in agricultural soils and are inherited from parent material that contains metals in forms which are not toxic to biota (Park et al., 2011). Anthropogenic activities are the most important sources of metals to agricultural soils and contain metals in forms that are highly mobile and toxic to living organisms (Lamb et al., 2009). Anthropogenic sources of metals include automobile vehicles through emissions (Sternbeck et al., 2002), industrial emissions, copper–nickel smelters, burning of diesel oil (Barałkiewicz and Siepak, 1999; Depledge and Fossi, 1994), waste disposal, fertilizers and pesticides, sewage sludge application and raw city effluent (Alloway and Jackson, 1991; Hussain et al., 2006). Phosphatic (P) fertilizers are the major source of metals particularly cadmium (Cd) in agricultural soils (Park et al., 2011). There is an approach using P fertilizer with low Cd level to decrease contamination of the soils. This is being achieved through use of selective rock phosphate having less Cd for manufacturing P fertilizers or through treatment of rock phosphate to remove Cd. Application of biosolids to agricultural soils is another important source of heavy metals to agricultural soils. In developing countries, use of raw city effluent for irrigation of crops in peri-urban areas is a major source of metal contamination of soils. Raw city effluent is a blend of municipal sewage with a significant share of industrial wastewater. Use of raw city effluent creates potential and risks as municipal waste water contains nutrients and industrial waste water potentially contains heavy metals, boron, pathogens and salts (Ensink et al., 2005) that could deteriorate soil productivity (Mapanda et al., 2005).

ROLE OF OM IN PHYTOAVAILABILITY OF METALS

Soil OM contains residues of plants and animals and their by-products at different stages of decomposition (You et al., 1999). By-products of decomposition of OM include high and low molecular weight organic molecules having different sizes and binding sites of varying binding energy (Stevenson and Fitch, 1986; Yin et al., 2002). Fulvic and humic acids and humin are the most stable by-products of OM decomposition (Stevenson and Cole, 1999). Plant residues are major source of OM in the soils and their amount, composition and properties are important for the decomposition process in the terrestrial environment (Kögel-Knabner, 2002). Microbial biomass makes a significant contribution to soil OM and subsequently serves as a parent material for humus formation (Haider, 1992). The OM in the soil comprises of particulate organic matter (POM) and dissolved organic matter (DOM). Dissolved organic matter is a fraction of OM that can pass through a 0.45-μm filter; however, some scientists have used finer filter paper (0.2-μm filter) for the separation of DOM (Dafner and Wangersky, 2002; Park et al., 2011; Zsolnay, 2003). Dissolved OM is a most reactive fraction of OM that influences different soil physical, chemical and biological properties (Park et al., 2011).

Organic matter in soils could retain metals by forming stable complexes with metals, and its effect on CEC and soil pH (Clemente and Bernal, 2006;

Elliott et al., 1986). The effect of organic materials in decreasing metal phytoavailability can not be isolated from its effects on other soil properties and soil pH. Addition of OM could decrease concentration of Ni and other metals in soils; however, decomposition of OM may alter soil pH thus affecting bioavailability of Ni and other metals (Arnesen and Singh, 1998; Karaca, 2004b). Organic matter affects the uptake of metals by plants depending upon its fraction in the soil viz. fulvic acid (soluble fraction) or humic acid (insoluble fraction). The OM can affect metal availability due to the formation of complexes with metals or competition for sorption sites on soil matrix; OM could enhance metal adsorption by forming ligands having strong affinity with soil particles or may have no clear effect due to weak complexation with metals (Huang and Lin, 1981). Application of OM to the soils in arid/semi-arid areas is crucial due to inherently low content of OM in the soils of these areas. Organic matter could immobilize Ni and other metals by transferring the soluble/exchangeable Ni to the forms that sorb with OM, carbonate and oxides of Fe and Mn (Walker et al., 2003). Sorption of metals depends upon type of metals, soil properties, heavy metal contents of soil, amount of OM, extent of humification of OM and its impact on soil pH (Almås et al., 1999; Ross, 1994; Shuman, 1999). The effect of OM added by different amendments on metal solubility depends upon the humification of OM, salt contents of amendments, pH, cation exchange capacity and redox conditions of soil (Almås et al., 1999; Ross, 1994; Shuman, 1999).

Soil pH is the most important property that controls mobility and phytoavailability metals in soils directly or indirectly. Solution metal speciation changes with changes in soil pH thereby affecting sorption or desorption of metals (Sauve et al., 2000). Adsorption of hydrolyzed metal ions is stronger than that of aqueous metal ions due to lower salvation energy (James and Healy, 1972; Yin et al., 2002). Metals may be precipitated at soil surface at high pH due to competition with protons and decrease in surface potential (Adriano, 2001).

Soil pH is greatly influenced by addition of OM through different organic amendments depending upon the nature of OM, release of basic cations and NH_3/NH_4^+ production during decomposition, adsorption of H^+ ions onto negatively charged soil surface, onset of reducing conditions due to enhanced microbial activity and displacement of OH^- from oxide surfaces by organic ligands (Pocknee and Sumner, 1997; Walker et al., 2004). Organic amendments (mushroom compost , grape marc and tobacco dust) changed the Ni partitioning between soil solution and solid phase due to the effect on soil pH (Karaca, 2004b). Cow manure, pig manure and peat soil decreased diethlyene triamine penta acetic acid (DTPA) extractable Ni in the soil by altering soil pH (Narwal and Singh, 1998).

The pH controls adsorption and desorption of Ni and other metals in soils due to changes in the metal species in solution (Casagrande et al., 2008). Metal retention generally increases with increasing soil pH (Bolan and Duraisamy,

2003; Harter, 1983; Hooda and Alloway, 1998; Naidu et al., 1994). Several mechanisms for immobilization of metals due to increase in soil pH were proposed including: (1) an increase in the negative charge on colloid surface leading to an increase in metal cation adsorption (James and Bartlett, 1983; Naidu et al., 1994); (2) formation of hydroxy species of metals having greater affinity for adsorption sites than just the metal cations (Naidu et al., 1994); and (3) the precipitation of metals as metal hydroxides and their adsorption in soil components having variable charges like oxides of Fe and Mn is greatly influenced by soil pH. The activity of commonly found ionic species of Ni in soil solution $Ni(H_2O)_6^{2+}$, decreases with increasing pH, and Ni adsorption is maximum at pH 6 (Moreira et al., 2008). This effect could be due to less numbers of available sites for cation adsorption at lower soil pH. However, Ni solublization at higher pH is attributed to the formation of Ni-OM complexes (Yin et al., 2002). At high pH, concentration of DOM is more, leading to the formation of soluble complexes. Metal concentration (Cu, Zn, Mn) in *Chenopodium album* L. were inversely related with soil pH indicating soil pH as the main determinant factor in controlling phytoavailability of metals (Walker et al., 2004; Yoo and James, 2002). Decrease in sorption of metals in soils may be due to protonation of OM and other charged sites caused by low pH (Clemente et al., 2005). Another reason of decreasing metal adsorption in soil may be the competition of metals with H^+ for sorption sites (Rooney et al., 2007; Weng and Huang, 2004). Conversely, high pH increases adsorption of metals due to hydrolysis of metal thus decreasing solvation energies of hydrolysed metal. Metals precipitate at soil surface due to a decrease in surface potential and proton competition at high pH which ultimately favor metal binding (McBride, 1989). Nickel and other metals' concentration in soil solution is determined by pH, total metal concentration and/or OM contents (Yin et al., 2002). At acidic pH, metals remain in cationic form while metals form stable complexes with humate at high pH (Morera et al., 2002). The pH controls behaviour of humic substances due to functional groups like carboxyl, phenolic and amine groups which may be protonated at low pH or de-protonated at high pH (Silveira et al., 2003). At high pH, Ni complexation with dissolved organic carbon was high due to more solubility of OM at high pH (Spark et al., 1995). Organic matter retains the metals by physical adsorption, electrostatic attraction, hydrogen bonding and coordination complexation (Stevenson, 1994). Complexation of OM with metals can occur in solution and solid phases of the soil due to the presence of specific functional groups which can bind Ni and other metals in both soil phases (Mellis et al., 2004). Metals can form complexes with different fractions of OM but fulvic and humic acids are important in metal complexation. Complexes of metals with fulvic acid are generally soluble while with humic acid, complexes are insoluble (Chirenje et al., 2002; Silveira et al., 2003). Complexation of metals with OM may be due to the direct coordination between metals and OM or through inner-sphere complex formation with acid functional groups (McBride, 1989). In metal-contaminated soils, metal sorption capacity and type of soil minerals are crucial to control availability of Ni and other metals (Massoura et al., 2006).

ORGANIC AMENDMENTS AND PHYTOAVAILABILITY OF METALS IN CONTAMINATED SOILS

Phytoavailability of metals is influenced by a number of soil properties and OM is the most important due its effect on other soil physical and chemical properties (Seregin and Kozhevnikova, 2006). Immobilization of Ni and other metals can decrease adverse effects on micro-organisms, plants, animals and humans. In situ immobilization of Ni and other metals with soil-applied amendments may decrease the solubility of metals. Additionally, organic amendments improve soil physico-chemical properties and improve plant growth and help the re-vegetation of contaminated soils. Different organic amendments can be used for immobilization of metals decreasing their phytoavailability, like farm yard manure, poultry manure, pressmud, tobacco dust, mushroom compost, biochar and pig manure (Karaca, 2004a; Sabir et al., 2008).

Manures

Farm yard manure (FYM) is the conventional manure and is most easily available to farmers. It is a decomposed mixture of farm animals' dung and urine with straw and litter used as bedding for animals and residues of fodder fed to the farm animals. It is a rich source of plant nutrients, cellulose and crude fibre (Sulieman and Hago, 2009) and could be used to immobilize Ni and other metals in contaminated soils (Paulose et al., 2007). In contaminated soil, application of FYM improved growth of *Chenopodium album* L. compared with that of compost (Walker et al., 2004). Similarly, increase in biomass of *Beta vulgaris* and *Beta maritima* was recorded with application of cow manure compared to the control and in olive husk treated contaminated soil (pH 7.7–8.1, EC 0.13 dS m^{-1}, OM 6.3 %) which was attributed to the decreased availability of metals in soils thus improving plant growth (Clemente et al., 2007). High biomass production by addition of FYM could be due to additional supply of macronutrients, increased OM content, biological activity, improved nutrient cycling, increased CEC and buffering capacity (Paulose et al., 2007; Stewart et al., 2000). Manure application decreased plant tissue concentration of metals (Cu, Zn and Pb) compared with those grown in compost treated soil or control soil and it was attributed primarily to increase in soil pH due to manure application (Walker et al., 2004).

Difference in pH of soil due to application of different organic amendments was possibly due to the differences in pH of applied amendments (Clemente et al., 2007; Narwal and Singh, 1998). Amendments may increase soil pH due to release of basic cations and NH$_3$ into soils during decomposition microbially induced reducing conditions and displacement of hydroxyls ions from sesquioxide surfaces by organic anions (Clemente et al., 2007; Pocknee and Sumner, 1997; Walker et al., 2004). Since the availability of Ni and other metals in soils is mainly controlled by the chemical equilibrium of metals in solid and solution phases, hence adsorption reactions are important to determine the availability of metals to plants and their mobility throughout the soil (Paulose et al., 2007). It was reported that cow manure decreased Zn availability in contaminated soils compared to compost and control treatments (Clemente and Bernal, 2006) that could be due to the precipitation of Zn (Clemente et al., 2003). Differential effect of farm manure compared to other amendments could be due to difference in rate of decomposition, and ensuing microbial activity, formation of inorganic salts and stable OM (Clemente et al., 2006; Karaca, 2004b; Walker et al., 2003). Vegetables (celeriac and leek) accumulated less Cd in farm yard manure treated soil (pH 5.7, OM 1.5%) compared to the control soil (Zaniewicz-Bajkowska et al., 2007). Addition of poultry litter reduced Ni concentration in shoots and grain of rice in Ni-contaminated soil (Chamon et al., 2005a).

Walker et al. (2003) recorded low DTPA-extractable metals (Fe, Zn, Pb, Cu) in fresh manure treated soil compared to the compost treated soil due to precipitation of metals with P and other salts released during decomposition of OM (Kabata-Pendias and Mukherjee, 2007). Complexation of OM with metals might be prevented due to high pH and calcareous nature of the soil thus decreasing DTPA-extractable metals (Jahiruddin et al., 1985). Addition of compost and manure increased concentration of exchangeable metals in soil by increasing CEC of soil by 38 and 21%, respectively (Shuman, 1999). Species may vary in their response to metal concentration and organic matter addition. Addition of manure decreased the growth of *Brassica juncea* due to toxicity caused by higher concentration of Mn shoots.(Walker et al., 2003). Cow and pig manure decreased DTPA-extractable Ni in soil due to formation of strong complexes with OM having strong affinity with soil solids. This led to decreased Ni concentration in wheat grains (Narwal and Singh, 1998). Cow manure decreased DTPA-extractable metals (Fe, Zn, Cu) in a calcareous soil and increased biomass production of *Beta vulgaris* (Clemente et al., 2007). Phosphate and soluble salt contents of cow manure might lead to low metal availability in soil and thus low metal concentration in plants (Walker et al., 2003). Farm yard manure enhanced spinach fresh weight by decreasing Zn concentration in spinach leaves in a contaminated soil. Decreased Zn in spinach leaves could be due to its decreased mobility in soil owing to its adsorption onto OM (Yassen et al., 2007).

Compost

Compost is a decomposed OM of plants and animals decomposed under aerobic conditions. Composting is a simple practice carried out by individuals in their homes, farmers on their land and commercially by industries and city administrations (Petruzzelli et al., 1989). Compost improves soil structure and soil fertility due to its organic matter contents that can counter natural decrease in soil fertility (Eghball et al., 2004). Use of composts and other organic by-products improves soil chemical properties like pH (Ouédraogo et al., 2001), increased plant nutrients availability and crop yields (Hartl and Erhart, 2005; Miyasaka et al., 2001). It could be used as an alternate for traditional manures in areas of intensive agriculture due to poor availability of traditional manures (Gigliotti et al., 1996). Effect of compost and other organic amendments on availability of Ni and other metals depends on amount and composition of OM present (Sabir et al., 2008; Unsal and Sozudogru Ok, 2001), heavy metal contents, salts and degree of stabilization (Murillo et al., 1995). Addition of sewage-sludge compost and green waste compost decreased available metals (Cd, Cu, Zn and Pb) in contaminated soil compared to non-amended soils (van Herwijnen et al., 2007). Biosolid compost and sugar beet lime enhanced the natural vegetative cover and biomass production in mine spill contaminated soil by decreasing metal concentration in plants, incremental nutrients due to decomposition of composts and improved soil physico-chemical properties (Madejón et al., 2006). The decrease in DTPA extractable Ni concentration in soil due to application of mushroom compost and other organic amendments was reported (Karaca, 2004b). Application of green waste compost decreased the uptake of Cu, Pb and Zn in Greek Cress by 21, 54 and 16%, respectively, in contaminated calcareous soils (van Herwijnen et al., 2007). Kiikkilä et al. (2002) concluded that exchangeable Cu decreased after mulched treatment with garden soil, sewage sludge, compost and compost mixed with bark chips and woodchips. The authors opined that Cu concentration may decrease due to formation of complexes with particulate organic matter contributed by mulch onto soil at pH > 6 (Kiikkilä et al., 2002). Biosolid compost decreased arsenic concentration in carrot (*Daucus carota* L.) and lettuce (*Lactuca sativa* L.) compared to the control (Cao and Ma, 2004). This could be due to adsorption of As by OM that is evidenced by decreased arsenic concentration in extractable fractions. Liu et al., (2009) reported that application of compost effectively reduced Cd toxicity to wheat by decreasing > 50% Cd uptake by wheat tissue and improving wheat growth. Alleviating effect of compost could be due to high soil pH, Cd complexation with OM and co-precipitation with P content (Liu et al., 2009).

Activated Carbon / Biochar

Activated carbon contains carbonaceous material derived from charcoal. Activated carbon is produced by pyrolysis of organic materials of plant origin. These materials include coal, coconut shells and wood, sugarcane bagasse, soybean hulls and nutshell (Dias et al., 2007; Paraskeva et al., 2008). On a limited scale, animal manures are also used for the production of activated carbon. Use of activated carbon is common to remove metals from waste waters, but its use for metal immobilization is not common in contaminated soils (Gerçel and Gerçel, 2007; Lima and Marshall, 2005b). Poultry manure derived activated carbon had excellent metal binding capacity (Lima and Marshall, 2005a). Activated carbon is often used for remediation of pollutants in soil and water due to porous structure, large surface area and high adsorption capacity (Üçer et al., 2006). Activated carbon removes metals (Ni, Cu, Fe, Co, Cr) from solution through precipitation as metal hydroxide, adsorption on activated carbon (Lyubchik et al., 2004). Almond husk derived AC effectively removed Ni from waste waters with and without H_2SO_4 treatment (Hasar, 2003).

Recently, biochar has been used as a soil amendment due to its beneficial effects on different soil physical and chemical properties (Beesley et al., 2010). Biochar contains very high contents (up to 90%) depending upon the parent material (Chan and Xu, 2009). Addition of biochar improves the adsorption of dissolved organic carbon, soil pH, decreases metals in the leachates and supplement macro nutrients (Novak et al., 2009; Pietikäinen et al., 2000). Long term persistence of biochar in soil decreases input of metals through repeated application of other amendments (Lehmann and Joseph, 2009). Beesley et al. (2010) concluded that biochar decreased water soluble Cd and Zn in the soils due to increase in organic carbon and pH. Activated carbon decreased metal concentration (Ni, Cu, Mn, Zn) in shoots of maize plants grown in contaminated soils compared to the un-amended soil (Sabir et al., 2013). Biochar decreased high concentrations of soluble Cd and Zn in a contaminated soil (Beesley and Marmiroli, 2011). They concluded that sorption is an important mechanism for retention of metals by soils. Biochar decreased concentration of Cd and Zn to a 300- and 45-fold decrease in their leachate concentrations, respectively (Beesley and Marmiroli, 2011).

Pressmud

Pressmud is a waste produced by the sugar cane industry which contains essential nutrients for plants. Its composition varies from 68–70% moisture, 24–28% combustible fraction and 6–8% ash. Disposal of pressmud along road sides or its use as a manure in salt-affected soils is common (Gangavati et al., 2005). Pressmud is a very good source of additional nutrients to soils and can decrease use of nitrogenous fertilizers in rice–wheat cropping systems and can add up to 27.3% carbon to the soil (Gangavati et al., 2005). Application of fly ash with pressmud and cow dung increased the growth of *Cassia siamea* (Tripathi and Bhardwaj, 2004). Pressmud increased Ni concentration in soil and *Cassia siamea* due to decrease in pH caused by the pressmud. Although, use of pressmud is common for amelioration of salt-affected soils and as organic fertilizer, its use for remediation of metal-contaminated soils is not common.

EFFECT OF TIME ON DECOMPOSITION OF ORGANIC AMENDMENTS AND METAL PHYTOAVAILABILITY

Heavy metals added to soils anthropogenically tend to accumulate in top soil (Baker and Walker, 1990; Samsøe-Petersen et al., 2002) and thus exert toxic effects on plants, animals and humans by entering the food chain (Berti and Jacobs, 1996), and disrupting the ecosystem (Stalikas et al., 1997). Traditionally, organic amendments are used in agriculture to enhance crop productivity on a sustainable basis, but recently they have been used for remediation of metal-contaminated soils (Clemente et al., 2007; Walker et al., 2004). These amendments could alter the availability of metals by changing chemistry of soil (pH, CEC, nutrient content) and chelation of metals with OM. Availability of metals in the soil depends upon their residence time in soils (Joner and Leyval, 2001); generally metal availability has an inverse relationship with residence time (McLaughlin, 2001). Residence time of metals in soils is affected by complexation of metals with organic substances, their adsorption and precipitation (McLaughlin, 2001). However, the effect of time on the fate of metals in soils is not well explored (Bataillard et al., 2003; Ma and Uren, 1998). Changes in OM content and composition with time due to decomposition may alter availability of Ni and other metals in soils (Clemente and Bernal, 2006; Karaca, 2004b).

Addition of grape marc and mushroom compost decreased DTPA extractable Ni concentration in soil and Ni was higher at the termination of 6 months of incubation compared to the start of incubation (Karaca, 2004b). During 6 months of incubation, composition and content of OM changed considerably (Yuan and Lavkulich, 1997). Oxidation of OM led to the changes in its contents and composition triggering the changes in in Ni adsorption on soil surfaces (Elliott and Denneny, 1982; Yuan and Lavkulich, 1997).

Manure application decreased the $CaCl_2$-extractable Zn in soil after 14, 28 and 56 days of incubation. During the decomposition of OM, Zn may be precipitated as $ZnCO_3$ thus decreasing Zn extractability (Clemente and Bernal, 2006; Walker et al., 2003). Conversion of soluble fraction of organic substances into stable fraction and inorganic compounds like nitrates, sulphates and phosphates may cause depression in extractability of Zn (Clemente and Bernal, 2006; Walker et al., 2003). The decomposition of OM was more in solid oil mill waste amended soil in comparison to un-amended soil after 56 days of incubation. Microbial mediated decomposition of OM, adsorption on clay surface and formation of polyphenols could lead to low dissolved carbon and phenol in the soil (De la Fuente et al., 2008). Extractability of Zn increased immediately after application of solid oil mill waste but this effect did not persist at the end of 56 days of incubation period. Decomposition of inherent carbon of humic acid in treated soil was lower compared to the control after 28 weeks of incubation (Clemente et al., 2006). This may be due to the presence of highly stable carbon of humic acids. The Zn availability decreased in soil with time due to decrease of dissolved organic carbon in the same period (Antoniadis et al., 2007). The DTPA extractable Ni decreased with incubation time which could be due to immobilization of Ni with humified OM in soil (Halim et al., 2003) and complexation of humus with clay caused by bridging of divalent cations or hydrophobic adsorption on clay surfaces (Churchman et al., 2006). Generally, stability constants of complexes of metal with DTPA are higher compared to the complexes with humic substances (Stevenson and Fitch, 1986). Generally, organic carbon increase DTPA-extractable Ni and other metals in soil (Halim et al., 2003).

Extractability and availability of metals (Cu, Zn Pb, Cd) from soil decreased after eight weeks of incubation (Lu et al., 2005). Extractability of Ni varied with incubation depending upon contents of OM, soil type, and transformation of OM with time (Martinez et al., 2003a). Nickel solubility increased with incubation in soil due to increase in concentration of DOC in soil solution (Martinez et al., 2003a). Interaction of different fractions of DOC with Ni and heavy metals is variable due to changes in properties of DOC with time. It has been reported that DOC in soil is positively correlated with soluble metals indicating existent control of DOC on metal solubility in soils (McBride et al., 1997; Römkens and Dolfing, 1998). With increasing time and temperature, decomposition of OM increased thereby increasing DOC and dissolution of adsorbed metals (Martinez et al., 2003a).

RESIDUAL EFFECT OF ORGANIC AMENDMENTS ON METAL PHYTOAVAILABILITY

Although, soil OM immobilizes with heavy metals depending upon contents and composition, its decomposition could alter its effect and thus could release metals. Nature and stability of OM is important to affect partitioning of metals between solution and solid phase and thus determine behaviour of metals in the environment (Muhammad, 2009). Hydrolysis, oxidation and/or depolymerization of OM may release the dissolved organic matter during decomposition of OM and resultantly bound metals are released with time after application of OM (Martinez et al., 2003a). Organic matter in the soil decreased after a third year wheat crop since its application in the form of cow manure, pig manure and peat soil compared to that after the harvest of the first crop of wheat (Arnesen and Singh, 1998), thus altering its composition and interaction with Ni and other metals (Yuan and Lavkulich, 1997). Application of cow manure and pig manure decreased Ni in wheat grains in the second and third years, respectively (Arnesen and Singh, 1998). Although, OM composition changes with time due to decomposition, it remain crucial in environmental studies whatever its quantity (Clemente et al., 2005; Han and Thompson, 1999).

Growth behaviour and uptake pattern of metals (Zn, Fe, Cu) by *Brassica juncea* was different in two crops in a metal-contaminated soil treated with manure and compost (Clemente et al., 2003). *Brassica juncea* contained higher concentration of metals (Zn, Cu, Pb) in the first year of crop compared to second year crop. It was recorded that concentration of metals (Cu, Fe and As) in *Brassica juncea* correlated positively with organic carbon during the first year and negatively in the second year suggesting conversion of metal solublizing organic compounds into metals retaining organic compounds in soils (Almås et al., 1999; Shuman, 1999).

Production of different organic substances, particularly organic acids, could alter availability of Ni and other metals (Misra and Pande, 1974a). Decomposition of OM was increased as evidenced by an increase in DOC that was positively correlated with release of metals (Martinez et al., 2003b). Manure and compost increased fixation of metals (Pb, Zn, Fe) in calcareous soil and fixation of metals was determined by mineralization of inherent OM of these amendments (Arnesen and Singh, 1998). Decomposition of OM decreases the surface area and CEC that may ultimately increase the release of metals with time and thus define a time bomb effect (McBride, 1995). Dissolution of metals in soil increases with time depending on type of metals (Martinez et al., 2003b).

ORGANIC ACIDS AND METAL PHYTOAVAILABILITY

Organic acids like citric/oxalic/formic/acconitic/acetic acid produced during the decomposition of OM or released by roots and microbial metabolites, can affect the availability of metals (Christensen et al., 1996; Strobel et al., 2001). The mobility and availability of metals depends on the affinity of metal–organic acid complex with the soil surface (Davis and Leckie, 1978). Organic acids have no important role in metal retention and release soil due to their easily degradable nature (Harter and Naidu, 1995).

Type, number and position of functional groups determines the interaction of organic acids with metals (Burckhard et al., 1995). Citrate decreased Ni adsorption on soil but arginine has no effect on Ni adsorption at high Ni concentration and this may be attributed to formation of bidentate complexes of citrate with Ni (Poulsen and Hansen, 2000). Barley plants absorbed Ni more from the forms applied as Ni-citrate, Ni-glutamate and Ni-EDTA as compared to inorganic forms applied as $NiSO_4$ (Molas and Baran, 2004). Zn leachability in soil columns increased citric acid application compared to formic acid, succinic acid and control in contaminated soil (Burckhard et al., 1995). Complexation of metals with organic acids is controlled by the diffusion of complexing agent from bulk solution to adsorption site, residence time of complexing with adsorption site, dissolution of organic acids, nature and position of functional groups and affinities of chelating agents for metals (McColl and Pohlman, 1986).

PHYTOREMEDIATION WITH ORGANIC AMENDMENTS: CONCLUSION AND FUTURE THRUST

Phytoremediation of metal-contaminated soils is an environmentally friendly technique which is non-disruptive. Metal tolerant plant species with high biomass production are the best choice for phytoremediation and initial re-vegetation of metal-contaminated soils. In metal-contaminated soils, plant growth is generally inhibited due to adverse physical and chemical properties along with limited availability of nutrients. Conditioning of metal-contaminated soils with organic amendments could enhance OM, nutrient availability, microbial activity, physical and chemical properties of the soils which consequently enhance plant growth and help re-vegetation of metal-contaminated soils. Different researchers have reported contrasting results. Organic matter can lower the availability of metals by transforming soluble metals into insoluble metals. Conversely, OM can enhance metal solubility by increasing dissolved organic matter in the soil due to decomposition of OM with time. Such type of behaviour is common when fresh OM is being applied to the soils that undergo rapid mineralization, thus releasing the adsorbed metals. Generally, fresh OM or partially decomposed OM is used in immobilization studies. Although some work on the use of humic substances is reported, there is still a lot to be investigated regarding the role of humified OM in remediating the metals in the soils. Organic amendment can play a significant role in phytoremediation of metal-contaminated soils due to the improvement of soil physico-chemical properties that improve plant growth. However, the composition and contents of OM, decomposition rate, metal contents and nature of organic acids produced during the decomposition of OM are important in affecting phytoavailability of metals and must be considered.

REFERENCES

Adesodun, J., Mbagwu, J., 2008. Distribution of heavy metals and hydrocarbon contents in an alfisol contaminated with waste-lubricating oil amended with organic wastes. Bioresour. Technol. 99, 3195–3204.

Adriano, D.C., 2001. Trace elements in terrestrial environments: Biogeochemistry, bioavailability, and risks of metals. Springer.

Alloway, B.J., Jackson, A.P., 1991. The behaviour of heavy metals in sewage sludge-amended soils. Sci. Total Environ. 100, 151–176.

Almås, Å., Singh, B., Salbu, B., 1999. Mobility of cadmium-109 and zinc-65 in soil influenced by equilibration time, temperature, and organic matter. J. Environ. Qual. 28, 1742–1750.

Antoniadis, V., Tsadilas, C.D., Ashworth, D.J., 2007. Monometal and competitive adsorption of heavy metals by sewage sludge-amended soil. Chemosphere 68, 489–494.

Appel, C., Ma, L., 2002. Concentration, pH, and surface charge effects on cadmium and lead sorption in three tropical soils. J. Environ. Qual. 31, 581–589.

Arnesen, A., Singh, B., 1998. Plant uptake and DTPA-extractability of Cd, Cu, Ni and Zn in a Norwegian alum shale soil as affected by previous addition of dairy and pig manures and peat. Can. J. Soil Sci. 78, 531–539.

Baker, A.J., Walker, P.L., 1990. Ecophysiology of metal uptake by tolerant plants. Heavy metal tolerance in plants: Evolutionary aspects, 155–177.

Barałkiewicz, D., Siepak, J., 1999. Chromium, nickel and cobalt in environmental samples and existing legal norms. Pol. J. Environ. Stud. 8, 201–208.

Bataillard, P., Cambier, P., Picot, C., 2003. Short-term transformations of lead and cadmium compounds in soil after contamination. E. J. Soil Sci. 54, 365–376.

Beesley, L., Marmiroli, M., 2011. The immobilisation and retention of soluble arsenic, cadmium and zinc by biochar. Environ. pollut. 159, 474–480.

Beesley, L., Moreno-Jiménez, E., Gomez-Eyles, J.L., 2010. Effects of biochar and greenwaste compost amendments on mobility, bioavailability and toxicity of inorganic and organic contaminants in a multi-element polluted soil. Environ. Pollut. 158, 2282–2287.

Berti, W., Jacobs, L., 1996. Chemistry and phytotoxicity of soil trace elements from repeated sewage sludge applications. J. Environ. Qual. 25, 1025–1032.

Bolan, N.S., Duraisamy, V., 2003. Role of inorganic and organic soil amendments on immobilisation and phytoavailability of heavy metals: A review involving specific case studies. Soil Res. 41, 533–555.

Burckhard, S., Schwab, A., Banks, M., 1995. The effects of organic acids on the leaching of heavy metals from mine tailings. J. Hazard. Mater. 41, 135–145.

Cao, X., Ma, L.Q., 2004. Effects of compost and phosphate on plant arsenic accumulation from soils near pressure-treated wood. Environ. Pollut. 132, 435–442.

Casagrande, J.C., Soares, M.R., Mouta, E.R., 2008. Zinc adsorption in highly weathered soils. Pesquisa agropecuária brasileira 43, 131–139.

Chamon, A., Gerzabek, M., Mondol, M., Ullah, S., Rahman, M., Blum, W., 2005a. Influence of cereal varieties and site conditions on heavy metal accumulations in cereal crops on polluted soils of Bangladesh. Commun. Soil. Sci. Plant. Anal. 36, 889–906.

Chamon, A., Gerzabek, M., Mondol, M., Ullah, S., Rahman, M., Blum, W., 2005b. Influence of soil amendments on heavy metal accumulation in crops on polluted soils of Bangladesh. Commun. Soil. Sci. Plant. Anal. 36, 907–924.

Chan, K.Y., Xu, Z., 2009. Biochar: Nutrient properties and their enhancement. Biochar for environmental management: Science and technology. Earthscan, London. 67–84.

Chirenje, T., Rivero, C., Ma, L.Q., 2002. Leachability of Cu and Ni in wood ash-amended soil as impacted by humic and fulvic acid. Geoderma 108, 31–47.

Christensen, J.B., Jensen, D.L., Christensen, T.H., 1996. Effect of dissolved organic carbon on the mobility of cadmium, nickel and zinc in leachate polluted groundwater. Water res. 30, 3037–3049.

Churchman, G.J., Gates, W.P., Theng, B., Yuan, G., 2006. Clays and Clay Minerals for Pollution Control. Dev. Clay Sci. 1, 625–675.

Clemente, R., Bernal, M.P., 2006. Fractionation of heavy metals and distribution of organic carbon in two contaminated soils amended with humic acids. Chemosphere 64, 1264–1273.

Clemente, R., Escolar, Á., Bernal, M.P., 2006. Heavy metals fractionation and organic matter mineralisation in contaminated calcareous soil amended with organic materials. Bioresour. Technol. 97, 1894–1901.

Clemente, R., Paredes, C., Bernal, M., 2007. A field experiment investigating the effects of olive husk and cow manure on heavy metal availability in a contaminated calcareous soil from Murcia (Spain). Agriculture, ecosystems & environ. 118, 319–326.

Clemente, R., Walker, D.J., Bernal, M.P., 2005. Uptake of heavy metals and As by *Brassica juncea* grown in a contaminated soil in Aznalcóllar (Spain): The effect of soil amendments. Environ. pollut. 138, 46–58.

Clemente, R., Walker, D.J., Roig, A., Bernal, M.P., 2003. Heavy metal bioavailability in a soil affected by mineral sulphides contamination following the mine spillage at Aznalcóllar (Spain). Biodegradation 14, 199–205.

Cunningham, S.D., Berti, W.R., Huang, J.W., 1995. Phytoremediation of contaminated soils. Trends. Biotechnol. 13, 393–397.

Dafner, E.V., Wangersky, P.J., 2002. A brief overview of modern directions in marine DOC studies Part II—Recent progress in marine DOC studies. J. Environ. Monit. 4, 55–69.

Davis, J.A., Leckie, J.O., 1978. Surface ionisation and complexation at the oxide/water interface II. Surface properties of amorphous iron oxyhydroxide and adsorption of metal ions. J. Environ. Qual. 67, 90–107.

De la Fuente, C., Clemente, R., Bernal, M., 2008. Changes in metal speciation and pH in olive processing waste and sulphur-treated contaminated soil. Ecotoxicol. environ. saf. 70, 207–215.

Depledge, M., Fossi, M., 1994. The role of biomarkers in environmental assessment (2). Invertebrates. Ecotoxicology 3, 161–172.

Dias, J.M., Alvim-Ferraz, M.C.M., Almeida, M.F., Rivera-Utrilla, J., Sánchez-Polo, M., 2007. Waste materials for activated carbon preparation and its use in aqueous-phase treatment: A review. J. Environ. Manage. 85, 833–846.

Eghball, B., Ginting, D., Gilley, J.E., 2004. Residual effects of manure and compost applications on corn production and soil properties. Agron. J. 96, 442–447.

Elliott, H., Denneny, C., 1982. Soil adsorption of cadmium from solutions containing organic ligands. J. Environ. Qual. 11, 658–663.

Elliott, H., Liberati, M., Huang, C., 1986. Competitive adsorption of heavy metals by soils. J. Environ. Qual. 15, 214–219.

Ensink, J.H., van der Hoek, W., Mukhtar, M., Tahir, Z., Amerasinghe, F.P., 2005. High risk of hookworm infection among wastewater farmers in Pakistan. Trans. R. Soc. Trop. Med. Hyg. 99, 809–818.

Facchinelli, A., Sacchi, E., Mallen, L., 2001. Multivariate statistical and GIS-based approach to identify heavy metal sources in soils. Environ. Pollut. 114, 313–324.

Gangavati, P., Safi, M., Singh, A., Prasad, B., Mishra, I., 2005. Pyrolysis and thermal oxidation kinetics of sugar mill press mud. Thermochimica acta 428, 63–70.

Gerçel, Ö., Gerçel, H.F., 2007. Adsorption of lead (II) ions from aqueous solutions by activated carbon prepared from biomass plant material of *Euphorbia rigida*. Chem. Eng. J. 132, 289–297.

Gigliotti, G., Businelli, D., Giusquiani, P.L., 1996. Trace metals uptake and distribution in corn plants grown on a 6-year urban waste compost amended soil. Agriculture, ecosystems & environ. 58, 199–206.

Haider, K., 1992. Problems related to the humification processes in soils of temperate climates. Soil biochem. 7, 55–94.

Halim, M., Conte, P., Piccolo, A., 2003. Potential availability of heavy metals to phytoextraction from contaminated soils induced by exogenous humic substances. Chemosphere 52, 265–275.

Han, N., Thompson, M.L., 1999. Copper-binding ability of dissolved organic matter derived from anaerobically digested biosolids. J. Environ. Qual. 28, 939–944.

Harter, R.D., 1983. Effect of soil pH on adsorption of lead, copper, zinc, and nickel. Soil. Sci. Soc. Am. J. 47, 47–51.

Harter, R.D., Naidu, R., 1995. Role of metal-organic complexation in metal sorption by soils. Adv. Agronomy 55, 219–263.

Hartl, W., Erhart, E., 2005. Crop nitrogen recovery and soil nitrogen dynamics in a 10-year field experiment with biowaste compost. J. Plant Nutr. Soil Sci. 168, 781–788.

Hasar, H., 2003. Adsorption of nickel (II) from aqueous solution onto activated carbon prepared from almond husk. J. Hazard. Mater. 97, 49–57.

Hooda, P., Alloway, B., 1998. Cadmium and lead sorption behaviour of selected English and Indian soils. Geoderma 84, 121–134.

Huang, C., Lin, Y., 1981. Specific adsorption of Co (II) and [Co (III) EDTA]-complexes on hydrous oxide surfaces, Adsorption from aqueous solutions. Springer. pp. 61–91.

Hussain, S.I., Ghafoor, A., Ahmad, S., Murtaza, G., Sabir, M., 2006. Irrigation of crops with raw sewage: Hazard assessment of effluent, soil and vegetables. Pak. J. Agric Sci. 43, 97–102.

Işıklı, B., Demir, T.A., Akar, T., Berber, A., Ürer, S.M., Kalyoncu, C., Canbek, M., 2006. Cadmium exposure from the cement dust emissions: A field study in a rural residence. Chemosphere 63, 1546–1552.

Jahiruddin, M., Livesey, N., Cresser, M., 1985. Observations on the effect of soil pH upon zinc absorption by soils. Commun. Soil Sci. Plant Anal. 16, 909–922.

James, B.R., Bartlett, R.J., 1983. Behavior of chromium in soils. VI. Interactions between oxidation-reduction and organic complexation. J. Environ. Qual. 12, 173–176.

James, R.O., Healy, T.W., 1972. Adsorption of hydrolyzable metal ions at the oxide–water interface. III. A thermodynamic model of adsorption. J. Colloid. Interface. Sci. 40, 65–81.

Joner, E., Leyval, C., 2001. Time-course of heavy metal uptake in maize and clover as affected by root density and different mycorrhizal inoculation regimes. Biol. Fertil. Soils 33, 351–357.

Kabata-Pendias, A., Pendias, H., 2001. Trace elements in soils and plants. CRC Press, Boca Raton, FL.

Kabata-Pendias, A., Mukherjee, A.B., 2007. Trace elements from soil to human. Springer.

Kadem, D., Rached, O., Krika, A., Gheribi-Aoulmi, Z., 2004. Statistical analysis of vegetation incidence on contamination of soils by heavy metals (Pb, Ni and Zn) in the vicinity of an iron steel industrial plant in Algeria. Environmetrics 15, 447–462.

Karaca, A., 2004a. Effect of organic wastes on the extractability of cadmium, copper, nickel, and zinc in soil. Geoderma 122, 297–303.

Karaca, A., 2004b. Effect of organic wastes on the extractability of cadmium, copper, nickel, and zinc in soil. Geoderma 122, 297–303.

Kiikkilä, O., Pennanen, T., Perkiömäki, J., Derome, J., Fritze, H., 2002. Organic material as a copper immobilising agent: A microcosm study on remediation. Basic. Appl. Ecol. 3, 245–253.

Kögel-Knabner, I., 2002. The macromolecular organic composition of plant and microbial residues as inputs to soil organic matter. Soil Biol. Biochem. 34, 139–162.

Lamb, D.T., Ming, H., Megharaj, M., Naidu, R., 2009. Heavy metal (Cu, Zn, Cd and Pb) partitioning and bioaccessibility in uncontaminated and long-term contaminated soils. J. Hazard. Mater. 171, 1150–1158.

Lehmann, J., Joseph, S., 2009. Biochar for environmental management: Science and technology. Earthscan.

Lima, I., Marshall, W.E., 2005a. Utilisation of turkey manure as granular activated carbon: Physical, chemical and adsorptive properties. Waste Manag. 25, 726–732.

Lima, I.M., Marshall, W.E., 2005b. Adsorption of selected environmentally important metals by poultry manure-based granular activated carbons. J. Chem. Technol. Biotechnol. 80, 1054–1061.

Liu, L., Chen, H., Cai, P., Liang, W., Huang, Q., 2009. Immobilisation and phytotoxicity of Cd in contaminated soil amended with chicken manure compost. J. Hazard. Mater. 163, 563–567.

Lu, A., Zhang, S., Shan, X.-q, 2005. Time effect on the fractionation of heavy metals in soils. Geoderma 125, 225–234.

Lyubchik, S.I., Lyubchik, A.I., Galushko, O.L., Tikhonova, L.P., Vital, J., Fonseca, I.M., Lyubchik, S.B., 2004. Kinetics and thermodynamics of the Cr (III) adsorption on the activated carbon from co-mingled wastes. Colloids and Surfaces A: Physicochemical and Engineering Aspects 242, 151–158.

Ma, Y., Uren, N., 1998. Transformations of heavy metals added to soil—application of a new sequential extraction procedure. Geoderma 84, 157–168.

Madejón, E., De Mora, A.P., Felipe, E., Burgos, P., Cabrera, F., 2006. Soil amendments reduce trace element solubility in a contaminated soil and allow regrowth of natural vegetation. Environ. Pollut. 139, 40–52.

Manta, D.S., Angelone, M., Bellanca, A., Neri, R., Sprovieri, M., 2002. Heavy metals in urban soils: A case study from the city of Palermo (Sicily), Italy. Sci. Total Environ. 300, 229–243.

Mapanda, F., Mangwayana, E., Nyamangara, J., Giller, K., 2005. The effect of long-term irrigation using wastewater on heavy metal contents of soils under vegetables in Harare, Zimbabwe. Agriculture, ecosystems & environ. 107, 151–165.

Martínez, C.E., Jacobson, A.R., McBride, M.B., 2003a. Aging and temperature effects on DOC and elemental release from a metal contaminated soil. Environ. pollut. 122, 135–143.

Martinez, F., Cuevas, G., Calvo, R., Walter, I., 2003b. Biowaste effects on soil and native plants in a semiarid ecosystem. J. Environ. Qual. 32, 472–479.

Massoura, S.T., Echevarria, G., Becquer, T., Ghanbaja, J., Leclerc-Cessac, E., Morel, J.-L., 2006. Control of nickel availability by nickel bearing minerals in natural and anthropogenic soils. Geoderma 136, 28–37.

McBride, M., 1989. Reactions controlling heavy metal solubility in soils, Advances in soil science. Springer. pp. 1–56.

McBride, M., Sauve, S., Hendershot, W., 1997. Solubility control of Cu, Zn, Cd and Pb in contaminated soils. Eur. J. Soil Sci. 48, 337–346.

McBride, M.B., 1995. Toxic metal accumulation from agricultural use of sludge: Are USEPA regulations protective? J. Environ. Qual. 24, 5–18.

McColl, J., Pohlman, A., 1986. Soluble organic acids and their chelating influence on Al and other metal dissolution from forest soils. Water. Air. Soil. Pollut. 31, 917–927.

McLaughlin, M.J., 2001. Ageing of metals in soils changes bioavailability.

Meharg, A.A., 2005. Mechanisms of plant resistance to metal and metalloid ions and potential biotechnological applications, Root Physiology: From Gene to Function. Springer. 163–174.

Mellis, E.V., Cruz, M.C.P.d., Casagrande, J.C., 2004. Nickel adsorption by soils in relation to pH, organic matter, and iron oxides. Scientia Agricola 61, 190–195.

Mielke, H.W., Powell, E.T., Gonzales, C.R., Mielke, P.W., Ottesen, R.T., Langedal, M., 2006. New Orleans soil lead (Pb) cleanup using Mississippi River alluvium: Need, feasibility, and cost. Environ. Sci. Technol. 40, 2784–2789.

Misra, S., Pande, P., 1974a. Effect of organic matter on availability of nickel. Plant and soil 40, 679–684.

Misra, S., Pande, P., 1974b. Evaluation of a suitable extractant for available nickel in soils. Plant and soil 41, 697–700.

Miyasaka, S., Hollyer, J., Kodani, L., 2001. Mulch and compost effects on yield and corm rots of taro. Field Crops Research 71, 101–112.

Molas, J., Baran, S., 2004. Relationship between the chemical form of nickel applied to the soil and its uptake and toxicity to barley plants (*Hordeum vulgare* L.). Geoderma 122, 247–255.

Moreira, C.S., Casagrande, J.C., Alleoni, L.R.F., de Camargo, O.A., Berton, R.S., 2008. Nickel adsorption in two Oxisols and an Alfisol as affected by pH, nature of the electrolyte, and ionic strength of soil solution. J. Soils and Sediments 8, 442–451.

Morera, M., Echeverria, J., Garrido, J., 2002. Bioavailability of heavy metals in soils amended with sewage sludge. Can. J. Soil Sci. 82, 433–438.

Muchuweti, M., Birkett, J., Chinyanga, E., Zvauya, R., Scrimshaw, M.D., Lester, J., 2006. Heavy metal content of vegetables irrigated with mixtures of wastewater and sewage sludge in Zimbabwe: Implications for human health. Agriculture, ecosystems & environ. 112, 41–48.

Muhammad, S., 2009. Phytoavailability of nickel (Ni) in contaminated soils in response to soil-applied amendments. University of Agriculture, Faisalabad.

Murillo, J., Cabrera, F., López, R., Martín-Olmedo, P., 1995. Testing low-quality urban composts for agriculture: Germination and seedling performance of plants. Agriculture, ecosystems & environ. 54, 127–135.

Murtaza, G., Ghafoor, A., Qadir, M., Owens, G., Aziz, M., Zia, M., 2010. Disposal and use of sewage on agricultural lands in Pakistan: A review. Pedosphere 20, 23–34.

Naidu, R., 1996. Contaminants and the soil environment in the Australasia-Pacific region: Proceedings of the First Australasia-Pacific Conference on Contaminants and Soil Environment in the Australasia-Pacific Region, held in Adelaide, Australia, 18–23 February 1996, 1st Australasia-Pacific Conference on Contaminants and Soil Environment in the Australasia-Pacific Region, Adelaide, S. Aust. (USA), 1996. Kluwer Academic.

Naidu, R., Bolan, N.S., Kookana, R.S., Tiller, K., 1994. Ionic-strength and pH effects on the sorption of cadmium and the surface charge of soils. Eur. J. Soil Sci. 45, 419–429.

Narwal, R., Singh, B., 1998. Effect of organic materials on partitioning, extractability and plant uptake of metals in an alum shale soil. Water. Air. Soil. Pollut. 103, 405–421.

Novak, J.M., Busscher, W.J., Laird, D.L., Ahmedna, M., Watts, D.W., Niandou, M.A., 2009. Impact of biochar amendment on fertility of a southeastern Coastal Plain soil. Soil science 174, 105–112.

Ouédraogo, E., Mando, A., Zombré, N., 2001. Use of compost to improve soil properties and crop productivity under low input agricultural system in West Africa. Agric. ecosystems environ. 84, 259–266.

Paraskeva, P., Kalderis, D., Diamadopoulos, E., 2008. Production of activated carbon from agricultural by-products. J. Chem. Technol. Biotechnol. 83, 581–592.

Park, J.H., Lamb, D., Paneerselvam, P., Choppala, G., Bolan, N., Chung, J.-W., 2011. Role of organic amendments on enhanced bioremediation of heavy metal (loid) contaminated soils. J. Hazard. Mater. 185, 549–574.

Paulose, B., Datta, S.P., Rattan, R.K., Chhonkar, P.K., 2007. Effect of amendments on the extractability, retention and plant uptake of metals on a sewage-irrigated soil. Environ. pollut. 146, 19–24.

Petruzzelli, G., Lubrano, L., Guidi, G., 1989. Uptake by corn and chemical extractability of heavy metals from a four year compost treated soil. Plant and soil 116, 23–27.

Pietikäinen, J., Kiikkilä, O., Fritze, H., 2000. Charcoal as a habitat for microbes and its effect on the microbial community of the underlying humus. Oikos 89, 231–242.

Pocknee, S., Sumner, M.E., 1997. Cation and nitrogen contents of organic matter determine its soil liming potential. Soil Sci. Soc. America J. 61, 86–92.

Poulsen, I.F., Hansen, H.C.B., 2000. Soil sorption of nickel in presence of citrate or arginine. Water, Air, and Soil Pollut. 120, 249–259.

Römkens, P.F., Dolfing, J., 1998. Effect of Ca on the solubility and molecular size distribution of DOC and Cu binding in soil solution samples. Environ. Sci. Technol. 32, 363–369.

Rooney, C.P., Zhao, F.-J., McGrath, S.P., 2007. Phytotoxicity of nickel in a range of European soils: Influence of soil properties, Ni solubility and speciation. Environ. Pollut. 145, 596–605.

Ross, S.M., 1994. Toxic metals in soil-plant systems. John Wiley & Sons ltd.

Sabir, M., Ghafoor, A., Saifullah, M.Z.-u.-R., Murtaza, G., 2008. Effect of organic amendments and incubation time on extractability of Ni and other metals from contaminated soils. Pak. J. Agri. Sci. 45, 1.

Sabir, M., Hanafi, M.M., Aziz, T., Ahmad, H.R., Zia-Ur-Rehman, M., Saifullah, G.M., Hakeem, K.R., 2013. Comparative effect of activated carbon, pressmud and poultry manure on immobilisation and concentration of metals in maize (Zea mays) grown on contaminated soil. Int. J. Agriculture and Biol. 15, 559–564.

Samsøe-Petersen, L., Larsen, E.H., Larsen, P.B., Bruun, P., 2002. Uptake of trace elements and PAHs by fruit and vegetables from contaminated soils. Environ. Sci. Technol. 36, 3057–3063.

Sauve, S., Hendershot, W., Allen, H.E., 2000. Solid-solution partitioning of metals in contaminated soils: Dependence on pH, total metal burden, and organic matter. Environ. Sci. Technol. 34, 1125–1131.

Schützendübel, A., Polle, A., 2002. Plant responses to abiotic stresses: Heavy metal–induced oxidative stress and protection by mycorrhization. J. Exp. Bot. 53, 1351–1365.

Seregin, I., Kozhevnikova, A., 2006. Physiological role of nickel and its toxic effects on higher plants. Russ. J. Plant Physiol. 53, 257–277.

Shuman, L.M., 1999. Organic waste amendments effect on zinc fractions of two soils. J. Environ. Qual. 28, 1442–1447.

Silveira, M.L.A., Alleoni, L.R.F., Guilherme, L.R.G., 2003. Biosolids and heavy metals in soils. Sci. Agricola 60, 793–806.

Singh, B.R., Oste, L., 2001. In situ immobilisation of metals in contaminated or naturally metal-rich soils. Environ. Rev. 9, 81–97.

Spark, K., Johnson, B., Wells, J., 1995. Characterising heavy–metal adsorption on oxides and oxyhydroxides. Eur. J. Soil Sci. 46, 621–631.

Stalikas, C., Mantalovas, A.C., Pilidis, G., 1997. Multielement concentrations in vegetable species grown in two typical agricultural areas of Greece. Soil. Sci. Soc. Am. J. 206, 17–24.

Sternbeck, J., Sjödin, Å., Andréasson, K., 2002. Metal emissions from road traffic and the influence of resuspension—results from two tunnel studies. Atmos. Environ. 36, 4735–4744.

Stevenson, F., Fitch, A., 1986. Chemistry of complexation of metal ions with soil solution organics. Interact. Soil Minerals Nat. Organics and Microbes. 29–58.

Stevenson, F.J., 1994. Humus chemistry: Genesis, composition, reactions. John Wiley & Sons.

Stevenson, F.J., Cole, M.A., 1999. Cycles of soils: Carbon, nitrogen, phosphorus, sulfur, micronutrients. Wiley. com.

Stewart, B., Robinson, C., Parker, D.B., 2000. Examples and case studies of beneficial reuse of beef cattle by-products. Land Appl. Agric. Ind. Municipal by-prod. 387–407.

Strobel, B.W., Hansen, H.C.B., Borggaard, O.K., Andersen, M.K., Raulund-Rasmussen, K., 2001. Cadmium and copper release kinetics in relation to afforestation of cultivated soil. Geochimica et Cosmochimica Acta 65, 1233–1242.

Sulieman, S.A., Hago, T.E., 2009. The effects of phosphorus and farmyard manure on nodulation and growth attributes of common bean (Phaseolus vulgaris L.) in Shambat soil under irrigation. Res. J. Agric. Biol. Sci. 5, 458–464.

Sun, B., Zhao, F., Lombi, E., McGrath, S., 2001. Leaching of heavy metals from contaminated soils using EDTA. Environ. Pollut. 113, 111–120.

Tripathi, G., Bhardwaj, P., 2004. Decomposition of kitchen waste amended with cow manure using an epigeic species (Eisenia fetida) and an anecic species (Lampito mauritii). Bioresour. Technol. 92, 215–218.

Üçer, A., Uyanik, A., Aygün, Ş., 2006. Adsorption of Cu (II), Cd (II), Zn (II), Mn (II) and Fe (III) ions by tannic acid immobilised activated carbon. Sep. Purif. Technol. 47, 113–118.

Unsal, T., Sozudogru Ok, S., 2001. Description of characteristics of humic substances from different waste materials. Bioresour. Technol. 78, 239–242.

van Herwijnen, R., Hutchings, T.R., Al-Tabbaa, A., Moffat, A.J., Johns, M.L., Ouki, S.K., 2007. Remediation of metal contaminated soil with mineral-amended composts. Environ. Pollut. 150, 347–354.

Walker, D.J., Clemente, R., Bernal, M.P., 2004. Contrasting effects of manure and compost on soil pH, heavy metal availability and growth of Chenopodium album L. in a soil contaminated by pyritic mine waste. Chemosphere 57, 215–224.

Walker, D.J., Clemente, R., Roig, A., Bernal, M.P., 2003. The effects of soil amendments on heavy metal bioavailability in two contaminated Mediterranean soils. Environ. Pollut. 122, 303–312.

Weng, C.-H., Huang, C., 2004. Adsorption characteristics of Zn (II) from dilute aqueous solution by fly ash. Colloids and Surf. A: Physicochemical and Eng. Aspects 247, 137–143.

Wilson, M.J., Maliszewska-Kordybach, B., 2000. Soil Quality, Sustainable Agriculture and Environmental Security in Central End Eastern Europe. Springer.

Yassen, A., Badran, N., Zaghloul, S., 2007. Role of some organic residues as tools for reducing heavy metals hazard in plants. World J. Agric. Sci. 3, 204–209.

Yin, Y., Impellitteri, C.A., You, S.-J., Allen, H.E., 2002. The importance of organic matter distribution and extract soil: Solution ratio on the desorption of heavy metals from soils. Sci. Total Environ. 287, 107–119.

Yoo, M.S., James, B.R., 2002. Zinc extractability as a function of pH in organic waste-amended soils. Soil sci. 167, 246–259.

You, S.-J., Yin, Y., Allen, H.E., 1999. Partitioning of organic matter in soils: Effects of pH and water/soil ratio. Sci. Total Environ. 227, 155–160.

Yuan, G., Lavkulich, L., 1997. Sorption behavior of copper, zinc, and cadmium in response to simulated changes in soil properties. Commun. Soil Sci. Plant Anal. 28, 571–587.

Zaniewicz-Bajkowska, A., Rosa, R., Franczuk, J., Kosterna, E., 2007. Direct and secondary effect of liming and organic fertilisation on cadmium content in soil and in vegetables. Plant Soil and Environ. 53, 473.

Zheng, N., Liu, J., Wang, Q., Liang, Z., 2010. Health risk assessment of heavy metal exposure to street dust in the zinc smelting district, Northeast of China. Sci. Total Environ. 408, 726–733.

Zhou, L., Wong, J., 2001. Effect of dissolved organic matter from sludge and sludge compost on soil copper sorption. J. Environ. Qual. 30, 878–883.

Zsolnay, A., 2003. Dissolved organic matter: Artefacts, definitions, and functions. Geoderma 113, 187–209.

Chapter 18

Soil Contamination, Remediation and Plants: Prospects and Challenges

M.S. Abdullahi
Department of Chemistry, Federal College of Education, Kontagora, Nigeria

INTRODUCTION

Vegetables constitute parts of the human diet since they contain carbohydrates, proteins, vitamins, minerals and trace elements. Their consumption is on the increase, particularly among the urban community, due to increased awareness of their food values as a result of exposure to other cultures and education. However, they contain both essential and toxic elements over a wide range of concentrations. Metals accumulation in vegetables may thus pose threat to human health. Heavy metals are contaminants in the tissues of vegetables. Heavy metals such as cadmium, copper, lead, chromium and mercury are pollutants, particularly in areas under irrigation using wastewater. Investigations on water, soil, sediments and vegetables irrigated with wastewater are available in literatures from different researchers of different parts of Africa.

Agriculture is one of the major sources of environmental pollution (Gucel et al., 2009; Ashraf et al., 2010 a, b, 2012; Ozturk et al., 2010; Hakeem et al., 2013). Efforts were made and are still being made to improve the productivity of the low-nutrient status of soil in tropical Africa so as to enhance food production in order to sustain the projected population growth necessitate irrigated farming system. To ensure the success of this objective, the use of inorganic fertilizers had tripled. Superphosphate fertilizers contain not only the necessary elements for plant nutrients and growth, but impurities such as Cd, Pb or Hg. Thus fertilizer application is a source of soil contamination. As such, the intake of these heavy metals by plants, especially the leafy vegetables, is an avenue of their entry into the human food chain with harmful effects on health. The level of heavy metals in soils through phosphate fertilizer application depends on the origin of phosphate rocks from which the fertilizer is manufactured.

Soil Remediation and Plants. http://dx.doi.org/10.1016/B978-0-12-799937-1.00018-8

Industrial or municipal wastewaters are used for the irrigation of crops in a particular ecosystem, due to their availability and due to the scarcity of fresh water (Ozturk et al., 2011). Irrigation with wastewater contributes to heavy metal contents in soil. Heavy metals are harmful because of their non-biodegradable nature, long biological half-lives and their potentials to accumulate in different body parts. Heavy metals are toxic because of their solubility in water. Low concentrations of heavy metals have damaging effects on humans and animals because there is no mechanism for their elimination from the body. Wastewater contains substantial amounts of toxic metals which create problems. A high accumulation of these metals in agricultural soils through wastewater irrigation does not only result in soil contamination, but also affects food quality and safety. Food and water are the sources of essential metals: these are also the media through which we are exposed to toxic metals and their accumulation in the edible parts of leafy vegetables.

Environmental pollution is the undesirable change in the atmosphere, hydrosphere and lithosphere. Advanced industrialization processes have provided a comfort to man but have also resulted in the indiscriminate release of gases and liquids which pollute the environment. Large amounts of untreated sewage and industrial water have been discharged into surface bodies of rivers for disposal and caused their contamination. With water shortage, farmers use wastewater to irrigate their fields. Such irrigation practices give good crop yields as the water contains organic material and some inorganic elements essential for their growth. It may also contain non-essential metals which, when present in large amounts, could be transferred to man and animals through the food chain (Saeed and Oladeji, 2013).

Long-term use of polluted water for cultivation of vegetables has resulted in the accumulation of trace metals in agricultural soils, affecting microbial activities as well as their transfer to various crops under cultivation with levels of contamination that exceed permissible levels. The uptake and bioavailability of trace metals to plants is controlled by factors associated with soil and climatic conditions, plant genotype, and agronomic management, activities/passive transfer processes, sequestration and speciation, soil PH, soil organic matter content, redox states, type of root system, the response of plants to the elements, seasonal cycles and competitive metal interactions (Tukura et al., 2013).

Similarly, the uptake and bioavailability of heavy metals in vegetables are influenced by many factors such as climate, atmosphere dispositions, the concentrations of heavy metals in soil, the nature of soil and the degree of maturity of the plant at harvest. Air pollution may pose a threat to post-harvest vegetables during transportation and marketing, causing elevated levels of heavy metals in vegetables (Sale et al., 2013).

Rapid urbanization and industrial development during the last decades have provoked some serious concerns for the environment (Ozturk et al.,

2010a, b). Heavy metal contaminations are a very serious health issues in many cities of the world especially the highly concentrated industrial areas. The sources of heavy metals in the environment are natural or anthropogenic. Usually the concentration of heavy metals in a natural environment is very low and is mostly derived from the mineralogy and the weathering material components. Main anthropogenic sources of heavy metal contamination are mining, disposal of untreated and partially treated effluents, indiscriminate use of heavy-metal-containing fertilizer and pesticides in agricultural fields. The natural sources are volcanic eruptions, erosion and weathering. Trace metal concentrations are important due to their potential toxicity to the environment and humans. Some of the metals like Cu, Fe, Mn, Ni and Zn are essential micronutrients for the life processes in animals and plants while many other metals such as Cd, Cr, Pb Hg and As have no known physiological activities. Metals are non-degradable and can accumulate in the human body causing damage to the nervous system and internal organs (Ibiam and Ekpe, 2013).

SOURCES OF HEAVY METALS IN SOIL

Heavy metals occur naturally in the soil environment from the pedogenetic processes of weathering of parent materials at levels that are regarded as trace ($< 1000 \, mg \, kg^{-1}$) and rarely toxic (Pierzynski et al., 2000; Kabata-Pendias and Pendias, 2001). Due to the disturbance and acceleration of nature's slowly occurring geochemical cycle of metals by man, most soils of rural and urban environments may accumulate one or more of the heavy metals above at defined background values high enough to cause risks to human health, plants, animals, ecosystems or other media (Raymond and Felix, 2011). The heavy metals essentially become contaminants in the soil environments because (1) their rates of generation via man-made cycles are more rapid relative to natural ones, (2) they become transferred from mines to random environmental locations where higher potentials of direct exposure occur, (3) the concentrations of the metals in discarded products are relatively high compared to those in the receiving environment and (4) the chemical form (species) in which a metal is found in the receiving environmental system may render it more bioavailable.

Heavy metals in the soil from anthropogenic sources tend to be more mobile, hence bioavailable than pedogenic or lithogenic ones (Kuo et al, 1993; Kaasalainen and Yli-Halla, 2003). Metal-bearing solids at contaminated sites can originate from a wide variety of anthropogenic sources in the form of metal mine tailings, disposal of high metal wastes in improperly protected landfills, leaded gasoline and lead-based paints, land application of fertilizer, animal manures, biosolids (sewage sludge), compost, pesticides, coal combustion residues, petrochemicals and atmospheric deposition are discussed below (Alloway, 1990; Adriano, 2001; Nicholson et al., 2003; Zhang, 2006; Huang et al., 2007; Raymond and Felix, 2011).

Fertilizers

Historically, agriculture was the first major human influence on the soil. To grow and complete the lifecycle, plants must acquire not only macronutrients (N, P, K, S, Ca and Mg), but also essential micronutrients. Some soils are deficient in the heavy metals (such as Co, Cu, Fe, Mn, Mo, Ni and Zn) that are essential for healthy plant growth, and crops may be supplied with these as an addition to the soil or as a foliar spray. Cereal crops grown on Cu-deficient soils are occasionally treated with Cu as an addition to the soil, and Mn may similarly be supplied to cereal and root crops. Large quantities of fertilizers are regularly added to soils in intensive farming systems to provide adequate N, P and K for crop growth. The compounds used to supply these elements contain trace amounts of heavy metals (e.g. Cd and Pb) as impurities, which, after continued fertilizer application, may significantly increase their content in the soil (Raymond and Felix, 2011). Metals, such as Cd and Pb, have no known physiological activity. Application of certain phosphatic fertilizers inadvertently adds Cd and other potentially toxic elements to the soil, including F, Hg and Pb.

Pesticides

Several common pesticides used fairly extensively in agriculture and horticulture in the past contained substantial concentrations of metals. For instance, in the recent past, about 10% of the chemicals approved for use as insecticides and fungicides in United Kingdom were based on compounds which contain Cu, Hg, Mn, Pb or Zn. Examples of such pesticides are copper-containing fungicidal sprays such as Bordeaux mixture (copper sulphate) and copper oxychloride. Lead arsenate was used in fruit orchards for many years to control some parasitic insects. Arsenic-containing compounds were also used extensively to control cattle ticks and to control pests in banana in New Zealand and Australia, timbers have been preserved with formulations of Cu, Cr and As (CCA), and there are now many derelict sites where soil concentrations of these elements greatly exceed background concentrations. Such contamination has the potential to cause problems, particularly if sites are redeveloped for other agricultural or nonagricultural purposes. Compared with fertilizers, the use of such materials has been more localized, being restricted to particular sites or crops (McLaughlin et al., 2000).

Biosolids and Manures

The application of numerous biosolids (e.g. livestock manures, composts and municipal sewage sludge) to land inadvertently leads to the accumulation of heavy metals such as As, Cd, Cr, Cu, Pb, Hg, Ni, Se, Mo, Zn, Tl, Sb and so forth, in the soil (Raymond and Felix, 2011). Certain animal wastes such as poultry, cattle and pig manures produced in agriculture are commonly applied to crops and pastures either as solids or slurries. Although most manures are seen as valuable fertilizers, in the pig and poultry industry, the Cu and Zn added to diets as growth promoters and As contained in poultry health products may also have the potential to cause metal contamination of the soil (Chaney and Oliver, 1996). The manures produced from animals on such diets contain high concentrations of As, Cu and Zn and, if repeatedly applied to restricted areas of land, can cause considerable build-up of these metals in the soil in the long run.

Biosolids (sewage sludge) are primarily organic solid products, produced by wastewater treatment processes that can be beneficially recycled (Raymond and Felix, 2011). Land application of biosolids materials is a common practice in many countries that allow the reuse of biosolids produced by urban populations (Weggler et al., 2004). The term sewage sludge is used in many references because of its wide recognition and its regulatory definition. However, the term biosolids is becoming more common as a replacement for sewage sludge because it is thought to reflect more accurately the beneficial characteristics inherent to sewage sludge (Raymond and Felix, 2011). It is estimated that in the United States, more than half of approximately 5.6 million dry tons of sewage sludge used or disposed of annually is land applied, and agricultural utilization of biosolids occurs in every region of the country. In the European community, over 30% of the sewage sludge is used as fertilizer in agriculture (Weggler et al., 2004). In Australia over 175,000 tons of dry biosolids are produced each year by the major metropolitan authorities, and currently most biosolids applied to agricultural land are used in arable cropping situations where they can be incorporated into the soil (McLaughlin et al., 2000).

There is also considerable interest in the potential for composting biosolids with other organic materials such as sawdust, straw or garden waste. If this trend continues, there will be implications for metal contamination of soils. The potential of biosolids for contaminating soils with heavy metals has caused great concern about their application in agricultural practices (Raymond and Felix, 2011). Heavy metals most commonly found in biosolids are Pb, Ni, Cd, Cr, Cu and Zn, and the metal concentrations are governed by the nature and the intensity of the industrial activity, as well as the type of process employed during the biosolids treatment (Mattigod and Page, 1983). Under certain conditions, metals added to soils in applications of biosolids can be leached downwards through the soil profile and can have the potential to contaminate groundwater. Recent studies on some New Zealand soils treated with biosolids have shown increased concentrations of Cd, Ni and Zn in drainage leachates (Raymond and Felix, 2011).

Wastewater

The application of municipal and industrial wastewater and related effluents to land dates back 400 years and is now a common practice in many parts of the world. Worldwide, it is estimated that 20 million hectares of arable land are irrigated with wastewater. In several Asian and African cities, studies suggest that agriculture based on wastewater irrigation accounts for 50% of the vegetable supply to urban areas (Bjuhr, 2007). Farmers generally are not bothered about environmental benefits or hazards and are primarily interested in maximizing their yields and profits. Although the metal concentrations in wastewater effluents are usually relatively low, long-term irrigation of land with such can eventually result in heavy metal accumulation in the soil.

Metal Mining, Milling Processes and Industrial Wastes

Mining and milling of metal ores coupled with industries have bequeathed many countries the legacy of wide distribution of metal contaminants in soil. During mining, tailings (heavier and larger particles settled at the bottom of the flotation cell during mining) are directly discharged into natural depressions, including on-site wetlands resulting in elevated concentrations (Raymond and Felix, 2011). Extensive Pb and Zn ore mining and smelting have resulted in contamination of soil that poses risk to human and ecological health. Many reclamation methods used for these sites are lengthy and expensive and may not restore soil productivity. Soil heavy metal environmental risk to humans is related to bioavailability. Assimilation pathways include the ingestion of plant material grown in (food chain), or the direct ingestion (oral bioavailability) of contaminated soil (Basta and Gradwohl, 1998).

Other materials are generated by a variety of industries such as textile, tanning, petrochemicals from accidental oil spills or utilization of petroleum-based products, pesticides, and pharmaceutical facilities and are highly variable in composition. Although some are disposed of on land, few have benefits to agriculture or forestry. In addition, many are potentially hazardous because of their contents of heavy metals (Cr, Pb and Zn) or toxic organic compounds and are seldom, if ever, applied to land. Others are very low in plant nutrients or have no soil-conditioning properties (Raymond and Felix, 2011).

Airborne Sources

Airborne sources of metals include stack or duct emissions of air, gas, or vapor streams, and fugitive emissions such as dust from storage areas or waste piles. Metals from airborne sources are generally released as particulates contained in the gas stream. Some metals such as As, Cd and Pb can also volatilize during high-temperature processing. These metals will convert to oxides and condense as fine particulates unless a reducing atmosphere is maintained (Smith et al., 1995). Stack emissions can be distributed over a wide area by natural air currents until dry and/or wet precipitation mechanisms remove them from the gas stream. Fugitive emissions are often distributed over a much smaller area because emissions are made near the ground. In general, contaminant concentrations are lower in fugitive emissions compared to stack emissions. The type and concentration of metals emitted from both types of sources will depend on site-specific conditions. All solid particles in smoke from fires and in other emissions from factory chimneys are eventually deposited on land or sea; most forms of fossil fuels contain some heavy metals and this is, therefore, a form of contamination which has been continuing on a large scale since the industrial revolution began. For example, very high concentrations of Cd, Pb and Zn have been found in plants and soils adjacent to smelting works. Another major source of soil contamination is the aerial emission of Pb from the combustion of petrol containing tetraethyl lead; this contributes substantially to the content of Pb in soils in urban areas and in those adjacent to major roads. Zn and Cd may also be added to soils adjacent to roads, the sources being tyres and lubricant oils (USEPA, 1996).

POTENTIAL RISK OF HEAVY METALS TO SOIL

The most common heavy metals found at contaminated sites, in order of abundance are Pb, Cr, As, Zn, Cd, Cu and Hg (USEPA, 1996). These metals are important since they are capable of decreasing crop production due to the risk of bioaccumulation and biomagnification in the food chain. There is also the risk of superficial and groundwater contamination. Knowledge of the basic chemistry, and environmental and associated health effects of these heavy metals is necessary in understanding their speciation, bioavailability and remedial options. The fate and transport of a heavy metal in soil depend significantly on the chemical form and speciation of the metal. Once in the soil, heavy metals are adsorbed by initial fast reactions (minutes, hours), followed by slow adsorption reactions (days, years) and are, therefore, redistributed into different chemical forms with varying bioavailability, mobility and toxicity. This distribution is believed to be controlled by reactions of heavy metals in soils such as (1) mineral precipitation and dissolution, (2) ion exchange, adsorption and desorption, (3) aqueous complexation, (iv) biological immobilization and mobilization, and (5) plant uptake (Raymond and Felix, 2011).

SOIL CONCENTRATION RANGES AND REGULATORY GUIDELINES FOR SOME HEAVY METALS

The specific type of metal contamination found in contaminated soil is directly related to the operation that occurred at the site. The range of contaminant concentrations and the physical and chemical forms of contaminants will also depend on activities and disposal patterns for contaminated wastes on the site. Other factors that may influence the form, concentration and distribution of metal contaminants include soil and ground-water chemistry and local transport mechanisms (GWRTAC, 1997).

Soils may contain metals in the solid, gaseous, or liquid phases, and this may complicate the analysis and interpretation of reported results. For example, the most common method for determining the concentration of metal contaminants in soil is via total elemental analysis (USEPA Method 3050). The level of metal contamination determined by this method is expressed as $mg\,metal\,kg^{-1}$ soil. This analysis does not specify requirements for the moisture content of the soil and may therefore include soil water. This measurement may also be reported on a dry soil basis. The level of contamination may also be reported as leachable metals as determined by leach tests, such as the toxicity characteristic leaching procedure (TCLP) (USEPA Method 1311) or the synthetic precipitation-leaching procedure, or SPLP test (USEPA Method 1312). These procedures measure the concentration of metals in leachate from soil contacted with an acetic acid solution (TCLP) (DPR-EGASPIN, 2002) or a dilute solution of sulfuric and nitric acid (SPLP). In this case, metal contamination is expressed in $mg\,l^{-1}$ of the leachable metal. Other types of leaching tests have been proposed including sequential extraction procedures (Tessier et al., 1979; Ure et al., 1993) and extraction of acid volatile sulfide (DiToro et al., 1992). Sequential procedures contact the solid with a series of extractant solutions that are designed to dissolve different fractions of the associated metal. These tests may provide insight into the different forms of metal contamination present. Contaminant concentrations can be measured directly in metal-contaminated water. These concentrations are most commonly expressed as total dissolved metals in mass concentrations ($mg\,l^{-1}$ or $g\,l^{-1}$) or in molar concentrations ($mol\,l^{-1}$). In dilute solutions, a $mg\,l^{-1}$ is equivalent to one part per million (ppm), and $g\,l^{-1}$ is equivalent to one part per billion (ppb).

    Riley et al. (1992) and NJDEP (1996) have reported soil concentration ranges and regulatory guidelines for some heavy metals (Table 18.1). In Nigeria, in the interim period, whilst suitable parameters are being developed, the Department of Petroleum Resources has recommended guidelines for remediation of contaminated land based on two parameters intervention values and target values (Table 18.2) below.

REMEDIATION OF CONTAMINATED SOIL BY HEAVY METALS

Remediation of soils contaminated by heavy metals is another attempt to clean-up the soils used for agricultural practices, reduce risk associated with heavy metals, improve soil nutrients and increase food production. Soil remediation is a necessary tool for the enhancement of food security. In general, it is very difficult to eliminate metals from the environment. Once metals get into the soil they remain there and do not degrade like organic molecules. However, metals like mercury and selenium are reported to be transformed and volatilized by microorganisms.

    The overall objective of any soil remediation approach is to create a final solution that is protective of human health and the environment. Remediation is generally subject to an array of regulatory requirements and can also be based on assessments of human health and ecological risks where no legislated standards exist or where standards are advisory. The regulatory authorities will normally accept remediation strategies that centre on reducing metal

TABLE 18.1 Soil Concentration Ranges and Regulatory Guidelines for Some Heavy Metals

Metal	Soil Concentration Range (mg/kg)	Regulatory Limits (mg/kg)
Pb	1.00–69,000	600
Cd	0.10–345	100
Cr	0.05–3950	100
Hg	<0.01–1800	270
Zn	150–5000	1500

Sources: Riley et al. (1992); NJDEP (1996).

TABLE 18.2 Target and Intervention Values for Some Metals for a Standard Soil

Metal	Target Value (mg/kg)	Intervention Value (mg/kg)
Ni	140.00	720.00
Cu	0.30	10.00
Zn	—	—
Cd	100.00	380.00
Pb	35.00	210
As	200	625
Cr	20	240
Hg	85	530

Source: DPR-EGASPIN (2002).

bioavailability only if reduced bioavailability is equated with reduced risk, and if the bioavailability reductions are demonstrated to be long term (Martin and Ruby, 2004). For heavy-metal-contaminated soils, the physical and chemical form of the heavy metal contaminant in soil strongly influences the selection of the appropriate remediation treatment approach. Information about the physical characteristics of the site and the type and level of contamination at the site must be obtained to enable accurate assessment of site contamination and remedial alternatives. The contamination in the soil should be characterized to establish the type, amount and distribution of heavy metals in the soil. Once the site has been characterized, the desired level of each metal in soil must be determined. This is done by comparison of observed heavy metal concentrations with soil quality standards for a particular regulatory domain, or by performance of a site-specific risk assessment. Remediation goals for heavy metals may be set as total metal concentration or as leachable metal in soil, or as some combination of these (Raymond and Felix, 2011).

Several technologies exist for the remediation of metal-contaminated soil. Gupta et al. (2000) classified remediation technologies of contaminated soils into three categories of hazard-alleviating measures:

1. gentle in situ remediation;
2. in situ harsh soil restrictive measures; and
3. in situ or ex situ harsh soil destructive measures.

The goal of the last two harsh alleviating measures is to avert hazards either to man, plant or animal while the main goal of gentle in situ remediation is to restore the malfunctionality of soil (soil fertility), which allows safe use of the soil. At present, a variety of approaches have been suggested for remediating contaminated soils. USEPA (2007) has broadly classified remediation technologies for contaminated soils into:

1. Source control and
2. Containment remedies.

Source control involves in situ and ex situ treatment technologies for sources of contamination. In situ or in place means that the contaminated soil is treated in its original place; unmoved, unexcavated; remaining at the site or in the subsurface. In situ treatment technologies treat or remove the contaminant from the soil without excavation or removal of the soil. Ex situ means that the contaminated soil is moved, excavated or removed from the site or subsurface. Implementation of ex situ remedies requires excavation or removal of the contaminated soil. Containment remedies involve the construction of vertical engineered barriers, caps and liners used to prevent the migration of contaminants.

Another classification places remediation technologies for heavy-metal-contaminated soils into five categories of general approaches to remediation (Table 18.3): isolation, immobilization, toxicity reduction, physical separation and extraction (USEPA 2007). In practice, it may be more convenient to employ a hybrid of two or more of these approaches for more cost effectiveness. The key factors that may influence the applicability and selection of any of the available remediation technologies are:

1. cost,
2. long-term effectiveness / permanence,
3. commercial availability,
4. general acceptance,
5. applicability to high metal concentrations,

TABLE 18.3 Technologies for Remediation of Heavy-Metal-Contaminated Soils

Category	Remediation Technologies
Isolation	(i) Capping (ii) subsurface barriers.
Immobilization	(i) Solidification/stabilization (ii) vitrification (iii) chemical treatment. Toxicity and/or mobility reduction; physical separation (i) Chemical treatment (ii) permeable treatment walls (iii) biological treatment bioaccumulation, phytoremediation (phytoextraction phytostabilization and rhizofiltration), bioleaching, biochemical processes.
Extraction	(i) Soil washing, pyrometallurgical extraction, in situ soil flushing and electrokinetic treatment.

Source: Raymond and Felix (2011).

6. applicability to mixed wastes (heavy metals and organics),
7. toxicity reduction,
8. mobility reduction, and
9. volume reduction.

Excess heavy metal accumulation in soils is toxic to humans and other animals. Exposure to heavy metals is normally chronic (exposure over a longer period of time), due to food chain transfer. Acute (immediate) poisoning from heavy metals is rare through ingestion or dermal contact, but is possible. Chronic problems associated with long-term heavy metal exposures are:

● Lead – mental lapse;
● Cadmium – affects kidney, liver and GI tract; and
● Arsenic – skin poisoning, affects kidneys and central nervous system.

Hazardous heavy metals such as lead, cadmium, chromium, mercury and arsenic are pervasive in much of the environment and it may not be possible to completely avoid exposure. There are many safe, organic and affordable ways that can help detoxify and eliminate these toxins from the body. One such way is the use of EDTA (ethylenediaminetetraacetic acid). Of all the metals, lead is the most common soil contaminant. Unfortunately, plants do not accumulate lead under natural conditions. A chelator such as EDTA has to be added to the soil as an amendment. The EDTA makes the lead available to the plant (USEPA, 1993). The most common plant used for lead extraction is Indian mustard (*Brassica juncea*). Phytotech (a private research company) has reported that they have cleaned up lead-contaminated sites in New Jersey to below the industrial standards in one to two summers using Indian mustard (Wantanabe, 1997).

Therefore, background knowledge of the sources, chemistry and potential risks of toxic heavy metals in contaminated soils is necessary for the selection of appropriate remedial options. Remediation of soil contaminated by heavy metals is necessary in order to reduce the associated risks, make the land resource available for agricultural production, enhance food security and scale down land tenure problems. Immobilization, soil washing and phytoremediation are frequently listed among the best available technologies for cleaning up heavy-metal-contaminated soils but have been mostly demonstrated in developed countries. These technologies are recommended for field applicability and commercialization in the developing countries also where agriculture, urbanization and industrialization are leaving a legacy of environmental degradation.

PREVENTION OF HEAVY METAL CONTAMINATION

Preventing heavy metal pollution is critical because cleaning contaminated soils is extremely expensive and difficult. Applicators of industrial waste or sludge must abide by the regulatory limits set by the US Environmental Protection Agency (EPA) (see Table 18.4).

Prevention is the best method to protect the environment from contamination by heavy metals. With the above table, a simple equation is used to show the maximum amount of sludge that can be applied. For example, suppose city officials want to apply the maximum amount of sludge ($kg\,ha^{-1}$) on some agricultural land. The annual pollutant-loading rate for zinc is $140\,kg\,ha^{-1}/yr$ (from Table 18.4). The laboratory analysis of the sludge shows a zinc concentration of $7500\,mg\,kg^{-1}$ ($mg\,kg^{-1}$ is the same as parts per million). How much can the applicator apply ($tons\,A^{-1}$) without exceeding the $140\,kg\,ha^{-1}/yr$?

Solution:

1. Convert mg to kg ($1,000,000\,mg = 1\,kg$) so all units are the same: $7500\,mg \times (1\,kg/1,000,000\,mg) = 0.0075\,kg$.
2. Divide the amount of zinc that can be applied by the concentration of zinc in the sludge: $(140\,kg\,Zn\,ha^{-1})/(0.0075\,kg\,Zn$ per kg sludge$) = 18,667\,kg$ sludge ha^{-1}.
3. Convert to $lb\,A^{-1}$: $18,667\,kg\,ha^{-1} \times 0.893 = 16,669\,lbs\,A^{-1}$. Convert lbs to tons: $16,669\,lb\,A^{-1}/2000\,lb\,T^{-1} = 8.3\,T$ sludge per acre.

TRADITIONAL REMEDIATION OF CONTAMINATED SOIL

According to Jacob et al. (2013), traditional treatments for metal contamination in soils are expensive and cost prohibitive when large areas of soil are contaminated. Treatments can be done in situ (on-site), or ex situ (removed and treated

TABLE 18.4 Regulatory Limits on Heavy Metals Applied to Soils

Heavy Metal	Maximum Concentration in Sludge (mg/kg)	Annual Pollutant Loading Rates		Cumulative Pollutant Loading Rates	
		(kg/ha/yr)	(lb/A/yr)	(kg/ha)	(lb/A)
Arsenic	75	21.9	1.8	41	36.6
Cadmium	85	1.9	1.7	39	34.8
Chromium	3000	150	134	3000	2679
Copper	4300	75	67	1500	1340
Lead	420	21	14	420	375
Mercury	840	15	13.4	300	268
Molybdenum	57	0.85	0.80	17	15
Nickel	75	0.90	0.80	18	16
Selenium	100	5	4	100	89
Zinc	7500	140	125	2800	2500

Adapted from USEPA (1993).

off-site). Different methods exist for the remediation of heavy metals in the environment. Some of these methods include mechanical, chemical and bioremediation. Some treatments that are available include:

1. High-temperature treatments (produce a vitrified, granular, non-leachable material).
2. Solidifying agents (produce cement-like material).
3. Washing process (leaches out contaminants).

MANAGEMENT OF CONTAMINATED SOIL

Soil and crop management methods can help prevent uptake of pollutants by plants, leaving them in the soil. The soil becomes the sink, breaking the soil–plant–animal or human cycle through which the toxin exerts its effects (Brady and Weil, 1999). The following management practices will not remove the heavy metal contaminants, but will help to immobilize them in the soil and reduce the potential for adverse effects from the metals. Note that the kind of metal (cation or anion) must be considered.

1. **Increasing the soil pH to 6.5 or higher.** Cationic metals are more soluble at lower pH level, so increasing the pH makes them less available to plants and therefore less likely to be incorporated in their tissues and ingested by humans. Raising the pH has the opposite effect on anionic elements.

2. **Draining wet soils.** Drainage improves soil aeration and will allow metals to oxidize, making them less soluble. Therefore, when aerated, these metals are less available. The opposite is true for chromium, which is more available in oxidized forms. Active organic matter is effective in reducing the availability of chromium.

3. **Applying phosphate.** Heavy phosphate applications reduce the availability of cationic metals, but have the opposite effect on anionic compounds like arsenic. Care should be taken with phosphorus applications because high levels of phosphorus in the soil can result in water pollution.

4. **Carefully selecting plants for use on metal-contaminated soils.** Plants translocate larger quantities of metals to their leaves than to their fruits or seeds. The greatest risk of food-chain contamination is in leafy vegetables like lettuce or spinach. Another hazard is forage eaten by livestock.

The Use of Plants for Environmental Clean-Up

Research has demonstrated that plants are effective in cleaning up contaminated soil (Wenzel et al., 1999). Phytoremediation is a general term for using plants to remove, degrade, or contain soil pollutants such as heavy metals, pesticides, solvents, crude oil, polyaromatic hydrocarbons and landfill leacheates. For example, prairie grasses can stimulate breakdown of petroleum products. Wildflowers were recently used to degrade hydrocarbons from an oil spill in Kuwait. Hybrid poplars can remove ammunition compounds such as TNT as well as high nitrates and pesticides (Brady and Weil, 1999).

The Use of Plants for Treating Metal-Contaminated Soils

Plants have been used to stabilize or remove metals from contaminated soil and water. The three mechanisms used are *phytoextraction*, *rhizofiltration* and *phytostabilization*. This technical note will define rhizofiltration and phytostabilization but will focus on phytoextraction.

Rhizofiltration

Rhizofiltration is the adsorption onto plant roots or absorption into plant roots of contaminants that are in solution surrounding the root zone (rhizosphere). Rhizofiltration is used to decontaminate groundwater. Plants are grown in greenhouses in water instead of soil. Contaminated water from the site is used to acclimatize the plants to the environment. The plants are then planted on the site of contaminated groundwater where the roots take up the water and contaminants. Once the roots are saturated with the contaminant, the plants are harvested including the roots. In Chernobyl, Ukraine, sunflowers were used in this way to remove radioactive contaminants from groundwater (USEPA, 1998).

Phytostabilization

Phytostabilization is the use of perennial, non-harvested plants to stabilize or immobilize contaminants in the soil and groundwater. Metals are absorbed and accumulated by roots, adsorbed on to roots, or precipitated within the rhizosphere. Metal-tolerant plants can be used to restore vegetation where natural vegetation is lacking, thus reducing the risk of water and wind erosion and leaching. Phytostabilization reduces the mobility of the contaminant and prevents further movement of the contaminant into groundwater or the air and reduces the bioavailability for entry into the food chain.

Phytoextraction

Phytoextraction is the process of growing plants in metal-contaminated soil. Plant roots then translocate the metals into above-ground portions of the plant. After plants have grown for some time, they are harvested and incinerated or composted to recycle the metals. Several crop growth cycles may be needed to decrease contaminant levels to allowable limits. If the plants are incinerated, the ash must be disposed of in a hazardous waste landfill, but the volume of the ash is much smaller than the volume of contaminated soil if dug out and removed for treatment. See the following example of disposal.

Example of Disposal

Excavating and landfilling a 10-acre contaminated site to a depth of 1 foot requires handling roughly 20,000 tons of soil. Phytoextraction of the same site would result in the need to handle about 500 tons of biomass, which is about {1/40} of the mass of the contaminated soil. In this example, if we assume the soil was contaminated with a lead concentration of 400 ppm, six to eight crops would be needed, growing four crops per season (Phytotech, 2000).

Phytoextraction is done with plants called hyperaccumulators, which absorb unusually large amounts of metals in comparison to other plants. Hyperaccumulators contain more than $1000 \, \text{mg} \, \text{kg}^{-1}$ of cobalt, copper, chromium, lead or nickel; or $10,000 \, \text{mg} \, \text{kg}^{-1}$ (1%) of manganese or zinc in dry matter (Baker and Brooks, 1989). One or more of these plant types was planted at a particular site based on the kinds of metals present and site conditions. Phytoextraction is easiest with metals such as nickel, zinc and copper because these metals are preferred by a majority of the 400 hyperaccumlator plants. Several plants in the genus *Thlaspi* (pennycress) have been known to take up more than 30,000 ppm (3%) of zinc in their tissues. These plants can be used as an ore because of the high metal concentration (Brady and Weil, 1999).

Preventive Steps

The best solution for combating heavy metal toxins is to avoid them in the first place, today's modern world makes this task almost impossible, but significant improvements can be accomplished using the following guidelines (Ibiam and Ekpe, 2013):

- Avoid mercury–silver dental amalgams;
- Avoid fish and shellfish;
- Clean water;
- Clean food;
- Natural products;
- Thimerosal-free vaccines; and
- Avoid industrial areas.

CLASSIFICATION OF HEAVY METALS

Heavy metals can be classified into four major groups based on their health implications:

- Essential: Cu, Zn, Cr (III), Mn and Fe. These metals are also called micronutrients and are toxic when taken in excess of requirements.
- Nonessential: Ba, Al and Zn.
- Less toxic: Sn and Al.
- Highly toxic: Pb, Cd, As, Cr(VI) and Hg.

SOURCES OF HEAVY METALS IN THE ENVIRONMENT

Heavy metals originate from either natural or anthropogenic sources. The primary anthropogenic sources of heavy metals are point sources such as mines, foundries, smelters and coal-burning power plants, as well as diffuse sources such as combustion by-products and vehicle emissions. Humans also affect the natural geological and biological redistribution of heavy metals by altering the chemical form of heavy metals released to the environment. Natural erosion and weathering of crusted materials take place over long periods of time and the amount of heavy metals released is small.

Major sources of toxic metals in the environment are by human activities both from domestic and industrial wastewaters, effluents and their associated solid waste; cadmium, chromium, copper, lead, and mercury are used extensively in industries and the effluents from these industries contribute to polluting the environment. Table 18.5 gives a list of some metals used in major industries.

Solid waste or sewage sludge is commonly disposed of in landfills or sold as fertilizers. Heavy metals can be released through leaching of sewage sludge in landfills. Sewage sludge also contains plant nutrients and compares favourably to other fertilizers in crop production. The amount of sewage sludge that can be applied to cropland is regulated by the USEPA and depends on the concentrations of heavy metals in the sludge and the soil chemistry of the cropland.

TABLE 18.5 Selected Heavy Metals Presently or Formerly Used in Major Industries

Industries	Cd	Cr	Cu	Pb	Hg
Machinery product	X	X	X	X	X
Paint pigment and ink	X	X	X	X	X
Electroplating	X	X	X	X	
Textile mill	X	X			
Wood, pulp and paper	X	X	X	X	
Organic chemicals	X	X	X	X	
Inorganic chemicals	X	X	X	X	X
Rubber manufacturing	X	X			
Iron and steel foundries	X		X		
Nonferrous metal foundries	X	X	X	X	
Leather processing	X				
Petroleum refining		X	X		
Steam-generating power plants	X	X			

X denotes the heavy metal that is used in the industry.

BENEFITS OF HEAVY METALS TO PLANTS

Heavy metals are required in trace amounts by plants, because they are essential for certain metabolic functions while others are essential components of enzymes and pigments in living systems. Such heavy metals include iron, cobalt, copper, manganese, molybdenum, vanadium, strontium and zinc; but in excess, they can be detrimental to organisms (Harada, 1994). Cobalt is essential for nitrogen fixation by rhizobium in legume noodles. In practice, it has the largest significance to animal nutrition and is observed to be a component of vitamin B12 (cobalamine) molecule.

Iron is an essential element in plants and a micronutrient required in a relatively small quantity. It has a major role in a host of biochemical reactions, particularly in connection with the electron transport chain (cytochromes). However, iron is not part of the chlorophyll molecule but is essential in plants for the synthesis of chlorophyll. Some of the enzymes and carriers that function in the respiratory and photosynthesis mechanisms of living cells are iron compounds, for example cytochromes and ferrodoxin. The principal iron-containing pigments in animals are the red hemoglobin of vertebrate blood, the red erythrocruorin found in many annelids and moulds, and the green chlorocruorin of certain polycheate worms.

Chromium is required for normal carbohydrate, lipid and nucleic acid metabolism. Insufficient dietary chromium leads to impaired glucose and lipid metabolism and may ultimately lead to maturity onset diabetes and/or cardio-vascular diseases. The essentiality of chromium is seen in mice, chicken and sheep. In humans, it acts as a defence mechanism against weight loss, glucose intolerance and impaired energy metabolism.

Nickel is required to maintain health in animals. Small amount of nickel are probably essential for humans, although a lack of nickel has not been found to have any effect.

Copper and zinc are micronutrients essential for the normal growth and development of the body. Copper is found to be essential in hemoglobin formation, production of RNA, cholesterol utilization among other processes.

Zinc is essential for protein synthesis, carbon dioxide transport, and sexual function, among other processes.

Arsenic, lead, mercury and cadmium have no known beneficial importance in living organisms.

FUTURE PROSPECTS

Phytoremediation has been studied extensively in research and at small-scale demonstrations, but in only a few at full-scale applications. Phytoremediation is moving into the realm of commercialization (Watanabe, 1997). It is predicted that the phytoremediation market will reach $214 to $370 million by the year 2015 (Environmental Science & Technology, 1998). Given the current effectiveness, phytoremediation is best suited for clean-up over a wide area in which contaminants are present at low to medium concentrations. Before phytoremediation is fully commercialized, further research is needed to ensure that tissues of plants used for phytoremediation do not have adverse environmental effects if eaten by wildlife or used by humans for things such as mulch or firewood (USEPA, 1998). Research is also needed to find more efficient bioaccumulators and hyperaccumulators that produce more biomass, and to further monitor current field trials to ensure a thorough understanding. There is the need for a commercialized smelting method to extract the metals from plant biomass so they can be recycled. However, phytoremediation is slower than traditional methods of removing heavy metals from soil but much less costly. Prevention of soil contamination is far less expensive than any kind of remediation and much better for the environment.

CHALLENGES

The high concentration of heavy metals in agricultural soils affects the crop output and quality. It also deteriorates the growth, morphology and metabolism of microorganisms in soils. These effects can be considered a real threat to human food safety. In recent years, rapid industrialization and urbanization have caused heavy metal contamination to become a serious concern in many countries (Hani and Pazira, 2011). For any African country to introduce and practice the remediation technologies widely employed in the developed countries, the background knowledge of the sources, chemistry, and potential risks of toxic heavy metals in contaminated soils is necessary for the selection of appropriate remedial options. However, the financial resources needed to effectively adopt these remediation technologies are lacking in Africa. Remediation of soil contaminated by heavy metals is necessary in order to reduce the associated risks, make the land resource available for agricultural production, enhance food security, and scale down land tenure problems. Immobilization, soil washing and phytoremediation are frequently listed among the best available technologies for cleaning up heavy-metal-contaminated soils but have been mostly demonstrated in developed countries. These technologies are recommended for field applicability and commercialization in the developing countries also where agriculture, urbanization and industrialization are leaving a legacy of environmental degradation. Developing countries of the world (mostly in Africa) need to experiment with some of these remediation options and find ways to subsidize the possible ones from peasant farmers in order to reduce poverty among the citizens who cannot afford to adopt the remediation technology in their farmlands which are already contaminated by heavy metals and the resultant effect is low yield.

CONCLUSIONS

Soils may become contaminated by the accumulation of heavy metals and metalloids through fertilizer application, use of pesticides, herbicides, biosolids and manures, wastewater, mining, milling processes, industrial effluents, airborne and many other sources not mentioned here. However, some of these heavy metals can be removed and prevented if remediation technologies are adopted and practiced. The benefits are that food security will be enhanced, farmers can be rest-assured of a bumper harvest each year and more revenue will be generated in the country. The literature reports on agricultural soil contamination by heavy metals in Africa and the world at large further suggests that baseline values for heavy metals should be proposed in Africa. This is necessary to facilitate the identification of soil contamination processes over the whole continent as a basis for undertaking appropriate action to protect the soil resource quality for improved food production and reduce the health risks associated with heavy metals.

REFERENCES

Adriano, D.C., 2001. Trace elements in terrestrial environments. Biogeochemistry, Bioavailability and Risks of Metals. Springer, New York.

Alloway, B.J., 1990. Heavy Metals in Soils. Blackie, London.

Ashraf, M., Ozturk, M., Ahmad, M.S.A. (Eds.), 2010a. Plant Adaptation and Phytoremediation. Series: Tasks for Vegetation Science. Springer Verlag. 481 pp.

Ashraf, M., Ozturk, M., Ahmad, M.S.A., 2010b. Toxins and their phytoremediation. In: Ashraf, et al. (Ed.), Plant Adaptation and Phytoremediation. Springer Verlag-Tasks for Vegetation Science, pp. 1–34.

Ashraf, M., Ozturk, M., Ahmad, M.S.A., Aksoy, A. (Eds.), 2012. Crop Production for Agricultural Improvement. Productivity: Springer Verlag. 796 pp.

Baker, A.J.M., Brooks, R.R., 1989. Terrestrial plants which hyperaccumulate metallic elements – a review of their distribution, ecology, and phytochemistry. Biorecovery 1, 81–126.

Basta, N.T., Gradwohl, R., 1998. Remediation of heavy metal-contaminated soil using rock phosphate. Better Crops vol. 82 (4), 29–31. 1998.

Brady, N.C., Weil, R.R., 1999. The Nature and Properties of Soils, twelfth ed. Prentice Hall, Upper Saddle River, NJ.

Bjuhr, J., 2007. Trace Metals in Soils Irrigated with Waste Water in a Periurban Area Downstream Hanoi City, Vietnam, Seminar Paper. Institutionenförmarkvetenskap, Sverigeslantbruksuniversitet (SLU), Uppsala, Sweden. 2007.

Chaney, R.L., Oliver, D.P., 1996. Sources, potential adverse effects and remediation of agricultural soil contaminants. In: Naidu, R. (Ed.), Contaminants and the Soil Environments in the Australia-Pacific Region. Kluwer Academic Publishers, Dordrecht, The Netherlands, pp. 323–359.

DiToro, D.M., Mahony, J.D., Hansen, D.J., Scott, K.J., Carlson, A.R., Ankley, G.T., 1992. Acid volatile sulfide predicts the acute toxicity of cadmium and nickel in sediments. Environ. Sci. Technol. vol. 26 (1), 96–101.

DPR-EGASPIN, 2002. Environmental Guidelines and Standards for the Petroleum Industry in Nigeria (EGASPIN). Department of Petroleum Resources, Lagos, Nigeria. 2002.

Environmental Science & Technology, 1998. Phytoremediation; forecasting. Environ. Sci. Technol. vol. 32 (17), 399A.

Gucel, S., Kocbas, F., Ozturk, M., 2009. Metal bioaccumulation by barley in Mesaoria plain alongside the Nicosiafamagusta highway, Northern Cyprus. Fresenius Environ. Bull. vol. 18 (11), 2034–2039.

Gupta, S.K., Herren, T., Wenger, K., Krebs, R., Hari, T., 2000. In situ gentle remediation measures for heavy metal-polluted soils. In: Terry, N., Bañuelos, G. (Eds.), Phytoremediation of Contaminated Soil and Water. Lewis Publishers, Boca Raton, Fla, USA, pp. 303–322.

GWRTAC, 1997. Remediation of metals-contaminated soils and groundwater. Tech. Rep. TE-97–01, GWRTAC, Pittsburgh, Pa, USA, 1997, GWRTAC-E Series.

Hakeem, K.R., Parvaiz, A., Ozturk, M. (Eds.), 2013. Crop Improvement—New Approaches and Modern Techniques. Springer Verlag, XXVII. 494 pp.

Hani, A., Pazira, E., 2011. Heavy metals assessment and identification of their sources in agricultural soils of Southern Tehran, Iran. Environ. Monit. Assess. vol. 176, 677–691.

Harada, M., 1994. Environmental contamination and human rights, case of Minamata disease. Ind. Environ. Crisis and Qual. 8, 141–154.

Huang, S.S., Liao, Q.L., Hua, M., Wu, X.M., Bi, K.S., Yan, C.Y., 2007. Survey of heavy metal pollutionand assessment of agricultural soil in Yangzhongdistrict, Jiangsu Province, China. Chemosphere 67, 2148–2155.

Ibiam, J.A., Ekpe, I.I., 2013. Remediation, danger and health implications of hazardous heavy metals. Proc. 36th Annu. Int. Conf. Chem. Soc. Nigeria, Minna, Niger State 1, 346–352. Held between 16th – 20th September, 2013.

Jacob, J.O., Ajai, A.I., Nmadi, A.N., 2013. Heavy metal toxicity: A review. Proc. 36th Annu. Int. Conf. Chem. Soc. Nigeria 1, 268–273. 16th – 20th September, 2013.

Kaasalainen, M., Yli-Halla, M., 2003. Use of sequential extraction to assess metal partitioning in soils. Environ. Pollut. vol. 126 (2), 225–233.

Kabata-Pendias, A., Pendias, H., 2001. Trace Metals in Soils and Plants, second ed. CRC Press, Boca Raton, Fla, USA.

Kuo, S., Heilman, P.E., Baker, A.S., 1993. Distribution and forms of copper, zinc, cadmium, iron, and manganese in soils near a copper smelter. Soil Sci. vol. 135 (2), 101–109.

Martin, T.A., Ruby, M.V., 2004. Review of in situ remediation technologies for lead, zinc and cadmium in soil. Remediation vol. 14 (3), 35–53.

McLaughlin, M.J., Hamon, R.E., McLaren, R.G., Speir, T.W., Rogers, S.L., 2000. Review: a bioavailability-based rationale for controlling metal and metalloid contamination of agricultural land in Australia and New Zealand. Aust. J. Soil Res. vol. 38 (6), 1037–1086.

Mattigod, S.V., Page, A.L., 1983. Assessment of metal pollution in soil. In: Applied Environmental Geochemistry. Academic Press, London, UK, pp. 355–394.

Nicholson, F.A., Smith, S.R., Alloway, B.J., Carlton-Smith, C., Chambers, B.J., 2003. An inventoryof heavy metals inputs to agricultural soils in England and Wales. Sci. Total Environ. 311, 205–219.

NJDEP, 1996. Soil Cleanup Criteria, New Jersey Department of Environmental Protection, Proposed Cleanup Standards for Contaminated Sites. NJAC 7 (26D), 1996.

Ozturk, M., Mermut, A., Celik, A., 2010a. Land Degradation, Urbanisation, Land Use & Environment. NAM S. & T, Delhi-India. 445 pp.

Ozturk, M., Sakcali, S., Gucel, S., Tombuloğlu, H., 2010b. Boron and Plants. In: Ashraf, et al. (Ed.), Plant Adaptation & Phytoremediation. Springer Verlag, Part 2, pp. 275–311.

Ozturk, M., Gucel, S., Sakcali, S., Guvensen, A., 2011. An overview of the possiblities for wastewater utilization for agriculture in Turkey. Isr. J. Plant Sci. vol. 59, 223–234.

Phytotech, 2000. Phytoremediation technology. http://cluin.org/PRODUCTS/SITE/ongoing/demoong/phytotec.htm.

Pierzynski, G.M., Sims, J.T., Vance, G.F., 2000. Soils and Environmental Quality, second ed. CRC Press, London, UK.

Raymond, A.W., Felix, E.O., 2011. Heavy metals in contaminated soils: A review of sources, chemistry, risks and best available strategies for remediation. ISRN Ecol. vol. 2011, 1–20.

Riley, R.G., Zachara, J.M., Wobber, F.J., 1992. Chemical contaminants on DOE lands and selection of contaminated mixtures for subsurface science research.US-DOE. Energy Resource Subsurface Science Program, Washington, DC, USA.

Saeed, M.D., Oladeji, S.O., 2013. Proximate and heavy metals content in vegetables grown along Kubanni stream channel in Zaria town, Kaduna State. Proc. 36th Annu. Int. Conf. Chem. Soc. Nigeria, Minna, Niger State 1, 242–246. Held between 16th – 20th September, 2013.

Sale, J.F., Yahaya, A., Umar, S.M., 2013. Determination of heavy metals in some vegetables sold in Dekina, Kogi State, Nigeria. Proc. 36th Annu. Int. Conf. Chem. Soc. Nigeria 1, 334–338. 16th –20th September, 2013.

Smith, L.A., Means, J.L., Chen, A., 1995. Remedial Options for Metals-Contaminated Sites. Lewis Publishers, Boca Raton, Fla, USA.

Tessier, A., Campbell, P.G.C., Blsson, M., 1997. Sequential extraction procedure for the speciation of particulate trace metal. Anal. Chem. vol. 51 (7), 844–851.

Tukura, B.W., Yahaya, M., Madu, P.C., 2013. Evaluation of physiochemical properties of irrigated soil. J. Nat. Sci. Res. 3 (9), 135–139.

USEPA, 1993. Clean Water Act, sec. 503. vol. 58, no. 32. U.S. Environmental Protection Agency, Washington, D.C.

USEPA, 1998. A Citizen's Guide to Phytoremediation. http://cluin.org/PRODUCTS/CITGUIDE/Phyto2.htm.

USEPA, 1996. Report: recent Developments for In Situ Treatment of Metals contaminated Soils. U.S. Environmental Protection Agency, Office of Solid Waste and Emergency Response. 1996.

Ure, A.M., Quevauviller, P.H., Muntau, H., Griepink, B., 1993. Speciation of heavy metals in soils and sediments. An account of the improvement and harmonization of extraction techniques undertaken under the auspices of the BCR of Commission of the European Communities. Int. J. Environ. Anal. Chem. vol. 51 (1), 35–151.

USEPA, 2007. Treatment Technologies for Site Cleanup: Annual Status Report, twelfth ed. Tech. Rep. EPA-542-R-07-012. Soil waste and Emergency Response (5203p), Washington, DC, USA.

Watanabe, M.E., 1997. Phytoremediation on the brink of commercialization. Environ. Sci. Technol. News 31, 182–186.

Weggler, K., McLaughlin, M.J., Graham, R.D., 2004. Effect of chloride in soil solution on the plant availability of biosolid-borne cadmium. J. Environ. Qual. vol. 33 (2), 496–504. 2004.

Wenzel, W.W., Adriano, D.C., Salt, D., Smith, R., 1999. Phytoremediation: A plant-microbe based remediation system. In: Adriano, D.C., et al. (Ed.), Bioremediation of Contaminated Soils. American Society of Agronomy, Madison, WI, pp. 457–508.

Zhang, C., 2006. Using multivariate analyses and GIS to identify pollutants and their spatial patterns in urban soils in Galway, Ireland. Environ. Pollut. 142, 501–511.

Chapter 19

Improving Phytoremediation of Soil Polluted with Oil Hydrocarbons in Georgia

Gia Khatisashvili,* Lia Matchavariani† and Ramaz Gakhokidze‡

*Durmishidze Institute of Biochemistry and Biotechnology at Agricultural University of Georgia, Laboratory of Biological Oxidation, Tbilisi, Georgia; †Department of Soil Geography Faculty of Exact & Natural Sciences, Tbilisi State University of Iv. Javakhishvili, Tbilisi, Georgia; ‡Department of Bioorganic Chemistry, Faculty of Exact & Natural Sciences, Tbilisi State University of Iv. Javakhishvili, Tbilisi, Georgia

INTRODUCTION

One of the main sources of environmental contamination is determined by petroleum production, transportation, refining and accidental waste (Korte et al., 1992). In Georgia, which is the transit route of different goods transportation between East and West, the Baku–Supsa and Baku–Tbilisi–Ceyhan oil pipelines in particular pose great danger of contamination with oil hydrocarbons and require the creation of a special ecological technology of environmental remediation and protection.

Due to the great power of natural detoxification processes, interest in the ecological potential of microorganisms and plants has increased in the last two decades (Arthur and Coats, 1998; Salt et al., 1998; Tsao, 2003b; Kvesitadze et al., 2006). Microorganisms that transform organics play an important role in maintaining the ecological balance in various ecosystems and, due to their high degradation and transformation powers, are successfully used for sewage and soil purification. Plants actively participate in soil and air remediation processes. Plants and microorganisms, together or individually, mainly through their powerful oxidative enzyme systems, are capable of remediating environments polluted by a wide spectrum of contaminants.

Phytoremediation is a unique clean-up strategy (Tsao, 2003b; Kvesitadze et al., 2006). The realization of phytoremediation technologies implies the planting of a contaminated area with one or more specific, previously selected species of plants having the potential to extract contaminants from the soil. The treatment continues by harvesting the plants and composting or incinerating them. To create a truly effective phytoremediation system, all components of the system should

Soil Remediation and Plants. http://dx.doi.org/10.1016/B978-0-12-799937-1.00019-X

be thoroughly analyzed. The major constitutive component of such a system is obviously the plant. The goal of plant selection is to choose a definite plant species with appropriate characteristics. A survey of site vegetation should be undertaken to determine which species of plants would have the best growth on the contaminated site, taking into account the ability of the plants to accumulate and degrade the contaminants (Korte et al., 2000; Kvesitadze et al., 2001, 2006).

Evidence that plant roots and the rhyzosphere-associated microbial community are capable of enhancing the degradation of petroleum chemicals in soils provides a potentially important approach for the in situ treatment of contaminated sites. Vegetation may act to immobilize water-soluble contaminants, increase their stability in soil structure, and create a favorable environment for degradative microorganisms. Before phytoremediation can be practically or efficiently employed, more research is needed to reveal the basic mechanisms involved.

The combined impact of plant and microbes of soil pollutants is numerous, and many attempts have been made to manipulate enhancing contaminant degradation. The approach we searched for in our investigation refers to rhizoaugmentation, which is the addition of hydrocarbon-degrading microorganisms to soil with the intention of them associating with rhizospheres of plants involved in remediation (Tsao, 2003a; Kirk et al., 2005; Shaw and Burns, 2007). However, we have enhanced the aims of such rhizoaugmentation by using not only simple hydrocarbon-degrading microorganisms, but with those which demonstrate an ability to produce biosurfactants (Shin et al., 2006).

The rhyzosphere is a unique environment where soil conditions are much different than in bulk soil (Burgmann et al., 2005). Loading of carbon substrates from root exudates can provide a potentially attractive environment for bacteria as an initiating agent among hardly consumed petroleum hydrocarbons. Understanding how root exudates influence soil populations that have the ability to degrade different petroleum hydrocarbons may help our understanding of the phytoremediation strategy.

The main idea of the work carried out is to create a database for development of a novel approach for the ecological safety of oil pipelines and the rehabilitation of oil-polluted sites through biological treatment. The approach is based on a combined application of plant and microbial hydrocarbons degradation potential for remediation and long-term protection of the environment.

The success of this work should promote the solution of existing problems of hydrocarbon contamination in soil and water reservoirs. First, an ecological preservation strategy for oil sources near oil reservoirs and along pipelines will be created, to prevent hydrocarbons spreading. Second, the fundamental problem of the metabolic mechanisms of hydrocarbon degradation (including that of aromatic compounds) by plants and microorganisms will be thoroughly investigated. Third, commercially valuable technologies and recommendations to preserve the environment from hydrocarbons will be created. The results will promote the preservation of the ecology along existing pipelines and those under construction in Georgia.

CHARACTERIZATION OF SOIL TYPES

The oil pipelines (Baku–Supsa and Baku–Tbilisi–Ceyhan), crossing over the most part of the territory of Georgia pass through 23 different types of soils, which are characterized by various physico-chemical and biological characteristics.

On development of an environmental protection strategy against contamination with petroleum hydrocarbons in the case of damage overflow, different adsorption features of soils should be taken into consideration. The risk level of total petroleum hydrocarbons (TPH) penetration into ground waters should also be estimated (Kvesitadze et al., 2008). To this end, the capability of soil samples to adsorb oil hydrocarbons was studied. The soil samples that represented 13 different types of soils were collected in Georgia from different sites along the Baku–Tbilisi–Ceyhan and Baku–Supsa Oil Pipelines (Figure 19.1).

The results showed that investigated soil types differently adsorb hydrocarbons. In particular, they may be divided into three groups:

1. Soils from which oil hydrocarbons are washed out rapidly. Oil contamination of such types of soils represents a danger to penetration of more than 70% of hydrocarbons into ground waters under the influence of atmospheric precipitation. Soils which belong to these types of soils along the pipelines passing through the territory of Georgia are as follows:
   - Alluvial calcareous (Calcaric fluvisols);
   - Yellow brown forest (Chromic cambisols and stagnic alisols);
   - Cinnamonic calcareous (Calcaric cambisols and Calcic kastanozems);
   - Raw humus calcareous (Rendzic leptosols);
   - Cinnamonic (Eutric cambisols and Calcic kastanozems).
2. Soils which have average ability of TPH absorption. From these soils 50–70% of oil hydrocarbons are washed out by water. Such types of soils are:
   - Meadow grey cinnamonic (Calcic vertisols);
   - Grey cinnamonic (Calcic kastanozems);
   - Chernozems (Chernozems);
   - Brown forest weakly unsaturated (Eutric cambisols);
   - Cinnamonic light (Calcic kastanozems);
   - Black calcareous (Calcic vertisosls).
3. Soils distinguished by strong TPH adsorption ability. Contamination in such types of soils is maintained over a long period of time at less than 50%. In the territory of Georgia such soil types are as follows:
   - Mountain meadow soddy (Leptosols, cambisols and cryosols);
   - Brown forest podzolized (Dystric cambisols).

Thus, during oil contamination, considerable amounts of hydrocarbons might appear in groundwaters from soils of types *Alluvial calcareous*, *Yellow brown forest*, *Cinnamonic calcareous*, *Raw humus calcareous* and *Cinnamonic* under the influence of atmospheric precipitation.

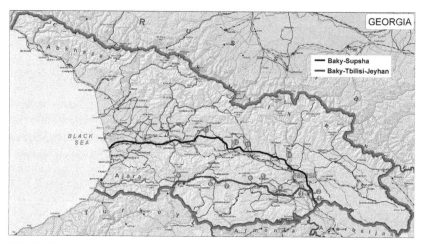

FIGURE 19.1   Baku–Supsa and Baku–Tbilisi–Ceyhan Oil Pipelines in the territory of Georgia. Sampling points of soils are indicated on the map. The types of soils are as follows: (1) Meadow grey cinnamonic (Calcic vertisols) (pH 5.2); (2) Grey cinnamonic (Calcic kastanozems) (pH 4.7); (3) Alluvial calcareous (Calcaric fluvisols) (pH 4.5); (4) Raw humus calcareous (Rendzic leptosols) (pH 6.0); (5) Chernozems (Chernozems) (pH 4.5); (6) Mountain meadow soddy (Leptosols, cambisols and cryosols) (pH = 4.0); (7) Brown forest podzolized (Dystric cambisols) (pH 4.5); (8) Brown forest weakly unsaturated (Eutric cambisols) (pH 5.0); (9) Cinnamonic (Eutric cambisols and Calcic kastanozems) (pH 5.5); (10) Cinnamonic calcareous (Calcaric cambisols and Calcic kastanozems) (pH 5.0); (11) Cinnamonic light (Calcic kastanozems) (pH 5.0); (12) Black calcareous (Calcic vertisosls) (pH 5.0); (13) Yellow brown forest (Chromic cambisols and stagnic alisols) (pH 5.0). For a colour version of this figure, please see the section at the end of this book.

SELECTION OF MICROORGANISMS

The use of microorganisms during soil bioremediation after oil spills and other contaminations is one of the most ecologically progressive approaches compared with physico-chemical treatments, which usually result in sharp disfunction of soil ecosystems. Plenty of bacteria, actinobacteria, microscopic fungi and yeasts are well-known degradants of petroleum hydrocarbons, At the present time the active search is on for the methods of intensification of biological degradation of oil hydrocarbons in ground and water. The microbiological method based on introduction of active oil-oxidizing microorganisms into contaminated environment is the most promising. For realization of this method it is necessary to select the most active microorganisms adapted to high concentrations of oil, and to develop the technology of their use by taking into account the complexity of soil properties and the level of pollution.

A large amount of hydrocarbon-degrading bacteria and fungi have been isolated from the contaminated and non-contaminated soils along the pipeline: 360 bacterial and 200 fungal strains were obtained and purified, and their taxonomic affiliations identified.

As a result of screening, 20 strains of microorganisms from different taxonomic groups (bacteria, microscopic fungi and yeast) were selected according to their abilities to effectively assimilate oil hydrocarbons when grown on media containing crude oil or separate individual hydrocarbons (alkanes: C_6H_{14}, C_8H_{18}, $C_{11}H_{24}$, $C_{13}H_{28}$, $C_{14}H_{30}$, $C_{15}H_{32}$ and $C_{16}H_{34}$, arenes: benzene and toluene). The capability of selected strains and their consortia to utilize oil hydrocarbons was determined by their submerged cultivation in nutrient media with crude oil (3%) as the sole source of carbon. The chromatographic analysis has shown that for oil degradation the most efficient combination of bacterial cultures is the consortium consisting of *Pseudomonas* and *Rhodococcus* strains. Chromatographic analysis also reveals that hydrocarbon fractions to C_{17} are assimilated completely and heavier hydrocarbons remain in the incubation medium in minimal amounts (Figure 19.2).

SELECTION OF PLANTS

The assessment of plant detoxification potential is determined by the rate and depth of contaminant uptake from the soil, how they accumulate in the plant cell, and the degree of their transformation to regular cell metabolites. To select the best plants for a particular phytoremediation task, ideally many plant characteristics should be made available. Firstly, the actual phytoremediation-related characteristics of the candidate plants should be established, notably (Kvesitadze et al., 2006):

- Their overall ability to take up and degrade contaminants existing in the soil or groundwater.
- Their ability to accumulate organic and inorganic contaminants in their cells and intracellular spaces.
- Their excretion of exudates to stimulate the multiplication of soil microorganisms and secretion of enzymes participating in the initial transformations of the contaminants.
- The existence, within the cells, of contaminant-degrading or conjugating enzymes (oxidases, reductases, transferases, esterases, etc.).
- Their high resistance against contaminants, that is that the plants' growth and metabolism is not adversely affected by the contaminants.
- Their root system (main and fibrous); the range of root depths of the plants.
- Whether the plants are endemic and non-agricultural.
- Their tolerance to salty soil (halophilicity).
- Their appropriate adaptediveness (to warm or cold conditions).
- Their growth rate.

Plants have the capacity to uptake organic pollutants and to subsequently metabolize or transform them into less toxic metabolites. Once taken up and translocated, the organic chemicals generally undergo three transformation stages: (1) chemical modification (oxidations, reductions, hydrolysis, etc.); (2)

(A)

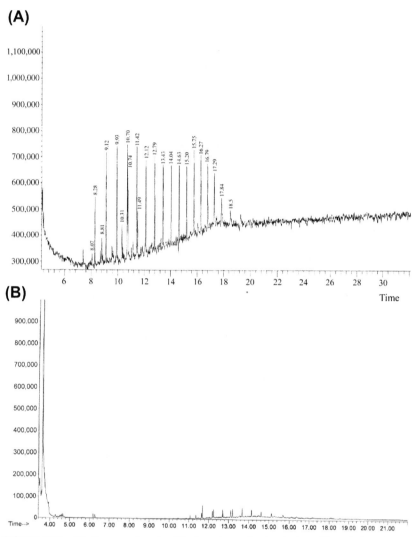

FIGURE 19.2   GS diagrams of crude oil hydrocarbons after incubation with strain from genera *Rhodococcus*. (A) Control variant: incubation without bacterial strain. (B) Test variant: incubation with bacterial strain.

conjugation (with glutathione, carbohydrates, amino acids, etc.); and (3) compartmentalization in vacuoles and/or the cell wall (Sandermann, 1994; Burken, 2003; Kvesitadze et al., 2006). Reactions occurring during all three detoxification processes are enzymatic in nature. In the absence of xenobiotics, these enzymes catalyze other reactions typical for plant cell regular metabolism (Kvesitadze et al., 2006). Phytoremediation abilities of plants, besides anatomical and physiological characteristics of plants, significantly depend on high activity of these enzymes and their induction by chemical agents (Sandermann, 1994; Kvesitadze et al., 2001).

Hydrocarbon degradation in plants proceeds mainly through a multistage oxidation pathway. The initial and rate-limiting stage of such transformation is hydroxylation. As a result of hydroxylation nonpolar hydrocarbon molecules get a polar functional (hydroxyl) group and become easily accessible for further oxidative degradation. The following oxidative metal-containing enzymes catalyze hydroxylation: cytochrome P450 containing monooxygenases (EC 1.14.14.1), peroxidases (EC 1.11.1.7) and phenoloxydases (EC 1.14.18.1) (Marrs, 1996; Kvesitadze et al., 2001, 2006; DeRidder et al., 2002).

Chemical modification of organic pollutants is a process whereby a molecule of a hydrophobic organic xenobiotic acquires a hydrophilic functional group (hydroxyl, amino, carboxyl, etc.) as a result of enzymatic transformations (Kvesitadze et al., 2006). Due to the functional group the polarity and reactivity of the toxicant molecule is enhanced. This promotes an increase of its affinity to enzymes, catalyzing further transformation (conjugation or further deep oxidation). Further, xenobiotic oxidative degradation proceeds to standard cell metabolites and mineralization to CO_2 (Chrikishvili et al., 2006). Through this pathway the plant cell not only fully detoxifies the xenobiotic but also utilizes its carbon atoms for intracellular biosynthetic and energetic needs. The totality of such transformations is the essence of the detoxification process.

Conjugation is a process where a xenobiotic is chemically coupled to cell endogenous compounds (proteins, peptides, amino acids, organic acids, mono-, oligo- and polysaccharides, lignin, etc.) by formation of peptide, ether, ester, thioether or other bonds of a covalent nature (Kvesitadze et al., 2006). Intermediates of xenobiotic transformations or xenobiotics already bearing functional groups capable of reacting with intracellular endogenous compounds are susceptible to conjugation. The formation of conjugates leads to the enhancement of the hydrophilicity of organic contaminants, and consequently to an increase in their mobility. Such characteristics further simplify compartmentation of the transformed toxic compounds. Being in conjugated form, a xenobiotic in the plant cell is kept apart from vital processes and is therefore rendered harmless for the plant (Burken et al., 2003).

A group of enzymes, called glutathione S-transferases (GSTs) (EC 2.5.1.18) have wide specificity and couple electrophilic xenobiotics and their metabolites with the reduced tripeptide glutathione (GSH) (Marrs, 1996; DeRidder et al., 2002). In plants, a large and diverse gene family encodes the GSTs. GSTs facilitate the reaction between the functional group of the contaminant intermediates and the SH-group of the glutathione cysteine residue. They participate in the conjugation of a wide spectrum of toxic compounds and their metabolites. As a consequence, the toxicant is bound to intracellular compounds via a covalent bond to the sulphur atom (Marrs, 1996).

Conjugation is not the most successful pathway of xenobiotic detoxification from the ecological point of view. Plant remains, containing the conjugated contaminants, actually become the toxicant carrier. Typically 70% or more of the absorbed xenobiotics accumulate in plants in the form of conjugates (Kvesitadze et al., 2001). This fact must be taken into account when considering the ultimate ecological fate of xenobiotics. Conjugates of toxic compounds are especially hazardous on insertion into the food chain: enzymes of the digestive tract of warm-blooded animals can hydrolyze conjugates and release the xenobiotics or products of their partial transformation, which in some cases, due to increased reactivity, are more toxic than the initial xenobiotics. Therefore, it is highly desirable that plants applied to phytoremediation have a phenomenal capability to accomplish deep enzymatic degradation of xenobiotics. The selection of such plants, or the promotion of gene expression of enzymes participating in plant detoxification processes are the basic strategies of modern phytoremediation technologies.

Thus, the main biocemical criteirion for selection of plant phytoremediators with regard to contamination with oil hydrocarbons is high activities of basic oxidative enzymes (cytochrome P450-containing monooxygenase, peroxidase and phenoloxidase) participating in oil hydrocarbons degradation. During the selection of such plants, 12 species (grasses, shrubs and trees) which are well adapted to almost every type of soil that exists along the oil pipelines in Georgia, are tested: evergreens – trifoliate orange (*Poncirus trifoliate*) and privet (*Ligustrum sempervirens*); deciduous plants – poplar (*Populus canadensis*) and mulberry (*Morus alba*); annual plants – maize (*Zea mays*), rye-grass (*Lolium multiflorum*), alfalfa (*Medicago sativa*) and soybean (*Glycine max*), chickpea (*Cicer arietinum*), chickling vetch (*Lathyrus sativum*), mung bean (*Vigna radiata*) and pea (*Pisum sativum*).

According to the results, the following conclusions can be drawn:

- Evergreens privet and trifoliate orange are respectively distinguished by high phenoloxidase and peroxidase activities. Highest monooxygenase activity is characteristic for rye-grass and poplar.
- Alfalfa and chickpea possess a stronger pool of oxidative enzymes, especially hemoproteins, and mung bean prevails in phenoloxidase activity.
- For oxidative enzymes benzene is a much stronger indicator (maximum level of induction degree is 185–200%) than octane (145–160%).
- Investigated oxidative enzymes were induced by the action of penetrated hydrocarbons and if the plant possesses a stronger pool of oxidative enzymes the induction degree of these enzymes is comparatively low.
- Two types of changes of oxidative enzyme activities are observed: (i) specific activation of certain oxidase (in alfalfa, mung bean and soybean) or (ii) equal induction of all oxidative enzymes (in rye-grass, chickling vetch, chickpea and pea). In the first case oxidative metabolism of hydrocarbons in roots is preferably realized via the monooxygenase pathway, while oxidation via phenoloxidase is more active in leaves. In the second case, the oxidation of hydrocarbons is carried out via inter-replacement of haemoproteins (cytochrome P450-containing monooxigenase and peroxidase) and copper-containing (phenoloxidase) oxidative enzymes.

- Oil hydrocarbons with open chains, as well as with aromatic ring, undergo oxidative degradation via participating of following plant enzymes: peroxidase, phenoloxidase, cytochrome P450-containing monooxygenase. Oxidative metabolism of benzene and octane in roots of tested plants is realized via participation of monooxygenase and peroxidase, and in leaves by phenoloxidase.
- Trifoliate orange, privet, rye-grass, alfalfa and soybean, according to content of oxidizing enzymes and their tolerance towards high concentrations of oil hydrocarbons, are serviceable for phytoremediation of soils polluted by oil spills.

In addition, the effects of petroleum hydrocarbons on plant cell ultrastructure and enzymes of basic metabolism, such as nitrogen assimilation and energy generation, were studied.

The understanding of the oil biodegradation process after a spill is becoming increasingly important. Bacterial pathways of hydrocarbon metabolism, for example, are reasonably well delineated (Ensley and Gibson, 1983; Bartha, 1986). A number of comprehensive reviews have addressed microbial-based bioremediation (Lee et al., 1988; Lee and Banks, 1993; Sims et al., 1989; Huesemann, 1994). Degradation of oil by bacteria and fungi may be limited by a lack of sufficient oxygen and nutrients, given that hydrocarbon metabolism is primarily an aerobic process (Dibble and Bartha, 1979; Bartha, 1986; Leahy and Colwell, 1990; Pritchard and Costa, 1991; Ferro et al., 1997; Siciliano and Germida, 1998).

Several studies have demonstrated that spilled oil degrades more rapidly on vegetated soils than on soils lacking vegetation (Lee and Banks, 1993; Schwab and Banks, 1994; Qiu et al., 1997; Lin and Mendelssohn, 1998; Wiltse et al., 1998; Banks et al., 2000), but the mechanisms of this acceleration have not been clearly defined. Organic pollutants penetrated into plant cells cause significant changes across a whole range of intracellular metabolic processes. This is first manifested in the activation of inductive processes directed to the synthesis of enzymes and enzymatic systems participating in xenobiotics detoxification. As a result of the progressive oxidation of the xenobiotics, standard cellular intermediates are formed (Kvesitadze et al., 2006).

Despite the fact that collateral biochemical processes accompanying the detoxification process in plants are not well investigated, there are quite a few examples in the literature indicating that the activities of the enzymes participating in different regular cellular processes are also influenced by xenobiotics that have penetrated into the cell. These xenobiotics may, furthermore, affect the activity of the regulatory enzymes involved in the tricarboxylic acid cycle and in the process of cell main metabolism (Kvesitadze et al., 2006).

The revelation of plant response to xenobiotics, expressed as cell structure–function deviations, characteristic to each plant species, enables plant resistance to contaminated environment to be revealed and allows the estimation of the prospects of their application in phytoremediation technologies.

The destructive changes in the ultrastructure of alfalfa and rye-grass root cells under the influence of octane as a typical oil hydrocarbon have been

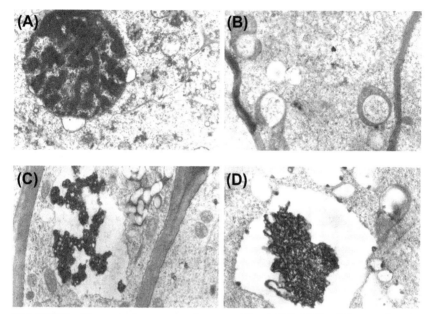

FIGURE 19.3 Changes in ultrastructure of rye-grass (A) and alfalfa (B, C and D) root cells under the action of 0.1 mM octane solution.

investigated. Under the influence of octane, the root cells of rye-grass were completely destroyed (Figure 19.3A).

To reveal measurable declines in plant cell homeostasis after the influence of oil hydrocarbons, activities of enzymes [glutamate dehydrogenase (GDH), malate dehydrogenase (MDH) and glutamine synthetase (GS)] were investigated. GDH catalyzing reversible reaction – oxidative deamination of L-glutamate and reductive amination of 2-oxoglutarate and thus connecting nitrogen metabolism with the tri-carboxylic acid cycle, occupies the central position in cell metabolism. GS plays an important role in the incorporation of inorganic nitrogen in a form of ammonium into amino acids and proteins. This enzyme catalyzes ammonium assimilation in the process of glutamine synthesis. The significance of GDH and GS is conditioned by the fact that the products of the reactions catalyzed by these enzymes are the donors of amino groups for biosynthesis of other amino acids. GDH and GS play key roles in maintaining the balance of carbon and nitrogen in plant cells (Miflin and Habash, 2002). MDH, participating in the tricarboxylic acid cycle, is involved in the processes of respiration and energy exchange.

The obtained results show that octane caused induction of GDH in all grassy plants; however, the rate of induction decreased parallel to the increase of the hydrocarbon concentration. Octane at a concentration of 1 mM caused induction of GDH in privet, trifoliate and white mulberry shoots. Higher concentration (10 mM) caused corresponding further increase in GDH activity in leaves of privet and white mulberry shoots but not in trifoliate orange. 100 mM octane was characterized by its inhibitory effect on GDH activity in all plants. Octane had an inductive effect on MDH in almost all plants and all tested concentrations, indicating the intensification of the tricarboxylic acid cycle.

Activation of GDH at low hydrocarbon concentrations indicates participation of the enzymes in plant defence mechanisms, namely, in intensification of amino acid catabolism, leading to energy generation which is in demand in a cell under contamination stress. The induction of MDH in plants exposed to increasing octane concentrations indicates the intensification of the tricarboxylic acid cycle, probably for further oxidation of xenobiotic oxidation intermediates.

Assessment of declinations from the normal range of metabolic processes under the influence of oil hydrocarbons allows elaboration of plant selection criteria for their application in phytoremediation of sites polluted by hydrocarbons.

DETERMINATION OF THE DEGREE OF OXIDATIVE DEGRADATION OF HYDROCARBONS

The metabolism of $[1-6\ ^{14}C]$ benzene in alfalfa and rye-grass has been investigated. For this aim, the radioactivity of low- and high-molecular compounds of plants after uptake of $[1-6\ ^{14}C]$ benzene by plants is detected. The results have shown that in both plants the main part of uptaken benzene (about 80–85%) is metabolized to low-molecular compounds and 6–9% undergoes deep oxidation with cleavage of aromatic ring to form $^{14}CO_2$. The rest part of ^{14}C label is detected in high-molecular compounds.

To determine the level of oxidative degradation of benzene, application of $[1-6\ ^{14}C]$ benzene was carried out. As seen from the obtained results among organic acids, fumaric, muconic and succinic acids are metabolites of oxidative degradation in plants. The presence of the acids, which are standard cell metabolites, indicates to the cleavage of the benzene aromatic ring in the process of oxidative transformation. Apart from organic acids, amino acids with aromatic rings, tyrosine and phenylalanine, are also dominantly formed in plants after benzene degradation. The identification of alanine, glycine and asparagine among ^{14}C-labelled amino acids suggest that uptaken benzene undergoes deep oxidation by oxidative enzymes of plant tissues.

According to the obtained data the inferred scheme of benzene metabolism in plants is the following:

Benzene   Phenol   Catechol   o-Quinone

Cis-cis-muconic acid   Fumaric acid   Tricarboxylic Acid Cycle → CO_2

As seen from the above scheme, benzene after gradual hydroxylation to phenol, catechol and o-quinone is oxidized with cleavage of the aromatic ring and forms cis-cis-muconic acid, which can form fumaric acid and, thus, be incorporated into the general metabolism of organic acids in plant cells.

Similarly, in the works to determine the level of oxidative degradation of cyclohexane, the experiments were carried out by applying of $[1\text{-}^{14}C]$ cyclohexane. The radioactive preparation of cyclohexane was added to seedlings of maize and after 48 h of incubation, the individual components of the fractions organic acids and amino acids were isolated. As a result, the insertion of cyclohexane radioactive label in organic acids is 10 times more intense than in amino acids. Succinic acid is the major metabolite of cyclohexane among organic acids, and in amino acids ^{14}C is distributed equally. The presence of the acids, which are standard cell metabolites, suggests cleavage of the cyclohexane ring in the process of oxidative transformation.

According to the obtained results and literature data (Wagner et al., 2002) the inferred scheme of cyclohexane metabolism in plants is the following:

As seen from the scheme, in the initial stage, cyclohexane undergoes oxidation to unsaturated cyclic intermediates with oxo- or hydroxyl-groups. In the next stage of metabolism these intermediates form cyclohexene-3-diol-1,2, which is oxidized with cleavage of the carbonic cycle and is transformed into adipinic acid. This metabolite can form fumaric acid and, thus, may be incorporated into the general metabolism of organic acids in plant cells.

Summing up the possible pathways of the studied hydrocarbons oxidative degradation, it can be concluded that hydroxylation is the initial stage of exogenous arenes and cycloalkanes oxidative degradation in plants, followed by deep oxidation of intermediates to standard endogenous compounds and carbon dioxide. It could be supposed that the oil hydrocarbons, penetrated into tested plants, preferably undergo oxidative degradation, which is realized via the cytochrome P450-containing monooxygenase and/or peroxidase pathways in roots, and via phenoloxidase in leaves. The products of bacterial oxidation of oil hydrocarbons, penetrated into tested plants undergo subsequent oxidative degradation preferably via cytochrome P450-containing monooxygenase and peroxidase, and they are also detoxified via conjugation with the reduced form of tripeptide glutathione, catalyzed by glutathione S-transferase.

REVELATION OF PLANT–MICROBIAL INTERACTION

Initially, suspension of microorganisms was inoculated in soils. Plants were sown after 35 days of incubation with bacterial strains (in the case of microscopic fungi, after 10 days). After 3 weeks from sowing plants, the seeds' germination abilities (the correlation between the number of germinated and sown seeds) and plants raw biomass were estimated.

The results show that contamination of soil decreased plant growth in all cases, but the presence of the bacterial consortium in contaminated soils significantly stimulates the germination of seeds and accumulation of plant biomass. In experiments where microscopic fungi were used, similar results were obtained, but in some cases fungi inhibited the growth of plants, in particular: maize did not germinate on the soil treated by selected strains from genera of *Chaetomium*; and rye-grass revealed weak growth on the soil treated by selected strains from genera of *Aspergillus versicolor.*

MODEL EXPERIMENTS

The joint action of the plant–microbial consortium for phytoremediation of soils polluted with oil hydrocarbons has been studied in model experiments. Four plant species (alfalfa, maize, rye-grass and soybean) together with a bacterial consortium (consisting of selected strains of *Rhodococcus, Pseudomonas* and *Mycobacterium*) and separate microscopic fungi (Strains of *Aspergillus, Chaetomium* and *Trichoderma viride*) were tested on soils contaminated by oil hydrocarbons with long chains (C_{30}–C_{60}). In tested variants, the levels of soil contamination with TPH were equal 44,000, 70,000, 96,000, 113,000 and 142,000 ppm.

Initially, the suspension of microorganisms was inoculated in soils. Plants were sown after 30–35 days of incubation with bacterial consortium (in the case of microscopic fungi, after 10 days). During the experiment the soil samples were analyzed for TPH content. Results are presented in Figures 19.4, 19.5, 19.6, 19.7, 19.8 and 19.9.

The obtained results show that the effect of soil cleaning depends on the initial contamination of soil. The use of microorganisms over 2 months without plants caused a decrease of TPH content in contaminated soil by:

- 60–75% in soil with 44,000 ppm initial contamination (Figures 19.6A, 19.7A, 19.8A and 19.9A).
- 40–50% in soil with 70,000 ppm initial contamination (Figures 19.6B, 19.7B, 19.8B and 19.9B).
- 40–45% in soil with 96,000 ppm initial contamination (Figures 19.4).
- 35–40% in soil with 142,000 ppm initial contamination (Figures 19.5).

The effect of soil cleaning of microorganisms is enhanced by 15–25% as a result of the use of plants as phytoremediators (Figures 19.4–19.9). In tested variants, soybean is revealed to have the highest phytoremediation ability. It has been established that rye-grass, soybean and alfalfa, together with the selected bacterial consortium and microscopic fungi, are the best tools for phytoremediation of soils, polluted with oil hydrocarbons. A bacterial consortium consisting of strains of *Rhodococcus* and *Pseudomonas* is the best remediation agent among tested microorganisms.

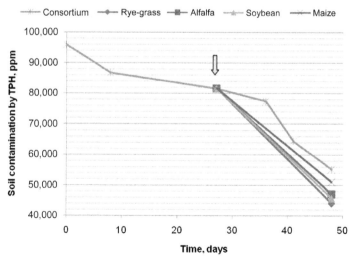

FIGURE 19.4 The dynamics of decrease of TPH content in soil contaminated with oil hydro-carbons during incubation with bacterial consortium (strains of *Rhodococcus, Pseudomonas* and *Mycobacterium*) and plants. The bacterial suspension was inoculated in the soil at the beginning of the experiment. On the 27th day of incubation, plants were sown in separate samples of soil (indi-cated by arrow). Initial degree of contamination, 96,000 ppm of TPH. Temperature, 20–25 C. For a colour version of this figure, please see the section at the end of this book.

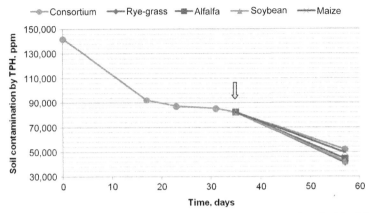

FIGURE 19.5   The dynamics of decreasing TPH content in soil contaminated with oil hydrocarbons during incubation with bacterial consortium (strains of *Rhodococcus, Pseudomonas* and *Mycobacterium*) and plants. The bacterial suspension was inoculated in the soil at the beginning of the experiment. On the 35th day of incubation, plants were sown in separate samples of soil (indicated by arrow). Initial degree of contamination: 142,000 ppm of TPH. Temperature: 20–25°C. For a colour version of this figure, please see the section at the end of this book.

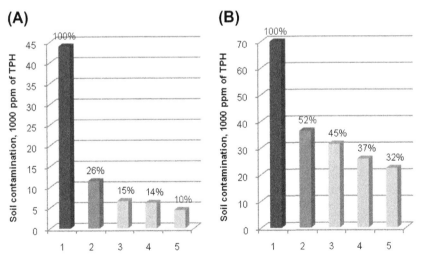

FIGURE 19.6   TPH content in soils contaminated with oil hydrocarbons after treating with bacterial consortium (strains of *Rhodococcus* and *Pseudomonas*) and sowing with plants. The suspension of bacteria was inoculated in the soil at the beginning of the experiment. On the 31st day of incubation, the plants were sown in separate samples of soil. Initial degree of contamination: (A) 44,000 ppm; (B) 70,000 ppm of TPH. Total time of incubation: 2 months; temperature: 20–25°C. Sample variants: (1) contaminated soil at the beginning of the experiment; (2) soil treated with bacterial consortium; (3) soil treated with bacterial consortium and sown with rye-grass; (4) soil treated with bacterial consortium and sown with alfalfa; (5) soil treated with bacterial consortium and sown with soybean.

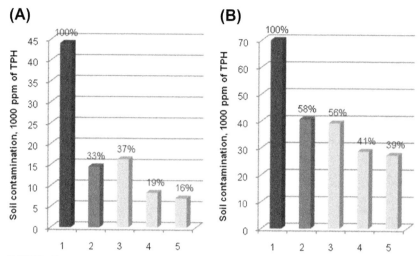

FIGURE 19.7 TPH content in soils contaminated with oil hydrocarbons after treating with a strain of *Aspergillus* and sowing with plants. The suspension of fungi was inoculated in the soil at the beginning of the experiment. On the 31st day of incubation, the plants were sown in separate samples of soil. Initial degree of contamination: (A) 44,000 ppm; (B) 70,000 ppm of TPH. Total time of incubation: 2 months; temperature: 20–25°C. Sample variants: (1) contaminated soil at the beginning of the experiment; (2) soil treated with a strain of *Aspergillus*; (3) soil treated with a strain of *Aspergillus* and sown with rye-grass; (4) soil treated with a strain of *Aspergillus* and sown with alfalfa; (5) soil treated with a strain of *Aspergillus* and sown with soybean.

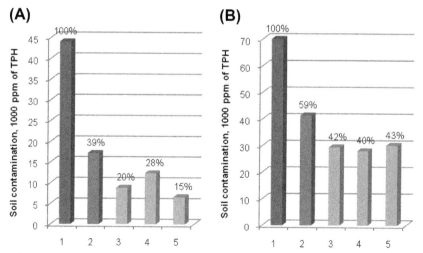

FIGURE 19.8 TPH content in soils contaminated with oil hydrocarbons after treating with a strain of *Trichoderma* and sowing with plants. The suspension of fungi was inoculated in the soil at the beginning of the experiment. On the 31st day of incubation, the plants were sown in separate samples of soil. Initial degree of contamination: (A) 44,000 ppm; (B) 70,000 ppm of TPH. Total time of incubation: 2 months; temperature: 20–25°C. Sample variants: (1) contaminated soil at the beginning of the experiment; (2) soil treated with a strain of *Trichoderma*; (3) soil treated with a strain of *Trichoderma* and sown with rye-grass; (4) soil treated with a strain of *Trichoderma* and sown with alfalfa; (5) soil treated with a strain of Trichoderma and sown with soybean.

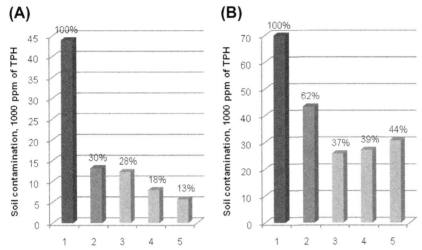

FIGURE 19.9   TPH content in soils contaminated with oil hydrocarbons after treating with a strain of *Chaetomium* and sowing with plants. The suspension of fungi was inoculated in the soil at the beginning of the experiment. On the 31st day of incubation, the plants were sown in separate samples of soil. Initial degree of contamination: (A) 44,000 ppm; (B) 70,000 ppm of TPH. Total time of incubation: 2 months; temperature: 20–25°C. Sample variants: (1) contaminated soil at the beginning of the experiment; (2) soil treated with a strain of *Chaetomium*; (3) soil treated with a strain of *Chaetomium* and sown with rye-grass; (4) soil treated with a strain of *Chaetomium* and sown with alfalfa; (5) soil treated with a strain of *Chaetomium* and sown with soybean.

The chromatographic analysis of soil samples (Figures 19.10 and 19.11) shows that in the case of 70,000 ppm initial contamination in the beginning of remediation the ratio between 'light' and 'heavy' petroleum hydrocarbons (C_{8-15} : C_{16-35}) equals 95:5 (Figure 19.10A); and after remediation this ratio is decreased to 53:47 (Figure 19.10B). In the case of 44,000 ppm initial contamination, this effect is expressed most clearly: the ratio is decreased from 87:13 (Figure 19.11A) to 6:94 (Figure 19.11B). These results indicate that the assimilation of C_{8-15} hydrocarbons takes place mostly during bioremediation.

Thus, conducted model experiments have shown that after the process of remediation implemented by applying plants and microorganisms, heavy fraction of oil hydrocarbons that do not undergo phytoremediation still reside in the soil. It can be concluded that efficiency of phytoremediation significantly decreases when utilization of contaminants by microorganisms and plants is complicated by small mobility, and, correspondingly, by low bioavailability of pollutant molecules. Similar cases arise during long-term contamination of soils with crude oil when light hydrocarbons become volatile, and components with a long-chain form a resinous mass and become extremely difficult to remove. In this case, it is necessary to use agents able to increase mobility and solubility of such hydrocarbons, which then will be effectively destroyed by microorganisms and easily extracted by plants. The feasible solution to this problem can be the application of surface-active substances, the most promising of which

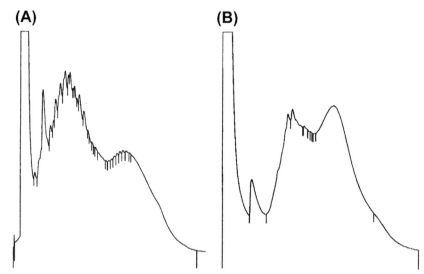

FIGURE 19.10   GC diagrams of hexane extracts of soil samples in beginning (A) and after (B) remediation process. Contaminated soil (initial contamination, 70,000 ppm of TPH) is treated by bacterial consortium (strains of *Rhodococcus* and *Pseudomonas*). Duration of remediation: 2 months; temperature: 20–25°C.

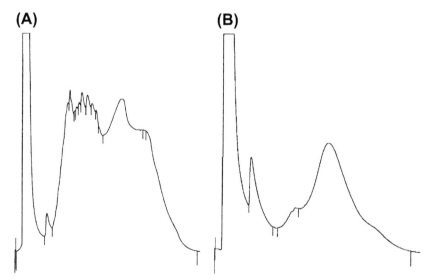

FIGURE 19.11   GC diagrams of hexane extracts of soil samples in beginning (A) and after (B) remediation process. Contaminated soil (initial contamination, 44,000 ppm of TPH) is treated by bacterial consortium (strains of *Rhodococcus* and *Pseudomonas*). Duration of remediation: 2 months; temperature: 20–25°C.

are biogenic surfactants (biosurfactants). They are not inferior to the synthetic ones by their efficiency and, at the same time, are environmentally friendly. Due to their physico-chemical properties (desorption of hydrophobic substances from the soil, their solubilization, reduction of surface and interfacial tension of solutions) and biological activity, biosurfactants may increase the degree of biodegradation of contaminants.

Application of natural biosurfactants – surface-active substances of microbiological origin – can become the real solution to this specified problem of phytoremediation. Microbial surfactants are biologically active compounds characterized by high efficiency, biodegradability and low toxicity. They are able to emulsify resin mass of heavy oil hydrocarbons, increase their bioavailability and improve their transport to plant cells. Parameters of their superficial and interphase tension of their solutions, ability to form stable fine-grained emulsion of water phase with hydrophobic compounds (hydrocarbons, including heavy fractions of oil, vegetable oils, fats) testify to high activity of biosurfactants (Desai and Banat, 1997; Bognolo, 1999). Due to all of the above, biosurfactants increase permeability of toxic compounds to plant cells (Vasileva-Tankova et al., 2001; Sotirova et al., 2008). In addition, surfactants are capable of raising bioavailability of pollutants and regulate their transport in plant cells. Some types of biosurfactants, for example, rhamnolipids, possess metal-chelating properties that enable their application in phytoextraction of heavy metals (Maier et al., 2001; Juwarkar et al., 2007). All these properties of biosurfactants should considerably increase the efficiency of application in phytoremediation technologies for cleaning soils contaminated with long-chain oil components.

Based on the above, we decided to test the use of various biosurfactants in phytoremediation technology. For this reason, model phytoremediation experiments were carried out.

In the model experiments the following objects were used:

- The soil artificially contaminated with raw oil (the level of initial contamination by TPH equals 27,500 ppm) – as the object of phytoremediation.
- Bacterial consortium that consist of strains of *Pseudomonas* and *Bacillus* – as bioremediation agents.
- Biosurfactants preparations: rhamnolipids, trehalose lipids and rhamnolipid biocomplex PS (they were used separately) – as agents to increase the efficiency of the phytoremediation process.
- Alfalfa – as plant phytoremediator.

The mass of the contaminated soil samples was 7.5 kg. Suspension of bacteria (1.21 in each soil sample) and solutions of biosurfactants (100 mg / 500 ml in each soil samples) were inoculated in the soil at the beginning of the experiment. On the 14th day from inoculation plants were sown in soil samples. The experiment continued for 3 months (from 21 June to 21 September, 2011) in greenhouse conditions.

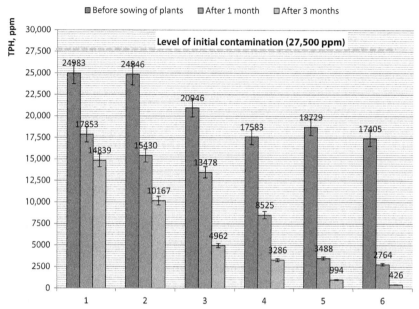

FIGURE 19.12 Phytoremediation of soil artificially contaminated with crude oil by using of plants (alfalfa), bacterial consortium (*Pseudomonas* sp. 6R67+*Bacillus* sp. 3Zu9) and biosurfactants (Rhamnolipids, complex of biosurfactants and Trehalose lipids; concentration – 0.01%). Sample variants: (1) control (contaminated soil without plants, bacteria and biosurfactants); (2) plants (contaminated soil with plants; without bacteria and biosurfactants); (3) plants + bacteria (contaminated soil with plants and bacteria; without biosurfactants); (4) plants + bacteria + complex of biosurfactants (contaminated soil with plants, bacteria and biosurfactants); (5) plants + bacteria + trehalose lipids (contaminated soil with plants, bacteria and biosurfactants); (6) plants + bacteria + rhamnolipids (contaminated soil with plants, bacteria and biosurfactants).

The obtained results (Figure 19.12) show that, as a result of abiotic and biotic factors action (volatilization of hydrocarbons and their assimilation by aboriginal microflora), TPH content in contaminated soil was decreased by 45% over 3 months. The use of plants (without bacterial consortium and biosurfactants) caused a decrease in contamination of 63%. Plants and bacteria jointly are capable of assimilating 82% of hydrocarbons. The application of biosurfactants raises the intensity of the bioremediation process additionally and in cases of biopreparations of trehalose lipids and thamnolipids, the maximum cleaning effects (correspondingly by 96 and 98%) are achieved.

Thus, as a result of research performed, effective phytoremediation technology for cleaning soils polluted with oil hydrocarbons has been developed. This technology can be used to eliminate the results of pollution caused by accidental oil spills that can be a result of oil transportation through the pipelines in Georgia.

REFERENCES

Arthur, E.L., Coats, J.R., 1998. Phytoremediation. In: Kearney, P., Roberts, T. (Eds.), Pesticide remediation in soils and water. Wiley, New York, pp. 251–283.

Banks, M.K., Govindaraju, R.S., Schwab, A.P., Kulakow, P., 2000. Part I: Field demonstration. In: Fiorenza, S., Oubre, C.L., Ward, C.H. (Eds.), Phytoremediation of Hydrocarbon-Contaminated Soil. Lewis Publishers, Baton Rouge, LA, pp. 3–88.

Bartha, R., 1986. Biotechnology of petroleum pollutant biodegradation. Microb. Ecol. 12, 155–172.

Bognolo, G., 1999. Biosurfactants as emulsifying agents for hydrocarbons. Colloids and Surf. A: Physicochemical and Eng. Aspects 152, 41–52.

Burgmann, H., Meier, S., Bunge, M., 2005. Effect of model root exudates on structure and activity of a soil community. Environ. Microbiol. 7, 1711–1724.

Burken, J.G., 2003. Uptake and metabolism of organic compounds: Green liver model. In: McCutcheon, S.C., Schnoor, J.L. (Eds.), Phytoremediation. Transformation and Control of Contaminants. Wiley-Interscience, Hoboken, New Jersey, pp. 59–84.

Chrikishvili, D., Sadunishvili, T., Zaalishvili, G., 2006. Benzoic acid transformation via conjugation with peptides and final fate of conjugates in higher plants. Ecotoxicol. Environ. Saf. 64, 390–399.

DeRidder, B.P., Dixon, D.P., Beussman, D.J., Edward,s, R., Goldsbrough, P.B., 2002. Induction of glutathione S-transferases in Arabidopsis by herbicide safeners. Plant. Physiol. 130, 1497–1505.

Desai, J.D., Banat, I.M., 1997. Microbial production of surfactants and their commercial potential. Microbiol. Mol. Biol. Rev. 61, 47–69.

Dibble, J.T., Bartha, R., 1979. Effect of enironmental parameters on the biodegradation of oil sludge. Appl. Environ. Microbiol. 37, 729–739.

Ensley, B.D., Gibson, D.T., 1983. Naphthalene dioxygenase: Purification and properties of a terminal oxygenase component. J. Bacteriol. 155, 505–511.

Ferro, A., Kennedy, J., Doucette, W., Nelson, S., Jauregui, G., McFarland, B., Bugbee, B., 1997. Fate of benzene in soils planted with alfalfa: Uptake, volatilization, and degradation. In: Kruger, E.L., Anderson, T.A., Coats, J.R. (Eds.), Phytoremediation of Soil and Water Contaminants. American Chemical Society, Washington, D.C., pp. 223–237.

Huesemann, M.H., 1994. Guidelines for landtreating petroleum hydrocarbon contaminated soils. J. Soil Contam. 3, 299–318.

Juwarkar, A.A., Nair, A., Dubey, K.V., Singh, S.K., Devotta, S., 2007. Biosurfactant technology for remediation of cadmium and lead contaminated soils. Chemosphere 68, 1996–2002.

Kirk, J., Klironomos, J., Lee, H., Trevors, J., 2005. The effect of perennial ryegrass and alfalfa on microbial abundance and diversity in petroleum contaminated soil. Environ. Pollut. 133, 455–465.

Korte, F., Behadir, M., Klein, W., Lay, J.P., Parlar, H., Scheunert, I., 1992. Lehrbuch der okologischen chemie. Grundlagen and Konzepte fur die Okologische Beureilung von Chemikalien. Georg Thieme Verlag, Suttgart.

Korte, F., Kvesitadze, G., Ugrekhelidze, D., Gordeziani, M., Khatisashvili, G., Buadze, O., Zaalishvili, G., Coulston, F., 2000. Review: Organic toxicants and plants. Ecotoxicol. Environ. Saf. 47, 1–26.

Kvesitadze, G., Gordeziani, M., Khatisashvili, G., Sadunishvili, T., Ramsden, J.J., 2001. Some aspects of the enzymatic basis of phytoremediation. J. Biol. Phys. Chem. 1, 49–57.

Kvesitadze, G., Khatisashvili, G., Sadunishvili, T., Gagelidze, N., Kharebashvili, M., Ramsden, J.J., 2008. On the strategy of ecological guarantee of oil pipelines. 10th International UFZ-Deltares / TNO Conference on Soil–Water Systems in cooperation with Provincia di Milano. ConSoil 2008, Milano, Italy.

Kvesitadze, G., Khatisashvili, G., Sadunishvili, T., Ramsden, J.J., 2006. Biochemical mechanisms of detoxification in higher plants. Basis of Phytoremediation. Springer, Berlin Heidelberg New York.

Leahy, J.G., Colwell, R.R., 1990. Microbial degradation of hydrocarbons in the environment. Microbiol. Mol. Biol. Rev. 54, 305–315.

Lee, E., Banks, M.K., 1993. Bioremediation of petroleum contaminated soil using vegetation: A microbial study. J. Environ. Sci. Health A28, 2187–2198.

Lee, M.D., Thomas, J.M., Morden, R.C., Bedient, P.B., Ward, C.H., Wilson, J.T., 1988. Biorestoration of aquifers contaminated with organic compounds. Crit. Rev. Environ. Control 18, 29–89.

Lin, Q., Mendelssohn, I.A., 1998. Phytoremediation for oil spill cleanup and habitat restoration in Louisiana coastal marshes: Effects of marsh plant species and fertilizer. Louisiana Applied Oil Spill Research and Development Program. OSRADP Techn. Rep. (p. 47), Ser. 97–0006.

Maier, R.M., Neilson, J.W., Artiola, J.F., Jordan, F.L., Glenn, E.P., Deschere, S.M., 2001. Remediation of metal-contaminated soil and sludge using biosurfactant technology. Int. J. Occup. Med. Environ. Health 14, 241–248.

Marrs, K.A., 1996. The function and regulation of glutathione S-transferases in plants. Annu. Rev. Plant. Physiol. Plant. Mol. Biol. 47, 127–158.

Miflin, B.J., Habash, D.Z., 2002. The role of glutamine synthetase and glutamate dehydrogenase in nitrogen assimilation and possibilities for improvement in the nitrogen utilization of crops. J. Exp. Bot. 53, 979–987.

Pritchard, P.H., Costa, C.F., 1991. EPA's Alaska oil spill bioremediation project. Environ. Sci. Technol. 25, 372–379.

Qiu, X., Leland, T.W., Shah, S.I., Sorensen, D.L., Kendall, E.W., 1997. Field study: Grass remediation for clay soil contaminated with polycyclic aromatic hydrocarbons. In: Kruger, E.L., Anderson, T.A., Coats, J.R. (Eds.), Phytoremediation of Soil and Water Contaminants. American Chemical Society, Washington, D.C., pp. 186–199.

Salt, D.E., Blaylock, M., Nanda Kumar, P.B.A., Dushenkov, V.P., Ensley, B.D., Chet, I., Raskin, I., 1998. Phytoremediation: A novel strategy for the environment using plants. Biotechnology 13, 468–474.

Sandermann, H., 1994. Higher plant metabolism of xenobiotics: The "green liver" concept. Pharmacogenetics 4, 225–241.

Schwab, A.P., Banks, M.K., 1994. Biologically mediated dissipation of polyaromatic hydrocarbons in the root zone. In: Anderson, T.A., Coats, J.R. (Eds.), Bioremediation through rhizosphere. Tech. American Chemical Society Maple Press, York, PA, pp. 132–141.

Shaw, L., Burns, R., 2007. Biodegradation of organic pollutants in the rhizosphere. Adv. Appl. Microbiol. 53, 1–60.

Shin, K., Kim, K., Ahn, Y., 2006. Use of biosurfactant to remediate phenanthrene-contaminated soil by combined solubilizaton–biodegradation process. J. Hazard. Mater. 137, 1831–1837.

Siciliano, S.D., Germida, J.J., 1998. Mechanisms of phytoremediation: Biochemical and ecological interactions between plants and bacteria. Environ. Rev. 6, 65–79.

Sims, J.L., Sims, R.C., Matthews, J.E., 1989. Bioremediation of contaminated surface soils. Robert S. Kerr Env. Research Laboratory, Office of Research and Development, US EPA. EPA/600/9–89/073.

Sotirova, A., Spasova, D., Galabova, D., Karpenko, E., 2008. Rhamnolipid–biosurfactant permeabilizing effects on Gram-positive and Gram-negative bacterial strains. Curr. Microbiol. DOI. 10.1007/s00284-008-9139-3.

Tsao, D.T., 2003a. Overview of phytotechnologies. Adv. Biochem. Eng. Biotechnol. 78, 4–50.

Tsao, D.T., 2003b. Phytoremediation. Advances in Biochemical Engineering and Biotechnology. Springer, Berlin Heidelberg New York.

Vasileva-Tankova, E., Galabova, D., Karpenko, E., 2001. Biosurfactants-rhamnolipid effects on yeast cells. Lett. Appl. Microbiol. 33, 280–284.

Wagner, M., Ma, J., Kale, Y., 2002. Biocatalysis / Biodegradation Database. Cyclohexane Pathway Map. http://umbbd.msi.umn.edu/chx/chx_map.html.

Wiltse, C.C., Rooney, W.L., Chen, Z., Schwab, A.P., Banks, M.K., 1998. Greenhouse evaluation of agronomic and crude oil-phytoremediation potential among alfalfa genotypes. J. Environ. Qual. 27, 169–173.

Chapter 20

Remediation of Cd-Contaminated Soils: Perspectives and Advancements

Syed Hammad Raza,* Fahad Shafiq,* Umer Rashid,[†] Muhammad Ibrahim[‡] and Muhammad Adrees[‡]

*Department of Botany, Government College University, Faisalabad, Pakistan, [†]Institute of Advanced Technology, Universiti Putra Malaysia, Serdang, Selangor, Malaysia, [‡]Department of Environmental Sciences, Government College University, Faisalabad, Pakistan

BACKGROUND AND INTRODUCTION

Cadmium metal is a persistent environmental toxicant (di Toppi and Gabbrielli, 1999; Zulfiqar et al., 2012) and its inclusion in terrestrial communities is increasing at alarming rates. Sustainability of our ecosystem is highly questionable due to rapid induction of high quantities of this toxic heavy metal since the human industrial revolution 200 years or so. Natural resources are also the major contributors in the emission of cadmium into terrestrial commodities. However, release of this metal from unnatural sources is commonly more mobile than those from native natural origins (Chlopecka et al., 1996; Nordic Council of Ministers, 2003). As the most toxic heavy metal among all the others, it has no known biological function in aquatic and terrestrial living organisms, including humans (Chen et al., 2007). The global annual production of cadmium was 22,200 metric tons in 2011 (Tolcin, 2012); and for humans, the weekly tolerable limit of Cd intake established by WHO is $7 \mu g\, kg^{-1}$ of total body weight (FAO / WHO, 2003).

CADMIUM EMISSIONS

Soil cadmium contamination is principally derived from natural and anthropogenic sources (De Meeus et al., 2002; Silvera and Rohan, 2007). Elemental forms of Cd with its principal minerals like octavite (CdSe), monteponite (CdO) and greenockite (CdS) are rare in nature (Kabata-Pendias and Mukherjee, 2007). Associations of cadmium with the sulphide ores of lead, copper and mainly with zinc which is the major contributor of its release, are also found

Soil Remediation and Plants. http://dx.doi.org/10.1016/B978-0-12-799937-1.00020-6

in nature (ECB, 2007; Atsdr, 2008; Cotuk, 2010). According to an estimate, 6 million tons of cadmium is released as a by-product of the zinc industry out of 1.9 billion tons of zinc resources worldwide holding 0.3% cadmium (Tolcin, 2009). Weathering, volcanic eruptions and forest fires are the significant contributing factors for natural cadmium emission, which fluctuates between 10 and 50% of the total emission (Nordic Council of Ministers, 1992; OECD, 1994; Van Assche, 1998).

However, major causative factors to environmental cadmium contamination are anthropogenic in origin. The most prominent release of cadmium into the environment is from metal-mining industries (Miura, 2009). It is extensively released as a by-product from metal industries (Wu et al., 2004; Tolcin, 2009).

Moreover, the use of phosphate fertilizers is also responsible for Cd release in agro-ecosystems, which in turn entirely depends on the parental rock used for the manufacture of the fertilizer (Yu-Jing et al., 2003). Various reports regarding release of cadmium from inorganic phosphate fertilizers are evident (Grant and Sheppard 2008, 2011). Approximately 43 million tons of P_2O_5 fertilizer was produced during 2011–2012, out of which 99% derived from phosphate rocks (FAO, 2012). The igneous origin of phosphate rocks used for fertilizer manufacturing contains $15\,mg\,kg^{-1}$ of cadmium while sedimentary phosphate rocks have higher cadmium $20–245\,mg\,kg^{-1}$ which is one of the major vehicles for depositing cadmium in our agro-ecosystems (Cotuk et al., 2010).

In addition to the above-mentioned sources, the excessive use of cadmium salts in different industries contributes to its release into the atmosphere. For instance, the use of cadmium chloride in electroplating, dyeing, photocopying, lubricants and as stabilizers; cadmium nitrate in coloured glass, porcelain and nuclear reactors; cadmium hydroxide in alkaline batteries (IARC, 1993; HSDB, 2009) is common. Furthermore, the deposition of industrial effluents and sewage water in agricultural lands has also resulted in cadmium build up. Presently, the irrigation of agricultural lands with sewage water and effluents as a source of nutrients is a common practice in third world countries, which leads to soil contamination due to elevated levels of these heavy metals, especially cadmium (Wu et al., 2004).

In the same way, the manufacture and use of Ni–Cd batteries occupies a significant portion of Cd use and release. At present, the use of cadmium-coated materials and pigments is continuing to decrease while the use of Ni–Cd batteries is increasing (Cotuk et al., 2010). To an estimate, approximately 85% of the global Ni–Cd batteries markets is located in Asia (Tolcin, 2008). About 80% of electric vehicles and a limited number of hybrid vehicles are powered by Ni–Cd batteries (Tolcin, 2008, 2009). Not only this, other sources of cadmium emission involve the use of cadmium oxide as ascaricide and nematocide (HSDB, 2009) pesticides (Papafilippaki et al., 2007), combustion of fossil fuels (De Rosa et al., 2003), power stations, cement industries, heavy road traffic (Wu et al., 2004), electroplating and stabilizers (di Toppi and Gabrielli, 1999).

SOIL DYNAMICS, RETENTION AND AVAILABILITY OF METALS

Metal are ubiquitous entities in soil modulating geochemistry by being important components of clay and mineral oxides of iron and manganese (Gadd, 2009). Commonly, sorption, ion-exchange capacities and binding energies of metals vary with respect to variation in soil components (Violante et al., 2010). The proportion of organic matter and clay in soil is one of the key factors governing the availability of metals in soil medium. Higher clay contents, organic matter and hydrous oxides adsorb metal ions and tend to immobilize these due to their interaction, thereby reducing metal bioavailability in soil (Miller et al., 1976). Clay possess a significant feature associated with its particles, owing to higher specific surface area, which is responsible for entrapping higher concentrations of metals ions in comparison with coarse-textured soils (Ahmad et al., 2011). Considering the toxic ions' entry into food chain, retention of metals on clayey soil minerals gives the advantage to sandy soils where these metals are far more easily available (Efremova and Izosimova, 2012). This retention of metals on clay particles can be a result of complexation, ion exchange, precipitation and adsorption (Zachara et al., 1993; Adriano et al., 2004). Conversely, organic matter decomposition may result in the recycling of organically bound metals (Tomáš et al., 2012).

Prime factors responsible for metal mobility and availability comprise pH, redox reactions, extent of organic and inorganic ligands, nature of the sorbent, charged mineral particles, nutrients and organic acids from root exudates (Efremova and Izosimova, 2012; Violante et al., 2010). To be more precise, the bioavailability and uptake of metals from soil is prominently dependent on soil physical and chemical characteristics along with climatic conditions (Tomáš et al., 2012).

The pH of the soil has a strong influence on solubility and speciation of metals from mineral sites, therefore it greatly affects the mobility and bioavailability of metals in soil medium (Zhao et al., 2010). Generally, solubility and availability of metallic species like Cr^{3+}, Cd^{2+}, Zn^{2+}, Hg^{2+}, Ni^{2+}, Fe^{2+}, Pb^{2+} and Mn^+ decrease to a greater extent as the pH of the soil increases (McLaughlin, 2007). However, it is not the sole factor controlling all this movement of metallic ions within the soil. The bioavailability and solubility of heavy metals are also reported to be linearly correlated with redox potential (Patrick et al., 1990; Masscheleyn et al., 1991). It has been shown that, at identical pH, a decrease in redox potential may increase solubility of metals (Yaron et al., 1996).

Not surprisingly, free metals present in the soil also determine the bioavailability of a particular metal and its subsequent plant uptake (Moffett and Brand, 1996). Apart from this, application of sewage can increase metal mobility in recipient soils by lowering pH through the process of nitrification and microbial carbon dioxide production (Smith, 1996). Therefore, the characterization of the factors responsible for adsorption, leaching, availability and toxicity of metals in soils is of significant importance (Tomáš et al., 2012).

DYNAMICS OF CADMIUM IN SOILS

Likewise other metals, complexation, adsorption, ion exchange and precipitation influence cadmium bioavailability in soils (Christensen and Tjell, 1990). Clay fraction of the soil is a significant component mediating bioavailability and leaching of cadmium (Allen, 1993). The sorption of Cd in soil achieves its equilibrium state within 1 h, however, this is a reversible phenomenon (Christensen and Tjell, 1990). Similarly, soil pH is a key factor driving cadmium (Cd) mobility and bioavailability (McBride, 1989; Cotuk et al., 2010). The mobility of Cd is greater in soils with low pH range, possessing fewer clay particles, low organic matter in soil, and low iron and manganese oxides, high Cd concentrations, lower cation-exchange capacities, zinc deficiency and at elevated temperatures (Kabata-Pendias, 2001). Generally, Cd mobility is greater at pH range 4.2–6.6, while it is moderate at pH range 6.7–8.8 (Schmitt and Sticher, 1991). Gray et al. (1999) showed that increasing the soil pH from 5.3 to 7.0 results in reduced Cd uptake in five different plant species. Thus, soil acidification phenomena like acid rain can positively affect Cd mobility and availability (Nigam et al., 2001). However, contrasting reports regarding pH effect on Cd mobility have also been documented (Eriksson, 1989).

It is not only the physicochemical properties of soil that modulate the Cd bioavailability, but it is also dependent on physiological attributes of roots, plant age and genetics of plant species (Harter and Naidu, 2001; Jung, 2008). The exchangeable, chelated and soluble Cd constituents are relatively more mobile in most of the soils with enhanced phytoavailability (Schmitt and Sticher, 1991). Conversely, carbonates present in soil can adsorb divalent cations like Pb^{2+}, Ba^{2+} and Cd^{2+} on their reactive surfaces (Ming, 2002).

INFLUENCE OF THE ASSOCIATED CATIONS AND ANIONS ON CADMIUM BIOAVAILABILITY IN SOIL

The existence of metals in specific forms influences their relative behaviour within soil systems (Ge et al., 2005). The presence of various cations and anions is a significant force driving metal availability and dynamics in soil. Within soil systems, anions contribute to several reactions including the desorption process of heavy metals and are considered in the following.

Cadmium availability is affected by the presence of certain ions in the soil owing to their role in complexation (Degryse et al., 2004), ionic strength (Gothberg et al., 2004) and competition of surface exchange sites (Tlustos, 2006). An inverse relationship between cadmium availability and ionic strength of the growing medium has been documented (Gothberg et al., 2004). Several anionic species have been reported to enhance the process of desorption of metals from soil particles. It is reported that presence of acetate and chloride ions increase mobility of metal ions through complexation. Extractability of Cd was 11 times higher with magnesium chloride (1 M $MgCl_2$) than magnesium nitrate [1 M $Mg(NO_3)_2$], as documented by Gommy et al. (1998). Likewise anions, monovalent and divalent cationic species take part in soil dynamic reactions with reference to metal and soil components.

Monovalent cations, for instance Na^+ and K^+, are useful for increasing extraction of metals while divalent cations like Ca^{2+}, Mg^{2+} can desorb metals and, therefore, extraction rates are far greater than for monovalent cations. Divalent cations including Ca^{2+}, Zn^{2+}, Mn^{2+} and Mg^{2+} compete with Cd^{2+} for metal exchange sites (Degryse, 2004) and for uptake by plants (Ramachandran and D'Souza, 2002; Tlustos et al., 2006). The competence for metal exchange is reported as: $Ba^{2+} > Ca^{2+} > Mg^{2+} > NH_4^+ > K^+ > Na^+$ with monovalent cations showing the least value (Gommy et al., 1998).

RESPONSE OF Cd TOWARDS NATURAL ELEMENTAL INORGANIC AMENDMENTS

Inorganic elements exhibit certain interactive effects in natural soil systems influencing Cd dynamics in soil. Many studies highlighted the effects of inorganic amendments on cadmium, either positive or negative, under different textured soils. Hence, these can be utilized in cadmium-contaminated soils depending on objectives either for reducing cadmium bioavailability or to immobilize it through various interactive properties. As stated above, the interactive effect of inorganic amendments in immobilizing Cd is not always positive. Furthermore, these inorganic amendments can primarily be nutrients because of their tendency to mitigate metal stress (Jalloh et al., 2009). In order to comprehensively elucidate these mechanisms, the effect of diverse inorganic amendments with subsequent response towards cadmium has been taken into account.

Calcium (Ca)

If the purpose is to limit or immobilize cadmium in soil, the primary step can be targeting the soil pH through liming. Such a process increases soil pH and ultimately reduces Cd mobility while entrapping it in soil particles (Reeves and Chaney, 2004; Tsadilas et al., 2005). This practice can be achieved using inorganic salts of calcium such as $CaCO_3$, $Ca(OH)_2$, CaO and $CaSO_4 \cdot 2H_2O$ (Filius et al., 1998). In addition to this, applying base-rich fertilizers can also be helpful in this respect (Sarwar et al., 2010). Liming is a cost-effective process for reducing metal transport in plants (Efremova and Izosimova, 2012). Moreover, calcium ion competes with cadmium ions for absorption sites that are available at plant roots (Cataldo et al., 1983). In a study conducted by Tlustos et al. (2006), application of calcium oxide and calcium carbonate resulted in a 50% reduction of Cd uptake in spring wheat, and this was explained on the basis of immobilization in response to soil liming. Furthermore, chelation of cadmium into Cd–Ca crystals has also been reported in tobacco plants (Choi et al., 2001).

Nitrogen (N)

Addition of nitrogen to cadmium-contaminated soil can dramatically modulate Cd availability. Nitrogen application in both nitrate (NO_3^-) and ammonium (NH_4^+) ion forms is effective for plant growth (Jalloh et al., 2009), but the effects of both the nitrogen forms on geochemistry and Cd dynamics in soil are through different mechanisms. Inorganic nitrogen applied in the form of ammonium ions can acidify soil (Landberg and Greger, 2003) either by a process of nitrification or by the release of protons (H^+) that can seriously affect the bioavailability of Cd (Loosemore et al., 2004; Zaccheo et al., 2006). This is one of the major reasons why nitrogen fertilizers like mono-ammonium phosphate, di-ammonium phosphate, urea and ammonium sulphate can result in cadmium build-up, due to their ability to decrease soil pH (Zaccheo et al., 2006). In addition, ammonium-based amine complexes have been regarded as heavy-metal-mobilizing mechanisms (Lebourg et al., 1996; Pueyo et al., 2004). The dissociation of ammonium ion can ultimately result in the formation of soluble amine metal exchangeable complexes (Gryschko et al., 2005); however, contrasting reports are also evident in the literature. In some recent studies, NH_4^+ supplementation resulted in lower pH, but increased Cd contents were recorded in response to the application of NO_3^- (Xie et al., 2009). Similarly, Jalloh et al. (2009) reported higher Cd content in response to NO_3^- and attributed this to the synergistic interactions of nitrate on cadmium. Therefore, cadmium mobility and availability in response to nitrogen is dependent on the type of nitrogen source applied.

Sulphur (S)

In relation to cadmium, the role of sulphur is very significant as sulphur amendments in the form of sulphated compounds can limit cadmium bioavailability in soils. The low solubility of metal sulphates is effective in limiting Cd (Violante et al., 2010). Sulphur application can initiate the formation of insoluble CdS complexes through production of H_2S, entrapping cadmium in soil (Hassan et al., 2005) by vulcanization (Bingham, 1979). Interestingly, higher Cd content has been reported in crop plants irrigated with water containing high sulphate content and this was attributed to sulphato complexes (McLaughlin et al., 2006). Sulphate reduction-based mobilization of metals like Pb, Cd and Cu through the formation of sulphide collide has also been reported (Violante et al., 2010).

Zinc (Zn)

Marked antagonistic interactions between Cd and Zn are documented in literature (Cataldo et al., 1983; McKenna et al., 1993; Wei et al., 2003). Both of these elements posses close similarity in their ionic structure, electronegativity (Moustakas et al., 2011), electronic configuration and reactivity towards most of the ligands (Smilde et al., 1992). It is reported that Zn suppresses Cd uptake

(Cataldo et al., 1983); therefore, soil Zn/Cd ratio determines cadmium uptake in plant (Cotuk et al., 2010). Zinc availability in soils even at concentration $0.3\,mg\,L^{-1}$ results in a marked competition with Cd for the sorption sites. The relationship between these two elements in soil media is significantly correlated (Eriksson, 1990) and various investigations reported inhibitory and antagonistic effects of Zn on cadmium uptake (McLaughlin and Singh, 1999; Long et al., 2003). Soil amendments with the application of Zn decreased Cd content of plants (Moustakas et al., 2011). It is documented that mass flow is also a factor contributing to plants' zinc and cadmium uptake (Mullins et al., 1986). Higher Cd accumulations have been associated with Zn-deficient soils (Adiloglu, 2002).

Contrary to this, synergistic interactions between Zn and Cd were also evident in studies where concentrations of both metals resulted in higher metal content of crops (Xue and Harrison, 1991; Nan et al., 2002; Kachenko and Singh, 2006).

Phosphorus (P)

Most agricultural soils have low concentration of phosphorus in comparison with nitrogen and potassium (Sarwar et al., 2010), that ranges to less than 0.15% (Havlin et al., 2007). Many studies have reported positive effects of phosphorus in reducing Cd (Haghiri, 1974; Smilde et al., 1992). Appliance of phosphate to the Cd-contaminated soils hinders the mobile form of Cd by making it immobile through the formation of insoluble cadmium phosphate (Dheri et al., 2007; Matusik et al., 2008). Application of phosphate fertilizers along with soil pH between 6.75 and 9.00 reduced the bioavailability of Cd up to 99% (Matusik et al., 2008). This prominent reduction in Cd bioavailability is dependent on several factors. Phosphate in soil media induces adsorption of Cd^{2+}, its precipitation as $Cd_3(PO_4)_2$, a rise in the pH and surface charge of the soil medium and co-adsorption of P and Cd as an ion pair (Bolan et al., 2003a, b). Similarly, Dheri et al. (2007) attributed the phosphorus-mediated decrease in DTPA (diethylentriamene pentaacetate)-extractable Cd from soil to in situ immobilization.

Iron (Fe)

In most neutral and alkaline soils, iron exists as Fe^{3+} in the form of insoluble compounds (Sheng et al., 2008). Iron can regulate pH and redox potential of the soil and its scarcity in the soil can result in enhanced Cd accumulation within plants (Bao et al., 2009). In addition, iron also competes with Cd for membrane transporters in plant roots (Sarwar et al., 2010). Application of chelated iron fertilizer (EDTA – Na_2Fe) to the Cd-contaminated soil resulted in decreased Cd content; however, $FeSO_4$ application resulted in the opposite behaviour of Cd (Sheng et al., 2008). Similarly, it is reported that iron plaque formed from the oxidation of Fe^{2+} to Fe^{3+} can adsorb Cd on its surface, therefore iron supplementation can mitigate Cd toxicity by reducing its bioavailability. Similar positive effects of iron supplementation on Cd immobilization in soil and its subsequent positive influence on rice were reported by Liu et al. (2008b).

Manganese, Silicon and Chloride

In the literature, contrasting reports are documented related to Mn and Cd interactions in soil medium. A synergistic effect between both these heavy metals has been cited by Chen et al. (2007), while these two heavy metals can also behave antagonistically in soil (Ramachandran and D'Souza, 2002). The opposite behaviour of Mn and Cd within soil was also affirmed by Baszynski et al. (1980) and was attributed to its competition with Cd in the soil.

Silicon exits in soil solution naturally at concentrations ranging between 0.1 and 0.6 mmol (Epstein, 1999). It has been reported to be beneficial in mitigation of Cd toxicity in several crops including wheat (Cocker et al., 1998), maize and rice (Liang et al., 2005). However, the Si-mediated reduction in cadmium uptake needs to be discussed with reference to soil dynamics. The primary effect mediated by Si application is the increase in soil pH resulting in immobilization of Cd (Liang et al., 2005). Reports related to cadmium complex formation with chloride in soil systems are evident. The Cd–Cl complex is highly soluble, and results in increased cadmium mobilization within soils (Degryse et al., 2004). That is why the application of KCl to Cd-contaminated soil resulted in increased Cd content in barley (Grant et al., 1999).

ORGANIC AMENDMENTS VERSUS CADMIUM-CONTAMINATED SOILS

In the early parts of this chapter, the significant influence of soil organic matter content to the bioavailability was stated. Organic matter (OM) content of soil and Cd availability are closely interlinked. In comparison to soils with high mineral contents, the soils higher in organic matter can retain metallic cations and exhibit about 30 times more Cd sorption affinity (Sarwar et al., 2010). Certain reactive functional groups like carboxylic, hydroxyl and phenoxy groups control the adsorption of metals very effectively (Alloway, 1995). Therefore, the application of organic matter and related amendments like farmyard manure, poultry manure, biosolids, compost effectively immobilize Cd from contaminated soils (Jamode et al., 2003; Puschenreiter et al., 2005; Sampanpanish and Pongpaladisai, 2011). In this way, it has been established that the addition of organic compounds to metal-contaminated soil reduces metal availability from the contaminated soils by complexation (Angelova et al., 2010) and solubility considerations (Ciecko et al., 2001).

The influence of some commonly used organic amendments, both natural and of synthetic origin, is grouped together for proper understanding in relation to soil Cd content.

NATURAL ORGANIC ADDITIVES

Commonly used organic additives include cow, farmyard and poultry manure, and partially decomposed matter termed humus. The organic amendments have been effective in improving soil physical characteristics (Bradshaw and Chadwick, 1980; Bouajila and Sanaa, 2011) and imparted positive influence on crops (Kaihura et al., 1999). Application of farmyard manure (FYM) can reduce the toxic effects of heavy metals on crops (Yassen et al., 2007). Increased soil organic fractions directly regulate its sorption and cation exchange for Cd affecting bioavailability (Alamgir et al., 2011). One unit pH decrease in soil is reported with the addition of $320\,g\,kg^{-1}$ organic matter to soil but it resulted in 1.5- to 6-fold increased cation exchange capacity (CEC) depending on soil texture (He and Singh, 1993). Therefore, the decrease in plant cadmium content was attributed to organic-matter-mediated increment in CEC (He and Singh, 1993).

Similar reports regarding FYM-induced increase in CEC of soil (Alamgir et al., 2011), and in combination with lime to reduce cadmium in rice grain (Kibria et al., 2011) are evident. The above-mentioned FYM-mediated decrease in Cd availability was confirmed by Pearson's correlation analysis. The concentrations of Cd were found to be negatively correlated (= −0.841 and −0.869 for the shoot and root, respectively) in response to the FYM application (Alamgir et al., 2011).

In addition to FYM, application of chicken manure (Li et al., 2006) and compost (Chiu et al., 2006; Pitchel and Bradway, 2008) to the contaminated soil decrease the extent of phyto- or bio-available Cd by immobilizing it. Another organic amendment is humus that acquires negatively charged sites on phenol and carboxylic groups capable of binding metals (Stevenson and Fitch, 1994). Marked proportions of humified OM were found to be associated with cow manure which also resulted in metal immobilization (Tordoff et al., 2000). Furthermore, humus acts as a multi-ligand component system of soil due to possession of large numbers of complex sites (Buffle, 1988). Therefore, the organic portion of the soil medium not only regulates CEC of soil but also modulates its geochemistry. It augments soil buffer capacity, minimizes soil compactness and acts as a nutrient pool by recycling nutrients naturally (Stewart et al., 2000). Moreover, immobilized Cd–organic matter complexes are also reported (Putwattana et al., 2010). However, a few contrasting reports are also available in the literature showing a positive relationship between OM application and Cd mobilization and its subsequent bio-sequestration (Narwal and Singh, 1998).

ROOT EXUDATES AND THE CONCEPT OF ORGANIC ACIDS AS NATURAL CHELATORS

Certain natural organic chelates including amino acids, polysaccharides, organic acids and related compounds are introduced into the rhizosphere by plants roots (Dong et al., 2007) to alter and modulate the solubility of metal ions. Organic acids (anions) comprise the major proportion of these exudates that possess tendencies to strongly bind with metals despite their physical state (Jones and Darrah, 1994; Jones et al., 1996). Interaction of Cd with organic ligands upon supplementation with organic acids has been proposed (Nigam et al., 2001). In addition, organic acids can affect sorption and desorption of Cd despite soil type and texture (Wang et al., 2013). Therefore, the detailed effect of various low-molecular-weight organic acids is considered below.

LOW-MOLECULAR-WEIGHT ORGANIC ACIDS AND CADMIUM CHELATION

Citric, malic, fumaric, succinic, aspartic, glutamic and related low-molecular-weight organic acids (LMWOAs) are proposed as potential metal chelators (Naidu and Harter, 1998). Most specifically, the carboxylic acids, malic and citric acid can result in stable complexes with divalent cations (Cieslinski et al., 1998) and result in Cd chelation (Dong et al., 2007). In many investigations, the chelated Cd species in turn were translocated through the xylem and were accumulated by plants more rapidly. Application of citric acid ($10\,mmol\,kg^{-1}$) resulted in 1.3- to 3-fold increased uptake of Cd from Cd-contaminated growth medium in *Brassica juncea* (Quartacci et al., 2005). Similar positive beneficial effects have been documented by Duarte et al. (2007) and Qu et al. (2011). Generally, the use of citric acid is greatly preferred over other organic acids as it does not impair growth and biomass of plants, biodegrades effortlessly, is environmentally friendly and cost effective (Melo et al., 2008; Smolinska and Krol, 2010; Wuana, et al., 2010). Cd–citric acid chelation is attributed to transformation of exchangeable Cd into residual form (Mojiri, 2011). Conversely, the application of citric acid to soil for Cd removal from the soil is not always effective and can be attributed to rapid degradation (Ström et al., 2001). A degradation of approximately 90% in applied citric acid was recorded within the 25 days (Jia et al., 2009). In addition to this fact, citric acid can be mineralized by soil microorganisms (Lesage et al., 2005) which cause re-adsorption of metal contaminant on soil particles (Nascimento et al., 2006). Moreover, inefficacy of lower concentrations of citric acid is unable to desorb soil Cd (Elkhatib et al., 2001; Turgut et al., 2004), therefore it is regarded as a weak chelating agent (Kirpichtchikova et al., 2006).

The role of malic acid in mobilizing soil entrapped/sequestered Cd is also documented (Zhang et al., 1997; Naidu and Harter, 1998; Cieslinski et al., 1998) but citric acid application is preferred over malic acid (White et al., 1981). Besides, along with citric acid, many studies report the beneficial interaction of organic acids like malic and oxalic acids on mobilization of metals by increasing solubility (Peters, 1999; Nigam et al., 2001; Chen et al., 2003; Evangelou et al., 2006).

EFFICACY OF SYNTHETIC ORGANIC CHELATING AGENTS TOWARDS CADMIUM

The manufactured organic chelates are most effective in chelating metal contaminants (Huang et al., 2005; Anwer et al., 2012). These include EDTA (ethylenediaminetetraacetic acid), DTPA, CDTA (1,2-cyclohexanediaminetetraacetic acid low sodium), EDDS (ethylenediaminedisuccinate), NTA (nitrilotetraaccetic acid), EGTA (ethylene glycol-O,O-bis-[2-amino-ethyl]-N,N,N,N,-tetraacetic acid) and related compounds (Quartacci et al., 2007; Pastor et al., 2007; Anwer et al., 2012) that have a strong potential for mobilizing heavy metals from the soil (Kayser et al., 2000). The appliance of chelating agents results in metal complexation and ultimately increased uptake of the chelated metal species (Nascimento et al., 2006; Engelen et al., 2007), but this is not the case all the time. Despite the increasing metal solubility and its subsequent mobilization, these compounds do not always result in the enhancement of metal uptake in plants (Liu et al., 2008a; Khan et al., 2000). The effects of various synthetically derived organic chelates towards cadmium are discussed below.

Predominantly, EDTA is the most widely used synthetic chelant under Cd contamination due to its strong affinity for Cd and because it is slowly biodegradable (Means et al., 1980; Saifullah et al., 2009). The extremely high binding affinity of EDTA for various metals enables the release of metals from the insoluble phase to the soluble phase, therefore, plays a significant part in the soil metal mobilization (Nowack, 2002). Several studies resulted in Cd release from sequestered portions in soil, affirming the positive effect of EDTA. In various studies, lower concentrations of EDTA resulted in Cd leachate from the soil (Römkens et al., 2002). Furthermore at elevated EDTA concentration, pH effect is masked showing not much influence on chelation (Ghestem and Bermond, 1998). Other than EDTA-assisted removal of Cd from soil, there are some limitations as well. Firstly, EDTA is a non-selective chelating agent with strong affinity for most of the divalent cations (Zeng et al., 2005). Secondly, it is not environmentally friendly as its application leads to a severe decline in soil nematodes (Wu et al., 2004). Most importantly, its application in soil results in marked reduction in soil essential micronutrients such as iron, magnesium and zinc (Wasay et al., 1998).

Another chelating agent purposely used for Cd complexation is DTPA. In many studies, applications of DTPA resulted in increased Cd solubility (Mehmood et al., 2012), therefore increased its uptake in plants (Kirkham, 2006; Engelen et al., 2007). Concentrations of Cd in the lechate were directly correlated with the DTPA dose applied (Wu et al., 2004) and this effect was prominent on Cd solubility even after the course of 90 days depicting high persistence of DTPA in soil (Wenzel et al., 2003). However in contrast, some severe limitations are associated with DTPA application in soil. It is less efficient in Cd solubilization than EDTA and is light sensitive (Metsärinne et al., 2004; Engelen et al., 2007; Evangelou et al., 2007). Furthermore, its application is toxic to plants which makes it unsuitable for phytoextraction (Nascimento et al., 2006).

Other than EDTA and DTPA, there are certain other synthetic chelating compounds that modulate metal mobility but extensive investigations are required to properly elucidate their role in terms of Cd dynamics in soil. NTA is one such compound that can bind with several metals and increase their solubility (Bolton et al., 1996) but it is reported to be very toxic to plants. Application of NTA at 10 mmol kg^{-1} resulted in death of mustard plants within 48 h (Quartacci et al., 2006). Similarly, CDTA and EGTA can significantly form complexes with higher proportions of Cd from the soil (Mehmood et al., 2012) but investigations are limited.

Therefore, variation is evident considering the influence of various amendments on cadmium in soil. A generalized effect of these amendments is presented in Figure 20.1.

RECENT PRESENTED REPORTS REGARDING GRAIN CROPS

Potentiality and feasibility of cereal crops to be used for remediating cadmium-contaminated soils cannot be neglected. Generally, Cd translocation from soil to grain component is largely dependent on soil and plant type along with agricultural practices (Rodda et al., 2011). The trend of Cd bioaccumulation in cereal crops from various studies is discussed in the following paragraph (see also Table 20.1).

Cadmium stress resulted in reduced growth rates of barley seedlings (Wu et al., 2007). In a study conducted by Wu et al. (2003), four barley genotypes responded differently to Cd in terms of biochemical attributes. The grain Cd concentration depends on root and shoot Cd content (Wu et al., 2007). Interestingly, the cadmium content of developing barley was maximum (51%) in grain as compared with the other parts which was in following order awn > stem > grain > rach is > glume (Chen et al., 2007). However, the Cd content of wheat grain was least in grain followed by straw and root, respectively (Wang et al., 2012). Cd concentration in wheat grain grown in non-contaminated soils in the United States ranged from 0.002 to 0.207 mg Cd kg^{-1} dry weight. In comparison with bread wheat genotypes, durum wheat tends to accumulate more Cd content in the grains (Li et al., 1997) but variations in Cd bioaccumulation in grains among durum wheat genotypes is also evident (Clarke et al., 2002). Sorghum is also an important cereal and a fodder crop but its biomass and productivity is affected by soil Cd contamination. Application of organic matter led to an increase in sorghum biomass under Cd toxicity; however, increased growth indirectly contributed to higher Cd accumulation (Pinto et al., 2005). The effect of nitrogen fertilizer in increasing the wheat grain Cd content has been documented (Perilli et al., 2010). High soil salinity resulted in increased Cd bioaccumulation in maize was attributed to Cd–Cl soluble complexes (Chuken, 2012). Many other contrasting reports among genotypes of a single species are documented in literature. Consequently, from a soil and plant perspective, exploration of the physiological mechanisms of Cd accumulation in grain crops need further exploration.

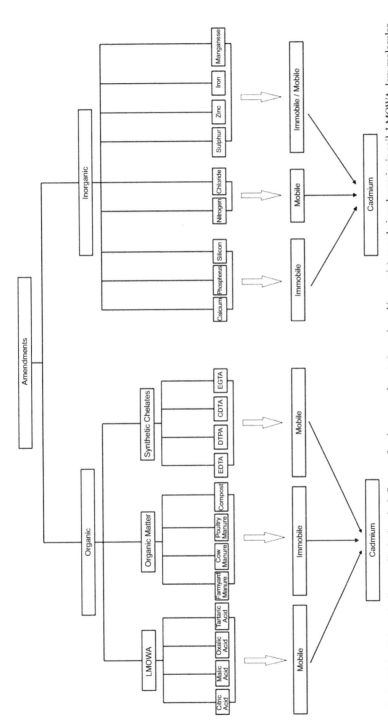

FIGURE 20.1   Schematic diagram displaying the influence of various amendments (organic and inorganic) on cadmium dynamics in soil. LMOWA, low-molecular-weight organic acids.

Soil Remediation and Plants

TABLE 20.1 Concentration of Cadmium in Grain of Some Cereal Crops from Recent Literature

| Sr. No | Crop | Grain Cadmium Concentration | | Experimental Soil / Growing Medium | Amendment Used | Reference |
		Min	Max			
1	Maize	3.5 mg kg^{-1}	12.3 mg kg^{-1}	Sandy Loam	Zinc	Akay and Koleli, 2007
2	Barley	0.01 mg kg^{-1}	0.07 mg kg^{-1}	—	—	Nakayama et al., 2009
3	Maize	26 ng g^{-1}	37 ng g^{-1}	—	Nitrogen	Kui et al., 2009
4	Sorghum	0.14 mg kg^{-1}	0.36 mg kg^{-1}	—	—	Faruruwa et al., 2013
5	Millet	0.11 mg kg^{-1}	0.22 mg kg^{-1}	—	—	Faruruwa et al., 2013
6	Wheat	0.015 mg kg^{-1}	0.040 mg kg^{-1}	Acidic (pH 5.1–5.8)	—	Wieczorek et al., 2005
7	Barley	0.020 mg kg^{-1}	0.064 mg kg^{-1}	Acidic (pH 5.1–5.8)	—	Wieczorek et al., 2005
8	Wheat	0.034 mg kg^{-1}	0.099 mg kg^{-1}	Loamy Soil	Inorganic Nitrogen	Li et al., 2011
9	Wheat	1.7 nmol plant^{-1}	5.1 nmol plant^{-1}	Loamy soil	Zinc	Herren and Feller, 1997
10	Rice	1.0 mg kg^{-1}	6.3 mg kg^{-1}	—	Farmyard manure	Mathew et al., 2002
11	Cow pea	1.2 mg kg^{-1}	2.8 mg kg^{-1}	—	Farmyard manure	Mathew et al., 2002
12	Sesame	3.2 mg kg^{-1}	13.7 mg kg^{-1}	—	Farmyard manure	Mathew et al., 2002
13	Rice	0.03 μg g^{-1}	0.12 μg g^{-1}	Paddy soils	—	Machiwa, 2010

TABLE 20.1 Concentration of Cadmium in Grain of Some Cereal Crops from Recent Literature—cont'd

14	Durum wheat	$0.08\,\mu g\,g^{-1}$	$0.32\,\mu g\,g^{-1}$	Five different soil types	—	Cifuentes et al., 2012
15	Rice	$0.0519\,mg\,kg^{-1}$	$0.0857\,mg\,kg^{-1}$	Polluted water irrigated soil	—	Bakhtiarian et al., 2001
16	Wheat	$2.99\,mg\,kg^{-1}$	$4.87\,mg\,kg^{-1}$	Hydroponics	—	Zhang et al., 2002
17	Durum wheat	$0.025\,mg\,kg^{-1}$	$0.359\,mg\,kg^{-1}$	Saline soil	—	Norvell et al., 2000
18	Rice (brown)	$0.1\,mg\,kg^{-1}$	$0.8\,mg\,kg^{-1}$	Hydroponics	—	Kukier and Chaney, 2002
19	Rice	$0.02\,mg\,kg^{-1}$	$5\,mg\,kg^{-1}$	Zn/Cd contaminated	—	Simmons et al., 2003
20	Soybean	$1.081\,mg\,kg^{-1}$	$1.71\,mg\,kg^{-1}$	Zn/Cd contaminated	—	Simmons et al., 2003
21	Wheat	$0.020\,mg\,kg^{-1}$	$0.049\,mg\,kg^{-1}$	Calcareous soils	Sewage Sludge	Qiong et al., 2012
22	Maize	$0.0019\,mg\,kg^{-1}$	$0.023\,mg\,kg^{-1}$	Calcareous soils	Sewage Sludge	Qiong et al., 2012

CONCLUSIONS AND THE CONCEPT OF COUPLED PHYTOREMEDIATION AS A FUTURE PERSPECTIVE

So far we have considered the efficacy of a diverse group of chelating agents vide inorganic and organic origins. Though most of these agents exhibited promising characteristics in affecting cadmium availability by altering its mobility in the soil system, the efficacy in terms of complete immobilization or mobilization is scarce.

Phytoremediation is an emergent and advanced methodology selected and executed recently for removing metal contaminants from the soil by growing plant species. Many studies emphasized the use of plants for remediation of Cd-contaminated soils but still certain limitations are associated with phytoremediation technology. This includes the inability of biomass production by plants or sometimes inefficient transfer of metal contaminant from the soil to aerial plant parts. The adverse effects of cadmium toxicity on biomass production ability and biochemical functioning (Raza and Shafiq, 2013; Raza et al., 2013) are important factors governing the selection of suitable plant species for phytoremediation technology. To tackle these limitations, coupled phytoremediation can be a comprehensive solution involving the use of two successive approaches. Coupled phytoremediation involves the selection of an efficient chelating agent along with suitable plant species. For instance, consider the chelating efficiency of ammonium sulphate. Ammonium ion can enhance Cd mobility while, in contrast, the acidic radical (sulphate) sometimes reduces its mobility within the soil system. In accordance with the concept of coupled phytoremediation, the use of ammonium nitrate will be far better for mobilizing Cd through the soil than ammonium sulphate as it liberates NH_4^+ and NO^{-3} ions. Both nitrate and ammonium ions (acidic and basic radicals) are pronounced in enhancing Cd mobility in the soils either by chelating processes or by affecting pH-related attributes. Furthermore the application of ammonium nitrate can be coupled with the cultivation of high-biomass-producing, metal-accumulating plant species. To be more precise, the removal of cadmium will be targeted by using two simultaneous approaches acting together. The use of inorganic salts having potential of mobilizing Cd or enhancing its bioavailability from the soil both through acidic and basic radicals can therefore be effective. Likewise, natural organic acids and synthetic organic chelates can also be coupled for phytoextraction of cadmium metal. However in contrast, if the purpose is to immobilize Cd in soil to limit its uptake by edible crop plant species, the use of inorganic and organic amendments effective in immobilizing soil Cd can be simultaneously coupled with growing genotypes exhibiting potentiality to exclude Cd and vice versa. In the latter context, the use of $CaSO_4$ or $ZnSO_4$ with the addition of organic matter will be beneficial in limiting Cd in soils.

REFERENCES

Adiloglu, A., 2002. The effect of zinc (Zn) application on uptake of cadmium (Cd) in some cereal species. Arch. Agronomy and Soil Sci. 48, 553–556.

Adriano, D.C., Wenzel, W.W., Vangronsveld, J., Bolan, N.S., 2004. Role of assisted natural remediation in environmental cleanup. Geoderma. 122, 121–142.

Ahmad, H.R., Ghafoor, A., Corwin, D.L., Aziz, M.A., Saifullah, Sabir, M., 2011. Organic and inorganic amendments affect soil concentration and accumulation of cadmium and lead in wheat in calcareous alkaline soils. Commun. Soil Sci. Plant Anal. 42, 111–122.

Akay, A., Koleli, N., 2007. Interaction between cadmium and zinc in barley (*Hordeum vulgare* L.) grown under field conditions. Bangladesh J. Bot. 36 (1), 13–19.

Alamgir, M., Kibria, M.G., Islam, M., 2011. Effects of farm yard manure on cadmium and lead accumulation in Amaranth (*Amaranthus oleracea* L.). J. Soil Sci. Environ. Manag. 2 (8), 237–240.

Allen, H.E., 1993. The significance of trace metal speciation for water, sediment and soil quality criteria and standards. Sci. Total Environ. 134 (1), 23–45.

Alloway, B.J., 1995. Cadmium. In: Alloway, B.J. (Ed.), Heavy Metals in Soils, second ed. Blackie Academic and Professional, London.

Angelova, V.R., Ivanov, K.I., Krustev, S.V., 2010. Effect of phosphorous, organic and sapropel amendments on lead, zinc and cadmium uptake by triticale from industrially polluted soils. Proc. Annual Int. Conf. on Soils, Sediments, Water and Energy 20 (13), 1–18.

Anwer, S., Ashraf, M.Y., Hussain, M., Ashraf, M., Jamil, A., 2012. Citric acid mediated phytoextraction of cadmium by maize (*Zea mays* L.). Pak. J. Bot. 44 (6), 1831–1836.

Atsdr, 2008. Draft toxicological profile for cadmium. Atlanta: US Department of Health and Human Services, Agency for Toxic Substances and Disease Registry. Retrieved April 21, 2013 from http://www.atsdr.cdc.gov/toxprofiles/tp5-p.pdf.

Bakhtiarian, A., Gholipour, M., Khansari, M.G.K., 2001. Lead and cadmium content of korbal rice in northern Iran. Iran. J. Public Health 30 (3–4), 129–132.

Bao, T., Sun, L., Sun, T., Zhang, P., Niu, Z., 2009. Iron-deficiency induces cadmium uptake and accumulation in *Solanum nigrum* L. Bull. Environ. Contam. Toxicol. 82, 338–342.

Baszynski, T., Wajda, L., Krol, M., Wolinska, D., Krupa, Z., Tukendorf, A., 1980. Photosynthetic activities of cadmium-treated tomato plants. Physiol. Plantarum. 63, 293–298.

Bingham, F.T., 1979. Bioavailability of Cd to food crops in relation to heavy metal content of sludge-amended soil. Environ. Health Perspect. 28, 39–43.

Bolan, N.S., Adriano, D.C., Duraisamy, P., Mani, P.A., Arulmozhiselvan, K., 2003a. Immobilization and phytoavailability of cadmium in variable charge soils. I. Effect of phosphate addition. Plant and Soil 250, 83–94.

Bolan, N.S., Adriano, D.C., Mani, P.A., Duraisamy, A., 2003b. Immobilization and phytoavailability of cadmium in variable charge soils. II. Effect of lime addition. Plant and Soil 251, 187–198.

Bolton, J.H., Girvin, D.C., Plymale, A.E., Harvey, S.D., Workman, D.J., 1996. Degradation of metal nitrilotriacetate complexes by Chelatobacter heintzii. Environ. Sci. Technol. 30 (3), 931–938.

Bouajila, K., Sanaa, M., 2011. Effects of organic amendments on soil physico-chemical and biological properties. J. Mater. Environ. Sci. 2 (1), 485–490.

Bradshaw, A.D., Chadwick, M.J., 1980. The restoration of land. Blackwell, Oxford.

Buffle, J., 1988. Complexation reactions in aquatic systems: An analytical approach. Ellis Horwood, Chicheste.

Cataldo, D.A., Garland, T.R., Wildung, R.E., 1983. Cadmium uptake kinetics in intact soybean plants. Plant. Physiol. 73, 844–848.

Chen, F., Dong, J., Wang, F., Wu, F., Zhang, G., Li, G., Chen, Z., Chen, J., Wei, K., 2007. Identification of barley genotypes with low grain Cd accumulation and its interaction with four microelements. Chemosphere 67, 2082–2088.

Chen, Y.X., Lin, Q., Luo, Y.M., He, Y.F., Zhen, S.J., Yu, Y.L., Tian, G.M., Wong, M.H., 2003. The role of citric acid on the phytoremediation of heavy metal contaminated soil. Chemosphere 50, 807–811.

Chiu, K.K., Ye, Z.H., Wong, M.H., 2006. Growth of Vetiveria zizanioides and Phragmities australis on Pb / Zn and Cu mine tailing amended with manure compost and sewage sludge: A greenhouse study. Bioresour. Technol. 97, 158–170.

Chlopecka, A., Bacon, J.R., Wilson, M.j., Kay, J., 1996. Forms of cadmium, lead, and zinc in contaminated soils from south-west Poland. J. Environ. Qual. 25 (1), 69–79.

Choi, Y.E., Harada, E., Wada, M., Tsuboi, H., Morita, Y., Kusano, T., Sano, H., 2001. Detoxification of cadmium in tobacco plant: Formation and active excretion of crystals containing cadmium and calcium through trichomes. Planta 213, 45–50.

Christensen, T., Tjell, J., 1984. Interpretation of experimental results on cadmium crop uptake from sewage sludge amended soil. In: L'Hermite, P.D., Ott, H. (Eds.), Processing and Use of Sewage Sludge. Reidel, Dodrecht, pp. 358–369.

Christensen, T.H., Tjell, J.C., 1990. Cadmium attenuation in soils. Department of Environmental Engineering 1865–1990. Technical University of Denmark. pp 30–31.

Chuken, U.J.L., Domınguez, U.L., Saldivar, R.R., Jiminez, E.M., Reyes, L.H., Mar, J.L.G., Saenz, E.O., 2012. Implications of chloride-enhanced cadmium uptake in saline agriculture: modeling cadmium uptake by maize and tobacco. Int. J. Environ. Sci. Technol. 9 (1), 69–77.

Ciecko, Z., Wyszkowski, M., Krajewski, W., Zabielska, J., 2001. Effect of organic matter and liming on the reduction of cadmium uptake from soil by Triticale and spring oilseed rape. Sci. Total Environ. 281, 37–45.

Cieslinski, G., Van Rees, K.C.J., Szmigielska, A.M., Krishnamurti, G.S.R., Huang, P.M., 1998. Low-molecular-weight organic acids in rhizosphere soils of durum wheat and their effect on cadmium bioaccumulation. Plant Soil 203, 109–117.

Cifuentes, G.A.R., Johnson, B.L., Berti, M.T., Norvell, W.A., 2012. Zinc fertilization effects on seed cadmium accumulation in oilseed and grain crops grown on North Dakota soils. Chilean J. Agricultural Res. 72 (1), 117–124.

Clarke, J.M., Norvell, W.A., Clarke, F.R., Buckley, W.T., 2002. Concentration of cadmium and other elements in the grain of near-isogenic durum lines. Can. J. Plant Sci. 82, 27–33.

Cocker, K.M., Evans, D.E., Hodson, M.J., 1998. The amelioration of aluminium toxicity by silicon in wheat (*Triticum aestivum* L.): malate exudation as evidence for an in plant mechanism. Planta 204, 318–323.

Cotuk, Y., Belivermis, M., Kilic, O., 2010. Environmental biology and pathophysiology of cadmium. IUFS J. Biol. 69 (1), 1–5.

De Meeus, C., Eduljee, G.H., Hutton, M., 2002. Assessment and management of risks arising from exposure to cadmium in fertilizers. Sci. Total Environ. 291 (1–3), 167–187.

De Rosa, M., Zarrilli, S., Paesano, L., Carbone, U., Boggia, B., Petretta, M., Masto, A., Cimmino, F., Puca, G., Colao, A., Lombardi, G., 2003. Traffic pollutants affect infertility in men. Humanit. Rep. 18, 1055–1061.

Degryse, F., Buekers, J., Smolders, E., 2004. Radio-labile cadmium and zinc in soils as affected by pH and source of contamination. Eur. J. Soil Sci. 55, 113–121.

Dheri, G.S., Brar, M.S., Malhi, S.S., 2007. Influence of phosphorus application on growth and cadmium uptake of spinach in two Cd-contaminated soils. J. Plant Nutr. Soil Sci. 170, 495–499.

di Toppi, L.S., Gabbrielli, R., 1999. Response to cadmium in higher plants. Environ. Exp. Bot. 41, 105–130.

Dong, J., Mao, W.H., Zhang, G.P., Wu, F.B., Cai1, Y., 2007. Root excretion and plant tolerance to cadmium toxicity – a review. Plant Soil and Environ. 53, (5), 193–200.

Duarte, B., Delgado, M., Cacador, I., 2007. The role of citric acid in cadmium and nickel uptake and translocation in Halimione portulacoides. Chemosphere 69 (5), 836–840.

ECB, 2007. European Union Risk Assessment Report. Cadmium oxide and cadmium metal, Part 1 – environment, Vol. 72. EUR 22919EN. Office for Official Publications of the European Communities, Luxembourg. Retrieved April 16, 2013 from http://ecb.jrc.ec.europa.eu/documents/ ExistingChemicals/RISK_ASSESSMENT/REPORT/cdmetal_cdoxideENVreport302.pdf [accessed June 2009].

Efremova, M., Izosimova, A., 2012. Contamination of Agricultural Soils with Heavy Metals. In: Jakobsson, C. (Ed.), Ecosystem Health and Sustainable Agriculture. Baltic University Press, Uppsala, pp. 250–252.

Elkhatib, E.A., Thabet, A.G., Mahdy, A.M., 2001. Phytoremediation of cadmium contaminated soils: Role of organic complexing agents in cadmium phytoextraction. Land Contamin. Reclamation 9 (4), 359–366.

Engelen, D.L.V., Sharpe-Pedler, R.C., Moorhead, K.K., 2007. Effect of chelating agents and solubility of cadmium complexes on uptake from soil by Brassica juncea. Chemosphere 68, 401–408.

Epstein, E., 1999. Silicon. Annu. Rev. Plant. Physiol. Plant. Mol. Biol. 50, 641–664.

Eriksson, J.E., 1989. The influence of pH, soil type and time on adsorption and uptake of cadmium. Water. Air. Soil. Pollut. 48, 317–335.

Eriksson, J.E., 1990. Water. Air. Soil. Pollut. 53, 69–78.

Evangelou, M.W.H., Ebel, M., Schaeffer, A., 2006. Evaluation of the effect of small organic acids on phytoextraction of Cu and Pb from soil with tobacco Nicotiana tabacum. Chemosphere 63, 996–1004.

Evangelou, M.W.H., Ebel, M., Schaeffer, A., 2007. Chelate assisted phytoextraction of heavy metals from sol. effect, mechanism, toxicity, and fate of chelating agents. Chemosphere 68, 989–1003.

Faruruwa, D.M., Birnin, U.A., Yauri, Dangoggo, S.M., 2013. Cadmium, copper, lead and zinc levels in sorghum and millet grown in the city of Kano and its environs. Global Adv. Res. J. Environ. Sci. Toxicol. 2 (3), 82–85.

Filius, A., Streck, T., Richter, J., 1998. Cadmium sorption and desorption in limed top soils as influenced by pH: Isotherms and simulate leaching. J. Environ. Qual. 27, 12–18.

Food and Agriculture Organization of the United Nations, 2012. Current world fertilizer trends and outlook to 2011 / 12. Electronic Publishing Policy and Support Branch, Communication Division, FAO, Rome.

Food and Agriculture Organization of the United Nations / World Health Organization, 2003. Joint FAO / WHO expert committee on food additives. Sixty-first meeting, Rome. 10–19 June, 2003.

Gadd, G.M., 2009. Heavy metal pollutants: environmental and biotechnological aspects. In: Schaechter, M. (Ed.), Encyclopedia of Microbiology. Elsevier, Oxford, pp. 321–334.

Ge, Y., Macdonald, D., Sauvoé, S., Hendershot, W., 2005. Modeling of Cd and Pb speciation in soil solutions by WinHumic V and NICA-Donnan model. Environ. Model. Software 20, 353–359.

Ghestem, J.P., Bermond, A., 1998. EDTA extractability of trace metals in polluted soils: A chemical–physical study. Environ. Technol. 19, 409–416.

Gommy, C., Perdrix, E., Galloo, J.C., Guillermo, R., 1998. Metal speciation in soil: Extraction of exchangeable cations from a calcareous soil with a magnesium nitrate solution. Int. J. Environ. Anal. Chem. 72, 27–45.

Gothberg, A., Greger, M., Holm, K., Bengtsson, B.E., 2004. Influence of nutrient levels on uptake and effects of mercury, cadmium, and lead in water spinach. J. Environ. Qual. 33, 1247–1255.

Grant, C.A., 2011. Influence of phosphate fertilizer on cadmium in agricultural soils and crops. Pedologist, 143–155.

Grant, C.A., Bailey, L.D., McLaughlin, M.J., Singh, B.R., 1999. Management factors which influence cadmium concentrations in crops, in Cadmium in Soils and Plants. In: McLaughlin, M.J., Singh, B.R. (Eds.). Kluwer Academic Publishers, Dordrecht, pp. 151–158.

Grant, C.A., Sheppard, S.C., 2008. Fertilizer impacts on cadmium availability in agricultural soils and crops. Hum. Ecol. Risk Assess. 14, 210–228.

Gray, C.W., Mclaren, R.G., Roberts, A.H.C., Condron, L.M., 1999. Effect of soil pH on cadmium phytoavailability in some new Zeland soils. N. Z. J. Crop and Horticultural Sci. 27, 169–179.

Gryschko, R., Kuhnle, R., Terytze, K., Breuer, J., Stahr, K., 2005. Soil extraction of readily soluble heavy metals and As with 1 M NH_4NO_3 solution. Evaluation of DIN 19730 J. Soils and Sediments 5, 101–106.

Haghiri, F., 1974. Plant uptake of cadmium as influenced by cation ex-change capacity, organic matter, zinc and soil temperature. J. Environ. Qual. 3 (2), 180–183.

Harter, R.D., Naidu, R., 2001. An assessment of environmental and solution parameter impact on trace-metal sorption by soil. Soil Sci. Soc. America J. 65, 597–612.

Hassan, M.J., Wang, F., Ali, S., Zhang, G., 2005. Toxic effects of cadmium on rice as affected by nitrogen fertilizer form. Plant Soil 277, 359–365.

Havlin, J.L., Tisdale, S.L., Nelson, W.L., Beaton, J.D., 2007. Soil Fertility and Fertilizer. In: An Introduction to Nutrient Management, seventh ed. Prentice Hall, New Jersey.

He, Q.B., Singh, B.R., 1993. Effect of organic matter on the distribution, extractability and uptake of cadmium in soils. J. Soil Sci. 44, 641–650.

Herren, F., Feller, U., 1997. Transport of cadmium via xylem and phloem in maturing wheat shoots: Comparison with the translocation of zinc, strontium and rubidium. Ann. Bot. 80, 623–628.

HSDB, 2009. Hazardous Substances Data Bank. National Librar y of Medicine. http://toxnet.nlm.nih.gov/cgi-bin/sis/htmlgen?HSDB. and search on C AS number. Last accessed: 10/26/09.

Huang, J.W., Chen, J.J., Berti, W.B., Cuningham, S.D., 2005. Phytoremediation of lead contaminated soils: Role of synthetic chelates in lead phytoextraction. Environ. Sci. Technol. 31, 800–805.

IARC., 1993. Cadmium and cadmium compounds. In Beryllium, Cadmium, Mercury and Exposures in the Glass Manufacturing Industry. IARC Monographs on the Evaluation of Carcinogenic Risk of Chemicals to Humans, Vol. 58. International Agency for Research on Cancer, Lyon, France. 119–239.

Jalloh, M.A., Chen, J., Zhen, F., Zhang, G., 2009. Effect of different N fertilizer forms on antioxidant capacity and grain yield of rice growing under Cd stress. J. Hazard Mater. 162, 1081–1085.

Jamode, A., Rao, M., Chandak, B.S., Jamode, V.S., Parwate, A.V., 2003. Applications of the inexpensive adsorbents for the removal of heavy metals from industrial wastewater: A brief review. J. Ind. Pollut. Control 19, 114–134.

Jia, W., Stacey, S.P., McLaughlin, M.J., Kirby, J.K., 2009. Biodegradation of rhamno lipid, EDTA and citric acid in cadmium and zinc contaminated soils. Soil Biol. Biochem. 41 (10), 2214–2221.

Jones, D.L., Darrah, P.R., 1994. Role of root derived organic acids in the mobilization of nutrients from the rhizosphere. Plant and Soil 166, 247–257.

Jones, D.L., Darrah, P.R., Kochian, V.L., 1996. Critical evaluation of organic acid mediated iron dissolution in the rhizosphere and its potential role in root iron uptake. Plant and Soil 180, 57–66.

Jung, M.C., 2008. Heavy metal concentration in soils and factors affecting metal uptake by plants in the vicinity of a Korean Cu–W mine. Sensors 8, 2413–2423.

Kabata-Pendias, A., 2001. Trace Elements in Soils and Plants, third ed. CRC Press, Boca Raton, Fl. p. 413.

Kabata-Pendias, A., Mukherjee, A.B., 2007. Trace Elements from Soil to Human. SpringerVerlag, Berlin.

Kachenko, A.G., Singh, B., 2006. Heavy metals contamination in vegetables grown in urban and metal smelter contaminated sites in Australia. Water. Air. Soil. Pollut. 169, 101–123.

Kaihura, B.S., Kullaya, I.K., Kilasara, M., Aune, J.B., Singh, B.R., Lal, R., 1999. Soil quality effects of accelerated erosion and management systems in three eco-regions of Tanzania. Soil and Tillage Res. 53, 59–70.

Kayser, A., Wenger, K., Attinger, W., Felix, H.R., Gupta, S.K., Schullin, R., 2000. Enhancement of phytoextraction of Zn, Cd and Cu from calcareous soil: The use of NTA and sulphur amendments. Environ. Sci. Technol. 24, 217–225.

Khan, A.G., Kuek, C., Chaudhry, T.M., Khoo, C.S., Hayes, W.J., 2000. Role of plants, mycorrhizae and phytochelators in heavy metal contaminated land remediation. Chemosphere 41, 197–207.

Kibria, M.G., Osman, K.T., Ahammad, M.J., Alamgir, M., 2011. Effects of farm yard manure and lime on cadmium uptake by rice grown in two contaminated soils of Chittagong. J. Agricultural Sci. Technol. 5 (3), 352–358.

Kirkham, M.B., 2006. Cadmium in plants on polluted soils: Effects of soil factors, hyperaccumulation, and amendments. Geoderma. 137, 19–32.

Kirpichtchikova, T.A., Manceau, A., Spadini, L., Panfili, F., Marcus, M.A., Jacquet, T., 2006. Speciation and solubility of heavy metals in contaminated soil using X-ray microfluorescence, EXAFS spectroscopy, chemical extraction and thermodynamic modeling. Geochimica et Cosmochimica Acta 70, 2163–2190.

Kui, R.Y., Suo, Z.F., Bo, S.J., 2009. Effects of nitrogen fertilization on heavy metal content of corn grains. Int. J. Exp. Bot. 78, 101–104.

Kukier, U., Chaney, R.L., 2002. Growing rice grain with controlled cadmium concentrations. J. Plant Nutr. 25 (8), 1793–1820.

Landberg, T., Greger, M., 2003. Influence of N and N supplementation on Cd accumulation in wheat grain. Conference conducted at the 7th International Conference on the Biogeochemistry of Trace Elements, Uppsala '03, Conference Proceedings 1: III, (pp. 90 – 91). Swedish University of Agricultural Sciences, Uppsala, Sweden.

Lebourg, A., Sterckeman, T., Ciesielski, H., Proix, N., 1996. Suitability of chemical extraction to assess risks of toxicity induced by soil trace metal bioavailability. Agronomie 16, 201–215.

Lesage, E., Meers, E., Vervaeke, P., Lamsal, S., Hopgood, M., Tack, F.M., Verioc, M.G., 2005. Enhanced phytoextraction: II. Effect of EDTA and citric acid on heavy metal uptake by *Helianthus annuus* from a calcareous soil. International Journal of Phytoremediation 7 (2), 143–152.

Li, S., Liu, R., Wang, M., Wang, X., Shen, H., Wang, H., 2006. Phytoavailability of cadmium to cherry-red radish in soils applied composed chicken or pig manure. Geoderma. 136, 260–271.

Li, X., Ziadi, N., Bélanger, G., Cai, Z., Xu, H., 2011. Cadmium accumulation in wheat grain as affected by mineral N fertilizer and soil characteristics. Can. J. Soil Sci. 91, 521–531.

Li, Y.M., Chaney, R.L., Schneiter, A.A., Miller, J.F., Elias, E.M., Hammond, J.J., 1997. Screening for low grain cadmium phenotypes in sunflower, durum wheat and flax. Euphytica 94, 23–30.

Liang, Y.C., Wong, J.W.C., Wei, L., 2005. Silicon-mediated enhancement of cadmium tolerance in maize (*Zea mays* L.) grown in cadmium contaminated soil. Chemosphere 58, 475–483.

Liu, D., Islam, E., Li, T., Yang, X., Jin, X., Mahmood, Q., 2008a. Comparison of synthetic chelators and low molecular weight organic acids in enhancing phytoextraction of heavy metals by two ecotypes of *Sedum alfredii* Hance. J. Hazard. Mater. 153, 114–122.

Liu, H., Zhang, J., Christie, P., Zhang, F., 2008b. Influence of iron plaque on uptake and accumulation of Cd by rice (*Oryza sativa* L.) seedlings grown in soil. Sci. Total. Environ. 394, 361–368.

Long, X.X., Yang, X.E., Ni, W.Z., Ye, Z.Q., He, Z.L., Calvert, D.V., Stoffela, J.P., 2003. Assessing zinc thresholds for phytotoxic and potential dietary toxicity in selected vegetable crops. Commun. Soil Sci. Plant Anal. 34, 1421–1434.

Loosemore, N., Straczek, A., Hinsinger, P., Jaillard, B., 2004. Zinc mobilisation from a contaminated soil by three genotypes of tobacco as affected by soil and rhizosphere pH. Plant and Soil 260, 19–32.

Machiwa, J.F., 2010. Heavy metal levels in paddy soils and rice (*Oryza sativa* L.) from wetlands of lake victoria basin, Tanzania. Tanzan. J. Sci. 36, 59–72.

Masscheleyn, P., Pardue, J., Delaune, R., Patrick, J., 1991. Effect of redox potential and pH on As speciation and solubility in a contaminated soil. Environ. Sci. Technol. 25, 1414–1419.

Mathew, U., VenugopaJ, V.K., Saraswathi, P., 2002. Cadmium content of plants as affected by soil application of cadmium and farm yard manure. J. Trop. Agriculture 40, 78–80.

Matusik, J., Bajda, T., Manecki, M., 2008. Immobilization of aqueous cadmium by addition of phosphates. J. Hazard. Mater. 152, 1332–1339.

Mcbride, M.B., 1989. Reactions controlling heavy metal solubility in soils. Adv. Soil Sci. 10, 1–56.

McKenna, I.M., Chaney, R.L., Williams, F.M., 1993. The effects of cadmium and zinc interactions on the accumulation and tissue distribution of zinc and cadmium in lettuce and spinach. Environ. Pollut. 79, 113–120.

McLaughlin, M., 2007. Heavy metals. In: Lal, R. (Ed.), Encyclopedia of Soil Science. Taylor and Francis Group, London, pp. 650–653.

McLaughlin, M.J., Singh, B.R., 1999. Cadmium in Soils and Plants. Kluwer Academic Publishers, Dordrecht. pp. 257–267.

McLaughlin, M.J., Whatmuff, M., Warne, M., Heemsbergen, D., Barry, G., Bell, M., Nash, D., Pritchard, D., 2006. A field investigation of solubility and food chain accumulation of bio solid-cadmium across diverse soil types. Environ. Chem. 3 (6), 428–432.

Means, J.L., Kucak, T., Crerar, D.A., 1980. Relative degradation rates of NTA, EDTA and DTPA and environmental applications. Environ. Pollut. B Chem. Phys. 1, 45–60.

Mehmood, F., Rashid, A., Mahmood, T., Dawson, L., 2012. Effect of DTPA on Cd solubility in soil – Accumulation and subsequent toxicity to lettuce. Chemosphere 90 (6), 1805–1810.

Melo, E.E.C., Nascimento, C.W.A., Accioly, A.M.A., Santos, A.C.Q., 2008. Phytoextraction and fractionation of heavy metals in soil after multiple applications of natural chelants. Sci. Agricola 65 (1), 61–68.

Metsärinne, S., Rantanen, P., Aksels, R., Tuhkanen, T., 2004. Biological and photochemical degradation rates of diethylenetriaminepentaacetic acid (DTPA) in the presence and absence of Fe (III). Chemosphere 55, 379–388.

Miller, J.E., Hassett, J.J., Koeppe, D.E., 1976. Uptake of cadmium by soybean as influenced by soil cation exchange capacity, pH and available phosphorus. J. Environ. Qual. 5, 157.

Ming, D., 2002. Carbonates. In: Lal, R. (Ed.), Encyclopaedia of Soil Science, pp. 139–142.

Miura, N., 2009. Individual susceptibility to cadmium toxicity and metallothionein gene polymorphisms: With references to current status of occupational cadmium exposure. In: Health 47, 487–494.

Moffett, J.W., Brand, L.E., 1996. The production of strong, extracellular Cu chelators by marine Cyanobacteria in response to Cu stress. Limnnology and Oceanography 41, 288–293.

Mojiri, A., 2011. The potential of corn (*Zea mays*) for phytoremediation of soil contaminated with cadmium and lead. J. Biol. Environ. Sci. 5 (13), 17–22.

Moustakas, N.K., Ioannidou, A.A., Barouchas, P.E., 2011. The effects of cadmium and zinc interactions on the concentration of cadmium and zinc in pot marigold (*Calendula officinalis* L.). Aust. J. Crop Sci. 5 (3), 277–282.

Mullins, G.L., Sommers, L.E., Barber, S.A., 1986. Modeling the plant uptake of cadmium and zinc from soils treated with sewage sludge. Soil Sci. Soc. America J. 50, 1245–1250.

Naidu, R., Harter, R.D., 1998. Effect of different organic ligands on Cadmium sorption and extractability from soils. Soil Sci. Soc. America J. 62, 644–650.

Nakayama, M., Kamewada, K., Kyoushima, R., 2009. Barley (*Hordeum vulgare*) cultivars that differ in cadmium accumulation. Bull. Tochigi Prefectural Agricultural Exp. Station 63, 17–25.

Nan, Z., Li, J., Zhang, J., Cheng, G., 2002. Cadmium and zinc interactions and their transfer in soil crop system under actual field conditions. Sci. Total Environ. 285, 187–195.

Narwal, R.P., Singh, B.R., 1998. Effect of organic materials on partitioning, extractability and plant uptake of metals in an alum shale soil. Water Air and Soil Pollut. 103, 405–421.

Nascimento, C.W.A., Amarasiriwardena, D., Xing, B., 2006. Comparison of natural organic acids and synthetic chelates at enhancing phytoextraction of metals from a multi-metal contaminated soil. Environ. Pollut. 140, 114–123.

Nigam, R., Srivastava, S., Prakash, S., Srivastava, M.M., 2001. Cadmium mobilization and plant availability – the impact of organic acids commonly exuded from roots. Plant and Soil 230, 107–113.

Nordic Council Ministers, 1992. Atmospheric heavy metal deposition in Northern Europe 1990. Nord 12, 41.

Nordic council of minister, 2003. Retrieved April, 18, 2013 from http://www.who.int/ifcs/documents/forums/forum5/nmr_cadmium.pdf.

Norvell, W.A., Wub, J., Hopkinsc, D.G., Welcha, R.M., 2000. Association of cadmium in durum wheat grain with soil chloride and chelate-extractable soil cadmium. Soil Sci. Soc. America J. 64 (6), 2162–2168.

Nowack, B., 2002. Environmental chemistry of aminopolycarboxylate chelating agents. Environ. Sci. Technol. 36, 4009–4016.

Organisation for Economic Co-operation and Development (OECD), 1994. Risk Reduction Monograph No. 5: Cadmium, OECD Environment Directorate. France, Paris. 1994.

Papafilippaki, A., Gasparatos, D., Haidouti, C., Stavroulakis, G., 2007. Total and bioavailable forms of Cu, Zn, Pb and Cr in agricultural soils. A study from the hydrological basin of Keritis, Chania, Greece. Global Nest J. 9, 201–206.

Pastor, J., Aparicio, A.M., Gutierrez-Maroto, A., Hernández, A.J., 2007. Effects of two chelating agents (EDTA and DTPA) on the autochthonous vegetation of a soil polluted with Cu, Zn and Cd. Sci. Total Environ. 378, 114–118.

Patrick, H., Ronald, D., Patrick, J., 1990. Transformations of Se, As affected by sediment oxidation–reduction potential and pH. Environ. Sci. Technol. 24, 91–96.

Perilli, P., Mitchell, L.G., Grant, C.A., Pisante, M., 2010. Cadmium concentration in durum wheat grain (*Triticum turgidum*) as influenced by nitrogen rate, seeding date and soil type. J. Sci. Food and Agriculture 90, 813–822.

Peters, R.W., 1999. Chelate extraction of heavy metals from contaminated soils. J. Hazard. Mater. 66, 151–210.

Pichtel, J., Bradway, D., 2008. Conventional crops and organic amendments for Pb, Cd and Zn treatment at a severely contaminated site. Bioresour. Technol. 99, 1242–1251.

Pinto, A.P., Vilar, M.T., Pinto, F.C., Mota, A.M., 2005. Organic Matter Influence in Cadmium uptake by sorghum. J. Plant. Nutr. 27 (12), 2175–2188.

Pueyo, M., Sanchez, J.F.L., Rauret, G., 2004. Assessment of $CaCl_2$, $NaNO_3$ and NH_4NO_3 extraction procedures for the study of Cd, Cu, Pb and Zn extractability in contaminated soils. Anal. Chim. Acta. 504, 217–226.

Puschenreiter, M., Horak, O., Friesl, W., Hartl, W., 2005. Low-cost agricultural measures to reduce heavy metal transfer into the food chain – a review. Plant Soil and Environ. 51, 1–11.

Putwattana, N., Kruatrachue, M., Pokethitiyook, P., Chaiyarat, P., 2010. Immobilization of cadmium in soil by cow manure and silicate fertilizer, and reduced accumulation of cadmium in sweet basil (Ocimum basilicum). ScienceAsia 36, 349–354.

Qiong, L., Yan, G.X., Hua, X.X., Bao, Z.Y., Pu, W.D., Bing, M., 2012. Phytoavailability of copper, zinc and cadmium in sewage sludge-amended calcareous Soils. Pedosphere 22 (2), 254–262.

Qu, J., Lou, C.Q., Yuan, X., Wang, X.H., Cong, Q., Wang, L., 2011. The effect of sodium hydrogen phosphate / citric acid mixtures on phytoremediation by alfalfa and metals availability in soil. J. Soil Sci. Plant Nutr. 11 (2), 85–95.

Quartacci, M.F., Baker, A.J.M., Navari-Izoo, F., 2005. Nitrioacetate and citric acid assisted phytoextraction of cadmium by Indian mustard (Barssica juncea). Chemosphere 59, 1249–1255.

Quartacci, M.F., Argilla, A., Baker, A.J.M., Navari-Izzo, F., 2006. Phytoextraction of metals from a multiply contaminated soil by Indian mustard. Chemosphere 63, 918–925.

Quartacci, M.F., Irtelli, B., Baker, A.J.M., Navari-Izzo, F., 2007. The use of NTA and EDDS for enhanced phytoextraction of metals from a multiply contaminated soil by Brassica carinata. Chemosphere 68, 1920–1928.

Ramachandran, V., D'Souza, T.J., 2002. Plant uptake of cadmium, zinc and manganese from four contrasting soils amended with Cd-enriched sewage sludge. J. Environ. Sci. Health A Tox. Hazard. Subst. Environ. Eng. 37, 1337–1346.

Raza, S.H., Shafiq, F., 2013. Exploring the role of salicylic acid to attenuate cadmium accumulation in radish (Raphanus sativus). Int. J. Agri. and Biol. 15, 547–552.

Raza, S.H., Shafiq, F., Tahir, M., 2013. Screening of cadmium tolerance in sugarcane using antioxidative enzymes as a selection criteria. Pak. J. Life and Soc. Sci. 11 (1), 8–13.

Reeves, P.G., Chaney, R.L., 2004. Marginal nutritional status of zinc, iron, and calcium increases cadmium retention in the duodenum and other organs of rats fed a rice-based diet. Environ. Res. 96, 311–322.

Rodda, M.S., Li, G., Reid, R.J., 2011. The timing of grain Cd accumulation in rice plants: The relative importance of remobilisation within the plant and root Cd uptake post- flowering. Plant and Soil 347, 105–114.

Römkens, P., Bouwman, L., Japenga, J., Draaisma, C., 2002. Potentials and drawbacks of chelate-enhanced phytoremediation of soils. Environ. Pollut. 116, 109–121.

Saifullah, Meers, E., Qadir, M., de Caritat, P., Tack, F.M.G., Du Laing, G.D., Zia, M.H., 2009. EDTA-assisted Pb Phytoextraction. Chemosphere 74, 1279–1291.

Sampanpanish, P., Pongpaladisai, P., 2011. Effect of Organic Fertilizer on Cadmium Uptake by Rice Grown in Contaminated Soil. 2011 International Conference on Environmental and Agriculture Engineering. IPCBEE. vol. 15(2011) ©(2011). IACSIT Press, Singapore.

Sarwar, N., Saifullah, Malhi, S.S., Zia, M.H., Naeem, A., Bibia, S., Farid, G., 2010. Role of mineral nutrition in minimizing cadmium accumulation by plants. J. Sci. Food. Agric. 90, 925–937.

Schmitt, H.W., Sticher, H., 1991. Heavy metal compounds in the soil. In: Merian, E. (Ed.), Metals and Their Compounds in the Environment – Occurrence, Analysis and Biological Relevance. VCH, New York, pp. 311–331.

Sheng, S.G., Xue, C.M., Ying, W.D., Mei, X.C., Yun, C.Z., Fu, Z.X., 2008. Using iron fertilizer to control Cd accumulation in rice plants: A new promising technology. Sci Sci. China C. Life Sci. 51, 245–253.

Silvera, S.A.N., Rohan, T.E., 2007. Trace elements and cancer risk: A review of the epidemiologic evidence. Cancer Causes and Control 18 (1), 7–27.

Simmons, R.W., Pongsakul, P., Chaney, R.L., Saiyasitpanich, D., Klinphoklap, S., Nobuntou, W., 2003. The relative exclusion of zinc and iron from rice grain in relation to rice grain cadmium as compared to soybean: Implications for human health. Plant and Soil 257, 163–170.

Smilde, B., Van Luit, K.W., Van Driel, W., 1992. The extraction by soil and absorption by plants of applied zinc and cadmium. Plant and Soil 143, 233–238.

Smith, S.R., 1996. Agricultural recycling of sewage sludge and the environment. Oxford University Press, New York.

Smolinska, B., Król, K., 2010. Phytoextraction of heavy metal contaminated soils. Food Chem. Biotechnol. 74, 89–98.

Stevenson, F.J., Fitch, A., 1994. Chemistry of complexation of metal ions with soil solution organics. In: Huang, P.M., Schnitzer, M. (Eds.), Interaction of Soil Minerals with Natural Organic and Microbes. Soil Science Society of America, Madison, WI.

Stewart, B.A., Robinson, C.A., Tarker, D.B., 2000. Examples and case studies of beneficial reuse of beef cattle by-product. In: Dick, W.A. (Ed.), Land Application of Agricultural, Industrial and Municipal By-product. Soil Science Society of America, Inc, Madison, pp. 387–407.

Ström, L., Owen, A.G., Godbold, D.L., Jones, D.L., 2001. Organic acid behaviour in a calcareous soil: Sorption reactions and biodegradation rates. Soil. Biol. Biochem. 33, 2125–2133.

Tlustos, P., Szakova, J., Korinek, K., Pavlikova, D., Hanc, A., Balik, J., 2006. The effect of liming on cadmium, lead and zinc uptake reduction by spring wheat grown in contaminated soil. Plant Soil and Environ. 2, 16–24.

Tolcin, A.C., 2008. U.S. Geological Survey, Mineral Commodity Summaries. http://minerals.usgs. gov/ds/2005/140/cadm ium.xls.

Tolcin, A.C., 2009. U.S. Geological Survey, Mineral Commodity Summaries. http://minerals.usgs. gov/ds/2005/140/cadmium.xls.

Tolcin, A.C., 2012. Cadmium, U.S. Geological Survey Minerals Yearbook—2011.

Tomáš, J., Árvay, J., Tóth, T., 2012. Heavy metals in productive parts of agricultural plants. J. Microbiol. Biotechnol. Food Sci. 1, 819–827.

Tordoff, G.M., Baker, A.J.M., Willis, A.J., 2000. Current approaches to the revegetation and reclamation of metalliferous mine wastes. Chemosphere 41, 219–228.

Tsadilas, C.D., Karaivazoglou, N.A., Tsotsolis, N.C., Stamatiadis, S., Samaras, V., 2005. Cadmium uptake by tobacco as affected by liming, N form and year of cultivation. Environ. Pollut. 134 (2), 239–246.

Turgut, C., Pepe, M.K., Cutright, T.J., 2004. The effect of EDTA and citric acid on phytoremediation of Cd, Cr, and Ni from soil using. *Helianthus annuus*. Environ. Pollut. 131 (1), 147–154.

Van Assche, F.J., 1998. A stepwise model to quantify the relative contribution of different environmental sources to human cadmium exposure. Proceedings of the 8th International Nickel–Cadmium Battery Conference, Prague, Czech Republic. September 21–22, 1998.

Violante, Cozzolino, V., Perelomov, L., Caporale, A.G., Pigna, M., 2010. Mobility and bioavailability of heavy metals and metalloids in soil environments. J. Soil Sci. Plant Nutr. 10 (3), 268–292.

Wang, C., Ji, J., Yang, Z.F., Chen, L., Browne, P., Yu, R., 2012. Effect s of Soil Properties on the Transfer of Cadmium from Soil to Wheat in the Yangtze River Delta Region, China—a Typical Industry–Agriculture Transition Area. Biol. Trace. Elem. Res. 148, 264–274.

Wang, J., Lv, J., Fu, Y., 2013. Effects of organic acids on Cd adsorption and desorption by two anthropic soils. Front. Environ. Sci. Eng. 7 (1), 19–30.

Wasay, S.A., Barrington, S.F., Tokunaga, S., 1998. Remediation of soils polluted by heavy metals using salts of organic acids and chelating agents. Environ. Technol. 19, 369–379.

Wei, L., Donat, J.R., Fones, G., Ahner, 2003. Interactions between Cd, Cu, and Zn influence particulate phytochelatin concentrations in marine phytoplankton: Laboratory results and preliminary field data. Environ. Sci. Technol. 37, 3609–3618.

Wenzel, W.W., Unterbrunner, R., Sommer, P., Sacco, P., 2003. Chelate-assisted hytoextraction using canola (*Brassica napus*) in outdoors pot and lysimeter experiments. Plant and Soil 249, 83–96.

White, M.C., Chaney, R.L., Decker, A.M., 1981. Metal complexation in xylem flfluid. III Electrophoretic evidence. Plant Physiol. 67, 311–315.

Wieczorek, J., Wieczorek, Z., Bieniaszewski, T., 2005. Cadmium and lead content in cereal grains and soil from cropland adjacent to roadways. Pol. J. Environ. Stud. 14 (4), 535–540.

Wu, Dong, J., Cai, Y., Chen, F., Zhang, G., 2007. Differences in Mn uptake and subcellular distribution in different barley genotypes as a response to Cd toxicity. Sci. Total Environ. 385, 228–234.

Wu, F., Zhang, G., Dominy, P., 2003. Four barley genotypes respond differently to cadmium: Lipid peroxidation and activities of antioxidant capacity. Environ. Exp. Bot. 50, 67–78.

Wu, F.B., Chen, F.K., Wei, K.G., Zhang, P., 2004. Effects of cadmium on free amino acids, glutathione, and ascorbic acid concentration in two barley genotypes (*Hordeum vulgare* L.) differing in cadmium tolerance. Chemosphere 57, 447–454.

Wuana, R.A., Okieimen, F.A., Imborvungu, J.A., 2010. Removal of heavy metals from a contaminated soil using organic chelating acids. Int. J. Environ. Sci. Technol. 7 (3), 485–496.

Xie, H.L., Jiang, R.F., Zhang, F.S., McGrath, S., PandZhao, F.J., 2009. Effectofnitrogenform on the rhizosphere dynamics and uptake of cadmium andzinc by the hyperaccumulator Thlaspi caerulescens. Plant Soil 318, 2.

Xue, Q., Harrison, H.C., 1991. Effect of soil zinc, pH and cultivar uptake in leaf lettuce (*Lactuca sativa* L. var. crispa). Commun. Soil. Sci. Plant. Anal. 22, 975–991.

Yaron, B., Calvet, R., Prost, R., 1996. Soil pollution processes and dynamics. Springer-Verlag, Berlin and Heidelberg. 313 p.

Yassen, A.A., Nadia, B.M., Zaghloul, M.S., 2007. Role of some organic residues as tools for reducing heavy metals hazards in plant. World J. Agricultural Sci. 3 (2), 204–207.

Yu-Jing, C., Zhong-Qui, Z., Wen-Ju, L., Shi-Bao, C., Yong-Guan, Z., 2003. Transfer of cadmium through soil–plant–human continuum and its affecting factors. Acta Ecologica Sinica 23 (10), 2133–2143.

Zaccheo, P., Laura, C., Valeria, D.M.P., 2006. Ammonium nutrition as a strategy for cadmium metabolisation in the rhizosphere of sunflower. Plant and Soil 283, 43–56.

Zachara, J.M., Smith, S.C., Resch, C.T., Cowan, C.E., 1993. Cadmium sorption on specimen and soil smectites in sodium and calcium electrolytes. Soil Sci. Soc. America J. 57 (6), 1491–1501.

Zeng, Q.R., Sauve, S., Allen, H.E., Hendershot, W.H., 2005. Recycling EDTA solutions used to remediate metal-polluted soils. Environ. Pollut. 133, 225–231.

Zhang, G., Fukamib, M., Sekimoto, H., 2002. Influence of cadmium on mineral concentrations and yield components in wheat genotypes differing in Cd tolerance at seedling stage. Field Crops Res. 77, 93–98.

Zhang, M., Alva, A.K., Li, Y.C., Calvert, D.V., 1997. Chemical association of Cu, Zn, Mn, and Pb in selected sandy citrus soils. Soil. Sci. 162, 181–188.

Zhao, K.L., Liu, X.M., Xu, J.M., Selim, H.M., 2010. Heavy metal contaminations in a soil–rice system: Identification of spatial dependence in relation to soil properties of paddy fields. J. Hazard. Mater. 181, 778–787.

Zulfiqar, S., Wahid, A., Farooq, M., Maqbool, N., Arfan, M., 2012. Phytoremediation of soil cadmium by using *Chenopodium* species. Pak. J. Agricultural Sci. 49, 435–445.

Chapter 21

Phytoremediation of Radioactive Contaminated Soils

Muhammad Ibrahim,* Muhammad Adrees,* Umer Rashid,[†]
Syed Hammad Raza[‡] and Farhat Abbas*

*Department of Environmental Sciences, Government College University, Faisalabad, Pakistan;
[†]Institute of Advanced Technology, Universiti Putra Malaysia, Serdang, Selangor, Malaysia;
[‡]Department of Botany, Government College University, Faisalabad, Pakistan

INTRODUCTION

The application of nuclear technology in various fields of basic and applied research has resulted in changing the habits and lifestyles of mankind. To a much greater extent, the transition towards radioactive sources has been responsible for improvements in health, agriculture, food and more importantly, in the energy sector. However in contrast, the uses of radioactive materials in industrial processes have resulted not only in the contamination of large areas of land but also the water resources. Once polluted, this type of contamination is very difficult to control and manage as it essentially requires a certain period of time for the decay of the particular radioactive substance. In addition to the previously mentioned limitations, radioactivity mostly spreads over a vast area so any outcomes of decontamination are very little or they are less effective and would be costly as well as time consuming. There is therefore a current need to develop realistic and efficient remediation methods or technologies to decontaminate these types of lands to benefit humanity. Phytoremediation and/or phytomanagement can be the best options for limiting or eradicating radioactivity. Phytoremediation employs the use of different plant species for in situ removal of radioactive contaminants from soil, water, sediments and air. Using plants for the removal of radioactive contamination is an efficient and pleasing technique of remediating sites having low to moderate degrees of radioactive contamination. Moreover, such techniques may be employed with other more conventional remedial techniques. Phytoremediation provides a permanent solution for in situ removal of radioactive contaminants rather than simply its translocation.

Soil Remediation and Plants. http://dx.doi.org/10.1016/B978-0-12-799937-1.00021-8

SCOPE AND LIMITATIONS

The research on radioactive materials and the remediation thereafter requires certain national and international protocols. Generally, this research has certain limitations for its acceptability and extension. The alliance by some European organizations has opened their doors to the international community for research in radio-ecology. The ultimate goal is to promote radio-linked research efficiency and to advance a science involving phytoremediation in order to decontaminate soils, if any.

The prime advantage of phytoremediation is its comparative economical standing compared to other remedial methods like excavation. In many cases it has been found to be less than 50% of the cost of alternative methods. However, phytoremediation carries its own faults, for example, it is dependent on specific plant species, depth of the roots and the tolerance of the plant to the radioactive contaminants. However, animal exposure to these plant species is becoming a major concern to environmentalists as herbivores may deposit these radioactive particles in different tissues which certainly have a major impact on the whole food web. Ecologists have devised some key questions for the exposure of plants, wildlife and humans. The researchers are particularly interested in knowing the nature of radioactive contaminants, and if censoriously addressed over the next 20 years, significant improvements in radio-ecology could be obtained. The three scientific challenges are presented in the form of schematic diagrams (Figures 21.1, 21.2 and 21.3).

MAJOR SOURCES OF RADIOACTIVE CONTAMINANTS TO SOIL AND ENVIRONMENT

A comprehensive overview considering the major sources of radioactive contaminants causing contamination to soil and the environment is presented.

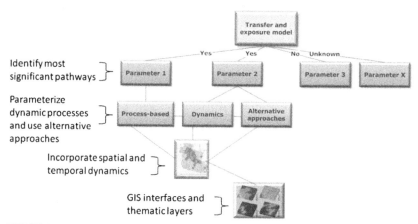

FIGURE 21.1 Scheme of key aspects to Challenge One: to predict human and wildlife exposure more robustly by quantifying key processes that influence radionuclide transfers, and incorporate the knowledge into new dynamic models. *With permission from Hinton et al. (2013).*

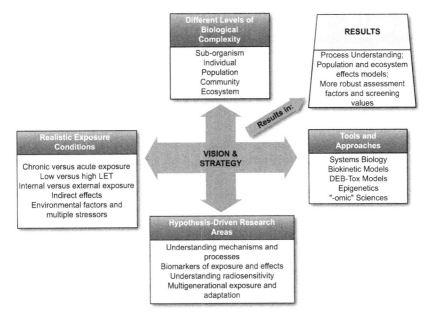

FIGURE 21.2 Schematic of the components and anticipated results of the strategic research agenda concerned with Challenge Two: to determine ecological consequences under the realistic conditions to which organisms are exposed. *With permission from Hinton et al. (2013).*

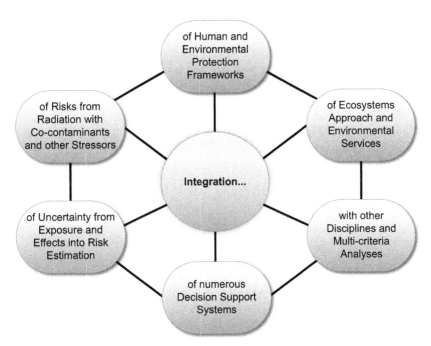

FIGURE 21.3 Six areas in Challenge Three: to improve human and environmental protection by integrating radioecology. *With permission from Hinton et al. (2013).*

Nuclear Weapons' Testing

This is the most important and most lethal of all the radioactive sources. Nuclear weapons' testing started in 1945 and continued through the years: from the Trinity test on 16 July 1945, the Operation Crossroads series in July 1946, the Operation Greenhouse shots of May 1951 which included the first boosted fission weapons test and many others dating back to the 1950s, to the late 1990s including the incident which took place during 1998 in the Indian Sub-Continent, and recent North Korean tests in 2006, 2009 and 2012. Most of the tests were performed during the cold war era by the USA and former USSR, UK and France in Europe, and China in the East. All these tests caused not only land contamination to the surroundings but also resulted in significant fallout as revealed and confirmed by the traces and radioactive remnants in the environment (Vandenhove, 2006). The doses for global fallout are much smaller and therefore remediation is not feasible but the local contamination is very severe. One such example is the Altai region (Russia) where atmospheric tests reported 10% activity close to the test site and likewise very significant land and population exposure was reported (Zeevaert et al., 1997).

Production of Nuclear Weapons

Information on the production of nuclear weapons and their subsequent effects on soil and environment is scanty. This is due to the secrecy of particular countries and related military operations and activities. The UK, France and the former USSR fabricated the nuclear weapons in Europe. Similarly USA, India and Pakistan have also produced nuclear weapons. Some of these installations to fabricate nuclear weapons produce a large amount of radioactive contamination and, therefore, generate huge environmental impacts. For instance, at Mayak (Russia), ^{137}Cs and ^{90}Sr contaminated the soil surface due to the discharge of the vast amount of medium- and high-level liquid waste. The waste was discharged to the Techa and Iset rivers aided by the explosion of a radioactive waste tank. The Government developed a rehabilitation plan in 1982 for farming and forest uses in contaminated areas (Romanov and Drozhko, 1996). Similarly, some other countries are also producing nuclear weapons in secret and the level of contamination and exposure of land, fauna and flora is still unknown.

The Nuclear Fuel Cycle

The nuclear fuel cycle is very important in the modern world and it involves the mining and milling of radioactive raw materials (e.g uranium ores), conversion of this raw material into fuel (in the reactor), reprocessing of used fuel, storage of fuel and waste, disposal and dispersal of waste materials. The removal and processing of large quantities of uranium (U) ores are a very important cause of radioactive pollution in the whole fuel cycle. For example, in Europe the major mining for U is carried out in France, Germany, Spain, Bulgaria, the Czech Republic and Hungary. Mostly, the mining sites are in France and Spain and the remediation process is ongoing as a matter of routine. In Eastern Europe, there are many abandoned sites which have never been remediated. These sites especially pose serious threats to terrestrial commodities like flora, fauna and the soils. Nuclides such as ^{230}Th, ^{222}Rn, ^{226}Ra, ^{210}Po, ^{238}U and ^{234}Th pose serious risks of atmospheric and soil contamination in the vicinity of the mining sites. The contamination from the sites can be in the form of ingestion, leaching to the groundwater or exposure to the land and degradation. The Chernobyl accident (in Ukraine, former USSR) in 1986 is the most dangerous and lethal accident that has ever occurred in these types of installations. Here the major contributor of contamination was radio-caesium. The total land contaminated area was estimated about $130,000\,km^2$ ($37\,kBq\,m^{-2}\,^{237}Cs$) (Vandenhove, 2006). The radioactivity of major nuclear pollutants recorded after some major nuclear accidents are given in Table 21.1, while ^{133}Xe and ^{131}I were recorded after the Three Mile Island accident.

Industrial Processes/Techniques Involving Radionuclides

There are a number of industrial processes which employ radionuclides such as uranium for mining and milling, smelting of metals and the phosphate-mining industry. ^{238}U and ^{232}Th are the most abundant commonly occurring radionuclides in Earth's crust, followed by a series of daughter products. The radionuclide levels in industry and the waste products of the oil and gas extraction industries, and the ceramics industry may be enhanced. The impact of the coal industry and production of electricity from coal is considered to be less important (Vandenhove 2000). The remediation of these sites may involve removal of sources through a multi-layer barrier (vegetation) or a sub-surface barrier (cement, etc.), separated by washing and flotation, and finally chemical extraction (Vandenhove et al., 2000).

Research Activities

The use of radioactive nuclides in basic and applied research has been increased to a great level during recent years. The use in agriculture (especially in

TABLE 21.1 Radionuclide Release from Major Nuclear Accidents in the World

	Kyshtym	Chernobyl (PBq)	Windscale
^{90}Sr	4	8	0.0002
^{137}Cs	0.03	85	0.05
^{239}Pu	—	0.03	0.002

Source: Values were taken from Vadenhove (2006).

breeding), food (in processing, packaging and storage), health and diagnosis, etc., has gained considerable success, and many researchers are using nuclear techniques even in hydrology and solute movement to map contaminant movement in a particular soil.

PHYTOREMEDIATION

Use of different crop species to remove hazardous materials from soils is much cheaper and has a significant impact as compared to the removal and replacement of contaminated soil altogether. Recently, various studies on different aspects of metal phytoremediation of contaminated soils are available (Raza et al., 2013; Raza and Shafiq, 2013). According to a definition stated by Salt et al. (1995) 'it is the use of plants to remove pollutants from the environment or to render them harmless'.

Phytoremediation can be subdivided into the following five groups:

- **Phytoextraction**: removal of metal contaminants using plants and accumulation of these metals in harvestable parts of crop species (Kumar et al., 1995).
- **Phytodegradation**: degradation of organic pollutants through different metabolic processes in plants and associated microbes (Burken and Schnoor, 1997).
- **Rhizofiltration**: a process through which metals are absorbed by plant roots from waste water current (Pulford and Watson 2003; Dushenkov et al., 1995).
- **Phytostabilization**: immobilization and prevention of pollutant migration in the environment by the use of different plant species which lead to the reduction in the mobility and bioavailability of these pollutants (Pulford and Watson, 2003).
- **Phytovolatilization**: different plant species are used for the volatilization of different pollutants into the atmosphere (Burken and Schnoor, 1999; Bañuelos and Ajwa, 1999).

POSSIBLE ROLES OF PHYTOREMEDIATION

Phytomanagement

Phytomanagement involves the use of plant species for managing target areas or contaminated soils. The term broadly considers various aspects and possibilities for using plant species to manage contaminated soils. Phytomanagment technology has already proven to be effective in restraining metal pollution from spreading in surrounding areas through percolation of water and wind erosion (Vangronsveld et al., 1995). Generally, the success of this technology is dependent upon the selection of suitable plant species that will, in turn, improve the effectiveness (Bidar et al., 2006). Therefore, the selection of suitable plant species will be dependent on its response to a particular contaminant and simultaneously on the physio-biochemical characteristics of the soil (Bidar et al.,

2006). Indeed, the information regarding the pollutant mobility, bioavailability and its chemical dynamics in the soil, other than their total contents, is of great value (Pueyo et al., 2004).

Phytoextraction

The use of plants to accumulate contaminants for a cleanup process is called phytoextraction. It requires target material (radionuclide), available to the plant root for absorption. Phytoextraction requires that the target metal (radionuclide) available for the plant root to be translated from the root to the shoot. The radionuclide will then be removed from the harvested biomass, and then subjected to either metal recovery or to concentrating metal by thermal, microbial and/or chemical treatments. Soil management practices such as soil pH, nutrients, addition of materials which may increase metal availability, efficient and hyperaccumulating plant genotype selection, translocation and tolerance ability of plants, may maximize the metal contents in the harvested biomass. The same is also true in case of plant nutrition on selected sites under normal and stress soil conditions (Ibrahim et al., 2011, 2012; Sarwar et al., 2010). The economic feasibility of phytoextraction depends on phytoextraction efficiency. This is also associated with costs used for crop management (described earlier), transport of postharvest biomass, biomass treatment and potential disposal (Vandenhove, 2006).

The Potential for Phytoextraction

The phytoextraction of radionuclides is different from the phytoextraction of heavy metals. Basic research in this scenario to check the phytoremediation (phytoextraction) was conducted in hydroponics or aqueous systems under controlled conditions by Vasudev et al. (1996) and Ensley et al. (1997), respectively.

A major process that determines the soil to plant transfer of metal contaminants is called *transfer factor* (TF) and it needs to be considered in the management of metal-contaminated soils. To minimize the exposure risks (to humans and food chain), strong measures should be taken. Conversely, remediation steps should be expedited resulting in soil phytomining and/or phytoextraction. Here, examples are given regarding three radiocontaminants having importance due to ubiquity and toxicity. These elements include natural radionuclide uranium (^{238}U, the predominant contaminant of U-mining and milling industry, and the phosphate industry), radio-caesium (^{137}Cs) and radio-strontium (^{90}Sr, long-lived fission products).

$$\text{Crop Removal} = \text{Yield} \times C_{\text{plant}} \qquad \text{(Eq. 21.1)}$$

The crop removal of contaminant from the soil with the harvested biomass (Bq ha^{-1}), is the product of the concentration in the plant (C_{plant}, in Bq kg^{-1}) and the yield of the harvested biomass (kg ha^{-1}). Soil type and concentration

TABLE 21.2 Transfer Factors for Three Elements

Elements	Range of Recommended Values	Total Range	Upper Limit
Uranium	0.00075–0.02	0.000006–21.13	21.13
Cesium	0.0038–0.29	0.00025–7.5	7.5
Strontium	0.017–3.2	0.0051–22	22

Source: Adapted from Vandenhove (2006).

TABLE 21.3 Phytoextraction Potential of Untreated and Treated (with Citric Acid) Soils

Plants	Annual Extraction %	Citric Acid	Soil
Rye-grass	0.007 ± 0.004	No	Control
	0.052 ± 0.008	No	Bicarb
	2.810 ± 0.689	Yes	Control
	3.477 ± 0.474	Yes	Bicarb
Mustard	0.010 ± 0.002	No	Control
	0.103 ± 0.030	No	Bicarb
	4.618 ± 0.384	Yes	Control
	3.284 ± 0.250	Yes	Bicarb
Amaranthus	0.0009 ± 0.0003	No	Control
	0.178 ± 0.058	Yes	Control

The data presented here are annual removal (%) within soil depth of 10 cm having bulk density of 1.5 kg dm⁻³.
Data derived with permission from Vandenhove (2006).

of a particular element in the soil regulate the plant concentration of that metal. The concentration of an element in the plant depends on its concentration in the soil, type of soil, plant type, etc. It has been reported that crop concentrations of non-essential radioactive elements (^{238}U, ^{90}Sr and ^{137}Cs) are proportional to the soil concentration of these elements for the same crop and soil type. This proportionality constant is termed as *Transfer Factor* (TF, which is dimensionless because it is a ratio).

$$TF = C_{plant}/C_{soil}$$

(Eq. 21.2)

where C_{plant} = the concentration of the radiocontaminant in the plant expressed as $Bq\,g^{-1}$; C_{soil} = concentration of the contaminant in the soil.

TABLE 21.4 Yearly Reduction (%) in Soil Contamination due to Phytoextraction

Yield (Mg ha^{-1})	5	10	15	20	30
TF (g g^{-1})					
Due to Phytoextraction (%)					
0.01	0.0033	0.0067	0.01	0.013	0.02
0.1	0.033	0.067	01	0.133	0.2
1	0.33	0.67	1.0	1.33	2.0
2	0.67	1.33	2.0	2.67	4.0
5	1.67	3.33	5.0	6.67	10.0
10	3.33	6.67	10.0	13.33	20.0
Due to Phytoextraction (%) and Radioactive Decay					
0.01	2.33	2.337	2.34	2.343	2.35
0.1	2.363	2.398	2.43	2.463	2.53
1	2.66	3	3.33	3.66	4.33
2	3	3.66	4.33	5	6.33
5	4	5.66	7.33	9	12.33
10	5.66	9	12.3	15.66	22.33

The values correspond to a soil depth of 10 cm and bulk density of 1.5 kg dm^{-3}.
Source: Values are derived from Vandenhove (2006).

The TF is considered to be an important parameter determining the potential of phytoextraction. The TF for these three elements and their recommended and total ranges are presented (Table 21.2).

The TF of radio-elements is very important to calculate the removal of a particular element on yearly bases. The following equation may be used to calculate the annual removal.

$$\text{Annual removal } (\%) = \text{TF} \times \text{Yield} \times 100/W_{\text{soil}} \qquad \text{(Eq. 21.3)}$$

where W_{soil} is the mass of contaminated soil layer expressed in kg ha^{-1}.

It is also clear from Eq. 21.3 that the annual removal of a particular element is directly proportional to the TF of the element and the yield. The TFs and yield are also interdependent because of their associations with growth of plants due to dilution effects. The phytoextraction potential and subsequent yearly reduction in contamination is given in Tables 21.3 and 21.4, respectively.

Associated Risks

In addition to the feasibility associated with using plant species on the radio-contaminated soils and recovering these soils, certain risks are also associated with it. Removal of the contaminants from the soil is the first step; however, the removed radiocontaminant still needs to be disposed of. The prime risk is the concern regarding the safe disposal of the extracted contaminant. There-fore, periodic harvesting and safe disposal is required. Moreover, the use of other inorganic amendments applied for increasing contaminant mobility can result in the leaching of the contaminant into groundwater. Therefore, soil profile should be taken into consideration while using chemically assisted phytoremediation.

IMPORTANT RADIONUCLIDES

Caesium (^{137}Cs)

Caesium Behaviour in Soil and Soil Factors Affecting Cs Availability

The tragic Chernobyl accident made the researchers and policy makers investigate the fate of radio-caesium in the environment because of its widespread contamination, relatively long half-life, and the similarity of ^{137}Cs with K (potassium) which favours its uptake by plants (Lembrechts, 1993; Nisbet, 1993; Shaw, 1993; Smolders and Merckx, 1993). The K in soil solution affects Cs availability and it is reported that higher K in soil lowers the Cs transfer to the plants grown in that particular soil. This may be explained in terms of cations (having positive charge and competing for the same uptake). Evans et al. (1983), Brouwer et al. (1983) and Cremers et al. (1988) are of the view that the existence of potassium and micaceous clay minerals govern the bio-availability of solid/liquid partitioning of radio-Cs in soil. Generally soils with higher clay contents are low in Cs-availability. In the investigations after the Chernobyl accident, $CaCO_3$ was used to reduce the radio-Cs TF, increasing the pH of the medium (Nisbet, 1993; Konoplev et al., 1993). At the lower pH the plant growth is affected and reduction in growth may be observed. The TF of caesium is reported to be higher on organic soils with substantial reduction in yield.

Potassium depletion and increased amounts of NH_4 in soil enhances Cs-uptake (Lembrechts, 1993; Bondar and Dutov, 1992; Cline and Hungate, 1960; Shaw et al., 1992; Smolders et al., 1997). Vandenhove (1999) reported a two-fold increase in Cs-TF by the addition of NH_4 along with increased production of plant biomass. These two favour phytoextraction and ulti-mately contribute in the remediation process. Considering soils with very low K, the TF of Cs may increase from 10 to 100 times with a significant reduction in yield. IUR (1989) reported a high TF for rye-grass (3.3 g g^{-1}) for a soil low in fertility. We may conclude that low pH, low K and high NH_4

may possibly enhance the radio-Cs TF and this increase in TF needs to be investigated in detail.

Phytoextraction Potential of Cs

The agricultural soils have a lower potential of ^{137}Cs flux compared to the grassland soils. The optimal plant growth needs high sorption of ^{137}Cs in soil and the availability of K levels in soil; both limit high off-take values. The values in the Table 21.5 describe higher radio-caesium TFs for sandy soils. Most of the crops have ranges of TFs, even for a specific soil type. Potential rye-grass yield was recorded from 15 to 20 Mg ha^{-1} with the supply of adequate fertilizers. This would correspond to phytoextraction (5.3–6.3%, decay included) (Vandenhove, 2006).

Strontium (^{90}Sr)

Factors Affecting the Strontium Transfer Factor

Radio-Cs and radio-Sr are among the most abundant radionuclides released routinely or accidently (as products of nuclear fission). Sr has relatively long half-life (~30 years) and is similar metabolically to calcium (Ca). The TF of Sr for different crops was reported to be the highest for sand and the lowest for clay and organic soils. The Sr-TF is affected by the soils contents of exchangeable Ca. A low Sr-TF was reported with higher soil solution Ca and higher cation exchange capacity (CEC) (Sauras et al., 1999).

Phytoextraction Potential of Sr

The availability of Sr is 10-fold greater than Cs and hence off-take of ^{90}Sr is higher than that of ^{137}Cs. The TFs of ^{90}Sr in green vegetables and Brassicas typically range between 1 and 10. If Sr contamination is in the upper 10 cm of the soil (a yield of 10 Mg ha^{-1} for leafy vegetables) a TF of 10 or 9% could be annually achieved. Under these conditions, a five-fold reduction in soil contamination would require almost 15 years. The phytoextraction of ^{90}Sr has not yet been investigated at large- or field-scale up until now. The Sr-TF having higher values suggest that their potential should be investigated on large scale and under agricultural crops. The values of Sr-TF for some plants have been given in Table 21.6.

Uranium (^{238}U)

Uranium usually occurs in soil in three isotopes, viz. ^{238}U, ^{235}U and ^{234}U. The relative abundance between these three isotopes is 9.27, 0.720 and 0.0055%, respectively. The behaviour of U in soil is similar to other heavy metals in soil and its toxicity mimics that of Pb. This element is toxic to kidneys and some insoluble U-compounds are carcinogenic (Hosner et al., 1998). The behaviour

TABLE 21.5 Soil-to-Plant Factor for Average Radio-Caesium

Crops	Plant Parts	Transfer Factor ($g\,g^{-1}$)	Reference
Sunflower	Seeds	0.15	GOPA, 1996
	Straw	0.59	
	Roots	2.88	
Summer wheat	Leaves	0.05	Lembrechts, 1993
Winter wheat	Leaves + grains	0.03	Lembrechts, 1993
Rye-grass	Whole	0.03	Lembrechts, 1993
Rye-grass	Whole	1.2	IUR, 1989
Rye-grass	Whole	1.87	Vandenhove et al., 1996
Potato	Tuber	0.09	Lembrechts, 1993
Lettuce	Whole	0.24	Lembrechts, 1993
Indian mustard	Whole	0.4–0.5	Lasat et al., 1998

of U in soil is complex and the metal speciation (especially pH-dependent) is very difficult to investigate. Uranium is present in soil (80–90%) in the +VI oxidation states as uranyl cations (UO_2^{-2}) (Ebbs, 1997). This is the predominant species of U in soil under acidic conditions. Generally, rare amounts of uranyl cations are present in available forms because of high solid–liquid distribution (K_d) as reported by Baes (1982). Scanty information is available on the uptake and translocation of uranium by the plants under diverse soil properties. Mostly researchers have investigated the U contents of plants growing under U-contaminated environments (Whicker and Ibrahim, 1984; Ibrahim and Whicker, 1988) or work has been reported by Sheppard et al. (1985, 1989) regarding the uptake of U by field and garden crop having importance to humans and animals.

The transfer of U from soil to plant, usually known as transfer factor (TF), varies with plant species and plant compartment. For example, roots have been shown to incorporate much more uranium than the stems, leaves and shoots (Apps et al., 1988; Ebbs et al., 1998a). Leafy vegetables recorded higher U-TFs followed by root, fruit and grain crop plants (data summarized in Tables 21.4, 21.5 and 21.6). It was to be noted that the TFs rarely exceed a value of 0.01 with the exception of some plants growing on very highly contaminated (acidic) U-mining soils. The TFs of different crops also depend on the soil pH and some are very sensitive to pH, for example, sage brush recorded the highest TF under

TABLE 21.6 Transfer Factor, Crop Yield and Crop Off-Take for Some Plants in the Case of Sr

Crop	TF $(g\,g^{-1})$	Crop Yield $(ton\,dm\,ha^{-1})$	Crop Off-Take (% of total in soil)
Vegetables (leafy)	0.45–9.1	5–10	0.07–3.0
Potato (tuber)	0.03–1.4	6–10	0.006–0.5
Cereals (grains)	0.02–0.94	5–7	0.0037–0.22

natural conditions rarely grown at pH below 4. The TF values for uranium to natural vegetation, vegetables and cereals are given in Tables 21.7, 21.8 and 21.9, respectively.

The uranium species (free UO_2^{+2}) is most readily taken up and translocated by plants (Ebbs, 1997). This U species is present at a pH of 5.5 or less, and acidification of uranium-contaminated sites is necessary for phytoextraction, as shown in Figure 21.4. This U species is also responsible for binding soil solids with organic matter; hence a reduction in plant uptake is recorded (Sheppard et al., 1984). Along with acidification of soil, some soil amendments may increase the availability of U by the formation of complexes.

The role of chelating agents and acidification on the solubilization of U was assessed by Ebbs et al. (1998a) and Huang et al. (1998). They tested organic acids (namely, acetic acid, citric acid, oxalic acid and malic acid). Citric acid was found to be the most effective for increasing U in the soil solution. Citric acid solubilized more uranium than simple soil acidification (HCl and HNO_3). They also reported a non-significant increase in U solublization with the use of chelating agents (EDTA and DTPA). The addition of 20–25 mmol citric acid per kg of soil, caused a decrease in soil pH (0.5 to 1.0 pH units), depending on initial soil pH. When *Brassica juncea* was subjected to citric acid treatment (20 mmol kg^{-1}) the U concentration was increased 1000-fold as reported by Huang et al. (1998). Encouraging results were also reported by Ebbs et al. (1998a) and Ebbs (1997). They found a 10 times increase in U-accumulation in *Beta vulgaris* with citric acid treatment of 25 or 20 mmol kg^{-1}. Citric acid was applied 1 week before harvest in the form of solution spray on the soil surface and its subsequent effect on U-accumulation was observed after 24 h, which reached the maximum after 3 days. There is no problem of residual citric acid because of its rapidly biodegradable nature. It is understood from the above discussion that addition of citric acid is required to get extraction levels high enough to make phytoextraction a feasible option for remediation, since some uncertainties exist affecting the phytoextraction potential. Keeping in view the

TABLE 21.7 Some Specific Uranium Transfer Factors (TF) to Natural Vegetation

	TF	Experimental Conditions	References
Sage brush	0.12	Back ground	Whicker and Ibrahim, 1984; Ibrahim and Whicker, 1992
	0.90	Edge of tailing pond	
Forbs	0.08	Edge of tailing pond	
	1.1	Exposed tailings	
Mixed grasses	0.07	Edge of tailing pond	
	0.69	Exposed tailings	
	0.16	background	
Grasses	0.003 – 0.18	Sandy and loamy soil	Frissel and van Bergeijck, 1989

global cost of phytoextraction in comparison with other remediation options, the following points must be considered:

1. plant growth stages affect the efficiency of citric acid addition, growth stage of maximal accumulation and expected yield;
2. the optimal level of citric acid addition is required to get the highest U-TF and the lowest impact on yield;
3. the effect of citric acid treatment on the soil.

Phytoextraction potential will largely depend on the selection of plants apart from the application of soil additives to maximize U-export with the plant biomass. Huang et al. (1998) investigated four varieties of *Brassica juncea* (Indian mustard) and reported a factor two difference in TF. The four varieties differed significantly in their abilities to export uranium under citric acid treatment.

The uptake of Sr by plants varies significantly between species and is correlated with the Ca uptake (Andersen, 1967). This ^{89}Sr/Ca ratio also varies between different parts of the same plants and may be attributed to discriminating against Sr relative to Ca during the translocation within the plants. The Sr/Ca ratio is usually high in the root (root crops) and low in case of fruits and seeds. The ^{89}Sr uptake is also dependent on soil type and the mount of exchangeable-Ca in the soils. As soil is a heterogeneous media, some other factors may influence the relative availability of strontium and calcium. The ^{137}Cs uptake also varies between plant species, but no relation could be found for the ^{137}Cs and K contents of plants grown in soils abundantly supplied with K (Anderson, 1967). The depletion of soil K after successive cropping increased the plant ^{137}Cs contents and a reduction in ^{137}Cs was observed

TABLE 21.8 Some Specific Uranium Transfer Factors (TF) to Vegetables

	TF	Experimental Conditions	References
Fruit Vegetables			
Tomato	0.0005	Lake sediments, sandy with pH 4	Frissel and van Bergeijck, 1989
Cucumber	0.0009	1.5% organic matter	
Leaf Vegetables			
Carrots	0.019	Sandy loam, pH 7	Frissel and van Bergeijck, 1989
Indian mustard	0.007	1.5% organic matter	Mortvedt, 1994
Lettuce	0.025	Loam, pH 7	Frissel and van Bergeijck, 1989
Turnip green	0.0058	Lake sediments, sandy with pH 4	Whicker and Ibrahim, 1984
Spinach	0.033	Loam, pH 7	Frissel and van Bergeijck, 1989
Root Crops			
Potato (fresh)	0.0009	Sand, pH 8.1	Lakshmnan and Venkteswarlu, 1988
Red beet	0.0024	Sandy or sandy loam, pH 7	Frissel and van Bergeijck, 1989
Turnip root	0.00099	Loam, pH 5.1–7.0	Whicker and Ibrahim, 1984
Sugar beet (fresh)	0.01–0.06	Sand, pH 8.1	Frissel and van Bergeijck, 1989

with an addition of K to the same soils. The type of soil, amount of clay and organic matter are also important factors in determining the uptake of these radio nuclides.

Risks and Potential

Terrestrial and aquatic communities can be exposed to radionuclides through soil, water or ingestion of the radio-contaminated food. Morris and VanHorn (1999) derived a model by which the associated risks at a given site can be calculated by monitoring Ecologically Based Screening Levels (EBSL). These EBSL can be calculated by determining the soil and/or water concentrations (C^E_{Si} and C^E_{Wi}) that would result in TQ=1. At TQ=1, D_T = TRV.

TABLE 21.9 Some Specific Uranium Transfer Factor (TF) to Cereals

Cereals	TF	Experimental Conditions	References
Corn	0.006–0.01	Lake sediments, organic matter, pH 4	Ebbs et al., 1998a
Corn	0.00021	Loam, pH 7.3	Frissel and van Bergeijck, 1989
Barley	0.0021	Loam, pH 7.3	
Rice	0.0005	1.5% organic matter	
Bush bean	0.0018	Lake sediments, organic matter, pH 4	Whicker and Ibrahim, 1984

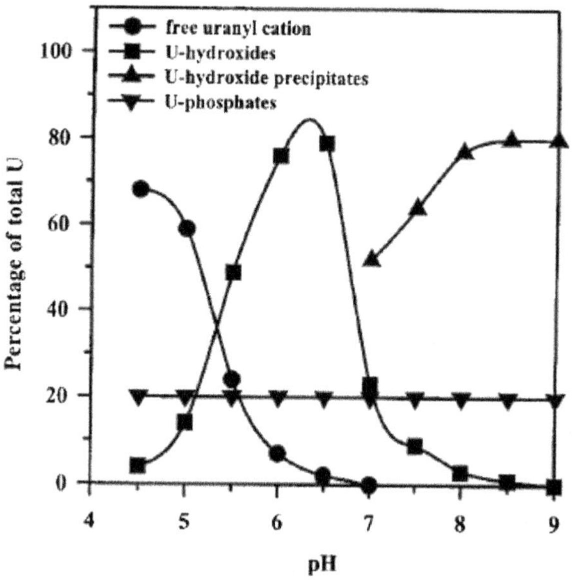

FIGURE 21.4 pH-dependent uranium speciation and uptake. *Source: Ebbs et al. (1998b).*

If a single radionuclide, i, is present in the contaminated soil and/or water, then:

$$TRV = J_{1i}C_{Si}^E + J_{2i}C_{Wi}^E$$

$$C_{Si}^E = \frac{TRV - J_{2i}C_{Wi}^E}{J_{1i}}$$

$$C_{Wi}^E = \frac{TRV - J_{1i}C_{Si}^E}{J_{2i}}$$

(Eq. 21.4)

Equation 21.4 is used to assess toxicity and risks with multi-radio nuclides.

For a given radionuclide and target organism, all parameters except C_{Si} and C_{Wi} are constant. While TRV refers to the Toxicity Reference Values. This model can also be used to assess the toxicity and associated risks with the multi-radio nuclides.

RHIZOFILTRATION

Possibilities and Prospects

Rhizofiltration is an important process going on in soil under various soil and environmental conditions. It corresponds to the accumulation of compounds from soil aqueous solutions by the adsorption on the surface of roots or assimilation through the roots and transmission to the aerial parts. Regarding radionuclides, rhizofiltration is searched for the removal of these nuclides from ground water and wastewater. There are certain factors which govern the removal of radionuclides from aqueous streams including plant dry weight and concentration in the medium called concentration factor (ratio of $Bq\,g^{-1}$ plant to $Bq\,g^{-1}$ water). Absorption in water (waste) per volume is lower than in soil and the CF is higher than TF. This is clear from the following relation:

$$\text{Soil} \overset{K_d}{\Leftrightarrow} \text{Solution} \overset{CF}{\Leftrightarrow} \text{Plant} \qquad \text{(Eq. 21.5)}$$

The relationship between these ratios may be expressed as follows:

$$TF = \frac{CF}{K_d} \qquad \text{(Eq. 21.6)}$$

where CF = concentration factor (ratio of radionuclide activity concentration in plant shoots to the soil solution), and K_d = soil liquid distribution coefficient of the radionuclide $(dm^3\,kg^{-1})$ indicating a ratio of radionuclide activity concentration in solid phase to that in solution phase.

Vasudev et al. (1996) reported the exclusion of U using sunflower (*Helianthus annuus*) from contaminated wastewater and ground water after root contact time of 24h. Chernobyl contaminated pond water was used in a similar experiment, and Cs and Sr levels decreased by 90% and 80%, respectively, after a contact time of 12h. Eight-week-old sunflower plants were used and these were replaced after every 48h. Sunflower showed higher removal rates than the competing crop plants (meadow foxtail, meadow peas and Indian mustard) (Vandenhove, 2006).

NON-FOOD CROPS / ALTERNATIVE CROPS

Since the Chernobyl incident, changed land use is a potentially sustainable remediation option to ensure an economic return so that the products of the land are radiologically acceptable. For example, Vandenhove et al. (2002) evaluated short rotation coppice (SRC) for energy production as an alternative land use in contaminated restricted sites at Chernobyl. The economic viability of the SRC wood production and conversion of the biofuel to energy is a prerequisite for promotion of this alternative land use with the condition that radiological and technical feasibility criteria are met. The parameters for profitable SRC production depended on total wood yield and harvesting methods, the market price of the biofuel; labour and machinery costs played a minor role. For appropriate soil conditions, potential SRC yields were calculated at about 10–12 Mg ha^{-1} per yr. In the case of sandy soils of Belarus, the yield calculations were low at about 5 Mg ha^{-1}. It was reported by Vandenhove et al. (2002) that the cost of production of SRC in Belarus is lower than in Western Europe owing to lower labour and machinery costs. But conversely, to get this advantage, the yield of SRC in Belarus (5 Mg ha^{-1}) must approach that in Western Europe (12 Mg ha^{-1} per yr). When these data sets were modeled, it was calculated that the process may still be profitable with yields of 6 Mg ha^{-1} per yr with optimal parameters. The cultivation of SRC for energy production on sandy soils (the Belarus continental climate conditions) is difficult to provide any profit since the yield is only about 5 Mg ha^{-1}.

Since sandy soils are not very productive in nature, these are low-fertility soils and have a number of other disadvantages. The potential of these soils to be used as alternate land use is reduced. Locally, the wood produced may be used for different purposes by the local community and the farmers. The production of electricity from woody biomass conversion is speculative and only feasible with price support or capital grants. The energy production from SRC is a potential ecologically and economically sustainable land-use option in Belarus (for example), if the land is contaminated or not, but the feasibility depends on a number of factors.

Forestry

Extensive tree plantation resulting in the establishment of artificial forests can also be a feasible step for rehabilitating radio-contaminated soils. Consequently, the area will be covered with vegetation resulting in less radioactive emissions from the contaminated soils. In addition to this, the plantation of metal-tolerant plant species will aid in soil clean-up. In the longer term, these forests will result in environment clean up, natural habitats for various insects, and eventually will result in the initiation of natural sustainable processes back on contaminated soil.

Biofuel / Biodiesel Crops

Phytoremediation and other methods and techniques used to clean-up contaminated sites largely depend on the following factors:

1. type of contamination, its deposition, etc.;
2. soil types;
3. land value (monetary terms);
4. alternative land use (if any);

5. population distribution;
6. extent of contamination (size of area); and
7. equipments available.

The main objective of any method tested is to restrict the entry of contaminant into the food chain and should have a minimum ecological impact. The selected methods and techniques must be safe, economically feasible, technically practical, and socially acceptable with no logistical problems and finally no problem of waste disposal.

STEPS INVOLVED IN REMEDIATION PROGRAMME MANAGEMENT

Recent advances have renewed interest in radiological assessment of discharges of naturally occurring radionuclides in the terrestrial environment. A major and important pathway for human exposure to radioactive materials is via ingestion of food crop materials and animal products. Soil-to-plant TFs of radioactive materials for food and fodder crops are an important aspect to investigate. These factors further need to be in line with the remediation management at a particular site. Here we will discuss the technologies being used in environmental remediation programmes, and not necessarily dealing with the emergency responses to the release of radioactive materials because emergency responses have a different character. When an emergency condition has prevailed, the very first attention is the safety of the affected people there and then to the action needed to control the accidental conditions. The remediation strategy should be aided with sound principles of project management and *as low as reasonably achievable* radiation protection principles according to the IAEA Basic Safety Standards (1996).

MAJOR STEPS IN THE MANAGEMENT OF A REMEDIATION PROGRAMME

Planning for Remediation

The following points are kept in mind for environmental remediation:

- ecological impacts on human health;
- risk of contamination;
- permanent effect of adverse contamination effects;
- community awareness and perception about the problem;
- protocols of radiological approach;
- chances for trans-boundary effects;
- technological advancements, solutions and capabilities, and
- affordability.

This preparation of a programme plan is linked to a number of other activities, which may include: preparation of programme plan; site characterization;

remediation criteria; suitable remediation approach and remediation activities; post-restoration activities; and finally considering any other special aspects. Sound planning is needed for these elements. Before specifying resources and committing efforts for this purpose, a detailed report on supporting activities related to these elements is necessary. The programme plan usually requires several iterations by taking a number of preliminary choices or strategic decisions to develop a plan. The supporting elements are further discussed below.

Site Characterization

This is the very first practical step needed to provide sufficient data to take early strategic decisions (environmental remediation activities). The site characterization will include the following aspects:

- characteristics, extent and distribution of radioactive/contamination sources, as well as their potential for future releases;
- exposure of humans to the environment and radioactive constituents;
- transport of radioactive constituents in the groundwater and/or surface water (any other pathways which may lead to exposure of humans).

The source characterization includes waste characterization and site characterization. It should provide reliable estimates for the rate of release of radioactive constituents and the constituent distribution. For example, in the case of rural zones, the transport of the radioactive material from soil to vegetation should also be estimated. An IAEA-TECDOC (1998) provides general information on the characterization of radioactively contaminated sites for remediation purposes.

Remediation Criteria

Any remediation programme must have clearly expressed objectives. If the programme is justified and clean-up action is optimized, then the criteria to target remediation activities are planned to assess performance. This is also important to verify that remediation has been achieved at its conclusion. The criteria may describe residual dose or concentration limits. It is also necessary to define re-entry criteria that will decide to allow the return of the population and/or reuse of the land for agriculture, and other activities (IAEA, 1997).

Remediation Strategy

Remediation strategies start just after the preliminary site characterization. They involve an engineering study to propose different remediation options capable of addressing specific contaminant problems and effective in minimizing the radiological and chemical exposure. The preliminary selection of options may include several factors, for example, technical and institutional considerations, future land use, social acceptability, economic viability, and regulatory requirements, etc. To start with remedy strategy, a bench scale and/or pilot scale tests of a specific technology may be conducted aiming at designing, to procure, and to operate a full-scale system. The final decision is made by the appropriate authority of the country. The remediation strategy after the onset of the engineering approach is aided with the use of plants for possible remediation called as phytoremediation. Different plant species may be used according to the agro-ecological conditions of the soil. For example, Gommers et al. (2005) described soil to wood transfer of radio-caesium in the case of willow SRC system. The radio-caesium soil-to-SRC wood TFs are in general very small. There is very little concern that the critical limits for fuel wood will be exceeded unless the radioactive contamination levels are extremely high. In another study Vandenhove et al. (2009) proposed new best estimates for soil-to-plant transfer of different radioactive materials. They presented TF estimates for major crop groups (e.g cereals, leafy and non-leafy vegetables, root crops, tubers, fruits, herbs, grasses and pastures, and fodder). Transfer factors were also influenced by soil texture and soil organic matter (SOM) level. The SOM, being the principal soil property controlling the soil behavior (Ibrahim et al., 2010, 2012). When the mechanism for clean-up is confirmed, then the preferred alternative should be selected (taking into consideration the future land-use constraints).

Duquene et al. (2006) conducted a greenhouse experiment to assess the differences of U-TFs by using pea, wheat, Indian mustard, rye-grass and maize, and concluded that TFs can be correlated to the changes in the rhizosphere and soil physical properties. Acidic and alkaline soils contaminated with ^{238}U were used and TFs ranged from 0.0005 to 0.021, and from 0.007 to 0.179 for acid and alkaline soils, respectively. Maize recorded the lowest U uptake while Indian mustard had the highest. Soil parameters (physical and chemical) influence the uptake and translocation of different chemicals from soil to plant. Different plants exhibit different tendencies to absorb and uptake different metals under a given set of conditions. Acidic and alkaline soils behave differently for U contamination and uptake by the plants (Duquene et al., 2006). Conversely, soil solution U concentration under the influence of soil parameters was assessed by Vandenhove et al. (2007). The maximum soil solution ^{238}U concentrations were observed at alkaline pH, high inorganic carbon content and low cation exchange capacity (CEC), organic matter content and clay content. The studies were conducted on pastures with a wide range of parameters important for U sorption. A significant correlation was found between solid–liquid distribution coefficients (K_d, $J kg^{-1}$) and organic matter contents ($R^2 = 0.70$).

Soil Solution Uranium (^{238}U)

Uranium sorption is strongly dependent on solution pH because of the changes in solution speciation, and surface species and surface charge as a function of pH. At acidic pH where the uranyl cation (UO_2^{+2}) predominates, sorption is weak. With an increase in pH, more negatively charged binding sites are available on mineral surfaces owing to release of protons. Increasing pH and increasing CO_3 concentrations increase complexing agents for uranium. At pH > 6, uranium complexing by carbonates increases and uranium mobility is enhanced (Langmuir, 1978; Waite et al., 1994; Zheng et al., 2003). The same is true for different other heavy metals and their sorption and transfer from soil to plant (Ahmad et al., 2011; Rehman et al., 2013a). In the higher pH range (pH ≥ 6), a linear decrease of log (K_d) with pH [log (K_d) = −1.18 pH + 10.8, R_2 = 0.65] was observed. This was attributed to the increased amount of soluble uranyl–carbonate complexes. A similar relationship was deduced by Echevarria et al. (2001) for French soils (pH ranged between 5.8 and 8.8). They found a relationship of [log (K_d) = −1.25 pH + 10.9, R_2 = 0.89]. When K_d values of eight Canadian soils were used in the regression analysis, the relationship was found to be [log(K_d) = −1.29 pH + 11.0, R_2 = 0.76]. It was concluded that the difference in behaviour of UO_2^{+2} and UO_2^{+2}–carbonate complexes is so high that it hides any other effect of soil properties on sorption. It is important to note that similar relationships between the K_d and pH of soils were obtained (with pH ≥ 6), despite the fact that the soils differed in contamination and incubation conditions (Echevarria et al., 2001). It showed that soil pH may be a relevant parameter to measure site-specific variation in uranium K_d values and this may reduce uncertainty associated with safety assessments. This is true for a pH value of approximately 6 and above, and when we consider K_d values at pH below 5.5 and lower, then this hypothesis of linearity is not valid.

Agricultural and Forested Zones and Their Remediation

If, at a particular site, agricultural lands have been contaminated by radioactive materials, special technology is selected capable of providing in situ, effective and economical remediation. It must also ensure ecological safety and respect of the environment. The areas after remediation must be utilized for agricultural production. In situ bioremediation and land farming have been reported but need development and improvements for optimal utilization to control contamination. Investigations have been made to remediate the forest area through the decontamination of wood cuttings and measures were taken to preserve forests (and natural radioactive was allowed). The forest was protected from insects/pests and diseases with improved fire-protection capabilities. These remediation strategies have had some success stories in Belarus, Russia and the Ukraine, where forests and agricultural areas were recovered using the following activities:

- control on living and economic activities of the dweller;
- self-cleansing;
- mitigating radioactively contaminated floods;
- depleting contamination from vadose zone through deep tillage of cultivable lands;

- supporting native vegetation;
- fertilizing soils for NPK (nitrogen, phosphorous and potassium);
- targeting and mitigating radionuclides from the soil profile;
- removal of contaminated vegetation;
- pollution-free feeding of cattle and poultry; and
- binding 137Cs with natural sorbents or substances in animal diets.

Numerous techniques have been tested since the Chernobyl accident (Hubert et al., 1996).

PHYTOSTABILIZATION OF RADIONUCLIDE CONTAMINATED SOILS

To remediate unvegetated or sparsely vegetated soil, phytostabilization is very important. This uses the plants and soil amendments for the establishment of vegetation on contaminated land. Phytostabilization does not achieve a soil clean-up, but it changes the mobility of potentially toxic substances by either reducing their concentrations in the soil water or exchange media *or* by reducing re-entrainment of toxic particulates following the development of a stable and permanent vegetation cover. These processes alter soil metal speciation and reduce potential environmental impact. These technologies employ fundamental soil and plant chemical processes and established agricultural and management practices (Ibrahim et al., 2011, 2012). Plant root activity may change metal speciation and also the behaviour of radionuclides in the soil. This phytostabilization may be more beneficial with changes in soil redox potential, secretion of protons or lowering of pH or chelating agents. The main benefit of this is that further spreading of the contamination from the site is prevented (Vandenhove, 2006).

REMEDIATION ACTIONS IMPLEMENTATION

Remediation actions include procurement of the selected technology, site preparation, health and safety plans, operational procedures, selection and training of staff and personnel involved, site clean-up and verification, waste disposal and release of site for any future use. After completion of the remediation activities, the site will have to meet the remediation objectives and long-term monitoring may be necessary in line with the quality assurance protocols.

Conducting Post-Remediation Activities

When the remediation activities have been completed and verified, the site may be released for restricted or unrestricted use. However, in most of the cases some post-remediation activities are required in the area of concern varying in comprehensiveness and duration according to the local conditions and the degree of remediation.

Post-institutional control necessary remedial activities are performed under controlled contexts including:

- monitoring barriers that isolate residual radioactively contaminated materials;
- environmental monitoring of the remediated site;
- maintenance of barriers and other protection systems;
- avoiding intrusion;
- adherence to licensing and the further terms and conditions;
- ensuring regulation and administrative controls; and
- securing, disseminating, and storage of project data.

Special Considerations

To accomplish the task of remediation of radionuclide-contaminated soils, special adaptations are required and may be included especially when the contaminated area is very large or the contamination is deep into the soil and difficult to access. Contaminated sites with smaller areas may benefit from removal or isolation approaches. These are not feasible for large areas.

Remediation of Areas of Extensive Surface Contamination

Nuclear explosive testing or nuclear accidents result in the dispersion of radioactive materials and environmental contamination. These may cover very large surface areas of hundreds of km^2. This may include yards, streets, walls and roofs (urban area), crops and grasslands, recreational parks, etc. (agricultural areas) and forested or undeveloped regions. For example, in 1967, 160-$km h^{-1}$ winds dispersed radioactive silts from the dried up Lake Karachai at Mayak, Russia over a region of approximately $3000 km^2$ (measured: $1000 km^2$ at a contamination level of $>7.4 \times 10^{10} Bq\,^{90}Sr\,km^{-2}$ and $2000 km^2$ at a contamination level of $>3.7 \times 10^9 Bq\,^{90}Sr\,km^{-2}$) (Fetisov et al., 1993). It is estimated that approximately $2 \times 10^{13} Bq$ of radionuclides (principally ^{90}Sr and ^{137}Cs) were spread over a distance of $75 km$ during this event. Also, years after the Chernobyl reactor accident, deposited radionuclides remain in the top 3 or 4 cm of the soil of fields in a wide zone around this site. This type of contamination is usually spread over a large area and radionuclides may be redistributed both vertically, and laterally and vertically with time. In the case of rainfall, the contaminants move deeper into soil sections and harm the water. Similarly, flood or surface run-off may also help in redistributing into the river flood plains, and may cause accumulation of nuclear elements in megastructures such as dams. Strong winds also help to spread the contamination. The remediation and the clean-up of radioactive contamination may result in secondary pathways of radioactive waste, which are impractical to recover. For example, municipal sanitary wastes contaminated by radioactivity, sludge from wastewater treatment, ash obtained by burning firewood and peat, and finally the most important: radioactively contaminated dredged soils.

REFERENCES

Ahmad, K., Ibrahim, M., Khan, Z.I., Rizwan, Y., Ejaz, A., Fardsous, A., Gondal, S., Lee, D.J., Al-Yemeni, M., 2011. Effect of sewage water on mineral nutritive potential of six fodder species grown under semiarid conditions. Saudi J. Biol. Sci. 18, 317–321.

Andersen, A.J., 1967. Investigations on the plant uptake of fission products from contaminated soils. I. Influence of Plant Species and Soil Types on the Uptake of Radioactive Strontium and Caesium. The Danish Atomic Energy Commission Research Establishment, RIBG Agricultural Research Department. Report No. 170.

Apps, M.J., Duke, M.J.M., Stephens-Newsham, L.G., 1988. A study of radionuclides in vegetation on abandoned uranium tailings. J. Radioanalytical Nuclear Chem. 123, 133–147.

Baes, C.F., 1982. Environmental transport and monitoring: Prediction of radionuclide K_d values from soil–plant concentration ratios. Trans. Am. Nucl. Soc. 42, 53–54.

Bañuelos, G.S., Ajwa, H.A., 1999. Trace elements in soils and plants: An overview. J. Environ. Sci. Health A. 34 (4), 951974.

Bidar, G., Garcon, G., Pruvot, C., Waterlot, C., Douay, F., Shirali, P., 2006. The phytomanagement of soils highly contaminated by metals: use of *Trifolium repens* and *Lolium perenne* as experimental model, Difpolmine Conference, December 12-14. Le Corum-Montpellier-France, 1–6.

Bondar, P.E., Dutov, A.I., 1992. Parameters of radiocaesium transfer into oats harvest on lime soils in connection with the application of mineral fertilizers and chemical ameliorants. In: Collections of Scientific Works – Problems of Agricultural Radiology. Kiev, Ukraine. 125–132.

Brouwer, E., Bayens, B., Maes, A., Cremers, A., 1983. Caesium and rubidium ion equilibria in illite clays. J. Phys. Chem. 87, 1213–1219.

Burken, J.G., Schnoor, J.L., 1997. Uptake and metabolism of atrazine by poplar trees. Environ. Sci. Technol. 31, 1399–1406.

Burken, J.G., Schnoor, J.L., 1999. Distribution and volatilisation of organic compounds following uptake by hybrid poplar trees. Int. J. Phytoremediation 1, 139151.

Cline, J.F., Hungate, F.P., 1960. Accumulation of potassium, caesium-137 and rubidium-86 in bean plants grown in nutrient solution. J. Plant. Physiol. 35, 826–829.

Cremers, A., Elsen, A., De Preter, P., 1988. Quantitative analysis of radiocesium retention in soils. Nature 335, 247–249.

Duquene, L., Vandenhove, H., Tack, F., Van der Avoort, E., Van Hees, M., Wannijn, J., 2006. Plant-induced changes in soil chemistry do not explain differences in uranium transfer. J. Environ. Radioact. 90, 1–14.

Dushenkov, V., Kumar, P.B.A.N., Motto, H., Raskin, I., 1995. Rhizofiltration: The use of plants to remove heavy metals from aqueous streams. Environ. Sci. Technol. 29, 1239–1245.

Ebbs, S.D., 1997. Identification of plant species and soil amendments that improve the phytoextraction of zinc and uranium from contaminated soil. Faculty of Graduate Studies. Cornell University, Michigan. 174.

Ebbs, S.D., Brady, D.J., Kochian, L.V., 1998b. Role of uranium speciation in the uptake and translocation of uranium by plants. J. Exp. Bot. 49, 1183–1190.

Ebbs, S.D., Norvell, W.A., Kochian, L.V., 1998a. The effect of acidification and chelating agents on the solubilisation of uranium from contaminated soil. J. Environ. Qual. 27, 1486–1494.

Echevarria, G., Sheppard, M.I., Morel, J.L., 2001. Effect of pH on the sorption of uranium in soils. J. Environ. Radioact. 53, 257–264.

Ensley, B.D., Raskin, I., Salt, D.E., 1997. Phytoremediation applications for removing heavy metals contamination from soil and water. Biotechnol. Sustainable Environ 6, 59–64.

Evans, D.W., Alberts, J.J., Clack, R.A., 1983. Reversible ion-exchange fixation of cesium-137 leading to mobilization from reservoir sediments. Geochimica et Cosmochimica Acta 47, 1041–1049.

Fetisov, V.I., Romanov, G.N., Drozhko, E.G., 1993. "Practice and problems of environment restoration at the location of the industrial association Mayak, 1994", European Commission Doc. XI-5027 / 94. Proc. Symp. Remediation and Restoration of Radioactive-contaminated Sites in Europe, Antwerp. 507–521.

Frissel, M.J., van Bergeijck, K.E. (Eds.), 1989. VIth Report of the IUR working group soil-to-plant transfer factors: Report of the working group meeting in Guttannen Switzerland. 198924–198926, 1989. Bilthoven, RIVM: 240.

Gommers, A., Gafvert, T., Smolders, E., Merckx, R., Vandenhove, H., 2005. Radiocaesium soil-to-wood transfer in commercial willow short rotation coppice on contaminated farm land. J. Environ. Radioact. 78, 267–287.

GOPA, 1996. Belarus: Study on alternative biodiesel sources in relation with soil decontamination. Project No. TACIS / REG93.

Hinton, T.G., Garnier-Laplace, J., Vandenhove, H., Dowdall, M., Adam-Guillermin, C., Alonzo, F., Barnett, C., Beaugelin-Seiller, K., Beresford, N.A., Bradshawe, C., Brownc, J., Eyrolle, F., Fevrier, L., Gariel, J.-C., Gilbin, R., Hertel-Aas, T., Horemans, N., Howard, B.J., Ikäheimonen, T., Mora, J.C., Oughton, D., Real, A., Salbu, B., Simon-Cornu, M., Steiner, M., Sweeck, L., Vives i Batlle, J., 2013. An invitation to contribute to a strategic research agenda in radioecology. J. Environ. Radioact. 115, 73–82.

Huang, J.W., Blaylock, M.J., Kapulnik, Y., Ensley, B.D., 1998. Phytoremediation of uranium contaminated soils: Role of organic acids in triggering hyperaccumulation in plants. Environ. Sci. Technol. 32, 2004–2008.

Hubert, P., et al., 1996. International scientific collaboration on the consequences of the chernobyl accident (1991–1995), Strategies of decontamination. Final report APASCOSU 1991–1995: ECP4 Project, EUR 16530 EN (1996).

Ibrahim, M., Yamin, M., Sarwar, G., Anayat, A., Habib, F., Ullah, S., Rehman, S., 2011. Tillage and farm manure affect root growth and nutrient uptake of wheat and rice under semi-arid conditions of Pakistan. Appl. Geochem. 26, S194–S197.

Ibrahim, M., Han, K.H., Ha, S.K., Zhang, Y.S., Hur, S.O., 2012. Physico-chemical characteristics of disturbed soils affected by accumulate of different texture in South Korea. Sains Malaysiana 41 (3), 285–291.

Ibrahim, M., Hassan, A., Arshad, M., Tanveer, A., 2010. Variation in root growth and nutrient element of wheat and rice: effect of rate and type of organic materials. Soil Environ. 29, 47–52.

Ibrahim, S.A., Whicker, F.W., 1988. Comparative uptake of U and Th by native plants at a U production site. Health Phys. 54 (4), 413–419.

Ibrahim, S.A., Whicker, F.W., 1992. Comparative plant uptake and environmental behavior of U-series radionuclides at a uranium mine-mill. J. Radio-anal. Nucl. Chem. 156 (2), 253–267.

International Atomic Energy Agency (IAEA), 1996. International basic safety standards for protection against ionizing radiation and for the safety of radiation sources. Safety Series No. 115. IAEA, Vienna (1996).

International Atomic Energy Agency, 1998. Characterization of radioactively contaminated sites for remediation purposes. IAEA-TECDOC-1017, Vienna.

International Atomic Energy Agency, 1997. Application of radiation protection principles to the cleanup of contaminated areas, Interim Report for Comment. IAEATECDOC-987, Vienna.

IUR, 1989. VIth Report of the IUR working group soil-to-plant transfer factors: Report of the working group meeting in Guttannen Switzerland. 24–26, 1989. Bilthoven, RIVM: 240.

Konoplev, A.V., Viktorova, N.V., Virchenko, E.P., Popov, V.E., Bulgakov, A.A., Desmet, G.M., 1993. Influence of agricultural countermeasures on the ratio of different chemical forms of radionuclides in soil and soil solution. Sci. Total. Environ. 137, 147–162.

Kumar, P.B.A.N., Dushenkov, V., Motto, H., Rasakin, I., 1995. Phytoextraction: The use of plants to remove heavy metals from soils. Environ. Sci. Technol. 29, 1232–1238.

Lakshmanan, A.R., Venkateswarlu, V.S., 1988. Uptake of uranium by vegetables and rice. Water. Air and Soil Pollut. 38, 151–155.

Langmuir, D., 1978. Uranium solution–mineral equilibria at low temperature with applications to sedimentary ore deposits. Geochimica et Cosmochimica Acta 42, 547–569.

Lasat, M.M., Fuhrmann, M., Ebbs, S.D., Cornish, J.E., Kochian, L.V., 1998. Phytoremediation of a radiocaesium contaminated soil: Evaluation of caesium-137 accumulation in the shoots of three plant species. J. Environ. Qual. 27, 165–169.

Lembrechts, J., 1993. A review of literature on the effectiveness of chemical amendments in reducing the soil-to-plant transfer of radiostrontium and radiocaesium. Sci. Total Environ. 137, 81–88.

Morris, R.C., VanHorn, R., 1999. Screening risks to terrestrial vertebrates from radionuclide contamination in soil and water. WM'99 Conference, February 28 – March 4, Assessed from http://www.wmsym.org/archives/1999/70/70-4.pdf.

Mortvedt, J.J., 1994. Plant and soil relationship of uranium and thorium decay series radionuclides – A review. J. Environ. Qual. 23, 643–650.

Nisbet, A.F., 1993. Effect of soil-based countermeasures on solid–liquid equilibria in agricultural soils contaminated with radiocaesium and radiostrontium. Sci. Total Environ. 137, 99–118.

Pueyo, M., López-Sánchez, J.F., Rauret, G., 2004. Assessment of $CaCl_2$, $NaNO_3$ and NH_4NO_3 extraction procedures for the study of Cd, Cu, Pb and Zn extractability in contaminated soils. Anal. Chimica Acta 504, 217–226.

Pulford, I.D., Watson, C., 2003. Phytoremediation of heavy metal-contaminated land by trees. Environ. Int. 29, 529540.

Raza, S.H., Shafiq, F., 2013. Exploring the role of salicylic acid to attenuate cadmium accumulation in radish (*Raphanus sativus*). Int. J. Agric. Biol. 15, 547–552.

Raza, S.H., Shafiq, F., Tahir, M., 2013. Screening of cadmium tolerance in sugarcane using antioxidative enzymes as a selection criteria. Pak. J. Life Soc. Sci. 11, 8–13.

Rehman, H., Ali, A., Tanveer, A., Hussain, M., 2013a. Agro-management practices for boosting yield and quality of hybrid maize (*Zea mays* L.). Pak. J. Life Soc. Sci. 11, 70–76.

Rehman, K., Ashraf, S., Rashid, U., Ibrahim, M., Hina, S., Iftikhar, T., Ramzan, S., 2013b. Comparison of proximate and heavy metal contents of vegetables grown with fresh and wastewater. Pak. J. Bot. 45, 391–400.

Romanov, G.N., Drozhko, Y.G., 1996. Ecological consequences of the activities at the Mayak plant. In: Proc. NATO Advanced Study Institute on Radio-active Contaminated Site Restoration. Zarechny, Russia, June 19-28, 1995, 45–56.

Salt, D.E., Smith, R.D., Raskin, I., 1995. Phytoremediation. Annu. Rev. Plant. Physiol. 49, 643–668.

Sarwar, M.A., Ibrahim, M., Tahir, M., Ahmad, K., Khan, Z.I., Valeem, E.E., 2010. Appraisal of pressmud and inorganic fertilizers on soil properties, yield and sugarcane quality. Pak. J. Bot. 42, 1361–1367.

Sauras, T.Y., Vallejo, V.R., Valcke, E., Colle, C., Förstel, H., Millan, R., Jouglet, H., 1999. 137 Cs and 90Sr root uptake prediction under close-to-real controlled conditions. J. Environ. Radioact. 45, 191–217.

Shaw, G., 1993. Blockade of fertilizers of caesium and strontium uptake into crops: Effects of root uptake process. Sci. Total Environ. 137, 119–133.

Shaw, G., Hewamanna, R., Lillywhite, J., Bell, J.N.B., 1992. Radiocaesium uptake and translocation in wheat with reference to the transfer factor concept and ion competition effects. J. Environ. Radioact. 16, 167–180.

Sheppard, M.I., Sheppard, S.C., Thibault, D.H., 1984. Uptake by plants and migration of uranium
and chromium in field lysimeters. J. Environ. Qual. 13, 357–361.

Sheppard, M.I., Thibault, D.H., Sheppard, S.C., 1985. Concentrations and concentration ratios of U,
As and Co in Scots Pine grown in waste site soil and an experimental contaminated soil. Water.
Air. Soil. Pollut. 26, 85–94.

Sheppard, S.C., Evenden, W.G., Pollock, R.J., 1989. Uptake of natural radionuclides by field and
garden crops. Can. J. Soil Sci. 69, 751–767.

Smolders, E., Merckx, R., 1993. Some principles behind the selection of crops to minimize radionuclide uptake from soil. Sci. Total Environ. 137, 135–146.

Smolders, E., Van den Brande, K., Merckx, R., 1997. Concentrations of 137Cs and K in soil solution
predict the plant availability of 137Cs in soils. Environ. Sci. Technol. 31, 3432–3438.

Vandenhove, H., 1999. Phytoextraction of low-level contaminated soil: Study of Feasibility of the
Phytoextraction Approach to Clean-up 137Cs Contaminated Soil from the Belgoprocess Site;
Part 2: Transfer factor screening test: Discussion of results. Internal SCK•CEN report R-3407.

Vandenhove, H., 2000. European sites contaminated by residues from ore extraction and processing
industries. In: Intl. Symp. Restoration of Environments with Radioactive Residues. Arlington,
Virginia, USA, November 29 to December 03, 1999. IAEA, STI/PUB/1092, Austria, Vienna,
pp. 61–89.

Vandenhove, H., 2006. Phytomanagement of radioactively contaminated sites. In: Morel,
J.-L., et al. (Eds.), Phytoremediation of Metal-Contaminated Soils. Springer, the Netherlands.
p. 191–228.

Vandenhove, H., Th. Zeevaert, A., Jackson, B.D., Lambers, B., Jensen, P.H., 2000. Investigation of
a possible basis for a common approach with regard to the restoration of areas affected by lasting radiation exposure as a result of past or old practice or work activity - CARE, Final report
for EC-DG XI-project 96-ET-006, Radiation Protection 115, Luxembourg, Office for Official
Publication by the European Communities, printed in Belgium. pp. 238.

Vandenhove, H., Goor, F., O'Brien, S., Grebenkov, A., Timofeyev, S., 2002. Economic viability of
short rotation coppice for energy production for reuse of caesium-contaminated land in Belarus.
Biomass and Bioenergy 22, 421–431.

Vandenhove, H., Olyslaegers, G., Sanzharova, N., Shubina, O., Reed, E., Shang, Z., Velasco, H.,
2009. Proposal for new best estimates of the soil-to-plant transfer factor of U, Th, Ra, Pb and
Po. J. Environ. Radioact. 100, 721–732.

Vandenhove, H., Van Hees, M., Wouters, K., Wannijn, J., 2007. Can we predict uranium bioavailability based on soil parameters? Part 1: Effect of soil parameters on soil solution uranium
concentration. Environ. Pollut. 145, 587–595.

Vandenhove, H., Van Hees, M., de Brouwer, S., Vandecasteele, C.M., 1996. Transfer of radiocaesium from podzol to ryegrass as affected by AFCF concentration. Sci. Total Environ. 187,
237–245.

Vangronsveld, J., van Assche, F., Clijsters, H., 1995. Reclamation of a bare industrial area contaminated by non-ferrous metals: in situ metal immobilization and revegetation. Environ. Pollut.
87, 51–59.

Vasudev, D., Ledder, T., Dushenkov, S., Epstein, A., Kumar, N., Kapulnic, Y., Ensleay, B.,
1996. Removal of radionuclide contamination from water by metalaccumulating terrestrial
plants. Prepared for presentation at Spring National Meeting, New Orleans, LA. In situ soil
and sediment remediation. Unpublished.

Waite, T.D., Davis, J.A., Payne, T.E., Waychunas, G.A., Xu, N., 1994. U (VI) sorption to ferrihydrite: Application of the surface complexation model. Geochimica et Cosmochimica Acta 58, 5465–5478.

Whicker, F.W., Ibrahim, S.A., 1984. Radioecological investigations of uranium mill tailings systems. Colorado State University, Fort Collins. 48.

Zeevaert, Th., Vanmarcke, H., Govaerts, P., 1997. Status of the restoration of contaminated sites in Europe. Radiation Protection 90. DGXI, Environment, Nuclear Safety and Civil Protection, Luxembourg.

Zheng, Z.P., Tokunaga, T.K., Wan, J.M., 2003. Influence of calcium carbonate on U (VI) sorption to soils. Environ. Sci. Technol. 37, 5603–5608.

Chapter 22

Heavy Metal Accumulation in Serpentine Flora of Mersin-Findikpinari (Turkey) – Role of Ethylenediamine Tetraacetic Acid in Facilitating Extraction of Nickel

Nurcan Koleli,* Aydeniz Demir,* Cetin Kantar,[†] Gunsu Altindisli Atag,[‡] Kadir Kusvuran[‡] and Riza Binzet*

*Mersin University, Faculty of Engineering, Department of Environmental Engineering, Mersin, Turkey; [†]Canakkale Onsekiz Mart University, Faculty of Engineering and Architecture, Department of Environmental Engineering, Canakkale, Turkey; [‡]Alata Horticultural Research Station Directorate, Erdemli, Turkey

INTRODUCTION

Soil and water contamination with Ni has become a worldwide problem (Chen et al., 2009). Nickel contamination mainly results from effluent disposal from mining and smelting, fossil-fuel burning, vehicle emissions, disposal of household, municipal and industrial wastes, fertilizer applications and organic manures (Memon et al., 1980; Alloway, 1995; Salt et al., 2000; Chen et al., 2009; McGrath et al., 2011; Ahmad and Ashraf, 2011; Sreekanth et al., 2013; Sabir et al., 2013). Nickel concentrations may reach $26,000 \, mg \, kg^{-1}$ in polluted soils and $0.2 \, mg \, l^{-1}$ in polluted surface waters: 20 to 30 times higher than found in unpolluted areas (Chen et al., 2009). Great efforts have been made in the last two decades to reduce pollution sources and remedy the polluted soil and water resources. Applications of phytoremediation phenomena to remove heavy metals from soils have been studied in recent years as one of the promising technologies (Bani et al., 2007; Ha et al., 2009; Sakakibara et al., 2011). Phytoextraction, a phytotechnology developed to extract heavy metals from contaminated soils using metal hyperaccumulator plants has received increasing attention due to its cost-effective, non-intrusive and aesthetically pleasing nature (Alkorta et al.,

Soil Remediation and Plants. http://dx.doi.org/10.1016/B978-0-12-799937-1.00022-X

2004a). Recently, the value of metal accumulating plants (hyperaccumulators) for environmental clean-up has been vigorously pursued, giving birth to the philosophy of 'phytoextraction' within a broader concept of phytoremediation (Ahmad et al., 2007; Purakayastha et al., 2008; Barbaroux et al., 2011). Some plants have demonstrated the ability to extract and accumulate high levels of heavy metals. These are, therefore, designated as hyperaccumulators.

Hyperaccumulator plants can accumulate metals in their shoots, regardless of soil metal concentrations with a metal accumulation capacity of $>10,000\,\mathrm{mg\,kg^{-1}}$ for Mn and Zn, $>1000\,\mathrm{mg\,kg^{-1}}$ for Cu, Co, Cr, Ni and Pb, and $>100\,\mathrm{mg\,kg^{-1}}$ for Cd (Baker and Brooks, 1989; Wenzel and Jockwer, 1999). More than 500 plant species have been identified to have potential for soil and water remediation (Sharma, 2011). Many Ni hyperaccumulator plants have been found to grow on serpentine soils (Reeves et al., 1999; Reeves and Adiguzel, 2004; Elena et al., 2005; Ghaderian et al., 2007, 2012; Kazakou et al., 2010; Tumi et al., 2012). Ultramafic rocks (also named as serpentines) and the soils derived from these, occur all over the globe, and may contain several thousand $\mathrm{mg\,kg^{-1}}$ of Ni (Shallari et al., 1998; Panwar et al., 2002; Ahmad et al.,2009; Kumar and Maiti, 2013).

Turkey is one of the rare places in the world with rocks containing ultramafic and serpentine soils. Mersin-Findikpinari, Turkey, located on the Bolkar Mountains, was selected as our study area due to the fact that the region is heavily covered with serpentine soils, and it is an interesting place from the point of endemism (Aytaç and Aksoy, 2000; Orcan et al., 2004). A perusal of the literature revealed that information on the type of native species and their Ni accumulation ability in the study area is scarce (Reeves, 1988; Chaney et al., 2008). Field studies carried out by several investigators on serpentine soils in Turkey have led to the discovery of new species including *Thlaspi elegans*, which can accumulate Ni to concentrations exceeding 0.1% of the dry weight (DW) of the plant (Reeves and Brooks, 1983; Kruckeberg et al., 1999; Reeves et al., 2001; Reeves and Adiguzel, 2004). *T. elegans*, based on the analysis of herbarium specimens, was first described as a hyperaccumulator of Ni by Reeves and Brooks (1983) in Mersin-Findikpinari, Turkey. It is known for its high concentration of Ni ($28600\,\mathrm{mg\,kg^{-1}}$ in leaves) (Prasad, 2005), and remarkable resistance to high levels of Ni in the soil. Nickel accumulation rates (in $\mathrm{mg\,kg^{-1}}$ of DW) and distribution of Ni hyperaccumulator in *Thlaspi* species and observed value in this study are listed in Table 22.1. To identify hyperaccumulate capacity of native heavy metal accumulator plants is necessary through further hydroponic and/or pot experiments. Very little information is available on the Ni accumulation ability of native Ni accumulator *T. elegans* under controlled conditions (Chaney et al., 2008) and no study has been performed till now to show whether the EDTA could efficiently improve its phytoextraction in Ni contaminated soils.

The use of hyperaccumulator species for phytoremediation on a commercial scale is limited due to its low biomass production and slow growth rate (Sun et al., 2009). In order to compensate for the low metal accumulation, many

TABLE 22.1 Nickel Hyperaccumulator *Thlaspi* Species

Ni Accumulation Rates (in mg kg^{-1} of dry weight)	Latin Name	Distribution
2000	*Thlaspi bulbosum* Spruner ex Boiss.	Greece
27300	Thlaspi caerulescens	Germany, Belgium
16200	*Thlaspi caerulescens* J. Presl	Cyprus
52120	*Thlaspi cypricum* Brnm.	Cyprus
20800	*Thlaspi elegans* Boiss.	Turkey
15693	*Thlaspi elegans* Boiss.	Findikpinari, Mersin-Turkey*
3000	*Thlaspi epirotum* Halacsy	Greece
12000	*Thlaspi goesingense* Halacsy	Greece
12400	*Thlaspi graecum* Jord	Greece
2440	*Thlaspi japonicum* H. Boissieu	Japan
26900	*Thlaspi jaubertii* Hedge	Turkey
13600	*Thlaspi kovatsii* Heuffel	Yugoslavia
5530	*Thlaspi montanum* L. var. *Montanum*	Turkey
4000	*Thlaspi ochroleucum* Boiss. and Heldr.	Greece
35600	*Thlaspi oxyceras* (Boiss.) Hedge	Turkey, Syria
18300	*Thlaspi rotundifolium* (L.) Gaudin var. corymbosum (Gay)	Central Europe
31000	*Thlaspi sylvium* (as *T. alpinim* subsp. *sylvium*)	Central Europe
11800	*Thlaspi tymphaeum* Hausskn.	Greece

*Observed value in this study.
Source: Prasad (2005).

workers have conducted studies using synthetic chelators such as EDTA (ethylendiaminetetraacetic acid) to increase the availability of heavy metals in soils and follow phytoextraction efficiency (Garbisu and Alkorta, 2001; Evangelou et al., 2007; Sun et al., 2009; Gupta et al., 2011; Chen et al., 2012). Enhanced phytoextraction of heavy metals using chelating agents and metal-hyperaccumulators has been proposed as an effective approach to remove heavy metals from contaminated soils (Sun et al., 2009). Chelating agents such as EDTA can strongly complex with metal ions, and formation of such organo–metal complexes may increase metal ion solubility, thereby enhancing metal uptake by plant roots (Panwar et al., 2002). Enhanced accumulation of metals by plant species with EDTA treatment is attributed to many factors working either singly or in combination. These factors include (1) an increase in the concentration of available metals, (2) enhanced metal–EDTA complex movement to roots, (3) less binding of metal–EDTA complexes with the negatively charged cell wall constituents, (4) damage to physiological barriers in roots either due to greater concentration of metals, EDTA or metal–EDTA complexes and (5) increased mobility of metals within the plant body when complexed with EDTA compared to free-metal ions facilitating the translocation of metals from roots to shoots (Wuana and Okieimen, 2011).

EDTA is very effective in dissolving metal ions (e.g. Ni) from contaminated soils, although the extraction efficiency depends on soil chemistry (e.g. pH, mineral content) and the strength of the metal–EDTA complex (Turgut et al., 2004; Turan and Ersingu, 2007). Haag-Kerwer et al. (1999) reported that about 80% of the total soil metal was solubilized, and became available for uptake by plants in the presence of EDTA.

The purposes of the current study are to identify possible native Ni hyperaccumulators growing on serpentine soils of Mersin-Findikpinari, Turkey, and to evaluate the impact of EDTA on the uptake and phytoextraction of Ni from contaminated agricultural soils by a native species of Ni hyperaccumulators. This chapter has been prepared from the project 'Selection of Heavy Metal Accumulator Plants Growing on Serpentin Soils in Mersin-Findikpinari and an Investigation of their Potential Use in Agricultural Areas' supported by the General Directorate of Agricultural Research (GDAR).

MATERIALS AND METHODS

Field Study: Site Description, Soil Characterization, Plant Analysis and Plant Selection

Surface soil samples (0–30cm) were collected from 11 different sites (such as agriculture soils, forest soils and sites near the Cr mining and processing unit) in Mersin-Findikpinari. The samples were transported to the laboratory in plastic bags. All samples after mixing and homogenizing in the laboratory were air-dried at room temperature, and passed through a 2-mm sieve. Samples were then preserved at room temperature in plastic bags for further analysis. The physical and chemical characteristics of the soil (organic matter, calcium carbonate, particle size, available P, pH, DTPA-extractable and total heavy metals concentration) were determined using standard soil analysis methods. Soil organic matter was determined using the Walkley and Black method (Nelson and Sommers, 1982), particle size distribution by the hydrometer method (Bouyoucos, 1962), the carbonate content volumetrically by calcimeter method, and the soil pH was measured at a 1 : 1 soil : water ratio (McLean, 1982). Available P was determined using the Olsen method (1952), the concentrations of DTPA-extractable heavy metals were estimated according to Lindsay and Norvell (1978). Soil samples (1 g) were digested in 5 ml of 69.5% HNO_3 and 2 ml 30% H_2O_2 using microwave digestion. Soil total and DTPA-extractable Cr, Ni, Cu, Zn, Cd, Pb, Co, Ca and Mg concentrations were determined using microwave digestion (DIN ISO 11466, 1995), inductively coupled plasma mass spectrophotometry (ICP-MS, Agilent 7500ce). The tests were performed in triplicate. Certified reference material (CRM 7003) was analyzed in order to control the data dependence. All results are given as oven-dried soil matter at $105°C$.

Plant samples were collected from the same sites as soils. A total of 123 plant species (members of 23 genera and 15 families) from five low to highly polluted different sampling locations (sites # 1, 5, 8, 9, 10) were collected during sampling periods over 24 months, and identified systematically. Plant sampling was done six times each year for a thorough survey, and samples identified taxonomically (Table 22.2). The shoots of identified plants were oven-dried at $70°C$ for dry matter (DM) determination, ground and digested in 2 ml 30% H_2O_2 and 5 ml 65% HNO_3 in sealed vessels of a microwave (MarsXpress). The Cr, Ni, Cu, Zn, Cd, Pb, Mg, Ca and Co concentrations in plants and soil samples were determined in triplicate, using ICP-MS. Certified reference materials (*SRM 1573A*, *SRM 1547*) were also analyzed in order to check the accuracy of the extraction technique used in the study. All experiments were carried out in triplicate, and the results are presented in terms of $mg\,kg^{-1}$ biomass (DW).

Greenhouse Study: Soil Characterization, Artificial Soil Contamination, Pot Experiment and Plant Analysis

A surface soil sample (0–30 cm) was collected from the experimental farm of the General Directorate of Agricultural Research, Tarsus-Mersin, Turkey. It was mixed and homogenized in the laboratory, air-dried at room temperature, and passed through a 2-mm sieve. The samples were preserved at room temperature in plastic bags for further evaluation. Some initial physical and chemical features of the soil used in pot experiments were measured using routine analytical methods. Two kilogrammes of air-dried soil was weighed and loaded into a plastic pot for artifical soil contamination and thoroughly mixed with deionized water containing 0, 500, 1000, 1500 and 2000 mg Ni^{2+}, which was in the form of dissolved salts of $Ni(NO_3)_2$. The slurry was then left to age at room temperature for almost three months with frequent and thorough mixing.

Thlaspi elegans Boiss. plant found as a Ni hyperaccumulator in the study area was grown under greenhouse conditions. At the begining, four seeds of the plant specimen were planted in plastic pots, each containing 2 kg soil supplemented with increasing supplies of Ni (0, 500, 1000, 1500 and 2000 mg kg^{-1} soil) in the absence or presence of 10 mg EDTA kg^{-1} soil, added in the form of Na_2EDTA ($C_{10}H_{14}N_2Na_2O_8.2H_2O$). A basal treatment of 200 mg kg^{-1} N as

TABLE 22.2 Species, Families and Sampled Plant Numbers of Identified Plants

Sample	Species	Family	Plant Number (n)
1	*Ajuga reptans* L.	Lamiaceae (Labiatae)	6
2	*Allium cepa* L.	Liliaceae	6
3	*Alyssum minus* (L.) Rothm.	Brassicaceae (Cruciferae)	2
4	*Alyssum murale* Waldst. & Kit.	Brassicaceae (Cruciferae)	10
5	*Anchusa granatensis* Boiss.	Boraginaceae	2
6	*Anthemis cretica* L.	Asteraceae (Compositae)	2
7	*Baptisia australis* (L.) R. Br. ex Ait. f.	Leguminoseae (Fabaceae)	4
8	*Campanula rapunculoides* L.	Campanulaceae	2
9	*Conium maculatum* L.	Umbelliferae (Apiaceae)	3
10	*Dianthus arpadianus* Ade & Born.	Caryophyllaceae	15
11	*Equisetum arvense* L.	Equisetaceae	2
12	*Asphodelus aestivus*	Liliaceae	8
13	*Euphorbia macrostegia* Boiss.	Euphorbiaceae	2
14	*Gladiolus italicus* Miller	Iridaceae	6
15	*Hordeum murinum* L.	Poaceae (Graminae)	8
16	*Onosma bracteosum* Hausskn. & Bornm.	Boroginaceae	3
17	*Podophyllum peltatum* L.	Berberidaceae	4
18	*Rumex obtusifolius* L.	Polygonaceae	5
19	*Sanicula europaea* L.	Umbelliferae (Apiaceae)	3
20	*Sideritis trojana* Bornm.	Lamiaceae (Labiatae)	2
21	*Thlaspi elegans* Boiss.	Brassicaceae (Cruciferae)	20
22	*Thymus pulvinatus* Celak	Lamiaceae (Labiatae)	4
23	*Vicia cassubica* L.	Leguminoseae (Fabaceae)	3

Source: Koleli et al. (2008a).

$Ca(NO_3)_2$ and $100 \, mg \, kg^{-1}$ P as KH_2PO_4 were applied to all pots. After germination, the seedlings were thinned to two plants per pot and left to grow for 60 days in the greenhouse. After this period only shoots of the plants were harvested, and dried at 70°C for the determination of DM production. During these 60 days, the soils in the pots were kept humid (~80% water holding capacity) throughout the experiment. The concentrations of Ni were determined using the dried and ground shoot samples with ICP-MS. All experiments were carried out in triplicate, and results are presented in terms of $mg \, Ni \, kg^{-1}$ biomass (DW) or µg Ni per plant. All results are given on the basis of dry plant matter at 70°C. Certified reference materials (*SRM 1573A, SRM 1547*) were also analyzed in order to check the accuracy of the extraction technique used in this study.

The shoots of the harvested plants were oven-dried at 70°C for DM determination. For Ni concentrations, the oven-dried shoot samples were first ground, and digested using $2 \, ml \, 30\% \, H_2O_2$ and $5 \, ml \, 65\% \, HNO_3$ in sealed vessels of a microwave (MarsXpress). The digested samples were subjected to analysis using ICP-MS for Ni. Certified reference materials (*SRM 1573A, SRM 1547*) were also used in order to check the accuracy of the extraction technique used in the study. All experiments were carried out in triplicate, and the results are presented in terms of $mg \, Ni \, kg^{-1}$ biomass (DW).

Statistical Analysis

All data collected was subjected to statistical Analysis of Variance (ANOVA), and two-tailed t-tests with p-values less than 0.05 were used for statistical comparison between averages. Nickel speciation was evaluated with FITEQL v. 4.0 (Herbelin and Westall, 1999) using solution phase reactions of Ni and EDTA.

RESULTS AND DISCUSSION

Field Study

Some Physical and Chemical Properties of the Serpentine Soils

Some physical and chemical features of the soils from 11 different sampling sites are summarized in Table 22.3. The soils in the area show local variations in their texture, pH, organic matter content and chemical composition depending on the sampling site. They are neutral to slighty alkaline in behaviour with pH ranging from 6.44 to 7.83. The organic matter contents of the soils varied between 1.06% and 4.89%. An analysis of all soils for their total metal contents indicates Ni concentrations ranging from 208 to $3615 \, mg \, kg^{-1}$. These are much higher than the typical concentration ranges observed for unpolluted soils (Turkish Soil Contamination and Control Legislation), and were above the critical values according to Kabata-Pendias and Pendias (2001). Like other soil characteristics, Ni concentrations also changed from one location to another.

TABLE 22.3 Some Physical and Chemical Properties of Soils Tested

Sample	Localization	pH	CaCO$_3$ %	Available P$_2$O$_5$ kg da^{-1}	Organic Matter %	Sand %	Silt %	Clay %	Texture Class
1	Akarca	6.90	4.12	9.87	3.70	43.53	31.38	25.09	Loam (L)
2	Findikpinari (Forest)	7.62	2.06	7.53	3.70	45.42	34.5	20.08	Loam (L)
3	Findikpinari–Agriculture Soil	7.30	2.06	>27.48	1.06	38.81	35.31	25.88	Loam (L)
4	Findikpinari (North)	7.74	2.06	11.24	2.64	50.21	22.66	27.13	Sandy Clay Loam (SCL)
5	Findikpinari-Bozon Güzlesi	7.83	4.12	7.99	1.85	54.85	23.03	22.12	Sandy Clay Loam (SCL)
6	Findikpinari–Kizilcamlik	7.72	4.12	24.46	3.70	52.99	28.96	18.05	Sandy Loam (SL)
7	Bozon Guzlesi–Agriculture Soil	7.72	8.24	>27.48	4.89	40.61	33.97	25.42	Loam (L)
8	Findikpinari (South)	7.20	2.06	4.69	4.49	48.64	23.89	27.47	Sandy Clay Loam (SCL)
9	Kuzucu–Mining I	7.12	2.06	4.53	1.58	47.76	33.51	18.73	Loam (L)
10	Kuzucu–Mining II	6.44	2.06	4.53	1.85	64.16	10.09	25.75	Sandy Clay Loam (SCL)
11	Kuzucu – Agriculture Soil	7.72	2.06	4.53	2.11	62.81	5.7	31.49	Sandy Clay Loam (SCL)

Source: Koleli et al. (2008b).

In general, the soils near mining areas exhibit much higher Ni concentrations as compared to less-contaminated agricultural areas.

Table 22.4 shows the lowest, highest and mean values for Cr, Ni, Cu, Zn, Cd, Pb, Mg, Ca and Co concentrations in the soils analyzed in the current study. As is apparent from Table 22.4, Ca concentration of all serpentine soils is less than Mg concentration for the soil sample numbers 1, 8, 9 and 10; a typical behaviour was observed for all serpentine soils as reported by Kazakou et al. (2010) too. The total concentrations of Cr, Ni, Cu, Zn, Cd and Co in the soils were higher than the typical concentration ranges observed for unpolluted soils as set by the Turkish Soil Contamination and Control Legislation and were above the critical value according to Kabata-Pendias and Pendias (2001), except for Pb. Our results also indicate that the heavy metal concentration of soils is highly dependent on sampling locations. The soil Ni and Cr concentrations were significantly higher than other metals. Nickel concentration ranged from $43\,mg\,kg^{-1}$ to $909\,mg\,kg^{-1}$ and Cr concentration ranged from $208\,mg\,kg^{-1}$ to $3615\,mg\,kg^{-1}$. The maximum concentrations of metals in soils (as DM) were $246\,mg\,Cu\,kg^{-1}$, $467\,mg\,Zn\,kg^{-1}$, $8.2\,mg\,Cd\,kg^{-1}$, $111\,mg\,Pb\,kg^{-1}$ and $214\,mg\,Co\,kg^{-1}$. These findings are in full agreement with the observations of earlier studies in Mersin (Koleli and Halisdemir, 2005).

The available concentration of heavy metals in the soils is shown in Table 22.5. The DTPA-extractable method provides evidence that the uptake of metals by plants decreases in the order of: $Cd > Pb > Zn > Ni > Cu > Cr$. The fate and transport behaviour of heavy metals in the soil highly depends on the soil pH, properties of metals, redox conditions, soil chemistry, organic matter content, clay content and cation exchange capacity (Kirkham, 1977).

Identification and Heavy Metal Concentrations of Sampled Plants in Mersin-Findikpinari

Table 22.6 shows the distribution of 123 plant species (from 23 genera and 15 families) collected from five different sampling locations (from the sites where soil samples were taken), with very high plant population in the study area. Sampling was carried out six times in one year.

The shoots of identifed plants were analyzed for their Cr, Ni, Cu, Zn, Cd, Pb, Mg, Ca and Co concentrations (Table 22.7). Among all the species collected in the area, *Thlaspi elegans* Boiss., *Alyssum murale* Waldst. & Kit., *Anthemis cretica* L. and *Sanicula europaea* L. exhibited excellent Ni uptake capacity with Ni concentrations of 15693, 13591, 7741 and $4247\,mg\,Ni\,kg^{-1}$ (DW), respectively. The Ni enrichment coefficients, commonly used to evaluate metal accumulating capacity of plants relative to the degree of soil contamination, are also shown in Table 22.7.

Two plant species collected from the study area exhibited excellent Ni uptake capacity (Table 22.6). These species are *T. elegans* ($15693\,mg\,Ni\,kg^{-1}$ DM), *A.murale* ($13591\,mg\,Ni\,kg^{-1}$ DM); followed by *A. cretica* ($7741\,mg\,Ni\,kg^{-1}$ DM) and *S.europaea* ($mg\,Ni\,kg^{-1}$ DM). Among these four different plant species,

TABLE 22.4 Concentration of Cr, Ni, Cu, Zn, Cd, Pb, Mg, Ca and Co in Soil (mg kg^{-1})

Sample	Localization	Cr	Ni	Cu	Zn	Cd	Pb	Mg	Ca	Co	Mg/Ca^{-1} Ratio
							(mg kg^{-1})				
1	Akarca	730	3605	29	208	4.1	42	30719	1897	209	16.2
2	Findikpinari (Forest)	43	208	31	242	8.2	36	2661	2298	14	1.2
3	Findikpinari – Agriculture Soil	98	277	70	284	6.6	50	4133	3286	26	1.3
4	Findikpinari (North)	126	222	60	467	3.1	109	3734	3070	36	1.2
5	Findikpinari – Bozon Güzlesi	115	239	93	332	1.6	59	3380	4147	32	0.8
6	Findikpinari – Kizilcamlik	291	774	49	253	4.5	61	4161	4618	46	0.9
7	Bozon Guzlesi – Agriculture Soil	151	699	79	257	2	53	8009	12821	41	0.6
8	Findikpinari (South)	909	3615	246	377	1.9	43	40195	1027	214	39.1
9	Kuzucu – Mining I	385	2289	12	274	4.7	60	14632	867	135	16.9

10	Kuzucu – Mining II	330	3334	55	423	7.8	111	21869	1168	181	18.7
11	Kuzucu – Agriculture Soil	151	464	37	358	1.1	73	2087	3567	39	0.6
	The lowest value in	43	208	12	208	0.1	36	2087	867	14	1
	The highest value	909	3615	246	467	8.2	111	40195	12821	214	39
	The mean value	303	1430	69	316	4.1	63	12325	3524	88	8.9
	Normal concentrations in soils*	5–1500	2–750	2–250	1–900	0.01–2	2–300			0.5–65	
	The critical concentration in the soils*	75–100	100	60–125	70–400	3.0–5.0	100–400			25–50	
	Soil Contamination Legistation value	100	75	140	300	3	300			—	

*Values according to Kabata-Pendias and Pendias (2001).
Source: Koleli et al. (2008b).

TABLE 22.5 DTPA-Extractable Cr, Ni, Cu, Zn, Cd, Pb and Co Concentrations in Soil (mg kg⁻¹ soil)

Sample	Localization	Cr	Ni	Cu	Zn	Cd	Pb	Co
					$(mg\,kg^{-1}\,soil)$			
1	Akarca	>b.d.	17.9 (%0.5)	8.4 (%29.1)	10.5 (%5.1)	0.4 (%9.9)	3.3 (%7.9)	22.0 (%10.5)
2	Findikpinari (Forest)	>b.d.	16.5 (%7.9)	3.6 (%11.5)	4.5 (%1.9)	0.4 (%4.8)	3.2 (%8.7)	11.5 (%34.1)
3	Findikpinari – Agriculture Soil	>b.d.	22 (%8.1)	1.5 (%2.1)	5.3 (%1.9)	0.02 (%0.4)	3.2 (%6.3)	74.8 (%18.9)
4	Findikpinari (North)	>b.d.	9.0 (%4.0)	4.6 (%7.6)	5.9 (%1.3)	0.4 (%13.3)	4.3 (%4.0)	4.0 (%10.8)
5	Findikpinari – Bozon Güzlesi	>b.d.	16.7 (%7.0)	7.4 (%8.0)	21.4 (%6.5)	1.1 (%71.4)	16.8 (%28.6)	8.0 (%24.9)
6	Findikpinari – Kizilcamlik	>b.d.	19.0 (%2.5)	7.9 (%15.9)	22.4 (%8.9)	1.3 (%29.8)	18.9 (%30.8)	9.4 (%20.7)
7	Bozon Guzlesi – Agriculture Soil	>b.d.	10.9 (%1.6)	3.4 (%4.2)	7.4 (%2.9)	0.4 (%20.5)	2.8 (%5.9)	5.7 (%14.2)

8	Findikpinari (South)	>b.d.	7.8 (%0.2)	3.9 (%1.6)	4.3 (%1.1)	0.4 (%20.0)	3.4 (%7.8)	3.3 (%1.6)
9	Kuzucu – Mining I	>b.d.	22.3 (%1.0)	2.5 (%21.5)	4.6 (%1.7)	0.3 (%5.5)	3.9 (%6.5)	2.7 (%2.0)
10	Kuzucu – Mining II	0.4 (% 0.1)	22.2 (%0.7)	3.3 (%6.0)	5.7 (%1.4)	0.1 (%0.7)	3.4 (%3.1)	64.9 (%35.9)
11	Kuzucu – Agriculture Soil	0.1 (% 0.1)	13.6 (%2.9)	4.3 (%11.7)	6.2 (%1.7)	0.6 (%27.3)	4.7 (%6.4)	147 (%14.0)
The lowest value in		0.11	7.78	1.50	4.30	0.02	2.83	62.5
The highest value		0.40	22	8.41	22	1.26	19	1423.7
The mean value		0.3	16.2	4.6	8.9	0.5	6.2	311.2

Source: Koleli et al. (2008b).

TABLE 22.6 Species, Families, Turkish Names and Sampled Plant Numbers of Identified Plants

Sample	Species	Family	Plant Number (n)
1	*Ajuga reptans* L.	Lamiaceae (Labiatae)	6
2	*Allium cepa* L.	Liliaceae	6
3	*Alyssum minus* (L.) Rothm.	Brassicaceae (Cruciferae)	2
4	*Alyssum murale* Waldst. & Kit.	Brassicaceae (Cruciferae)	10
5	*Anchusa granatensis* Boiss.	Boraginaceae	2
6	*Anthemis cretica* L.	Asteraceae (Compositae)	2
7	*Baptisia australis* (L.) R. Br. ex Ait. f.	Leguminoseae (Fabaceae)	4
8	*Campanula rapunculoides* L.	Campanulaceae	2
9	*Conium maculatum* L.	Umbelliferae (Apiaceae)	3
10	*Dianthus arpadianus* Ade & Born.	Caryophyllaceae	15
11	*Equisetum arvense* L.	Equisetaceae	2
12	*Asphodelus aestivus*	Liliaceae	8
13	*Euphorbia macrostegia* Boiss.	Euphorbiaceae	2
14	*Gladiolus italicus* Miller	Iridaceae	6
15	*Hordeum murinum* L.	Poaceae (Graminae)	8
16	*Onosma bracteosum* Hausskn. & Bornm.	Boroginaceae	3
17	*Podophyllum peltatum* L.	Berberidaceae	4
18	*Rumex obtusifolius* L.	Polygonaceae	5
19	*Sanicula europaea* L.	Umbelliferae (Apiaceae)	3
20	*Sideritis trojana* Bornm.	Lamiaceae (Labiatae)	2
21	*Thlaspi elegans* Boiss.	Brassicaceae (Cruciferae)	20
22	*Thymus pulvinatus* Celak	Lamiaceae (Labiatae)	4
23	*Vicia cassubica* L.	Leguminoseae (Fabaceae)	3

Source: Koleli et al. (2008a).

TABLE 22.7 Concentration of Cr, Ni, Cu, Zn, Cd, Pb, Mg, Ca and Co in Plants (mg kg⁻¹)

No	Species	Sample No.	Cr	Ni	Cu	Zn	Cd	Pb	Ca	Mg	Co	Mg/Ca Ratio
						(mg kg⁻¹)						
1	Ajuga reptans L.	6	>b.d.	34	4.5	140	0.1	1.8	1804	1926	0.9	1.07
2	Allium cepa L.	6	3.63	221	1.7	107	0.3	2.4	622	2369	3.5	3.81
3	Alyssum minus (L.) Rothm.	2	>b.d.	68	4.6	49	0.4	2	763	2045	1.2	2.68
4	Alyssum murale Waldst. & Kit.	10	27.16	13512	9.4	260	1.2	12.8	1292	3793	7.7	2.94
5	Anchusa granatensis Boiss.	2	6.56	210	13.5	86	0.2	4.2	1245	741	0.9	0.60
6	Anthemis cretica L.	2	>b.d.	7741	0.1	76	0.5	1.7	419	3909	8.9	9.34
7	Baptisia australis (L.) R. Br. ex Ait. f.	4	>b.d.	12	3.1	56	0.1	2	1980	3682	0.5	1.86
8	Campanula rapunculoides L.	2	>b.d.	23	12.5	93	0.1	1.9	848	835	0.7	0.98
9	Conium maculatum L.	3	1.48	22	7.4	71	0.1	2.3	2225	1290	1.1	0.58
10	Dianthus arpadianus Ade & Born.	15	>b.d.	40	4.7	51	0	2.4	985	979	0.3	0.99
11	Equisetum arvense L.	2	7.88	188	>b.d.	46	0.1	1.6	445	674	0.8	1.52
12	Asphodelus aestivus	8	0.05	106	3.1	100	0.1	1.7	3780	928	0.8	0.25
13	Euphorbia macrostegia Boiss.	2	0.7	8	12.9	204	0.2	16.3	2499	1359	0.9	0.54

Continued

TABLE 22.7 Concentration of Cr, Ni, Cu, Zn, Cd, Pb, Mg, Ca and Co in Plants (mg kg⁻¹)—cont'd

No	Species	Sample No.	Cr	Ni	Cu	Zn	Cd (mg kg⁻¹)	Pb	Ca	Mg	Co	Mg/Ca Ratio
14	Gladiolus italicus Miller	6	2.22	73	3.6	102	0.1	2	3021	4362	1.5	1.44
15	Hordeum murinum L.	8	6.97	99	5.2	59	0.3	3.8	1956	8253	0.8	4.22
16	Onosma bracteosum Hausskn. & Bornm.	3	16.72	107	7.8	155	0.1	2.5	3117	2113	6.6	0.68
17	Podophyllum peltatum L.	4	>b.d.	25	14	86	0.1	2	1274	2203	0.6	1.73
18	Rumex obtusifolius L.	5	0.63	50	10.1	76	0.1	2	2246	1627	0.8	0.72
19	Sanicula europaea L.	3	>b.d.	4247	0.7	67	0.1	2.4	1421	3604	5.4	2.54
20	Sideritis trojana Bornm.	2	6.56	210	13.5	86	0.2	4.2	1245	741	0.9	0.60
21	Thlaspi elegans Boiss.	20	3.88	15693	9.4	95	1.1	13.7	1389	2628	6.4	1.89
22	Thymus pulvinatus Celak	4	>b.d.	7	7.3	57	0.1	2	897	1155	0.2	1.29
23	Vicia cassubica L.	3	3.24	139	6.3	122	0.4	1.9	1267	1060	5.5	0.84
	Normal value*		0.03–14	0.02–5	5–20	1–400	0.1–2.4	0.2–20			0.02–1	
	Critical value*		5–30	10–100	20–100	100–400	5–30	30–300			15–50	

*Values according to Kabata-Pendias and Pendias (2001).
Source: Koleli et al. (2008a).

T. elegans is endemic to Mersin-Findikpinari, and other species of this genus are well established as Ni hyperaccumulators in the literature (Reeves, 1988) while *Alyssum murale* Waldst. & Kit. is an endemic species for serpentine soils (Chaney et al., 2008).

Enrichment Coefficient and Bioaccumulation Factor of Ni in Plants

Table 22.8 shows enrichment coefficients which define the heavy metal concentration in shoots divided by the heavy metal concentration in soil. These coefficients were used to evaluate metal-accumulating capacity of plants relative to the degree of soil contamination. The enrichment coefficients of Ni by plants decreased in the order of: *T. elegans* (10.97) > *A. murale* (9.45) > *A. cretica* (5.41) > *S.europaea* (2.97). The bioaccumulation factor, defined as a ratio of the heavy metal concentration in plant shoots to the extractable concentration of metal in the soil, was used for quantitative expression of accumulation. *T. elegans* plants showed the highest bioaccumulation efficiency for Ni compared to other species (Table 22.8). The bioaccumulation factor of Ni by plants decreased in the order of: *T. elegans* (807.17) > *A. murale* (937.46) > *A. cretica* (462.43) > *S. europaea* (253.70). The enrichment coefficients and bioaccumulation factor values are indicative of the accumulation potential of the heavy metals in plants, which can increase the possibility of toxic metals entering the food chain through the consumption of plants on the sites by either animals or humans (Chin, 2007) if consumed (Table 22.8).

Greenhouse Study

Greenhouse experiments were performed to determine the phytoremediation potential of *T. elegans* (15693 mg kg^{-1} DM) for Ni, since field studies indicated that it was an excellent Ni hyperaccumulator. The main goal was to determine the potential of this plant for the treatment of agricultural areas contaminated with Ni.

Some Initial Physical and Chemical Properties of the Soil Used in Pot Experiment

Some physical and chemical characteristics of the soil used in pot experiments are summarized in Table 22.9. The greenhouse pot experiments were performed using unpolluted agricultural soils with a clayey loam texture, pH of 8.1, 26.2% $CaCO_3$ and 1.3% organic matter contents. The analysis of soils for total and DTPA-extractable Ni contents indicates Ni concentrations of 57.0 and 2.2 mg kg^{-1} soil, respectively. The total metal contents are much lower than the typical concentration ranges observed for unpolluted soils as set by the 'Turkish Soil Contamination and Control Legislation', and were under the critical value (100 mg kg^{-1}) according to Kabata-Pendias (2001).

TABLE 22.8 Enrichment Coefficient and Bioaccumulation Factor of Ni in the Collected Plants

No	Species	Sample Number (n)	Mean Ni Concentration (mg Nikg^{-1})	Enrichment Coefficient of Ni*	Bioaccumulation Factor of Ni**
1	Ajuga reptans L.	6	34	0.02	2.03
2	Allium cepa L.	6	221	0.15	13.20
3	Alyssum minus (L.) Rothm.	2	68	0.05	4.06
4	Alyssum murale Waldst. & Kit.	10	13512	9.45	807.17
5	Anchusa granatensis Boiss.	2	210	0.15	12.54
6	Anthemis cretica L.	2	7741	5.41	462.43
7	Baptisia australis (L.) R. Br. ex Ait. f.	4	12	0.01	0.72
8	Campanula rapunculoides L.	2	23	0.02	1.37
9	Conium maculatum L.	3	22	0.02	1.31
10	Dianthus arpadianus Ade & Born.	15	40	0.03	2.39
11	Equisetum arvense L.	2	188	0.13	11.23
12	Asphodelus aestivus	8	106	0.07	6.33
13	Euphorbia macrostegia Boiss.	2	8	0.01	0.48
14	Gladiolus italicus Miller	6	73	0.05	4.36
15	Hordeum murinum L.	8	99	0.07	5.91

16	Onosma bracteosum Hausskn. & Bornm.	3	107	0.07	6.39
17	Podophyllum peltatum L.	4	25	0.02	1.49
18	Rumex obtusifolius L.	5	50	0.03	2.99
19	Sanicula europaea L.	3	4247	2.97	253.70
20	Sideritis trojana Bornm.	2	210	0.15	12.54
21	Thlaspi elegans Boiss.	20	15693	10.97	937.46
22	Thymus pulvinatus Celak	4	7	0.00	0.42
23	Vicia cassubica L.	3	139	0.10	8.30

*Enrichment coefficient (EC): the heavy metal concentration in plant above ground part / the heavy metal concentration in soil (at five stations, mean Ni concentration: 2313 mg/kg soil).
**Bioaccumulation factor (BAF): the heavy metal concentration in plant shoot / the extractable concentration of metal in the soil (at five stations, mean available Ni concentration: 16.74 mg/kg soil).
Source: Koleli et al. (2008a).

TABLE 22.9 Some Initial Physical and Chemical Properties of the Soil Used in Pot Experiment

Parameters	Tarsus soil
PH (1:2)	8.1
Organic matter (%)	1.3
$CaCO_3$ (%)	26.2
Particle Size Distribution	
Sand (%)	41
Silt (%)	36
Clay (%)	23
Texture class	Clay Loam (CL)
Total Ni (mg/kg)	57
DTPA extractable Ni (mg/kg)	2.2

Effect of EDTA on Morphology and Plant Growth

Following Ni application, visual symptoms were also monitored throughout the experiments. Although *T. elegans* did not exhibit any significant symptoms as observed in a typical Ni toxicity, the toxic effects of high concentrations of Ni in plants have been frequently reported leading to reductions in plant growth and adverse effects on fruit yield and quality (Chen et al., 2009). Plants grown in Ni-contaminated soil and media show various responses and toxicity symptoms including retardation of germination, inhibition of growth, reduction of yield, induction of leaf chlorosis and wilting, disruption of photosynthesis, inhibition of CO_2 assimilation (Chen et al., 2009).

As is clear from the Figure 22.1, after 60 days of growth, addition of EDTA increased the growth of plants as well as dry biomass yields as compared to the plants in the control. A two-sided t-test with p-values less than 0.05 confirmed that the EDTA application had a positive impact on plant growth. Increasing Ni application led to a significant decrease in the growth of plants (Figure 22.1), especially at Ni concentrations $>500\,mg\,kg^{-1}$. Kopittke et al. (2007) found that the addition of Ni^{2+} to cowpea significantly reduced plant growth, and the degree of Ni toxicity varied with Ni concentration applied. However, Sabir et al. (2013) demonstrated in a recent study that high concentrations could modify the growth performance and mineral status of wheat cultivars under some conditions. Outridge and Scheuhammer (1993) suggest that adverse effects on chlorophyll metabolism and growth occur at soil water concentrations as low as $1\,mg\,l^{-1}$ Ni. In the presence of complexing ligands

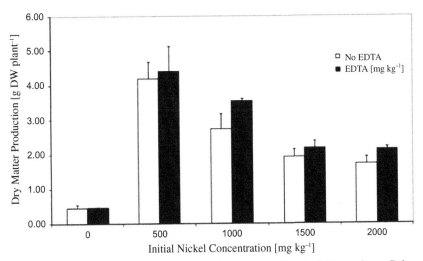

FIGURE 22.1   Effect of Ni and EDTA on dry matter production (shoot) of *Thlaspi elegans* Boiss.

such as EDTA, on the other hand, the toxic effect of Ni^{2+} on plant growth decreased significantly, most probably due to the formation of less toxic Ni–EDTA complexes. A two-sided t-test with *p*-values less than 0.05 confirmed that the EDTA application had a positive impact on plant growth. Numerous reports suggest that addition of EDTA significantly increases plant growth (Gupta et al., 2011; Chen and Cutright, 2001; Lai and Chen, 2005; Meers et al., 2004; Alkorta et al., 2004b).

Effect of EDTA on Ni Concentration in Plant Shoots

Figure 22.2 shows Ni concentrations in shoots of *T. elegans* and the effects of EDTA on it. The addition of EDTA to the soils increased Ni uptake by *T. elegans* as compared to the control (Figure 22.2). An application of $10\,mg\,kg^{-1}$ EDTA resulted in an increase in Ni accumulation. These results demonstrate the strong dependence of Ni concentration on initial Ni dose and EDTA concentration. *T. elegans* accumulated more Ni in the presence of EDTA. The addition of EDTA ($10\,mg\,kg^{-1}$ soil) to pots significantly increased Ni accumulation relative to non-EDTA-containing systems. While *T. elegans* accumulated $13216\,mg\,kg^{-1}$ Ni in the absence of EDTA, the addition of $10\,mg\,kg^{-1}$ EDTA increased Ni accumulation to $16632\,mg\,kg^{-1}$ Ni at a Ni application dose of $2000\,mg\,kg^{-1}$ (Figure 22.2). Nickel concentrations $>1000\,mg\,kg^{-1}$ in shoots confirm Ni hyperaccumulation for *T. elegans* both the absence or presence of $10\,mg\,kg^{-1}$ EDTA.

In general, critical toxicity levels are $>10\,mg\,kg^{-1}$ DW in sensitive plant species, $50\,mg\,kg^{-1}$ DW in moderately tolerant plant species, and $>1000\,mg\,kg^{-1}$ DW in Ni hyperaccumulator plants, such as *Alyssum* and *Thlaspi* species (Chen et al., 2009). Wenzel et al. (2003) reported that average

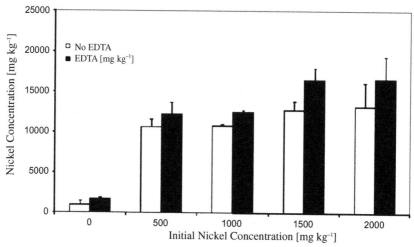

FIGURE 22.2 Effect of Ni and EDTA on shoot Ni concentration ($mg\,kg^{-1}$) of *Thlaspi elegans* Boiss.

Ni concentrations were $3180\,mg\,kg^{-1}$ in shoots for the Ni hyperaccumulator *Thlaspi goesingense*. Our data is about four times higher than this value, even in the absence of EDTA.

Statistical analysis (two-sided t-test) ($p < 0.05$) indicates that the differences obtained in Ni concentration in the absence and presence of EDTA are statistically significant. The increase in Ni accumulation by plants can be explained through the formation of highly soluble and less toxic Ni–EDTA complexes. Water solubility of Ni in soils and its bioavailability to plants are highly influenced by soil pH. Nickel forms less soluble immobile species under alkaline pH conditions (Anonymous, 1993). With the addition of complexing agents such as EDTA, Ni forms highly soluble Ni–EDTA complexes, making Ni available for uptake by plants. In a soil contaminated with Cd, Cr and Ni, Chen and Cutright (2001) found that the application of EDTA to soils at a rate of $500\,mg\,kg^{-1}$ resulted in a significant increase in the metal uptake of *Helianthus annuus* plants. Jean et al. (2008) observed that the enhanced solubiliziation of Ni from contaminated soils using EDTA as a leaching agent led to a significant increase in Ni uptake by *Datura innoxia*. Panwar et al. (2002) found that metal uptake by *Brassica* species could be significantly enhanced in the presence of EDTA due to formation of highly soluble metal–ligand complexes.

Effect of EDTA on Ni Content in Plant Shoots

The nickel content of the shoots was calculated by multiplying shoot Ni concentration by shoot DWs. Figure 22.3 shows the data presented in terms of Ni accumulated per plant as obtained by multiplying shoot Ni concentration by

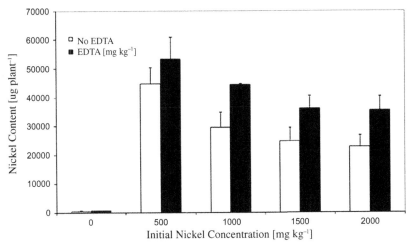

FIGURE 22.3   Effect of Ni and EDTA on Ni uptake by *Thlaspi elegans* Boiss., plotted in terms of initial Ni concentration versus shoot Ni content (μg Ni plant^{-1}).

shoot DW (μg kg^{-1} Ni plant). Nickel uptake by plants was at its maximum at an initial Ni dosage of 500 mg kg^{-1}, and then decreased sharply with increasing initial Ni dose due to the fact that Ni toxicity reduced the biomass of plants (Figure 22.1), thereby decreasing the amount of Ni accumulation. A correlation analysis was performed between Ni uptake and the plant growth. The correlation between plant growth and Ni uptake in *T. elegans* was much stronger ($r > 0.95$, $p < 0.01$). This can be explained through much higher affinity of *T. elegans* to accumulate Ni as is clear from Figure 22.3. The Ni uptake by plants significantly increased in the presence of 10 mg kg^{-1} EDTA compared to non-EDTA containing systems. For instance, while the Ni content of *T. elegans* was 44831 μg per plant in the absence of EDTA, the addition of 10 mg kg^{-1} EDTA increased Ni content to 53740 μg per plant at an initial Ni application dose of 500 mg kg^{-1}. The enhanced Ni accumulation in the presence of EDTA indicates the formation of less toxic Ni–EDTA complexes, which led to an increase in plant growth (Figure 22.1), thereby increasing both Ni tolerance and capacity for Ni transport to the shoot. An addition of EDTA to the soil can change the concentration and relative abundance of different Ni species. Kramer et al. (1996) found that the production of complexing agents such as amino acid histidine during Ni exposure to plants was responsible for the Ni hyperaccumulation phenotype in *Alyssum*.

According to Laurie et al. (1991) and Molas and Baran (2004), there are mainly two different pathways for the uptake of metal ions from contaminated soils. The first involves the transport of free metal ion (M^{+2}) to cell root across the plasmalemma following the dissociation of metal–ligand complex in the diffuse layer. The second, conversely, involves the absorption of metal–ligand complex by the root cell membrane where the metal–ligand complex is either transported to root cells across plasmalemma, or dissociated in the cell membrane; free metal is then transported to the cell, and the ligand goes back to the solution. The increase in Ni uptake in the presence of EDTA in our case may be explained through a mechanism similar to the second pathway mentioned above. Molas and Baran (2004), however, suggests that Ni is absorbed by plants in the form of a free ion rather than a Ni–ligand complex. The data by several authors shows that in response to excess Ni exposure, the concentration of complexing ligands such as citrate and amino acid histidine increases in hyperaccumulator plants (Kramer et al., 1996; Yang et al., 1997).

While Ni^{2+} and $NiOH^+$ are the dominant Ni species in alkaline soils, the main solution species in acidic soils includes Ni^{2+}, $NiSO_4$ and $NiHPO_4$ (Anonymous, 1993). Kopittke et al. (2007) found that the addition of Ni^{2+} to cowpea significantly reduced plant growth, and the degree of Ni toxicity varied with the Ni concentration applied. Outridge and Scheuhammer (1993) suggest that adverse effects on chlorophyll metabolism and growth occur at soil water concentrations as low as $1\,mg\,Ni\,l^{-1}$. In the presence of complexing ligands such as EDTA, however, the toxic effect of Ni^{2+} on plant growth decreases significantly, most probably due to the formation of less toxic Ni–EDTA complexes (Table 22.10). A two-sided t-test with p-values

TABLE 22.10 Thermodynamic Data for Aqueous Phase Reactions of Ni and EDTA

Reaction	log K $(I=0)$[a]
$Ni^{2+} + H_2O = Ni(OH)^+ + H^+$	−9.86
$Ni^{2+} + 2H_2O = Ni(OH)_2 + 2H^+$	−19.00
$Ni^{2+} + H_2CO_3 = NiCO_3 + 2H^+$	−9.73
$Ni^{2+} + 2H_2CO_3 = Ni(CO_3)_2^{2-} + 4H^+$	−27.8
$EDTA^{4-} + H^+ = HEDTA^{3-}$	10.95
$EDTA^{4-} + 2H^+ = H_2EDTA^{2-}$	17.11
$EDTA^{4-} + 3H^+ = H_3EDTA^-$	20.36
$EDTA^{4-} + 4H^+ = H_4EDTA$	22.58
$Ni^{2+} + EDTA^{4-} = NiEDTA^{2-}$	19.41
$Ni^{2+} + EDTA^{4-} + H^+ = NiHEDTA^-$	22.96
$Ni^{2+} + EDTA^{4-} + H_2O = NiOHEDTA^{3-} + H^+$	6.84

[a]Values from Smith and Martell (2004), corrected to ionic strength $I=0$ using the Davies equation if necessary.

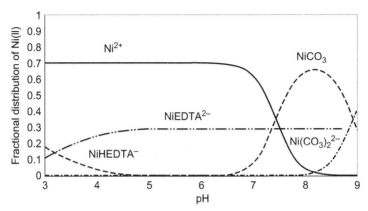

FIGURE 22.4 Distribution of solution phase Ni(II) species in equilibrium with atmospheric CO_2 as a function of pH, including Ni–EDTA complexes for $Ni_T = 1.7 \times 10^{-2}$ and EDTA $= 5 \times 10^{-2}$ M. Equilibrium constants for Ni (II) and EDTA complexes are given in Table 22.10.

less than 0.05 confirmed that the EDTA application had a positive impact on plant growth of the species. Figure 22.4 shows the distribution of aqueous-phase Ni species in the presence of EDTA. The Ni forms strong complexes with EDTA (e.g. $NiEDTA^{2-}$) under acidic to alkaline pH conditions. Molas and Baran (2004) found that Ni toxicity to barley plants is highly dependent on the type of complex formed, with toxicity decreasing in the order: $NiSO_4 > Ni$–citrate $> Ni$–glutamate $> Ni$–EDTA.

Distribution of solution phase Ni (II) species in equilibrium with atmospheric CO_2 as a function of pH, including Ni-EDTA complexes (for $Ni_T = 1.7 \times 10^{-2}$ and EDTA $= 5 \times 10^{-2}$ M) showed equilibrium constants for Ni (II) and EDTA complexes as given in Table 22.10.

Biological Accumulation Coefficient (BAC)

The bioaccumulation factor (BAC) was used to evaluate the effectiveness of a plant in metal accumulation and translocation (Sun et al., 2009). Adding $10 \, mg \, kg^{-1}$ EDTA significantly increased the bioaccumulation coefficient of Ni in *T. elegans* shoots (Table 22.11). As listed in Table 22.10, the BAC values increased with the application of EDTA and ranged from 8.3 to 24.4.

According to the criteria described by Brooks (1998) to define hyper-accumulators, bioaccumulation coefficient (ratio of metal concentration in plant to soil) is greater than 1. Biological accumulation coefficient (BAC) was defined as the concentration of heavy metals in plant shoots divided by the heavy metal concentration in soil [BAC = (Metal) shoot / (Metal) soil] (Zu et al., 2005) and indicate the ability of plants to tolerate and accumulate heavy metals.

TABLE 22.11 Bioaccumulation Coefficient of Ni in *Thlaspi elegans* Boiss. Growing in Artifically Contaminated Soil

Application (mg kg^{-1})		Bioaccumulation Coefficient
Ni	EDTA	
500	0	3.6
1000	0	2.8
1500	0	2.1
2000	0	1.9
500	10	5.6
1000	10	3.6
1500	10	2.7
2000	10	2.0

CONCLUSION

Field and laboratory experiments performed to determine the effects of EDTA on Ni uptake by *T. elegans* for Ni removal from agricultural soils have provided encouraging results for phytoremediaiton. The field tests used to evaluate Ni accumulation capability of the native plant species growing on serpentine soils of Mersin-Findikpinari, Turkey, as well as their tolerance to Ni clearly indicated that among all native species tested, *T. elegans* came out as the most suitable for the phytoremediation of soils because of its higher biomass production and tolerance to Ni.

The laboratory greenhouse experiments performed with the experimental plant indicated that *T. elegans* was more tolerant to high concentrations of Ni. An increasing initial Ni dose (0, 500, 1000, 1500 mg kg^{-1}) adversely affected the growth. The results also indicated that the addition of EDTA at an initial dose of 10 mg kg^{-1} significantly increased Ni uptake in the *T. elegans* plants as compared to the non-EDTA-containing systems. This was most probably due to the formation of highly soluble and less toxic Ni–EDTA complexes, which led to an increase in plant biomass, thereby increasing both Ni tolerance and capacity for Ni transport to the shoot.

The results showed that EDTA increased the solubility of Ni in soil and it can be a good chelator candidate for *T.elegans* when used for environmentally safe phytoextraction of Ni in soils. The results of this study also highlighted further studies with regard to the maximum concentrations of Ni uptake by *Thlaspi elegans* and its potential use as Ni hyperaccumulator from contaminated soils.

Thlaspi species are the hyperaccumulators most extensively studied by the scientific community and are known to hyperaccumulate more than one metal. The species used are *T. caerulescens* for Cd, Ni, Pb and Zn, *T. goesingense* for Ni and Zn, *T. ochroleucum* for Ni and Zn, and *T. rotundifolium* for Ni, Pb and Zn (Prasad and Freitas, 2003). Future studies should include work on the hyperaccumulation of more than one metal by *T. elegans*. The plant could be used for the phytoextraction of Ni-contaminated soils.

Recently some drawbacks of chelator-induced phytoextraction have been reported that most synthetic chelators, such as EDTA, due to potential risk of leaching of metals into groundwater or uptake of crop plants could prove problematic. The application of chelating agents to soil can cause prolonged negative effects upon the growth of plants and soil microfauna and they can be also persistent in the environment due to their poor photo-, chemo- and biodegradability in soil environments (Sun et al., 2009). Most important is that the use of EDTA may result in potential risks of surface and groundwater pollution through uncontrolled solubilization and migration of metals (Quartacci et al., 2006; Sun et al., 2009). Therefore, potential environmental risk should be considered when chelate enhancement is used to improve photoremediation efficiency (Sun et al., 2009). The restrictions concerning chelator-assisted phytoextractionmay can be overcome only when easily biodegradable and low phytotoxic chelating agents are applied. Various strategies have been suggested to raise the efficiency of metal uptake and, at the same time, to reduce the risk of environmental pollution. The approaches to obtain those effects include the use of natural, easily biodegradable compounds such as low-molecular-weight organic acids or EDDS, lowering the concentrations of chelators or splitting dosages, addition of acrylamide hydrogels, clay minerals or apatite mixtures in order to improve soil sorption properties, as well as application of slow-release coated EDTA granules and leachate recirculation (Karczewska et al., 2009).

ACKNOWLEDGEMENTS

This research was supported by a grant from the General Directorate of Agricultural Research (GDAR), Project No: TAGEM/TA/05320C01. We thank Dr Ayse Everest, Emre Ozer and Ersin Ozturk for their assistance in plant identification. We also thank to Dr M.N. V. Prasad, Dr Hatice Dağhan and Abdi Kurt for critical reading of the manuscript and helpful comments.

REFERENCES

Ahmad, M.S.A., Hussain, M., Saddiq, R., Alvi, A.K., 2007. Mungbean: A nickle indicator, accumulator or excluder? Bull. Environ. Contam. Toxicol. 78 (5), 319–324.

Ahmad, K., Khan, Z.I., Ashraf, M., Valeem, E.E., Shah, Z.A., Mcdowell, L.R., 2009. Determination of forage concentrations of lead, nickle, and chromium in relation to the requirements of grazing ruminants in the salt range of Pakistan. Pak. J. Bot. 41 (1), 61–65.

Ahmad, M.S.A., Ashraf, M., 2011. Essential Roles and Hazardous Effects of Nickel in Plants. Rev. Environ. Contam. Toxicol. 214, 125–167.

Alkorta, I., Hernandez-Allica, J., Becerril, J.M., Amezaga, I., Albizu, I., Garbisu, C., 2004a. Recent findings on the phytoremediation of soils contaminated with environmentally toxic heavy metals and metalloids such as zinc, cadmium, lead, and arsenic. Rev. Environ. Sci. Biotechnol. 3, 71–90.

Alkorta, I., Hernández-Allica, J., Becerril, J.M., Amezaga, I., Albizu, I., Onaindia, M., Garbisu, S., 2004b. Chelate-enhanced phytoremediation of soils polluted with heavy metals. Rev. Environ. Sci. Biotechnol. 3, 55–70.

Alloway, B.J., 1995. Soil processes and the behavior of heavy metals. In: Alloway, B.J. (Ed.), Heavy Metal in Soils. Blackie Academic and Professional, London, pp. 25–34.

Anonymous, 1993. Toxicological Profile for Nickel. U.S. Public Health Service, Agency for Toxic Substances and Disease Registry. Report TP-92/14. pp.158, Atlanta, Georgia.

Aytaç, Z., Aksoy, A., 2000. A new Sideritis species (S. ozturkii-Labiatae) from Turkey. Flora Mediterranea I0, 181–184.

Baker, J.M., Brooks, R.R., 1989. Terrestrial higher plants which hyperaccumulate metallic elements—a review of their distribution, ecology and phytochemistry. Biorecovery 1, 81–126.

Bani, A., Echevarria, G., Sulçe, S., Morel, J.L., Mullai, A., 2007. In-situ phytoextraction of Ni by a native population of Alyssum murale on an ultramaphic site (Albania). Plant Soil 293, 79–89.

Barbaroux, R., Mercier, G., Blais, J.F., Morel, J.L., Simonnot, M.O., 2011. A new method for obtaining nickel metal from the hyperaccumulator plant Alyssum murale. Separation Purif. Technol. 83, 57–65.

Bouyoucos, G.J., 1962. Hydrometer method improved for making particle size analysis of soils. Agron. J. 54, 464–465.

Brooks, R.R., 1998. Plants that hyperaccumulate heavy metals: Their role in phytoremediation, microbiology, archaeology mineral exploration and phytomining. CAB International, Wallingford, UK and New York., 380.

Chaney, R.L., Chen, K.Y., Li, Y.M., Angle, J.S., Baker, A.J.M., 2008. Effects of calcium on nickel tolerance and accumulation in Alyssum species and cabbage grown in nutrient solution. Plant Soil 311, 131–140.

Chen, H., Cutright, T., 2001. EDTA and HEDTA effects on Cd, Cr, and Ni uptake by Helianthus annuus. Chemosphere 45, 21–28.

Chen, C., Huang, D., Liu, J., 2009. Functions and toxicity of nickel in plants: Recent advances and future prospects. Clean – Soil Air Water 37 (4–5), 304–313.

Chen, K.F., Yeh, T.Y., Lin, C.F., 2012. Phytoextraction of Cu, Zn, and Pb Enhanced by Chelators with Vetiver (Vetiveria zizanioides): Hydroponic and Pot Experiments. ISRN Ecol. Vol. Article ID 729693, 12 pages http://dx.doi.org/10.5402/2012/729693.

Chin, L., 2007. Investigations into Lead (Pb) Accumulation in Symphytum officinale L.: A Phytoremediation Study. Ph.D. Thesis in Plant Biotechnology, University of Canterbury, ABD.

DIN ISO 11466, 1995. Soil Quality, Extraction of Trace Elements Soluble in Aqua Regia. Berlin, Germany.

Elena, C., Whiting, S.N., Neumann, P.M., Baker, A.J.M., 2005. Salt (NaCl) tolerance in the Ni hyperaccumulator Alyssum murale and the Zn hyperaccumulator Thlaspi caerulescens. Plant Soil 270 (1), 91–99.

Evangelou, M.W.H., Ebel, M., Schaeffer, A., 2007. Chelate assisted phytoextraction of heavy metals from soil. Effect, mechanism, toxicity, and fate of chelating agents. Chemosphere 68 (6), 989–1003.

Garbisu, C., Alkorta, I., 2001. Phytoextraction: A cost-effective plant-based technology for the removal of metals from the environment. Bioresour. Technol. 77 (3), 229–236.

Ghaderian, S.M., Mohtadi, A., Rahiminejad, R., Reeves, R.D., Baker, A.J.M., 2007. Hyperaccumulation of nickel by two Alyssum species from the serpentine soils of Iran. Plant Soil 293, 91–97.

Ghaderian, S.M., Mohtadi, A., Rahiminejad, R., Reeves, R.D., Baker, A.J.M., 2012. Hyperaccumulation of nickel by two *Alyssum* species from the serpentine soils of Iran. Plant Soil 293 (1), 91–97.

Gupta, A., Shaw-Wei, S., Zueng-Sang, C., 2011. Heavy-metal bioavailability and chelate mobilization efficiency in an assisted phytoextraction process by *Sesbania sesban* (L.) Merr. Commun. Soil Sci. Plant Anal. 42 (2), 231–245.

Ha, N.T.H., Sakakibara, M., Sano, S., Hori, R.S., Sera, K., 2009. The Potential of *Eleocharis acicularis* for Phytoremediation: Case Study at an Abandoned Mine Site. Clean – Soil Air Water 37 (3), 203–208.

Haag-Kerwer, A., Schafer, H.J., Heiss, S., Walter, C., Rausch, T., 1999. Cadmium exposure in *Brassica juncea* causes a decline in transpiration rate and leaf expansion without effect on photosynthesis. J. Exp. Bot. 50 (341), 1827–1835.

Herbelin, A.L., Westall, J.C., 1999. FITEQL. A computer program for determination of chemical equilibrium constants (OR). Oregon State University. Department of Chemistry, Corvallis. Version 4.0, Report 99–01.

Jean, L., Bordas, F., Gautier-Moussard, C., Vernay, P., Hitmi, A., Bollinger, J.C., 2008. Effect of citric acid and EDTA on chromium and nickel uptake and translocation by *Datura innoxia*. Environ. Pollut. 153, 555–563.

Kabata-Pendias, A., Pendias H., 2001. Trace Elements in Soils and Plants, third ed. CRC Press, Boca Raton, USA.

Karczewska, A., Galka, B., Kabala, C., Szopka, K., Kocan, K., Dziamba, K., 2009. Effects of arious chelators on the uptake of Cu, Pb, Zn and Fe by maize and Indian mustard from silty loam soil polluted by the emissions from copper smelter. Fresenius Env. Bull. 18 (10a), 1967–1974.

Kazakou, E., Adamidis, G.C., Baker, A.J.M., Reeves, R.D., Godino, M., Dimitrakopoulos, P.G., 2010. Species adaptation in serpentine soils in Lesbos Island (Greece) metal hyperaccumulation and tolerance. Plant Soil 332, 369–385.

Kirkham, M.B., 1977. Trace elements sludge on land: Effect on plants, soils, and ground water. In: Laehr, R.C. (Ed.), Land as a Waste Management Alternative. Ann Arbor Science Publishers, Ann Arbor, MI, pp. 209–247.

Koleli, N., Halisdemir, B., 2005. Distribution of chromium, cadmium, nickel and lead in agricultural soils collected from Kazanli-Mersin. Int. J. Environ Pollut. 23, 409–415.

Koleli, N., Altındisli Atag, G., Kusvuran, K., Kantar, C., Demir, A., Binzet, R., Eke, M., 2008a. The study of metal hyperaccumulator in Mersin-Findikpinarii. Mersin Symposium, Mersin, Turkey. 171–179.

Koleli, N., Altındisli Atag, G., Kusvuran, K., Kantar, C., Demir, A., Seyhanlı, İ., 2008b. Heavy metal contents of soils in Mersin-Findikpinari. Mersin Symposium, Mersin. pp. 160–170, Turkey.

Kopittke, P.M., Asher, C.J., Menzies, N.W., 2007. Toxic effects of Ni^{2+} on growth of cowpea (*Vigna unguiculata*). Plant Soil 292, 283–289.

Kramer, U., Cotter-Howells, J.D., Charnock, J.M., Baker, A.J.M., Smith, J.A.C., 1996. Free histidine as a metal chelator in plants that accumulate nickel. Nature 379, 635–638.

Kruckeberg, A.R., Adiguzel, N., Reeves, R.D., 1999. Glimpses of the flora and ecology of Turkish (Anatolian) serpentines. The Karaca Arboretum Magazine 5 (2), 67–86.

Kumar, A., Maiti, S.K., 2013. Availability of chromium, nickel and other associated heavy metals of ultramafic and serpentine soil/rock and in plants. Int. J. Emerg. Technol. Adv. Eng. 3 (2), 256–268.

Lai, H.Y., Chen, Z.S., 2005. The EDTA effect on phytoextraction of single and combined metals-contaminated soils using rainbow pink (*Dianthus chinensis*). Chemosphere 60, 1062–1071.

Laurie, S.H., Tancock, N.P., McGrath, S.P., Sanders, J.R., 1991. Influence of complexation on the uptake by plants of iron, manganese, cooper and zinc: II. Effect of DTPA in a multi-metal and computer stimulation study. J. Exp. Bot. 42 (4), 515–519.

Lindsay, W.L., Norvell, W.A., 1978. Development of a DTPA Test for Zinc, Iron, Manganese, and Copper. Soil Sci. Soc. Am. J. 42, 421–428.

McGrath, S.P., Zhao, F.J., Lombi, E., 2001. Plant and rhizosphere processes involved in phytoremediation of metal-contaminated soils. Plant Soil 232, 207–214.

McLean, E. O. (1982) in Methods of Soil Analysis (Eds.: A. L. Page, R. H. Miller, D. R. Keeney), Agronomy Monograph No. 9. second ed., American Society of Agronomy, Madison, USA. 595-624

Meers, E., Hopgood, M., Lesage, E., Vervaeke, P., Tack, F.M.G., Verloo, M.G., 2004. Enhanced Phytoextraction: In Search of EDTA Alternatives. Int. J. Phytoremediation 6 (2), 95–109.

Memon, A.R., Ito, S., Yatazawa, M., 1980. Taxonomic characteristics in accumulating cobalt and nickle in temperate forest vegetation of central Japan. Soil Sci. Plant Nutr. 26, 271–280.

Molas, J., Baran, S., 2004. Relationship between the chemical form of nickel applied to the soil and its uptake and toxicity to barley plants (Hordeumvulgare L.). Geoderma 122, 247–255.

Nelson, D. W., and Sommers, L. E. (1982) In "Methods of soil analysis" (A. L. Page, R. H. Miller, D. R. Keeney, Ed.), Agronomy Monograph No. 9. second ed., American Society of Agronomy, Madison, USA. 539-577

Orcan, N., Binzet, R., Yaylalioglu, E., 2004. The flora of Findikpinari (Mersin-Turkey) plateau. Flora Medit., 309–345.

Outridge, P.M., Scheuhammer, A.M., 1993. Bioaccumulation and toxicology of nickel: Implications for wild mammals and birds. Environ. Rev. 1, 172–197.

Panwar, B.S., Ahmed, K.S., Mittal, S.B., 2002. Phytoremediation of nickel-contaminated soils by Brassica species. Environ. Dev. Sustain. 4, 1–6.

Prasad, M.N.V., Freitas, H., 2003. Metal hyperaccumulation in plants – Biodiversity prospecting for phytoremediation technology. Electron. J. Biotechnol. 6, 275–321.

Prasad, M.N.V., 2005. Nickelophilous plants and their significance in phytotechnologies. Braz. J. Plant Physiol. 17 (1), 113–128.

Purakayastha, T.J., Viswanath, T., Bhadraray, S., Chhonkar, P.K., Adhikari, P.P., Suribabu, K., 2008. Phytoextraction of zinc, copper, nickel and lead from a contaminated soil by different species of Brassica. Int J. Phytoremed. 10, 61–72.

Quartacci, M.F., Argilla, A., Baker, A.J.M., Navari-Izzo, F., 2006. Phytoextraction of metals from a multiply contaminated soil by Indian mustard. Chemosphere 63, 918–925.

Reeves, R.D., 1988. Nickel and zinc accumulation by species of Thlaspi L., Cochlearia L. and other genera of the Brassicaceae. Taxon 37, 309–318.

Reeves, R.D., Adiguzel, N., 2004. Rare plants and nickel accumulators from Turkish serpentine soils, with Special Reference to Centaurea Species. Turk. J. Bot. 28, 147–153.

Reeves, R.D., Baker, A.J.M., Borhidi, A., Berazaín, R., 1999. Nickel hyperaccumulation in the serpentine flora of Cuba. Ann. Bot. 83, 29–38.

Reeves, R.D., Brooks, R.R., 1983. European species of Thlaspi L. (Cruciferae) as indicators of nickel and zinc. J. Geochem. Explor. 18, 275–283.

Reeves, R.D., Kruckeberg, A.R., Adiguzel, N., Kramer, U., 2001. Studies on the flora of serpentine and other metalliferous areas of western Turkey. S. Afr. J. Bot. 97, 513–517.

Sabir, M., Hakeem, K.R., Aziz, T., Zia-ur-Rehman, M., Ozturk, M., 2013. High Ni levels in soil can modify growth performance and mineral status of wheat cultivars. Clean – Soil Air Water. http://dx.doi.org/10.1002/clen.201300352.

Sakakibara, M., Ahmori, Y., Ha, N.T.H., Sano, S., Sera, K., 2011. Phytoremediation of heavy metal-contaminated water and sediment by Eleocharis acicularis. Clean – Soil Air Water 39 (8), 735–741.

Salt, D.E., Kato, N., Krämer, U., Smith, R.D., Raskin, I., 2000. The role of root exudates in nickel hyperaccumulation and tolerance in accumulator and nonaccumulator species of *Thlaspi*. In: Terry, N., Banuelos, G. (Eds.), Phytoremediation of Contaminated Soil and Water. Lewis Publishers, Boca Raton, FL, pp. 189–200.

Shallari, S., Schwartz, C., Hasko, A., Morel, J.L., 1998. Heavy metals in soils and plants of serpentine and industrial sites of Albania. Sci. Total. Environ. 209 (2–3), 133–142.

Sharma, H., 2011. Metal Hyperaccumulation in plants: A review focusing on phytoremediation technology. J. Environ. Sci. Technol. 4 (2), 118–138.

Sreekanth, T.V.M., Nagajyothi, P.C., Lee, K.D., Prasad, T.N.V.K.V., 2013. Occurrence, physiological responses and toxicity of nickel in plants. Int. J. Environ. Sci. Technol. 2013 (10), 1129–1140.

Smith, R.M., Martell, A.E., 2004. NIST critical stability constants of metal complexes database. NIST Standard Reference Database 46. U.S. 891 Department of Commerce, Gaithersburg, MD.

Sun, Y.B., Zhou, Q.X., An, J., Liu, W.T., Liu, R., 2009. Chelator-enhanced phytoextraction of heavy metals from contaminated soil irrigated by industrial wastewater with the hyperaccumulator plant (*Sedum alfredii* Hance). Geoderma 150 (1–2), 106–112.

Tumi, A.F., Mihailović, N., Gajić, B.A., Marjan, N., Gordana, T., 2012. Comparative study of hyperaccumulation of nickel by *Alyssum murale* s.l. Populations from the Ultramafics of Serbia. Polish J. Environ. Stud. 21 (6), 1855–1866.

Turan, M., Esringu, A., 2007. Phytoremediation based on canola (*Brassica napus* L.) and Indian mustard (*Brassica juncea* L.) planted on spiked soil by aliquot amount of Cd, Cu, Pb, and Zn. Plant Soil Environ. 53 (1), 7–15.

Turgut, C., Pepe, M.K., Cutright, T.J., 2004. The effect of EDTA and citric acid on phytoremediation of Cd, Cr, and Ni from soil using *Helianthus annuus*. Environ. Pollut. 131, 147–154.

Wenzel, W.W., Jockwer, F., 1999. Accumulation of heavy metals in plants grown on mineralized soils of the Austrian Alps. Environ. Pollut. 104, 145–155.

Wenzel, W.W., Unterbrunner, R., Sommer, P., Sacco, P., 2003. Chelate assisted phytoextraction using canola (*Brassica napus* L.) in outdoors pot and lysimeter experiments. Plant Soil 249, 83–89.

Wuana, R.A., Okieimen, F.E., 2011. Heavy metals in contaminated soils: A review of sources, chemistry, risks and best available strategies for remediation. ISRN Ecol. http://dx.doi.org/10.5402/2011/402647.

Yang, X.E., Baligar, V.C., Foster, J.C., Martens, D.C., 1997. Accumulation and transport of nickel in relation to organic acids in ryegrass and maize grown with different nickel levels. Plant Soil 196, 271–276.

Zu, Y.Q., Li, Y., Chen, J.J., Chen, H.Y., Qin, L., Schvartz, C., 2005. Hyperaccumulation of Pb, Zn and Cd in herbaceous grown on lead–zinc mining area in Yunnan, China. Environ. Int. 31, 755–762.

Chapter 23

Phytomanagement of Padaeng Zinc Mine Waste, Mae Sot District, Tak Province, Thailand

M.N.V. Prasad,* Woranan Nakbanpote,† Abin Sebastian,* Natthawoot Panitlertumpai† and Chaiwat Phadermrod‡

*Department of Plant Sciences, University of Hyderabad, Hyderabad, India; †Department of Biology, Faculty of Science, Mahasarakham University, Khamriang, Kantarawichi, Mahasarakham, Thailand; ‡Padaeng Industry Public Co. Ltd, Phratad Padaeng, Mae Sot, Tak, Thailand

INTRODUCTION

Thailand is rich in metallic minerals of zinc (Zn), lead (Pb), tin (Sn) and gold (Au). Rich sources (20–30%) of zinc have been found since 1947 in Doi Phadaeng, Phatat Phadaeng sub-district, Mae Sot district, Tak Province, northwest of Thailand (Figure 23.1). Padaeng Industry was established on April 10, 1981 and mining activities started on April 30, 1982. Padaeng Industry is engaged in zinc mining and smelting with the primary objective of producing zinc metal (67,000 tons per year based on the last 10 years average) and zinc alloys (101,100 tons per year based on the last 10 years average). The most common uses of zinc are shown in Figure 23.2 (Prasad, 2002). Zinc is extracted electrolytically from zinc-containing solutions prepared by crushing and leaching zinc ores. The zinc production process is shown in Figure 23.3. High levels of Cd and Zn have been reported in sediment samples from the creek sand, paddy field areas and rice grain in the vicinity of the mining during surveys in 2001–2004 (National Research for Environmental and Hazardous Waste Management, 2005; Pollution Control Department, 2004; Simmons et al., 2005). The average levels of both Cd and Zn in the soil and rice posed a health risk to the public. Cd was higher along the shore of the Mae Tao creek than further inland. The amount of Cd in the soil ranged from 3.4–284 mg kg^{-1} of soil, which was between 1.13 and 94 times higher than allowed under the European Community's regulations (3 mg kg^{-1} of soil). The total of Zn and Cd in agricultural soil in the vicinity of the zinc mine and tailing soil are shown in Table 23.1. The range of Cd that was found in rice grain was 0.1–0.4 mg kg^{-1}

Soil Remediation and Plants. http://dx.doi.org/10.1016/B978-0-12-799937-1.00023-1

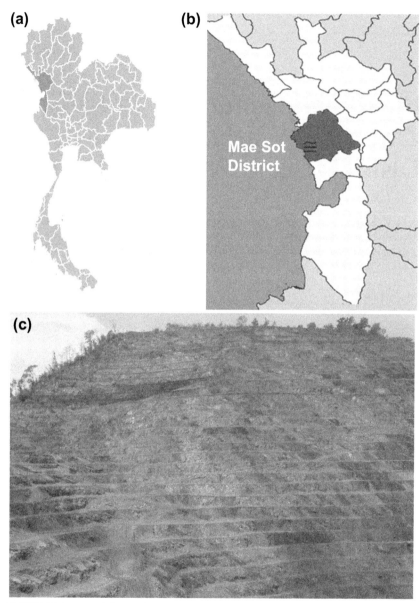

FIGURE 23.1 (a) and (b) Map showing the location of Padaeng zinc mine in Mae Sot district of the Tak provice of Thailand's north-west region; (c) Zinc-rich ore removed from mine area—a small hillock of terraces prone to soil erosion due to rains.

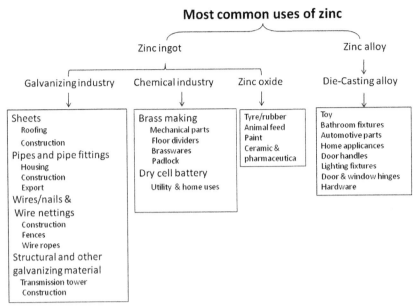

FIGURE 23.2   Most common uses of zinc. (Prasad, 2002)

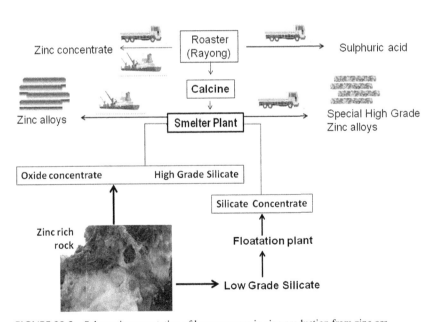

FIGURE 23.3   Schematic presentation of key processes in zinc production from zinc ore.

TABLE 23.1 Zn and Cd Contamination in and around Mae Sot Mine during 2013

Site	Cd mg kg⁻¹	Zn mg kg⁻¹
Agricultural soil in the vicinity of zinc mine	3–85	306–4883
Tailing soil	1660–2777	39,062–89,881

See also Phaenark et al. (2009) for additional information.

(Simmon et al., 2005, 2009). The results were also over the limit for Cd in rice grain which is $0.2 \, mg \, kg^{-1}$ (Codex, 2005; Ikeda et al., 2004). The health impacts of Cd overexposure on the Mae Sot population have been reported since 2007 (Swaddiwudhipong et al., 2007, 2010, 2012; Teeyakasem et al., 2010; Limpatanachote et al., 2009). Cultivation of various crops such as rice, corn, soy beans, garlic and shallots is the livelihood for nearby villagers. About 5295 acres of land in the 12 rural villages near the mines have been declared as unsafe for growing food crops. The villagers were advised not to cultivate rice as the arable land contained high levels of cadmium, a 'guest' metal in exploration of Zn. During 2007–2009, the government purchased the rice grown in these contaminated areas for destroying and supported the production of non-food crops (Svasti, 2007).

Prolonged excessive oral exposure can cause chronic cadmium poisoning such as anemia, hypertension, cancer, cardiac failure, cerebrovascular infarction, emphysema, proteinuria, serious damage to lungs, renal dysfunction, cataract formation in the eyes and osteoporosis. Thus, the arable lands in the vicinity of the mineral-rich areas of Phatat Phadaeng sub-district, Mae Sot, Tak Province posed serious health implications to humans. The government has supported the production of non-food crops, and heaps of mine tailings have to be decontaminated to recover land and to prevent erosion and leaching of Zn and Cd from the contaminated soil. The chemical–physical technologies for the decontamination of contaminated ecosystems are complex and expensive and may cause undesirable side effects for the environment that then must in turn be decontaminated. Thus, phytomanagement and related agronomic practices are considered to be cost effective for this problem. Several of the small holding farmers in the vicinity of zinc mines cultivate 'Thai jasmine rice'. Therefore, phytomanagement and phytotechnologies are considered to be the feasible options for minimization of Cd in rice and are presented in this chapter with specific references to Phadaeng zinc mine. Padaeng industry also procures raw materials from other countries (silicate ore, oxide and sulphide concentrate) to produce zinc ingots and zinc alloys (Figure 23.3).

PHYTOMANAGEMENT OF A ZINC-MINE-INDUSTRY-RAVAGED ECOSYSTEM

Metallophytes are metal-tolerant plants that successfully complete their life-cycle in metal-rich soils. The tolerance to extreme metal-rich concentrations makes them ideal for a variety of applications including mine reclamation and mine waste stabilization (Prasad, 2013). Thus, metallophytes form a key resource for the mine industry, in mineral exploration as well as reclamation (Prasad, 2013).

Phadaeng zinc mine is located in a mixed deciduous reserved forest in Thailand, one of the hot-spots for tropical biodiversity. The ore from this mine is rich in zinc silicate [Hemimorphite Zn_4 $(Si_2O_7)(OH)_2.2H_2O$] with a marginal zinc carbonate (Smithsonite $ZnCO_3$) (Pollution Control Department, 2004). However, the top soil (20–30 cm) contains high levels of cadmium along with zinc (Table 23.1) Therefore, plants that are suitable for phytomanagement should be selected from the native flora (Figure 23.4). During phytosociological investigations of the mine area, we have observed native trees, shrubs, herbs and grasses growing luxuriantly as shown in Table 23.2.

Vertiveria zizanioides (vertiver grass) was planted in 2003 on mine waste heaps as a cover crop to prevent soil erosion and improve soil quality (Figures 23.5 and 23.6). The root of *V. zizanioides* can deeply plunge into soil. (Chiu et al., 2006; Chen et al., 2004; Roongtanakiat et al., 2001, 2008, 2011; Truong, 1999). It is known to have multiple uses. Vetiver grass was first introduced as an ideal grass for the Mae Sot zinc mine rehabilitation programme as it had excellent capability to control soil erosion prevention and soil quality improvement. More than 17.5 million Vetiver plants have been planted at the Mae Sot zinc site, an example of Thailand's single largest concentration of vetiver plantation. This extraordinary grass is adaptable to multiple environmental conditions and is globally recognized as an easy and economical alternative to control soil erosion and to solve a variety of environmental problems. The grass has been successfully used to stabilize mining overburden and highly saline, sodic, magnesic and alkaline (pH 9.5) tailings of coalmines, as well as highly acidic (pH 2.7) arsenic tailings of gold mines. It is one of the best choices for revegetation of Pb/Zn mine tailings in China due to its high metal tolerance (Chen et al., 2004; Chiu et al., 2006; Chong and Chu, 2007; Rotkittikhun et al., 2007a,b; Makris et al., 2007; Pang et al., 2003; Singh et al., 2007; Truong, 2000; Wilde et al., 2005).

Legumes are well suited to planting on harsh, degraded sites as they are fast-growing, readily established and managed and often nitrogen fixing (Hughes and Styles, 1987). The heap of mine tailing and mined area was successfully rehabilitated with perennial and fast-growing trees native to the neighborhood, such as *Acacia mangium* (Family: Fabaceae), *Leucaena leucocephala* (Family: Fabaceae), *Xylia xylocarpa* (Family: Fabaceae), *Gmelina arborea* (Family: Lamiaceae), *Bauhinia variegata* (Family: Fabaceae),

FIGURE 23.4   (a) Zinc-tolerant native plant and (b) cryptogams on zinc-rich rock. For a colour version of this figure, please see the section at the end of this book.

Duabanga grandiflora (Family: Lythraceae), *Azadirachta indica* (Family: Meliaceae), *Alstonia scholaris* (Family: Apocynaceae), *Musa acuminate* (Family: Musaceae) and *Bambusa bambos* (Family: Poaceae).

   Gynura pseudochina (Figure 23.7a–e) is a perennial herbaceous native Cd/Zn hyperaccumulator (Phaenark et al., 2009; Panitlertumpai et al., 2013). Its capability to accumulate high Zn and Cd concentration is attributed to sulphur-rich proteins. Zn and Cd accumulation in this plant are dependent on the metal concentrations and exposure time (Panitlertumpai et al., 2008, 2013).

   Murdannia spectabilis (Figure 23.7f) is frequently found near *G. psudochina* and accumulates high quantities of Zn and Cd (Panitlertumpai et al., 2003; Rattanapolsan et al., 2013). It adapted leaf surface anatomy when grown in Zn/Cd-contaminated soil, which might be an adaptive mechanism of Zn and Cd detoxification (Rattanapolsan et al., 2013).

   Cyperus rotundus was found growing over the zinc mine (Panitlertumpai et al, 2003). Sao et al. (2007) reported that *C. rotundus* accumulated cadmium in root and shoot. The mechanism of cadmium accumulation by grasses involves cadmium precipitation in the stable form of cadmium silicate. This plant could be grown to prevent soil erosion and to remediate cadmium-contaminated soil.

   Siam weed (*Chromolaena odorata*) and vetiver grass (*Vetiveria zizanioides*) are applied as cover plants at Mae Sot mine (Figure 23.6). Although vetiver grass performed Cd accumulation in its roots, siam weed accumulated Cd especially in the aboveground parts (Sampanpanish et al., 2008). Therefore, native

TABLE 23.2 Predominant Native Plants in Mae Sot Zinc Mine Waste That Serve as Raw Materials for Phytomanagement and Rehabilitation

Genus	Scientific Name
Acanthaceae	*Justicia procumbens* Herb
	Justicia sp. Herb
Amaranthaceae	*Aerva sanguinolenta* Herb
Aracae	*Colocasia esculenta* Herb
Asteraceae	*Ageratum conyzoides* Shrub
	Bidens biternata Herb
	Blumea mollis Herb
	Blumea napifolio Herb
	Chromolaena odoratum Shrub
	Conyza sumatrensis Herb
	Crassocephalum crepidioides Herb
	Grangea maderaspatina Herb
	Gynura pseudochina Herb
	Laggera pteradonta Herb
	Sonchus arvensis Herb
	Spilanthes iahadicensis Herb
	Wedelia trilobata Herb
Balsaminaceae	*Impatiens violaeflora* Herb
Boraginaceae	*Heliotropium indicum* Herb
Buddlejaceae	*Buddleja asiatica* Shrub
Cyperaceae	*Kylinga brevifolia* Sedge
	Cyperus rotundus Sedge
Commelinaceae	*Cyanotis tuberosa* Herb
	Murdannia spectabilis Herb
Euphorbiaceae	*Euphorbia hirta* Herb
Fabaceae	*Aeschynomene americana* Herb
	Crotalaria montana Herb
Lamiaceae	*Hyptis suaveolens* Shrub

Continued

TABLE 23.2 Predominant Native Plants in Mae Sot Zinc Mine Waste That Serve as Raw Materials for Phytomanagement and Rehabilitation—cont'd

Genus	Scientific Name
Onagraceae	*Ludwigia hyssopifolia* Herb
Poaceae	*Brachiaria sp.* Grass
	Imperata cylindrica Grass
	Neyraudia arundinacea Grass
	Thysanolaena maximaremota Grass
	Vertiveria zizanioides Shrub
Rubiaceae	*Spermacoce remota* Herb
	Rubia sp. Herb
Scrophulariaceae	*Lindenbergia philippensis* Under Shrub
	Scoparia dulcis Herb
Zigiberaceae	*Kaempferia marginata* Herb

Also see Phaenark et al. (2009) for additional information.

flora such as *G. pseudochina*, *M. spectabilis*, *V. zizanioides* and *Chromolaena odorata*, (Asteraceae), are ideal for the Mae Sot rehabilitation programme. A metal-tolerant weed has the capability to phytoremediate metal-contaminated soil (Atagana, 2011, Tanhan et al., 2007).

Additionally tree, legume, bamboo and grass association is ideal for stabilization of mine waste allowing the growth of the understorey flora (Figures 23.8 and 23.9).

The success of large-scale mine waste reclamation depends upon in-house nursing of native plants capable of growing on mine waste. The Padaeng mine industry therefore established a large nursery with huge stocks of seedlings. Furthermore, the nursery has its own vermicompost and biofertilizer units to meet demand (Figure 23.10a–f).

PHYTOMANAGEMENT FOR SUSTAINABLE AGRICULTURE IN THE VICINITY OF MAE SOT ZINC MINE

Phytoremediation is the process through which contaminated land is ameliorated by growing plants that have the ability to remove the contaminating chemicals. Although relatively slow, phytoremediation is environmentally friendly, cheap, requires little equipment or labour, is easy to perform, and sites can be cleaned without removing the polluted soil; it is an in situ method. In addition,

FIGURE 23.5   (a) Zinc mine waste as over burden; (b) Vetiveria species being planted on zinc-rich soils to control soil erosion.

precious metals, such as gold, zinc and chromium, collected by the hyperaccumulator can be harvested and extracted as phytoextraction. Also, plant design for successful phytoremediation in any contaminated area should be concerned with ground water, run off, geology and the effect of growing the plant on

FIGURE 23.6 (a-e) Vetiveria grass stem stumps intermingled with *Chromolena odorata* association; (f) *Chromolena odorata* growing in metal-rich environment. For a colour versiosn of this figure, please see the section at the end of this book.

biodiversity. In particular, harvesting management and by-product utilization should be studied and investigated to convince local people and government of the usefulness of phytoremediation (Raskin and Ensley, 2000).

Cd contamination in arable land and agricultural products has been reported (Simmons et al., 2005, 2009). An alternative promotional campaign was instituted for rice farmers to cultivate sugarcane [*Saccharum officinarum* (L.)] in place of paddy cultivation. Sugar cane produce was diverted for ethanol production by Maesot Clean Energy Co. Ltd. The factory requirement is 5000 tons of sugar cane per day (Svasti, 2007).

FIGURE 23.7   (a-e) *Gynura pseudochina* (L.) DC. (Wan-Maha-Kan), a tuber bearing Asteraceae. It grows luxuriantly in high zinc and cadmium-contaminated soils. (f) *Murdannia spectabilis*, a tuber bearing Commelinaceae. They also accumulate zinc and cadmium. For a colour version of this figure, please see the section at the end of this book.

FIGURE 23.8   (a) *Doubanga grandiflora* (Native tree); (b) *Phyllostachys mannii* Bamboo; and (c) *Leucaena leucocephala* stands on zinc-rich mine soils. For a colour version of this figure, please see the section at the end of this book.

Nakbanpote et al. (2010) reported that polyphenolic compounds containing antioxidant properties, especially rutin and caffeine, could be extracted from the leaves of *Gynura pseudochina*, a Zn/Cd hyperaccumulative plant in the Astera-ceae family. The zinc concentration in the extracts were within the legal level

Vetiveria zizinoides
Imperata cylindrica
Thysanolaena maximarenmota

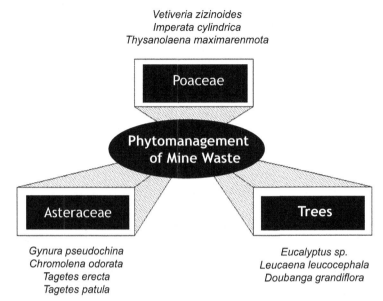

Poaceae

Phytomanagement
of Mine Waste

Asteraceae

Trees

Gynura pseudochina
Chromolena odorata
Tagetes erecta
Tagetes patula

Eucalyptus sp.
Leucaena leucocephala
Doubanga grandiflora

FIGURE 23.9   Tree (preferably legume), grass and asteraceae association, which allows the growth of the understorey flora, is ideal for stabilization of Padaeng zinc mine waste.

for human consumption. Then unneeded biomass could be turned to ash in an incinerator before zinc extraction using sulphuric acid as in the mining process. Plants that can be used for renewable energy or biodiesel that can also be grown in highly Cd-contaminated areas could be an alternative solution to these problems. Physic nut [*Jatropha curcas* (L.)] can be used in oil production because its oil content is 34% and the extraction process is not complicated (Suppadit et al., 2012). Kitisin et al. (2013) investigated the potential antioxidative activity of crude rice oil extracted from cadmium-contaminated rice as an alternative for utilizing cadmium-contaminated rice without compromising public health with hazard risks. This study demonstrated that crude rice oil extracted from cadmium-contaminated rice exhibited antioxidative activity, which can be used for pharmaceutical and cosmeceutical use.

Some rice farmers, however, were reluctant to follow this transition and insisted on growing their rice for consumption and income (Swaddiwudhipong et al., 2010). Others whose land plots were not fit for sugarcane continued growing rice crops. Therefore, many Thai scientists have carried out research to decrease the cadmium accumulated in rice growing in Cd-contaminated land. Sumpanpanish and Pongpaladisai (2011) studied organic fertilizer effects on Cd uptake by rice grown in contaminated soil. The four Thai rice varieties viz. Khao Dawk Mali 105, RD6, Phitsanulok and Niaw San-Pa-Tong were cultivated in contaminated soil using cow dung as organic fertilizer. Cow dung supplementation reduced Cd uptake. Sukreeyapongse

FIGURE 23.10 (a, b) Locally available degradable waste is used for vermin-compost production; (c–f) stocks of seedlings of different plant species being maintained in a nursery on a mine site for use in phytomanagement of zinc-contaminated soil.

et al. (2010) grew Khao Dawk Mali 105 by the irrigation and soil washing method. Soil management methods can reduce Cd concentration in rice grains. Rice accumulation of less Cd can be explained by the following hypotheses: (1) in submerged paddy soil sulphate ions are reduced to sulphide resulting in the precipitation of the Cd minerals and rice will uptake less Cd; (2) after soil washing, Cd concentration in soil will decrease and rice will uptake less Cd; and (3) biochar and agricultural residues such as corncob and coir pith, can reduce Cd accumulation in the shoot part of rice (Siswanto et al., 2013; Suppadit et al., 2012).

FEASIBLE OPTIONS FOR THE MANAGEMENT OF ARABLE LANDS MINE TAILING WATER

Since 2007, Mae Sot mine have had to reform their environmental management according to the resolution of the Office of National Environment Board. A series of fourteen sedimentation ponds, check dams and widening canals were built to protect irrigation and run-off of soil from the mining area to Mae Tow Creek. In addition, some flocculants have been applied for increased precipitation during the raining season. The water in the flotation process has been reused in the process, and not drained into the sedimentation ponds. Heaps of soil obtained from opening pit and mine tailing have mine water leach being rich in heavy metals such as Cd. Removal of heavy metal from mine water outlets is the most important step in minimizing heavy metals in food crops irrigated with this water. Mine water leachate must be subjected to treatments that will either precipitate heavy metal or filter out the metals. The very first step to cut down heavy metal load in the water reservoir and canal-based irrigation method must be followed in these regions. This will allow monitoring of the water bodies used for irrigation purposes for heavy metal pollution. This method also allows natural precipitation reactions of heavy metals and hence decreases the amount of these metals in the irrigation water. The physiochemical properties of heavy metal allows various precipitation reactions (Figure 23.11). Once the precipitate is formed, these metal salts can be used either for recycling of individual metals or discarded from the precipitation canal for environmental friendly disposal methods such as land filling (Diels et al., 2002; Wuana and Okieimen, 2011). Thus the addition of chemical species that react with heavy metal is promising to remove these metals from irrigation water. Once the precipitate/suspension containing these precipitates is formed, it can be treated for sedimentation or filtering using a membrane system. The precipitation reaction can also be performed with organic materials such as powder prepared from biomass of weeds or charcoal made from weeds which will be able to bind the metal; and after a period of sedimentation these materials can be utilized for land filling (Lalhruaitluanga et al., 2010; Meitei and Prasad, 2013). This method is the cheapest because of the plentiful availability of the raw material for metal absorption. It must be noted that application of organic material needs to be monitored because the degradation of the material leads to loss of metal holding capacity. Electrochemistry-based methods are permissible with regard to separation of metal species from mine lechate (Meunier et al., 2006). But these methods are not economically feasible and hence rarely followed. Another approach is the usage of columns having adsorbents which are able to bind heavy metals. Once the water flows to these columns, it will be able to bind the heavy metals in the water. But efficiency of this method is low compared to a water bund barrier system that can be applied in the water channels. It has been reported that charcoal is efficient at binding with heavy metals (Lalhruaitluanga et al., 2010). Production of

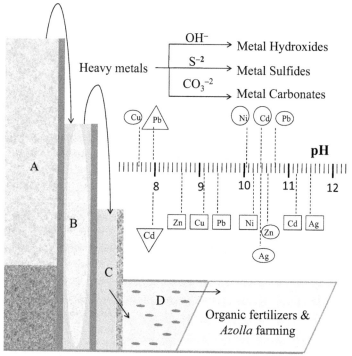

A – Natural metal precipitation, B – Chemical precipitation,
C – Filter bunds, D – Rhizofiltration, E – In situ field remediation

FIGURE 23.11 Barrier model for irrigation of crops near mine tailings: Bunds (A, B, C) act as metal precipitation or filtration chambers where (A) represents blocking of irrigation water without any chemical treatments that allows natural precipitation of heavy metals along with organic matter sedimentation; (B) represents metal precipitation by chemical treatments; and (C) represents the barrier that filters water to remove heavy metals. Rhizofiltration (D) further ensures purity of water in a sustainable and economically feasible method. Along with these practices, addition of organic matters which will biosorb heavy metals and co-cultivation of azolla which will accumulate metals in the plant body ensures reduction in heavy metal uptake by crop plants such as rice. Heavy metal undergoes precipitation reaction with hydroxide, sulfide and carbonate ions and is pH dependant as indicated by the pH scale in the figure. Thus addition of chemical compounds which supply these ions promise heavy metal precipitation in the irrigation water. NB: Circle, square and triangle represents hydroxides, sulphides and carbonates.

charcoal from low cost woody species is economically feasible too. A sand–charcoal combination is ideal to create a charcoal–sand filtering bund system which will control the flow of water into the field as well as filtering heavy metals (Figure 23.11). Other metal binding compounds such as perlite or diatomaceous earth which are known to bind heavy metals also can be utilized for making this kind of bund system (Torab-Mostaedi et al., 2010; Khraisheh et al., 2004). Once the barrier is saturated with metals it must be removed and the metals eluted from it using eluting agents and hence these barrier filters will be reuseable. A method which combines metal precipitation reactions and barrier filtration is highly efficient in removing heavy metals from mine water leachate. Traditional rice growing belts are characterized by heavy rain fall. Hence rain fall harvesting technology and utilization of rain water for irrigation of fields completely eliminates the need for mine water leachate for irrigation purposes. This will cut down the cost of water purification practices to a large extent. Since the mining industry also produces atmospheric emissions of heavy metals, the rain water collected must be checked for metal content (Pacyna et al., 2007). Apart from water purification methods, water management in rice fields is highly efficient at eliminating heavy metal accumulation in rice. Maintaining the water level in the field has been described as a crucial step in minimizing heavy metal uptake. In fact management of water brings control of redox potential in the field and hence control of redox dependent metal uptake by the plant (Sebastian and Prasad, 2014a). Flooding of the field ensures sorption of heavy metal to Al/Fe/Mn oxy-hydroxides or precipitation by forming carbonates or sulphide compounds (Arao et al., 2009; Chlopecka, 1996). In short, purification of mine water leachate and control of water input in the field play significant roles in the minimization of heavy metals in food crops.

SOIL REMEDIATION

Soil remediation allows crops to grow in metal-polluted sites free from metal accumulation. Soil remediation may use either ex situ or in situ methods (Boopathy, 2000). Soil dressing is an ex situ method of soil remediation where a new layer of soil is deposited on the top of a contaminated layer of soil. This method can be further improved by removing the contaminated soil layer and layering the unpolluted soil. This method requires huge investment and hence its application is limited. Compared with ex situ methods, in situ methods such as soil washing, phytoremediation and application of soil amendments are economically feasible. Soil washing is performed by subjecting soil to various treatments such as washing with metal chelators that wash out metal from the soil. The hard pan beneath the sub surface layer of paddy fields which is impermeable to water makes soil washing successful in rice fields because this prevents ground water pollution during the washing process. Chloride, nitrate or sulfate salts of calcium, iron, magnesium, manganese and potassium are the major soil washing chemicals (Bolan et al., 2013). These washing agents not only wash out toxic heavy metals such as Cd but also replenish the pool of essential nutrients in the field. Addition of these metal salts brings chemical changes in the soil by forming metal hydroxides that release hydrogen ions in the soil and hence reduction of soil pH. Hence after the process of soil washing, an alkaline treatment with calcium hydroxide is preferred. A typical in situ soil washing procedure involves chemical washing using micro or macronutrient salts mixed into the soil followed by washing with water and waste water treatment. The first step removes Cd from the soil by cation exchange process and the latter two steps remove Cd from the field and purify the released water, respectively. Application of metal chelators such as EDTA (Ethylenediaminetetraacetic acid) is also useful in the soil washing process but the lack of biodegradability is a disadvantage of this chemical compound. Organic acids such as malic acid, citric acid and tartaric acid are also metal complexing agents but field level application needs field trails since these acids releases hydrogen ions in the field (Wasay et al., 2001). In the case of semi-aquatic crop cultivating systems, cultivation of aquatic plants such as azolla together with the crop further reduces the chance of metal accumulation in the crop (Sood et al., 2012). It must be noted that chemical washing may cause reduction in micronutrients too. Thus micronutrients must be replenished by fertilizer application.

Phytoremediation has been successfully employed as a soil remediation step, especially in areas contaminated due to mining processes (Prasad et al., 2010). Phytoremediation using hyperaccumulator plants brings down soil heavy metal concentration and allows cultivation of crops in metal-contaminated areas. The time period required for remediating soil with metal accumulating plants stands as a major drawback in the process of phytoremediation. But the incorporation of metal-mobilizing microbes speeds up the process of phytoremediation substantially. Potential phytosiderophore-producing microbes not only help to mobilize metal for plant uptake but also enhance plant nutrition (Rajkumar et al., 2010). Application of economically important crops rather than food crops in phytoremediation makes for a successful remediation strategy in metal-contaminated soils. Many of the plants mentioned in the previous sections are promising native plants that are useful for phytoremediation. Many of the agroecosytems contaminated by heavy metals are characterized by suitability of the field for a particular crop. In such cases the phytoremediation can be done with the crop being cultivated. For example, semi-aquatic rice fields contaminated with heavy metals need to be remediated with rice plants having higher metal accumulation capacity. This process can be further accelerated by the addition of fertilizers that enhance metal accumulation. Another important aspect in phytoremediation is the introduction of crops having mycorrhizae associations. Extended absorptive surfaces of root systems due to the presence of this symbiotic association leads to higher uptake of heavy metal in the plant. Thus the presence of mycorrhizae leads to phytoextraction of toxic metals from the soil which may either lead to higher root to shoot translocation of heavy metals in plants or retention of heavy metals in roots, called phytostabilization, that immobilizes heavy metals in the soil (Gohre and Paszkowski, 2006). The phytostabilization process essentially prevents leaching of heavy metals in the soil.

Soil amendment is another promising method that brings management of metal polluting fields by adding chemical or organic material that can bind available Cd in the soil and hence lessen the chance of metal accumulation by plants (Sebastian and Prasad, 2014a). Addition of organic amendments often brings retention of heavy metals in the soil by preventing metal leaching and plant uptake of the metal. Studies revealed that organic amendments work better in acidic soil having low cation exchange capacity. Application of farmyard manure and straw as organic amendments showed that these materials are efficient at preventing metal accumulation in crops by 25%–45% (Puschenreiter et al., 2005). Organic amendments are prone to faster degradation than chemical amendments. Thus degradation of material which uses amendments brings down efficiency of amendments chosen to adsorb or absorb the metal. Organic amendments can be spiked to the soil at regular intervals unlike chemical amendments and are feasible for annual crops. Addition of biochar, as well as charcoal, is reported to reduce plant available heavy metals (Park et al., 2011a). Both these materials are created by pyrolysis of biomass. Hence utilization of weeds for the production of these materials cuts down cost for soil remediation steps. Addition of biochar also has the benefit of increasing soil pH and conductivity of soil solution in the soil carbon pool. It had been reported that addition of biochar brings down cadmium accumulation in rice and the effect prevails for up to 2 years (Cui et al., 2011). This indicates longetivity in the effect of biochar application compared with organic manure treatments such as farmyard manure etc., whose effects persist for short-term periods only. Conversely, studies with the wetland species *Juncus subsecundus* indicate that biochar immobilizes Cd in soil but does not support plant growth (Zhang et al., 2013). Thus application of biochar must compare with sustainable soil remediation practices such as biodynamic farming which have advantages in crop productivity too. The use of cow dung, poultry manure or pig manure during biodynamic farming replenishes soil organic matter content which binds with heavy metals as well as increase nitrogen content in the field. Since these methods use by-products of livestock, the remediation process is economically feasible. Papermill sludge and water treatment sludge are also low-cost ecofriendly amendments that bring down soil heavy metals in the soil (Park et al., 2011b; Puschenreiter et al., 2005). Studies with application of paper mill sludge indicated that the effect of these materials had a long-term effect and these materials bring down the amount of heavy metals in the leachate of acidic soil. The presence of carbonate appears to be a crucial factor that limits heavy metal load, apart from silicate and organic matter, in the papermill sludge. Xylogen is also a by-product from paper mill industrial waste which was shown to immobilize Zn, Pb and Hg (Guo et al., 2006). Bark saw dust from the timber industry, chitosan from the crab meat canning industry, baggase from sugar cane, rice hulls from rice cultivation, leaves of senna, redwood, pine and straws of cotton, rice and maize are also reported as promising organic amendments that bring down plant available heavy metals in the field (Guo et al., 2006). It is the abundance of tannins rich in polyhydroxyl groups which ensures immobilization of metals by these materials. Chemical amendments are also known to reduce plant available heavy metals, as well as leaching heavy metals in the soil. These compounds either adsorb heavy metals or form hydroxides that precipitate metals in the soil and hence lower the availability of heavy metals to plants. Application of calcium carbonate is a traditional practice in acidic soil to increase soil pH. But this practice not only increases soil pH but also reduces plant available heavy metals by a cation exchange process. Chemical compounds such as CaO, $Ca(OH)_2$, $CaMgCO_3$, $CaHPO_3$, $Ca(H_2 PO_3)_2$, K_2HPO_4, H_3PO_4 and $(NH_4)HPO_4$ are efficient at immobilizing heavy metals in soil (Guo et al., 2006). The high solubility of calcium hydroxide makes this compound effective in a deeper as well as wider radius of soil compared with other chemical compounds. Oxides of iron and manganese are also reported to reduce plant uptake of heavy metals such as Cd (Gadde and Laitinen, 2006; Hua et al., 2012). Hence, salts of these micronutrients are efficient at preventing heavy metal uptake in plants. It has been demonstrated that oxides of these salts form plaque over the root surface which adsorbs toxic heavy metal with higher efficiency. Since iron and manganese are limiting plant nutrients, this approach also could be a solution for iron and manganese deficiency in plants. The industrial waste product fly ash is another promising amendment proven to be useful in soil remediation. Occasionally flyash contains toxic heavy metals such as Cd, Pb, etc., even though the main components of flyash are SiO_2, Al_2O_3, Fe_2O_3 and CaO. Hence the usage of flyash must come after monitoring heavy metal composition. Zeolites derived from flyash were also found to be effective in immobilization of heavy metals in the soil (Ojha et al., 2004; Querol et al., 2006). It is the sorption reaction to soil clay mineral called illite and increase of soil pH that makes zeolite immobilize heavy metals in the soil. Hydroxyapetite, slag, portland cement, gravel sludge, beringite, bauxite, bentionite, Ca-montmorillonite, ettringite and rock phosphate are also potential agents that help to remediate heavy metal polluted areas. A major drawback of chemical amendments is the saturation of metal binding properties. Thus areas continue to irrigate with mine water and the efficiency of these materials to bind heavy metals decreases and soil washing must be performed. It is beneficial to apply a combination of soil amendments with fertilizer application that decreases transfer of toxic metals into plants by operating metal binding processes as well as competition with micronutrient uptake. It has been reported that micronutrients such as zinc cause reduction in Cd translocation in rice plants (Sebastian and Prasad, 2013). Hence, application of zinc fertilizers will be feasible to reduce transfer of toxic metal from root to edible parts. Nitrogen supply is found to enhance Cd chelation with the aid of metal chelators in roots (Sebastian and Prasad, 2014b). Thus, fertilizer mix containing nitrogen is also beneficial to reduce cadmium translocation into the aerial part of the plants. The usage of ammonium fertilizer agents must be performed along with soil acidity regulators because of the chance of soil acidification processing that will further enhance metal uptake into plants.

REDUCTION OF Cd IN CROP PRODUCE

Crop cultivation in metal polluted areas must be carefully performed, and edible crops that have the ability to accumulate toxic heavy metals must be avoided. Apart from this, management of the field on the basis of sustainable agriculture patterns ensures preservation of soil structure in the area. It is worthy to note that toxic heavy metals such as Cd accumulate in crops in the increasing order of leaf vegetables > root vegetables > grain crops (Singh et al., 2010, 2012). Leafy vegetables such as lettuce, spinach, chinese cabbage, onion, and underground edible crops such as radish, carrot, red beet, potato, etc. are known to accumulate metals in edible parts. Hence these crops must be avoided in metal polluted sites. Among other leafy crops, tobacco is well known for Cd accumulation ability in its leaves. Plant families such as brassicaceae and asteraceae are also well known for metal accumulation ability (Prasad and Frietas, 2002). Cereals, legumes, melons, pepper and tomato accumulate comparatively low amounts of heavy metals and hence these crops are ideal for metal polluted sites. But choosing of crops must be done on the basis of level of field contamination and polluting sources. For example, even though staple foods such as rice accumulate lower amounts of Cd compared with other cereals such as wheat, daily consumption of rice leads to human health implications. If the field is massively contaminated as well as continuously irrigated with metal-contaminated water, cultivation of above low level Cd accumulating edible crops must be prohibited. The vast availability of germplasm in cases of low cadmium accumulating crop is promising to screen low Cd up taking cultivars. It has been reported that a natural variation of 12–23-fold variation exists in diverse japonica rice cultivars (Arao and Ae, 2003; Ueno et al., 2009). This indicates the potential of germplasm that allow selection and breeding of low Cd accumulating rice. Tissue dilution effect also must be considered as a factor during crop management. This factor is crucial in such a way that cultivars which are able to produce higher biomass create a dilution effect of metal in the plant tissue. If the biomass produced differs mainly in non-edible regions while edible regions such as biomass of grain do not have much variation, plants with higher biomass production must be chosen for cultivation. Existence of difference in biomass ratio of root and shoot also is an important factor that determines accumulation of metals in edible parts. For example, rice plants show highest accumulation of heavy metals such as Cd in root followed by shoot and grain. By considering shoot to grain translocation of heavy metals, soghum and maize, which produce more shoot and root biomass, have an advantage over rice and wheat which produce lower biomass. Even though areas heavily polluted with heavy metals are not suitable for edible crops, ornamental plants as well as plants that are commercially important for fibre production, secondary metabolite production, etc. can be used. Cotton, flax, hemp and energy crops such as corn, jatropha, salix trees, reed canary grass and gliricidia are examples of such non edible crops that can be utilized for large scale field remediation. Crop rotation strategy is also beneficial in areas polluted with heavy metals. It is important to alternate non-edible crops of higher metal accumulation capacity with edible crops of low metal accumulation capacity. Dicots like vigna, lupins, etc. and monocots viz. grasses are characterized by the ability to secrete citric acid and phytosiderphores which can mobilize heavy metal in the soil to plants (Jakkeral and Kajjidon, 2011; Meda et al., 2007). Growth of legumes is associated with nitrification-related reduction in soil pH (Yan et al., 1996). Thus alteration of crops has the advantage of removing heavy metal from soil as well as enrichment of soil with nitrogen. An ideal crop rotation pattern could start with cultivation of non edible crops that help mobilize metal followed by legumes and ornamental plants. If it is found that plant available metal reduced to a significant amount, these areas can be further utilized for cereals being used for staple food production. Because of the variation observed among cereals where wheat accumulates more heavy metals compared with maize and rice, crop ration is preferrably started with rice, followed by maize and wheat.

CONCLUSIONS

Thailand is rich in metallic minerals of zinc, lead, tin and gold. Doi Phadaeng, Phatat Phadaeng sub-district and Mae Sot district are rich in zinc ores. Cadmium is mostly a 'guest' metal in Zn mineralization. Although contamination of Cd has been reported in the vicinity of a mine near Mae Tow creek, the sediment in Mae Ku creek, which did not pass through the zinc mine, contains Zn and Cd. Therefore, Cd contamination in Phatat Phadaeng sub-district resulted from natural activity, agricultural activity and also careless mining in the past. This problem of heavy metal contamination from old mining has been reported in many countries. The chemical–physical technologies for the decontamination of contaminated ecosystems are complex and expensive and may cause undesirable side effects for the environment that then must in turn be decontaminated. Thus, phytoremediation and related agronomic practices are considered to be cost effective for this problem. This chapter showed Padaeng zinc mine in phytomanagement to recover the ecosystem and showed feasible options for the management of mine tailing water to prevent erosion and leaching of Zn and Cd from contaminated soil. Phytomanagement and soil remediation can be applied for sustainable agriculture. Several of the small hold farmers in the vicinity of zinc mines cultivate 'Thai jasmine rice'. Therefore, bioremediation of zinc mine soil waste and Cd minimization strategies in rice and critical multidisciplinary perspectives were discussed in this chapter.

ACKNOWLEDGEMENTS

The authors gratefully acknowledge the receipt of financial support under the auspices of India–Thailand bilateral scientific cooperation ref. DST/INT/THAI/P-02/2012 dated 31-1-13. Thanks are also due to Mahasarakham

University for logistical support for the visit to Mae Sot zinc mine area, and Padaeng Industry Public Co. Ltd. (Mae Sot Office) for permission to conduct this phytomanagement study.

REFERENCES

Arao, T., Ae, N., 2003. Genotypic variations in cadmium levels of rice grain. Soil Sci. Plant Nutr. 287, 223–233.

Arao, T., Kawasaki, A., Baba, K., Mori, S., Matsumoto, S., 2009. Effects of water management on cadmium and arsenic accumulation and dimethyl arsinic acid concentrations in Japanese rice. Environ. Sci. Technol. 43, 9361–9367.

Atagana, H.I., 2011. Bioremediation of co-contamination of crude oil and heavy metals in soil by phytoremediation *using Chromolaena odorata* (L) King & H.E. Robinson. Water. Air. Soil. Pollut. 215, 261–271.

Bolan, N.S., Makino, T., Kunhikrishnan, A., Kim, P.J., Ishikawa, S., Murakami, M., Naidu, R., Kirkham, M.B., 2013. Cadmium contamination and its risk management in rice ecosystems. Adv. Agronomy 119, 183–273.

Boopathy, R., 2000. Factors limiting bioremediation technologies. Bioresour. Technol. 74, 63–67.

Chen, Y., Shen, Z., Li, X., 2004. The used of vetiver grass (*Veiveria zizaniodes*) in thephytoremediation of soil contaminated with heavy metals. Appl. Geochem. 19, 1553–1565.

Chiu, K.K., Ye, Z.H., Wong, M.H., 2006. Growth of *Vetiveria zizanioides* and *Phragamities australis* on Pb/Zn and Cu mine tailings amended with manure compost and sewage sludge: A greenhouse study. Bioresour. Technol. 97, 158–170.

Chlopecka, A., 1996. Forms of Cd, Cu, Pb, and Zn in soil and their uptake by cereal crops when applied jointly as carbonates. Water. Air. Soil. Pollut. 87, 297–309.

Chong, C.W., Chu, L.M., 2007. Growth of vetivergrass for cutslope landscaping: Effects of container size and watering rate. Urban Forestry and Urban Greening 6, 135–141.

Codex, 2005. Report of the 37th Session of the Codex Committee on Food Additives and Contaminants. Alinorm 05/28/12. Codex Alimentarius Comm. 1–189.

Cui, L., Li, L., Zhang, A., Pan, G., Bao, D., Chang, A., 2011. Biochar amendment greatly reduces rice Cd uptake in a contaminated paddy soil: A two-year field experiment. BioResources 6, 2605–2618.

Diels, L., van der Lelie, N., Bastiaens, L., 2002. New developments in treatment of heavy metal contaminated soils. Rev. Environ. Sci. Biotechnol. 1, 75–82.

Gadde, R.R., Laitinen, H.A., 2006. Heavy metal adsorption by hydrous iron and manganese oxides. Environ. Sci. Technol. 40, 2213–2218.

Gohre, V., Paszkowski, 2006. Contribution of the arbuscular mycorrhizal symbiosis to heavy metal phytoremediation. Planta 223, 1115–1122.

Guo, G., Zhou, Q., Ma, L.Q., 2006. Availability and assessment of fixing additives for the *in situ* remediation of heavy metal contaminated soils: A review. Environ. Monit. Assess. 116, 513–528.

Hua, M., Zhang, S., Pan, B., Zhang, W., Lv, L., Zhang, Q., 2012. Heavy metal removal from water/wastewater by nanosized metal oxides: A review. J. Hazard. Mater. 211–212, 317–331.

Hughes, C.E., Styles, B.T., 1987. The benefits and potential risks of woody legume introductions. Int. Tree Crops J. 4, 209–248.

Ikeda, M., Ezaki, T., Tsukahara, T., Moriguchi, J., 2004. Dietary cadmium intake in polluted and non-polluted areas in Japan in the past and in the present. Int. Arch. Occup. Environ. Health 77, 227–234.

Jakkeral, S.A., Kajjidon, S.T., 2011. Root exudation of organic acids in selected genotypes under phosphorus deficient condition in black gram (*Vigna mungo* L. Hepper). Karnataka J. agricultural sci. 24, 316–319.

Khraisheh, M.A.M., Al-degsb, Y.S., Mcminn, W.A.M., 2004. Remediation of wastewater containing heavy metals using raw and modified diatomite. Chem. Eng. J. 99, 177–184.

Kitisin, T., Kosiyachinda, P., Luplertuop, N., 2013. Potential anti-oxidative activity of crude rice oil extracted from cadmium-contaminated rice as determined using an *in vitro* primary human fibroblast cell model. J. Agricultural Sci. 5, 104–121.

Lalhruaitluanga, H., Jayaram, K., Prasad, M.N.V., Kumar, K.K., 2010. Lead(II) adsorption from aqueous solutions by raw and activated charcoals of *Melocanna baccifera* Roxburgh (bamboo) — a comparative study. J. Hazard. mater. 175, 311–318.

Limpatanachote, P., Swaddiwudhipong, W., Mahasakpan, P., Krintratun, S., 2009. Cadmium-exposed population in Mae Sot District, Tak Province: 2. Prevalence of renal dysfunction in the adults. J. Med. Assoc. Thai. 92, 1345–1353.

Makris, K.C., Shskya, K.M., Datta, R., Sarkar, D., Pachanoor, D., 2007. High uptake of 2,4,6-trinitrotoluene by vetiver grass- potential for phytoremediation? Environ. Pollut. 146, 1–4.

Meda, A.R., Scheuermann, E.B., Prechsl, U.E., Erenoglu, B., Schaaf, G., Hayen, H., Weber, G., von Wiren, N., 2007. Iron acquisition by phytosiderophores contributes to cadmium tolerance. Plant. Physiol. 143, 1761–1773.

Meitei, M.D., Prasad, M.N.V., 2013. Lead (II) and cadmium (II) biosorption on *Spirodela polyrhiza* (L.) Schleiden biomass. J. Environ. Chem. Eng. doi.org/10.1016/j.jece.2013.04.016.

Meunier, N., Drogui, P., Montané, C., Hausler, R., Blais, J., Mercier, G., 2006. Heavy metals removal from acidic and saline soil leachate using either electrochemical coagulation or chemical precipitation. J. Environ. Eng. 132, 545–554.

Nakbanpote, W., Panitlertumpai, N., Sukadeetad, K., Meesungneon, O., Noisa-nguan, W., 2010. Advances in phytoremediation research: A case study of *Gynura pseudochina* (L.) DC. In: Fürstner, I. (Ed.), Advanced Knowledge Application in Practice. Sciyo, Croatia, pp. 353–378.

National Research for Environmental and Hazardous Waste Management, Chula longkorn University, 2005. Distribution of Cadmium and Absorption by Rice Plants in Areas Nearby the Zinc Mine in Mae Sot District. Chulalongkorn University, Bangkok.

Ojha, K., Pradhan, N.C., Samanta, A., 2004. Zeolite from fly ash: Synthesis and characterisation. Bull. Mater. Sci. 27, 555–564.

Pacyna, E.G., Pacyna, J.M., Fudala, J., Strzelecka-Jastrzab, E., Hlawiczka, S., Panasiuk, D., Nitter, S., Pregger, T., Pfeiffer, H., Friedrich, R., 2007. Current and future emissions of selected heavy metals to the atmosphere from anthropogenic sources in Europe. Atmos. Environ. 41, 8557–8566.

Pang, J., Chan, G.S.Y., Zhang, J., Liang, J., Wong, M.H., 2003. Physiological aspects of vetiver grass for rehabilitation in abandoned metalliferous mine wastes. Chemosphere 52, 1559–1570.

Panitlertumpai, N., Nakbanpote, W., Thiravetyan, P., Surarungchai, W., 2003. "The exploration of zinc-hyperaccumulative plants from mining area of Tak province in Thailand", 29th Congress on Science and Technology of Thailand, 20–22 October. Khon Kaen University, Khon Kaen, Thailand.

Panitlertumpai, N., Mongkhonsin, B., Nakbanpote, W., Jitto, P., 2008. Zinc hyperaccumulation by *Gynura pseudochina* (L.) DC. Zinc process. 08, 25–26. Aug, Stamford Plaza Hotel, Brisbane, Australia.

Panitlertumpai, N., Nakbanpote, W., Sangdee, A., Thumanu, K., Nakai, I., Hokura, A., 2013. Zinc and/or cadmium accumulation in *Gynura pseudochina* (L.) DC. studied *in vitro* and the effect on crude protein. J. Mol. Struct. 1036, 279–291.

Park, J.H., Choppala, G.K., Bolan, N.S., Chung, J.W., Chuasavathi, T., 2011a. Biochar reduces the bioavailability and phytotoxicity of heavy metals. Plant Soil 348, 439–451.

Park, J.H., Lamb, D., Paneerselvam, P., Choppala, G., Bolan, N., Chung, J., 2011b. Role of organic amendments on enhanced bioremediation of heavy metal(loid) contaminated soils. J. Hazard. Mater. 185, 549–574.

Phaenark, C., Pokethitiyook, P., Kruatrachue, M., Ngernsansaruay, C., 2009. Cd and Zn accumulation in plants from the Padaeng zinc mine area. Int. J. Phytoremediation 11, 479–495.

Pollution Control Department, 2004. Cadmium Contamination in Mae Tao Creek, Mae Sot District, Tak Province. Thailand Ministry of Natural Resources and Environment, Bangkok.

Prasad, M.N.V., 2002. Zinc is the friend and foe of life. Zeszyty Naukowe PAN 33, Com. In: Kabata-Pendias, A., Szteke, B., Warsaw (Eds.), *Man and Biosphere*, pp. 49–54.

Prasad, M.N.V., 2012. Exploitation of weeds and ornamentals for bioremediation of metalliferous substrates in the era of climate change. In: Ahmad, P., Prasad, M.N.V. (Eds.), Environmental Adaptations and Stress Tolerance of Plants in the Era of Climate Change. Springer, pp. 487–508. http://dx.doi.org/10.1007/978-1-4614-0815-4.

Prasad, M.N.V., 2013. Metallophytes — Properties, Functions and Applications.The Botanica 62 & 63, 17–26.

Prasad, M.N.V., Frietas, H., 2002. Metal hyperaccumulation in plants—biodiversity prospecting for phytoremediation technology. Electron. J. Biotechnol. 6, 285–321.

Prasad, M.N.V., Freitas, H., Fraenzle, S., Wuenschmann, S., Markert, B., 2010. Knowledge explosion in phytotechnologies for environmental solutions. Environ. Pollut. 158, 18–23.

Puschenreiter, M., Horak, O., Friesl, W., Hartl, W., 2005. Low-cost agricultural measures to reduce heavy metal transfer into the food chain — a review. Plant Soil and Environ. 51, 1–11.

Querol, X., Alastuey, A., Moreno, N., Alvarez-Ayuso, E., García-Sánchez, A., Cama, J., Ayora, C., Simón, M., 2006. Immobilisation of heavy metals in polluted soils by the addition of zeolitic material synthesized from coal fly ash. Chemosphere 62, 171–180.

Rajkumar, M., Ae, N., Prasad, M.N.V., Freitas, H., 2010. Potential of siderophore-producing bacteria for improving heavy metal phytoextraction. Trends. Biotechnol. 28, 142–149.

Raskin, I., Ensley, B.D., 2000. Phytoremediation of Toxic Metals: Using Plants to Clean up the Environment. John Wiley & Sons. ISBN 0-471-19254-6, New York.

Rattanapolsan, L., Nakbanpote, W., Saensouke, P., 2013. Metals accumulation and leaf anatomy of *Murdannia spectabilis* growing in Zn/Cd contaminated soil. Environ. Asia 6, 71–82.

Roongtanakiat, N., Chairoj, P., 2001. Vetiver grass for the remediation of soil contaminated with heavy metals. Kasetsart J. (Natural Science) 35, 433–440.

Roongtanakiat, N., Sanoh, S., 2011. Phytoextraction of zinc, cadmium and lead from contaminated soil by vetiver grass. Kasetsart J. (Natural Science) 45, 603–612.

Roongtanakiat, N., Tangruangkiat, S., Meesat, R., 2007. Utilisation of vetiver grass (*Vetiveria zizanioides*) for removal of heavy metals from industrial wastewater. Sci. Asia 33, 397–403.

Rotkittikhun, P., Chaiyarat, R., Kruatreehue, M., Pokethitiyook, P., Baker, A., 2007a. Growth and lead accumulation by grasses *Vetiveria zizanioides* and *Thysanolaena maxima* in lead-contaminated soil amended with pig manure and fertilizer: A glasshouse study. Chemosphere 66, 45–53.

Rotkittikhun, P., Chaiyarat, R., Kruatreehue, M., Pokethitiyook, P., Baker, A., 2007b. Growth and lead accumulation by grasses *Vetiveria zizanioides* and *Thysanolaena maxima* in lead-contaminated soil amended with pig manure and fertiliser: A glasshouse study. Chemosphere 66, 45–53.

Roongtanakiat, N., Yongyuth, O., Charoen, Y., 2008. Effect of soil amendment on growth and-heavy metals content in vetiver grown in iron ore tailings. Kasetsart J. (Natural Science) 42, 397–406.

Sampanpanish, P., Chaengcharoen, W., Tongcumpou, C., 2008. Heavy metals removal from contaminated soil by siam weed (*Chromolaena odorata*) and vetiver grass (*Vetiveria zizanioides*). Res. J. Chem. Environ. 12, 23–34.

Sao, V., Nakbanpote, W., Thiravetyan, P., 2007. Cadmium accumulation by *Axonopus compressus* (Sw.) P. Beauv and *Cyperus rotundas* Linn growing in cadmium solution and cadmium-zinc contaminated soil. Songklanakarin. J. Sci.Technol. 29, 881–892.

Sebastian, A., Prasad, M.N.V., 2013. Cadmium accumulation retard activity of functional components of photo assimilation and growth of rice cultivars amended with vermicompost. Int. J. Phytoremediation 15, 965–978.

Sebastian, A., Prasad, M.N.V., 2014a. Cadmium minimisation in rice. A review. Agronomy for Sustainable Development 34, 155–173. http://dx.doi.org/10.1007/s13593-013-0152-y.

Sebastian, A., Prasad, M.N.V., 2014b. Photosynthesis mediated decrease in cadmium translocation protect shoot growth of *Oryza sativa* seedlings up on ammonium phosphate– sulfur fertilisation. Environ. Sci. Pollut. Res. 21, 986–997. http://dx.doi.org/10.1007/s11356-013-1948-7.

Simmons, R.W., Pongsakul, P., Saiyasitpanich, D., Klinphoklap, S., 2005. Elevated levels of Cd and zinc in paddy soils and elevated levels of Cd in rice grain downstream of a zinc mineralised area in Thailand: Implications for Public Health. Environ. Geochem. Health 27, 501–511.

Simmons, R.W., Noble, A.D., Pongsakul, P., Sukreeyapongse, O., Chinabut, N., 2009. Cadmium-hazard mapping using a general linear regression model (Irr-Cad) for rapid risk assessment. Environ. Geochem. Health 31, 71–79.

Singh, S.K., Juwarkar, A.A., Kumar, S., Meshram, J., Fan, M., 2007. Effect of amendment on phyto-extraction of arsenic by *Vetiveria zizanioides* from soil. Int. j. Environ. Sci. Technol. 4, 339–344.

Singh, A., Sharma, R.K., Agrawal, M., Marsha, F.M., 2010. Health risk assessment of heavy metals via dietary intake of food from the wastewater irrigated site of a dry tropical area of India. Food. Chem. Toxicol. 48, 611–619.

Singh, S., Zacharias, M., Kalpana, S., Mishra, S., 2012. Heavy metals accumulation and distribution pattern in different vegetable crops. J. Environ. Chem. Ecotoxicol. 4, 170–177.

Siswanto, D., Suksabye, P., Thiravetyan, P., 2013. Reduction of cadmium uptake of rice plants using soil amendments in high cadmium contaminated soil: A pot experiment. J. Trop. Life Sci. 3, 132–137.

Sood, A., Uniyal, P.L., Prasanna, R., Ahluwalia, A.S., 2012. Phytoremediation potential of aquatic macrophyte, azolla. AMBIO: A J. Hum. Environ. 41, 122–137.

Sukreeyapongse, O., Srisawat, L., Chomsiri, O., Notesiri, N., Aug 1–6 2010. Soil management for reduce Cd concentration in rice grains, 19th World Congress of Soil Science, Soil Solutions for a Changing World. Brisbanem, Austrlia.

Sumpanpanish, P., Pongpaladisai, P., 2011. Effect of organic fertiliser on cadmium uptake by rice grown in contaminated soil. International Conference on Environmental and Agriculture Engineering, IPCBEE' 2011, vol. 15, IACSIT Press, Singapore. 103–109.

Suppadit, T., Kitikoon, V., Phubphol, A., Neunoi, P., 2012. Effect of quail litter biochar on productivity of four new physic nut varieties planted in cadmium-contaminated soil. Chilean J. Agricultural Res. 72, 125–132.

Svasti, P., 2007. Poisoned livelihoods. Bangkok Post, May 14, 2007.

Swaddiwudhipong, W., Limpatanachote, P., Mahasakpan, P., Krintratun, S., Padungtod, C., 2007. Cadmium-exposed population in Mae Sot District, Tak Province: 1. Prevalence of high urinary cadmium levels in the adults. J. Med. Assoc. Thai. 90, 143–148.

Swaddiwudhipong, W., Mahasakpan, P., Funkhiew, T., Limpatanachote, P., 2010. Changes in cadmium exposure among persons living in cadmium-contami nated areas in northwestern Thailand: A five-year follow-up. J. Med. Assoc. Thai. 93, 1217–1222.

Swaddiwudhipon, W., Limpatanachote, P., Mahasakpan, P., Krintratun, S., Punta, B., Funkhiew, T., 2012. Progress in cadmium-related health effects in persons with high environmental exposure in northwestern Thailand : A five-year follow up. Environ. Res. 112, 194–198.

Tanhan, P., Kruatrachue, M., Pokethitiyook, P., Chaiyarat, R., 2007. Uptake and accumulation of cadmium, lead and zinc by Siam weed [*Chromolaena odorata* (L.) King & Robinson]. Chemosphere 68, 323–329.

Teeyakasem, W., Nishijo, M., Honda, R., Satarug, S., Swaddiwudhipong, W., Ruangyuttikarn, W., 2007. Monitoring of cadmium toxicity in a Thai population with high-level cadmium exposure. Toxicol. Lett. 169, 185–195.

Torab-Mostaedi, M., Ghassabzadeh, H., Ghannadi-Maragheh, M., Ahmadi, S.J., Taheri, H., 2010. Removal of cadmium and nickel from aqueous solution using expanded perlite. Braz. J. Chem. Eng. 27, 299–308.

Truong, P.N., 1999. Vetiver grass technology for mine rehabilitation. Pacific Rim Vetiver Network. Office of Royal Development Projects Board, Bangkok. Tech. Bull. 12. No. 1999/2.

Truong, P.N., Mason, F., Waters, D., Moody, P., 2000. Application of vetiver grass technology in off-site pollution control. I. Trapping agrochemicals and nutrients in agricultural lands. In: Proceeding of the Second International Vetiver Conference Thailand, January 2000.

Ueno, D., Koyama, E., Kono, I.O.T., Yano, M., Ma, J.F., 2009. Identification of a novel major quantitative trait locus controlling distribution of Cd between roots and shoots in rice. Plant. Cell. Physiol. 50, 2223–2233.

Wasay, S.A., Barrington, S., Tokunaga, S., 2001. Organic Acids for the *in situ* remediation of soils polluted by heavy metals: Soil flushing in columns. Water. Air. Soil. Pollut. 27, 301–314.

Wilde, E.W., Brigmon, R.L., Dunn, D.L., Heitkamp, Dagnan, D.C., 2005. Phytoextraction of lead from firing range soil by Vetiver grass. Chemosphere 61, 1451–1457.

Wuana, R.A., Okieimen, F.E., 2011. Heavy metals in contaminated soils: A review of sources, chemistry, risks and best available strategies for remediation. ISRN Ecol. doi:10.5402/2011/402647.

Yan, F., Schubert, S., Mengel, K., 1996. Soil pH changes during legume growth and application of plant material. Biol. Fertil. Soils 23, 236–242.

Zhang, Z., Solaiman, Z.M., Meney, K., Murphy, D.V., Rengel, Z., 2013. Biochars immobilise soil cadmium, but do not improve growth of emergent wetland species *Juncus subsecundus* in cadmium-contaminated soil. J. Soils and Sediments 13, 140–151.

Chapter 24

Effect of Pig Slurry Application on Soil Organic Carbon

Ibrahim Halil Yanardağ,* Asuman Büyükkılıç Yanardağ,* Angel Faz Cano* and Ahmet Ruhi Mermut†

*Sustainable Use, Management and Reclamation of Soil and Water Research Group, Agrarian Science and Technology Department, Technical University of Cartagena, Cartagena, Murcia, Spain; †Harran University, Agriculture Faculty, Soil Science Department, Şanlıurfa, Turkey

INTRODUCTION: IMPORTANCE OF SOIL ORGANIC MATTER

Some soil properties such as water-holding capacity, water infiltration rate, erodibility, nutrient cycling and pesticide adsorption are strongly related to soil organic matter (SOM) (Francioso et al., 2000; Wander and Yang, 2000). Information on the relationship between the stability and chemical structure of organic matter is essential. Although it is generally thought that humic substances are non-labile, have various chemical structures and consequently their degradation rates also vary with initial organic substances. They may become refractory by losing their aliphatic moieties, during polymerization–condensation and oxidation processes (Watanabe et al., 2006). Humic acids are a major fraction of humic substances that account for up to 40% of total SOM (Watanabe et al., 2006).

The animal manure application increases SOM quantity, nutrient availability, soil aggregation and other soil functions (Hatfield and Stewart, 1998). Therefore, management of manure is very important to SOM content and consequently to the fate of agriculture. Many studies have also recognized SOM as a central indicator of soil quality and health (Soil and Water Conservation Society, 1995).

Composition of SOM is very complex, and also includes a complex mixture of living organisms, dead organic debris and anthropogenic inputs. During the decay of plants, a major part of plant carbon (~99%) is biodegraded and recycled to the atmosphere as CO_2. The organic matter remaining in soils is thus a carbon pool exhibiting high resistance to biodegradation. Also some SOM may be in the minor classes: carbohydrates, amino acids, lipids and phenols have so far been identified (Lichtfouse et al., 1998).

Soil Remediation and Plants. http://dx.doi.org/10.1016/B978-0-12-799937-1.00024-3

Characterization of organic matter in soils and sediments is mostly determined by solid-state ^{13}C nuclear magnetic resonance (NMR) spectroscopy which provides detailed information (Ivanova and Randall, 2003; Mao et al., 2000). Ding et al. (2000) demonstrated that both DRIFT and ^{13}C NMR techniques are suitable for examining the effects of agricultural management on SOM (Ding et al., 2002). Solid-state ^{13}C cross polarization (CP) NMR was used to find out that mineral fertilizer and manure differed little in their effects on the chemical composition of SOM associated with clay- and silt-sized fractions. Several investigators concluded that the chemical composition of SOM was determined primarily by the interaction between the organisms responsible for decomposition and the mineral soil matrix, rather than by the nature of the substrate input (Mao et al., 2008).

PIG SLURRY APPLICATION

Pig slurry (PS) is not a waste product: it is a valuable plant nutrient resource with the added benefit of improving soil quality. However, because of the complex nature of the material, it can result in negative environmental impact when it is mismanaged. Spain is the second leading European country with regard to pig population (22.4 million head in 2000), representing 18% of the total production within the European Union, with a substantial increase over the last 10 years (38% from 1990 to 2000) (Daudén and Quílez, 2004). Therefore, sustainable PS application can play a very important role in the environment. It is known that the pig slurry applications can directly or indirectly affect the soil physical, chemical and biological properties.

In two different studies, we aimed to determine PS effects on soil properties and quality. In the first study, we applied three different doses on silty loam soils: Single (D1), Double (D2), Triple (D3); and unfertilized plots (C) served as controls. Samples were collected at two different levels, surface (0–30 cm) and subsurface (30–60 cm). Using PS as an organic fertilizer, we tried to determine the optimum amount that can be added to the soil, and the effect on soil properties and quality. The results demonstrated that the D1 application dose, which is the agronomic rate of N-requirement (170 kg N ha^{-1} per year) (European Directive 91/676/CEE), is very appropriate in terms of sustainable agriculture and can also improve physical, chemical and biological soil properties (Büyükkılıç Yanardağ, 2013a,b).

In second study, we applied PS from physical, chemical and biological separation processes and different pig diets. The different sources of PS [raw pig slurry (RPS) and treated pig slurry (TPS) from liquid and solid feeding diets] were applied on sandy loam soil under barley cultivation to determine its effects, in terms of environment, pollution, soil quality and soil C dynamics. The study suggested that RPS applications are appropriate in terms of soil carbon cycle and quality (Yanardağ et al., 2013). There are many factors to be considered when attempting to assess the overall net impact of a management practice on productivity. Additions of pig manure to soils at agronomic rates

(170 kg N ha^{-1} per year) to match crop nutrient requirements are expected to have a positive impact on soil productivity. Therefore, the benefits from the use of application depend on the management of PS, carbon and environmental quality. However, PS has high micronutrient contents, and for this reason the application of high doses can pollute soils and damage human, animal and plant health, which is not suitable in terms of sustainable agriculture. Therefore, long-term use of PS with low dose may necessarily enhance soil quality in the long term (Büyükkılıç Yanardağ, 2013a,b). Yanardag et al. (2013) observed that separation processes declined bioavailable metal concentration of PS with both liquid and solid breeding diets.

There are still a lack of information about the agronomic benefits and disadvantages of the organic matter quality from anaerobically digested pig slurries, although they are widely used in agricultural soils. This is of particular interest since SOM is decreasing in cultivated soils, which is a matter currently receiving much attention in Europe (Marcato et al., 2009). In the context of climate change and sustainable agricultural development, C sequestration in the topsoil attracts much of the attention, as C sinks in national greenhouse gas inventories (Saby et al., 2008).

EFFECT OF PIG SLURRY APPLICATION ON SOIL ORGANIC CARBON

Organic manures can improve the nutrient status of the soil and maintain high crop yields and high levels of residues that can be returned to the soil to increase the soil organic carbon (SOC) concentration (Holeplass et al., 2004), especially in semiarid agro-ecosystems where the low organic matter content may favour soil degradation and erosion (Garcia-Gil et al., 2000). Increased crop yields and SOM in the soil result in higher SOM content and biological activity (Haynes and Naidu, 1998). The role of manures in maintaining and increasing levels of SOC in agricultural soils has been well documented (Rudrappa et al. 2006).

Qualities of organics (i.e. lignin, C/N, etc.) also influence the quantity of SOC development in the soil. In intensive cropping systems, burning or removal of crop residues is a common practice, where combine harvesters are used. Thus, only the roots are recycled or retained under intensive cultivation (Manna et al., 2005).

Manure is more resistant to microbial decomposition than plant residues are. Consequently, for the same carbon input, carbon storage is higher with manure application than with plant residues (Feng and Li, 2001). In organic degradation, manure and other sources of organic matter are utilized as a source of energy by a succession of living microbial organisms.

Farage et al. (2003) observed that only 1 ton of carbon would actually be available for incorporating into the soil, if manure contains 25% carbon (Figure 24.1). Clearly, much less carbon is available for sequestering into the soil when it passes through the heterotrophic chain (Farage et al., 2003). Therefore, there is a need

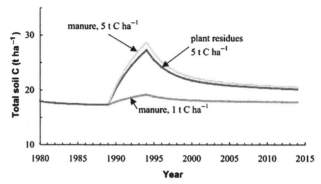

FIGURE 24.1 Change in organic soil carbon for a Vertisol in semi-arid Andhra Pradesh, India modeled with the Roth C soil carbon model. Five tons of plant residue carbon and 5 tons of cattle manure carbon were applied in 1989 for five consecutive years. Application of 1 ton of carbon in cattle manure represents the amount of manure carbon that would be available for incorporation into the soil if the plant residues had been directly fed to the cattle. For a colour version of this figure, please see the section at the end of this book. *Source: Farage et al. (2003).*

to quantify the C input returned back to the soil, and to determine whether it is sufficient to improve the SOC restorative process in the long run. We also need to know whether total organic matter and things related to it are the major constraints for yield improvement (Manna et al., 2005). Some studies indicated that manure-bound nutrients are gradually released from the organic component of the manure (Bouldin and Klausner, 1998; Mao et al., 2008).

Pig slurry contains relatively small amounts of C compared to N (between 3 and 6%), and applied slurry provides minerals to native SOM, rather than incorporating applied commercial N fertilizers to the soil (Plaza et al., 2004). Effects of manure addition on increasing SOM content are more pronounced on soils of lower organic matter content and low fertility (Loro et al., 1997). The application of manure can increase the SOM, and SOM contributes to nutrient supply, improving soil physical and chemical properties (Jimenez et al., 2002). Many researchers have observed increases in SOM following repeated applications of manure (Campbell et al., 1997; Sommerfeldt et al., 1988; Larney et al., 2000).

Black Carbon

Black carbon (BC) is a generic term used to refer to highly condensed organic C structures and has been insignificant in the global carbon cycle. Black carbon is a *climate forcing* agent formed through the incomplete combustion of fossil fuels, biofuel and biomass, and is emitted in both anthropogenic and naturally occurring soot (fine carbon particles adhering to the side of the chimney or pipe) (Rovira et al., 2009). Global production rates of black carbon are relatively low, $0.05–0.27\,Pg\,yr^{-1}$, compared to terrestrial net primary productivity of about $60\,Pg\,yr^{-1}$ (Forbes et al., 2006; Denman et al., 2007).

The charred plant residues contribute to SOM formation in volcanic ash soils in Japan (Kumada, 1983). However, that charcoal may represent the inert carbon pool used in SOM models and recalcitrant carbon may be an indicator, on a geological time scale in sediments (Glaser et al., 1998).

Black carbon is generally aromatic in nature, and is therefore thought to be stable with regard to microbial decay, which shows high chemical resistance and a high survival potential in sediments and some soils. It increases the passive SOM pool and, thus the passive SOM pool represents an important sink in the global C cycle. In order to elucidate its role in geobiochemical cycles and soil ecological functions, its quantification is essential (Schmidt and Noack, 2000; Knicker, 2007).

The interest in furthering our knowledge of the different fractions of C, depending on its strength and ease of degradation, is important and lies in the characterization of C cycle and soil as a sink for C (Kuzyakov et al., 2009). The most recalcitrant carbon, or black carbon, undergoes microbiological and chemical changes very slowly; it is one of the most stable pools of soil organic carbon, contributing to the stability of humus in the soil (Brodowski et al., 2007). In Mediterranean countries, slash-and-burn agriculture or even the simple burning of harvest residues are widespread practices. Thus, charred organic carbon (OC) may account for up to 45% of the total OC in the Chernozemic soils of Germany (Schmidt et al., 1999).

The average amount of black carbon ranged between 7.9 (1.8) and 29.2 (11.8) $mg\,C\,kg^{-1}$ in surface, and 3.2 (0.9) to 20.0 (10.7) $mg\,C\,kg^{-1}$ in subsurface horizon (Table 24.1). The results of studies showed that soils had similar black carbon concentrations in the surface horizons for agricultural soils (Currie et al., 2002; Skjemstad et al., 2002; Masiello and Druffel, 2003). There were significant differences in these microbiological properties between control and the amended soils. The results showed that TPS applications increased black carbon content, while RPS did not affect the surface horizon significantly in 2010 ($P>0.05$); even so, black carbon was increased with all PS applications except for the RPS treatment from liquid feeding diet and that ratios increased in the solid feeding diet applications in 2011. All PS applications decreased black carbon content except for the RPS from the solid feeding diet in the subsurface horizon in 2011 (Buyukkilic Yanardag, 2013 a,b).

Following PS application, mineralization of approximately one-third of C added as PS was observed within 3 months, providing evidence that C is retained in the soil. Soil retention of C is related to the presence of recalcitrant compounds or to the immobilization of C by microbial biomass (Balota et al., 2011).

Soluble Carbon

Soluble organic C is not only a C source for microorganisms, but its production is also believed to be microbially mediated (Christ and David, 1996). McGill et al. (1986) suggested that the flow of C through water-soluble components

TABLE 24.1 Evolution of Soluble C, MBC and Black Carbon by PS Application in Surface and Subsurface Soils, Treatment n:3, Control n:12

| Year | Treatments | Soluble C (mg C kg^{-1}) | | MBC (mg C kg^{-1}) | | Black Carbon (mg C kg^{-1}) | |
		As (0–30 cm)	Ap (30–60 cm)	As (0–30 cm)	Ap (30–60 cm)	As (0–30 cm)	Ap (30–60 cm)
2009	Control	248.19ab (52.6)	149.1ab (42.6)	433.72abc (71.7)	244.7ab (84.1)	12.01a (4.1)	5.8 (2.9)
	RPS LFD	375.96b (67.5)	141.0ab (38.7)	497.42bc (91.9)	161.0a (62.9)	11.30a (1.0)	5.0 (2.9)
	TPS LFD	348.54b (59.5)	169.6ab (23.2)	482.34abc (128.4)	204.7ab (66.0)	15.66ab (3.4)	6.2 (4.1)
	RPS SFD	344.00ab (64.4)	265.2b (36.7)	511.15bc (131.8)	125.7a (45.7)	22.75ab (6.4)	8.4 (0.9)
	TPS SFD	353.04b (48.8)	165.0ab (88.9)	619.90c (118.2)	167.0ab (22.0)	20.74ab (4.7)	3.7 (2.2)
2010	Control	189.88ab (58.3)	99.6a (24.0)	270.75ab (85.6)	138.5a (68.7)	21.40ab (5.4)	11.9 (7.2)
	RPS LFD	184.60a (10.5)	133.9ab (59.3)	643.96c (88.2)	342.0b (120.5)	21.48ab (6.3)	10.3 (6.9)
	TPS LFD	269.46ab (78.6)	137.9ab (41.3)	252.42a (14.5)	187.8ab (15.0)	29.20b (11.8)	11.8 (4.6)
	RPS SFD	262.88ab (71.3)	158.6ab (18.8)	325.55ab (57.9)	69.4a (2.3)	19.90ab (1.9)	10.7 (2.6)
	TPS SFD	331.59ab (50.4)	155.6ab (65.7)	217.83a (63.4)	115.5a (1.6)	25.30ab (4.5)	9.0 (1.6)
2011	Control	164.01a (84.0)	93.7a (59.9)	341.61ab (92.5)	155.6a (62.9)	18.03ab (4.8)	11.6 (3.2)
	RPS LFD	244.27ab (29.8)	137.6ab (16.6)	447.39abc (45.5)	102.3a (60.0)	14.86a (7.8)	9.2 (3.7)
	TPS LFD	230.06ab (27.2)	115.2ab (15.7)	315.04ab (24.0)	201.4ab (78.3)	18.58ab (3.8)	10.4 (3.6)
	RPS SFD	183.55a (58.4)	40.0a (6.2)	535.99bc (65.2)	227.8ab (0.8)	20.52ab (7.2)	12.4 (4.5)
	TPS SFD	151.49a (11.3)	25.6a (6.0)	395.67abc (107.5)	141.8a (49.1)	24.84ab (5.3)	11.4 (5.2)
	f	6.385****	3.360****	9.166****	3.875****	3.837****	1.525ns

RPS, Raw Pig Slurry; TPS, Treated Pig Slurry; LFD, Liquid Feeding Diet; SFD, Solid Feeding Diet; As, Surface Horizon; Ap, Subsurface Horizon; MBC, Microbial Biomass C; Sub: *$P < 0.05$, **$P < 0.01$, ***$P < 0.001$, ns, not significant.

supplies substrate for microbial biomass turnover (Gregorich et al., 2000). Generally, water-soluble organic carbon (WSOC) concentrations are affected by different land uses, vegetation types and also by different fertilizers (Haynes, 2000).

The soluble carbon concentrations in different treatments and controls in surface and subsurface horizons are given in Table 24.1. The average of soluble C ranged from 65.3 (11.1) to 376.0 (67.5) mg C kg^{-1} in surface, and 14.9 (6.2) to 265.2 (150.5) mg C kg^{-1} in subsurface horizon, where soils had normal soluble C concentrations in the surface horizon for agricultural soil compared with previous studies (Cosandey et al., 2003; Fierer and Schimel, 2002). There were significant differences in the microbiological properties between the control and the amended soil ($p < 0.05$). All PS applications increased soluble carbon content except for the RPS from liquid feed diet in surface horizon in 2010, and it was increased with all PS applications except for TPS from solid feed diet in 2011. PS applications increased soluble C values in the subsurface in 2010, whereas it was decreased with solid feed diet applications in 2011.

Correlation coefficient analysis results are given in Table 24.2. There are high positive correlations between soluble carbon and TOC, TN, P, EC, Mg, Ca, β-glucosidase, Arylesterase enzyme activity, MBC and black carbon. Conversely, there were negative correlations between pH water, pH KCl and C:N ratio.

Other studies observed that the use of various manures significantly increased water-soluble carbon contents of soil (Rochette et al., 2000; Bol et al., 2003), whereas these changes were not permanent over time. According to Chantigny (2003), this increase is attributed to the presence of soluble compounds in the amendment which is rapidly degraded and the soil recovers in a relatively short period of time to the initial level of soluble organic matter. Furthermore, Ma et al. (2010) found that WSOC concentrations were decreased with increased soil depth and there were positive correlations between WSOC and SOC.

Microbial Biomass Carbon

The world soil is a rich habitat for numerous creatures including microbes that convert complex organic compounds such as proteins, carbohydrates, cellulose, etc., into a useable form that plants can incorporate for growth. Soil microorganisms help to create a healthy productive soil. Thus, soil microbial activity is a potential indicator of soil quality (Chen et al., 2003). Microbial biomass, composition of microflora, mineralization and synthesizing processes are ecologically important soil characteristics that play a key role in different ecologically important soil functions. The fallow is also very often the subject of discussion; some authors treatise biological activities of such soils (Růžková et al., 2008).

The MBC concentrations in different treatments and controls in the surface and subsurface horizons are given in Table 24.1. The mean of MBC ranged between 179.7 (15.0) and 644.0 (88.2) mg C kg^{-1} in surface, and 69.4 (19.0)

TABLE 24.2 Correlation between Different Soil Properties Determined

	Total N	P_Olsen	pH	EC	CaCO₃	C:N	CEC	K	Mg	Ca	C_sol	C_mic	C_Bl
TOC	0.937**	0.754**	−0.719**	0.148	−0.02	−0.216	0.369**	0.183	0.712**	0.682**	0.538**	0.606**	0.704**
Total N	*	0.815**	−0.740**	0.270*	−0.061	−0.466**	0.301*	0.101	0.828**	0.694**	0.652**	0.638**	0.753**
P_Olsen		*	−0.539	0.521**	−0.254*	−0.302**	0.103	0.164	0.802**	0.591**	0.641**	0.587**	0.442**
pH			*	−0.191	−0.058	0.272	−0.295**	0.011	−0.575**	−0.553**	−0.505**	−0.458**	−0.629**
EC				*	−0.243*	−0.318**	−0.117	0.331**	0.562**	0.313**	0.520**	0.154	0.034
CaCO₃					*	0.131	0.062	0.003	−0.224	−0.138	−0.144	−0.266*	−0.036
C:N						*	0.077	0.156	−0.487**	−0.327**	−0.460**	−0.300*	−0.378**
CEC							*	0.234*	0.169	0.216	−0.176	0.129	0.391**
K								*	0.263*	0.313**	0.004	0.035	−0.04
Mg									*	0.756**	0.746**	0.588**	0.597**
Ca										*	0.614**	0.520**	0.511**
C_sol											*	0.448**	0.473**
C_mic												*	0.459**
C_Bl													*

Sample size=48. TOC, total organic C; T Nitrogen, total Nitrogen; C:N, C:N Ratio; EC, electrical conductivity; C_{mic}, microbial biomass C; C_{sol}, Soluble Carbon; C_{Bl}, Black Carbon; Sub: *P<0.05; **P<0.01; ***P<0.001.
Source: Yanardag et al. (2013).

to 342.0 (120.5) mg C kg^{-1} in subsurface soil. The result showed that MBC content was slightly high for semi-arid climate condition soils (Debosz et al., 1999; Plaza et al., 2004; Tripathi et al., 2007). The MBC contents were significantly affected by all PS applications (Table 24.1) ($p < 0.05$): while decreased with TPS application, it was increased with RPS applications in the surface horizon in 2010. The organic matter content in the MBC was increased with all PS applications except for the TPS from liquid feeding diet in the surface in 2011. The Liquid feeding diet (LFD) applications increased the MBC content in the subsurface horizon in 2010. However, it was increased by applications of the TPS in liquid feeding diet and RPS from solid feeding diet in 2011 in the subsurface horizon.

Correlation coefficient analysis results (Table 24.2) have shown that there were high positive correlations between MBC and TOC, TN, P, Mg, Ca, β-glucosidase, β-glucosidase, Arylesterase enzyme activity and black carbon. However, there were negative correlations between pH water, pH KCl, CaCO$_3$ and C:N ratio.

The application of PS for 9 years tended to increase total organic C content and the MBC, but this positive effect was limited and occurred only in the surface layer (Dambreville et al., 2006). The C provided by liquid slurry is considered to be rapidly mineralized and able to develop a short-term effect on soil respiration and MBC, whereas any existing long-term effect is limited (Rochette et al., 2000; Dambreville et al., 2006).

Balota et al. (2011) found that an increase in pig slurry application resulted in a decrease in the ratio of soil enzyme activities per unit of MBC. In this study, soil samples were taken after application of PS in 6th month; previous studies also showed that after the PS application MBC content rapidly increased in the short term, but in the long term the effects were not found to be significant. TPS applications had no effect on the MBC content while, RPS applications increased it in the long term.

Soil Respiration

Soil respiration is very important to C cycling in terrestrial ecosystems. Soil respiration in terrestrial ecosystems is estimated to be 50–75 Pg C year^{-1}. While microbial decomposition of organic debris is essential to SOM formation, the process also represents a major mechanism of C loss from the soil systems (Raich and Schlesinger., 1992). However, respiration of the soil can be partitioned into two processes: metabolic activity of plant roots (autotrophic respiration) and the decomposition of dead organic material (heterotrophic respiration) (Bernhardt et al., 2006). Autotrophic respiration ranges from 10% to 90% of total soil CO$_2$ efflux (Hanson et al., 2000). Both autotrophic and heterotrophic respiration can respond independently to a number of factors, such as climate, forest type, soil type and human disturbances (Figure 24.2) (Luo and Zhou, 2006).

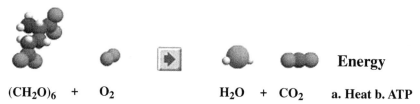

$$(CH_2O)_6 \quad + \quad O_2 \qquad\qquad H_2O \quad + \quad CO_2 \qquad \textbf{a. Heat b. ATP}$$

FIGURE 24.2 Basic soil CO_2 generation processes: oxidation of organic matter due to aerobic microbial respiration (Available on-line, http://www.oocities.org/b1gcee/fessay/biol/ res_pho.htm, 2013). For a colour version of this figure, please see the section at the end of this book.

The effect of temperature is described as an exponential function, although there is a discussion as to which exponential formulation is the best (Lloyd and Taylor, 1994). Soil respiration and photosynthetic activity are closely related (Moyano et al., 2008). Elevated CO_2 can increase photosynthetic assimilation rates, therefore increasing the above- and below-ground biomass production. The increase in below-ground biomass increases C loss from the soil (Edwards and Norby, 1999). Release of CO_2 from the soil, due to production of CO_2 by roots, and soil organisms to a lesser extent, is commonly referred to as soil respiration (Deng et al., 2009).

Many studies have shown that there is a positive correlation between soil respiration and temperature (Raich and Schlcsinger; 1992). The amount of soil respiration and the microbial metabolic quotient reflect the soil microbial activities, which are affected by the eco-environment changes (Alvarez et al., 1995). Soil microbial activity is an index of the actual (basal respiration) and the potential microbial respirometric activity (Bailey et al., 2002). Several incubation studies have shown that mineralizable C is directly related to soluble C, which is an index of the immediate availability of C to soil microorganisms (Rochette and Gregorian, 1998). The positive correlation between soluble organic C and soil temperature suggests that the production of soluble C increases more than the mineralization of soluble C, as air temperature increases (Rochette and Gregorian, 1998).

The result show that soil respiration in the RPS from liquid feeding diet 1.02 (0.14) $\mu mol\, m^{-2} per\, s$ (the data not shown), TPS from liquid feeding diet 1.68 (0.80) $\mu mol\, m^{-2} per\, s$, RPS from solid feeding diet 1.88 (0.99) $\mu mol\, m^{-2} per\, s$, TPS from SFD 1.83 (1.07) $\mu mol\, m^{-2} per\, s$ and control 1.76 (0.81).

In previous studies some researchers studied with barley and McGinn et al. (1998) found 30–40 $\mu mol\, C\, m^{-2} per\, s$, Eriksen and Jensen (2001) found 3–4 $\mu mol\, m^{-2} per\, s$ and Moyano et al. (2007) found 2 $\mu mol\, m^{-2} per\, s$ soil respiration ratios.

Carbon Functional Groups

For the characterization of organic matter in soils and sediments, solid-state ^{13}C-NMR spectroscopy is providing more detailed information (Ivanova and Randall, 2003). Ding et al. (2000) demonstrated that both DRIFT and ^{13}C-NMR techniques are useful and suitable for examining the effects of agricultural management on SOM. Using solid-state ^{13}C cross polarization (CP) NMR, Mao et al. (2008) found that mineral fertilizer and manure differed little in their effects on the chemical composition of SOM associated with clay- and silt-sized fractions. The authors concluded that the chemical composition of SOM was determined primarily by the interaction between the organisms responsible for decomposition and the mineral soil matrix rather than by the nature of substrate input.

In several recent studies, the ^{13}C-NMR spectra were divided into the seven chemical shift regions: alkyl-C (0–50 ppm); methoxyl (50–60 ppm); O-alkyl C (60–98 ppm); di-O-alkyl (98–112 ppm); aromatic C (112–145 ppm); phenolic C (145–163 ppm); and carbonyl C (163–190 ppm). Table 24.3 presents typical chemical shift ranges for constituents found in SOM. Areas of the chemical shift regions were measured by integration and were described as a percentage of the total area (relative intensity). Sidebands contribution generated by the carbonyl region was corrected by multiplying the area under the peak to the low field sideband (215–230 ppm) by a factor of 2 and adding it to the carboxyl area (163–190 ppm) (Faz Cano et al., 2002; Kavdır et al., 2005).

One-way repeated analyses of variance (ANOVA) showed that there were differences between pre- and post-PS application among the SOM functional groups (data not shown). However, alkyl, methoxyl, phenolic and carboxyl C functional groups were statistically significantly different ($p < 0.05$). Alkyl C groups predominated the functional groups of OM and they were decreased with liquid feeding diet applications. Treatment with treated solid feeding diet caused a statistically significant increase in alkyl C group, except the RPS treatment. Even in the control plot, the slight changes in alkyl groups suggest that this carbon oxidizes with time in arid and semi-arid areas. Differences in O-alkyl groups, however, under different treatments were not statistically significant.

Stable Carbon Isotope

^{13}C natural abundance technique was first used to study the dynamics of SOC, and has been widely used in the field ever since (Nissenbaum and Shallinger, 1974; Balesdent et al., 1993). Natural abundance of ^{13}C (stable C isotope ratio, δ^{13}C) in SOM depends on the migration and decomposition dynamics of organic matter itself and on plant type, and δ^{13}C can be used in ascertaining the rate of turnover of total SOC and that of migration or decomposition of some of its fractions (Haefelea et al., 2004; Ma et al., 2011).

The utility of ^{13}C isotopic tracers for SOM studies derives from the fact that all plants discriminate against ^{13}C during photosynthesis, but to different degrees depending on their photosynthetic pathways (Farquhar et al., 1989). Terrestrial plants with the C3 (Calvin cycle) pathway have δ^{13}C mean values of −27‰. Plants with the C4 (Hatch–Slack) pathway have higher δ^{13}C mean values of −12‰ (Stevenson et al., 2005). The isotopic composition of SOM

TABLE 24.3 Typical Chemical Shift Ranges for Constituents Found in SOM

Chemical Shift Range (ppm)	C Structure Characteristics
0–50	Unsubstituted alkyl C: includes straight-chain methylene C (30–35 ppm) and terminal methyl groups (15 ppm). Branched methylene C is found more downfield (35–50 ppm) and Substituted alkyl C such as that found in amines (45–46 ppm)
50–60	Methoxyl groups
60–98	Oxygen-substituted C, ring C in carbohydrates and C in ethers
98–112	Dioxygen-substituted aliphatic C and anomeric C in carbohydrates (105 ppm)
112–145	Aromatic C
145–163	Phenolic C
163–190	Carboxylic, amide and ester C

Source: Adapted from Faz et al (2002) and Simpson and Preston (2006).

closely resembles the isotopic composition of the vegetation from which it was derived, because the fractionation during decomposition is small relative to the original fractionation during C fixation (Nadelhoffer and Fry, 1988). When one type of vegetation is replaced with another, $\delta^{13}C$ values can be used to identify SOM whether derived from residues of the original vegetation or from the new vegetation and added residues (Bernoux et al., 1998).

The results showed that $\delta^{13}C$ contents after 3 years, in D2 and D3 plots had higher $\delta^{13}C$ values. D3 plots had higher $\delta^{13}C$ values than D1, D2 plots and C in the subsurface horizon (data not shown). This applied PS has high OC content, therefore application of high doses of PS can increase the soil OC content which can increase the $\delta^{13}C$ value of the soil. The $\delta^{13}C$ value of the surface soil and subsurface soil ranged between −23.63‰ and −25.44‰, and between −23.30‰ and −25.10‰ during the experiment.

Wang and Anderson (1998) observed the average $\delta^{13}C$ value of the SOM to be about −25.5‰ for the whole profile. An analysis of $\delta^{13}C$ value of the SOM of the Amulet transect indicates that a lower $\delta^{13}C$ (−27.5‰) occurs in the surface horizon in the slough area, and becomes less negative (−25.5‰) downward to a depth of 75 cm (Nadelhoffer and Fry, 1988; Gregorich et al., 1995).

The C enrichment with depth is thought to be the result of overall discrimination against ^{13}C during organic matter decomposition or due to the preferential preservation of litter components or organic matter which is enriched in ^{13}C (Nadelhoffer and Fry, 1988). However, in the upper slope soil, the $\delta^{13}C$ value of SOM is higher (-22.5‰) in the surface horizon, and becomes more negative (-25.5‰) at 75 cm. The lower $\delta^{13}C$ values for the upper slope soils appear to be a likely consequence of a recent colonization of this site by blue grama grass, with the change to C plants showing up mainly in the surface horizons where organic additions are greater (Desjardins et al., 1994).

CONCLUSIONS

The different sources of PS addition had very significant effects on the enzyme activities, microbial biomass, soluble and black carbon, compared with controls under barley monoculture in Mediterranean soil. This effect may originate from the organic C, N, P and S compounds added with PS application. The highest enzyme activities, microbial biomass, black and soluble carbon were observed in the soil samples from the RPS treatment. There may have been a transient positive effect of the RPS treatments on these measured soil biochemical parameters. RPS from solid feeding diet significantly increased β-glucosidase and β-galactosidase enzyme activities; however, changes in β-galactosidase enzyme activity were not statistically significant. The beneficial effects of the PS additions were less pronounced in the 0- to 30-cm soil layer. The study suggests that TPS applications were not beneficial in terms of MBC, β-glucosidase and β-galactosidase enzyme activities.

PS applications increased the SOC in the terrestrial system; however, applications may increase the micronutrient contents in some areas, therefore the benefits from the use of application depend on management of PS, carbon and environmental quality. The application of manure directly influenced the SOC pools, plant productivity, and SOC turnover compared to control and treatments. Application to silty loam soils of the Mediterranean ecosystem has positive effects on carbon to the atmosphere and groundwater. Certain doses seem to be very suitable in terms of sustainable agriculture and can also improve physical, chemical and biological soil properties.

$\delta^{13}C$ NMR proved to be useful to track the fate of applied organic residues. From the data available, if there is a distinct difference between the $\delta^{13}C$ values of original SOC and added C, then calculations can be made regarding the fate of added organic matter. This is a completely new technology for tracing the oxidation process in the soil. Our stable isotope data have proven that there is a high correlation between carbon isotopic values and the amount of PS added to the soil.

REFERENCES

Alvarez, R., Diaz, R.A., Barbero, N., Santanatoglia, O.J., Blotta, L., 1995. Soil organic carbon, microbial biomass and C–CO$_2$ production from three tillage systems. Soil Till. Res. 33, 17–28.

Bailey, V.L., Smith, J.L., Bolton Jr, H., 2002. Fungal to bacterial biomass ratios in soil investigated for enhanced carbon sequestration. Soil Biol. Biochem. 34, 997–1007.

Balesdent, J., Girardin, C., Mariotti, A., 1993. Site-related ^{13}C of tree leaves and soil organic matter in a temperate forest. Ecology 74, 1713–1721.

Balota, E.L., Machineski, O., Truber, V.P., Auler, P.M.A., 2011. Effect of tillage systems and permanent groundcover intercropped with orange trees on soil enzyme activities. Braz. Arch. Biol. Technol. 54 (2), 221–228.

Bernhardt, E.S., Barber, J.J., Pippen, J.S., Taneva, L., Andrews, J.A., Schlesinger, W.H., 2006. Long-term effects of Free Air CO_2 Enrichment (FACE) on soil respiration. Biogeochemistry 77, 91–116.

Bernoux, M., Cerri, C.C., Neill, C., de Moraes, J.F.L., 1998. The use of stable isotopes for estimating soil organic matter turnover rates. Geoderma. 82 (1–3), 43–58.

Bol, R., Kandeler, E., Amelung, W., Glaser, B., Marx, M.C., Preedy, N., Lorenz, K., 2003. Short-term effects of dairy slurry amendment on carbon sequestration and enzyme activities in temperate grassland. Soil Biol. Biochem. 35, 1411–1421.

Bouldin, D.R., Klausner, S.D., 1998. Managing nutrients in manure: General principles and applications to dairy manure in New York. In: Hatfield, J.L., Stewart, B.A. (Eds.), Animal Waste Utilization: Effective Use of Manure as a Soil Resource. Lewis Publishers, Boca Raton, FL, pp. 65–88.

Brodowski, S., Amelung, W., Haumaier, L., Zech, W., 2007. Black carbon contribution to stable humus in German arable soils. Geoderma.139 (1–2), 220–228.

Büyükkılıç Yanardağ, A., 2013a. Long-term Effect of Pig Slurry Application on Soil Carbon Storage, Quality and Yield Sustainability in Murcia Region, Spain. Geophys. Res. Abstr. 15. EGU2013-7336.

Büyükkılıç Yanardağ, A., 2013b. Management of Soil Organic Carbon using Farmyard Manure in Murcia, Spain. Ph. D. Thesis, Technical University of Cartagena, Murcia, Spain.

Campbell, A.J., Mac Leod, J.A., Stewart, C., 1997. Nutrient characterization of stored liquid hog manure. Can. Agricultural Eng. 39, 43–48.

Chantigny, M.H., 2003. Dissolved and water-extractable organic matter in soils: A review on the influence of land use and management practices. Geoderma.113, 357–380.

Chen, G., Zhu, H., Zhang, Y., 2003. Soil activities and carbon and nitrogen fixation. Res. Microbiol. 154, 393–398.

Christ, M., David, M.B., 1996. Dynamics of extractable organic carbon in Spodosol forest floors. Soil Biol. Biochem. 28, 1171–1179.

Cosandey, A.C., Maitre, V., Guenat, C., March 2003. Temporal denitrification pattern of different horizon of two riparian soils. Eur. J. Soil Sci. 54, 25–37.

Currie, L.A., Benner, Jr. B.A., Kessler, J.D., Klinedinst, D.B., Klouda, G.A., Marolf, J.V., Slater, J.F., Wise, S.A., Cachier, H., Cary, R., Chow, J.C., Watson, J., Druffe, E.R.M., Masiello, C.A., Eglinton, T.I., Pearson, A., Reddy, C.M., Gustafsson, Ö., Quinn, J.G., Hartmann, P.C., Hedges, J.I., Prentice, K.M., Kirchstetter, T.W., Novakov, T., Puxbaum, H., Schmid, H., 2002. A critical evaluation of interlaboratory data on total, elemental, and isotopic carbon in the carbonaceous particle reference material, NIST SRM 1649a. J. Res. Natl. Inst. Stand. Technol. 107, 279–298.

Dambreville, C., Hallet, S., Nguyen, C., Morvan, T., Germon, J.C., Philippot, L., 2006. Structure and activity of the denitrifying community in a maize-cropped field fertilized with composted pig manure or ammonium nitrate. FEMS Microbiol. Ecol. 56, 119–131.

Daudén, A., Quílez, D., 2004. Pig slurry versus mineral fertilization on corn yield and nitrate leaching in a Mediterranean irrigated environment. Eur. J. Agron. 21, 7–19.

Debosz, K., Rasmussen, P.H., Pedersen, A.R., 1999. Temporal variations in microbial biomass C and cellulolytic enzyme activity in arable soils: Effects of organic matter input. Appl. Soil Ecol. 13, 209–218.

Deng, Q., Zhou, G., Liu, J., Liu, S., Duan, H., Zhang, D., 2009. Responses of soil respiration to elevated carbon dioxide and nitrogen addition in subtropical forest ecosystems in China. Biogeosciences Discuss 6, 8359–8391.

Denman, K.L., Brasseur, G., Chidthaisong, A., Ciais, P., Cox, P.M., Dickinson, R.E., Hauglustaine, D., Heinze, C., Holland, E., Jacob, D., Lohmann, U., Ramachandran, S., da Silva Dias, P.L., Wofsy, S.C., Zhang, X., 2007. Couplings between changes in the climate system and biogeochemistry. In: Solomon, S., Qin, D., Manning, M., Chen, Z., Marquis, M., Averyt, K.B., Tignor, M., Miller, H.L. (Eds.), Climate Change 2007: The Physical Science Basis. Contribution of Working Group I to the Fourth Assessment Report of the Intergovernmental Panel on Climate Change. Cambridge University Press, Cambridge, United Kingdom and New York, NY, USA, pp. 499–587.

Desjardins, T., Andreux, F., Volkoff, B., Cerri, C.C., 1994. Organic carbon and ^{13}C contents in soils and soil size-fractions, and their changes due to deforestation and pasture installation in eastern Amazonia. Geoderma 61, 103–118.

Ding, G., Novak, J.M., Amarasiriwardena, D., Hunt, P.G., Xing, B., 2002. Soil organic matter characteristic as affected by Tillage Management. Soil Sci. Soc. Am. J. 66, 421–429.

Ding, G., Amarasiriwardena, D., Herbert, S., Novak, J., Xing, B., 2000. Effect of cover crop systems on the characteristics of soil humic substances. p. 53–61. In: Ghabbour, E.A., Davis, G. (Eds.), Humic Substances: Versatile Components of Plants, Soil and Water. The Royal Society of Chemistry, Cambridge.

Edwards, N.T., Norby, R.J., 1999. Below-ground respiratory responses of sugar maple and red maple saplings to atmospheric CO2 enrichment and elevated air temperature. Plant Soil 206, 85–97.

Eriksen, J., Jensen, L.S., 2001. Soil respiration, nitrogen mineralization and uptake in barley following cultivation of grazed grasslands. Biol. Fertil. Soils 33, 139–145.

Farage, P., Pretty, J., Ball, A., Feb 2003. Biophysical Aspects of Carbon Sequestration in Drylands. University of Essex.

Farquhar, G.D., Ehleringer, J.R., Kubick, K.T., 1989. Carbon isotope discrimination and photosynthesis. Annu. Rev. Plant. Physiol. Plant. Mol. Biol. 40, 503–537.

Faz Cano, A., Mermut, A.R., Ortiz, R., Benke, M.B., Chatson, B., 2002. ^{13}C CP/MAS-NMR spectra of organic matter as influenced by vegetation, climate, and soil characteristics in soils from Murcia. Spain. Can. J. Soil Sci., 403–411.

Feng, Y.S., Li, X.M., 2001. An analytical model of soil organic carbon dynamics based on a simple "hockey stick" function. Soil Sci. 166, 431–440.

Fierer, Schimel, 2002. Noah Fierer and Joshua P. Schimel. Effects of drying–Rewetting frequency on soil carbon and nitrogen transformations. Soil Biol. Biochem. 34, 777–787.

Forbes, M.S., Raison, R.J., Skjemstad, J.O., 2006. Formation, transformation and transport of black carbon (charcoal) in terrestrial and aquatic ecosystems. Sci. Total. Environ. 370, 190–206.

Francioso, O., Ciavatta, C., Sanchez-Cortes, S., Tugnoli, V., Sitti, L., Gessa, C., 2000. Spectroscopic characterization of soil organic matter in long-term amendment trials. Soil Sci. 165, 495–504.

García, C., Hernández, T., Pascual, J.A., Moreno, J.L., Ros, M., 2000. Microbial activity in soils of SE Spain exposed to degradation and desertification processes. Strategies for their Rehabilitation. In: García, C., Hernández, M.T. (Eds.), Research and perspectives of soil enzymology in Spain. Consejo Superior de Investigaciones Científicas, Madrid, pp. 93–143.

Glaser, B., Haumaier, L., Guggenberger, G., Zech, W., Jan 1, 1998. Black carbon in soils: The use of benzenecarboxylic acids as specific markers. Org. Geochem. 29 (4), 811–819.

Gregorich, E.G., Ellert, B.H., Monreal, C.M., 1995. Turnover of soil organic matter and storage of corn residue carbon estimated from natural ^{13}C abundance. Can. J. Soil Sci. 75, 161–167.

Gregorich, E.G., Liang, B.C., Drury, C.F., Mackenzie, A.F., McGill, W.B., 2000. Elucidation of the source and turnover of water soluble and microbial biomass carbon in agricultural soils. Soil Biol. Biochem. 32, 581–587.

Haefelea, S.M., Wopereis, M.C.S., Schloebohm, A.M., and Wiechmann, H., Jan 8, 2004. Long-term fertility experiments for irrigated rice in the West African Sahel: Effect on soil characteristics. Field Crops Res. 85 (1), 61–77.

Hanson, P.J., Edwards, N.T., Garten, C.T., Andrews, J.A., 2000. Separating root and soil microbial contributions to soil respiration: A review of methods and observations. Biogeochemistry 48, 115–146.

Hatfield, J.L., Stewart, B.A. (Eds.), 1998. Animal waste utilization: Effective use of manure as a soil resource. Lewis Publishers, Boca Raton, FL.

Haynes, R.J., Naidu, R., 1998. Influence of lime, fertilizer and manure applications on soil organic matter content and soil physical conditions: A review. Nutrient Cycling in Agroecosystems 51, 123–137.

Haynes, R.J., 2000. Labile organic matter as an indicator of organic matter quality in arable and pastoral soils in New Zealand. Soil. Biol. Biochem. 32, 211–219.

Holeplass, H., Singh, B.R., Lal, R., 2004. Carbon sequestration in soil aggregates under different crop rotations and nitrogen fertilization in an Inceptisol in southeastern Norway. Nutr. Cycl. Agroecosyst. 70, 167–177.

Ivanova, G., Randall, E.W., 2003. ^{13}C NMR and Mass Spectrometry of Soil Organic Matter. Cent. Eur. J. Chem. 1, 10–27.

Jimenez, M.P., Horra, A.M., Pruzzo, L., Palma, R.M., 2002. Soil quality: A new index based on microbiological and biochemical parameter. Biol. Fertil. Soils 35, 302–306.

Kavdır, Y., Ekinci, H., Yüksel, O., Mermut, A.R., Dec, 2005. Soil aggregate stability and ^{13}C CP/MAS-NMR assessment of organic matter in soils influenced by forest wildfires in Çanakkale, Turkey. Geoderma.129 (3–4), 219–229.

Knicker, H., 2007. How does fire affect the nature and stability of soil organic nitrogen and carbon? A review. Biogeochem 85, 91–118.

Kumada, K., 1983. Carbonaceous materials as a possible source of soil humus. Soil Sci. Plant Nutr. 29 (1983), 383–386.

Kuzyakov, Y., Subbotina, I., Chen, H., Bogomolova, I., Xu, X., 2009. Black carbon decomposition and incorporation into soil microbial biomass estimated by ^{14}C labeling. Soil Biol. Biochem. 41, 210–219.

Larney, F.J., Olson, B.M., Janzen, H.H., Lindwall, C.W., 2000. Early impact of topsoil removal and soil amendments on crop productivity. Agron. J. 92, 948–956.

Lichtfouse, É., Chenu, C., Baudin, F., Leblond, C., Da Silva, M., Behar, F., Derenne, S., Largeau, C., Wehrung, P., Albrecht, P., 1998. A novel pathway of soil organic matter formation by selective preservation of resistant straight-chain biopolymers: Chemical and isotope evidence. Org. Geochem. 28, 411–415.

Lloyd, J., Taylor, J.A., 1994. On the temperature dependence of soil respiration. Funct. Ecol. 8, 315–323.

Loro, P.J., Bergstrom, D.W., Beauchamp, E.G., 1997. Intensity and duration of denitrification following application of manure and fertilizer to soil. J. Environ. Qual. 26, 706–713.

Luo, Y., Zhou, X., 2006. Soil respiration and the environment. Elsevier, San Diego, CA.

Ma, Li, Yang, L-Z., Xia, L-Z., Shen, M-X., Yin, S-X., Li, Y-D., April, 2011. Long-Term effects of inorganic and organic amendments on organic carbon in a paddy soil of the Taihu Lake Region, China. Pedosphere 21 (2), 186–196.

Ma, X.Z., Chen, L.J., Chen, Z.H., Wu, Z.J., Zhang, L.L., Zhang, Y.L., 2010. Soil glycosidase activities and water soluble organic carbon under different land use types. J. Soil Sci. Plant Nutr. 10 (2), 1–9.

Manna, M.C., Swarup, A., Wanjari, R.H., Ravankar, H.N., Mishra, B., Saha, M.N., Singh, Y.V., Sahi, D.K., Sarap, P.A., Sep 14, 2005. Long-term effect of fertilizer and manure application on soil organic carbon storage, soil quality and yield sustainability under sub-humid and semi-arid tropical India. Field Crops Res. 93 (2–3), 264–280.

Mao, Jingdong, Olk, Dan C., Fang, Xiaowen, He, Zhongqi, Schmidt-Rohr, Klaus, 2008. Influence of animal manure application on the chemical structures of soil organic matter as investigated by advanced solid-state NMR and FT-IR spectroscopy. Geoderma.146, 353–362.

Mao, J.D., Hu, W., Schmidt-Rohr, K., Davies, G., Ghabbour, E.A., Xing, B., 2000. Quantitative characterization of humic substances with solid state NMR. Soil Sci. Soc. Am. J. 64, 873–884.

Marcato, C.E., Mohtar, R., Revel, J.-C., Pouech, P., Hafidi, M., Guiresse, M., 2009. Impact of anaerobic digestion on organic matter quality in pig slurry. Int. Biodeterioration & Biodegradation 63 (3), 260–266. ISSN 0964-8305.

Masiello, C.A., Druffel, E.R.M., 2003. Organic and black carbon 13C and 14C through the Santa Monica Basin oxic–anoxic transition. Geophys. Res. Lett. 30 (4).

McGill, W.B., Cannon, K.R., Robertson, J.A., Cook, F.D., 1986. Dynamics of soil microbial biomass and water-soluble organic C in Breton L after 50 years of cropping to two rotations. Can. J. Soil Sci. 66, 1–19.

McGinn, S.M., Akinremi, O.O., McLean, H.D.J., Ellert, B., 1998. An automated chamber system for measuring soil respiration. Can. J. Soil Sci. 78, 573–579.

Moyano, F.E., Kutsch, W.L., Schulze, E.-D., 2007. Response of mycorrhizal, rhizosphere and soil basal respiration to temperature and photosynthesis in a barley field. Soil Biol. Biochem. 39, 843–853.

Moyano, F.E., Kutsch, W.L., Rebmann, C., 2008. Soil respiration fluxes in relation to photosynthetic activity in broad-leaf and needle-leaf forest stands. Agricultural and forest meteorol. 148 (1), 135–143.

Nadelhoffer, K.J., Fry, B., 1988. Controls on natural nitrogen-15 and carbon-13 abundance in forest soil organic matter. Soil Sci. Soc. Am. J. 52, 1633–1640.

Nissenbaum, A., Schallinger, K.M., 1974. The distribution of the stable carbon isotope ($^{13}C/^{12}C$) in fractions of soil organic matter. Geoderma.11, 137–145.

Plaza, C., Hernández, D., García-Gil, J., Polo, A., 2004. Microbial activity in pig slurry-amended soils under semiarid conditions. Soil Biol. Biochem. 36, 1577–1585.

Raich, J.W., Schlesinger, W.H., 1992. The global carbon dioxide flux in soil respiration and its relationship to vegetation and climate. Tellus 44B, 81–99.

Rochette, P., Gregorian, E.G., 1998. Dynamics of soil microbial biomass carbon, soluble organic carbon and CO_2 evolution after three years of manure application. Can. J. Soil Sci. 78, 283–290.

Rochette, P., van Bochove, E., Prevost, D., Angers, D.A., Cote, D., Bert rand, N., 2000. Soil carbon and nitrogen dynamics following application of pig slurry for the 19th consecutive year: II: N2O fluxes and mineral nitrogen. Soil. Sci. Soc. Am. J. 64, 1396–1403.

Rovira, Pere, Duguy, Beatriz, Ramon Vallejo, V., 2009. Black carbon in wildfire affected shrubland Mediterranean soils. J. Plant Nutr. Soil Sci. 2009 (172), 43–52.

Rudrappa, L., Purakayastha, T.J., Singh, D., Bhadraray, S., 2006. Long-term manuring and fertilization effects on soil organic carbon pools in a Typic Haplustept of semi-arid sub-tropical India. Soil Till. Res. 88, 180–192.

Růžková, M., Růžek, L., Voříšek, K., 2008. Soil biological activity of mulching and cut/harvested land set aside. Plant. Soil and Environ. 54, 204–211.

Saby, N., Bellamy, P.H., Morvan, X., Arrouays, D., Jones, R.J.A., Verheijen, F.G.A., Kibblewhite, M.G., Verdoodt, A., Berenyi, U., veges, J., Freudenschu, A., Simota, C., 2008. Will European soil-monitoring networks be able to detect changes in topsoil organic carbon content? Global Change Biol. 14 (10), 2432–2442.

Schmidt, M.W.I., Noack, A.G., 2000. Black carbon in soils and sediments: Analysis, distribution, implications, and current challenges. Global. Biogeochem. Cycles. 14, 777–793.

Schmidt, M.W.I., Skjemstad, J.O., Gehrt, E., Kögel-Knabner, I., 1999. Charred organic carbon in German chernozemic soils. Eur. J. Soil Sci. 50, 351–365.

Simpson and Preston, 2006.

Skjemstad, J.O., Reicosky, D.C., Wilts, A.R., McGowan, J.A., 2002. Charcoal carbon in US agricultural soils. Soil Sci. Soc. Am. J. 66, 1249–1255.

Soil and Water Conservation Society, 1995. Farming for a better environment – A white paper. Soil Water Conserv. Soc. Ankeny, IA.

Sommerfeld, T.G., Chang, C., Entz, T., 1988. Long term annual manure applications increase soil OM and nitrogen, decrease C:N ratio. Soil Soc. Am. J. 52, 1668–1672.

Stevenson, B.A., Kelly, E.F., McDonald, E.V., Busacca, A.J., 2005. The stable carbon isotope composition of soil organic carbon and pedogenic carbonates along a bioclimatic gradient in the Palouse region, Washington State, USA. Geoderma.124, 37–47.

Tripathi, R.P., Sharma, P., Singh, S., 2007. Influence of tillage and crop residue on physical properties and yields of rice and wheat under shallow water table conditions. Soil Tillage Res. 93, 221–226.

Wander, M.M., Yang, X., 2000. Influence of tillage on the dynamics of loose and occluded particulate and humified organic matter fractions. Soil Biol. Biochem. 32, 1151–1160.

Wang, D., Anderson, D., 1998. Stable carbon isotopes of carbonate pendants from Chernozemic soils of Saskatchewan, Canada. Geoderma. 84, 309–322.

Watanabe, O., Osaki, M., Yano, H., Rao, I., 2006. Internal mechanisms of plant adaptation to aluminum toxicity and phosphorus starvation in three tropical forages. J. Plant Nutr. 29, 1243–1255.

Yanardag, I.H., Mermut, A.R., Cano, A., Faz, Yanardag, A.B., 2013. Effect of pig manure on organic carbon functional groups in soils from Southeast of Spain. Unpublished paper.

Index

Note: Page numbers with "f" denote figures; "t" tables.

A

Acacia mangium (*A. mangium*), 177–179, 178t
ACC. *See* 1-aminocyclopropane-1-carboxylic acid
Accumulator plants, 256
Activated carbon, 511–512
Adenosine triphosphate (ATP), 20
Aerial emissions, 41–42
Agricultural activities in Turkey. *See also* Industrial activities in Turkey
    agricultural areas, 296–298
    contaminated waters, 296–298
    fertilizer impurities, 294–295
    input amounts of heavy metal, 296t
    pesticides, 295–296
    stubble burning, 298–299
Agricultural field, 482
Agricultural-based sources. *See also* Industrial-based sources
    cadmium contents, 41t
    chemical compositions of soils, 40t
    fertilizers, 40–41
    pesticides, 39–40
    zinc fertilizer materials, 42t
AHLS. *See* N-acyl-homoserine lactones
Airborne sources, 530–531
ALA. *See* Aminolevulinic acid
ALAD. *See* δ-aminolevulinic acid dehydratase
Alcaligin E, 271
Alfisols, 321
Allelopathy, 10
Alluvial soil, 316
AM. *See Arbuscular mycorrhizae*
Amaranthus, 351
1-aminocyclopropane-1-carboxylic acid (ACC), 263
Aminolevulinic acid (ALA), 65
Ammonium containing amendments, 384
Analysis of variance (ANOVA), 164, 635, 699
Antioxidant defence in plants, 461–462, 462f
    antioxidant enzyme activities, 467t
    LARG and HARG responses, 465
    MDA levels, 466t

    O. sativa, 462
    oxidative stress, 466–468
    regulation, 463t–464t
APX. *See* Ascorbate peroxidase
Aquatic ecosystem, 437f, 439f
AR. *See* Arsenate reductase
Arabidopsis, 31
Arable lands mine tailing water, 675–677
    barrier model for irrigation of crops, 676f
Arbuscular mycorrhizae (AM), 265–266, 482
Aridosols, 321–322
Arsenate (As^{5+}), 65
Arsenate reductase (AR), 450–451
Arsenic (As), 16, 47, 65. *See also* Zinc (Zn)
    carcinogenic effect, 47
    hazard, 441
        awareness build-up poster, 448f
        As contamination in crop, 445–446
        extensive study, 447
        As in groundwater, 443f–444f
        irrigated area, 445f
        Public Health and Engineering Department, 448
        Tubewell water, 443–445
    health hazards of, 16–17
    historical use, 48
    leading sources, 16t
    optimized rhizofiltration process, 18f
    phytoremediation use, 17, 48
    soil and water contamination, 47
Arsenic toxicity (As toxicity), 433
    approaches, 434–435
    arsenic hazards
        agronomic management, 468–470
        arsenic-contaminated soils phytoremediation, 470–482, 471t
        Bangladesh perspective, 441–448
        phytoextraction mechanism, 472f
        remediation, 468
    articles publishing, 434f
    As-hyperaccumulators, 473t–476t, 477f–478f
    As-tolerant plant species, 481t

Arsenic toxicity (As toxicity) *(Continued)*
  biotechnology role, 485
  PC biosynthesis, 483f
  presumed phytochelatin-mediated
    detoxification, 483–484
  contamination in soil, 436–438
    dynamics, 439f
    sources and routes, 437f
    Worldwide input, 437f
  contents in soils, 442t
  environmental chemistry, 435–436, 435f
  plant responses, 451
    growth, 455–456
    growth, physiology and productivity,
      452t–453t
    oxidative stress, 460–461
    photosynthesis, 457–459
    in rice plants, 455f
    seed germination, 451–455
    water relations, 456–457
    yield, 459–460
  soil threshold values, 442t
  status in world, 438–441
    documented world problem distribution,
      440f
    Worldwide emission of heavy metals,
      440f
  transportation in plants, 448–451
  As uptake, 448–451, 449f
  routes, 449f
Arsenic-contaminated soils phytoremediation,
  470–482, 471t
  As-hyperaccumulators, 473t–476t,
    477f–478f
  As-tolerant plant species, 481t
  phytoextraction mechanism, 472f
As^{5+}. *See* Arsenate
Ascorbate peroxidase (APX), 461–462
Asphalt batching, 6
ATP. *See* Adenosine triphosphate

B

BAC. *See* Biological absorption coefficient;
    Biological accumulation coefficient
Bacterial strains, 264
Baku–Supsa and Baku–Tbilisi–Ceyhan oil
    pipelines, 549, 550f
Bangladesh
  arsenic hazard, 441–448
  data analysis
    DIM in vegetables, 340
    HRI of heavy metal pollution, 340–341
    MPI in food crops, 340

P_{CLI} for soil, 340
P_{SLI} of soil, 338
environmental pollution in, 331–332
geology and geomorphology, 333
heavy metal transfer, 350–354
  for vegetables and plants, 352t–353t
heavy metals
  comparative study in soil and plants,
    349–350
  concentrations in top soil, 341–342
  levels in plants and vegetables, 344–348
  mean concentrations, 339t
  measurement, 337
  sample preparation, 335–336
  vegetables and plants for analysis, 335t
HRI of heavy metals in vegetables, 354–360,
  358t–359t
nuclear power plant in, 332
pollution index assessment
  of different crops, 348–349
  of soil, 342–344
PTF, 338
radioactivity
  measurement, 337–338
  sample preparation, 337
  in soil, 360
  in vegetables, 361–362
sampling, 334–335
  areas, 336f
toxic bio-accumulative metal pollutants, 332
Basella alba, 351
BC. *See* Black carbon
BCF. *See* Bioconcentration factor
Bio-piles, 7
Biochar, 53–54, 512
  and heavy metal pollution, 54
Bioconcentration factor (BCF), 89–91, 163,
  192–194, 480, 481t
Biodiesel crops, 616–617. *See also* Non-food
  crops
Biofuel crops, 616–617. *See also* Non-food
  crops
Biological absorption coefficient (BAC), 12
Biological accumulation coefficient (BAC),
  653, 654t
Biomolecules, 96–97
Bioremediation, 325
Biosolids, 528–529
Biosurfactants, 548, 563–565
Black carbon (BC), 692–693
Black gram (*Vigna mungo*), 469–470
Blastofiltration, 136–137
Brassica juncea (*B. juncea*), 514

Brookhaven National Laboratory (BNL), 233
Buckwheat (*Fagopyrum esculentum*), 113–115
Buffer strips, 234–235
Burdur Lake basin, 297

C

Cadmium (Cd), 23, 44–45, 295, 367, 571
  accumulation effect in plants, 368
  availability, 573
  bioassessment in soil, 370
  bioavailability in soil, 574–575
  chelation, 580
  contamination sources of agricultural soils,
    369–370
  coupled phytoremidiation, 586
  dynamics, 574, 583f
  emissions, 571–572
  factors influencing accumulation, 370
    climatic factors, 372
    microorganisms, 372
    plant factors, 371–372
    soil factors, 370–372
  grain crops, reports regarding, 582,
    584t–585t
  health hazards, 23
  response towards natural elemental inorganic
    amendments, 571
    Ca, 575
    chloride, 578
    iron, 577–578
    manganese, 578
    nitrogen, 576
    phosphorus, 577
    silicon, 578
    sulphur, 576
    Zn, 576–577
  levels in soil, 368–369
  LMWOA, 580
  natural organic additives, 579
  organic acids, 579–580
  organic amendments *vs.*, 578
  plant response to Cd concentrations,
    377–378
  reduction in crop production, 681–682
  retention, 573
  root exudates, 579–580
  soil dynamics, 573
  synthetic organic chelating agents efficacy,
    581–582
  threshold bio-available concentration,
    378–379
  toxic effects in plants, 374f
  uptake and accumulation in plants, 372–378

Cd induced oxidative stress in plants,
    377
  mineral nutrients effect, 376
  miscellaneous toxic effects, 377–378
  photosynthetic pigments effect, 376–377
  plant growth and biomass effect, 375
  seed germination effect, 374–375
  use of phytoremediation, 23–24
Caesium (^{137}Cs), 608
  behaviour in soil and soil factors, 608–609
  phytoextraction potential, 609
  soil-to-plant factor, 610t
Calcium (Ca), 575
  carbonate, 678–680
  containing inorganic amendments, 379–381
Carbon functional groups, 698–699
Carrot (*Daucus carota* L.), 510–511
Catalase (CAT), 461–462
Cation Diffusion Facilitator (CDF), 86–89,
    119–121
Cation exchange capacity (CEC), 115–117,
    371, 503–504, 579, 619
CAtion eXchanger (CAX), 137–141
CCA. *See* Copper-chromium arsenate
Cd-contaminated soils remediation, 379
  metals behaviour
    in response to ammonium containing
      amendments, 384
    in response to sulphur-containing
      amendments, 384–385
  metals response
    to calcium containing inorganic
      amendments, 379–381
    to phosphorus containing amendments,
      381–383
  techniques, 368
CDC. *See* Center for Disease Control
CDF. *See* Cation Diffusion Facilitator
CDTA. *See* 1, 2-cyclohexanediaminetetraacetic
    acid
CEC. *See* Cation exchange capacity
Cellular detoxification of metals, 89–91
Center for Disease Control (CDC), 39–40
Certified reference material (CRM), 632–633
CF. *See* Concentration factor
CGP. *See* Consolidated Growers and
    Processors
Chelate-assisted phytoextraction, 71–72
Chelating agents, 399–400
  comparison of synthetic, 407
  EDDS, 403–406
  EDTA, 400–403
  NTA, 406–407

Chelator assisted heavy metals
  phytoremeidation, 52–53
Chemical analysis, 162–163
Chimney gases, 300–301
Chloride, 578
Chromium, 542
Cobalt, 541
Colocasia, 351
Complex feedback mechanisms, 263
Composite Pollution Load Index (P$_{CLI}$), 340,
  342
Composite Pollution Load Index (P$_{CLS}$), 344
Compost, 510–511
Concentration factor (CF), 615
Concentration technology, 132–134
Consolidated Growers and Processors (CGP),
  243–244
Constructed wetlands, 237–238
Contaminated soil. *See also* Heavy metal(s)
  management, 537
    example of disposal, 539
    preventive steps, 539–540
    use of plants, 538–539
    traditional remediation, 536–537
Contaminated waters, 296–298
Continuous phytoextraction, 72
Continuous ponding (CP), 469
Copper, 21. *See also* Lead (Pb)
  accumulation, 22f
  health hazards, 21
  hyperaccumulation, 31
  use of phytoremediation
    candidates for, 22f
    to clean up, 21
    copper-tolerant species, 21–22
Copper concentration
  Acacia mangium, 184–185, 185t
  Dyera costulata, 185–187, 186t
  growth media and dry biomass, 189–190,
    190t
  in growth media
    after harvest, 164–168, 168f
    before planting, 164, 166t
    soil pH, 168, 169t
  Hopea odorata, 187–188, 188t
  J. curcas, 183–184, 183t
  phytoremediation potential of tested species,
    191–195
  in plant species, 188–189
Copper-chromium arsenate (CCA), 477–479
Copper-contaminated soil, 151
  background, 147–148
  literature review

environmental pollution, 149
  heavy metals, 149–151
  metal accumulation in plants, 156–159
  remediation of heavy metals, 153–156
  sources of contamination, 149
  toxicity of heavy metals in plants,
    151–153
  uptake and translocation, 153
materials and methods
  data collection, 160–161
  description of study area, 159
  dry biomass of plants, 161
  experimental design and treatments, 160
  greenhouse, 161f
  heavy metals uptake evaluation, 163–164
  laboratory analysis, 161–163
  plant species and planting, 160, 160t
  planting materials, 159
  statistical analysis, 164
objectives, 148–149
problem statement, 148
results and discussion, 164–195
CP. *See* Continuous ponding; Cross
  polarization
CRM. *See* Certified reference material
Cross polarization (CP), 690, 698–699
Cryoborolls, 321
Cryochrepts, 320
Cryumbrepts, 319
CS. *See* Cysteine synthase
1, 2-cyclohexanediaminetetraacetic acid
  (CDTA), 398–399, 581
Cynodon dactylon (*C. dactylon*), 94–96
Cyperrus rotundus (*C. rotundus*), 666
Cysteine synthase (CS), 462

D
Daily metal intake (DIM), 340, 354
  for individual heavy metals, 355t–357t
DAP. *See* Diammonium phosphates
Database file (dbf), 423
DCP. *See* Dicalcium phosphates
Decomposition, 259
Dehydroascorbate reductase (DHAR), 461–462
δ-aminolevulinic acid (δ-ALA), 457–458
δ-aminolevulinic acid dehydratase (ALAD), 65
Department of Environment (DOE), 150
Deposition hypothesis, 111
Depth of roots, 9
Dhaka Export Processing Zone, 331–332
DHAR. *See* Dehydroascorbate reductase
Diammonium phosphates (DAP), 381
Dicalcium phosphates (DCP), 382

Diethylenetrinitrilopentaacetic acid (DTPA), 53
DIM. *See* Daily metal intake
Dimethylarsinic acid (DMA), 448–451
Dimethylselenide (DMSe), 134–135
Dissolved organic carbon (DOC), 371
Dissolved organic matter (DOM), 506
DMA. *See* Dimethylarsinic acid
DMRT. *See* Duncan's Multiple Range test
DMSe. *See* Dimethylselenide
DOC. *See* Dissolved organic carbon
DOE. *See* Department of Environment
DOM. *See* Dissolved organic matter
Dry biomass production
    Acacia mangium, 177–179, 178t
    Dyera costulata, 179–180, 179t
    Hopea odorata, 180–181, 181t
    J. curcas, 176–177, 176t
    plant species, 182, 182f
Dry weight (DW), 113–115, 633, 649–650
DTPA. *See* Diethylenetrinitrilopentaacetic acid
Duncan's Multiple Range test (DMRT), 164
DW. *See* Dry weight
Dyera costulata, 179–180, 179t
Dystrochrepts, 319

E

Eber Lake, 297–298
EBSL. *See* Ecologically Based Screening Levels
Ecologically Based Screening Levels (EBSL),
    613
Economical Cooperation and Development
    Organisation (OECD), 298
EDDHA. *See* Etylenediamine-di
    (o-hydroxyphenylacetic acid)
EDDS. *See* Ethylene diamine disuccinic acid
EDTA. *See* Ethylene diamine tetra acetic acid
EDXRF. *See* Energy Dispersive X-ray
    Fluorescence
EGTA. *See* Ethyleneglycol-bis-tetraacetic acid
Electroremediation, 5
Elemental allelopathy, 110–111
Endophytes, 266–267
Energy Dispersive X-ray Fluorescence
    (EDXRF), 337
Engineering-based technologies, 220–221
Enhydra fluctuans, 351
Entisols, 318
Environment Protection Agency (EPA), 243
Environmental cleanup, 261
    use of plants for, 538
Environmental pollution, 149
    by heavy metals and radionuclides, 331–332
    study areas, 334

Environmental Protection Agency (EPA),
    536
Environmental remediation, 255
EPA. *See* Environment Protection Agency;
    Environmental Protection Agency
Erosion, 292–293
Ethylene diamine disuccinate. *See* Ethylene
    diamine disuccinic acid (EDDS)
Ethylene diamine disuccinic acid (EDDS),
    71–72, 398–399, 403, 581
    degradation, 405–406
    drawbacks, 406
    Pb accumulation by plants, 404–405
    Pb solubilization in soil, 404
Ethylene diamine tetraaceticacid (EDTA), 19,
    53, 398–400, 630–632
    effect
        distribution of solution phase, 653f
        morphology, 648–649
        Ni concentration, 649–653, 650f
    for lead extraction, 535–536
    metal chelator, 115–117
    nickel and
        on dry matter production, 649f
        thermodynamic data, 652t
    Pb solubilization in soil, 400–401
    Pb translocation to shoot tissues, 401–402
    Pb uptake by plants, 401
    risks association with, 402–403
Ethyleneglycol-bis-tetraacetic acid (EGTA),
    398–399, 581
Etylenediamine-di(o-hydroxyphenylacetic
    acid) (EDDHA), 398–399
Eutroboralfs, 321
Eutrochrepts, 320
EW. *See* Exempt Waste
Ex situ method, 14
Exclusion process, 94–96
Exempt Waste (EW), 218

F

Farm yard manure (FYM), 509, 579
FDWT. *See* Fractal dimension of reflectance
    with wavelet transform
Fertile plain soils, 303–304
Fertilizers, 1–2, 40–41, 528
    impurities, 294–295
    management, 469–470
Fibrous roots, 9
Field study
    nickel
        bioaccumulation factor, 645, 646t–647t
        enrichment coefficient, 645, 646t–647t

Field study *(Continued)*
  plant
    analysis, 632–633
    sampling, 634t
    selection, 632–633
  serpentine soils properties, 635–637, 636t
  site description, 632–633
  soil characterization, 632–633
Fluvaquents, 318
Flying ash, 2
Forensic phytoremediation, 236
Fractal dimension of reflectance with wavelet
    transform (FDWT), 417–418
Full width half maximum (FWHM), 337–338
FYM. *See* Farm yard manure

G

g-EC enzymes, 21
γ-aminobutyric acid (GABA), 65
Gastrointestinal tract (GI tract), 1–2
GDAR. *See* General Directorate of Agricultural
    Research
GDH. *See* Glutamate dehydrogenase
General Directorate of Agricultural Research
    (GDAR), 632
Genetic engineering
  enzyme activities improvement, 77–78
  to improving phytoremediation, 137–141,
    140t
  metal sequestering proteins and peptides,
    75–77
Genetically modified organism (GMO), 68–69
Geobotany, 236–237
Geographic information system (GIS), 416,
    418–421
GI tract. *See* Gastrointestinal tract
GIS. *See* Geographic information system
Glacial soil, 317–318
Global Positioning System (GPS), 416–417
Glutamate dehydrogenase (GDH), 556
Glutamine synthetase (GS), 377, 556
Glutathione (GSH), 89–91, 450–451, 461–462
Glutathione peroxidase (GPX), 461–462
Glutathione reductase (GR), 461–462
Glutathione S-transferase (GST), 461–462, 553
GMO. *See* Genetically modified organism
Goktas-Artvin copper factory, 300
GPS. *See* Global Positioning System
GPX. *See* Glutathione peroxidase
GR. *See* Glutathione reductase
Grain crops, reports regarding, 582, 584t–585t
Gravelly soil, 317
"Green" approach, 69

Greenhouse study
  artificial soil contamination, 633–635
  BAC, 653, 654t
  EDTA effect
    on morphology, 648–649
    on Ni concentration, 648–653
  plant analysis, 633–635
  pot experiment, 633–635, 645, 648t
  soil characterization, 633–635
Ground water, 239
GS. *See* Glutamine synthetase
GSH. *See* Glutathione
GST. *See* Glutathione S-transferase
Gynura pseudochina (*G. pseudochina*), 666,
    671f

H

Haplaquents, 318–319
Haplumbrepts, 320
Haplustalfs, 321
Haplustolls, 320
HARG. *See* High-As-accumulating
Health Risk Index (HRI), 340–341, 354
  for heavy metals in vegetables, 354–360,
    358t–359t
Heavy metal accumulation in serpentine flora
  EDTA, 630–632
  field study
    Ni bioaccumulation factor, 645, 646t–647t
    Ni enrichment coefficient, 645, 646t–647t
    plant analysis, 632–633
    plant sampling, 634t
    plant selection, 632–633
    serpentine soil properties, 635–637, 636t
    site description, 632–633
    soil characterization, 632–633
  greenhouse study
    artificial soil contamination, 633–635
    BAC, 653, 654t
    EDTA effect on morphology, 648–649
    EDTA effect on Ni concentration,
      648–653
    plant analysis, 633–635
    pot experiment, 633–635, 645, 648t
    soil characterization, 633–635
  hyperaccumulator plants, 630
  and identification, 637–645, 642t
  nickel contamination, 629–630
  nickel hyperaccumulator, 631t
  soil concentrations, 638t–639t
  statistical analysis, 635
Heavy Metal transporting ATPase (HMAs),
    119–121

Heavy metal(s), 2, 39, 149–150, 525.
  See also Radionuclides
  agriculture, 525
  benefits, 541–542
  bioavailability, 526
  biochar application, 53–55
  challenges, 542–543
  chelator assisted, 52–53
  classification, 540
  comparative study in soil and plants,
    349–350
  concentrations in top soil, 341–342
  contaminated soil remediation, 532–536,
    535t
  contamination prevention, 536
  environmental pollution, 331–332, 526
  heavy metal contamination prevention, 536
  hyperaccumulation, 4f
  input amounts, 296t
  levels in plants and vegetables, 344–348
  mean concentrations, 339t
  measurement, 337
  phytoextraction, 51–52
  pollution, 325
    HRI from vegetables, 340–341
    MPI in food crops, 340, 348
    P_{SLI}, 342
    in studied soil, 342
  potential risk, 531
  regulatory guidelines, 531–532, 533t
  removal, 51
  sample preparation, 335–336
  in soil, 39
    agricultural-based sources, 39–41
    build-up of, 43
    industrial-based sources, 41–43
  soil concentration ranges, 531–532, 533t
  soil pollutants, 288
    approach to removal of, 14
    ex situ method, 14
    in situ method, 14–15
  sources in environment, 540
  sources in soil, 527
    airborne sources, 530–531
    biosolids, 528–529
    fertilizers, 528
    industrial wastes, 530
    manures, 528–529
    metal mining, 530
    milling processes, 530
    pesticides, 528
    wastewater, 529–530
  status in Nepal, 322–324

  toxicity, 151–153
  toxicity symptoms in plants, 43–44
    As, 47–48
    Cd, 44–45
    Ni, 45–46
    Pb, 46–47
  uptake by plants, 332–333
  used in major industries, 541t
  vegetables and plants for analysis, 335t
Helencha, 347, 360
Helianthus annuus. *See* Sunflower
High Level Waste (HLW), 218–219
High molecular weight (HMW), 267
High-As-accumulating (HARG), 465
Highways, 303
Histidine (His), 77
Histogram, 419–420
HLW. *See* High Level Waste
HMAs. *See* Heavy Metal transporting ATPase
HMW. *See* High molecular weight
Hopea odorata, 180–181, 181t
"Hormesis" effect, 44–45
HRI. *See* Health Risk Index
Human Developed Index, 307
Hydraulic control, 235, 260
Hydraulic plume control. *See* Hydraulic control
Hydrocarbons oxidative degradation, 557–559
  benzene metabolism, 557f–558f
  cyclohexane metabolism, 558f
Hyperaccumulation, 15, 221–223
Hyperaccumulator(s), 51, 156, 539
  for phytoremediation of metal, 108–109
  plants, 256, 629–630
    accumulated heavy metal /metalloids, 257
    accumulating metals, 630
    heavy metal, 257
    minimum amount and metals /metalloids
      relationship, 256t
    nickel hyperaccumulator, 631t
    for phytoremediation, 630–632
    Thlaspi species, 655
  species of metals, 158t
Hypertolerance, 89–91

I

IAA. *See* Indoleacetic acid
IAEA. *See* International Atomic Energy
    Agency
ICP-MS. *See* Inductively coupled plasma mass
    spectrophotometry
IDW. *See* Inverse distance weighted
Illite reaction, 678–680
ILW. *See* Intermediate Level Waste

In situ biological remediation, 221
In situ method, 14–15
In-place inactivation. *See* Phytostabilization
Inceptisols, 318
Indoleacetic acid (IAA), 263, 270
Inductively coupled plasma mass
    spectrophotometry (ICP-MS),
    632–633
Industrial activities in Turkey
    industrial waste, 299–300
    OIZ, 299
    petroleum pollution, 301
    thermal power plants, 300–301
Industrial process, 603
Industrial wastes, 299–300, 530
Industrial-based sources
    aerial emissions, 41–42
    raw effluents, 42–43
Inorganic pollutants, 8
Intermediate Level Waste (ILW), 218
Intermittent ponding (IP), 469
International Atomic Energy Agency (IAEA),
    213–215
Inverse distance weighted (IDW), 421
IP. *See* Intermittent ponding
Iron, 541, 577–578

K

"King of poisons". *See* Arsenic (As)
Krigging, 421–423. *See also* Inverse distance
    weighted (IDW)
    database file creation, 423
        Arc Map GIS window, 425f
        geostatistical analyst in ARC Map, 427f
        heavy metal concentration, 426f
        spatial variation in nickel, 425f

L

Ladies' fingers, 351
Land degradation, 316
Land pollution. *See* Soil pollution
LARG. *See* Low-As-accumulating
Lead (Pb), 18, 46–47, 65. *See also* Mercury
    (Hg)
    concentration in potatoes, 19f
    health hazards, 18
    phytoremediation use, 19–20
Lead sulphide (PbS), 46
Lettuce (*Lactuca sativa* L.), 510–511
Liming process, 575
Lipopolysaccharide (LPS), 271
LLW. *See* Low level waste
LMW. *See* Low molecular weight

LMWOA. *See* Low-molecular-weight organic
    acids
Low level waste (LLW), 218
Low molecular weight (LMW), 267
Low-As-accumulating (LARG), 465
Low-molecular-weight organic acids
    (LMWOA), 53, 580
LPS. *See* Lipopolysaccharide

M

Mae Sot zinc mine, 668–670
    Cd contamination, 670
    polyphenolic compounds, 672–673
    Thai scientists, 673–674
Malate dehydrogenase (MDH), 556
Malondialdehyde (MDA), 460, 466t
Manganese, 578
Manganism, 347
Manures, 509–510, 528–529
MAP. *See* Monoammonium phosphate
MAPK. *See* Mitogen-activated protein kinase
Marigold (*Tagetes patula*), 479–480
MATE. *See* Multidrug and toxin extrusion
MCA. *See* Multichannel analyzer
MDA. *See* Malondialdehyde
MDH. *See* Malate dehydrogenase
MDHAR. *See* Monodehydroascorbate
    reductase
Mercury (Hg), 24, 64–65
    health hazards, 24
    use of phytoremediation
        to clean up, 24
        merA gene, 25, 25f
        merB gene, 25, 26f
        merC gene, 25, 26f
        methyl mercury, 25
Mersin-Findikpinari (Turkey), 630
    identification and heavy metal concentration,
        637–645
    soil plants concentration, 643t–644t
    soil plants distribution, 642t
    surface soil samples, 632–633
Metal
    accumulating plant species, 110–111
        compartmentation, 111
        complex formation, 111
        deposition hypothesis, 111
        hyperaccumulation, 112–113
    chelators, 118–119
    detoxification, 119–121, 121f
    excluders, 94–96, 110
    hyperaccumulators, 398, 470–472
    indicators, 110

intoxication, 66
mining, 530
phytoavailability, 515
sequestering proteins and peptides, 75–77
tolerance, 77–78
Metal pollution index (MPI), 340, 348
among different species of crops, 349
in different vegetables and plants, 348f
Metal-contaminated soils
geophysical techniques
GIS, 418–421
GPS, 417
histogram, 419–420
QQplot technique, 420
remote sensing, 417–418
semivariogram, 420–421, 423f
spatial distribution of lead, 420f
spatial distribution of zinc, 421f, 424f
heavy metals, 416
IDW, 421
immobilization of metals, 504–505
industrialization, 415
krigging, 421–423
database file for GIS environment,
423–426
metal phytoavailability, 515
OM, 503–504
in metals phytoavailability, 506–509
organic acids, 515
phytoremediation with organic amendments
activated carbon, 511–512
biochar, 512
compost, 510–511
manures, 509–510
and phytoavailability, 509
phytoremediation with, 515–516
pressmud, 512
residual effect, 514–515
effect of time, 512–514
plants use for, 538
phytoextraction, 539
phytostabilization, 539
rhizofiltration, 538
pollutants, 503
sources of metals, 505–506
urban agricultural soils, 415–416
Metallophytes, 15
Metallothioneins (MT), 75–76, 89–91
Metalloids, 435–436, 483
Methyl mercury, 25
Microbial
biomass carbon, 695–697
colonization, 266–267

inhabitants, 266
Microbial based remediation. See also
Rhizoremediation
bacteria resist toxicity of metals, 66–67
natural selection process, 67
toxic effects, 66
Microorganisms, 550–551
Milling processes, 530
Mimulus guttatus (M. guttatus), 31
Mine tailings, 219
Mine water, 675–677
Mineral deficiency, 323
Mineral nutrients effect, 376
Mining, 303–304
Mitogen-activated protein kinase (MAPK), 465
MMA. See Momethylarsonic acid
Model experiments, 559–566
decrease of TPH content, 560f–561f
GC diagrams of hexane, 564f
phytoremediation of soil, 566f
TPH content in soils, 561f–563f
Mollisols, 320
Molybdenum deficiency (Mo deficiency), 324
Momethylarsonic acid (MMA), 448–451
Monoammonium phosphate (MAP), 381, 576
Monodehydroascorbate reductase (MDHAR),
461–462
MPI. See Metal pollution index
MT. See Metallothioneins
Multichannel analyzer (MCA), 337–338
Multidrug and toxin extrusion (MATE),
137–141
Municipal waste, 2
Municipal wastewater irrigation, 332
Murdannia spectabilis (M. spectabilis), 666,
671f
Mycorrhizae, 94–96

N
N-acyl-homoserine lactones (AHLS), 268
National Research Council (NRC), 418–419
Natural organic additives, 579
Natural selection process, 67
NDVI. See Normalized difference vegetation
index
Nepal, 313
divisions, 313
generalized geological cross section, 315f
improper soil nutrient management, 322
nutrient and heavy metal status, 322–323
geology and soil type, 324
paddy crop rotations, 323
reports from Terai plains, 323

Nepal *(Continued)*
    soil fertility, 323–324
    studies in Arun Valley, 324
  physiographical division, 314t, 315f
  remediation of toxicity from soil, 325
  remediation studies in soil of Nepal, 326
  soils, 316, 317f
    alluvial soil, 316
    characteristics, 315–316
    fertility loss, 322
    glacial soil, 317–318
    gravelly soil, 317
    orders, 317–318
    residual soil, 317
    sandy and alluvial soil, 316–317
Nickel (Ni), 45–46
  bioaccumulation factor, 645, 646t–647t
  contamination, 629–630
  enrichment coefficient, 645, 646t–647t
  hyperaccumulator, 631t
  spatial variation in, 425f
Nitrilotriacetate. *See* Nitrilotriacetic acid (NTA)
Nitrilotriacetic acid (NTA), 53, 398–399, 406, 581
  Pb phytoremediation, 406–407
Nitrogen (N), 384, 576
Nitrogen, phosphorus and potassium (NPK), 316
Nitrophos (NP), 381
NMR. *See* Nuclear magnetic resonance
Non-food crops, 615–616
  biodiesel crops, 616–617
  biofuel crops, 616–617
  forestry, 616
Non-plant methods, 220–221
"Nor Westers", 333
Normalized difference vegetation index (NDVI), 417–418
NP. *See* Nitrophos
NPK. *See* Nitrogen, phosphorus and potassium
NRC. *See* National Research Council
NTA. *See* Nitrilotriacetic acid
Nuclear fuel cycle, 602–603
Nuclear magnetic resonance (NMR), 698–699
Nuclear weapons
  production, 602
  testing, 602
"Nugget", 420–421

O

OECD. *See* Economical Cooperation and Development Organisation
Oil hydrocarbons, 547, 555, 559. *See also* Phytoremediation of soil
  GS diagrams, 552f
  TPH content, 560f–562f
OIZ. *See* Organized Industrial Zones
OM. *See* Organic matter
Organic acids, 515, 579–580
Organic amendments, 504–505, 507
  in contaminated soils, 509–512
  residual effect on metal phytoavailability, 514–515
  time effect on decomposition, 512–514
Organic matter (OM), 503–505
  compost, 510–511
  role in phytoavailability of metals, 506–509
Organic pollutants, 8. *See also* Inorganic pollutants
Organized Industrial Zones (OIZ), 299

P

Padaeng industry, 661–664
Para phytoremediation, 223
Parent material, 293–294
Particulate organic matter (POM), 506
Pb. *See* Lead
Pb-contaminated soils, 397
  chelating agents, 399–407
  metal hyperaccumulators, 398
  phytoextraction approach, 398–399
  problem of, 399
PbS. *See* Lead sulphide
PC. *See* Phytochelatins
P_{CLI}. *See* Composite Pollution Load Index
P_{CLS}. *See* Composite Pollution Load Index
PCS. *See* Phytochelatin synthase
Pearl millet (*Pennisetum typhoides*), 451
Permanent raised bed (PRB), 469
Peroxidases (POX), 461–462
Pesticides, 1–2, 39–40, 295–296, 528. *See also* Fertilizers
Petroleum pollution, 301
PGPR. *See* Plant growth promoting rhizobacteria
Phosphate rocks (PR), 381
Phosphorus (P), 381, 577
  containing amendments, 381–383
Photosynthesis, 457–459
Photosynthetic pigments effect, 376–377
Photosystem II (PS II), 376–377
Physic nut (*Jatropha curcas* L.), 176–177, 176t, 672–673
Physical analysis, 161
Physico-chemical properties of control media, 164–195, 165t
Phytoabsorption. *See* Phytoextraction

Phytoaccumulation. *See* Phytoextraction
Phytoavailability of metals
  in aquatic and terrestrial environments,
    504–505
  in contaminated soils, 503–504, 509
    activated carbon, 511–512
    biochar, 512
    compost, 510–511
    manures, 509–510
    phytoremediation, 515–516
    pressmud, 512
    residual effect, 514–515
    effect of time, 512–514
  OM role, 506–509
  organic acids and, 515
  residual effect of organic amendments,
    514–515
  time effect on decomposition, 512–514
Phytochelatin synthase (PCS), 450–451
Phytochelatins (PC), 32, 89–91, 118–119, 373,
    378, 450–451
Phytodegradation, 13, 155–156, 231, 259, 604
Phytoextraction, 12, 69–70, 86–89, 132–134,
    156, 604–605, 629–630
  cellular detoxification of metals, 89–91
  chelate-assisted phytoextraction, 71–72
  continuous, 72
  environmental aspects, 13
  heavy metals, 12
  metal-contaminated soil, 539
  metals and metalloids, 71
  mobilization into soil solution, 70–71
  plant species, 12
  potential for, 605–607
    of Cs, 609
    of Sr, 609
    TF, 606t
    untreated and treated soils, 606t
    yearly reduction, 607t
  rhizosphere changes, 91–93
  roles, 69–70
  root adaptations, 91–93
  slow removal process, 223–231
Phytoimmobilization. *See* Phytostabilization
Phytoinvestigation, 236–237
Phytomanagement
  for agriculture in Mae Sot zinc mine,
    668–670
    Cd contamination, 670
    polyphenolic compounds, 672–673
    Thai scientists, 673–674
  arable lands mine tailing water, 675–677
    barrier model for irrigation of crops, 676f

Cd reduction, 681–682
Padaeng industry, 661–664
Phatat Phadaeng sub-district, 661–664, 662f
and phytotechnologies, 664
soil remediation, 677–680
zinc-mine-industry-ravaged ecosystem,
    665–668
zinc
  and Cd contamination, 664t
  production process, 661–664, 663f
  uses, 663f
Phytomining, 27, 28f, 108–109
Phytoremediation of soils, 1, 7–8, 63, 107–108,
    255–257, 599, 604, 668–670
  accessibility of sites for, 30
  accumulator plants, 256
  associated risks, 608
  background, 108–109
  Baku–Supsa and Baku–Tbilisi–Ceyhan oil
    pipelines, 549, 550f
  cap, 234–235
  challenges, 28–29
    accumulation of contaminants, 30
    availability of space, 29
    climatic conditions, 29
    genetically modified plants, 30
    less tolerance in plants, 29–30
    plants' roots, 29
    slow growth cycle, 29
  economics of, 11
  efficiency, 261
  enhancing bioremediation
    through genetic engineering, 67
    novel metal binding peptides and proteins,
      68
  environmental clean-up, 538
  factors affecting phytoremediation, 115
    availability of metals in soil, 115–117
    metal chelators, 118–119
    plant microbe interactions, 117–118
    plant uptake, 117
    translocation, 117
  genetic engineering, 75, 137–141, 140t
    enzyme activities improvement, 77–78
    metal sequestering proteins and peptides,
      75–77
  genetic improvement of plants, 30–31
    hyperaccumulation, 31
    oxidation state of heavy metals, 32
    phytochelatins, 32
    recombinant DNA technology, 31
  in Georgia, 547
  global scenario of soil pollution

Phytoremediation of soils *(Continued)*
  America, 3
  Asia, 3
  Europe, 2–3
  Fiji, 4
  Kiribati, 4
  Marshall Islands, 4
  Nauru, 4
  Niue, 4
  Pacific Islands, 3–4
  GMO, 68–69
  "green" approach, 69
  groups, 604
  heavy metal soil pollutants, 14–16
  hydraulic control technique, 260
  hydrocarbons oxidative degradation,
    557–559
  hyperaccumulator plants, 256
    accumulated heavy metal /metalloids, 257
    heavy metal, 257
    minimum amount and metals /metalloids
      relationship, 256t
  ideal plant characteristics for, 15
  and mechanisms, 86–97
  metal detoxification, 119–121
  metal pollutants and human health, 64
    As, 65
    Hg, 64–65
    metal intoxication, 66
    Pb, 65
  metal-contaminated soils, 87t–88t
  methods, 257, 258f
  microbial based remediation
    bacteria resist toxicity of metals, 66–67
    natural selection process, 67
    toxic effects, 66
  microorganisms selection, 550–551
  model experiments, 559–566
  phytodegradation, 259
  phytoextraction, 69–72, 257–258, 605–607
  phytomanagement, 604–605
  phytostabilization, 74–75, 259
  phytovolatilization, 260
  plant
    growth stimulation, 268–271
    and microbes interactions, 260–266
    plant–microbial interaction revelation, 559
    response to heavy metals, 109–115
  prospects for, 26
    cleans up contaminants, 27–28
    environmentally friendly, 27
    genetically engineered, 28
    help mining industries, 27

    phytomining, 27, 28f
    reduction of noise pollution, 28
    solar energy driven and cost effective,
      27
  recent trends and approaches, 131–132
  rhizodegradation, 259–260
  rhizofiltration, 258–259
  rhizoremediation, 73–74
  rhizosphere microbiome, 266–268
  rhyzosphere, 548
  selection of plants, 551
    activation of GDH, 557
    biocemical criteirion, 554
    GS diagrams, 552f
    GSTs, 553
    hydrocarbon degradation, 553
    metabolic process, 557
    oil biodegradation process, 555
    ryegrass ultrastructure, 556f
  site remediation
    bio-piles, 7
    electroremediation, 5
    soil flushing, 5
    soil vapour extraction, 5
    soil washing, 7
    stabilization, 6
    technologies for, 5
  soil, plant and green energy recovery system,
    132f
  soil pollutant types
    fertilizers, 1–2
    flying ash, 2
    heavy metals, 2
    municipal waste, 2
    pesticides, 1–2
    wastewater irrigation, 2
  soil types characterization, 549
  system design, 8
    allelopathy, 10
    climate considerations, 10–11
    concentration of pollutants, 8
    depth of roots, 9
    inorganic pollutants, 8
    organic pollutants, 8
    plant growth rate, 9
    plant species, 9
    plant type, 10, 11t
    pollutant characteristics, 8–9
    seed and plant source, 9–10
    transpiration rate, 9
    type of root, 9
  technologies, 12–14, 132–137
  unique clean-up strategy, 547–548

use of, 14–16
waste disposal, considerations for, 12
Phytorestoration. *See* Phytostabilization
Phytosequestration. *See* Phytoextraction
Phytostabilization
  applications, 135–136
  goal, 74–75
  mechanisms, 94–96
  phytoextraction of metals, 93–94
  plants' ability, 259
  pollutant migration, 604
  purpose, 13
  using vegetation, 232
Phytotoxicity, 29–30
Phytotransformation. *See* Phytodegradation
Phytovolatilization, 30
  biomethylation, 134–135
  dispose of metal ions, 32
  heavy metals and organic solvents, 14
  organic and inorganic contaminants, 75
  toxic metals and metaloids tranformation,
    96–97
  tritium, 232–233
  uptake of pollutants, 260, 604
Pig slurry (PS), 690
  application, 690–691
  effect on SOC, 691
    BC, 692–693
    carbon functional groups, 698–699
    change in organic soil carbon, 692f
    CO_2 generation process, 698f
    manure application, 692
    microbial biomass carbon, 695–697
    soil properties, 696t
    soil respiration, 697–698
    soluble carbon, 693–695, 694t
    SOM chemical shift ranges, 700t
    stable carbon isotope, 699–701
Plant analysis
  field study, 632–633
  greenhouse study, 633–635
Plant growth promoting rhizobacteria (PGPR),
    269
Plant growth stimulation, 85–86, 268
  ACC-deaminase activity, 269–270
  biotic or abiotic stress, 268–269
  Fe source for plants, 270
  IAA, 270
  Kluyvera ascorbata SUD165, 270
  PGPR, 269, 271
  plant growth rate, 9
  toxic heavy metal constituents, 271
Plant nutrient matter (PNM), 295

plant species, growth performance of
  Acacia mangium, 171, 172f
  Dyera costulata, 171–174, 173f
  Hopea odorata, 174–175, 175f
  J. curcas, 169–171, 170f
Plant transfer factor (PTF), 332–333,
    350–351
  for heavy metal, 338
  for radionuclides, 338
Plant(s)
  antioxidant defence, 461–462, 462f
    antioxidant enzymes activities, 467t
    LARG and HARG responses, 465
    MDA levels, 466t
    O. sativa, 462
    oxidative stress, 466–468
    regulation, 463t–464t
  cadmium
    accumulation effect, 368
    induced oxidative stress in, 377
    uptake and accumulation in, 372–378
  comparative study of heavy metal contents,
    349–350
  for heavy metal analysis, 335t
  heavy metal transfer, 352t–353t
  levels of heavy metals in, 344–348
  microbe interactions, 117–118
  and microbes interactions, 261
    accumulation, 260–261
    bacterial strains, 264
    complex feedback mechanisms, 263
    endophytic-and rhizobacteria, 265
    metal immobilization, 264
    metal mobilization, 264
    mutualistic relationship, 262
    plant exudates and survival, 263
    plant–endophyte associations, 265
    rhizospherial strains, 263
    rhizospheric control over pollutant
      toxicity, 261–262
    roots, 262
    soil uses, 260
    soil-borne microbial communities,
      262
    transpiration rate and concentrations,
      265–266
  MPI in, 348f, 349
  organic pollutants, 266
  Pearson correlation, 350t
  response to Cd concentrations, 377–378
  response to heavy metals, 109–110
    buckwheat (*Fagopyrum esculentum*),
      113–115

Plant(s) *(Continued)*
  metal accumulating plant species,
    110–115
  metal excluders, 110
  metal indicators, 110
  MTs action in plants, 115f
  sampling, 633, 634t
  selection, 551, 632–633
    activation of GDH, 557
    biocemical criteirion, 554
    GS diagrams, 552f
    GSTs, 553
    hydrocarbon degradation, 553
    metabolic process, 557
    oil biodegradation process, 555
    ryegrass ultrastructure, 556f
  source, 9–10
  tissue analysis, 163
  toxic effects of Cd in, 374f
  uptake, 117
Planting materials
  growth media for copper contamination, 159
  seedlings, 159
Plant–microbial interaction revelation, 559
PNM. *See* Plant nutrient matter
Point of zero charge (PZC), 382
"Poison of kings". *See* Arsenic (As)
Pollutants
  characteristics, 8–9
  concentration of, 8
  inorganic, 8
  organic, 8
Pollution index assessment
  of different crops, 348–349
  of soil
    P_{CLS}, 344
    P_{SLI}, 342
POM. *See* Particulate organic matter
Population, 302
Pot experiment, 645, 648t
Potassium, 361–362
POX. *See* Peroxidases
PR. *See* Phosphate rocks
PRB. *See* Permanent raised bed
Pressmud, 512
PS. *See* Pig slurry
PS II. *See* Photosystem II
P_{SLI}. *See* Single pollution load index
PT50, 378
PTF. *See* Plant transfer factor
Public Health and Engineering Department,
  448
PZC. *See* Point of zero charge

Q
Quality Criteria For Continental Water resource
  Classesî and Water Pollution Control
  Regulations, 297
Quantile–Quantile plot (QQplot), 420, 422f

R
Radio-ecology, 600
Radioactive contaminated soils
  areas in challenge, 601f
  components and anticipated results, 601f
  key aspects, 600f
  non-food crops, 615–617
  phytoremediation, 599, 604
    associated risks, 608
    groups, 604
    phytoextraction, 605–607
    phytomanagement, 604–605
  radionuclide contaminated soils
    phytostabilization, 621
  radionuclides
    Caesium, 608–609
    risks and potential, 613–615
    Strontium, 609
    Uranium, 609–613
  remediation programme management, 617
    agricultural and forested zones, 620–621
    planning for, 617–618
    remediation actions implementation,
      621–622
    remediation criteria, 618
    remediation strategy, 618–619
    site characterization, 618
    soil solution uranium, 619–620
  rhizofiltration, 615
  sources and environment, 600
    industrial process, 603
    nuclear fuel cycle, 602–603
    nuclear weapons production, 602
    nuclear weapons' testing, 602
    radionuclides, 603, 603t
    research activities, 603–604
Radioactive pollution, 307
Radioactive waste treatment
  classification and categories, 217–219,
    217f
  costs and economics, 240, 241t–242t
  isotopes and radionucleotides, 209t–212t
  management and disposal, 219–220
  phytoremediation, 207
    advantages and disadvantages, 214t
    combinations, 238
    and hyperaccumulation, 221–223

methods in, 223–235
and non-plant methods, 220–221
plant species, 224t–230t
radioactive material and safety, 215–217
tolerance and extraction, 235–236
transgenic phytoremediation, 244–245
transportation and responsibility, 220
treatment, evaluation and objectives,
238–239
air, 240
ground water, 239
soil, sediment and sludge, 240
surface water, 239
waste water, 239
uptake and distribution, 236–237
wetlands and aquatic phytoremediation,
237–238
Worldwide radioactive waste inventory, 216f
Radioactivity, 599, 602–603
measurement, 337–338
sample preparation, 337
in soil, 360
in vegetables, 361–362
Radionuclides, 603, 603t
activity concentrations, 361f
Caesium, 608–609
comparison of activity concentration, 361f
environmental pollution by, 331–332
phytostabilization, 621
risks and potential, 613–615
strontium, 609
transfer factors, 362, 362f
uptake by plants, 332–333
uranium, 609–613
Randomized complete block design (RCBD),
160
Raw effluents, 42–43
Raw pig slurry (RPS), 690–691. *See also*
Treated pig slurry (TPS)
RBC. *See* Red blood cells
RCBD. *See* Randomized complete block
design
RE. *See* Removal efficiency
Re-vegetation, 509, 515–516
Reactive nitrogen species (RNS), 66
Reactive oxygen species (ROS), 66, 433–434
Red blood cells (RBC), 65
Red edge position (REP), 417–418
Reference oral dose (R_fD), 340
Rehabilitation, 665
Relative growth rate (RGR), 455–456
Relative water content (RWC), 457
Remediation actions implementation, 621

extensive surface contamination areas, 622
post-remediation activities, 621–622
special considerations, 622
Remediation programme management, 617
agricultural and forested zones, 620–621
criteria, 618
planning for, 617–618
remediation actions implementation,
621–622
site characterization, 618
soil solution uranium, 619–620
strategy, 618–619
Remediation studies in Turkey, 305
contaminated soil acreage estimation, 306
Human Developed Index, 307
metallothionein gene expression effect, 306
phytoextraction, 306
soil pollution
control administration, 307
problems, 305–306
Remote sensing, 417–418
Removal efficiency (RE), 163, 191–192
REP. *See* Red edge position
Research activities, 603–604
Residual soil, 317
R_fD. *See* Reference oral dose
RGR. *See* Relative growth rate
Rhizobacteria, 325
Rhizodegradation, 73–74, 234, 259–260
Rhizofiltration, 13, 73, 136–137, 233–234,
258, 538, 615
Rhizoplane, 266–267
Rhizoremediation
rhizodegradation, 73–74
rhizofiltration, 73
rhizosphere effect, 73
Rhizosphere
colonization, 266–267
microbiome, 266
bioremediation, 267–268
composition, 267
degradative microbes, 268
exudates, 267
microbial activity, 266–267
microbial cells, 268
Rhizospherial strains, 263
Rhodustalfs, 321
Rhyzosphere, 548
Ribulose-1, 5-bisphosphate carboxylase/
oxygenase (RuBisCO), 457–458
RNS. *See* Reactive nitrogen species
RO. *See* Root oxidizability
Root colonization, 266–267

Root exudates, 579–580
Root oxidizability (RO), 460
ROS. *See* Reactive oxygen species
RPS. *See* Raw pig slurry
RuBisCO. *See* Ribulose-1, 5-bisphosphate
    carboxylase/oxygenase
RWC. *See* Relative water content

S

S-adenosylmethionine synthetase (SAMS), 462
Sandy and alluvial soil, 316–317
Savannah River Site (SRS), 233
SeCys. *See* Selenocysteine
Seed germination, 451–455
    effect, 374–375
    growth, physiology and productivity,
        452t–453t
    in rice plants, 455f
    yield reduction of crop plants, 454t
Seed source, 9–10
Selenocysteine (SeCys), 134–135
Selenomethionine (SeMet), 134–135
Semivariogram, 420–421, 423f
Serpentine soils, 635–637
    DTPA-extractable, 640t–641t
    properties, 636t
    soil concentrations, 638t–639t
Serpentines. *See* Ultramafic rocks
Sewage sludge, 302
Shallow tube wells (STW), 445–446
Shoot:root quotient (SRQ), 480
Short rotation coppice (SRC), 615–616,
    618–619
Siam weed (*Chromolaena odorata*), 666–668
Silicon, 578
Single pollution load index (P_{SLI}), 338
    heavy metals in soil, 342, 343t
Single super phosphate (SSP), 381
SIT. *See* Statistical Institute of Turkey
SITE. *See* Superfund Innovative Technology
    Evaluation
Site remediation
    bio-piles, 7
    electroremediation, 5
    soil flushing, 5
    soil vapour extraction, 5
    soil washing, 7
    stabilization, 6
    technologies for, 5
SOC. *See* Soil organic carbon
SOD. *See* Superoxide dismutase
Soil contamination with metals, 37
    control measures, 51–55

heavy metals, 39–43
    sources, 49t–50t
metal concentrations, 38
risk assessment, 48–51
urban and industrial wastes, 37–38
Soil organic carbon (SOC), 691
pig slurry
    application, 690–691
    effect, 691–701
SOM, 689–690
for Vertisol, 692f
Soil organic matter (SOM), 54, 618–619, 689
    chemical shift ranges, 700t
Soil pollution, 1, 288–289
global scenario
    America, 3
    Asia, 3
    Europe, 2–3
    Fiji, 4
    Kiribati, 4
    Marshall Islands, 4
    Nauru, 4
    Niue, 4
    Pacific Islands, 3–4
on human health and environment, 4
land use of Turkish soils, 289
    arable lands, 289
    lack of environmental measures, 290
    SIT data report, 290–291
    soil contamination effect, 291
radioactive pollution, 307
remediation methods, 304
    phytoremediation, 305
    plants for in situ cleaning, 304–305
remediation of toxicity in Nepal, 325
remediation studies in Turkey, 305–307
sources in Turkey, 291
    agricultural activities, 294–299
    erosion, 292–293
    industrial activities, 299–301
    mining, 303–304
    non-agricultural use, 292
    parent material, 293–294
    preventive measures against soil
        contamination, 291–292
    soil sources, 290f
    ultramafic geology areas, 294f
    urbanization, 302–303
status in Nepal
    nutrient and heavy metal status, 322–324
    physiographical division, 314t, 315f
    soil orders, 317–318
    soils, 316–322, 317f

Soil-borne microbial communities, 262
Soil(s), 1
    amendment, 678–680
    cadmium bioassessment in, 370
    cadmium contamination sources, 369–370
    characterization
        field study, 632–633
        greenhouse study, 633–635
    dressing, 677–678
    fertility, 316
    flushing, 5, 6f
    heavy metals polluting, 288
    layer formation period, 287–288
    natural cadmium levels in, 368–369
    in Nepal, 316, 317f
        alluvial soil, 316
        characteristics, 315–316
        glacial soil, 317–318
        gravelly soil, 317
        remediation studies on toxicity removal,
            326
        residual soil, 317
        sandy and alluvial soil, 316–317
    P_{CLI} for, 340
    pH, 507
    P_{SLI} of, 338
    radioactivity in, 360
    remediation, 677–678
        calcium carbonate application, 678–680
        phytoremediation, 678
        soil amendment, 678–680
        soil dressing, 677–678
    respiration, 697–698
    vapour extraction, 5
    washing, 7
Soil–water–air relations, 287
Soluble carbon, 693–695, 694t
SOM. See Soil organic matter
Sorghum, 582
SPLP. See Synthetic precipitation-leaching
        procedure
Spodosols, 320
Spring wheat (Triticum aestivum), 446
SRC. See Short rotation coppice
SRQ. See Shoot:root quotient
SRS. See Savannah River Site
SSP. See Single super phosphate
Stabilization, 6. See also Phytostabilization
    asphalt batching, 6
    vitrification, 6
Stable carbon isotope, 699–701
Statistical analysis, 164
Statistical Institute of Turkey (SIT), 290–291

Strontium (^{90}Sr), 609, 611t
Stubble burning, 298–299
STW. See Shallow tube wells
Sulphur, 576
Sulphur-oxidizing bacteria (Thiobucillus),
        384–385
Sunflower (Helianthus annuus), 615
Superfund Innovative Technology Evaluation
        (SITE), 132–134
Superoxide dismutase (SOD), 460–461
Surface expression of novel metal binding
        peptides and proteins, 68
Surface water, 239
Synthetic organic chelating agents, 581–582
Synthetic precipitation-leaching procedure
        (SPLP), 531–532

T
Tap roots, 9
TCE. See Trichloroethylene
TCLP. See Toxicity characteristic leaching
        procedure
Terrestrial radiation, 331
TF. See Transfer factor; Translocation factor
Thermal power plants, 300–301
Thlaspi elegans Boiss., 633–635
    Ni and EDTA effect, 649f–651f
    Ni bioaccumulation coefficient, 654t
TNT. See Trinitrotoluene
Tolerable weekly intake (TWI), 367
Total petroleum hydrocarbons (TPH), 549, 565
Toxicity characteristic leaching procedure
        (TCLP), 531–532
TPH. See Total petroleum hydrocarbons
TPS. See Treated pig slurry
Transfer factor (TF), 605, 606t, 610–611, 615
Transgenic phytoremediation, 244–245
Translocation, 117
Translocation factor (TF), 163, 480
    As-tolerant plant species values, 481t
    heavy metals uptake evaluation, 163
    phytoremediation potential evaluation,
        194–195
    of plant species at harvest, 194f
Transpiration rate, 9
Treated pig slurry (TPS), 690–691
Trichloroethylene (TCE), 107–108
Trinitrotoluene (TNT), 13, 107–108
Triple super phosphate (TSP), 381
Tritium (^{3}H), 232–233
TSP. See Triple super phosphate
Tubewell water, 443–445
Tulatoli village of Teknaf, 334

Turkey
  land use of Turkish soils, 289–291
  radioactive pollution, 307
  remediation methods, 304–305
  remediation studies in, 305–307
  soil pollution sources in, 291–304
  soil sources in, 290f
TWI. *See* Tolerable weekly intake

U

Ultisols, 321
Ultramafic rocks, 630
United States Environmental Action Group
  (USEAG), 149–150
United States Environmental Protection
  Agency (USEPA), 41–42
Uranium (^{238}U), 609–613
  pH-dependent uranium speciation, 614f
  transfer factor
    cereals, 614t
    natural vegetation, 612t
    vegetables, 613t
Urbanization
  highways, 303
  population, 302
  wastes, 302
USEAG. *See* United States Environmental
  Action Group
USEPA. *See* United States Environmental
  Protection Agency
Ustifluvents, 318
Ustochrepts, 319
Ustorthents, 318

V

Vegetables
  DIM in, 340, 355t–357t
  gardens, 334
  health risk assessment, 354–360
  for heavy metal analysis, 335t
  heavy metal transfer, 352t–353t
  HRI of heavy metal, 340–341, 358t–359t
  leafy and non-leafy, 334, 340, 347, 360
  levels of heavy metals in, 344–348
  MPI in, 348f, 349

Pearson correlation, 350t
  radioactivity in, 361–362
Vegetation cap, 234–235
Vertical engineered barriers (VEB), 534
Vertiver grass (*Vertiveria zizanioides*), 665,
  670f
Very Low Level Waste (VLLW), 218
Very Short Lived Waste (VSLW), 218
Vitrification, 6

W

Waste disposal, considerations for, 12
Waste water, 239, 529–530
  irrigation, 2
Wastes, 302
Water-insoluble fertilizers, 381
Water-soluble organic carbon (WSOC),
  693–695
Winter wheat (*T. aestivum*), 451
World soil, 695

X

Xylem loading, 373
Xylogen, 678–680

Z

Zea mays, 351, 354
Zinc (Zn), 20, 576–577
  health hazards of, 20
  hyperaccumulation, 31
  phytoremediation use, 20–21
Zinc transporter of Arabidopsis Thaliana
  (ZAT), 137–141
Zinc-mine-industry-ravaged ecosystem
  phytomanagement, 665
  legumes, 665–666, 672f
  Padaeng mine industry, 668
  plants in Mae Sot zinc mine waste,
    667t–668t
  Siam weed, 666–668
  tree, grass and asteraceae association, 673f
  vermin-compost production, 674f
  vertiver grass, 665, 670f
  zinc mine waste, 669f
  zinc tolerant native plant, 666f

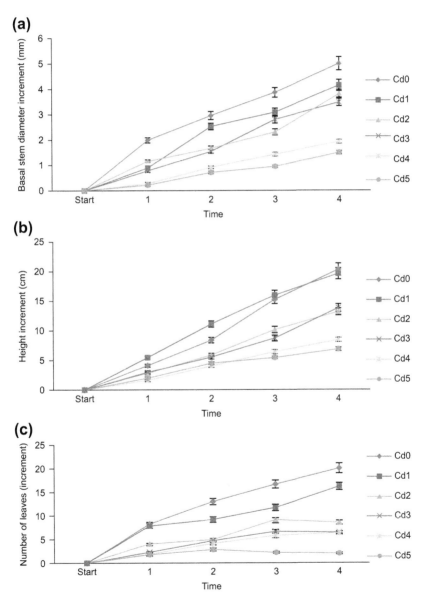

FIGURE 7.4 Increment (increase per month) of basal stem diameter (a) and height (b) and the total number of leaves (c) of *Jatropha curcas*, as influenced by different copper levels.

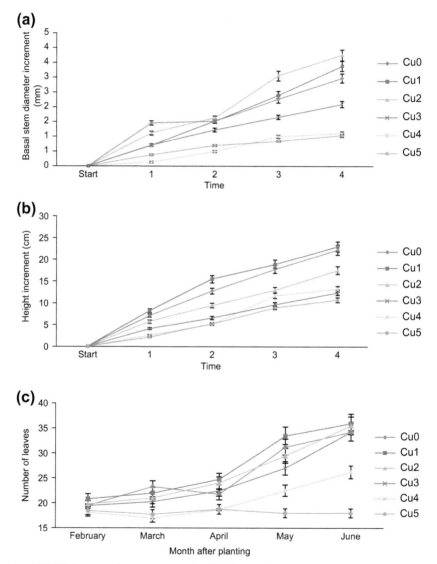

FIGURE 7.5 Increment (increase per month) of basal stem diameter (a) and height (b) and the total number of leaves (c) of *Acacia mangium*, as influenced by different copper levels.

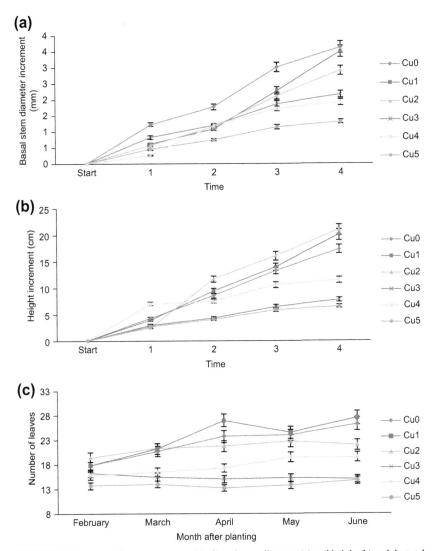

FIGURE 7.6 Increment (increase per month) of basal stem diameter (a) and height (b) and the total number of leaves (c) in *Dyera costulata*, as influenced by different copper levels.

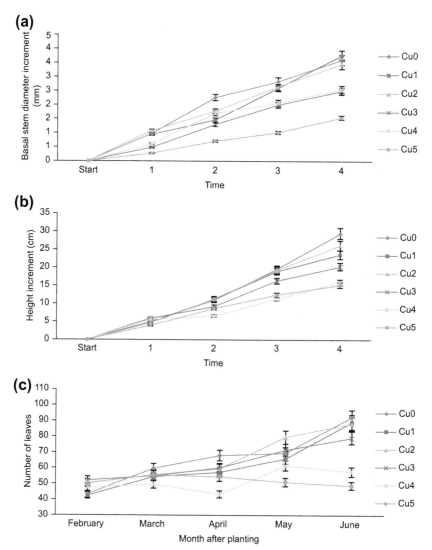

FIGURE 7.7 Increment (increase per month) of basal stem diameter (a), height (b) and the total number of leaves (c) of *Hopea odorata*, as influenced by different copper levels.

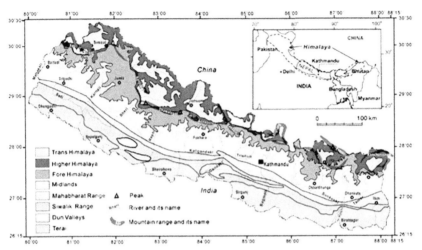

FIGURE 11.1 Physiography of the Nepal Himalaya. *Source: Dahal, 2006.*

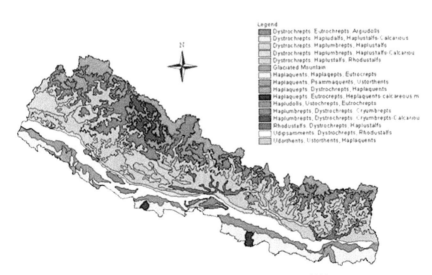

FIGURE 11.3 Soils of Nepal. *Source: Pariyar, 2008.*

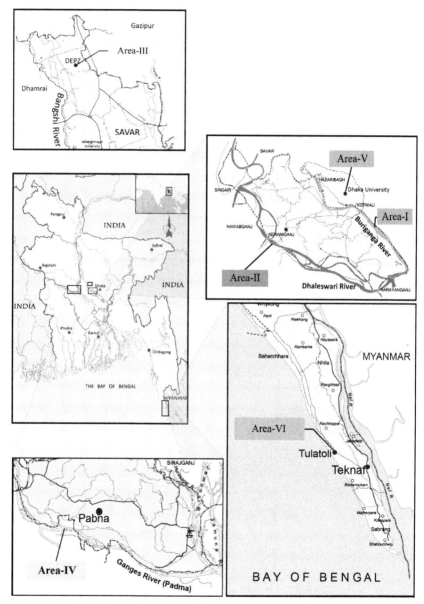

FIGURE 12.1 Map showing different sampling areas of Bangladesh.

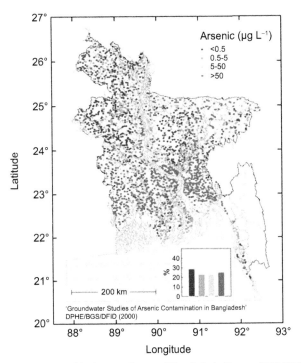

FIGURE 16.8 Presence of As in groundwater of Bangladesh. *Source: BGS / BGS (2001).*

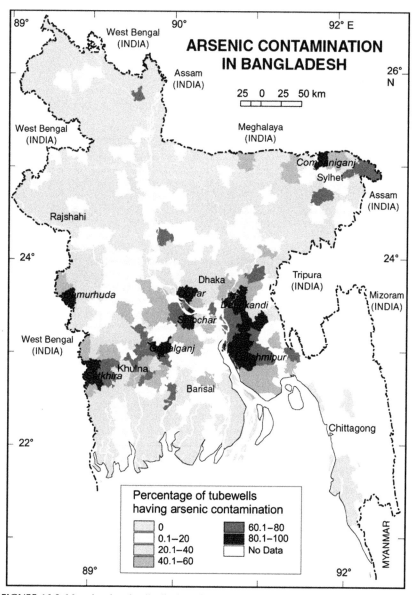

FIGURE 16.9 Map showing the distribution of As in groundwater from shallow (<150 m) tube-wells in Bangladesh. *Source: Banglapedia (2013).*

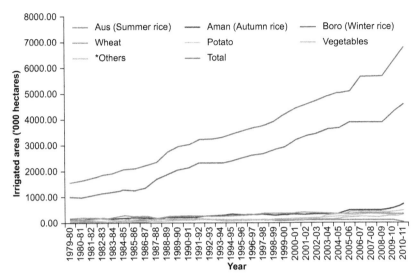

FIGURE 16.10 Irrigated Area under Different Crops, 1978–79 to 2010–11, Bangladesh. *Others means others cereals, pulses, oilseeds, sugarcane, cotton and other minor crops. *Source: BRRI (2013).*

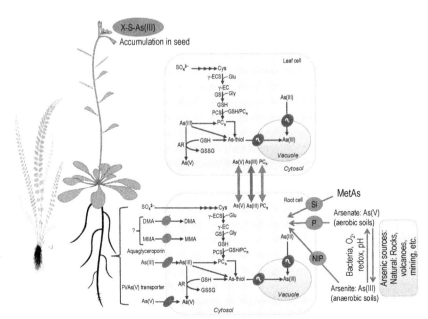

FIGURE 16.13 Mechanism of Arsenic uptake, translocation and detoxification in plants. *Sources: Tripathi et al. (2007), Briat (2010).*

FIGURE 16.14 Toxicity symptoms in rice plants exposed to different levels (mg l⁻¹) of arsenic. *Photo Courtesy of Prof. Dr Md. Asaduzzaman Khan, SAU, Dhaka.*

FIGURE 16.17 Some As hyperaccumulator ferns widely grown in As-contaminated soils of Bangladesh. (a) fronds of *Pteris vittata;* (b) whole plant of *Pteris cretica;* (c) frond of *Pteris cretica;* (d) whole plant; (e) frond; and (f) root mass of *Nephrodium molle* (Dryopteridaceae). *Photo credit: Nafia Jahan Rashmi.*

FIGURE 16.18 Some As-hyperaccumulators in terrestrial and aquatic higher plants. (a) *Colocasia esculanta*; (b). *Azolla* sp., (c) *Pistia stratiotes*; (d) *Hydrilla verticillata*; (e) *Monochoria hastata*; (f) *Eichhornia crassipes*; (g) *Echinochola cruss-galli*; (h) *Cyperus rotundus*; (i) *Lemma minor*; (j) *Xanthium italicum*; (k) *Typha latifolia*; (l) *Leucaena leucocephala*.

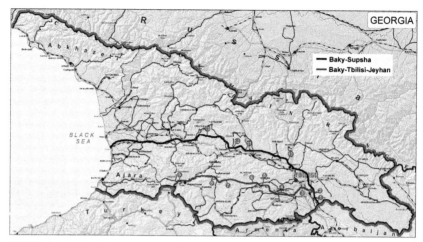

FIGURE 19.1 Baku–Supsa and Baku–Tbilisi–Ceyhan Oil Pipelines in the territory of Georgia. Sampling points of soils are indicated on the map. The types of soils are as follows: (1) Meadow grey cinnamonic (Calcic vertisols) (pH 5.2); (2) Grey cinnamonic (Calcic kastanozems) (pH 4.7); (3) Alluvial calcareous (Calcaric fluvisols) (pH 4.5); (4) Raw humus calcareous (Rendzic leptosols) (pH 6.0); (5) Chernozems (Chernozems) (pH 4.5); (6) Mountain meadow soddy (Leptosols, cambisols and cryosols) (pH = 4.0); (7) Brown forest podzolized (Dystric cambisols) (pH 4.5); (8) Brown forest weakly unsaturated (Eutric cambisols) (pH 5.0); (9) Cinnamonic (Eutric cambisols and Calcic kastanozems) (pH 5.5); (10) Cinnamonic calcareous (Calcaric cambisols and Calcic kastanozems) (pH 5.0); (11) Cinnamonic light (Calcic kastanozems) (pH 5.0); (12) Black calcareous (Calcic vertisosls) (pH 5.0); (13) Yellow brown forest (Chromic cambisols and stagnic alisols) (pH 5.0).

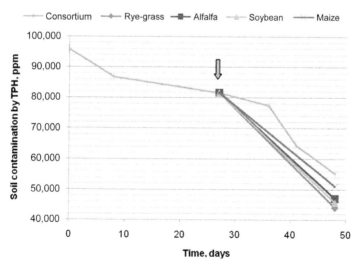

FIGURE 19.4 The dynamics of decrease of TPH content in soil contaminated with oil hydrocarbons during incubation with bacterial consortium (strains of *Rhodococcus*, *Pseudomonas* and *Mycobacterium*) and plants. The bacterial suspension was inoculated in the soil at the beginning of the experiment. On the 27th day of incubation, plants were sown in separate samples of soil (indicated by arrow). Initial degree of contamination, 96,000 ppm of TPH. Temperature, 20–25 C.

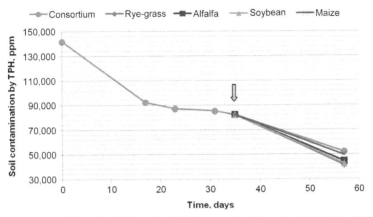

FIGURE 19.5 The dynamics of decreasing TPH content in soil contaminated with oil hydrocarbons during incubation with bacterial consortium (strains of *Rhodococcus, Pseudomonas* and *Mycobacterium*) and plants. The bacterial suspension was inoculated in the soil at the beginning of the experiment. On the 35th day of incubation, plants were sown in separate samples of soil (indicated by arrow). Initial degree of contamination: 142,000 ppm of TPH. Temperature: 20–25°C.

FIGURE 23.4 (a) Zinc-tolerant native plant and (b) cryptogams on zinc-rich rock.

FIGURE 23.6 (a–e) Vetiveria grass stem stumps intermingled with *Chromolena odorata* association; (f) *Chromolena odorata* growing in metal-rich environment.

FIGURE 23.7 (a-e) *Gynura pseudochina* (L.) DC. (Wan-Maha-Kan), a tuber bearing Asteraceae. It grows luxuriantly in high zinc and cadmium-contaminated soils. (f) *Murdannia spectabilis*, a tuber bearing Commelinaceae. They also accumulate zinc and cadmium.

FIGURE 23.8 (a) *Doubanga grandiflora* (Native tree); (b) *Phyllostachys mannii* Bamboo; and (c) *Leucaena leucocephala* stands on zinc-rich mine soils.

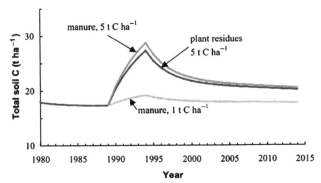

FIGURE 24.1 Change in organic soil carbon for a Vertisol in semi-arid Andhra Pradesh, India modeled with the Roth C soil carbon model. Five tons of plant residue carbon and 5 tons of cattle manure carbon were applied in 1989 for five consecutive years. Application of 1 ton of carbon in cattle manure represents the amount of manure carbon that would be available for incorporation into the soil if the plant residues had been directly fed to the cattle. *Source: Farage et al. (2003).*

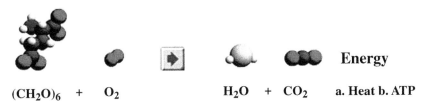

$(CH_2O)_6$ + O_2 H_2O + CO_2 a. Heat b. ATP

FIGURE 24.2 Basic soil CO_2 generation processes: oxidation of organic matter due to aerobic microbial respiration (Available on-line, http://www.oocities.org/b1gcee/fessay/biol/ res_pho.htm, 2013).

Printed in the United States
By Bookmasters